Earth Materials

Kevin Hefferan was born and raised in Jersey City, NJ to parents originating from Kiltimagh, County Mayo, Ireland. Kevin received his geological training at New Jersey City State University, Bryn Mawr College and Duke University. Kevin is married to Sherri (Cramer) Hefferan and is the proud father of Kaeli, Patrick, Sierra, Keegan and Quintin of Stevens Point, WI. Kevin is a professor of geology at the University of Wisconsin–Stevens Point Department of Geography and Geology.

John O'Brien is married (to Anita) with two sons (Tyler and Owen). He was born (on December 10, 1941) in Seattle, Washington, and was raised there and in Ohio and southern California. His parents were teachers, so summers were spent with the family traveling throughout the west, imbuing him with a passion for the natural world. He discovered an enthusiasm for working with students as a teaching assistant at Miami University (Ohio) and combined the two interests in a career teaching geological sciences at New Jersey City University. A sedimentologist by training, he took over responsibility for the mineralogy, petrology and structure courses when a colleague departed. The *Earth Materials* text is in part the result of that serendipitous occurrence.

Companion website

A companion website for this book, with resource materials for students and instructors is available at: www.wiley.com/go/hefferan/earthmaterials

Earth Materials

Kevin Hefferan and John O'Brien

WILEY-BLACKWELL

A John Wiley & Sons, Ltd., Publication

This edition first published 2010, © 2010 by Kevin Hefferan and John O'Brien

Blackwell Publishing was acquired by John Wiley & Sons in February 2007. Blackwell's publishing program has been merged with Wiley's global Scientific, Technical and Medical business to form Wiley-Blackwell.

Registered office: John Wiley & Sons Ltd, The Atrium, Southern Gate, Chichester, West Sussex, PO19 8SQ, UK

Editorial offices: 9600 Garsington Road, Oxford, OX4 2DQ, UK
The Atrium, Southern Gate, Chichester, West Sussex, PO19 8SQ, UK
111 River Street, Hoboken, NJ 07030-5774, USA

For details of our global editorial offices, for customer services and for information about how to apply for permission to reuse the copyright material in this book please see our website at www.wiley.com/wiley-blackwell

The right of the author to be identified as the author of this work has been asserted in accordance with the Copyright, Designs and Patents Act 1988.

Wiley also publishes its books in a variety of electronic formats. Some content that appears in print may not be available in electronic books.

Designations used by companies to distinguish their products are often claimed as trademarks. All brand names and product names used in this book are trade names, service marks, trademarks or registered trademarks of their respective owners. The publisher is not associated with any product or vendor mentioned in this book. This publication is designed to provide accurate and authoritative information in regard to the subject matter covered. It is sold on the understanding that the publisher is not engaged in rendering professional services. If professional advice or other expert assistance is required, the services of a competent professional should be sought.

Library of Congress Cataloguing-in-Publication Data

Hefferan, Kevin.
    Earth materials / Kevin Hefferan and John O'Brien.
        p. cm.
    Includes bibliographical references and index.
    ISBN 978-1-4051-4433-9 (hardcover : alk. paper) – ISBN 978-1-4443-3460-9 (pbk. : alk. paper)   1. Geology–Textbooks.   I. O'Brien, John, 1941–   II. Title.
    QE26.3.H43 2010
    550–dc22
                                        2009050260

A catalogue record for this book is available from the British Library.

Set in 11 on 12 pt Sabon by Toppan Best-set Premedia Limited

Printed and bound in Malaysia by Vivar Printing Sdn Bhd

4   2016

Contents

Color plate sections between pp. 248 and 249, and pp. 408 and 409

Companion website for this book: wiley.com/go/hefferan/earthmaterials

Preface

Particularly since the 1980s, Earth science at the undergraduate level has experienced fundamental changes with respect to curricula and student goals. Many traditional geology and Earth science programs are being revamped in response to evolving employment and research opportunities for Earth science graduates.

As a result, many colleges and universities have compressed separate mineralogy, optical mineralogy, petrology and sedimentology courses into a one- or two-semester Earth materials course or sequence. This in part reflects the increasing demand on departments to serve students in environmental sciences, remote imaging and geographical information systems and science education. This change has occurred at an accelerating pace over the last decade as departments have adjusted their course offerings to the new realities of the job market. At present, a glaring need exists for a textbook that reflects these critical changes in the Earth science realm.

No book currently on the market is truly suitable for a one- or two-semester Earth materials course. Currently available texts are restricted to specific topics in mineralogy, sedimentology or petrology; too detailed because they are intended for use in traditional mineralogy, sedimentology or petrology course sequences; or not appropriately balanced in their coverage of the major topic areas. This book is intended to provide balanced coverage of all the major Earth materials subject areas and is appropriate for either a one-semester or two-semester mineralogy/petrology or Earth materials course.

The chapters that follow illuminate the key topics involving Earth materials, including:

- Their properties, origin and classification.
- Their associations and relationships in the context of Earth's major tectonic, petrological, hydrological and biogeochemical systems.
- Their uses as resources and their fundamental role in our lives and the global economy.
- Their relation to natural and human-induced hazards.
- Their impact on health and on the environment.

This *Earth Materials* text provides:

- A comprehensive descriptive analysis of Earth materials.
- Graphics and text in a logical and integrated format.
- Both field examples and regional relationships with graphics that illustrate the concepts discussed.
- Examples of how the concepts discussed can be used to answer significant questions and solve real-world problems.
- Up-to-date references from current scientific journals and review articles related to new developments in Earth materials research.
- A summative discussion of how an Earth materials course impacts both science and non-science curricula.

Chapter 1 contains a brief introduction to Earth materials and an overview of system Earth, including a discussion of Earth's interior and global tectonics. This introductory chapter provides a global framework for the discussions that follow.

A minerals section begins with Chapter 2, which addresses necessary background chemistry and mineral classification. Chapter 3 examines the fundamentals of crystal chemistry, phase diagrams and stable and unstable isotopes. Chapter 4 reviews the basic principles of crystallography. Chapter 5 examines mineral formation, macroscopic mineral properties and the major rock-forming minerals. Chapter 6 focuses on the microscopic optical properties of minerals and petrographic microscope techniques.

The igneous rocks section begins with Chapter 7, which discusses the composition, texture and classification of igneous rocks. Chapter 8 addresses the origin and evolution of magmas and plutonic structures. Chapter 9 focuses on volcanic structures and processes. In Chapter 10, the major igneous rock associations are presented in relation to plate tectonics.

The sedimentary rock section begins with Chapter 11, which is concerned with the sedimentary cycle and sedimentary environments. This chapter also focuses on sediment entrainment, transport and deposition agents and the sedimentary structures produced by each. Chapter 12 addresses weathering and soils and the production of sedimentary materials. Chapter 13 examines the composition, textures, classification and origin of detrital sedimentary rocks. Chapter 14 focuses on the composition, texture, classification and origin of carbonate sedimentary rocks, while providing coverage of evaporites, siliceous, iron-rich and phosphatic sedimentary rocks. It ends with a brief synopsis of carbon-rich sedimentary materials, including coal, petroleum and natural gas.

The metamorphic rock section begins with Chapter 15, which introduces metamorphic agents, processes, protoliths and types of metamorphism. Chapter 16 addresses metamorphic structures in relationship to stress and strain. Chapter 17 investigates rock textures and the classification of metamorphic rocks. Chapter 18 concentrates on metamorphic zones, metamorphic facies and metamorphic trajectories in relationship to global tectonics. Lastly, Chapter 19 addresses ore minerals, industrial minerals, gems and environmental and health issues related to minerals.

In addition to information contained in the book, graphics, links and resources for instructors and students are available on the website that supports the text: www.wiley.com/go/hefferan/earthmaterials.

Our overall goal was to produce an innovative, visually appealing, informative textbook that will meet changing needs in the Earth sciences. *Earth Materials* provides equal treatment to minerals, igneous rocks, sedimentary rocks and metamorphic rocks and demonstrates their impact on our personal lives as well as on the global environment.

Acknowledgments

We are indebted to Wiley-Blackwell publishers for working with us on this project. We are especially indebted to Ian Francis, who accepted our proposal for the text in 2005 and worked with us closely over the last 4 years, offering both guidance and support. Kelvin Matthews, Jane Andrew, Rosie Hayden, Delia Sandford, Camille Poire and Catherine Flack all made significant contributions to this project.

We gained much useful input from our mineralogy and petrology students at the University of Wisconsin-Stevens Point (UWSP) and New Jersey City University (NJCU). UWSP and NJCU provided sabbatical leave support for the authors that proved essential to the completion of the text, given our heavy teaching loads. We are also particularly thankful to the excellent library staffs at these two institutions.

We are truly appreciative of the many individuals and publishers who generously permitted reproduction of their figures and images from published work or from educational websites such as those created by Stephen Nelson, Patrice Rey and Steve Dutch.

Several reviewers provided critical feedback that greatly improved this book. Reviews by Malcolm Hill, Stephen Nelson, Lucian Platt, Steve Dutch, Duncan Heron, Jeremy Inglis, Maria Luisa Crawford, Barbara Cooper, Alec Winters, David H. Eggler, Cin-Ty Lee, Samantha Kaplan and Penelope Morton were particularly helpful.

Lastly we would like to thank our families, to whom we dedicate this text. Kevin's family includes his wife Sherri and children Kaeli, Patrick, Sierra, Keegan and Quintin. John's family includes his wife Anita and sons Tyler and Owen.

Chapter 1

Earth materials and the geosphere

1.1 EARTH MATERIALS

This book concerns the nature, origin, evolution and significance of Earth materials. Earth is composed of a variety of naturally occurring and synthetic materials whose composition can be expressed in many ways. Solid Earth materials are described by their chemical, mineral and rock composition. Atoms combine to form minerals and minerals combine to form rocks. Discussion of the relationships between atoms, minerals and rocks is fundamental to an understanding of Earth materials and their behavior.

The term mineral is used in a number of ways. For example, elements on your typical breakfast cereal box are listed as minerals. Oil and gas are considered mineral resources. All these are loose interpretations of the term mineral. In the narrowest sense, minerals are defined by the following five properties:

1   Minerals are **solid**, so they do not include liquids and gases. Minerals are solid

Earth Materials, 1st edition. By K. Hefferan and J. O'Brien. Published 2010 by Blackwell Publishing Ltd.

because all the atoms in them are held together in fixed positions by forces called chemical bonds (Chapter 2).

2   Minerals are **naturally occurring**. This definition excludes synthetic solids produced through technology. Many solid Earth materials are produced by both natural and synthetic processes. Natural and synthetic diamonds are a good example. Another example is the solid materials synthesized in high temperature and high pressure laboratory experiments that are thought to be analogous to real minerals that occur only in the deep interior of Earth.

3   Minerals usually form by **inorganic** processes. Some solid Earth materials form by both inorganic and organic processes. For example, the mineral calcite ($CaCO_3$) forms by inorganic processes (stalactites and other cavestones) and is also precipitated as shell material by organisms such as clams, snails and corals.

4   Each mineral species has a **specific chemical composition** which can be expressed by a chemical formula. An example is common table salt or halite which is

composed of sodium and chlorine atoms in a 1 : 1 ratio (NaCl). Chemical compositions may vary within well-defined limits because minerals incorporate impurities, have atoms missing, or otherwise vary from their ideal compositions. In addition some types of atoms may substitute freely for one another when a mineral forms, generating a well-defined range of chemical compositions. For example, magnesium (Mg) and iron (Fe) may substitute freely for one another in the mineral olivine whose composition is expressed as $(Mg,Fe)_2SiO_4$. The parentheses are used to indicate the variable amounts of Mg and Fe that may substitute for each other in olivine group minerals (Chapter 3).

5 Every mineral species possesses a **long-range, geometric arrangement of constituent atoms or ions**. This implies that the atoms in minerals are not randomly arranged. Instead minerals crystallize in geometric patterns so that the same pattern is repeated throughout the mineral. In this sense, minerals are like three-dimensional wall paper. A basic pattern of atoms, a motif, is repeated systematically to produce the entire geometric design. This long-range pattern of atoms characteristic of each mineral species is called its **crystal structure**. All materials that possess geometric crystal structures are **crystalline** materials. Solid materials that lack a long-range crystal structure are **amorphous** materials, where amorphous means without form; without a long-range geometric order.

Over 3500 minerals have been discovered to date (Wenk and Bulakh, 2004) and each is distinguished by a unique combination of crystal structure and chemical composition. Strictly speaking, naturally-occurring, solid materials that lack one of the properties described above are commonly referred to as **mineraloids**. Common examples include amorphous materials such as volcanic glass and organic crystalline materials such as those in organic sedimentary rocks such as coal.

Most of the solid Earth is composed of various types of rock. A rock is an **aggregate of mineral crystals and/or mineraloids**. A **monominerallic** rock consists of multiple crystals of a single mineral. Examples include the sedimentary rock quartz sandstone, which may consist of nothing but grains of quartz held together by quartz cement, and the igneous rock dunite, which can consist entirely of olivine crystals. Most rocks are **polymineralllic**; they are composed of many types of mineral crystals. For example, granite commonly contains quartz, potassium feldspar, plagioclase, hornblende and biotite and may include other mineral species.

Mineral composition is one of the major defining characteristics of rocks. Rock textures and structures are also important defining characteristics. It is not surprising that the number of rock types is very large indeed, given the large number of different minerals that occur in nature, the different conditions under which they form, and the different proportions in which they can combine to form aggregates with various textures and structures. Helping students to understand the properties, classification, origin and significance of rocks is the major emphasis of this text.

1.2   THE GEOSPHERE

Earth materials can occur anywhere within the geosphere, whose radius is approximately 6380 km (Figure 1.1). In static standard models of the geosphere, Earth is depicted with a number of roughly concentric layers. Some of these layers are distinguished primarily on the basis of differences in composition and others by differences in their state or mechanical properties. These two characteristics by which the internal layers of Earth are distinguished are not totally independent, because differences in chemical, mineralogical and/or rock composition influence mechanical properties and state.

1.2.1   Compositional layers

The layers within Earth that are defined largely on the basis of chemical composition (Figure 1.1; left side) include: (1) the **crust**, which is subdivided into **continental** and **oceanic** crust, (2) the **mantle**, and (3) the **core**. Each of these layers has a distinct combination of chemical, mineral and rock compositions that distinguishes it from the others as

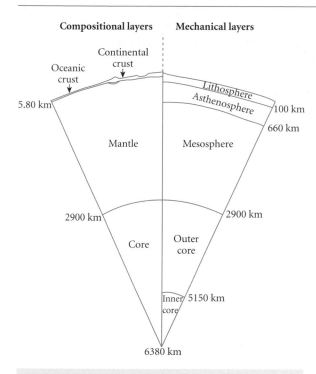

Figure 1.1 Standard cross-section model of the geosphere showing the major compositional layers on the left and the major mechanical layers on the right.

described in the next section. The thin crust ranges from 5 to 80 km thick and occupies <1% of Earth's volume. The much thicker mantle has an average radius of ~2885 km and occupies ~83% of Earth's volume. The core has a radius of ~3480 km and comprises ~16% of Earth's volume.

1.2.2 Mechanical layers

The layers within Earth defined principally on the basis of mechanical properties (Figure 1.1; right side) include: (1) a strong lithosphere to an average depth of ~100 km that includes all of the crust and the upper part of the mantle; (2) a weaker asthenosphere extending to depths ranging from 100 to 660 km and including a transition zone from ~400 to 660 km; and (3) a mesosphere or lower mantle from ~660 to 2900 km. The core is divided into a liquid outer core (~2900–5150 km) and a solid inner core, below ~5150 km to the center of Earth. Each of these layers is distinguished from the layers above and below by its unique mechanical properties. The major

features of each of these layers are summarized in the next section.

1.3 DETAILED MODEL OF THE GEOSPHERE

1.3.1 Earth's crust

The outermost layer of the geosphere, Earth's crust, is extremely thin; in some ways it is analogous to the very thin skin on an apple. The crust is separated from the underlying mantle by the **Mohorovičić (Moho) discontinuity**. Two major types of crust occur.

Oceanic crust

Oceanic crust is composed largely of dark-colored, mafic rocks enriched in oxides of magnesium, iron and calcium (MgO, FeO and CaO) relative to average crust. The elevated iron (Fe) content is responsible for both the dark color and the elevated density of oceanic crust. Oceanic crust is thin; the depth to the Moho averages 5–7 km. Under some oceanic islands, its thickness reaches 18 km. The elevated density and small thickness of oceanic crust cause it to be less buoyant than continental crust, so that it occupies areas of lower elevation on Earth's surface. As a result, most oceanic crust of normal thickness is located several thousand meters below sea level and is covered by oceans. Oceanic crust consists principally of rocks such as basalt and gabbro, composed largely of the minerals pyroxene and calcic plagioclase. These mafic rocks comprise layers 2 and 3 of oceanic crust and are generally topped with sediments that comprise layer 1 (Table 1.1). An idealized stratigraphic column (see Figure 1.8) of ocean crust consists of three main layers, each of which can be subdivided into sublayers.

Oceanic crust is young relative to the age of the Earth (~4.55 Ga = 4550 Ma). The oldest ocean crust, less than 180 million years old (180 Ma), occurs along the western and eastern borders of the north Atlantic Ocean and in the western Pacific Ocean. Older oceanic crust has largely been destroyed by subduction, but fragments of oceanic crust, perhaps as old as 2.5 Ga, may be preserved on land in the form of ophiolites. **Ophiolites** may be slices of ocean crust thrust onto continental margins and, if so, provide evidence for the existence of Precambrian oceanic crust.

Table 1.1 A comparison of oceanic and continental crust characteristics.

Properties	Oceanic crust	Continental crust
Composition	Dark-colored, mafic rocks enriched in MgO, FeO and CaO Averages ~50% SiO_2	Complex; many lighter colored felsic rocks Enriched in K_2O, Na_2O and SiO_2 Averages ~60% SiO_2
Density	Higher; less buoyant Average 2.9–3.1 g/cm^3	Lower; more buoyant Average 2.6–2.9 g/cm^3
Thickness	Thinner; average 5–7 km thickness Up to 15 km under islands	Thicker; average 30 km thickness Up to 80 km under mountains
Elevation	Low surface elevation; mostly submerged below sea level	Higher surface elevations; mostly emergent above sea level
Age	Up to 180 Ma for in-place crust ~3.5% of Earth history	Up to 4000 Ma 85–90% of Earth history

Continental crust

Continental crust has a much more variable composition than oceanic crust. Continental crust can be generalized as "granitic" in composition, enriched in K_2O, Na_2O and SiO_2 relative to average crust. Although igneous and metamorphic rocks of granitic composition are common in the upper portion of continental crust, lower portions contain more rocks of dioritic and/or gabbroic composition. Granites and related rocks tend to be light-colored, lower density felsic rocks rich in quartz and potassium and sodium feldspars. Continental crust is generally much thicker than oceanic crust; the depth to the Moho averages 30 km. Under areas of very high elevation, such as the Himalayas, its thickness approaches 80 km. The greater thickness and lower density of continental crust make it more buoyant than oceanic crust. As a result, the top of continental crust is generally located at higher elevations and the surfaces of the continents tend to be above sea level. The distribution of land and sea on Earth is largely dictated by the distribution of continental and oceanic crust. Only the thinnest portions of continental crust, most frequently along thinned continental margins and rifts, occur below sea level.

Whereas modern oceans are underlain by oceanic crust younger than 180 Ma, the oldest well-documented continental crust includes 4.03 Ga rocks from the Northwest Territories of Canada (Stern & Bleeker, 1998). Approximately 4 Ga rocks also occur in Greenland and Australia. Greenstone belts (Chapter 18) may date back as far as 4.28 Ga (O'Neill et al., 2008) suggesting that crust began forming within 300 million years of Earth's birth. Individual zircon grains from metamorphosed sedimentary rocks in Australia have been dated at 4.4 Ga (Wilde et al., 2001). The great age of some continental crust results from its relative buoyancy. In contrast to ocean crust, continental crust is largely preserved as its density is too low for it to be readily subducted. Table 1.1 summarizes the major differences between oceanic and continental crust.

1.3.2 Earth's mantle

The mantle is thick (~2900 km) relative to the radius of Earth (~6370 km) and constitutes ~83% of Earth's total volume. The mantle is distinguished from the crust by being very rich in MgO (30–40%) and, to a lesser extent, in FeO. It contains an average of approximately 40–45% SiO_2 which means it has an **ultrabasic composition** (Chapter 7). Some basic rocks such as eclogite occur in smaller proportions. In the upper mantle (depths to 400 km), the silicate minerals olivine and pyroxene are dominant; spinel, plagioclase and garnet are locally common. These minerals combine to produce dark-colored **ultramafic rocks** (Chapter 7) such as peridotite, the dominant group of rocks in the upper mantle. Under the higher pressure conditions deeper in the mantle similar chemical components combine to produce dense minerals with tightly packed structures. These high pressure mineral transformations are largely indicated

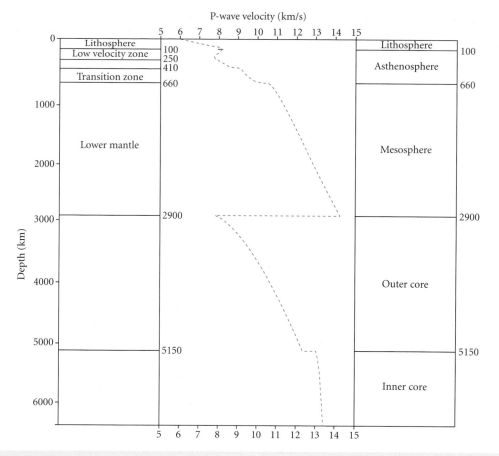

Figure 1.2   Major layers and seismic (P-wave) velocity changes within Earth, with details of the upper mantle layers.

by changes in seismic wave velocity, which reveal that the mantle contains a number of sublayers (Figure 1.2) as discussed below.

Upper mantle and transition zone

The uppermost part of the mantle and the crust together constitute the relatively rigid **lithosphere**, which is strong enough to rupture in response to stress. Because the lithosphere can rupture in response to stress, it is the site of most earthquakes and is broken into large fragments called plates, as discussed later in this chapter.

A discrete **low velocity zone** (LVZ) occurs within the upper mantle at depths of ~100–250 km below the surface. The top of LVZ marks the contact between the strong lithosphere and the weak asthenosphere (Figure 1.2). The **asthenosphere** is more plastic and

flows slowly, rather than rupturing, when subjected to stress. The anomalously low P-wave velocity of the LVZ has been explained by small amounts of partial melting (Anderson et al., 1971). This is supported by laboratory studies suggesting that peridotite should be very near its melting temperature at these depths due to high temperature and small amounts of water or water-bearing minerals. Below the base of the LVZ (250–410 km), seismic wave velocities increase (Figure 1.2) indicating that the materials are more rigid solids. These materials are still part of the relatively weak asthenosphere which extends to the base of the transition zone at 660 km.

Seismic discontinuities marked by increases in seismic velocity occur within the upper mantle at depths of ~410 and ~660 km (Figure 1.2). The interval between the depths of 410 and 660 km is called the **transition zone** between the upper and lower mantle. The sudden jumps in seismic velocity record sudden increases in rigidity and incompressibility. Laboratory studies suggest that the minerals in peridotite undergo transformations into new minerals at these depths. At approximately 410 km depth (~14 GPa), olivine (Mg_2SiO_4) is transformed to more rigid, incompressible beta spinel (β-spinel), also known as wadleysite (Mg_2SiO_4). Within the transition zone, wadleysite is transformed into the higher pressure mineral ringwoodite (Mg_2SiO_4). At ~660 km depth (~24 GPa), ringwoodite and garnet are converted to very rigid, incompressible perovskite [$(Mg,Fe,Al)SiO_3$] and oxide phases such as periclase (MgO). The mineral phase changes from olivine to wadleysite and from ringwoodite to perovskite are inferred to be largely responsible for the seismic wave velocity changes that occur at 410 and 660 km (Ringwood, 1975; Condie, 1982; Anderson, 1989). Inversions of pyroxene to garnet and garnet to minerals with ilmenite and perovskite structures may also be involved. The base of the transition zone at 660 km marks the base of the asthenosphere in contact with the underlying mesosphere or lower mantle (see Figure 1.2).

Lower mantle (mesosphere)

Perovskite, periclase [$(Mg,Fe)O$], magnesiowustite [$(Mg,Fe)O$], stishovite (SiO_2), ilmenite [$(Fe,Mg)TiO_2$] and ferrite [$(Ca,Na,Al)Fe_2O_4$] are thought to be the major minerals in the lower mantle or **mesosphere**, which extends from depths of 660 km to the mantle–core boundary at ~2900 km depth. Our knowledge of the deep mantle continues to expand, largely based on anomalous seismic signals deep within Earth. These are particularly common in a complex zone near the core–mantle boundary called the **D″ layer**. The D″ discontinuity ranges from ~130 to 340 km above the core–mantle boundary. Williams and Garnero (1996) proposed an **ultra low velocity zone** (ULVZ) in the lowermost mantle on seismic evidence. These sporadic ULVZs may be related to the formation of deep mantle plumes within the lower mantle. Other areas near the core–mantle boundary are characterized by anomalously fast velocities. Hutko et al.

(2006) detected subducted lithosphere that had sunk all the way to the D″ layer and may be responsible for the anomalously fast velocities. Deep subduction and deeply rooted mantle plumes support the concept of whole mantle convection and may play a significant role in the evolution of a highly heterogeneous mantle, but these concepts are highly controversial (Foulger et al., 2005).

1.3.3 Earth's core

Earth's core consists primarily of **iron** (~85%), with smaller, but significant amounts of nickel (~5%) and lighter elements (~8–10%) such as oxygen, sulfur and/or hydrogen. A dramatic decrease in P-wave velocity and the termination of S-wave propagation occurs at the 2900 km discontinuity (Gutenberg discontinuity or core–mantle boundary). Because S-waves are not transmitted by non-rigid substances such as fluids, the **outer core** is inferred to be a liquid. Geophysical studies suggest that Earth's outer core is a highly compressed liquid with a density of ~10–12 g/cm³. Circulating molten iron in Earth's outer core is responsible for the production of most of Earth's magnetic field.

The outer/inner core boundary, the Lehman discontinuity, at 5150 km, is marked by a rapid increase in P-wave velocity and the occurrence of low velocity S-waves. The solid inner core has a density of ~13 g/cm³. Density and magnetic studies suggest that Earth's **inner core** also consists largely of iron, with nickel and less oxygen, sulfur and/or hydrogen than in the outer core. Seismic studies have shown that the inner core is seismically anisotropic; that is, seismic velocity in the inner core is faster in one direction than in others. This has been interpreted to result from the parallel alignment of iron-rich crystals or from a core consisting of a single crystal with a fast velocity direction.

In this section, we have discussed the major layers of the geosphere, their composition and their mechanical properties. This model of a layered geosphere provides us with a spatial context in which to visualize where the processes that generate Earth materials occur. In the following sections we will examine the ways in which all parts of the geosphere interact to produce global tectonics. The ongoing story of global tectonics is one of the most

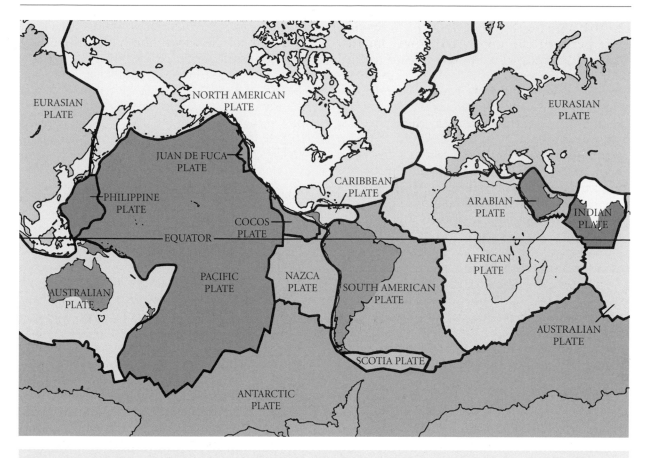

Figure 1.3 World map showing the distribution of the major plates separated by boundary segments that end in triple junctions. (Courtesy of the US Geological Survey.)

fascinating tales of scientific discovery in the last century.

1.4 GLOBAL TECTONICS

Plate tectonic theory has profoundly changed the way geoscientists view Earth and provides an important theoretical and conceptual framework for understanding the origin and global distribution of igneous, sedimentary and metamorphic rock types. It also helps to explain the distribution of diverse phenomena that include faults, earthquakes, volcanoes, mountain belts and mineral deposits.

The fundamental tenet of plate tectonics (Isacks et al., 1968; Le Pichon, 1968) is that the lithosphere is broken along major fault systems into large pieces called **plates** that move relative to one another. The existence of the strong, breakable lithosphere permits plates to form. The fact that they overlie a weak, slowly flowing asthenosphere permits them to move. Each plate is separated from adjacent plates by plate boundary segments ending in **triple junctions** (McKenzie and Morgan, 1969) where three plates are in contact (Figure 1.3).

The relative movement of plates with respect to the boundary that separates them defines three major types of plate boundary segments (Figure 1.4) and two hybrids: (1) divergent plate boundaries, (2) convergent plate boundaries, (3) transform plate boundaries, and (4) divergent–transform and convergent–transform hybrids (shown).

Each type of plate boundary produces a characteristic suite of features composed of a characteristic suite of Earth materials. This relationship between the kinds of Earth materials formed and the plate tectonic settings in

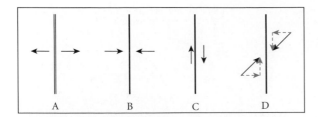

Figure 1.4 Principal types of plate boundaries: A, divergent; B, convergent; C, transform; D, hybrid convergent–transform boundary. Thick black lines represent plate boundaries and arrows indicate relative motion between the plates; blue dashed arrows show components of convergent and transform relative motion.

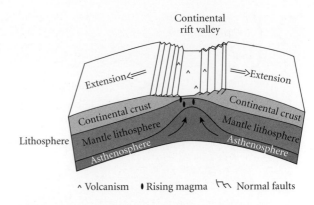

Figure 1.5 Major features of continental rifts include rift valleys, thinned continental crust and lithosphere and volcanic–magmatic activity from melts generated in the rising asthenosphere.

which they are produced provides a major theme of the chapters that follow.

1.4.1 Divergent plate boundaries

Divergent plate boundaries occur where two plates are moving apart relative to their boundary (Figure 1.4a). Such areas are characterized by horizontal extension and vertical thinning of the lithosphere. Horizontal extension in continental lithosphere is marked by continental rift systems and in oceanic lithosphere by the oceanic ridge system.

Continental rifts

Continental rift systems form where horizontal extension occurs in continental lithosphere (Figure 1.5). In such regions, the lithosphere is progressively stretched and thinned, like a candy bar being stretched in two. This stretching occurs by brittle, normal faulting near the cooler surface and by ductile flow at deeper, warmer levels. Extension is accompanied by uplift of the surface as the hot asthenosphere rises under the thinned lithosphere. Rocks near the surface of the lithosphere eventually rupture along normal faults to produce continental rift valleys. The East African Rift, the Rio Grande Rift in the United States and the Dead Sea Rift in the Middle East are modern examples of continental rift valleys.

If horizontal extension and vertical thinning occur for a sufficient period of time, the continental lithosphere may be completely rifted into two separate continents. Complete **continental rifting** is the process by which supercontinents such as Pangea and Rodinia were broken into smaller continents such as those we see on Earth's surface at present. When this happens, a new and growing ocean basin begins to form between the two continents by the process of **sea floor spreading** (Figure 1.6). The most recent example of this occurred when the Arabian Peninsula separated from the rest of Africa to produce the Red Sea basin some 5 million years ago. Older examples include the separation of India from Africa to produce the northwest Indian Ocean basin (~115 Ma) and the separation of North America from Africa to produce the north Atlantic Ocean basin (~180 Ma). Once the continental lithosphere has rifted completely, the divergent plate boundary is no longer situated within continental lithosphere. Its position is instead marked by a portion of the oceanic ridge system where oceanic crust is produced and grows by sea floor spreading (Figure 1.6).

Oceanic ridge system

The **oceanic ridge system** (ridge) is Earth's largest mountain range and covers roughly 20% of Earth's surface (Figure 1.7). The ridge is >65,000 km long, averages ~1500 km in width, and rises to a crest with an average

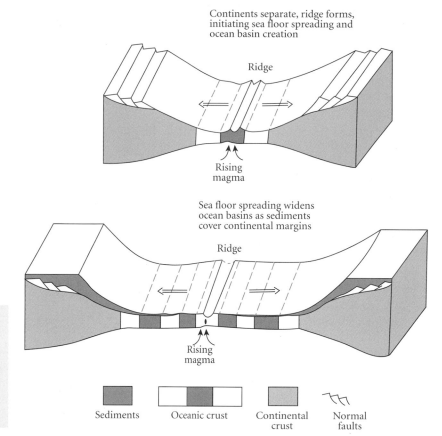

Continents separate, ridge forms, initiating sea floor spreading and ocean basin creation

Ridge

Rising magma

Sea floor spreading widens ocean basins as sediments cover continental margins

Ridge

Rising magma

Sediments   Oceanic crust   Continental crust   Normal faults

Figure 1.6 Model showing the growth of ocean basins by sea floor spreading from the ridge system following the complete rifting of continental lithosphere along a divergent plate boundary.

Figure 1.7 Map of the ocean floor showing the distribution of the oceanic ridge system. (Courtesy of Marie Tharp, with permission of Bruce C. Heezen and Marie Tharp, 1977; © Marie Tharp 1977/2003. Reproduced by permission of Marie Tharp Maps, LLC, 8 Edward Street, Sparkhill, NT 10976, USA.) (For color version, see Plate 1.7, opposite p. 248.)

elevation of ~3 km above the surrounding sea floor. A moment's thought will show that the ridge system is only a broad swell on the ocean floor, whose slopes on average are very gentle. Since it rises only 3 km over a horizontal distance of 750 km, then the average slope is 3 km/750 km which is about 0.4%; the average slope is about 0.4°. We exaggerate the vertical dimension on profiles and maps in order to make the subtle stand out. Still there are differences in relief along the ridge system. In general, warmer, faster spreading portions of the ridge such as the East Pacific Rise (~6–18 cm/yr) have gentler slopes than colder, slower spreading portions such as the Mid-Atlantic Ridge (~2–4 cm/yr). The central or axial portion of the ridge system is marked by a rift valley, especially along slower spreading segments, or other rift features, and marks the position of the divergent plate boundary in oceanic lithosphere.

One of the most significant discoveries of the 20th century (Dietz, 1961; Hess, 1962) was that oceanic crust and lithosphere form along the axis of the ridge system, then spreads away from it in both directions, causing ocean basins to grow through time. The details of this process are illustrated by Figure 1.8. As the lithosphere is thinned, the asthenosphere rises toward the surface generating basaltic–gabbroic melts. Melts that crystallize in magma bodies well below the surface form plutonic rocks such as gabbros that become layer 3 in oceanic crust. Melts intruded into near-vertical fractures above the chamber form the basaltic–gabbroic sheeted dikes that become layer 2b. Lavas that flow onto the ocean floor commonly form basaltic pillow lavas that become layer 2a. The marine sediments of layer 1 are deposited atop the basalts. In this way layers 1, 2 and 3 of the oceanic crust are formed. The underlying mantle consists of ultramafic rocks (layer 4). Layered ultramafic rocks form by differentiation near the base of the basaltic–gabbroic magma bodies, whereas the remainder of layer 4 represents the unmelted, refractory residue that accumulates below the magma body.

Because the ridge axis marks a divergent plate boundary, the new sea floor on one side moves away from the ridge axis in one direction and the new sea floor on the other side moves in the opposite direction relative to the ridge axis. More melts rise from the astheno-

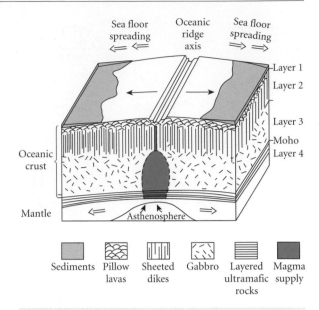

Figure 1.8 The formation of oceanic crust along the ridge axis generates layer 2 pillow basalts and dikes, layer 3 gabbros of the oceanic crust and layer 4 mantle peridotites. Sediment deposition on top of these rocks produces layer 1 of the crust. Sea floor spreading carries these laterally away from the ridge axis in both directions.

sphere and the process is repeated, sometimes over >100 Ma. In this way ocean basins grow by sea floor spreading as though new sea floor is being added to two conveyor belts that carry older sea floor in opposite directions away from the ridge where it forms (Figure 1.8). Because most oceanic lithosphere is produced along divergent plate boundaries marked by the ridge system, they are also called **constructive** plate boundaries.

As the sea floor spreads away from the ridge axis, the crust thickens from above by the accumulation of additional marine sediments and the lithosphere thickens from below by a process called **underplating**, which occurs as the solid, unmelted portion of the asthenosphere spreads laterally and cools through a critical temperature below which it becomes strong enough to fracture. As the entire lithosphere cools, it contracts, becomes denser and sinks so that the floors of the ocean gradually deepen away from the thermally elevated ridge axis. As explained in the next section, if the density of oceanic litho-

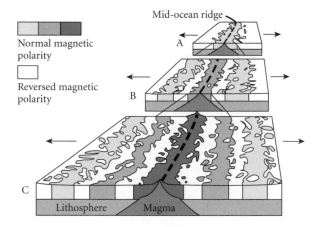

Normal magnetic
polarity

Reversed magnetic
polarity

Mid-ocean ridge

A

B

C

Lithosphere   Magma

Figure 1.9 Model depicting the production of alternating normal (colored) and reversed (white) magnetic bands in oceanic crust by progressive sea floor spreading and alternating normal and reversed periods of geomagnetic polarity (A through C). The age of such bands should increase away from the ridge axis. (Courtesy of the US Geological Survey.)

sphere exceeds that of the underlying asthenosphere, subduction occurs.

The formation of oceanic lithosphere by sea floor spreading implies that the age of oceanic crust should increase systematically away from the ridge in opposite directions. Crust produced during a period of time characterized by normal magnetic polarity should split in two and spread away from the ridge axis as new crust formed during the subsequent period of reversed magnetic polarity forms between it. As indicated by Figure 1.9, repetition of this splitting process produces oceanic crust with bands (linear magnetic anomalies) of alternating normal and reversed magnetism whose age increases systematically away from the ridge (Vine and Matthews, 1963).

Sea floor spreading was convincingly demonstrated in the middle to late 1960s by paleomagnetic studies and radiometric dating that showed that the age of ocean floors systematically increases in both directions away from the ridge axis, as predicted by sea floor spreading (Figure 1.10).

Hess (1962), and those who followed, realized that sea floor spreading causes the outer layer of Earth to grow substantially over time.

If Earth's circumference is relatively constant and Earth's lithosphere is growing horizontally at divergent plate boundaries over a long period of time, then there must be places where it is undergoing long-term horizontal shortening of similar magnitude. As ocean lithosphere ages and continues to move away from ocean spreading centers, it cools, subsides and becomes more dense over time. The increased density causes the ocean lithosphere to become denser than the underlying asthenosphere. As a result, a plate carrying old, cold, dense oceanic lithosphere begins to sink downward into the asthenosphere, creating a convergent plate boundary.

1.4.2   Convergent plate boundaries

Convergent plate boundaries occur where two plates are moving toward one another relative to their mutual boundary (Figure 1.11). The scale of such processes and the features they produce are truly awe inspiring.

Subduction zones

The process by which the leading edge of a denser lithospheric plate is forced downward into the underlying asthenosphere is called **subduction**. The downgoing plate is called the subducted plate or downgoing slab; the less dense plate is called the overriding plate. The area where this process occurs is a subduction zone. The subducted plate, whose thickness averages 100 km, is always composed of oceanic lithosphere. Subduction is the major process by which oceanic lithosphere is destroyed and recycled into the asthenosphere at rates similar to oceanic lithosphere production along the oceanic ridge system. For this reason, subduction zone plate boundaries are also called **destructive** plate boundaries.

The surface expressions of subduction zones are **trench–arc systems** of the kind that encircle most of the shrinking Pacific Ocean. Trenches are deep, elongate troughs in the ocean floors marked by water depths that can approach 11 km. They are formed as the downgoing slab forces the overriding slab to bend downward forming a long trough along the boundary between them.

Because the asthenosphere is mostly solid, it resists the downward movement of the

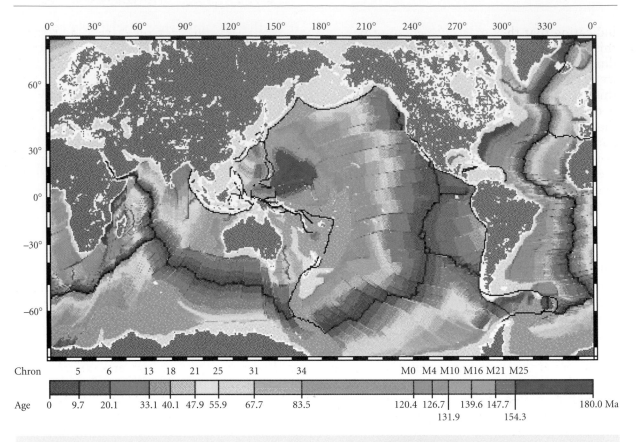

Figure 1.10   World map showing the age of oceanic crust; such maps confirmed the origin of oceanic crust by sea floor spreading. (From Muller et al., 1997; with permission of the American Geophysical Union.) (For color version, see Plate 1.10, opposite p. 248.)

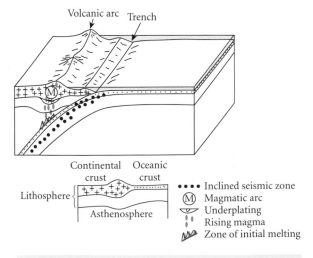

Figure 1.11   Convergent plate boundary, showing a trench–arc system, inclined seismic zone and subduction of oceanic lithosphere.

subducted plate. This produces stresses in the cool interior of the subducted lithosphere that generate earthquakes (Figure 1.11) along an **inclined seismic (Wadati–Benioff) zone** that marks the path of the subducted plate as it descends into the asthenosphere. The three largest magnitude earthquakes in the past century occurred along inclined seismic zones beneath Chile (1909), Alaska (1964) and Sumatra (2004). The latter event produced the devastating Banda Aceh tsunami which killed some 300,000 people in the Indian Ocean region.

What is the ultimate fate of subducted slabs? Earthquakes occur in subducted slabs to a depth of 660 km, so we know slabs reach the base of the asthenosphere transition zone. Earthquake records suggest that some slabs flatten out as they reach this boundary, indicating that they may not penetrate below this. Seismic tomography, which images three-dimensional variations in seismic wave velocity within the mantle, has shed some light on this question, while raising many questions. A consensus is emerging (Hutko et al., 2006) that some subducted slabs become dense enough to sink all the way to the core–mantle boundary where they contribute material to the D″ layer. These recycled slabs may ultimately be involved in the formation of mantle plumes, as suggested by Jeanloz (1993).

Subduction zones produce a wide range of distinctive Earth materials. The increase in temperature and pressure within the subducted plate causes it to undergo significant metamorphism. The upper part of the subducted slab, in contact with the hot asthenosphere, releases fluids as it undergoes metamorphism which triggers partial melting. A complex set of melts rise from this region to produce **volcanic–magmatic arcs**. These melts range in composition from basaltic–gabbroic through dioritic–andesitic and may differentiate or be contaminated to produce melts of granitic–rhyolitic composition. Melts that reach the surface produce volcanic arcs such as those that characterize the "ring of fire" of the Pacific Ocean basin. Mt St. Helens in Washington, Mt Pinatubo in the Philippines, Mt Fuji in Japan and Krakatau in Indonesia are all examples of composite volcanoes that mark the volcanic arcs that form over Pacific Ocean subduction zones.

When magmas intrude the crust they also produce plutonic igneous rocks that add new continental crust to the Earth. Most of the world's major **batholith belts** represent plutonic magmatic arcs, subsequently exposed by erosion of the overlying volcanic arc. In addition, many of Earth's most important ore deposits are produced in association with volcanic–magmatic arcs over subduction zones.

Many of the magmas generated over the subducted slab cool and crystallize at the base of the lithosphere, thickening it by underplating. Underplating and intrusion are two of the major sets of processes by which new continental crust is generated by the solidification of melts. Once produced, the density of continental crust is generally too low for it to be subducted. This helps to explain the great age that continental crust can achieve (>4.0 Ga).

Areas of significant relief, such as trench–arc systems, are ideal sites for the production and accumulation of detrital (epiclastic) sedimentary rocks. Huge volumes of detrital sedimentary rocks produced by the erosion of volcanic and magmatic arcs are deposited in forearc and backarc basins (Figure 1.12). They also occur with deformed abyssal sediments in the forearc subduction complex. As these sedimentary rocks are buried and deformed, they are metamorphosed.

Continental collisions

As ocean basins shrink by subduction, portions of the ridge system may be subducted. Once the ridge is subducted, growth of the ocean basin by sea floor spreading ceases, the ocean basin continues to shrink by subduction, and the continents on either side are brought closer together as subduction proceeds. Eventually they converge to produce a continental collision.

When a **continental collision** occurs (Dewy and Bird, 1970), subduction ceases, because continental lithosphere is too buoyant to be subducted to great depths. The continental lithosphere involved in the collision may be part of a continent, a microcontinent or a volcanic–magmatic arc. As convergence continues, the margins of both continental plates are compressed and shortened horizontally and thickened vertically in a manner analogous to what happens to two vehicles in a head-on collision. In the case of continents colliding at a convergent plate boundary, however, the convergence continues for millions of years resulting in a severe horizontal shortening and vertical thickening which results in the progressive uplift of a mountain belt and/or extensive elevated plateau that mark the closing of an ancient ocean basin (Figure 1.13).

Long mountain belts formed along convergent plate boundaries are called **orogenic belts**. The increasing weight of the thickening orogenic belt causes the adjacent continental lithosphere to bend downward to produce **foreland basins**. Large amounts of detrital sediments derived from the erosion of the mountain belts are deposited in such basins. In addition, increasing temperatures and pressures within the thickening orogenic belt cause regional metamorphism of the

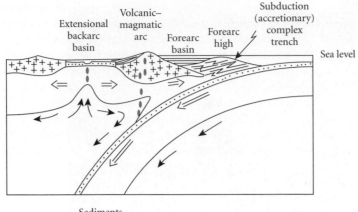

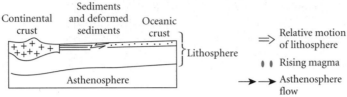

Figure 1.12 Subduction zone depicting details of sediment distribution, sedimentary basins and volcanism in trench–arc system forearc and backarc regions.

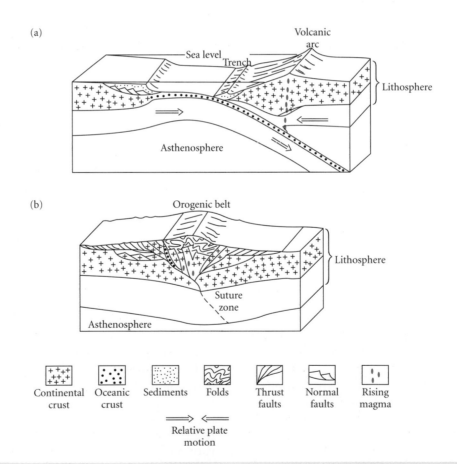

Figure 1.13 (a) Ocean basins shrink by subduction, as continents on two plates converge. (b) Continental collision produces a larger continent from two continents joined by a suture zone. Horizontal shortening and vertical thickening are accommodated by folds and thrust faults in the resulting orogenic belt.

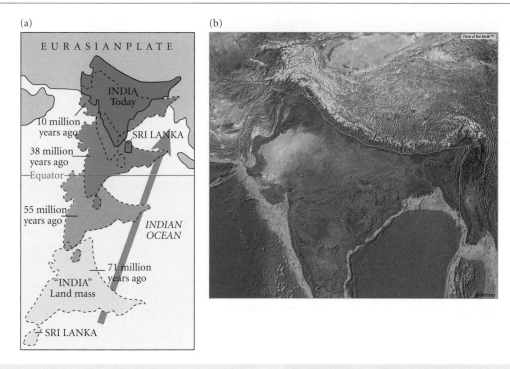

Figure 1.14 (a) Diagram depicting the convergence of India and Asia which closed the Tethys Ocean. (Courtesy of NASA.) (b) Satellite image of southern Asia showing the indentation of Eurasia by India, the uplift of Himalayas and Tibetan Plateau and the mountains that "wrap around" India. (Courtesy of UNAVCO.)

rocks within it. If the temperatures become high enough, partial melting may occur to produce melts in the deepest parts of orogenic belts that rise to produce a variety of igneous rocks.

The most striking example of a modern orogenic belt is the Himalayan Mountain range formed by the collision of India with Eurasia over the past 40 Ma. The continued convergence of the Indian microcontinent with Asia has resulted in shortening and regional uplift of the Himalayan mountain belt along a series of major thrust faults and has produced the Tibetan Plateau. Limestones near the summit of Mt Everest (Chomolungma) were formed on the floor of the Tethys Ocean that once separated India and Asia, and were then thrust to an elevation of nearly 9 km as that ocean was closed and the Himalayan Mountain Belt formed by continental collision. The collision has produced tectonic indentation of Asia, resulting in mountain ranges that wrap around India (Figure 1.14). The Ganges River in northern India flows approximately west–east in a trough that represents a modern foreland basin.

Continental collision inevitably produces a larger continent. It is now recognized that supercontinents such as Pangea and Rodinia were formed as the result of collisional tectonics. **Collisional tectonics** only requires converging plates whose leading edges are composed of lithosphere that is too buoyant to be easily subducted. In fact all the major continents display evidence of being composed of a collage of terranes that were accreted by collisional events at various times in their histories.

1.4.3 Transform plate boundaries

In order for plates to be able to move relative to one another, a third type of plate boundary is required. **Transform plate boundaries** are characterized by horizontal motion, along transform fault systems, which is parallel to the plate boundary segment that separates two plates (see Figure 1.4c). Because the rocks

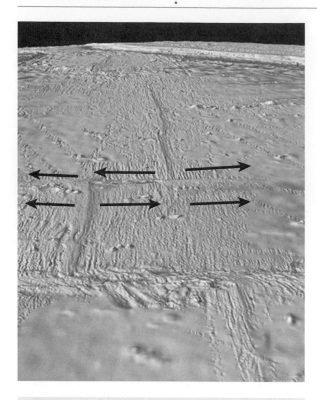

Figure 1.15 Transform faults offsetting ridge segments on the eastern Pacific Ocean floor off Central America. Arrows show the directions of sea floor spreading away from the ridge. Portions of the fracture zones between the ridge segments are transform plate boundaries; portions beyond the ridge segments on both sides are intraplate transform scars. (Courtesy of William Haxby, LDEO, Columbia University.) (For color version, see Plate 1.15, between pp. 248 and 249.)

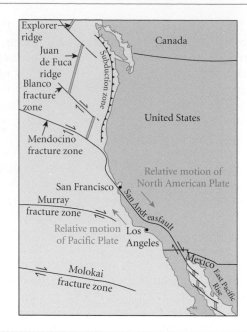

Figure 1.16 Fracture zones, transform faults and ridge segments in the eastern Pacific Ocean and western North America. The San Andreas Fault system is a continental transform fault plate boundary. (Courtesy of the US Geological Survey.)

on either side slide horizontally past each other, transform fault systems are a type of strike-slip fault system.

Transform faults were first envisioned by J. T. Wilson (1965) to explain the seismic activity along fracture zones in the ocean floor. **Fracture zones** are curvilinear zones of intensely faulted, fractured oceanic crust that are generally oriented nearly perpendicular to the ridge axis (Figure 1.15). Despite these zones having been fractured by faulting along their entire length, earthquake activity is largely restricted to the transform portion of fracture zones that lies between offset ridge segments. Wilson (1965) reasoned that if sea

floor was spreading away from two adjacent ridge segments in opposite directions, the portion of the fracture zone between the two ridge segments would be characterized by relative motion in opposite directions. This would produce shear stresses resulting in strike-slip faulting of the lithosphere, frequent earthquakes and the development of a transform fault plate boundary. The exterior portions of fracture zones outside the ridge segments represent oceanic crust that was faulted and fractured when it was between ridge segments, then carried beyond the adjacent ridge segment by additional sea floor spreading. These portions of fracture zones are appropriately called healed transforms or **transform scars**. They are no longer plate boundaries; they are intraplate features because the sea floor on either side is spreading in the same direction (Figure 1.15).

Transform plate boundaries also occur in continental lithosphere. The best known modern examples of continental transforms include the San Andreas Fault system in California (Figure 1.16), the Alpine Fault system in New Zealand and the Anatolian Fault systems in Turkey and Iran. All these are characterized by active strike-slip fault systems of the type that characterize transform plate boundaries. In places where such faults bend or where their tips overlap, deep **pull-apart basins** may develop in which thick accumulations of sedimentary rocks accumulate rapidly.

Plates cannot simply diverge and converge; they must be able to slide past each other in opposite directions in order to move at all. Transform plate boundaries serve to accommodate this required sense of motion. Small amounts of igneous rocks form along transform plate boundaries, especially those hybrids that have a component of divergence or convergence as well. They produce much smaller volumes of igneous and metamorphic rocks than are formed along divergent and convergent plate boundaries.

1.5 HOTSPOTS AND MANTLE CONVECTION

Hotspots (Wilson, 1963) are long-lived areas in the mantle where anomalously large volumes of magma are generated. They occur beneath both oceanic lithosphere (e.g., Hawaii) and continental lithosphere (e.g., Yellowstone National Park, Wyoming) as well as along divergent plate boundaries (e.g., Iceland). Wilson pointed to linear seamount chains, such as the Hawaiian Islands (Figure 1.17), as surface expressions of hotspots. At any one time, volcanism is restricted to that portion of the plate that lies above the hotspot. As the plate continues to move, older volcanoes are carried away from the fixed hotspot and new volcanoes are formed above it. The age of these seamount chains increases systematically away from the hotspot in the direction of plate motion. For the Hawaiian chain, the data suggest a west–northwest direction of plate motion for the last 45 Ma. However, a change in orientation of the seamount chain to just west of north for older volcanoes suggests that the seafloor may have spread over the hotspot in a more northerly direction prior to 45 Ma. A similar trend of volcanism of increasing age extends southwestward from the Yellowstone Caldera.

In the early 1970s, Morgan (1971) and others suggested that hotspots were the surface expression of fixed, long-lived mantle

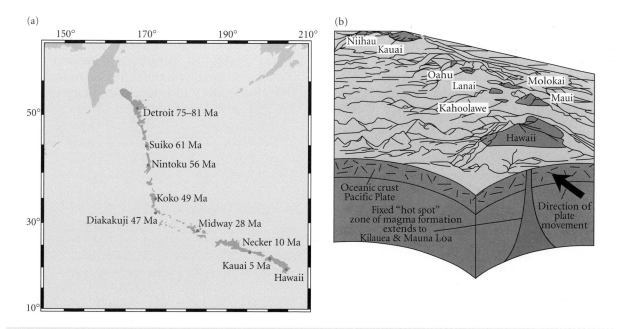

Figure 1.17   (A) Linear seamount chain formed by plate movement over the Hawaiian hotspot and/or hotspot motion. (After Tarduno et al., 2009; with permission of *Science Magazine*.) (B) "Fixed" mantle plume feeding the surface volcanoes of the Hawaiian chain. (Courtesy of the US Geological Survey.)

plumes. **Mantle plumes** were hypothesized to be columns of warm material that rose from near the core–mantle boundary. Later workers hypothesized that deep mantle plumes originate in the ULVZ of the D″ layer at the base of the mantle and may represent the dregs of subducted slabs warmed sufficiently by contact with the outer core to become buoyant enough to rise. Huge **superplumes** (Larson, 1991) were hypothesized to be significant players in extinction events, the initiation of continental rifting, and in the **supercontinent cycle** (Sheridan, 1987) of rifting and collision that has caused supercontinents to form and rift apart numerous times during Earth's history. Eventually most intraplate volcanism and magmatism was linked to hotspots and mantle plumes.

The picture has become considerably muddled over the past decade. Many Earth scientists have offered significant evidence that mantle plumes do not exist (Foulger et al., 2005). For example, there is no seismic velocity evidence for a deep plume source beneath the Yellowstone hotspot. Others have suggested that mantle plumes exist, but are not fixed (Nataf, 2000; Koppers et al., 2001; Tarduno et al., 2009). Still others (Nolet et al., 2006) suggest on the basis of fine-scale thermal tomography that some of these plumes originate near the core–mantle boundary, others at the base of the transition zone (660 km) and others at around 1400 km in the mesosphere. They suggest that the rise of some plumes from the deep mantle is interrupted by the 660 km discontinuity, whereas other plumes seem to cross this discontinuity. This is reminiscent of the behavior of subducted slabs, some of which spread out above the 660 km discontinuity, whereas other penetrate it and apparently sink to the core–mantle boundary. It is likely that hotspots are generated by a variety of processes related to mantle convection patterns that are still not well understood. Stay tuned; this will be an exciting area of Earth research over the coming decade.

We have attempted to provide a spatial and tectonic context for the processes that determine which Earth materials will form where. One part of this context involves the location of compositional and mechanical layers within the geosphere where Earth materials form. Ultimately, however, the geosphere cannot be viewed as a group of static layers. Plate tectonics implies significant horizontal and vertical movement of the lithosphere with compensating motion of the underlying asthenosphere and deeper mantle. Global tectonics suggests significant lateral heterogeneity within layers and significant vertical exchange of material between layers caused by processes such as convection, subduction and mantle plumes.

Helping students to understand how variations in composition, position within the geosphere and tectonic processes interact on many scales to generate distinctive Earth materials is the fundamental task of this book. We hope you will find what follows is both exciting and meaningful.

Chapter 2

Atoms, elements, bonds and coordination polyhedra

If we zoom in on any portion of Earth, we will see that it is composed of progressively smaller entities. At very high magnification, we will be able to discern very small particles called atoms. Almost all Earth materials are composed of atoms that strongly influence their properties. Understanding the ways in which these basic chemical constituents combine to produce larger scale Earth materials is essential to understanding our planet.

In this chapter we will consider the fundamental chemical constituents that bond together to produce Earth materials such as minerals and rocks. We will discuss the nucleus and electron configuration of atoms and the role these play in determining both atomic and mineral properties and the conditions under which minerals form. This information will provide a basis for understanding how and why minerals, rocks and other Earth materials have the following characteristics:

1 They possess the properties that characterize and distinguish them.
2 They provide benefits and hazards through their production, refinement and use.
3 They form in response to particular sets of environmental conditions and processes.
4 They record the environmental conditions and processes that produce them.
5 They permit us to infer significant events in Earth's history.

2.1 ATOMS

Earth materials are composed of smaller entities. At very high magnification, we are able to discern particles called atoms whose effective diameters are a few angstroms (1 angstrom = 10^{-10} m). These tiny entities, in turn, consist of three main particles – electrons, protons and neutrons – which were discovered between 1895 and 1902. The major properties of electrons, protons and neutrons are summarized in Table 2.1.

Protons (p^+) and **neutrons** (n^0) each have a mass of ~1 amu and are clustered together in a small, positively-charged, central region

Earth Materials, 1st edition. By K. Hefferan and J. O'Brien. Published 2010 by Blackwell Publishing Ltd.

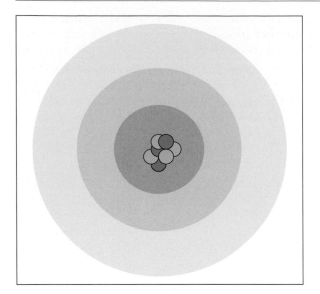

Figure 2.1 Model atom with nucleus that contains positively-charged protons (dark blue) and electrically neutral neutrons (light blue) surrounded by an electron cloud (gray shades) in which negatively-charged electrons move in orbitals about the nucleus.

Table 2.1 The major properties of electrons, protons and neutrons.

Particle type	Electric charge	Atomic mass (amu)*
Proton (p^+)	+1	1.00728
Neutron (n^0)	0	1.00867
Electron (e^-)	−1	0.0000054

* amu = atomic mass unit = 1/12 mass of an average carbon atom.

called the **nucleus** (Figure 2.1). Protons possess a positive electric charge and neutrons are electrically neutral. The nucleus is surrounded by a vastly larger, mostly "empty", region called the **electron cloud**. The electron cloud represents the area in which the **electrons** (e^-) in the atom move in orbitals about the nucleus (Figure 2.1). Electrons have a negative electric charge and an almost negligible mass of 0.0000054 amu. Knowledge of these three fundamental particles in atoms is essential to understanding how minerals and other materials form, how they can be used as resources and how we can deal with their sometimes hazardous effects.

2.1.1   The nucleus, atomic number and atomic mass number

The nucleus of atoms is composed of positively-charged protons and uncharged neutrons bound together by a strong force. Ninety-two fundamentally different kinds of atoms called elements have been discovered in the natural world. More than 20 additional elements have been created synthetically in laboratory experiments during the past century. Each **element** is characterized by the number of protons in its nucleus. The number of protons in the nucleus, called the **atomic number**, is symbolized by the letter **Z**. The atomic number (Z) is typically represented by a subscript number to the lower left of the element symbol. The 92 naturally occurring elements range from hydrogen (Z = 1) through uranium (Z = 92). Hydrogen ($_1$H) is characterized by having one proton in its nucleus. Every atom of uranium ($_{92}$U) contains 92 protons in its nucleus. The atomic number is what distinguishes the atoms of each element from atoms of all other elements.

Every atom also possesses mass that largely results from the protons and neutrons in its nucleus. The mass of a particular atom is called its **atomic mass number**, and is expressed in atomic mass units (amu). As the mass of both protons and neutrons is ~1 amu, the atomic mass number is closely related to the total number of protons plus neutrons in its nucleus. The simple formula for atomic mass number is:

$$\text{Atomic mass number} = \text{number of protons}$$
$$\text{plus number of neutrons} = p^+ + n^0$$

The atomic mass number is indicated by a superscript number to the upper left of the element symbol. For example, most oxygen atoms have eight protons and eight neutrons so the atomic mass number is written ^{16}O.

Isotopes

Although each element has a unique atomic number, many elements are characterized by atoms with different atomic mass numbers. Atoms of the same element that possess different atomic mass numbers are called **isotopes**. For example, three different isotopes of hydrogen exist (Fig. 2.2A). All hydrogen ($_1$H)

(a)

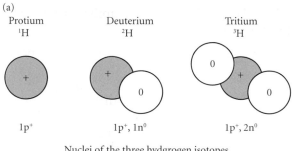

Nuclei of the three hydgrogen isotopes

(b)

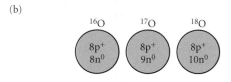

Nuclei of the three oxygen isotopes

Figure 2.2 (a) Nuclear configurations of the three common isotopes of hydrogen. (b) Nuclear configurations of the three common isotopes of oxygen.

trons to the lower right and the atomic mass number (number of protons + number of neutrons) to the upper left. For example, the most common isotope of uranium has the symbolic nuclear configuration of 92 protons +146 neutrons and an atomic mass number of 238:

$$^{238}_{92}U_{146}$$

Stable isotopes have stable nuclei that tend to remain unchanged; they retain the same number of protons and neutrons over time. On the other hand, **radioactive isotopes** have unstable nuclear configurations (numbers of protons and neutrons) that spontaneously change over time via radioactive decay processes, until they achieve stable nuclear configurations. Both types of isotopes are extremely useful in solving geological and environmental problems, as discussed in Chapter 3. Radioactive isotopes also present serious environmental hazards.

2.1.2   The electron cloud

Electrons are enigmatic entities, with properties of both particles and wave energy, that move very rapidly around the nucleus in ultimately unpredictable paths. Our depiction of the electron cloud is based on the probabilities of finding a particular electron at a particular place. The wave-like properties of electrons help to define the three-dimensional shapes of their electron clouds, known as **orbitals**. The size and shape of the electron cloud defines the chemical behavior of atoms and ultimately the composition of all Earth materials they combine to form. Simplified models of the electron cloud depict electrons distributed in spherical orbits around the nucleus (Figure 2.3); the reality is much more complex. Because the electron cloud largely determines the chemical behavior of atoms and how they combine to produce Earth materials, it is essential to understand some fundamental concepts about it.

Every electron in an atom possesses a unique set of properties that distinguishes it from all the other electrons in that atom. An individual electron's identity is given by properties that include its principal quantum number, its azimuthal quantum number, its

atoms have an atomic number of 1. The common form of hydrogen atom, sometimes called **protium**, has one proton and no neutrons in the nucleus; therefore protium has an atomic mass number of 1, symbolized as ^1H. A less common form of hydrogen called **deuterium**, used in some nuclear reactors, has an atomic mass number of 2, symbolized by ^2H. This implies that it contains one proton and one neutron in its nucleus ($1p^+ + 1n^0$). A rarer isotope of hydrogen called **tritium** has an atomic mass number of 3, symbolized by ^3H. The nucleus of tritium has one proton and two neutrons. Similarly oxygen occurs in three different isotopes: ^{16}O, ^{17}O and ^{18}O. All oxygen atoms contain eight protons but neutron numbers vary between ^{16}O, ^{17}O and ^{18}O, which contain eight, nine and 10 neutrons, respectively (Fig. 2.2B). The average atomic mass for each element is the weighted average for all the isotopes of that element. This helps to explain why the listed atomic masses for each element do not always approximate the whole numbers produced when one adds the number of protons and neutrons in the nucleus.

The general isotope symbol for the nucleus of an atom expresses its atomic number to the lower left of its symbol, the number of neu-

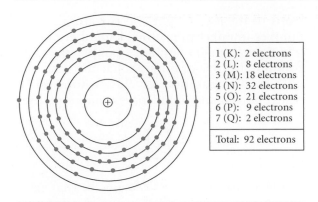

1 (K): 2 electrons
2 (L): 8 electrons
3 (M): 18 electrons
4 (N): 32 electrons
5 (O): 21 electrons
6 (P): 9 electrons
7 (Q): 2 electrons

Total: 92 electrons

Figure 2.3 Distribution of electrons in the principal quantum levels ("electron shells") of uranium.

Table 2.2 Quantum designations of electrons in the 92 naturally occurring elements. The numbers refer to the principal quantum region occupied by the electrons within the electron cloud; small case letters refer to the subshell occupied by the electrons.

Principal quantum number	Subshell description	Number of electrons
1 (K)	1s	2
2 (L)	2s	2
	2p	6
3 (M)	3s	2
	3p	6
	3d	10
4 (N)	4s	2
	4p	6
	4d	10
	4f	14
5 (O)	5s	2
	5p	6
	5d	10
	5f	14
6 (P)	6s	2
	6p	6
7 (Q)	7s	2
		Total = 92

magnetic quantum number and its spin number. The **principal quantum number (n)** signifies the **principal quantum energy level** or "**shell**" in which a particular electron occurs. Principle quantum regions are numbered in order of increasing electron energies **1, 2, 3, 4, 5, 6 or 7** or alternatively lettered **K, L, M, N, O, P or Q**. These are arranged from low quantum number for low energy positions closer to the nucleus to progressively higher quantum number for higher energy positions farther away from the nucleus.

Each principal quantum level contains electrons with one or more **azimuthal quantum numbers** which signify the directional quantum energy region or "**subshell**" in which the electron occurs. This is related to the angular momentum of the electron and the shape of its orbital. Azimuthal quantum numbers or subshells are labeled **s, p, d and f**. The number of electrons in the highest principal quantum level s and p subshells largely determines the behavior of chemical elements. Table 2.2 summarizes the quantum properties of the electrons that can exist in principle quantum shells 1 through 7.

Atomic nuclei were created largely during the "big bang" and subsequently by fusion reactions between protons and neutrons in the interior of stars and in supernova. When elements are formed, electrons are added to the lowest available quantum level in numbers equal to the number of protons in the nucleus. Electrons are added to the atoms in a distinct

sequence, from lowest quantum level electrons to highest quantum level electrons. The relative quantum energy of each electron is shown in Figure 2.4.

The **diagonal rule** is a simple rule for remembering the sequence in which electrons are added to the electron cloud. The order in which electrons are added to shells is depicted by a series of diagonal lines in Figure 2.5, each from 1s upper right to 7p lower left.

Table 2.3 shows the ground state electron configurations for the elements. One can write the electron configuration of any element in a sequence from lowest to highest energy electrons. For example, calcium (Z = 20) possesses the electron configuration $1s_2$, $2s_2$, $2p_6$, $3s_2$, $3p_6$, $4s_2$. Elements with principal quantum levels (shells) or azimuthal quantum s– and p– subshells that are completely filled (that is they contain the maximum number of electrons possible) possess very stable electron configurations. These elements include the noble gas elements such as helium (He), neon (Ne), argon (Ar) and krypton (Kr) which,

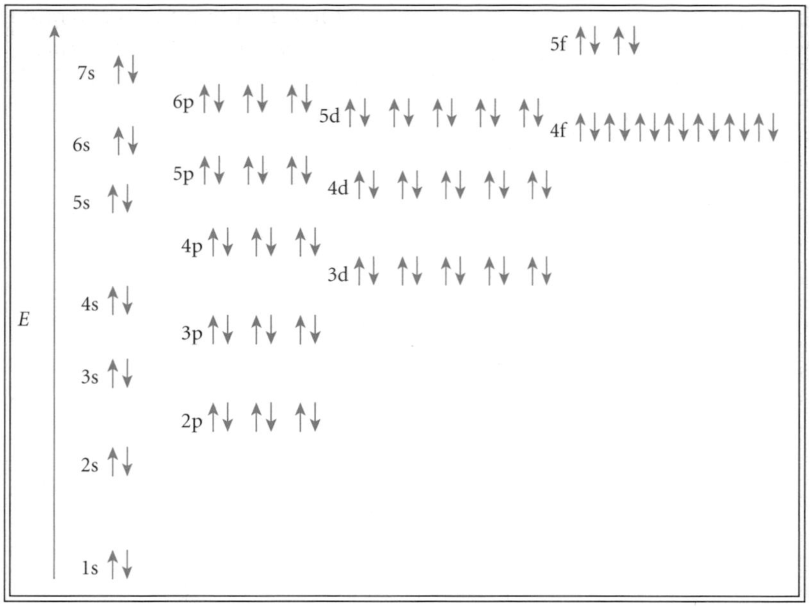

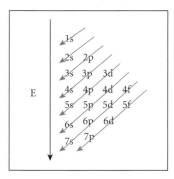

Figure 2.4 The quantum properties of electrons in the 92 naturally occurring elements, listed with increasing quantum energy (E) from bottom to top.

Figure 2.5 The diagonal rule for determining the sequence in which electrons are added to the electron cloud.

because of their stable configurations, tend not to react with or bond to other elements. For elements other than helium, the highest quantum level stable configuration is s_2, p_6, sometimes referred to as the "**stable octet**".

2.2 THE PERIODIC TABLE

The naturally occurring and synthetic elements discovered to date display certain periodic traits; that is several elements with different atomic numbers display similar chemical behavior. Tables that attempt to portray the periodic behavior of the elements are called **periodic tables**. It is now well known that the periodic behavior of the elements is related to their electron configurations. In most modern periodic tables (Table 2.3) the elements are arranged in **seven rows or periods** and **eighteen columns or groups**. Two sets of elements, the **lanthanides** and the **actinides**, which belong to the sixth and seventh rows, respectively, are listed separately at the bottom of such tables to allow all the elements to be shown conveniently on a printed page of standard dimensions. For a rather different approach to organizing the elements in a periodic table for Earth scientists, readers are referred to Railsbach (2003).

2.2.1 Rows (periods) on the periodic table

On the left-hand side of the periodic table the row numbers 1 to 7 indicate the highest principle quantum level in which electrons occur in the elements in that row. Every element in a given horizontal row has its outermost electrons in the same energy level. Within each row, the number of electrons increases with

Table 2.3 Periodic table of the naturally-occurring elements displaying atomic symbols, atomic number (Z), average mass, ground state electron configuration, common valence states and electronegativity of each element.

IA 1	IIA 2	IIIB 3	IVB 4	VB 5	VIB 6	VIIB 7	VIIIB 8	9	10
1 2.20 H 1.008 $(1s_1)$ (±1)									
3 0.98 Li 6.941 $(He + 2s_1)$ $(+1)$	**4** 1.57 Be 9.012 $(He + 2s_2)$ $(+2)$								
11 0.93 Na 22.990 $(Ne + 3s_1)$ $(+1)$	**12** 1.31 Mg 24.305 $(Ne + 3s_2)$ $(+2)$								
19 0.82 K 39.098 $(Ar + 4s_1)$ $(+1)$	**20** 1.00 Ca 40.080 $(Ar + 4s_2)$ $(+2)$	**21** 1.36 Sc 44.956 $(Ar + 4s_2 + 3d_1)$ $(+3)$	**22** 1.54 Ti 47.900 $(Ar + 4s_2 + 3d_2)$ $(+4,+2)$	**23** 1.63 V 50.942 $(Ar + 4s_2 + 3d_3)$ (many)	**24** 1.66 Cr 51.996 $(Ar + 4s_1 + 3d_5)$ $(+3,+6)$	**25** 1.55 Mn 54.938 $(Ar + 4s_2 + 3d_5)$ (many)	**26** 1.63 Fe 55.847 $(Ar + 4s_2 + 3d_6)$ $(+2,+3)$	**27** 1.85 Co 58.933 $(Ar + 4s_2 + 3d_7)$ $(+2,+3)$	**28** 1.91 Ni 58.700 $(Ar + 4s_2 + 3d_8)$ $(+2)$
37 0.82 Rb 85.468 $(Kr + 5s_1)$ $(+1)$	**38** 0.95 Sr 87.620 $(Kr + 5s_2)$ $(+2)$	**39** 1.22 Y 88.906 $(Kr + 5s_2 + 4d_1)$ $(+3)$	**40** 1.33 Zr 91.220 $(Kr + 5s_2 + 4d_2)$ $(+4,+3)$	**41** 1.60 Nb 92.906 $(Kr + 5s_1 + 4d_4)$ (many)	**42** 2.16 Mo 95.940 $(Kr + 5s_1 + 4d_5)$ (many)	**43** 1.90 Tc (98) $(Kr + 5s_2 + 4d_5)$ (many)	**44** 2.20 Ru 101.07 $(Kr + 5s_1 + 4d_7)$ (many)	**45** 2.28 Rh 102.91 $(Kr + 5s_1 + 4d_8)$ (many)	**46** 2.20 Pd 106.40 $(Kr + 5s_2 + 4d_8)$ $(+2,+4)$
55 0.79 Cs 132.91 $(Xe + 6s_1)$ $(+1)$	**56** 0.89 Ba 137.33 $(Xe + 6s_2)$ $(+2)$	**57** 1.10 La 138.91 $(Xe + 6s_2 + 5d_1)$ $(+3)$	**72** 1.30 Hf 178.48 $(Xe + 6s_2 + 4f_{14}5d_2)$ $(+4)$	**73** 1.50 Ta 180.95 $(Xe + 6s_2 + 4f_{14}5d_3)$ $(+5)$	**74** 2.36 W 183.85 $(Xe + 6s_2 + 4f_{14}5d_4)$ (many)	**75** 1.90 Re 186.21 $(Xe + 6s_2 + 4f_{14}5d_5)$ (many)	**76** 2.12 Os 190.20 $(Xe + 6s_2 + 4f_{14}5d_6)$ (many)	**77** 2.20 Ir 192.22 $(Xe + 6s_2 + 4f_{14}5d_7)$ (many)	**78** 2.28 Pt 195.09 $(Xe + 6s_1 + 4f_{14}5d_9)$ $(+2,+4)$
87 0.70 Fr (223) $(Rn + 7s_1)$ $(+1)$	**88** 0.87 Ra 226.03 $(Rn + 7s_2)$ $(+2)$	**89** 1.10 Ac 227.03 $(Rn + 7s_2 + 6d_1)$ $(+3)$							

	58 1.12 Ce 140.12 $(Xe + 6s_2 + 5d_14f_1)$ $(+3,+4)$	59 1.13 Pr 140.91 $(Xe + 6s_2 + 5d_14f_2)$ $(+3,+4)$	60 1.14 Nd 144.24 $(Xe + 6s_2 + 5d_14f_3)$ $(+3)$	61 1.13 Pm (145) $(Xe + 6s_2 + 5d_14f_4)$ $(+3)$	62 1.17 Sm 150.40 $(Xe + 6s_2 + 5d_14f_5)$ $(+3,+2)$	63 1.20 Eu 151.96 $(Xe + 6s_2 + 5d_14f_6)$ $(+3,+2)$	64 1.20 Gd 157.25 $(Xe + 6s_2 + 5d_14f_7)$ $(+3)$	65 1.20 Tb 158.93 $(Xe + 6s_2 + 5d_14f_8)$ $(+3,+4)$
Lanthanides								
Actinides	90 1.30 Th 232.04 $(Rn + 7s_2 + 6d_15f_1)$ $(+4)$	91 1.30 Pa 231.04 $(Rn + 7s_2 + 5d_14f_2)$ $(+5,+4)$	92 1.38 U 238.03 $(Rn + 7s_2 + 5d_14f_3)$ (many)					

IB 11	IIB 12	IIIA 13	IVA 14	VA 15	VIA 16	VIIA 17	VIIIA 18
							2 ---- He 4.003 $(1s_2)$ (0)
		5 2.04 B 10.810 $(He + 2s_2 2p_1)$ (+3)	**6** 2.55 C 12.011 $(He + 2s_2 2p_2)$ (+4, 0)	**7** 3.04 N 14.007 $(He + 2s_2 2p_3)$ (many)	**8** 3.44 O 15.999 $(He + 2s_2 2p_4)$ (-2)	**9** 3.95 F 18.998 $(He + 2s_2 2p_5)$ (-1)	**10** ---- Ne 20.179 $(He + 2s_2 2p_6)$ (0)
		13 Al Al 26.962 $(Ne + 3s_2 3p_1)$ (+3)	**14** Si Si 28.086 $(Ne + 3s_2 3p_2)$ (+4)	**15** P P 30.974 $(Ne + 3s_2 3p_3)$ (many)	**16** S S 32.060 $(Ne + 3s_2 2p_4)$ (-2, +6)	**17** Cl Cl 35.453 $(Ne + 3s_1 3p_5)$ (-1)	**18** Ar Ar 39.948 $(Ne + 3s_2 3p_6)$ (0)
29 1.90 Cu 63.546 $(Ar + 4s_1 + 3d_{10})$ (+1,+2)	**30** 1.65 Zn 65.380 $(Ar + 4s_2 + 3d_{10})$ (+2)	**31** 1.81 Ga 69.720 $(Ar + 4s_2 4p_1 + 3d_{10})$ (+3)	**32** 2.01 Ge 72.590 $(Ar + 4s_2, 4p_2 + 3d_{10})$ (+4)	**33** 2.18 As 74.922 $(Ar + 4s_2, 4p_3 + 3d_{10})$ (many)	**34** 2.55 Se 78.960 $(Ar + 4s_2, 4p_4 + 3d_{10})$ (-2,+6)	**35** 2.96 Br 79.904 $(Ar + 4s_2, 4p_5 + 3d_{10})$ (-1)	**36** ---- Kr 83.800 $(Ar + 4s_2, 4p_6 + 3d_{10})$ (0)
47 1.93 Ag 107.87 $(Kr + 5s_1 + 4d_{10})$ (+1)	**48** 1.69 Cd 112.41 $(Kr + 5s_2 + 4d_{10})$ (+2)	**49** 1.78 In 114.82 $(Kr + 5s_2 5p_1 + 4d_{10})$ (+3)	**50** 1.96 Sn 118.69 $(Kr + 5s_2 5p_2 + 4d_{10})$ (+4,+2)	**51** 2.05 Sb 121.75 $(Kr + 5s_2 5p_3 + 4d_{10})$ (+5,+3)	**52** 2.10 Te 127.60 $(Kr + 5s_2 5p_4 + 4d_{10})$ (-2,+6)	**53** 2.66 I 126.90 $(Kr + 5s_2 5p_5 + 4d_{10})$ (-1)	**54** ---- Xe 131.30 $(Kr + 5s_2 5p_6 + 4d_{10})$ (0)
79 2.54 Au 196.97 $(Xe + 6s_1 + 4f_{14} 5d_{10})$ (+1,+3)	**80** 2.00 Hg 200.59 $(Xe + 6s_2 + 4f_{14} 5d_{10})$ (+2,+1)	**81** 2.04 Tl 204.37 $(Xe + 6s_2 6p_1 + 4f_{14} 5d_{10})$ (+3,+1)	**82** 2.33 Pb 208.98 $(Xe + 6s_2 6p_2 + 4f_{14} 5d_{10})$ (+4,+2)	**83** 2.02 Bi 208.98 $(Xe + 6s_2 6p_3 + 4f_{14} 5d_{10})$ (+5,+3)	**84** 2.00 Po (209) $(Xe + 6s_2 6p_4 + 4f_{14} 5d_{10})$ (+4,+2)	**85** 2.20 At (210) $(Xe + 6s_2 6p_5 + 4f_{14} 5d_{10})$ (many)	**86** ---- Rn (222) $(Xe + 6s_2 6p_6 + 4f_{14} 5d_{10})$ (0)

66 1.22 Dy 162.50 $(Xe + 6s_2 + 5d_1 4f_9)$ (+3)	**67** 1.23 Ho 164.93 $(Xe + 6s_2 + 5d_1 4f_{10})$ (+3)	**68** 1.24 Er 167.29 $(Xe + 6s_2 + 5d_1 4f_{11})$ (+3)	**69** 1.25 Tm 168.94 $(Xe + 6s_2 + 5d_1 4f_{12})$ (+3,+2)	**70** 1.10 Yb 173.04 $(Xe + 6s_2 + 5d_1 4f_{13})$ (+3,+2)	**71** 1.27 Lu 174.97 $(Xe + 6s_2 + 5d_1 4f_{14})$ (+3)

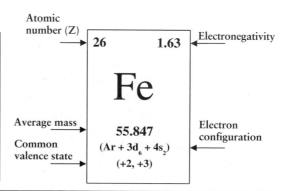

Atomic number (Z) → **26**   **1.63** ← Electronegativity

Fe

Average mass → **55.847**   Electron configuration

Common valence state → $(Ar + 3d_6 + 4s_2)$ (+2, +3)

the atomic number from left to right. The number of elements in each row varies, and reflects the sequence in which electrons are added to various quantum levels as the atoms are formed. For example, row 1 has only two elements because the first quantum level can contain only two 1s electrons. The two elements are hydrogen ($1s_1$) and helium ($1s_2$). Row 2 contains eight elements that reflect the progressive addition of 2s, then 2p electrons during the formation of lithium (helium + $2s_1$) through neon (helium + $2s_2$, $2p_6$). Row 3 contains eight elements that reflect the filling of the 3s and 3p quantum regions respectively during the addition of electrons in sodium (neon + $3s_1$) through argon (neon + $3s_2$, $3p_6$) as indicated in Table 2.3. The process continues through rows 6 and 7 ending with uranium. In summary, elements are grouped into rows on the periodic table according to the highest ground state quantum level (1–7)

occupied by their electrons. Their position within each row depends on the distribution and numbers of electrons within the principle quantum levels.

2.2.2 Ionization

The periodic table not only organizes the elements into rows based on their electron properties, but also into vertical columns based upon their tendency to gain or lose electrons in order to become more stable, thereby forming atoms with a positive or negative charge (Table 2.3). Ideal atoms are electrically neutral because they contain the same numbers of positively charged protons and negatively charged electrons ($p^+ = e^-$). Many atoms are not electrically neutral; instead they are electrically charged particles called **ions**. The process by which they acquire their charge is called **ionization** (Box 2.1). In order for

---

Box 2.1 Ionization energy

Ionization energy (IE) is the amount of energy required to remove an electron from its electron cloud. Ionization energies are periodic as illustrated for 20 elements in Table B2.1.

Table B2.1 Ionization energies for hydrogen through calcium (units in kjoules/mole).

Element	Ionization energy							
	1st	2nd	3rd	4th	5th	6th	7th	8th
H	1312							
He	2372	5250						
Li	520	7297	11810					
Be	899	1757	14845	21000				
B	800	2426	3659	25020	32820			
C	1086	2352	4619	6221	37820	47260		
N	1402	2855	4576	7473	9452	53250	64340	
O	1314	3388	5296	7467	10987	13320	71320	84070
F	1680	3375	6045	8408	11020	15150	17860	91010
Ne	2080	3963	6130	9361	12180	15240	–	–
Na	496	4563	6913	9541	13353	16610	20114	26660
Mg	737	1451	7733	10540	13630	17995	21703	25662
Al	578	1817	2745	11575	14830	18376	23292	–
Si	787	1577	3231	4356	16091	19784	23783	–
P	1012	1903	2912	4956	6273	22233	25397	–
S	1000	2251	3361	4564	7012	8495	27105	–
Cl	1251	2297	3822	5160	6540	7458	11020	–
Ar	1520	2665	3931	5570	7238	8781	11995	–
K	418	3052	4220	5877	7975	9590	11343	14944
Ca	590	1145	4912	6491	8153	10496	12270	14206

Box 2.1 *Continued*

The first ionization energy is the amount of energy required to remove one electron from the electron cloud; the second ionization energy is the amount required to remove a second electron and so forth. Ionization energies are lowest for electrons that are weakly held by the nucleus and higher for electrons that are strongly held by the nucleus or are in stable configurations. Ionization energies decrease down the periodic table because the most weakly held outer electrons are shielded from the positively-charged nucleus by a progressively larger number of intervening electrons. Elements with relatively low first ionization energies are called **electropositive elements** because they tend to lose one or more electrons and become positively-charged cations. Most elements with high first ionization energies are **electronegative elements** because they tend to add electrons to their electron clouds and become negatively-charged anions. Since opposite charges tend to attract, you can imagine the potential such ions have for combining to produce other Earth materials. The arrangement of elements into vertical columns or groups within the periodic table helps us to comprehend the tendency of specific atoms to lose, gain or share electrons. For example, on the periodic table (see Table 2.3), column 2 (IIA) elements commonly exist as divalent (+2) cations because the first and second ionization energies are fairly similar and much lower than the third and higher ionization energies. This permits two electrons to be removed fairly easily from the electron cloud, but makes the removal of additional electrons much more difficult. Column 13 (IIIA) elements commonly exist as trivalent (+3) cations (see Table 2.4). These elements have somewhat similar first, second and third ionization energies, which are much smaller than the fourth and higher ionization energies. The transfer of electrons is fundamentally important in the understanding of chemical bonds and the development of mineral crystals.

an ion to form, the number of positively-charged protons and negatively-charged electrons must become unequal. **Cations** are positively-charged ions because they have more positively-charged protons than negatively-charged electrons ($p^+ > e^-$). Their charge is equal to the number of excess protons ($p^+ - e^-$). Cations form when electrons are lost from the electron cloud. Ions that have more negatively-charged electrons than positively-charged protons, such as the ion chlorine (Cl–), will have a negative charge and are called **anions**. The charge of an anion is equal to the number of excess electrons ($e^- - p^+$). Anions form when electrons are added to the electron cloud during ionization.

2.2.3 Ionization behavior of columns (groups) on the periodic table

The elements in every column or group on the periodic table (see Table 2.3) share a similarity in electron configuration that distinguishes them from elements in every other column. This shared property causes the elements in each group to behave in a similar manner during chemical reactions. As will be seen

later in this chapter and throughout the book, knowledge of these patterns is fundamental to understanding and interpreting the formation and behavior of minerals, rocks and other Earth materials. The tendency of atoms to form cations or anions is indicated by the location of elements in columns of the periodic table.

Metallic elements have relatively low first ionization energies (<900 kJ/mol) and tend to give up one or more weakly held electrons rather easily. Column 1 and 2 (group IA and group IIA) elements, the alkali metals and alkali earths, respectively, tend to display the most metallic behaviors. Many of the elements in columns 3–12 (groups IIIB through IIB), called the transition metals, also display metallic tendencies.

Non-metallic elements have high first ionization energies (>900 kJ/mol) and tend not to release their tightly bound electrons. With the exception of the stable, non-reactive, very non-metallic noble elements in column 18 (group VIIIA), non-metallic elements tend to be electronegative and possess high electron affinities. Column 16–17 (group VIA and group VIIA) elements, with their

Table 2.4   Common ionization states for common elements in columns on the periodic table.

Column (group)	Ionic charge	Description	Examples
1 (IA)	+1	Monovalent cations due to low first ionization energy	Li^{+1}, Na^{+1}, K^{+1}, Rb^{+1}, Cs^{+1}
2 (IIA)	+2	Lose two electrons due to low first and second ionization energy	Be^{+2}, Mg^{+2}, Ca^{+2}, Sr^{+2}, Ba^{+2}
3–12 (IIIB–IIB)	+1 to +7	Transition elements; lose variable numbers of electrons depending upon environment	Cu^{+1}, Fe^{+2}, Fe^{+3}, Cr^{+2}, Cr^{+6}, W^{+6}, Mn^{+2}, Mn^{+4}, Mn^{+7}
13 (IIIA)	+3	Lose three electrons due to low first through third ionization energy	B^{+3}, Al^{+3}, Ga^{+3}
14 (IVA)	+4	Lose four electrons due to low first through fourth ionization energy; may lose a smaller number of electrons	C^{+4}, Si^{+4}, Ti^{+4}, Zr^{+4}, Pb^{+2}, Sn^{+2}
15 (VA)	+5 to −3	Lose up to five electrons or capture three electrons to achieve stability	N^{+5}, N^{-3}, P^{+5}, As^{+3}, Sb^{+3}, Bi^{+4}
16 (VIA)	−2	Generally gain two electrons to achieve stability; gain six electrons in some environments	O^{-2}, S^{-2}, S^{+6}, Se^{-2},
17 (VIIA)	−1	Gain one electron to achieve stable configuration	Cl^{-1}, F^{-1}, Br^{-1}, I^{-1}
18 (VIIIA)	0	Stable electron configuration; neither gain nor lose electrons	He, Ne, Ar, Kr

especially high electron affinities, display a strong tendency to capture additional electrons to fill their highest principle quantum levels. They provide the best examples of highly electronegative, non-metallic elements. A brief summary of the characteristics of the columns and elemental groups is presented below and in Table 2.4.

- **Column 1 (IA)** metals are the only elements with a single s-electron in their highest quantum levels. Elements in column 1 (IA) achieve the stable configuration of the next lowest quantum level when they lose their single s-electron from the highest principal quantum level. For example, if sodium (Na) with the electron configuration ($1s_2$, $2s_2$, $2p_6$, $3s_1$) loses its single 3s electron (Na^{+1}), its electron configuration becomes that of the stable noble element neon ($1s_2$, $2s_2$, $2p_6$) with the "stable octet" in the highest principle quantum level.
- **Column 2 (IIA)** metals are the only elements with only two electrons in their highest quantum levels in their electrically neutral states. Column 2 (IIA) elements achieve stability by the removal of two s-electrons from the outer electron shell to become +2 citations.
- **Columns 3–12 (IIIB through IIB)** transition elements are situated in the middle of the periodic table. Column 3 (IIIB) elements tend to occur as trivalent cations by giving up three of their electrons (s_2, d_1) to achieve a stable electron configuration. The other groups of transition elements, from column 4 (IVB) through column 12 (IIB) are cations that occur in a variety of ionization states. Depending on the chemical reaction in which they are involved, these elements can give up as few as one s-electron as in Cu^{+1}, Ag^{+1} and Au^{+1} or as many as six or seven electrons, two s-electrons and four or five d-electrons as in Cr^{+6}, W^{+6} and Mn^{+7}. An excellent example of the variable ionization of a transition metal is iron (Fe). In environments where oxygen is relatively scarce, iron commonly gives up two electrons to become Fe^{+2} or ferrous iron. In other environments, especially those where oxygen is abundant, iron gives up three electrons to become smaller Fe^{+3} or ferric iron.
- **Column 13 (IIIA)** elements commonly exist as trivalent (+3) cations by losing 3 electrons (s_2, p_1).

- **Column 14 (IVA)** elements commonly exist as tetravalent (+4) cations by losing four electrons. The behavior of the heavier elements in this group is somewhat more variable than in those groups discussed previously. It depends on the chemical reaction in which the elements are involved. Tin (Sn) and lead (Pb) behave in a similar manner to silicon and germanium in some chemical reactions, but in other reactions they only lose the two s-electrons in the highest principal quantum level to become divalent cations.
- **Column 15 (VA)** elements commonly have a wide range of ionization states from tetravalent (+5) cations through trivalent (−3) anions. These elements are not particularly electropositive, nor are they especially electronegative. Their behavior depends on the other elements in the chemical reaction in which they are involved. For example, in some chemical reactions, nitrogen attracts three additional electrons to become the trivalent anion N^{-3}. In other chemical reactions, nitrogen releases as many as five electrons in the second principal quantum level to become the tetravalent cation N^{+5}. In still other situations, nitrogen gives up or attracts smaller numbers of electrons to form a cation or anion of smaller charge. All the other elements in group VA exhibit analogous situational ionization behaviors. Phosphorous, arsenic, antimony and bismuth all have ionic states that range from +5 to −3.
- **Column 16 (VIA)** non-metallic elements commonly exist as divalent (−2) anions. These elements attract two additional electrons into their highest principal quantum levels to achieve a stable electron configuration. For example, oxygen adds two electrons to become the divalent anion O^{-2}. With the exception of oxygen, however, the column 16 elements display other ionization states as well, especially when they react chemically with oxygen, as will be discussed later in this chapter. Sulfur and the other VIA elements are also quite electronegative, with strong electron affinities, so that they tend to attract two electrons to achieve a stable configuration and become divalent anions such as S^{-2}; but in the presence of oxygen these elements may lose electrons and become cations such as S^{+6}.
- **Column 17 (VIIA)** non-metallic elements commonly exist as monovalent (−1) anions. Because electrons are very difficult to remove from their electron clouds, these elements tend to attract one additional electron into their highest principal quantum level to achieve a stable electron configuration.
- **Column 18 (VIIIA)** noble gas elements contain complete outer electron shells (s_2, p_6) and do not commonly combine with other elements to form minerals. Instead, they tend to exist as monatomic gases.

The periodic table is a highly visual and logical way in which to illustrate patterns in the electron configurations of the elements. Elements are grouped in rows or classes according to the highest principal quantum level in which electrons occur in the ground state. Elements are grouped into columns or groups based on similarities in the electron configurations in the higher principal quantum levels; those with the highest quantum energies are farthest from the nucleus. A more thorough explanation of the periodic table and the properties of elements is available in a downloadable file on the website for this text.

From the discussion above, it should be clear that during the chemical reactions that produce Earth materials, elements display behaviors that are related to their electron configurations. Group 18 (VIIIA) elements in the far right column of the periodic table have stable electron configurations and tend to exist as uncharged atoms. Metallic elements toward the left side of the periodic table are strongly electropositive and tend to give up one or more electrons to become positively-charged particles called cations. Non-metallic elements toward the right side of the periodic table, especially in groups 16 (VIA) and 17 (VIIA), are strongly electronegative and tend to attract electrons to become negatively-charged particles called anions. Elements toward the middle of the periodic table are somewhat electropositive and tend to lose various numbers of electrons to become cations with various amounts of positive charge. These tendencies are summarized in Table 2.4.

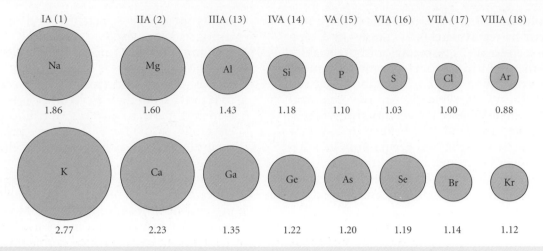

Figure 2.6   Trends in variation of atomic radii (in angstroms; 1 angstrom = 10^{-10} m) with their position on the periodic table, illustrated by rows 3 and 4. With few exceptions, radii tend to decrease from left to right and from bottom to top.

2.2.4   Atomic and ionic radii

Atomic radii are defined as half the distance between the nuclei of bonded identical neighboring atoms. Because the electrons in higher quantum levels are farther from the nucleus, the effective radius of electrically neutral atoms generally increases from top to bottom (row 1 through row 7) in the periodic table (see Table 2.3). However, atomic radii generally decrease within rows from left to right (Figure 2.6). This occurs because the addition of electrons to a given quantum level does not significantly increase atomic radius, while the increase in the number of positively-charged protons in the nucleus causes the electron cloud to contract as electrons are pulled closer to the nucleus. Atoms with large atomic numbers and large electron clouds include cesium (Cs), rubidium (Rb), potassium (K), barium (Ba) and uranium (U). Atoms with small atomic numbers and small electron clouds include hydrogen (H), beryllium (Be) and carbon (C).

Electrons in the outer electron levels are least tightly bound to the positively-charged nucleus. This weak attraction results because these electrons are farthest from the nucleus and because they are shielded from the nucleus by the intervening electrons that occupy lower quantum level positions closer to the nucleus.

These outer electrons or **valence electrons** are the electrons that are involved in a wide variety of chemical reactions, including those that produce minerals, rocks and a wide variety of synthetic materials. The loss or gain of these valence electrons produces anions and cations, respectively.

Atoms become ions through the gain or loss of electrons. When atoms are ionized by the loss or gain of electrons, their **ionic radii** invariably change. This results from the electrical forces that act between the positively-charged protons in the nucleus and the negatively-charged electrons in the electron clouds. The ionic radii of cations tend to be smaller than the atomic radii of the same element (Figure 2.7). As electrons are lost from the electron cloud during cation formation, the positively-charged protons in the nucleus tend to exert a greater force on each of the remaining electrons. This draws electrons closer to the nucleus, reducing the effective radius of the electron cloud. The larger the charge on the cation, the more its radius is reduced by the excess positive charge in the nucleus. This is well illustrated by the radii of the common cations of iron (Figure 2.7). Ferric iron (Fe^{+3}) has a smaller radius (0.64 angstroms) than does ferrous iron (Fe^{+2} = 0.74 angstroms) because greater excess positive charge in the nucleus draws the electrons

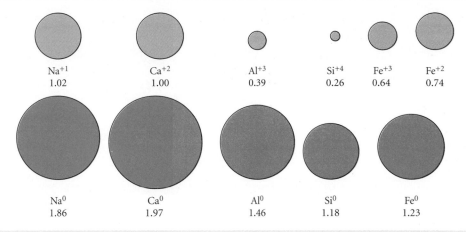

Figure 2.7   Radii (in angstroms) of some common cations in relationship to the atomic radius of the neutral atoms.

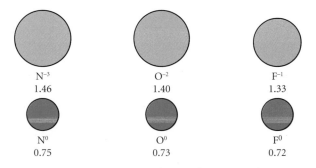

Figure 2.8   Radii (in angstroms) of some common anions in relationship to the atomic radius of the neutral atoms.

closer to the nucleus which causes the electron cloud to contract. Both iron cations possess much smaller radii than neutral iron (Fe^0 = 1.23 angstroms) in which there is no excess positive charge in the nucleus.

The ionic radii of anions are significantly larger than the atomic radii of the same element (Figure 2.8). When electrons are added to the electron cloud during anion formation, the positively-charged protons in the nucleus exert a smaller force on each of the electrons. This allows the electrons to move farther away from the nucleus, which causes the electron cloud to expand increasing the effective radius of the anion. The larger the charge on the anion, the more its effective radius is increased.

The expansion of anions and the contraction of cations are well illustrated by the

common ions of sulfur (Figure 2.9). The divalent sulfur (S^{-2}) anion possesses a relatively large average radius of 1.84 angstroms. In this case, the two electrons gained during the formation of a divalent sulfur anion produce a large deficit between positive charges in the nucleus and negative charges in the electron cloud. This leads to a significant increase in the effective ionic radius. The anion is considerably larger than electrically neutral sulfur (S^0 = 1.03 angstroms) which in turn is larger than the divalent sulfur cation (S^{+2} = 0.37 angtsroms) and the very small, highly charged hexavalent sulfur cation (S^{+6} = 0.12 angstroms). Keep in mind that the effective radius of a particular anion does vary somewhat. As we will see in the following sections, this depends on the environment in which bonding occurs, the number of nearest neighbors and the type of bond that forms.

2.3   CHEMICAL BONDS

2.3.1   The basics

Atoms in minerals and rocks are held together by forces or mechanisms called **chemical bonds**. The nature of these bonds strongly influences the properties and behavior of minerals, rocks and other Earth materials. The nature of the bonds is, in turn, strongly influenced by the electron configuration of the elements that combine to produce the material.

Five principle bond types and many hybrids occur in minerals. The three most common

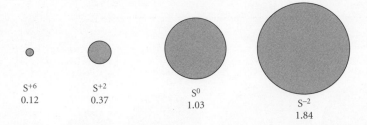

Figure 2.9   Radii (in angstrom units) of some common anions and cations of sulfur in relationship to the neutral atom radius.

bond types are (1) ionic, (2) covalent, and (3) metallic. They can be modeled based on the behavior of valence electrons. Valence electrons, which occur in the outer shells or quantum levels of atoms, display varying tendencies to change position based on their periodic properties. In discussing chemical bonds, it is useful to divide elements into those that are metallic and those that are non-metallic.

Ionic, covalent and metallic bonds involve the linking together of (1) metallic and non-metallic elements, (2) non-metallic and non-metallic elements, and (3) metallic and metallic elements, respectively. Hybrids between these bond types are common and minerals with such hybrid or transitional bonds commonly possess combinations of features characteristic of each bond type. Other bond types include van der Waals and hydrogen bonds. Chemical bonding is a very complicated process; the models used below are simplifications designed to make this complex process easier to understand.

Another useful concept, developed originally by Linus Pauling (1929), is the concept of electronegativity (see Table 2.3). **Electronegativity** (En) is an empirical measure that expresses the tendency of an element to attract electrons when atoms bond. Highly electronegative elements (En >3.0) have a strong tendency to become anions during bonding. Many column 16 (group VIA) and column 17 (group VIIA) elements are highly electronegative, requiring capture of two or one electrons, respectively, to achieve a stable electron configuration. Elements with low electronegativity (En <1.5) are electropositive, metallic elements with a tendency to give up electrons during bonding to become positively charged cations. Highly electropositive ele-

ments include column 1 (group IA) and column 2 (group IIA) elements that tend to release one or two electrons, respectively, to achieve a stable electron configuration. Electronegativity is a very helpful concept in discussions of how atoms bond to produce larger molecules and minerals.

2.3.2   Ionic (electrostatic) bonds

When very metallic atoms bond with very non-metallic atoms, an **ionic bond**, also called an **electrostatic bond**, is formed. Because the very metallic atoms (e.g., columns 1 and 2) are electropositive, they have a strong tendency to give up one or more electrons to achieve a stable configuration in their highest principal quantum level. In doing so, they become positively-charged cations with a charge equal to the number of electrons each has lost. At the same time, very non-metallic atoms (columns 16 and 17) are electronegative and have a strong tendency to gain one or more electrons in order to achieve a stable configuration in their highest principal quantum level. In doing so, they become negatively-charged anions with a charge equal to the number of electrons each has gained. When very metallic and very non-metallic atoms bond, the metallic atoms give up or donate their valence electrons to the non-metallic atoms that capture them. It is like a tug-of-war in which the electronegative side always wins the battle for electrons. In the electron exchange process, the atoms of both elements develop stable noble element electron configurations while becoming ions of opposite charge. Because particles of opposite charge attract, the cations and anions are held together by the electrostatic attraction between

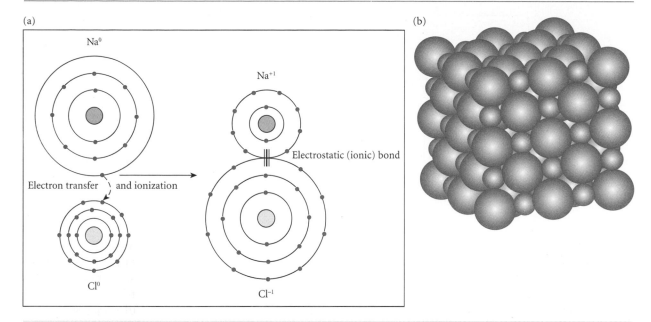

Figure 2.10 (a) Ionic bonding develops between highly electronegative anions and highly electropositive cations. When neutral sodium (Na^0) atoms release an electron to become cations (Na^{+1}) their ionic radius decreases. When neutral chlorine (Cl^0) atoms capture an electron to become anions (Cl^{-1}) their ionic radius increases. (b) Ions of opposite charge attract to form crystals such as sodium chloride (NaCl).

them that results from their opposite charges. Larger clusters of ions form as additional ions exchange electrons and are bonded and crystals begin to grow.

The most frequently cited example of ionic bonding is the bonding between sodium (Na^{+1}) and chloride (Cl^{-1}) ions in the mineral halite (NaCl) (Figure 2.10). As a column 1 (group IA) element, sodium is very metallic and electropositive, with a rather low electronegativity (1.0). Sodium has a strong tendency to give up one electron to achieve a stable electron configuration. On the other hand, chlorine, as a column 17 (group VIIA) element, is very non-metallic, has a strong affinity for electrons and has a high electronegativity (3.5). It has a strong tendency to gain one electron to achieve a stable electron configuration. When sodium and chlorine atoms bond, the sodium atoms release one electron to become smaller sodium cations (Na^{+1}) with the "stable octet" electron configuration (neon), while the chlorine atoms capture one electron to become larger chloride anions (Cl^{-1}) with "stable octet" electron configurations. The two atoms are then

joined together by the electrostatic attraction between particles of opposite charge to form the compound NaCl. In macroscopic mineral specimens of halite, many millions of sodium and chloride ions are bonded together, each by the electrostatic or ionic bond described above. Note that the numbers of chloride anions and sodium cations must be equal if the electric charges are to be balanced so that the mineral is electrically neutral. Other group IA (1) and group VIIA (17) elements bond ionically to produce minerals such as **sylvite** (KCl).

Ionic bonds also form when group IIA and group VIA elements combine. In the mineral periclase (MgO), magnesium (Mg^{+2}) and oxygen (O^{-2}) ions are bonded together to form MgO. In this case, electropositive, metallic magnesium atoms from group IIA tend to donate two valence electrons to become stable, smaller divalent magnesium cations (Mg^{+2}) while highly electronegative, non-metallic oxygen atoms from group VIA capture two valence electrons to become stable, larger divalent oxygen anions (O^{-2}). The two oppositely charged ions are then held

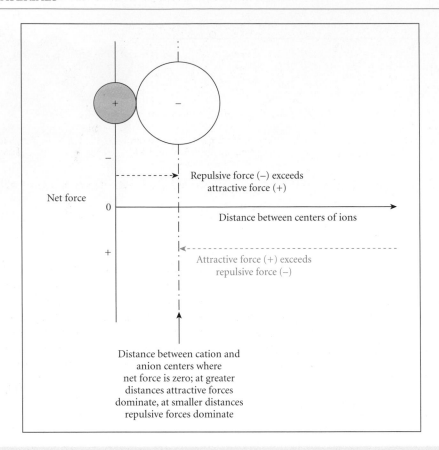

Figure 2.11   The relationship between attractive and repulsive forces between ions produces a minimum net force when the near spherical surfaces of the ions are in contact.

together by virtue of their opposite charges by an electrostatic or ionic bond. Once again, the number of magnesium cations (Mg^{+2}) and oxygen anions (O^{-2}) in periclase (MgO) must be the same if electrical neutrality is to be conserved. A slightly more complicated example of ionic bonding involves the formation of the mineral fluorite (CaF_2). In this case, electropositive, metallic calcium atoms from class IIA release two electrons to become stable divalent cations (Ca^{+2}). At the same time, two non-metallic, strongly electronegative fluorine atoms from class VIIA each accept one of these electrons to become stable univalent anions (F^{-1}). Pairs of F^{-1} anions bond to each Ca^{+2} cation to form ionic bonds in electrically neutral fluorite (CaF_2).

In ideal ionic bonds, ions can be modeled as spheres of specific ionic radius in contact with one another (Figure 2.10), as though they were ping-pong balls or marbles in contact with each other. This approximates

real situations because the attractive force between ions of opposite charge (Coulomb attraction) and the repulsive force (Born repulsion) between the negatively-charged electron clouds are balanced when the two ions approximate spheres in contact (Figure 2.11). If they were moved farther apart, the electrostatic attraction between ions of opposite charge would move them closer together. If they were moved closer together, the repulsive forces between the negatively-charged electron clouds would move them farther apart. It is when they approximate spheres in contact that these attractive and repulsive forces are balanced.

Bonding mechanisms play an essential role in contributing to material properties. Crystals with ionic bonds are generally characterized by the following:

1   Variable hardness.
2   Brittle at room temperatures.

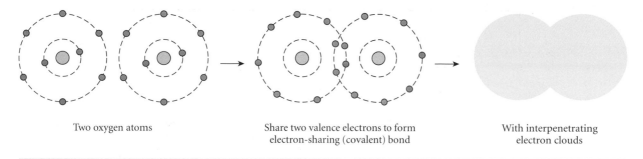

Two oxygen atoms   Share two valence electrons to form   With interpenetrating
                     electron-sharing (covalent) bond   electron clouds

Figure 2.12   Covalent bonding in oxygen (O_2) by the sharing of two electrons from each atom.

3   Quite soluble in polar substances (such as water).
4   Intermediate melting temperatures.
5   Do not absorb much light, producing translucent to transparent minerals with light colors and vitreous to sub-vitreous luster in macroscopic crystals.

2.3.3   Covalent (electron-sharing) bonds

When non-metallic atoms bond with other non-metallic atoms they tend to form **covalent bonds**. Because the elements involved are electronegative they each tend to attract electrons; neither gives them up easily because of their high first ionization potentials and electron affinities. This is a little bit like a tug-of-war in which neither side can be moved so neither side ends up with sole possession of the electrons needed to achieve stable electron configurations. In simple models of covalent bonding, the atoms involved share valence (thus covalent) electrons (Figure 2.12). By sharing electrons, each atom gains the necessary electrons to achieve a stable electron configuration in its highest principal quantum level.

A relevant example is oxygen gas (O_2) molecules. Each oxygen atom requires two electrons to achieve a stable electron configuration in its highest principal quantum level. Since both oxygen atoms have an equally large electron affinity and electronegativity, they tend to share two electrons in order to achieve the stable electron configuration. This sharing is modeled as interpenetration or overlapping of the two electron clouds (Figure 2.12). Interpenetration of electron clouds due to the sharing of valence electrons forms a strong covalent or electron-sharing bond. Because the bonds are localized in the region where the electrons are "shared" each atom has a larger probability of electrons in the area of the bond than it does elsewhere in its electron cloud. This causes each atom to become electrically polarized with a more negative charge in the vicinity of the bond and a less negative charge away from the bond. Polarization of atoms during covalent bonding is accentuated when covalent bonds form between different atoms with different electronegativities, which causes the electrons to be more tightly held by the more electronegative atom which distorts the shape of the atoms so that they cannot be as effectively modeled as spheres in contact.

Other diatomic gases with covalent bonding mechanisms similar to oxygen include the column 17 (group VIIA) gases chlorine (Cl_2), fluorine (F_2) and iodine (I_2) in which single electrons are shared between the two atoms to achieve a stable electron configuration. Another gas that possesses covalent bonds is nitrogen (N_2) where three electrons from each atom are shared to achieve a stable electron configuration. Nitrogen is the most abundant gas (>79% of the total) in Earth's lower atmosphere. Because the two atoms in nitrogen and oxygen gas are held together by strong electron-sharing bonds that yield stable electron configurations, these two molecules are the most abundant constituents of Earth's lower atmosphere.

The best known mineral with covalent bonding is **diamond**, which is composed of carbon (C). Because carbon is a column 14 atom, it must either lose four electrons or gain four electrons to achieve a stable electron

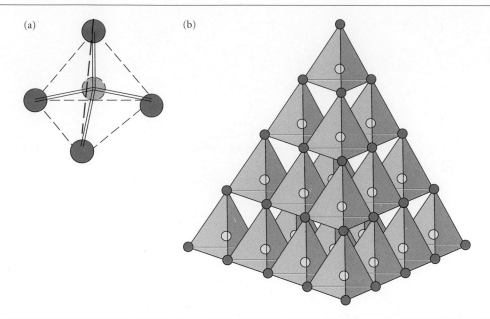

Figure 2.13 (a) Covalent bonding (double lines) in a carbon tetrahedron with the central carbon atom bonded to four carbon atoms that occupy the corners of a tetrahedron (dashed lines). (b) A larger scale diamond structure with multiple carbon tetrahedra. (Courtesy of Steve Dutch.)

configuration. In diamond, each carbon atom in the structure is bonded to four nearest neighbor carbon atoms that share with it one of their electrons (Figure 2.13). In this way, each carbon atom attracts four additional electrons, one from each of its neighbors, to achieve the stable noble electron configuration. The long-range crystal structure of diamond is a pattern of carbon atoms in which every carbon atom is covalently bonded to four other carbon atoms.

Covalently bonded minerals are generally characterized by the following:

1 Hard and brittle at room temperature.
2 Insoluble in polar substances such as water.
3 Crystallize from melts.
4 Moderate to high melting temperatures.
5 Do not absorb light, producing transparent to translucent minerals with light colors and vitreous to sub-vitreous lusters in macroscopic crystals.

2.3.4 Metallic bonds

When metallic atoms bond with other metallic atoms, a **metallic bond** is formed. Because very metallic atoms have low first ionization energies, are highly electropositive and possess low electronegativities they do not tend to hold their valence electrons strongly. In such situations, each atom releases valence electrons to achieve a stable electron configuration. The positions of the valence electrons fluctuate or migrate between atoms. Metallic bonding is difficult to model, but is usually portrayed as positively-charged partial atoms (nuclei plus the strongly held inner electrons) in a matrix or "gas" of delocalized valence electrons that are only temporarily associated with individual atoms (Figure 2.14). The weak attractive forces between positive partial atoms and valence electrons bond the atoms together. Unlike the strong electron-sharing bonds of covalently bonded substances, or the sometimes strong electrostatic bonds of ionically bonded substances, metallic bonds are rather weak, less permanent and easily broken and reformed. Because the valence electrons are not strongly held by any of the partial atoms, they are easily moved in response to stress or in response to an electric field.

Many good examples of metallic bonding exist in the native metals such as native gold (Au), native silver (Ag), and native copper (Cu). When subjected to an electric potential or field, delocalized electrons flow toward

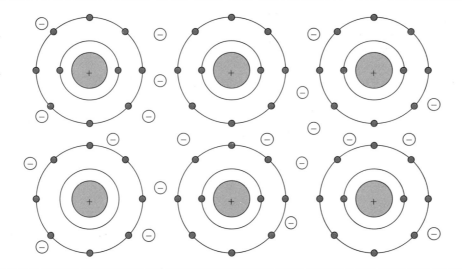

Figure 2.14 A model of metallic bonds with delocalized electrons surrounding positive charge centers that consist of tightly held lower energy electrons (blue dots) surrounding individual nuclei (blue).

the positive anode, creating and maintaining a strong electric current. Similarly, when a thermal gradient exists, thermal vibrations are transferred by delocalized electrons, making such materials excellent heat conductors. When metals are stressed, the weakly held electrons tend to flow, which helps to explain the ductile behavior that characterizes copper and other metallically bonded substances.

Minerals containing metallic bonds are generally characterized by the following features:

1 Fairly soft to moderately hard minerals.
2 Plastic, malleable and ductile.
3 Excellent electrical and thermal conductors.
4 Frequently high specific gravity.
5 Excellent absorbers and reflectors of light; so are commonly opaque with a metallic luster in macroscopic crystals.

2.3.5 Transitional (hybrid) bonds

Transitional or **hybrid bonds** display combinations of ionic, covalent and/or metallic bond behavior. Some transitional bonds can be modeled as ionic–covalent transitional, others as ionic–metallic or covalent–metallic

transitional. A detailed discussion of all the possibilities is beyond the scope of this book, but because most bonds in Earth materials are transitional, it is a subject worthy of mention. The following discussion also serves to illustrate once again the enigmatic behavior of which electrons are capable.

Earlier in this chapter, we defined electronegativity in relation to the periodic table. Linus Pauling developed the concept of electronegativity to help model transitional ionic–covalent bonds. In models of such bonds, electrons are partially transferred from the less metallic, more electronegative element to the more metallic, less electronegative element to produce a degree of ionization and electrostatic attraction typical of ionic bonding. At the same time, the electrons are partially shared between the two elements to produce a degree of electron sharing associated with covalent bonding. Such bonds are best modeled as hybrids or transitions between ionic and covalent bonds. Materials that possess such bonds commonly display properties that are transitional between those of ionically bonded substances and those of covalently bonded substances. Using the **electronegativity difference** – the difference between the electronegativities of the two elements sharing the bond – Pauling was able to predict the percentages of covalent and ionic bonding,

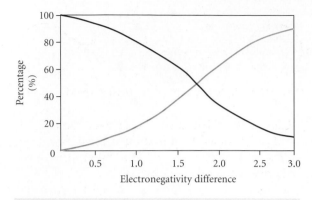

Figure 2.15 Graph showing the electronegativity difference and bond type in covalent–ionic bonds. Percent covalent bonding is indicated by the black line and percent ionic bonding by the blue line.

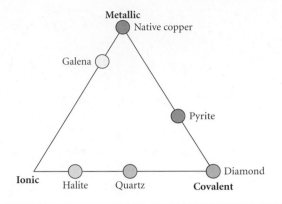

Figure 2.16 Triangular diagram representing the bond types of some common minerals.

that is, the percentages of electron sharing and electron transfer that characterize ionic–covalent transitional bonds. Figure 2.15 illustrates the relationship between electronegativity difference and the percentages of ionic and covalent bond character that typify the transitional ionic–covalent bonds.

Where electronegativity differences in transitional ionic–covalent bonds are smaller than 1.68, the bonds are primarily electron-sharing covalent bonds. Where electronegativity differences are larger than 1.68, the bonds are primarily electron-transfer ionic bonds. Calculations of electronegativity and bond type lead to some interesting conclusions. For example, when an oxygen atom with En = 3.44 bonds with another oxygen atom with En = 3.44 to form O_2, the electronegativity difference (3.44 − 3.44 = 0.0) is zero and the resulting bond is 100% covalent. The valence electrons are completely shared by the two oxygen atoms. This will be the case whenever two highly electronegative, non-metallic atoms of the same element bond together. On the other hand, when highly electronegative, non-metallic atoms bond with strongly electropositive, metallic elements to form ionically bonded substances, the bond is never purely ionic. There is always at least a small degree of electron sharing and covalent bonding. For example, when sodium (Na) with En = 0.93 bonds with chlorine

(Cl) with En = 3.6 to form sodium chloride (NaCl), the electronegativity difference (3.6 − 0.93 = 2.67) is 2.67 and the bond is only 83% ionic and 17% covalent. Although the valence electrons are largely transferred from sodium to chloride and the bond is primarily electrostatic (ionic), a degree of electron sharing (covalent bonding) exists. Even in this paradigm of ionic bonding, electron transfer is incomplete and a degree of electron sharing occurs. The bonding between silicon (Si) and oxygen (O), so important in silicate minerals, is very close to the perfect hybrid since the electronegativity difference is 3.44 − 1.90 = 1.54 and the bond is 45% ionic and 55% covalent.

This simple picture of transitional ionic–covalent bonding does not hold in bonds that involve transition metals. For example, the mineral galena (PbS) has properties that suggest its bonding is transitional between metallic and ionic. In this case some electrons are partially transferred from lead (Pb) to sulfur (S) in the manner characteristic of ionically bonded substances, but some electrons are weakly held in the manner characteristic of metallic bonds. As a result, galena displays both ionic properties (brittle and somewhat soluble) and metallic properties (soft, opaque and a metallic luster). Figure 2.16 utilizes a triangle, with pure covalent, ionic and metallic bonds at the apices, to depict the pure and transitional bonding characteristic of selected minerals, including those discussed above.

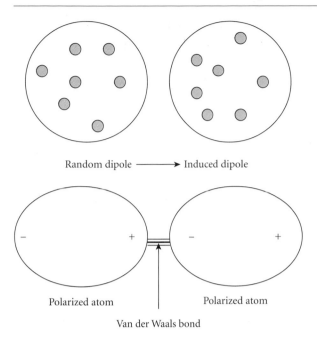

Random dipole ⟶ Induced dipole

Polarized atom   Polarized atom

Van der Waals bond

Figure 2.17 Van der Waals bonding occurs when one atom becomes dipolar as the result of the random concentration of electrons in one region of an atom. The positively-charged region of the atom attracts electrons in an adjacent atom causing it to become dipolar. Oppositely charged portions of adjacent dipolar atoms are attracted creating a weak van der Waals bond. Larger structures result from multiple bonds.

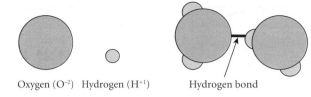

Oxygen (O^{-2}) Hydrogen (H^{+1})   Hydrogen bond

Figure 2.18 Diagram showing two water molecules joined by a hydrogen bond that links the hydrogen in one molecule to the oxygen in the other molecule.

2.3.6   Van der Waals and hydrogen bonds

Because the distributions of electrons in the electron cloud are probabilistic and constantly changing, they may be, at any moment, asymmetrically distributed within the electron cloud. This asymmetry gives rise to weak electric dipoles on the surface of the electron cloud; areas of excess negative charge concentration where the electrons are located and areas of negative charge deficit (momentary positive charge) where they are absent. Areas of momentary positive charge on one atom attract electrons in an adjacent atom, thus inducing a dipole in that atom. The areas of excess negative charge on one atom are attracted to the areas of positive charge on an adjacent atom to form a very weak bond that holds the atoms together (Figure 2.17). Bonds that result from weak electric dipole forces

that are caused by the asymmetrical distribution of electrons in the electron cloud are called **van der Waals bonds**. The presence of very weak van der Waals bonds explains why minerals such as graphite and talc are extremely soft.

The **hydrogen bond** is a bond that exists between hydrogen in a molecule such as water or hydroxyl ion and an electronegative ion such as oxygen. Because of the profound importance of water (H_2O) and hydroxyl ion (OH^{-1}), in both organic and inorganic compounds, this type of bond has been given its own separate designation (Figure 2.18). Hydrogen bonds are relatively weak bonds that occur in hydrated or hydroxyl minerals.

Atoms are held together by a variety of chemical bonds. The type of bond that forms depends largely on the electron configurations of the combining elements as expressed by their electronegativities. Each bond type imparts certain sets of properties to Earth materials that contain those bonds. In the following section we will discuss factors that determine the three-dimensional properties of the molecular units that result from such bonding. In Chapter 4 we will elaborate the long-range crystalline structures that form when these molecular units combine to produce crystals. Remember: it all starts with atoms, their electron properties and the way they bond together to produce crystals.

2.4   PAULING'S RULES AND COORDINATION POLYHEDRA

2.4.1   Pauling's rules and radius ratio

Linus Pauling (1929) established five rules, now called **Pauling's rules**, describing cation–

anion relationships in ionically bonded substances, which are paraphrased below:

- **Rule 1:** A polyhedron of anions is formed about each cation with the distance between a cation and an anion determined by the sum of their radii (radius sum). The number of coordinated anions in the polyhedron is determined by the cation:anion radius ratio.
- **Rule 2:** An ionic structure is stable when the sum of the strengths of all the bonds that join the cation to the anions in the polyhedron equals (balances) the charge on the cation and the anions. This rule is called the **electrostatic valency rule.**
- **Rule 3:** The sharing of edges and particularly faces by adjacent anion polyhedral elements decreases the stability of an ionic structure. Similar charges tend to repel. If they share components, adjacent polyhedra tend to share corners, rather than edges.
- **Rule 4:** Cations with high valence charges and small coordination number tend not to share polyhedral elements. Their large positive charges tend to repel.
- **Rule 5:** The number of different cations and anions in a crystal structure tends to be small. This is called the **rule of parsimony.**

Pauling's rules provide a powerful tool for understanding crystal structures. Especially important are the rules concerning radius ratio and coordination polyhedra. Coordination polyhedra provide a powerful means for visualizing crystal structures and their relationship to crystal chemistry. In fact, they provide a fundamental link between the two. When atoms and ions combine to form crystals, they bond together into geometric patterns in which each atom or ion is bonded to a number of nearest neighbors. The number of nearest neighbor ions or atoms is called the **coordination number (CN).** Clusters of atoms or ions bonded to other coordinating atoms produce **coordination polyhedron** structures. Polyhedrons include cubes, octahedrons and other geometric forms.

When ions of opposite charge combine to form minerals, each cation attracts as many nearest neighbor anions as can fit around it "as spheres in contact". In this way, the basic units of crystal structure are formed which grow into crystals as multiples of such units are added to the existing structure. One can visualize crystal structures in terms of different coordinating cations and coordinated anions that together define a simple three-dimensional polyhedron structure. As detailed in Chapters 4 and 5, complex polyhedral structures develop by linking of multiple coordination polyhedra.

The number of nearest neighbor anions that can be coordinated with a single cation "as spheres in contact" depends on the **radius ratio (RR = Rc/Ra)** which is the radius of the smaller cation (Rc) divided by the radius of the larger anion (Ra). For very small, highly charged cations coordinated with large, highly charged anions, the radius ratio (RR) and the coordination number (CN) are small. This is analogous to fitting basketballs as spheres in contact around a small marble. Only two basketballs can fit as spheres in contact with the marble. For cations of smaller charge coordinated with anions of smaller charge, the coordination number is larger. This is analogous to fitting golf balls around a larger marble. One can fit a larger number of golf balls around a large marble as spheres in contact because the radius ratio is larger.

The general relationship between radius ratio, coordination number and the type of coordination polyhedron that results is summarized in Table 2.5. For radius ratios less than 0.155, the coordination number is 2 and the "polyhedron" is a line. The appearance of these coordination polyhedra is summarized in Figure 2.19.

When predicting coordination number using radius ratios, several caveats must be kept in mind.

1 The ionic radius and coordination number are not independent. As illustrated by Table 2.6, effective ionic radius increases as coordination number increases.
2 Since bonds are never truly ionic, models based on spheres in contact are only approximations. As bonds become more covalent and more highly polarized, radius ratios become increasing less effective in predicting coordination numbers.
3 Radius ratios do not successfully predict coordination numbers for metallically bonded substances.

Table 2.5 Relationship between radius ratio, coordination number and coordination polyhedra.

Radius ratio (Rc/Ra)	Coordination number	Coordination type	Coordination polyhedron
<0.155	2	Linear	Line
0.155–0.225	3	Triangular	Triangle
0.225–0.414	4	Tetrahedral	Tetrahedron
0.414–0.732	6	Octahedral	Octahedron
0.732–1.00	8	Cubic	Cube
>1.00	12	Cubic or hexagonal closest packed	Cubeoctahedron complex

The great value of the concept of coordination polyhedra is that it yields insights into the fundamental patterns in which atoms bond during the formation of crystalline materials. These patterns most commonly involve three-fold (triangular), four-fold (tetrahedral), six-fold (octahedral), eight-fold (cubic) and, to a lesser extent, 12-fold coordination polyhedra or small variations of such basic patterns. Other coordination numbers and polyhedron types exist, but are rare in inorganic Earth materials.

Another advantage of using spherical ions to model coordination polyhedra is that it allows one to calculate the size or volume of the resulting polyhedron. In a coordination polyhedron of anions, the cation–anion distance is determined by the **radius sum (R_Σ)**. The radius sum is simply the sum of the radii of the two ions (Rc + Ra); that is, the distance between their respective centers. Once this is known, the size of any polyhedron can be calculated using the principles of geometry. Such calculations are beyond the scope of this book but are discussed in Klein and Dutrow (2007) and Wenk and Bulakh (2004).

2.4.2 Electrostatic valency

An important concept related to the formation of coordination polyhedra is **electrostatic valency (EV)**. In a stable coordination structure, the total strength of all the bonds that reach a cation from all neighboring anions is equal to the charge on the cation. This is another way of saying that the positive charge on the cation is neutralized by the electrostatic component of the bonds between it and its nearest neighbor anions. Similarly, every anion in the structure is surrounded by some number of nearest neighbor cations to which

it is bonded, and the negative charges on each anion are neutralized by the electrostatic component of the bonds between it and its nearest neighbor cations. For a cation of charge Z bonded to a number of nearest neighbor anions, the electrostatic valency of each bond is given by the charge of the cation divided by the number of nearest neighbors to which it is coordinated:

$$EV = Z/CN$$

For example, in the case of the silica tetrahedron $(SiO_4)^{-4}$ each Si^{+4} cation is coordinated with four O^{-2} anions (Figure 2.20). The electrostatic valency of each bond is given by $EV = Z/CN = +4/4 = +1$. What this means is that each bond between the coordinating silicon ion (Si^{+4}) and the coordinated oxygen ions (O^{-2}) balances a charge of +1. Another way to look at this is to say that each bond involves an electrostatic attraction between ions of opposite charge of one charge unit. Since there are four Si–O bonds, each balancing a charge of +1, the +4 charge on the silicon ion is fully neutralized by the nearest neighbor anions to which it is bonded. However, although the +4 charge on the coordinating silicon ion is fully satisfied, the −2 charge on each of the coordinated ions is not. Since each has a −2 charge, a single bond involving an electrostatic attraction of one charge unit neutralizes only half their charge. They must attract and bond to one or more additional cations, with an additional total electrostatic valency of one, in order to have their charges effectively neutralized. So it is that during mineral growth, cations attract anions and anions attract additional cations of the appropriate charge and radius which in turn attract additional anions of the

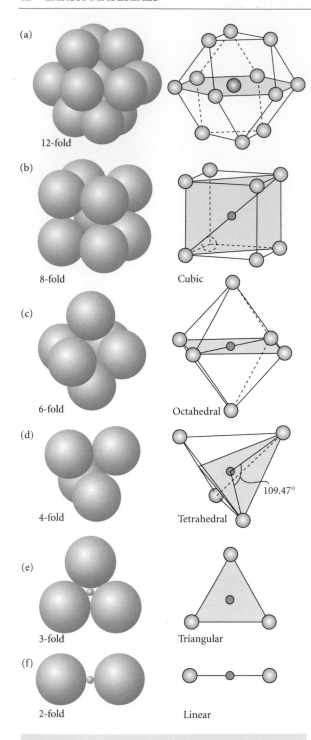

Figure 2.19 Common coordination polyhedra: (a) cubic closest packing, (b) cubic, (c) octahedral, (d) tetrahedral, (e) triangular, (f) linear. (From Wenk and Bulakh, 2004; with permission of Cambridge University Press.)

Table 2.6 Variations in ionic radius (in angstroms) with coordination number (CN) for some common cations.

Ion	CN = 4	CN = 6	CN = 8
Na^{+1}	0.99	1.02	1.18
K^{+1}		1.38	1.51
Rb^{+1}		1.52	1.61
Cs^{+1}		1.67	1.74
Mg^{+2}	0.57	0.72	
Al^{+3}	0.39	0.48	
Si^{+4}	0.26	0.40	
P^{+5}	0.17	0.38	
S^{+6}	0.12	0.29	

appropriate charge and radius as the mineral grows. In this manner minerals retain their essential geometric patterns and their ions are neutralized as the mineral grows. In the following section we will introduce the major mineral groups and see how their crystal chemistry forms the basis of the mineral classification.

2.5 CHEMICAL CLASSIFICATION OF MINERALS

The formation and growth of most minerals can be modeled by the attractive forces between cations and anions, the formation of coordination polyhedra with unsatisfied negative charges and the attraction of additional ions *ad infinitum*, until the conditions for growth cease to exist. It is useful to visualize minerals in terms of major anions and anion groups and/or radicals bonded to various cations that effectively neutralize their charge during the formation and growth of minerals. One common way to group or classify minerals is to do so in terms of the major anion group in the mineral structure. Those that contain $(SiO_4)^{-4}$ silica tetrahedra, discussed in the previous section, are silicate minerals, by far the most common minerals in Earth's crust and upper mantle. Those that do not contain silica tetrahedra are non-silicate minerals and are further subdivided on the basis of their major anions. Table 2.7 summarizes the common mineral groups according to this major anion group classification system.

Oxygen (O) and silicon (Si) are the two most abundant elements in Earth's continen-

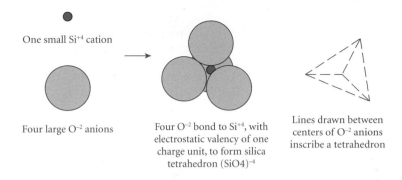

One small Si^{+4} cation

Four large O^{-2} anions

Four O^{-2} bond to Si^{+4}, with electrostatic valency of one charge unit, to form silica tetrahedron (SiO4)$^{-4}$

Lines drawn between centers of O^{-2} anions inscribe a tetrahedron

Figure 2.20 A silica tetrahedron is formed when four oxygen ions (O^{-2}) bond to one silicon ion (Si^{+4}) in the form of a tetrahedron. The electrostatic valency of each silicon–oxygen bond in the silica tetrahedron is one charge unit, which fully neutralizes the charge on the central silicon ion (four = four), while leaving the charge on the oxygen ions only partially neutralized (one is one-half of two).

Table 2.7 Mineral classification based on the major anion groups.

Mineral group	Major anion groups	Mineral group	Major anion groups
Native elements	None	Nitrates	$(NO_3)^{-1}$
Halides	F^{-1}, Cl^{-1}, Br^{-1}	Borates	$(BO_3)^{-3}$ and $(BO_4)^{-5}$
Sulfides	S^{-2}, S^{-4}	Sulfates	$(SO_4)^{-2}$
Arsenides	As^{-2}, As^{-3}	Phosphates	$(PO_4)^{-3}$
Sulfarsendies	As^{-2} or As^{-3} and S^{-2} or S^{-4}	Chromates	$(CrO4)^{-5}$
Selenides	Se^{-2}	Arsenates	$(AsO_4)^{-3}$
Tellurides	Te^{-2}	Vanadates	$(VO_4)^{-3}$
Oxides	O^{-2}	Molybdates	$(MO_4)^{-2}$
Hydroxides	$(OH)^{-1}$	Tungstates	$(WO_4)^{-2}$
Carbonates	$(CO_3)^{-2}$	Silicates	$(SiO_4)^{-4}$

tal crust, oceanic crust and mantle. Under the relatively low pressure conditions that exist in the crust and the upper mantle, the most abundant rock-forming minerals (Chapter 5) are silicate minerals. **Silicate minerals**, characterized by the presence of silicon and oxygen that have bonded together to form silica tetrahedra, are utilized here to show how coordination polyhedra are linked to produce larger structures with the potential for the long-range order characteristic of all minerals.

2.5.1 The basics: silica tetrahedral linkage

Silica tetrahedra are composed of a single, small, tetravalent silicon ion (Si^{+4}) in four-fold, tetrahedral coordination with four larger, divalent oxygen ions (O^{-2}). These silica

tetrahedra may be thought of as the basic building blocks, the LEGO®, of silicate minerals. Because the electrostatic valency of each of the four Si–O bonds in the tetrahedron is one (EV = 1), the +4 charge of the silicon ion is effectively neutralized. However, the -2 charges on the oxygen (O^{-2}) ions are not neutralized. Each oxygen ion possesses an unsatisfied charge of -1 which it can only neutralize by bonding with one or more additional cations in the mineral structure. Essentially, as a crystal forms, oxygen anions can bond to another silicon (Si^{+4}) ion to form a second bond with an electrostatic valency of 1 or it can bond to some other combination of cations (e.g., Al^{+3}, Mg^{+2}, Fe^{+2}, Ca^{+2}, K^{+1}, Na^{+1}) with a total electrostatic valency of 1.

Many factors influence the type of silica tetrahedral structure that develops when

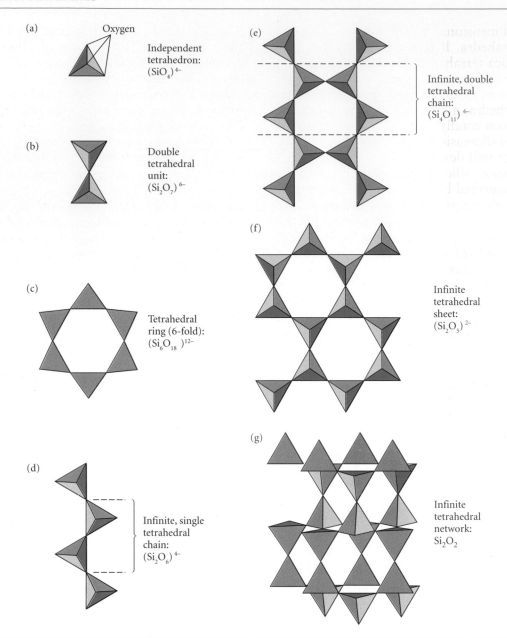

Figure 2.21 Major silicate structures: (a) nesosilicate, (b) sorosilicate, (c) cyclosilicate, (d) single-chain inosilicate, (e) double chain inosilicate, (f) phyllosilicate, (g) tectosilicate. (From Wenk and Bulakh, 2004; with permission of Cambridge University Press.)

silicate minerals form; the most important is the relative availability of silicon and other cations in the environment in which the mineral crystallizes. Environments that are especially enriched in silicon (and therefore in silica tetrahedra) tend to favor the linkage of silica tetrahedra through shared oxygen ions. Environments that are depleted in silicon tend to favor the linkage of the oxygen ions in silica tetrahedra to cations other than silicon.

In such situations, silica tetrahedra tend to link to coordination polyhedral elements other than silica tetrahedra.

If none of the oxygen ions in a silica tetrahedron bond to other silicon ions in adjacent tetrahedra, the silica tetrahedron will occur as an isolated tetrahedral unit in the mineral structure. If all the oxygen ions in a silica tetrahedron bond to other silicon ions of adjacent tetrahedra, the silica tetrahedra form a three-dimensional framework structure of silica tetrahedra. If some of the oxygen ions in the silica tetrahedra are bonded to silicon ions in adjacent tetrahedra and others are bonded to other cations in adjacent coordination polyhedra, a structure that is intermediate between totally isolated silica tetrahedra and three-dimensional frameworks of silica tetrahedra will develop.

Six major silicate groups (Figure 2.21) are distinguished based upon the linkage patterns of silica tetrahedra and consist of the following: (1) nesosilicates, (2) sorosilicates, (3) cyclosilicates, (4) inosilicates, (5) phyllosilicates, and (6) tectosilicates. Nesosilicates ("island" silicates) are characterized by isolated silica tetrahedra that are not linked to other silica tetrahedra through shared oxygen ions. Sorosilicates ("bow-tie" silicates) contain pairs of silica tetrahedra linked through shared oxygen ions. In cyclosilicates ("ring" silicates), each silica tetrahedron is linked to two others through shared oxygen ions into ring-shaped structural units. In single-chain inosilicates each silica tetrahedron is linked

through shared oxygen anions to two other silica tetrahedra in the form of a long, one-dimensional chain-like structure. When two chains are linked through shared oxygen anions a double-chain inosilicate structure is formed. When chains are infinitely linked to one another through shared oxygen anions, a two-dimensional sheet of linked silica tetrahedra is formed which is the basic structural unit of phyllosilicates ("sheet" silicates). Finally, when silica tetrahedra are linked to adjacent silica tetrahedra by sharing all four oxygen anions, a three-dimensional framework of linked silica tetrahedra results, which is the basic structure of tectosilicates ("framework" silicates).

Because these constitute the most significant rock-forming minerals in Earth's crust and upper mantle they are discussed more fully in Chapter 5. In Chapter 3, we will further investigate significant aspects of mineral chemistry, including substitution solid solution and the uses of isotopes and phase stability diagrams in understanding Earth materials.

Chapter 3

Atomic substitution, phase diagrams and isotopes

3.1   ATOMIC (IONIC) SUBSTITUTION

Minerals are composed of atoms or ions that occupy structural sites in a crystal structure. Different ions can occupy the same structural site if they have similar size, have similar charge and are available in the environment in which the mineral is forming. This process of one ion replacing another ion is called **substitution**. In mineral formulas, ions that commonly substitute for one another are generally placed within a single set of parentheses. In the olivine group, iron and magnesium can freely substitute for one another in the sixfold, octahedral site. As a result, the formula for olivine is commonly written as $(Mg,Fe)_2SiO_4$.

Substitution is favored for ions of similar ionic radius. In general, cation substitution at surface temperatures and pressures is limited when differences in cation radii exceed 10–15% and becomes negligible for differences greater than 30%. Such ions are "too big" or "too small" to easily substitute for one another (Figure 3.1a), while ions of similar size are "just right". Substitution of ions of significantly different radii distorts coordination polyhedra and decreases the stability of crys-

tals. However, at higher temperatures, ions with larger differences in radius may substitute for one another.

Substitution is favored for ions of similar charge. Where substitutions occur in only one coordination site, substitution is largely limited to ions with the same charge (Figure 3.1b). This enables the mineral to remain electrically neutral, which increases its stability. However, where substitution can occur in multiple coordination sites, ions of different charge may substitute for one another in one site so long as this charge difference is balanced by a second substitution of ions of different charge in a second coordination site.

Substitution is favored for ions that are widely available in the environment in which the mineral is growing (Figure 3.1). As minerals grow, coordination sites will preferentially select ions with the appropriate radii and charge that are available in the vicinity of the growing crystal. The ions that occupy a coordination site in a mineral provide vital clues to the composition of the system, and the environmental conditions under which crystallization occurred.

3.1.1   Simple complete substitution

Simple complete substitution exists when two or more ions of similar radius and the same

Earth Materials, 1st edition. By K. Hefferan and J. O'Brien. Published 2010 by Blackwell Publishing Ltd.

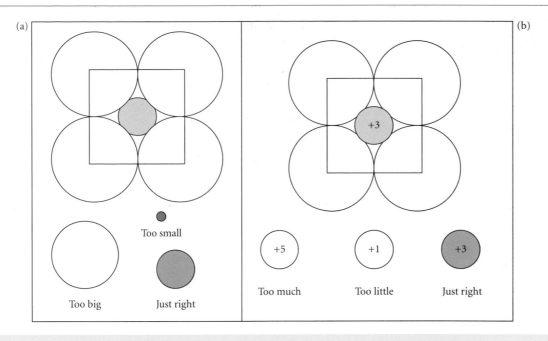

Figure 3.1 Criteria for substitution are similar size (a), similar charge (b) and availability.

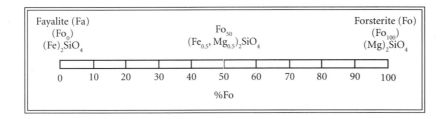

Figure 3.2 Olivine complete substitution solid solution series.

charge substitute for one another in a coordination site in any proportions. In such cases, it is convenient to define end members or **components** that have only one type of ion in the structural site in question. The olivine group illustrates complete substitution. In the olivine group, $(Mg,Fe)_2SiO_4$, Mg^{+2} (radius = 0.66 angstroms) and Fe^{+2} (radius = 0.74 angstroms) can substitute for one another in the octahedral site in any proportion. The two end members are the pure magnesium silicate component called forsterite $[(Mg)_2SiO_4]]$ and the pure iron silicate component called fayalite $[(Fe)_2SiO_4]$. Since these two end members can substitute for one another in any proportion in olivine, a **complete solid solution series** exists between them. As a result, the composition of any olivine can

be expressed in terms of the proportions of forsterite (Fo) and/or fayalite (Fa). Simple two-component, complete solid solution series are easily represented by a number line called a **tie line** between the two end members (Figure 3.2).

Compositions of any olivine can be represented in a number of different ways. For example, pure magnesium olivine can be represented by a formula (Mg_2SiO_4), by a name (forsterite), by its position on the tie line (far right) or by the proportion of either end member (Fo_{100} or Fa_0). Similarly, pure iron olivine can be represented by a formula (Fe_2SiO_4), by a name (fayalite), by its position on the tie line (far left) or by the proportion of either end member (Fo_0 or Fa_{100}). Any composition in the olivine complete solid solution

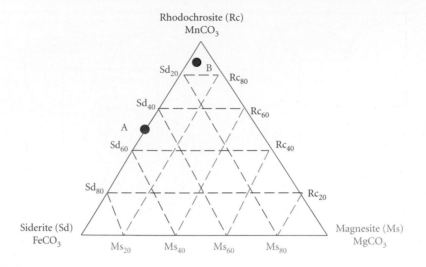

Figure 3.3   Compositions of carbonate minerals expressed in terms of the proportions of iron, magnesium and manganese; that is of the three components: siderite (Sd), magnesite (Ms) and rhodochrosite (Rc) plotted on a ternary diagram.

series can be similarly represented. For example, an olivine with equal amounts of the two end member components can be represented by the formula $[(Mg_{0.5},Fe_{0.5})_2SiO_4]$, by its position on the line (halfway between the ends) or by the proportions of either end member (Fo_{50} or Fa_{50}). Typically the forsterite component is used (e.g., Fo_{50}).

In cases where three ions substitute freely for one another in the same coordination site, it is convenient to define three end member components. Each of these end member components contains only one of the three ions in the structural site in which substitution occurs. For example, ferrous iron (Fe^{+2}), magnesium (Mg^{+2}) and manganese (Mn^{+2}) can all substitute for one another in any proportions in the cation site of rhombohedral carbonates. The general formula for such carbonate minerals can be written as $(Fe,Mg,Mn)CO_3$. The three end member components are the "pure" minerals siderite ($FeCO_3$), magnesite ($MgCO_3$) and rhodochrosite ($MnCO_3$). On a three-component diagram, the three pure end member components are plotted at the three apices of a triangle (Figure 3.3).

Points on the apices of the triangle represent "pure" carbonate minerals with only one end member component. Percentages of any component decrease systematically from 100% at the apex toward the opposite side of

the triangle where its percentage is zero. Each side of the triangle is a line connecting two end members. Points on the sides represent carbonate solid solutions between two end member components. Point A on Figure 3.3 lies on the side opposite the magnesite apex and so contains no magnesium. Because it lies halfway between rhodochrosite and siderite, its composition may be written as $Rc_{50}Sd_{50}$ or as $(Mn_{0.5},Fe_{0.5})CO_3$. Any point that lies within the triangle represents a solid solution that contains all three end member components. The precise composition of any three-component solid solution can be determined by the distance from the point to the three apices of the triangle. Point B in Figure 3.3 lies closest to the rhodochrosite apex and farthest from the magnesite apex and so clearly contains more Mn than Fe and more Fe than Mg. Its precise composition can be expressed as $Sd_{10}Ms_2Rc_{88}$ or as $(Fe_{0.10},Mg_{0.02},Mn_{0.88})CO_3$. Many other examples of three-component systems with complete solid solution exist. All may be represented in a similar fashion by their position on a triangular diagram.

3.1.2   Coupled (paired) ionic substitution

Coupled (paired) ionic substitution involves the simultaneous substitution of ions of different charges in two different structural sites

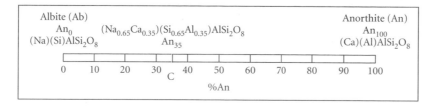

Figure 3.4 Coupled ionic substitution in the plagioclase solid solution series.

that preserves the electrical neutrality of the crystal lattice (Figure 3.4). The substitution of ions of different charge in one structural site changes the electric charge and requires a second set of substitutions of ions in a second structural site to balance that change in charge. Many examples of coupled ionic substitution exist; none are more important than those that occur in the plagioclase feldspars, the most abundant mineral group in Earth's crust.

In the plagioclase feldspars, similar size ions of sodium (Na^{+1}) and calcium (Ca^{+2}) can substitute for one another in any proportion in the large cation coordination site. However, when calcium (Ca^{+2}) substitutes for sodium (Na^{+1}), the positive charge of the crystal lattice is increased, and when the reverse occurs, the positive charge of the lattice is decreased. These changes in charge are balanced by a second set of substitutions. This second set of substitutions occurs in the small tetrahedral cation coordination site where aluminum (Al^{+3}) and silicon (Si^{+4}) substitute for one another. When a sodium (Na^{+1}) ion is added to the large cation coordination site, a silicon (Si^{+4}) ion is added to the small cation structural site. The two sites together contain a total charge of +5 that is balanced by the anions in the plagioclase structure. When a calcium (Ca^{+2}) is added to the large cation site, an aluminum (Al^{+3}) is added to the small cation site. Once again, the two sites together contain a total charge of +5 which is balanced by the anions in the plagioclase structure. If a sodium (Na^{+1}) ion replaces a calcium (Ca^{+2}) ion in the first coordination site, a silicon (Si^{+4}) ion must simultaneously replace an aluminum (Al^{+3}) ion in the second structural site for the two sites to total +5 and for the electrical neutrality of the crystal lattice to be maintained. Therefore all substitutions are paired and any change in the proportion of sodium

to calcium (Na/Ca) in the large ion site must be balanced by a similar change in the proportion of silicon to aluminum (Si/Al) in the small ion site. As a result, the general composition of plagioclase can be represented by the formula $(Na,Ca)(Si,Al)AlSi_2O_8$ to emphasize the nature of coupled ionic substitutions.

The plagioclase group can be represented as a two-component system with coupled ionic substitution (Figure 3.4). The two components are the "pure" sodium plagioclase called **albite** (**Ab**), whose formula can be written as $NaSiAlSi_2O_8$ (or $NaAlSi_3O_8$), and the "pure" calcium plagioclase called **anorthite** (**An**), whose formula can be written as $CaAlAlSi_2O_8$ (or $CaAl_2Si_2O_8$). Since a complete solid solution series exists between these two end members, any plagioclase composition can also be represented by its position on a tie line between the end member components or by the proportions of albite (Ab) and/or anorthite (An). In Figure 3.4, the composition of "pure" sodium plagioclase can be represented by (1) its position on the left end of the tie line, (2) Ab_{100}, (3) An_0, or (4) the formula $[(Na_{1.0},Ca_{0.0})(Si_{1.0},Al_{0.0})AlSi_2O_8]$ = $NaAlSi_3O_8$. Similarly, the composition of "pure" calcium plagioclase can be represented by (1) its position on the right end of the tie line, (2) Ab_0, (3) An_{100}, or (4) the formula $[(Na_{0.0},Ca_{1.0})(Si_{0.0},Al_{1.0})AlSi_2O_8]$ = $CaAl_2Si_2O_8$. Plagioclase compositions are generally expressed in terms of anorthite proportions, with the implication that the proportion of albite is (100 − An). An intermediate plagioclase solid solution such as the composition marked C on the tie line in Figure 3.4 has a composition that is represented by its position on the tie line, which can be represented as An_{35} (= Ab_{65}). An_{35} can also be expressed as $(Na_{0.65},Ca_{0.35})(Si_{0.65},Al_{0.35})AlSi_2O_8$ and indicates that 65% of the large cation site is occupied by sodium

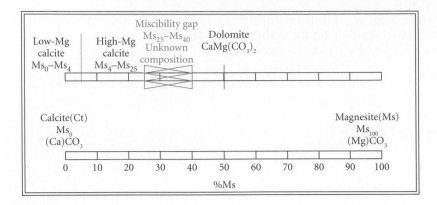

Figure 3.5 Limited substitution and miscibility gap in calcium–magnesium carbonates with compositional ranges of low-Mg calcite, high-Mg calcite and dolomite.

(Na^{+1}) ions and 35% by calcium (Ca^{+2}) ions with a coupled substitution of 65% silicon (Si^{+4}) and 35% aluminum (Al^{+3}) existing in the small cation site.

3.1.3 Limited ionic substitution

Substitution is limited by significant differences in the ionic radii or charge of substituting ions. Ions of substantially different size limit the amount of substitution so that only a **limited solid solution** can exist between end member components. This situation can be illustrated in the rhombohedral carbonates by the limited solid solution series that exists between calcite ($CaCO_3$), the major mineral in limestone and the raw material from which cement is refined, and magnesite ($MgCO_3$). Once again, the potential solid solution series can be represented as a line between the two end members, and the composition of any calcium–magnesium-bearing, rhombohedral carbonate may be represented by a formula, by its position on the tie line or by the proportion of an end member component (calcite = Ct or magnesite = Ms). However, because calcium cations (Ca^{+2}) are more than 30% larger than magnesium (Mg^{+2}) cations, the substitution between the two end members is limited. Because the amount of substitution is limited, many potential compositions do not exist in nature. Such gaps in a solid solution series are called **miscibility gaps** by analogy with immiscible liquids that do not mix in certain proportions. In this series, a miscibility gap exists between approximately $Ms_{25} = Ct_{75}$ and $Ms_{40} = Ct_{60}$ (Figure 3.5). To the left of this miscibility gap, a partial solid solution series exists between $Ms_0 = (Ca_{1.0},Mg_{0.0})CO_3$ and $Ms_{25} = (Ca_{0.75},Mg_{0.25}) CO_3$. Many organisms secrete shells in this compositional range. Within this range, we can define **low magnesium calcite** and **high magnesium calcite** in terms of their proportions of calcite (Ct) and magnesite (Ms) end members. Low magnesium calcites generally contain less than 4% magnesium (Mg^{+2}) substituting for calcium (Ca^{+2}) in this structural site and so have compositions in the range $Ct_{96-100} = Ms_{0-4}$ (Figure 3.5). High magnesium calcites have more than 4% magnesium substituting for calcium and therefore have compositions in the range $Ct_{75-96} = Ms_{4-25}$. Compositions from $Ms_{40-55} = Ct_{45-60}$ actually have a different structure – that of the double carbonate mineral **dolomite** whose average composition is $CaMg(CO_3)_2$. Many other examples exist of limited substitution series with miscibility gaps. The importance of mineral compositional variations that result from variations in substitution can be more fully understood in the context of phase diagrams, as discussed in the following section.

3.2 PHASE DIAGRAMS

The behavior of materials in Earth systems can be modeled using thermodynamic calculations and/or laboratory investigations. The results of such calculations and/or investigations are commonly summarized on phase stability diagrams. A **phase** is a mechanically separable part of the system. **Phase stability diagrams** display the stability fields for various phases separated by lines representing conditions under which phase changes occur. Phase stability diagrams related to igneous rocks and processes summarize relationships between liquids (melts) and solids (crystals) in a system. Such diagrams usually have temperature increasing upward on the vertical axis and composition shown on the horizontal axis. At high temperatures the system is completely melted. The stability field for 100% liquid is separated from the remainder of the phase diagram by a phase boundary line called the **liquidus** that represents the temperature above which the system exists as 100% melt and below which it contains some crystals. The low temperature stability field for 100% solid is separated from higher temperature conditions by a phase boundary line called the **solidus**. At intermediate temperatures between the solidus and liquidus, the system consists of two types of stable phases in equilibrium, both liquid and solid crystals. Phase equilibrium diagrams, based on both theoretical and laboratory analyses, exist for a variety of multicomponent systems. A one-component and five representative two-component systems related to the discussion of igneous rocks and processes (Chapters 7–10) are discussed below. Metamorphic phase diagrams are discussed in Chapter 18. For discussions of systems that are beyond the scope of this text, including three- and four-component systems, the reader is referred to mineralogy books by Dyar et al. (2008), Klein and Dutrow (2007), Wenk and Bulakh (2004) and Nesse (2000). Some of the more important terms you will encounter in this discussion are defined in Table 3.1.

3.2.1   The phase rule

The **phase rule** (Gibbs, 1928) governs the number of phases that can coexist in equilibrium in any system and can be written as:

$$P = C + 2 - F$$

where:

- P represents the number of phases present in a system. Phases are mechanically sepa-

Table 3.1   A list of some common terms used in phase diagrams.

Terms	Definitions
Liquidus	Phase boundary (line) that separates the all-liquid (melt) stability field from stability fields that contain at least some solids (crystals)
Solidus	Phase boundary (line) that separates the all-solid (crystal) stability field from stability fields that contain at least some liquid (melt)
Eutectic	Condition under which liquid (melt) is in equilibrium with two different solids
Peritectic	Condition under which a reaction occurs between a pre-existing solid phase and a liquid (melt) to produce a new solid phase
Phase	A mechanically separable part of the system; may be a liquid, gas or solid with a discrete set of mechanical properties and composition
Invariant melting	Occurs when melts of the same composition are produced by melting rocks of different initial composition
Incongruent melting	Occurs when a solid mineral phase melts to produce a melt and a different mineral with a different composition from the initial mineral
Discontinuous reaction	Mineral crystals and melt react to produce a completely different mineral; negligible solid solution exists between the minerals
Continuous reaction	Mineral crystals and melt react to continuously and incrementally change the composition of both; requires a mineral solid solution series
Solvus	Phase boundary (line) that separates conditions in which complete solid solution occurs within a mineral series from conditions under which solid solution is limited

rable varieties of matter that can be distinguished from other varieties based on their composition, structure and/or state. Phases in igneous systems include minerals of various compositions and crystal structures, amorphous solids (glass) and fluids such as liquids or gases. All phases are composed of one or more of the components used to define the composition of the system.

- C designates the minimum number of chemical components required to define the phases in the system. These chemical components are usually expressed as proportions of oxides. The most common chemical components in igneous reactions include SiO_2, Al_2O_3, FeO, Fe_2O_3, MgO, CaO, Na_2O, K_2O, H_2O and CO_2. All phases in the system can be made by combining components in various proportions.

- F refers to the number of degrees of freedom or variance. Variance means the number of independent factors that can vary, such as temperature, pressure and the composition of each phase, without changing the phases that are in equilibrium with one another. We will use the first phase diagram in the next section to show how the phase rule can be applied to understanding phase diagrams. A discussion of the phase rule and of phase diagrams related to metamorphic processes is presented in Chapter 18.

3.2.2 One-component phase diagram: silica polymorphs

Pure silica (SiO_2) occurs as a number of different mineral phases, each characterized by a different crystal structure. These silica minerals include low quartz (alpha quartz), high quartz (beta quartz), tridymite, cristobalite, coesite and stishovite. Each polymorph of silica is stable under a different set or range of temperature and pressure conditions. A phase stability diagram (Figure 3.6), where pressure increases upward and temperature increases to the right, shows the stability fields for the silica minerals. The stability fields represent the temperature and pressure conditions under which each mineral is stable. Each stability field is bounded by phase boundaries, lines that define the limits of the stability field as well as the conditions under which phases in adjoining fields can coexist. Where three phase boundaries intersect, a unique set of conditions is defined under which three stable phases can coexist simultaneously.

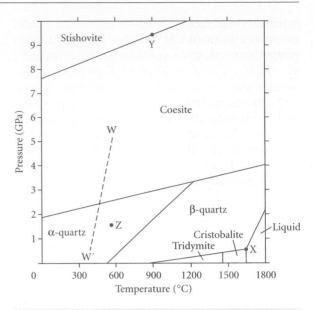

Figure 3.6 Phase diagram for silica depicting the temperature–pressure stability fields for the major polymorphs. (After Wenk and Bulakh, 2004.)

The positions of the stability fields show that **stishovite** and **coesite** are high pressure varieties of silica and that **tridymite** and **cristobalite** are high temperature/low pressure minerals. The high pressure polymorphs, coesite and stishovite, occur in association with meteorite impact and thermonuclear bomb sites, and stishovite is a likely constituent of the deep mantle. The diagram shows that quartz is the stable polymorph of silica over a broad range of temperature–pressure conditions common in Earth's crust. This wide stability range and an abundance of silicon and oxygen help to explain why quartz is such an abundant rock-forming mineral in the igneous, sedimentary and metamorphic rocks of Earth's crust. Figure 3.6 also shows that alpha quartz (low quartz) is generally more stable at lower temperatures than beta quartz (high quartz). Lastly, the diagram is bounded on the right by a phase stability boundary (melting curve) that separates the lower temperature conditions under which silica is solid from the higher temperature conditions under which it is a liquid.

The phase rule permits a deeper understanding of the relationships portrayed in the diagram. Places where three phase boundaries intersect represent unique temperature and pressure conditions where three stable mineral phases can coexist. For example, at point X at ~1650°C and ~0.4 GPa, high quartz, tridymite and cristobalite coexist because the high quartz/tridymite, high quartz/cristobalite and tridymite/cristobalite phase boundaries intersect. Because there are three phases and one component, the phase rule ($P = C + 2 - F$) yields $3 = 1 + 2 - F$, so that F must be 0. This simply means that the temperature and pressure cannot be varied if three phases are to coexist. There are no degrees of freedom.

Figure 3.6 shows that the three phases coexist only under this unique set of temperature and pressure conditions. If either temperature or pressure is varied, the system moves to a place on the diagram where one or more phases are no longer stable. There are no degrees of freedom; the system is invariant. As stated previously, two phases coexist under the conditions marked by phase stability boundaries. The phase rule ($P = C + 2 - F$) yields $2 = 3 - F$, so that F must be 1. For example, under the conditions at point Y (900°C, 9.2 GPa), both coesite and stishovite can coexist. If the temperature increases the pressure must also increase, and vice versa, in order for the system to remain on the phase stability boundary where these two phases coexist. There is only one independent variable or 1 degree of freedom. The temperature and pressure cannot be changed independently. In a similar vein, two phases, one solid and one liquid, can coexist anywhere above or below X on the melting curve that separates the liquid and solid stability fields.

However, for any point within a phase stability field (e.g., point Z) only one phase is stable (low quartz). The phase rule ($P = C + 2 - F$) yields $1 = 1 + 2 - F$, so that F must be 2. This means that the temperature and the pressure can change independently without changing the phase composition of the system. For point Z, the temperature and pressure can increase or decrease in many different ways without changing the phase that is stable. There are two independent variables and 2 degrees of freedom. Points to the right of the melting curve in the liquid field represent the stability conditions for a single phase, liquid silica.

One can also use this diagram to understand the sequence of mineral transforma-tions that might occur as Earth materials rich in silica experience different environmental conditions. A system cooling at a pressure of 0.3 bar will begin as melt. At ~1650°, cristobalite will begin to crystallize. As it cools, the system will reach the cristobalite/tridymite phase boundary (1470°C), where cristobalite will be transformed into tridymite and continue to cool until it reaches the tridymite/high quartz phase boundary. Here it will be transformed into high quartz, then cool through the high quartz field until it reaches the low quartz/high quartz phase boundary, where it will be converted to low quartz and continue to cool. Two phases will coexist only at phase boundaries during phase transformations that take finite amounts of time to complete (Chapter 4).

Similarly, a system undergoing decompression and cooling as it slowly rises toward the surface might follow line W–W′ on the phase diagram. It will start as coesite and be converted into alpha quartz (low quartz) as it crosses the phase boundary that separates them. Note that low quartz is the common form of quartz in low temperature, low pressure Earth materials.

3.2.3 Two-component phase diagram: plagioclase

Figure 3.7 is a phase stability diagram for plagioclase, the most abundant mineral group in Earth's crust. One critical line on the phase stability diagram is the high temperature, convex upward liquidus line, above which is the all liquid (melt) stability field that comprises the conditions under which the system is 100% liquid (melt). A second critical line is the lower temperature, convex downward solidus line, below which is the all solid stability field that comprises the conditions under which the system is 100% solid (plagioclase crystals). A third stability field occurs between the liquidus and the solidus. In this melt plus solid field, conditions permit both crystals and liquid to coexist simultaneously.

To examine the information that can be garnered from the plagioclase phase stability diagram, let us examine the behavior of a system with equal amounts of the two end member components albite and anorthite whose composition can be expressed as An_{50} (Figure 3.7). On the phase diagram, the system

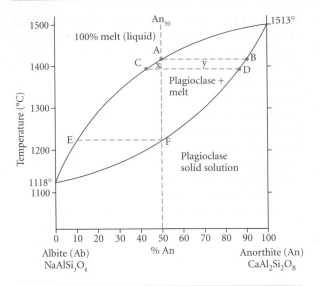

Figure 3.7 Plagioclase phase stability diagram with a complete solid solution between the two end member minerals albite (Ab) and anorthite (An).

is located on the vertical An_{50} composition line that is above the liquidus (100% liquid) at high temperatures, between the liquidus and solidus (liquid + solid) at intermediate temperatures and below the solidus (100% solid) at low temperatures. If this system is heated sufficiently, it will be well above the liquidus temperature for An_{50} and will be 100% melt, much like an ideal magma. Now let us begin to cool the system until the system reaches the liquidus (1420°C) at point A. Once the system moves incrementally below A, it moves into the melt plus solid field, so that crystallization of the melt begins at point A. To determine the composition of the first crystals, a horizontal line (A–B), called a tie line, is constructed between the liquidus and the solidus. The tie line represents the composition of the two phases (liquid and solid solution) in equilibrium with each other at that temperature. The intersection of the tie line with the liquidus (point A) represents the composition of the liquid ($\sim An_{50}$), because the melt has just begun to crystallize. The tie line intersection with the solidus (point B) represents the composition of the first solid solution mineral ($\sim An_{90}$) to crystallize from the melt.

As the system continues to cool, the composition of the melt continues to change incrementally down the liquidus line (e.g., to point C) while the composition of the crystalline solid solution simultaneously changes incrementally down the solidus line (e.g., to point D). This process continues as liquid compositions evolve down the liquidus and solid compositions evolve down the solidus until the vertical system composition line intersects the solidus at point F. Any further cooling brings the system into the 100% solid field. The tie line E–F at this temperature indicates that the last drops of liquid in the system have the composition $\sim An_{10}$, whereas the final solid crystals will be the same as the system composition ($\rightarrow An_{50}$).

Clearly the percentage of crystals must increase (from 0 to 100) and the percentage of melt must decrease (from 100 to 0) as cooling proceeds, while the composition of the melt and the crystals continuously changes down the liquidus and solidus lines, respectively. How does this happen? As the system cools through the melt plus solid field, two phenomena occur simultaneously. First, the melt and the existing crystals continuously react with one another so that crystal compositions are progressively converted into more albite-rich crystals (lower An) stable at progressively lower temperatures. Second, newly formed (lower An) crystals of the stable composition form and earlier formed crystals continue to grow as they react with the melt, so that the percentage of crystalline material increases progressively at the expense of melt. Crystal compositions evolve down the solidus line toward more albite-rich compositions (decreasing An) as temperature decreases. Liquid compositions evolve down the liquidus, also toward more albite-rich composition (decreasing An), as temperature decreases because the additional crystals that separate from the melt are always enriched in anorthite relative to the melt composition.

The precise proportion of melt and solid at any temperature can be determined by the lever rule. The **lever rule** states that the proportion of the tie line on the solidus side of the system composition represents the proportion of liquid in the system, whereas the proportion of the tie line on the liquidus side of the system composition represents the proportion of crystals (solid solution) in the system. In Figure 3.7, the proportion of tie line A–B on the solidus side of the system composition line is ~100% and the proportion on the liquidus side of the system composition line is ~0%. This makes sense because crystallization has just begun. So tie line A–B indicates that ~0% solids of composition An_{90} coexist with ~100% of composition An_{50}, just as crystallization begins. As the system cools (1) the percentage of crystals increases at the expense of the melt; (2) crystal composition evolves down the solidus; and (3) liquid composition evolves down the liquidus during continuous melt–crystal reaction and additional crystallization.

We can check this by drawing tie lines between the liquidus and the solidus for any temperature in which melt coexists with solids. Tie line C–D provides an example. In horizontal (An) units, this tie line is 45 units long (An_{86}–An_{41} = 45). The proportion of the tie line on the liquidus side of the system composition (x) that represents the percentage of crystals is 20% (9/45), whereas the proportion of the tie line on the solidus side (y) that represents the percentage of liquid is 80% (36/45). The system is 20% crystals of composition An_{86} and 80% liquid of composition An_{41}. As the system cooled from temperature A–B to temperature C–D, existing crystals reacted continuously with the melt and new crystals continued to separate from the melt. Therefore, the percentage of crystals progressively increases as crystal composition evolved incrementally down the solidus line and melt composition evolved incrementally down the liquidus line. When the system has cooled to the solidus temperature (1225°C), the proportion of the tie line (E–F) on the liquidus side approaches 100% indicating that the system is approaching 100% solid and the proportion of the solidus side approaches 0%, implying that the last drop of liquid of composition An_{10} is reacting with the remaining solids to convert them into An_{50}. We can use the albite–anorthite phase diagram to trace the progressive crystallization of any composition in this system. The lever rule can be used for compositions and temperatures other than those specifically discussed in this example.

The crystallization behavior of plagioclase in which An-rich varieties crystallize at high temperatures and react with the remaining melt to form progressively lower temperature Ab-rich varieties forms the basis for understanding the meaning of the continuous reaction series (Chapter 8) of Bowen's reaction series. Phase stability diagrams summarize what happens when equilibrium conditions are obtained. In the real world, disequilibrium conditions are common so that incomplete reaction between crystals and magmas occurs. These are discussed in the section of Chapter 8 that deals with fractional crystallization.

In addition, phase diagrams permit the melting behavior of minerals to be examined by raising the temperature from below the solidus. Let us do this with the same system we examined earlier (An_{50}). As the system is heated to the solidus temperature (1225°C), it will begin to melt. The lever rule (line E–F) indicates that the first melts (An_{10}) will be highly enriched in the albite component. As the temperature increases, the percentage of melt increases and the percentage of remaining crystals decreases as the melt and crystals undergo the continuous reactions characteristic of systems with complete solid solution. The melt continues to be relatively enriched in the albite (lower temperature) component, but progressively less so, as its composition evolves incrementally up the liquidus. Simultaneously, the remaining solids become progressively enriched in the anorthite (higher temperature) component as the composition of the solids evolves up the solidus. The lever rule allows us to check this at 1400°C where tie line C–D provides an example. The proportion of the tie line on the liquidus side of the system composition that represents the percentage of crystals is 20% (9/45), whereas the proportion of the tie line on the solidus side that represents the percentage of liquid is 80% (36/45). The system is 20% crystals of composition An_{86} and 80% liquid of composition An_{41}. Complete equilibrium melting of the system occurs at 1420°C (point A), where the last crystals of An_{90} melt to produce 100% liquid with the composition of the original system (An_{50}).

Why are phase diagrams important in understanding igneous processes? Several important concepts concerning melting in igneous systems are illustrated in the plagioclase phase diagram. All partial melts are enriched in low temperature components such as albite relative to the composition of the original rock. The smaller the amount of partial melting that occurs in a system, the more enriched are the melts in low temperature constituents such as albite. Progressively larger percentages of partial melting progressively dilute the proportion of low temperature constituents. In addition, if melts separate from the remaining solids, the solids are enriched in high temperature, refractory constituents. During crystallization, the liquidus indicates two conditions: (1) the temperature at which a system of a given composition (An content) begins to crystallize; and (2) the stable composition of any liquid in contact with crystals in the melt plus solid field. During crystallization, the solidus represents the stable composition of any solid crystals that are in contact with liquid in the melt plus solid field as crystallization continues and the temperature of crystallization.

Phase stability diagrams deliver quantitative information regarding the behavior of melts and crystals during both melting and crystallization. This provides simple models for understanding such significant processes as anatexis (partial melting) and fractional crystallization, which strongly influence magma composition and the composition of igneous rocks. All these topics are explored in the context of igneous rock composition, magma generation and magma evolution in Chapters 7 and 8. Let us now consider two-component systems with distinctly different end members, between which no solid solution exists, using the diopside–anorthite binary phase diagram.

3.2.4 Two-component phase diagram: diopside–anorthite

Figure 3.8 illustrates a simple type of two-component or binary phase stability diagram in which the two end members possess entirely different mineral structures so that there is no solid solution between them. The two components are the calcic plagioclase anorthite ($CaAl_2Si_2O_8$), a tectosilicate mineral, and the calcium-magnesium clinopyroxene diopside ($CaMgSi_2O_6$), a single-chain inosilicate mineral. The right margin of the diagram represents 100% anorthite component and the left margin represents 100% diopside component. Compositions in the system are expressed as weight % anorthite component; the weight

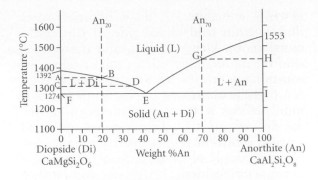

Figure 3.8 Diopside–anorthite phase diagram at atmospheric pressure.

% diopside component is 100% minus the weight % anorthite component. Temperature (°C) increases upward on the vertical axis.

The diopside–anorthite phase stability diagram illustrates the temperature–composition conditions under which systems composed of various proportions of diopside and anorthite end member components exist as 100% melt, melt plus solid crystals and 100% solid crystals. At high temperatures all compositions of the system are completely melted. The stability field for 100% liquid is separated from the remainder of the phase diagram by the liquidus. The liquidus temperature increases in both directions away from a minimum value of An_{42} (Di_{58}), showing that a higher anorthite (An) or a higher diopside (Di) component requires higher temperatures for complete melting. The phase diagram also shows that at low temperatures the system is completely crystallized. The stability field for 100% solid is separated from the remainder of the phase diagram by the solidus. For compositions of An_{100} (Di_0) and Di_{100} (An_0), the solidus temperature is the same as the liquidus temperature so that the solidus and liquidus intersect at 1553°C and 1392°C, respectively. For all intermediate compositions, the solidus temperature is a constant 1274°C.

The liquidus and solidus lines define a third type of stability field that is bounded by the two lines. This stability field represents the temperature–composition conditions under which both melt and crystals coexist so that a liquid of some composition coexists with a plagioclase solid solution of some composition. Two melt plus solid fields are defined: (1) a melt plus diopside field for compositions of <42% anorthite by weight, and (2) a melt plus anorthite field for compositions of >42% anorthite by weight. The liquidus and the solidus intersect where these two fields meet at a temperature of 1274°C and a composition of 42% anorthite by weight (An_{42}). This point defines a temperature trough in the liquidus where it intersects the solidus and is called a **eutectic point** (E in Figure 3.8). Let us use a couple of examples, one representative of compositions of <42% anorthite by weight and the other of compositions of >42% anorthite by weight, to illustrate how this works.

To investigate crystallization behavior in this system, we start with a system rich in diopside component with a composition of An_{20} (Di_{80}), at a temperature above the liquidus temperature for this composition (Figure 3.8). As the system cools it will eventually intersect the liquidus at a temperature of ~1350°C (point B in Figure 3.8). To determine the composition of the first crystals, a horizontal tie line (A–B) is drawn between the liquidus and the solidus. The intersection of the tie line with the liquidus (point B) represents the composition of the liquid (~An_{20}) because the melt has just begun to crystallize and its intersection with the solidus (point A) indicates the composition of the first crystals (diopside). As the system continues to cool, diopside crystals continue to form and grow, increasing the percentage of solid diopside crystals in the system while incrementally increasing its proportion of anorthite component as the percentage of melt decreases. As the system continues to cool to 1315°C (tie line C–D), the composition of the melt continues to change incrementally down the liquidus line (to point D) while the composition of the crystalline solid remains pure diopside (point C). As cooling continues, liquid compositions evolve down the liquidus and solid compositions evolve down the solidus until the vertical system composition line intersects the solidus at point E, after which any further cooling brings the system into the 100% solid (diopside plus anorthite) field. As the system approaches 1274°C (tie line E–F), it contains a large proportion of diopside crystals and a smaller proportion of melt with the composition ~An_{42}. When the system reaches the eutectic point at 1274°C, where the liquidus and solidus intersect, the remaining melt crys-

tallizes completely by isothermal eutectic crystallization of diopside and anorthite until all the melt has been crystallized. Cooling of the system below 1274°C causes it to enter the all-solid diopside plus anorthite field.

The percentage of crystals must increase (from 0 to 100) and the percentage of melt must decrease (from 100 to 0) as cooling proceeds, while the composition of the melt continuously changes down the liquidus and the solids are crystallized in the sequence all diopside prior to the eutectic and diopside plus anorthite at the eutectic. Can we quantify these processes? In Figure 3.8, the proportion of tie line A–B on the solidus side of the system composition line is ~100% and the proportion on the liquidus side of the system composition line is ~0%. This makes sense because crystallization has just begun. So tie line A–B indicates that ~0% solid diopside coexists with ~100% melt of composition An_{20}, at the moment crystallization begins. As the system cools, the percentage of crystals should increase at the expense of the melt as liquid composition evolves down the liquidus during continuous separation and growth of diopside crystals. We can check this by drawing tie lines between the liquidus and the solidus for any temperature in which melt coexists with solids. Tie line C–D provides an example. In horizontal (An) units, this tie line is 35 units long (An_{35}–An_0 = 35). The proportion of the tie line on the liquidus side of the system composition that represents the percentage of crystals is 43% (15/35), whereas the proportion of the tie line on the solidus side is 57% (20/35). The system is 43% diopside crystals (An_0) and 57% liquid of composition An_{35}. As the system cools from temperature A–B to temperature C–D, existing diopside crystals grow and new crystals continue to separate from the melt so that the percentage of crystals progressively increases as melt composition evolves incrementally down the liquidus line toward more anorthite-rich compositions. When the system approaches the eutectic temperature, the tie line (E–F) is 42 An units long and proportion of the tie line (E–F) on the liquidus side approaches 52% (22/42), indicating that the system contains 52% diopside crystals, and the proportion on the solidus side is 48% (20/42) liquid of composition An_{42}. At the eutectic temperature, diopside and anorthite simultaneously crystallize isothermally until the remaining melt is depleted. The proportion of crystals that form during eutectic crystallization of the remaining melt (48% of the system) is given by the lever rule as 42% (42/100) anorthite crystals and 58% (58/100) diopside crystals. The composition of the final rock is given by the proportions of the tie line between the solid diopside and solid anorthite that lie to the right and left of the system composition line. For this system, with a composition of An_{20}, the lever rule yields a final rock composition of 20% anorthite and 80% diopside.

The specific example related above is representative of the behavior of all compositions in this system between An_0 and An_{42}. When the system cools to the liquidus, diopside begins to crystallize, and as the system continues to cool, diopside continues to crystallize, so that the composition of the remaining melt evolves down the liquidus toward the eutectic. Separation of crystals from the melt causes melt composition to change. When the system reaches the eutectic composition, isothermal crystallization of diopside and anorthite occurs simultaneously until no melt remains.

For compositions between An_{42} and An_{100} (e.g., An_{70}), the system diopside–anorthite behaves differently. For these compositions, when the system cools to intersect the liquidus, the first crystals formed are anorthite crystals (tie line G–H). Continued separation of anorthite crystals from the cooling magma causes the melt to be depleted in anorthite component (and enriched in diopside component) so that the melt composition evolves down the liquidus line to the left. Tie lines can be drawn and the lever rule can be used for any temperature in the anorthite plus liquid field. When the system cools to approach the eutectic temperature (tie line E–I), it contains a proportion of anorthite crystals in equilibrium with a liquid of composition $\sim An_{42}$. At the eutectic temperature, both diopside and anorthite simultaneously crystallize isothermally in eutectic proportions (58% diopside, 42% anorthite) until no melt remains.

Several important concepts emerge from studies of the equilibrium crystallization of two-component eutectic systems such as diopside–anorthite: (1) which minerals crystallize first from a magma depends on the specifics of melt composition, (2) separation of crystals from the melt generally causes melt composition to change, and (3) multiple minerals can crystallize simultaneously from a magma. This means that no standard reaction series, such as Bowen's reaction series (Chapter 8), can be applicable to all magma compositions because the sequence in which minerals crystallize or whether they crystallize at all is strongly dependent on magma composition, as well as on other variables. It also means that the separation of crystals from liquid during magma crystallization generally causes magma compositions to change or evolve through time. These topics are discussed in more detail in Chapter 8, which deals with the origin, crystallization and evolution of magmas.

Phase diagrams can also provide simple models for rock melting and magma generation. To do this, we choose a composition to investigate starting at subsolidus temperatures low enough to ensure that the system is 100% solid, and then gradually raise the temperature until the system reaches the solidus line where partial melting begins. As temperature continues to rise, we can trace the changes in the composition and proportions of melts and solids, using the lever rule, until the system composition reaches the liquidus, which implies that it is 100% liquid. Let us examine such melting behavior, using the two compositions previously used in the discussion of crystallization. A solid system of composition 20% anorthite (An_{20}) and 80% diopside (Di_{80}) will remain 100% solid until it has been heated to a temperature of 1274°C, where it intersects the solidus. Further increase in temperature causes the system to enter the melt plus diopside field as indicated by tie line E–F. The composition of the initial melt is given by the intersection of the tie line with the liquidus (point E), so that first melts have the eutectic composition (An_{42}), and the composition of the remaining, unmelted solids is indicated by the intersection of the tie line with the solidus (point F = An_0 = Di_{100}). As the system is heated incrementally above the eutectic, the tie line (E–F) is 42 An units long and the proportion of the tie line on the liquidus side is 52% (22/42) indicating that the system contains 52% diopside crystals, and the proportion on the solidus side is 48% (20/42), indicating that all the anorthite and some of the diopside have melted at the eutectic to produce a liquid of composition An_{42}. At the eutectic temperature, both diopside and anorthite simultaneously melt isothermally until the remaining anorthite is completely melted. The proportion of crystals that melt during eutectic melting (48% of the system) is given by the lever rule and is 42% anorthite crystals and 58% diopside crystals as reflected in the melt composition. Further increases in temperature cause more diopside to melt, which increases the amount of melt and changes the melt composition toward less An-rich compositions as melt composition evolves up the liquidus toward progressively diopside-enriched, anorthite-depleted compositions. When the temperature approaches the liquidus temperature for the bulk composition (An_{20}) of the system, the lever line (A–B) clearly indicates that the system consists of nearly 100% melt (An_{20}) and nearly 0% diopside (An_0) as the last diopside is incorporated into the melt.

Several important concepts emerge from an examination of melting behavior in two-component systems such as diopside–anorthite: (1) the composition of first melts in such systems is the same – is invariant – for a wide variety of system compositions; (2) melt compositions depend on the proportion of melting so that increasing degrees of partial melting cause liquid compositions to change; and (3) changes in liquid composition depend on the composition of the crystals being incorporated into the melt. **Invariant melting** helps to explain why some magma compositions (e.g., basaltic magmas) are more common than others because some magma compositions can be generated by partial melting of a wide variety of source rock compositions. The dependence of melt composition on the degree of partial melting suggests that it might be an important influence on melt composition. The ways in which magma composition depends on the incorporation of constituents from crystals in contact with the melt is also discussed in Chapter 8 in conjunction with a discussion of magma origin and evolution.

3.2.5   Two-component phase diagram: albite–orthoclase

Mineral compositions may offer vital clues to the conditions under which they were pro-

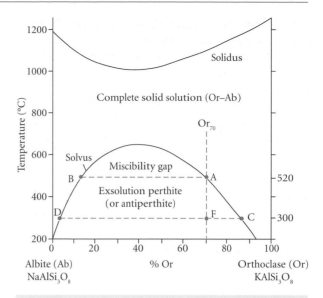

Figure 3.9   Albite–orthoclase phase diagram at atmospheric pressure.

duced. This is well illustrated by the temperature-dependent substitution of potassium (K^{+1}) and sodium (Na^{+1}) in the alkali feldspars $(Na,K)AlSi_3O_8$, as illustrated by the albite–orthoclase phase diagram (Figure 3.9).

At high temperatures (>~670°C) a complete substitution solid solution series exists between the two end members: (1) the potassium feldspar orthoclase ($KAlSi_3O_8$) and (2) the sodium plagioclase feldspar, albite ($NaAlSi_3O_8$). Crystals that form at high temperatures can have any proportions of orthoclase (Or) or albite (Ab) end member. Actual proportions will depend largely on the composition of the system; that is, the availability of potassium and sodium ions. Because a complete solid solution exists between the two end members, crystallization and melting in this system share many similarities with the albite–anorthite system (see Figure 3.7) discussed earlier. For systems with <40% Or, initial crystals are rich in the albite plagioclase component. As plagioclase crystals continue to separate on cooling, they react continuously with the melt so that crystal composition changes down the solidus as the remaining liquid changes composition down the liquidus, both toward increasing Or content until no melt remains.

The result is a rock composed of sodic plagioclase, with a potassic orthoclase component in solid solution. For systems with >40% Or, initial crystals are enriched in the potassium feldspar (orthoclase) component. As such crystals continue to separate on cooling, they react continuously with the melt so that crystal composition changes down the solidus as the remaining liquid changes composition down the liquidus, both toward decreasing Or content until all the melt is used up. The result is a rock composed of a feldspar solid solution. For systems with >40% Or, these crystals are potassic orthoclase crystals with an albite component in solid solution. All solid solutions between the two end members are stable at high temperatures after they begin to cool below the solidus temperature.

However, at lower temperatures (<~620°C), the solid solution between orthoclase and albite becomes limited and a miscibility gap exists in which the solid solution between the two end members is limited. The lower the temperature, the more limited the solid solution and the larger the miscibility gap becomes. As high temperature potassium-sodium feldspar solid solutions cool, they may eventually reach the **solvus** (Figure 3.9), a phase stability boundary that separates the conditions under which a complete solid solution is stable from conditions under which solid solutions are limited. The temperature at which a complete solid solution becomes unstable is called the solvus temperature and is generally highest for compositions with large amounts of both end members.

Let us examine a potassium-rich feldspar (line Or_{70} in Figure 3.9) that is a complete solid solution of composition Or_{70} (Ab_{30}) as it cools below the solidus temperature. As this feldspar cools it eventually intersects the solvus at point A, at a temperature of 520°C, below which the solid solution becomes unstable. At temperatures below the solvus, the original solid solution unmixes or **exsolves** into two stable, but distinctly different, feldspars whose compositions lie on the solvus line that borders the miscibility gap. In this case, plagioclase of composition Ab_{70} (Or_{30}) begins to unmix (exsolve) from the potassic feldspar as the solid solution becomes limited and a miscibility gap is created. Because the solid solution becomes increasingly limited and the miscibility gap widens as the temperature decreases, more plagioclase exsolves from the potassic feldspar and becomes increasingly sodic (Ab rich). As a result, the composition of the exsolved plagioclase evolves down the solvus to the left toward increasing Ab enrichment. Because so much albite component is exsolving from the potassic feldspar, its orthoclase content progressively increases as its composition evolves down the solvus to the right. The lever rule can be used to trace the proportions and the composition of the exsolved plagioclase and the potassic feldspar at any temperature. Tie line C–D (85 Or units long) between the two feldspar compositions on the solvus can be used for this purpose. For example, upon cooling to 300°C, the potassic feldspar component (point C) is Or_{95} and the plagioclase component (point D) is Or_3. The percentage of exsolved Ab-rich plagioclase, given by line segment C–F, is 21% (18/85), and the percentage of potash feldspar, given by line segment D–F, is 79% (67/85). Progressive unmixing produces one feldspar increasingly enriched in potassium (Or) and another feldspar increasingly enriched in sodium (Ab). For initially potassium-rich feldspar solid solutions, the result is a specimen of potassium-rich feldspar that contains sodium-rich feldspar blebs, stringers or patches. A potassium feldspar crystal that contains sodium feldspar blebs, stringers or patches produced by the unmixing or **exsolution** of two distinct feldspars is called **perthite**. Look closely at most potash feldspar crystals (e.g., orthoclase or microcline) and you will see the generally less transparent blebs and stringers of plagioclase produced by exsolution. For initially albite-rich compositions, the result can be plagioclase crystals called **antiperthite** that contain blebs, patches and/or stringers of orthoclase in albite. Antiperthite is less common than perthite because calcium-rich plagioclase does not form a solid solution series with orthoclase or any other potassic feldspar.

3.2.6 Two-component phase diagram: nepheline–silica

The nepheline–silica phase diagram (Figure 3.10) illustrates a type of two-component system in which there is an intermediate compound whose composition can be produced by combining the compositions of the two end member components. In this case silica

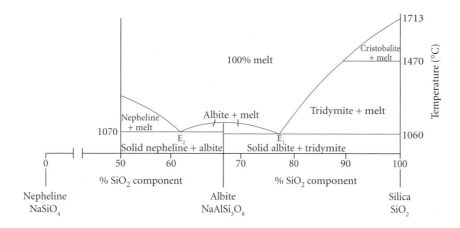

Figure 3.10 Phase diagram for the system nepheline–silica with the intermediate compound albite, at atmospheric pressure.

(SiO_2) and nepheline ($NaAlSiO_4$) are the two end member components. The intermediate compound formed by combining one molecular unit of nepheline and two of silica [$NaAlSiO_4 + 2(SiO_2)$] is the plagioclase mineral albite ($NaAlSi_3O_8$). No solid solution exists between nepheline, albite and silica minerals. Compositions are expressed on the horizontal axis in terms of molecular percent silica (SiO_2) component, so that the percentage of nepheline component is %Ne = 100% − %SiO_2 component. The composition of the intermediate compound albite is two-thirds SiO_2 component. Temperature increases on the vertical axis. The polymorphs of silica (see Figure 3.6) that crystallize in this system are cristobalite and tridymite, but the more common polymorph of silica is quartz. All these minerals have the same chemical composition but different crystal structures.

Several important concepts are illustrated and reinforced by the nepheline–silica phase diagram (Figure 3.10). The most important is the notion of **silica saturation**, which is fundamental to igneous rock classification. When there is sufficient silica component (more than two-thirds) so that each molecular unit of nepheline component can be converted into albite by adding a molecular unit of silica with an additional silica component remaining, the system is said to be oversaturated with respect to silica. Evidence for silica oversaturation is the presence of a silica mineral such as tridymite or quartz along with plagioclase feldspar in the final rock. When there is insufficient silica component (less than two-

thirds) to convert each molecular unit of nepheline into albite by adding a molecular unit of silica, the system is said to be undersaturated with respect to silica. Evidence for silica undersaturation is the presence of a low silica feldspathoid mineral such as nepheline in the final rock. Only when the silica component is exactly two-thirds is there precisely the amount of silica component required to convert each molecular unit of nepheline into albite. Such systems are said to be exactly saturated with respect to silica. Evidence for exact silica saturation is the presence of feldspar and the absence of both silica and feldspathoids from the final rock, which in the system nepheline–silica consists of 100% albite. The International Union of Geological Sciences (IUGS) classification of igneous rocks (Chapter 7) is largely based on the concept of silica saturation; rocks in the upper triangle contain quartz and feldspar, whereas those in the lower half contain feldspathoids and feldspar. Rocks that lie on the join between the two triangles contain feldspar but neither quartz nor feldspathoids and are approximately saturated with respect to silica.

The nepheline–silica phase diagram shows some similarity to the diopside–anorthite phase diagram. The most significant difference is the presence of two eutectic points where troughs in the liquidus intersect the solidus. For many purposes, this diagram may be interpreted as two side-by-side eutectic diagrams: one diagram for undersaturated compositions (less than two-thirds silica component), with a eutectic point at 1070°C and 62% silica component, and a second diagram for oversaturated compositions (over two-thirds silica component), with a eutectic point at 1060°C and 77% silica component. A brief discussion of the crystallization and melting behaviors for these two compositional ranges follows.

Let us first examine the behavior of silica oversaturated systems with between 78 and 100% silica component. If a cooling melt with a silica content between 89 and 100% intersects the liquidus, the first crystals to separate are composed of the silica polymorph called cristobalite. With continued cooling (Figure 3.10), more cristobalite separates from the melt, causing its composition to evolve down the liquidus toward lower percentages of silica as the proportion of melt decreases. As the system reaches a temperature of 1470°C, cristobalite becomes unstable and inverts isothermally to the stable polymorph of silica called tridymite. This ideal inversion temperature is shown by the phase boundary between cristobalite and tridymite in the silica plus melt field. With continued cooling below 1470°C, more tridymite separates from the melt, and melt composition continues to evolve down the liquidus toward the eutectic (E$_1$) at 1060°C. Upon reaching the eutectic, both albite and tridymite crystallize simultaneously until the melt is used up and the system enters the albite plus tridymite field. For compositions between 78 and 89% silica component, the behavior is similar except that the first crystals to form are tridymite.

For compositions between 67 and 78% SiO$_2$ (Figure 3.10), albite crystallizes when the system cools to the liquidus temperature. Continued separation of albite on cooling causes the liquid composition to move down the liquidus toward increasing silica content. As the system cools to the eutectic temperature of 1060°C, albite and tridymite crystallize simultaneously and isothermally until the melt is used up. The final rock contains both albite and a silica mineral in proportions that can be determined by the lever rule.

Let us now examine the behavior of so-called silica undersaturated systems with between 0 and 50% silica component. For compositions between 62 and 67% silica, cooling of the system to the liquidus temperature causes albite crystals to separate from the melt (Figure 3.10). Continued cooling below the liquidus temperature causes further separation of albite from the melt; this causes melt compositions to change down the liquidus toward decreasing silica content. As the eutectic temperature (E$_2$) is reached at 1070°C, both albite and nepheline crystallize isothermally until the melt is used up. The final rock contains percentages of both albite and nepheline that can be determined by the lever rule. Lastly, for those compositions with 50–62% silica component covered in the diagram (complexities exist for systems with lower amounts of silica component), the first crystals to separate are nepheline crystals. Continued separation of silica-poor nepheline causes melt compositions to change down the liquidus toward the eutectic at 1070°C. At the eutectic, both albite and nepheline crystallize isothermally until the melt is used up. Once again the final rock is composed of albite and nepheline, and their percentages can be calculated using the lever rule.

3.2.7 Two-component phase diagram: forsterite–silica

Another set of mineral relationships is well illustrated by the two-component system forsterite–silica (Figure 3.11). Forsterite

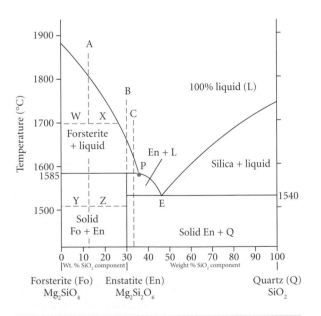

Figure 3.11 Phase diagram for the system forsterite–silica with the intermediate compound enstatite, at atmospheric pressure.

(Mg_2SiO_4) is the magnesium end member of the olivine solid solution series, and the silica mineral is commonly quartz (SiO_2). As in the nepheline–silica system discussed above, this system contains an intermediate compound, in this case the orthopyroxene mineral enstatite ($MgSi_2O_6$), that can be thought of as being composed of one molecular unit of each of the two end member components. The horizontal axis in this phase diagram is weight % silica end member component (%SiO_2), rather than the molecular proportion used in the nepheline–silica diagram. Temperature increases on the vertical axis. Six phase stability fields are defined: (1) 100% melt, (2) melt + quartz, (3) melt + enstatite, (4) melt + forsterite, (5) forsterite + enstatite, and (6) enstatite + quartz. No solid solution exists between the three minerals in this system (forsterite, enstatite and quartz). Instead, **discontinuous reactions** occur between forsterite and enstatite in which early formed minerals react with the melt to produce different minerals. These reactions occur when the system reaches point P on the liquidus line, the **peritectic point** at 1585°C and 35% silica component. There is also a eutectic point (E) located in the trough in the liquidus where it intersects the solidus at 1540°C and 46% silica. Let us examine four selected compositions in this system during crystallization, each of which demonstrates different behaviors and/or results.

For compositions of >46% silica component by weight, the system behaves as a simple eutectic system. As melts cool to the liquidus, silica (quartz) begins to separate from the melt and continues to separate as the system cools further (Figure 3.11). This causes the composition of the melt to evolve down the liquidus toward lower silica contents. Upon reaching the eutectic at 1540°C, both quartz and enstatite crystallize simultaneously. For compositions of 35–46% silica that are richer in silica component than the peritectic (P) composition, the system also behaves as a simple eutectic. The only change is that enstatite crystallizes first, causing the liquid to evolve down the solidus toward increasing silica content until it reaches the eutectic, where enstatite and quartz crystallize simultaneously until the melt is used up. The lever rule can be used to analyze phase percentages

and compositions for any composition in which two phases coexist.

Systems of between 0 and 35% silica component behave somewhat differently because they pass through the peritectic point where reactions occur between forsterite olivine, enstatite and melt. During cooling and crystallization, three fundamentally different situations can be recognized. For example, with a composition of 12% silica component (dashed line A, Figure 3.11), the system cools to the liquidus at 1810°C, where forsterite begins to separate. Continued cooling causes additional forsterite to separate from the melt, which causes the melt composition to evolve down the liquidus toward the peritectic. At 1700°C, the system consists of 50% forsterite crystals (line segment x) and 50% melt (line segment w) with a composition of 24% silica component as inferred from the lever rule. Further cooling and separation of forsterite crystals cause the melt composition to approach the peritectic point at 1580°C, where the lever rule shows that the system contains 66% forsterite and 34% melt of the peritectic composition (35% silica component). Below this temperature the system enters the 100% solid forsterite plus enstatite field with 60% forsterite olivine (line segment z) and 40% enstatite (line segment y). So what happens when the melt reaches the peritectic? The percentage of solids increases as the melt is used up, and the percentage of solid forsterite decreases while the percentage of solid enstatite increases dramatically. The percentage of forsterite decreases because some of the forsterite reacts with some of the remaining melt to produce enstatite. The percentage of enstatite increases dramatically because some olivine is converted to enstatite, while new enstatite crystallizes simultaneously from the remaining melt until it is used up. More generally, for all compositions of <30% silica component (enstatite composition), the equilibrium behavior is (1) forsterite crystallization as the melt cools below the liquidus; (2) increasing proportions of olivine and decreasing proportions of melt as the melt cools; (3) evolution of the remaining silica-enriched melt down the liquidus toward the peritectic; and (4) isothermal conversion of some forsterite to enstatite by discontinuous reaction with the remaining melt at the peritectic accompanied by additional isothermal crystallization of enstatite until the melt is used up. Some forsterite always remains because there is insufficient silica component to convert all of it into the intermediate compound enstatite. The peritectic reaction that converts the olivine mineral forsterite to the pyroxene mineral enstatite is called a **discontinuous reaction** as it is an excellent example of how early formed crystals can react with remaining melt to produce an entirely different mineral. These reactions are characteristic of the minerals in the discontinuous reaction series of Bowen's reaction series (Chapter 8).

Systems with exactly 30% silica component have the composition of the intermediate compound enstatite. They are precisely saturated with respect to silica. Line B (Figure 3.11) indicates the cooling behavior of such a system. Initial crystallization at the liquidus produces forsterite crystals. Continued cooling, accompanied by addition, separation and growth of forsterite crystals, causes the remaining melt to evolve down the liquidus toward the peritectic with forsterite and melt percentages and melt compositions given by the lever rule. When this system reaches the peritectic, it contains 14% forsterite crystals and 86% melt with 35% silica component. Discontinuous reaction between the remaining melt and the forsterite crystals converts all the forsterite crystals to enstatite, while simultaneous crystallization of enstatite causes the melt to be used up. The final rock is 100% enstatite and is neither undersaturated (lacks forsterite) nor oversaturated (lacks quartz) with respect to silica.

Systems with silica component contents of 30–35% silica are oversaturated with respect to silica and thus exhibit another set of behaviors. Line C (33% silica component) is representative of these behaviors (Figure 3.11). The system cools to the liquidus at 1650°C where forsterite crystals begin to separate. Continued separation of forsterite causes melt composition to evolve down the liquidus toward the peritectic. As the melt composition reaches the peritectic, the lever rule shows that the system contains 6% (2/35) forsterite crystals and 94% (33/35) melt with 35% silica component. At the peritectic, all the forsterite is converted to enstatite with the remaining melt and additional enstatite crystallizes, but additional melt remains. The lever rule shows that as the system leaves the peritectic and enters

the enstatite plus liquid stability field it contains 40% (2/5) enstatite and 60% (3/5) melt. Further cooling leads to additional crystallization of enstatite (30% silica component), which causes the remaining melt to evolve down the liquidus toward the eutectic (at 46% silica component). As the system reaches the eutectic, it contains 81% enstatite (13/16) and 19% (3/16) melt with 46% silica component. At the eutectic, enstatite and quartz crystallize simultaneously until the melt is used up. The final rock contains 96% (67/70) enstatite and 4% (3/70) quartz.

Compositions in this system between 30 and 35% silica component clearly show that minerals such as forsterite that are undersaturated with respect to silica can crystallize from magmas that are oversaturated with respect to silica. If equilibrium conditions between these crystals and the melt are maintained, they will eventually be converted to the intermediate compound and therefore will not be preserved. However, if disequilibrium conditions occur, of the kind that commonly occur during fractional crystallization, early formed crystals may well be preserved in the final rock. In addition, because silica in the remaining melt was not used to convert forsterite to enstatite, the melt will be more enriched in silica than would otherwise be the case. As discussed in Chapter 8, such concepts are very important in understanding the evolution of magma composition.

The system forsterite–silica (Figure 3.11) clearly illustrates the concept of silica saturation. Compositions of >30% silica (SiO_2) end member component by weight are oversaturated with respect to silica, so that there is sufficient silica to convert all the forsterite into enstatite. Equilibrium crystallization in such silica-rich systems produces the intermediate compound enstatite with excess silica to form quartz. As discussed in connection with the nepheline–silica diagram (see Figure 3.10), quartz forms by equilibrium crystallization of melts that are oversaturated with respect to silica. On the other hand, compositions of <30% silica component by weight are undersaturated with respect to silica, so that there is insufficient silica to convert all the forsterite into enstatite. Equilibrium crystallization in such silica-poor systems produces forsterite plus as much of the intermediate compound enstatite as can be converted by discontinuous reaction at the peritectic by the available silica. Forsterite forms by equilibrium crystallization of melts that are undersaturated with respect to silica. As detailed in Chapter 7, both forsterite-rich olivine and feldspathoids suggest crystallization from systems undersaturated with respect to silica. Systems with exactly 30% silica component are exactly saturated with respect to silica because they possess precisely the amount of silica component required to convert forsterite into enstatite without excess silica remaining.

One can also investigate melting behaviors in this system. For compositions of >35% silica, the system behaves as a simple eutectic, producing first melts with a composition of 46% silica. Melts possess this composition until either quartz (for systems 35-46% silica component) or enstatite (for systems >46% silica component) is completely melted. Subsequent melting of the remaining mineral causes the liquid to change composition up the liquidus. These behaviors once again demonstrate the ways in which melt compositions depend both on the percentage of partial melting and on the composition of the original rock. For compositions of <35% silica, melting involves peritectic reactions. In these systems, whenever the system reaches the peritectic, some or all of the enstatite remaining in the solid fraction melts to produce both forsterite and melt at the peritectic (35% silica component). This behavior is essentially the reverse of what happens when systems cool through the peritectic, and melt plus olivine yields enstatite. Such behavior, in which the melting of one crystalline material produces both a new crystalline material and a melt of different compositions, is called **incongruent melting**. It also illustrates how silica oversaturated melts might be obtained from the partial melting of silica undersaturated, forsterite-rich rocks such as the ultramafic peridotites in the mantle.

3.3   ISOTOPES

This section provides a brief introduction to the uses of radioactive isotopes and stable isotopes (Chapter 2) in understanding Earth materials and processes. Isotope studies provide powerful insights concerning the age, behavior and history of Earth materials. In geology, a thorough understanding of both stable and radioactive isotopes is essential in determining the ages and origin of minerals and rocks. Isotope ratios, determined by mass spectroscopy, are also instrumental in understanding a variety of other phenomena discussed in this book, including the determination of:

1   Source rocks from which magmas are derived.
2   Origin of water on Earth's surface.
3   Source rocks for petroleum and natural gas.
4   Changes in ocean water temperatures, biological productivity and circulation.
5   History of ice age glacial expansions and contractions.
6   Timing of mountain building events involving igneous intrusions and metamorphism.

3.3.1   Stable isotopes

Stable isotopes contain nuclei that do not tend to change spontaneously. Instead, their nuclear configurations (number of protons and neutrons) remain constant over time. Many elements occur in the form of multiple stable isotopes with different atomic mass numbers. In many cases, these isotopes exhibit subtly different behaviors in Earth environments. These differences in behavior are recorded as differences in the ratios between isotopes that can be used to infer the conditions under which the isotopes were selectively incorporated into Earth materials. We will use oxygen and carbon isotopes to illustrate the uses of stable isotope ratios in increasing our understanding of Earth materials and processes.

Oxygen isotopes

Three isotopes of oxygen occur in Earth materials (Chapter 2): oxygen-18 (^{18}O), oxygen-17 (^{17}O) and oxygen-16 (^{16}O). Each oxygen isotope contains eight protons in its nucleus; the remaining mass results from the number of neutrons in the nucleus. ^{16}O constitutes >99.7% of the oxygen on Earth, ^{18}O constitutes ~0.2% and ^{17}O is relatively rare. The ratio $^{18}O/^{16}O$ provides important information concerning Earth history.

During evaporation, water with lighter ^{16}O is preferentially evaporated relative to water with heavier ^{18}O. During the evaporation of ocean water, water vapor in the atmosphere is enriched in ^{16}O relative to ^{18}O (lower $^{18}O/^{16}O$) while the remaining ocean water is preferentially enriched in ^{18}O relative to ^{16}O (higher $^{18}O/^{16}O$). Initially, these ratios were related to temperature because evaporation rates are proportional to temperature. Higher $^{18}O/^{16}O$ ratios in ocean water record higher temperatures, which cause increased evaporation and increased removal of ^{16}O. It was quickly understood that organisms using oxygen to make calcium carbonate ($CaCO_3$) shells could preserve this information as carbonate sediments accumulated on the sea floor over time. Such sediments would have the potential to record changes in water temperature over time, especially when the changes are large and the signal is clear (see Box 3.1).

However, it was soon realized that small, short-term temperature signals could be largely obliterated by a second set of processes that involve changes in global ice volumes associated with the expansion and contraction of continental glaciers, e.g., during ice ages. Glaciers expand when more snow accumulates each year than is ablated, leading to a net growth in glacial ice volume. Because atmospheric water vapor largely originates by evaporation, it is enriched in ^{16}O and has a low $^{18}O/^{16}O$ ratio. As glaciers expand, they store huge volumes of water with low $^{18}O/^{16}O$ ratios, causing the $^{18}O/^{16}O$ ratio in ocean water to progressively expand. As a result, periods of maximum glacial ice volume correlate with global periods of maximum $^{18}O/^{16}O$ in marine sediments. Prior to the use of oxygen isotopes, the record of Pleistocene glaciation was known largely from glacial till deposits on the continent, and only four periods of maximum Pleistocene glaciation had been established. Subsequently, the use of oxygen isotope records from marine sediments and ice (H_2O) cores in Greenland and Antarctica has established a detailed record that involves dozens of glacial ice volume expansions and contractions during the Pliocene and Pleistocene.

$^{18}O/^{16}O$ ratios are generally expressed with respect to a standard in terms of $\delta^{18}O$. The standard is the $^{18}O/^{16}O$ ratio in a belemnite from the Cretaceous Pee Dee Formation of South Carolina, called PDB. $\delta^{18}O$ is usually expressed in parts per thousand (mils) and calculated from:

$$\delta^{18}O = \frac{[^{18}O/^{16}O_{SAMPLE} - {}^{18}O/^{16}O_{PDP}]}{^{18}O/^{16}O_{PDB}} \times 1000$$

Because the Cretaceous was an unusually warm period in Earth's history, with high evaporation rates, PDB has an unusually high $^{18}O/^{16}O$ ratio. As a result, most Pliocene–Pleistocene samples have a negative $\delta^{18}O$, with small negative numbers recording maximum glacial ice volumes and larger negative numbers recording minimum glacial ice volumes. Because different organisms selectively fractionate ^{18}O and ^{16}O, a range of organisms must be analyzed and the results averaged when determining global changes in $^{18}O/^{16}O$. Nothing is ever as easy as it first seems.

It should be noted that many $\delta^{18}O$ analyses have used a different standard. This standard is the average $^{18}O/^{16}$ ratio in ocean water known as standard mean ocean water (SMOW). Because the original SMOW and PDB standards have been used up in comparative analyses, yet another standard, Vienna standard mean ocean water (VSMOW), is also used. This name is misleading as the Vienna standard is actually a pure water sample with no dissolved solids. There is currently much discussion concerning the notion of which standards are most appropriate and how $\delta^{18}O$ and other isotope values should be reported.

Carbon isotopes

Three isotopes of carbon occur naturally in Earth materials: carbon-12 (^{12}C), carbon-13 (^{13}C) and the radioactive carbon-14 (^{14}C). Each carbon isotope contains six protons in its nucleus; the remaining mass results from the number of neutrons (six, seven or eight) in the nucleus. ^{12}C constitutes >98.9% of the stable carbon on Earth, and ^{13}C constitutes most of the other 1.1%.

When organisms synthesize organic molecules, they selectively utilize ^{12}C in preference to ^{13}C so that organic molecules have lower than average $^{13}C/^{12}C$ ratios. Enrichment of the organic material in ^{12}C causes the $^{13}C/^{12}C$ in the water column to increase. Ordinarily, there is a rough balance between the selective removal of ^{12}C from water during organic synthesis and its release back to the water column by bacterial decomposition. Mixing processes produce a relatively constant $^{13}C/^{12}C$ ratio in the water column. During periods of stagnant circulation in the oceans or other water bodies, disoxic–anoxic conditions develop in the lower part of the water column and/or in bottom sediments, which inhibit bacterial decomposition (Chapters 12 and 14) and lead to the accumulation of ^{12}C-rich organic sediments. These sediments have unusually low $^{13}C/^{12}C$ ratios; as they accumulate, the remaining water column, depleted in ^{12}C, is progressively enriched in $^{13}C/^{12}C$. However, any process, such as the return of vigorous circulation and oxidizing conditions, that rapidly releases the ^{12}C-rich carbon from organic sediments is associated with a rapid decrease in $^{13}C/^{12}C$. By carefully plotting changes in $^{13}C/^{12}C$ ratios over time, paleo-oceanographers have been able to document both local and global changes in oceanic circulation. In addition, because different organisms selectively incorporate different ratios of ^{12}C to ^{13}C, the evolution of new groups of organisms and/or the extinction of old groups of organisms can sometimes be tracked by rapid changes in the $^{13}C/^{12}C$ ratios of carbonate shells in marine sediments or organic materials in terrestrial soils.

$^{13}C/^{12}C$ ratios are generally expressed with respect to a standard in terms of $\delta^{13}C$. The

Box 3.1   The Paleocene–Eocene thermal maximum

In the mid-19th century, scientists recognized a rapid change in mammalian fossils that occurred early in the Tertiary era. The earliest Tertiary epoch, named the Paleocene (early life), was dominated by archaic groups of mammals that had mostly been present in the Mesozoic. The succeeding period, marked by the emergence and rapid radiation of modern mammalian groups, was called the Eocene (dawn of life). The age of the Paleocene–Eocene boundary is currently judged to be 55.8 Ma. Later workers noted that the boundary between the two epochs was also marked by the widespread extinction of major marine groups, most prominently deep-sea benthic foraminifera (Pinkster, 2002). The cause of these sudden biotic changes remained unknown. Oxygen and carbon isotope studies have given us some answers.

Kennett and Stott (1991) reported a rapid rise in $\delta^{18}O$ at the end of the Paleocene, which they interpreted as resulting from a rapid rise in temperature, since they believed that no prominent ice sheets existed at this time. Subsequent work (e.g., Zachos et al., 1993; Rohl et al., 2000) has confirmed that temperatures rose between 6 and 8°C at high latitudes and ~2°C at low latitudes over a time interval not longer than 20,000 years. Rapid global warming, in this case the Paleocene–Eocene thermal maximum (PETM), has apparently occurred in the past, with significant implications for life on Earth. Researchers have also shown that the higher temperatures lasted for approximately 100 ka (Pinkster, 2002). How long will the current period of global warming last?

What caused the rapid global warming? Researchers studying carbon isotopes have shown that the sudden increase in temperature implied by rising $\delta^{18}O$ values corresponds with sudden decreases in $\delta^{13}C$. Several hypotheses have been suggested, most of which involve the release of large quantities of ^{12}C from organic carbon reservoirs. Two rapid spikes in negative $\delta^{13}C$, each occurring over time periods of less than 1000 years, suggest that some releases were extremely rapid. The currently favored hypothesis involves the melting of frozen clathrates in buried ocean floor sediments. Clathrates consist of frozen water in which methane, methanol and other organic carbon molecules are trapped. The hypothesis is that small amounts of warming cause clathrates to melt, releasing large volumes of methane to the atmosphere in sudden bursts. This would account for the sudden negative $\delta^{13}C$ spikes. Because methane (CH_4) is a very effective greenhouse gas (10–20 times more effective than CO_2), this theory also accounts for the sudden warming of Earth's surface and the extinction and mammalian radiation events that mark the Paleocene–Eocene boundary. Of course scientists wonder if the current episode of global warming might be accelerated by the sudden release of clathrates, and how long the effects of such releases might linger.

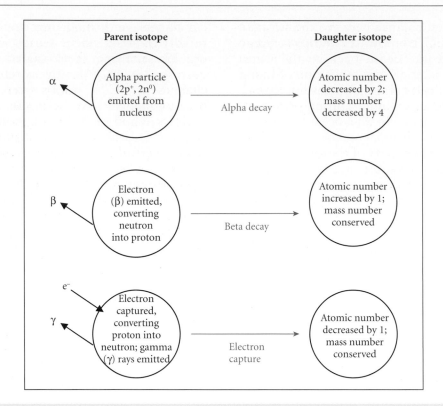

Figure 3.12 The three types of radioactive decay: alpha decay, beta decay and electron capture (gamma decay) and the changes in nuclear configuration that occur as the parent isotope decays into a daughter isotope.

standard once again is the $^{13}C/^{12}C$ ratio Pee Dee Belemnite, or PDB. $\delta^{13}C$ is usually expressed in parts per thousand (mils) and calculated from:

$$\delta^{13}C = \frac{[^{13}C/^{12}C_{SAMPLE} - ^{13}C/^{12}C_{PDP}]}{^{13}C/^{12}C_{PDB}} \times 1000$$

Box 3.1 illustrates an excellent example of how oxygen and carbon isotope ratios can be used to document Earth history, in this case a period of sudden global warming that occurred 55 million years ago.

3.3.2 Radioactive isotopes

Radioactive isotopes possess unstable nuclei whose nuclear configurations tend to be spontaneously transformed by **radioactive decay**. Radioactive decay occurs when the nucleus of an unstable **parent isotope** is transformed into that of a **daughter isotope**. Daughter isotopes have different atomic numbers and/or differ-

ent atomic mass numbers from the parent isotope. Three major radioactive decay processes (Figure 3.12) have been recognized: alpha (α) decay, beta (β) decay and electron capture.

Alpha decay involves the ejection of an alpha (α) particle plus gamma (γ) rays and heat from the nucleus. An alpha particle consists of two protons and two neutrons, which is the composition of a helium (^4He) nucleus. The ejection of an alpha particle from the nucleus of a radioactive element reduces the atomic number of the element by two ($2p^+$) while reducing its atomic mass number by four ($2p^+ + 2n^0$). The spontaneous decay of uranium-238 (^{238}U) into thorium-234 (^{234}Th) is but one of many examples of alpha decay.

Beta decay involves the ejection of a beta (β) particle plus heat from the nucleus. A beta particle is a high-speed electron (e^-). The ejection of a beta particle from the nucleus of a radioactive element converts a neutron into a proton ($n^0 - e^- = p^+$) increasing the atomic number by one while leaving the atomic mass number unchanged. The spontaneous decay of radioactive rubidium-87 (Z = 37) into stable strontium-87 (Z = 38) is one of many examples of beta decay. **Electron capture** involves the addition of a high-speed electron to the nucleus with the release of heat in the form of gamma rays. It can be visualized as the reverse of beta decay. The addition of an electron to the nucleus converts one of the protons into a neutron ($p^+ + e^- = n^0$). Electron capture decreases the atomic number by one while leaving the atomic mass number unchanged. The decay of radioactive potassium-40 (Z = 19) into stable argon-40 (Z = 18) is a useful example of electron capture. It occurs at a known rate, which allows the age of many potassium-bearing minerals and rocks to be determined. Only about 9% of radioactive potassium decays into argon-40; the remainder decays into calcium-40 (^{40}Ca) by beta emission.

The time required for one half of the radioactive isotope to be converted into a new isotope is called its **half-life** and may range from seconds to billions of years. Radioactive decay processes continue until a stable nuclear configuration is achieved and a stable isotope is formed. The radioactive decay of a parent isotope into a stable daughter isotope may involve a sequence of decay events. Table 3.2 illustrates the 14-step process required to convert radioactive uranium-238 (^{238}U) into the stable daughter isotope lead-206 (^{206}Pb).

All of the intervening isotopes are unstable, so that the radioactive decay process continues. The first step, the conversion of ^{238}U into thorium-234 by alpha decay, is slow, with a half-life of 4.47 billion years (4.47 Ga). Because many of the remaining steps are relatively rapid, the half-life of the full sequence is just over 4.47 Ga. As the rate at which ^{238}U atoms are ultimately converted into ^{206}Pb atoms is known, the ratio $^{238}U/^{206}Pb$ can be used to determine the crystallization ages for minerals, especially for those formed early in Earth's history, as explained below.

Those isotopes that decay rapidly, beginning with protactinium-234 and radon-222, produce large amounts of decay products in short amounts of time. Radioactive decay products can produce significant damage to crystal structures and significant tissue damage in human populations (Box 3.2). Radioactive isotopes also have significant applications in medicine, especially in cancer treatments. Radioactive materials are fundamentally important global energy resources, even though the radioactive isotopes in spent fuel present significant long-term hazards with respect to its disposal. Radioactive decay provides a major energy source through nuclear fission in reactors. Radioactive decay is also the primary heat engine within Earth and is partly responsible for driving plate tectonics and core–mantle convection. Without radioactive heat, Earth would be a very different kind of home.

Table 3.2 The 14-step radioactive decay sequence that occurs in the conversion of the radioactive isotope ^{238}U into the stable isotope ^{206}Pb.

Parent isotope	Daughter isotope	Decay process	Half-life
Uranium-238	Thorium-234	Alpha	4.5×10^9 years
Thorium-234	Protactinium-234	Beta	24.5 days
Protactinium-234	Uranium-234	Beta	1.1 minutes
Uranium-234	Thorium-230	Alpha	2.3×10^5 years
Thorium-230	Radium-226	Alpha	8.3×10^4 years
Radium-226	Radon-222	Alpha	1.6×10^3 years
Radon-222	Polonium-218	Alpha	3.8 days
Polonium-218	Lead-214	Alpha	3.1 minutes
Lead-214	Bismuth-214	Beta	26.8 minutes
Bismuth-214	Polonium-214	Beta	19.7 minutes
Polonium-214	Lead-210	Alpha	1.5×10^{-4} seconds
Lead-210	Bismuth-210	Beta	22.0 years
Bismuth-210	Polonium-210	Beta	5.0 days
Polonium-210	**Lead-206**	Alpha	140 days

Box 3.2 Radon and lung cancer

Inhalation of radon gas is the second largest cause of lung cancer worldwide, second only to cigarette smoking. In the 1960s, underground uranium miners began to show unusually high incidences of lung cancer. The cause was shown to be related to the duration of the miner's exposure to radioactive materials. To cause lung cancer, the radioactive material must enter the lungs as a gas, which then causes progressive damage to the bronchial epithelium or lining of the lungs. What is the gas and how does it originate? Table 3.2 shows the many radioactive isotopes that are produced by the decay of the common isotope of uranium (^{238}U). Uranium miners would be exposed to all these, but which one would they inhale into their lungs? Because radon possesses a stable electron configuration, it tends not to combine with other elements. Like most noble elements, under normal near surface conditions, it tends to exist as separate atoms in the form of a gas. In the confined space of poorly ventilated underground mines, radioactive decay in the uranium series produces sufficient concentrations of radon to significantly increase the incidence of lung cancer. The other property that makes radon-222 so dangerous is its short half-life (3.825 days). Within days, most of the radon inhaled by miners has decayed into polonium-218 with the emission of alpha particles. Subsequently, most of the radioactive ^{218}Po decays within hours into lead-210 with the release of more alpha particles (^{4}He nuclei). Lung damage leading to lung cancer largely results from this rapid release of alpha particles over long periods of exposure. Scientific studies on radon exposure have been complicated by the fact that many miners were also smokers.

Is the general public at risk of radon exposure? Uranium is ubiquitous in the rocks of Earth's crust, and so therefore is radon. Potassium feldspar-bearing rocks such as granites and gneisses, black shales and phosphates contain higher uranium concentrations (>100 ppm) than average crustal rocks (<5 ppm) and therefore pose a larger threat. Radon gas occurs in air spaces and is quite soluble in water; think of the dissolved oxygen that aqueous organisms use to respire or the carbon dioxide dissolved in carbonate beverages. Groundwater circulating through uranium-rich rocks can dissolve substantial amounts of radon gas and concentrate radium, another carcinogenic isotope. Ordinarily, this is not a problem; the gas rises and is released to the atmosphere, where it is dispersed and diluted to very low levels. But if radon gas is released into a confined space such as a home, especially one that is well insulated and not well ventilated, radon gas concentrations can reach hazardous levels. Most radon gas enters the home through cracks in the walls and foundations, either as gas or in water from which it is released. Most of the remainder is released when water from radon-contaminated wells is used, again releasing radon into the home atmosphere. The problem is especially bad in winter and spring months when homes are heated, basements flooded and ventilation poor. As warm air in the home rises, air is drawn from the soil into the home, increasing radon concentrations. The insulation that increases heat efficiency also increases radon concentrations. What can be done to reduce the risk? Making sure that basements and foundation walls are well sealed and improving ventilation can reduce radon concentrations to acceptable levels, even in homes built on soils with high concentrations of uranium. Be aware of uranium concentrations in rock types underlying your neighborhood.

In the following sections we have chosen a few examples, among the many that exist, to illustrate the importance of radioactive isotopes and decay series in the study of Earth materials.

Age determinations using radioactive decay series

Table 3.3 lists several radioactive parent to stable daughter transformations that can be used to determine the formation ages of Earth materials. All these are based on the principle that after the radioactive isotope is incorporated into Earth materials, the ratio of radioactive parent isotopes to stable daughter isotopes decreases through time by radioactive decay. The rate at which such parent:daughter ratios decrease depends on the rate of decay, which is given by the **decay constant** (λ), the proportion of the remaining

Table 3.3   Systematics of radioactive isotopes important in age determinations in Earth materials.

Decay series	Decay process	Decay constant (λ)	Half-life	Applicable dating range
$^{14}C \rightarrow {}^{14}N$	Beta decay	1.29×10^{-4}/year	5.37 Ka	<60 Ka
$^{40}K \rightarrow {}^{40}A$	Electron capture	4.69×10^{-10}/year	1.25 Ga	500 Ka to >4.5 Ga
$^{87}Rb \rightarrow {}^{87}Sr$	Beta decay	1.42×10^{-11}/year	48.8 Ga	10 Ma to >4.5 Ga
$^{147}Sm \rightarrow {}^{143}Nd$	Alpha decay	6.54×10^{-12}/year	106 Ga	200 Ma to >4.15 Ga
$^{232}Th \rightarrow {}^{208}Pb$	Beta and alpha decays	4.95×10^{-11}/year	14.0 Ga	10 Ma to >4.5 Ga
$^{235}U \rightarrow {}^{207}Pb$	Beta and alpha decays	9.85×10^{-10}/year	704 Ma	10 Ma to >4.5 Ga
$^{238}U \rightarrow {}^{206}Pb$	Beta and alpha decays	1.55×10^{-10}/year	4.47 Ga	10 Ma to >4.5 Ga

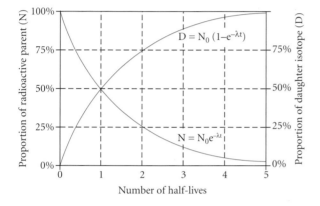

Figure 3.13   Progressive change in the proportions of radioactive parent (N) and daughter (D) isotopes over time, in terms of number of half-lives.

radioactive atoms that will decay per unit of time. One useful formula that governs decay series states that the number of radioactive atoms remaining at any given time (N) is equal to the number of radioactive atoms originally present in the sample (N_0) multiplied by a negative exponential factor ($e^{-\lambda t}$) that increases with the rate of decay (λ) and the time since the sample formed (t), that is, its age. These relationships are given by:

$$N = N_0 e^{-\lambda t} \qquad \text{(equation 3.1)}$$

It should be clear from the formula that when t = 0, N = N_0, and that N becomes smaller through time as a function of the rate of decay given by the decay constant; rapidly for a large decay constant, more slowly for a small one. Figure 3.13 illustrates a typical decay curve, showing how the abundance of the radioactive parent isotope decreases exponentially over time while the abundance of the

daughter isotope increases in a reciprocal manner. The two curves cross where the number of radioactive parent and stable daughter atoms is equal. The time required for this to occur is called the half-life of the decay series and is the time required for one half of the radioactive isotopes to decay into stable daughter isotopes.

More generally, the age of any sample may be calculated from the following equation:

$$t = (1/\lambda) \ln (d/p + 1) \qquad \text{(equation 3.2)}$$

where t = age, λ = the decay constant, d = number of stable daughter atoms and p = number of radioactive parent atoms. Where stable daughter atoms were present in the original sample, a correction must be made to account for them, as explained below.

Many radioactive to stable isotope decay series, each with a unique decay constant, can be used, often together, to determine robust formation ages for Earth materials, especially for older materials. Table 3.3 summarizes some common examples. Two of these are discussed in more detail in the sections that follow.

Uranium–lead systematics

Uranium (U) occurs in two radioactive isotopes, both of which decay in steps to different stable isotopes of lead (Pb). The closely related actinide element thorium (Th) decays in a similar fashion to yet another stable isotope of lead. The essential information on the radioactive and stable isotopes involved for these three decay series ($^{238}U \rightarrow {}^{206}Pb$, $^{235}U \rightarrow {}^{207}Pb$ and $^{232}Th \rightarrow {}^{208}Pb$) is summarized in Table 3.3. This includes the types of decay processes, their half-lives and the useful age range for dating by each of these methods.

Of the three decay series, the most commonly used is $^{238}U \rightarrow {}^{206}Pb$. This is because ^{238}U is much more abundant than the other two radioactive isotopes and thus easier to measure accurately. Typically, heavy minerals, such as zircon, sphene and/or monazite, are used for analyses because they contain substantial actinides, are relatively easy to separate and resist chemical alteration.

A problem arises because some ^{206}Pb occurs naturally in Earth materials, without radioactive decay, so that measured $^{238}U/{}^{206}Pb$ ratios include both daughter and non-daughter ^{206}Pb. This must be corrected if an accurate age is to be determined. The correction can be accomplished because yet another lead isotope, ^{204}Pb, is not produced by radioactive decay. The ratio of $^{204}Pb/{}^{206}Pb$ in meteorites, formed at the time Earth formed, is known and is taken to be the original ratio in the rock whose age is being determined. The correction is then accomplished by measuring the amount of ^{204}Pb in the sample and subtracting the meteoritic proportion of ^{206}Pb from total ^{206}Pb to arrive at the amount of ^{206}Pb that is the daughter of ^{238}U.

In most analyses, the ratio of $^{235}U/{}^{207}Pb$ is also determined. The $^{235}U/{}^{207}Pb$ decay series has a much shorter half-life than the $^{238}U/{}^{206}Pb$ series, so that the U/Pb ratios are much smaller for a given sample age. Once the ratios of both parent uranium isotopes to the corrected daughter isotopes ($^{238}U/{}^{206}Pb$, $^{235}U/{}^{207}Pb$) have been determined, sample ages are generally determined on a concordia plot (Figure 3.14) that displays the expected relationship between the reciprocal ratios $^{206}Pb/{}^{238}U$ and $^{207}Pb/{}^{235}U$ as a function of sample age.

Either decay series is capable of producing a sample age on its own. For example, a $^{206}Pb/{}^{238}U$ ratio of 0.5 indicates an age of approximately 2.6 Ga (Figure 3.14), while a $^{207}Pb/{}^{235}U$ ratio of 40 indicates an age of approximately 3.7 Ga. One advantage of these decay series is that the measurement of multiple, closely related isotopes permits ages to be checked against each other and provides robust sample ages when both methods yield a closely similar age. For example, if a sample possesses a $^{206}Pb/{}^{238}U$ ratio of 0.7 and a $^{207}Pb/{}^{235}U$ ratio of 32, a sample age near 3.5 Ga (Figure 3.14) is highly likely. When two age determinations are in agreement, they are said to be concordant. Another way to state this is that samples that plot on the concordia line when their $^{206}Pb/{}^{238}U$ versus $^{207}Pb/{}^{235}U$ ratios are plotted yield concordant ages. Such dating techniques are especially powerful in determining robust crystallization ages of igneous rocks that crystallized from magmas and lavas (Chapters 8 and 9).

A significant problem arises from the fact that subsequent events can alter the geochemistry of rocks, so that age determinations are no longer concordant. For example, when rocks undergo metamorphism, lead is commonly mobilized and lost from the minerals or rocks in question. This results in lower than expected amounts of daughter lead isotopes, lower Pb/U ratios and points that fall below the concordant age line on $^{206}Pb/{}^{238}U$ versus $^{207}Pb/{}^{235}U$ diagrams. When samples of similar real ages, affected differently during metamorphism, are plotted on a concordia diagram, they fall along a straight line below the concordant age line. One interpretation of such data is that the right end of the dashed straight line intersects the concordant age line at the original age of the sample (4.0 Ga in Figure 3.14) and that the left end intersects the concordant age line at the time of metamorphism (2.5 Ga in Figure 3.14). Wherever possible, such interpretations should be tested against other data so that a truly robust conclusion can be drawn. One such radiometric decay series, with widespread application, is described in the section that follows.

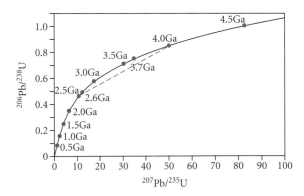

Figure 3.14 Uranium–lead concordia plot showing sample ages as a function $^{206}Pb/{}^{238}U$ and $^{207}Pb/{}^{235}U$ ratios and samples with lower $^{206}Pb/{}^{238}U$ and $^{207}Pb/{}^{235}U$ ratios, reset during a 2.5 Ga metamorphic event.

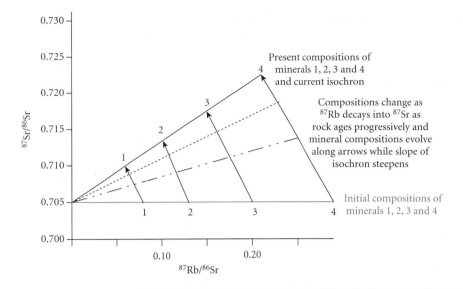

Figure 3.15   Rubidium–strontium systematics, showing evolution in the composition of four representative minerals (1–4) from initial composition to current composition as ^{87}Rb decays into ^{87}Sr over time. Whole rock compositions would lie somewhere between minerals 1 and 4 depending on the specifics of mineral composition and their proportions in the rock.

Rubidium–strontium systematics

Several isotopes exist of the relatively rare element rubidium (Rb), of which some 27% are radioactive ^{87}Rb. Most Rb^{+1} is concentrated in continental crust, especially in substitution for potassium ion (K^{+1}) in potassium-bearing minerals such as potassium feldspar, muscovite, biotite, sodic plagioclase and amphibole. Radioactive ^{87}Rb is slowly (half-life = 48.8 Ga) transformed by beta decay into strontium-87 (^{87}Sr). Unfortunately, the much smaller strontium ion (Sr^{+2}) does not easily substitute for potassium ion (K^{+1}) and therefore tends to migrate into minerals in the rock that contain calcium ion (Ca^{+2}) for which strontium easily substitutes. This makes using $^{87}Rb/^{86}Sr$ for age dating a much less accurate method than the U/Pb methods explained previously.

One basic concept behind rubidium–strontium dating is that the original rock has some initial amount of ^{87}Rb, some initial ratio of $^{87}Sr/^{86}Sr$ and some initial ratio of $^{87}Rb/^{86}Sr$. These ratios evolve through time in a predictable manner. Over time, the amount of ^{87}Rb decreases and the amount of ^{87}Sr increases by radioactive decay so that the $^{87}Sr/^{86}Sr$ ratio increases and the $^{87}Rb/^{86}Sr$ ratio decreases by amounts proportional to sample age. A second

basic concept is that the initial amount of ^{87}Rb varies from mineral to mineral, being highest in potassium-rich minerals. As a result, the rate at which the $^{87}Sr/^{86}Sr$ ratio increases depends on the individual mineral. For example in a potassium-rich (rubidium-rich) mineral, the $^{87}Sr/^{86}Sr$ ratio will increase rapidly, whereas for a potassium-poor mineral it will increase slowly. For a mineral with no K or ^{87}Rb, the $^{87}Sr/^{86}Sr$ ratio will not change; it will remain the initial $^{87}Sr/^{86}Sr$ ratio. However, the $^{87}Rb/^{86}Sr$ ratio decreases at a constant rate that depends on the decay constant.

In a typical analysis, the amounts of ^{87}Rb, ^{87}Sr and ^{86}Sr in the whole rock and in individual minerals are determined by mass spectrometry, and the $^{87}Rb/^{86}Sr$ and $^{87}Sr/^{86}Sr$ ratios are calculated for each. These are plotted on an $^{87}Sr/^{86}Sr$ versus $^{87}Rb/^{86}Sr$ diagram (Figure 3.15). At the time of formation, assuming no fractionation of strontium isotopes, the $^{87}Sr/^{86}Sr$ in each mineral and in the whole rock was a constant initial value, while the $^{87}Rb/^{86}Sr$ values varied from relatively high for rubidium-rich minerals such as biotite and potassium feldspar to zero for minerals with no rubidium. These initial $^{87}Sr/^{86}Sr$ and $^{87}Rb/^{86}Sr$ values are shown by the horizontal line in Figure 3.15. As the rock ages, ^{87}Rb

progressively decays to ^{87}Sr, which causes the $^{87}Sr/^{86}Sr$ ratio to increase at rates proportional to the initial amount of ^{87}Rb, while the $^{87}Rb/^{86}Sr$ ratio decreases at a constant rate. Over time, the $^{87}Sr/^{86}Sr$ ratios and $^{87}Rb/^{86}Sr$ ratios for each mineral and the whole rock evolve along paths shown by the arrowed lines in Figure 3.15. If each minerals acts as a closed system, points representing the current $^{87}Sr/^{86}Sr$ versus $^{87}Rb/^{86}Sr$ ratios will fall on a straight line whose slope increases through time (Figure 3.15). The slope of the best-fit line, called an **isochron** (line of constant age), yields the age of the sample. The y-intercept of any isochron yields the initial $^{87}Sr/^{86}Sr$ ratio, which is unchanging for a theoretical sample that contains no ^{87}Rb. The **initial** $^{87}Sr/^{86}Sr$ is especially important in identifying the source regions from which magmas are derived in the formation of igneous rocks (Chapter 8).

There are many other isotope series utilized in determining rock ages, the history of magmatic source rocks and/or the age of metamorphic events. Some of these are discussed in the chapters that follow in the contexts where they are especially important.

Chapter 4

Crystallography

4.1   CRYSTALLINE SUBSTANCES

Crystallography emphasizes the long-range order or crystal structure of crystalline substances. It focuses on the symmetry of crystalline materials and on the ways in which their long-range order is related to the three-dimensional repetition of fundamental units of pattern during crystal growth. In minerals, the fundamental units of pattern are molecular clusters of coordination polyhedra or stacking sequences (Chapter 2). The ways in which these basic units can be repeated to produce crystal structures with long-range order are called symmetry operations. In addition, crystallography focuses on the description and significance of planar features in crystals including planes of atoms, cleavage planes, crystal faces and the forms of crystals. Crystallography is also concerned with crystal defects, local imperfections in the long-range order of crystals. Given the broad scope of

Earth Materials, 1st edition. By K. Hefferan and
J. O'Brien. Published 2010 by Blackwell Publishing Ltd.

this text, a more detailed treatment of crystallography cannot be provided.

4.1.1   Crystals and crystal faces

Mineral crystals are one of nature's most beautiful creations. Many crystals are enclosed by flat surfaces called crystal faces. **Crystal faces** are formed when mineral crystals grow, and enclose crystalline solids when they stop growing. Perfectly formed crystals are notable for their remarkable symmetry (Figure 4.1). The external symmetry expressed by crystal faces permits us to infer the geometric patterns of the atoms in mineral crystal structures as well. These patterns inferred from external symmetry have been confirmed by advanced analytical techniques such as X-ray diffraction (XRD) and atomic force microscopy (AFM).

Mineralogists have developed language to describe the symmetry of crystals and the crystal faces that enclose them. Familiarizing students with the concepts and terminology of crystal symmetry and crystal faces is one of the primary goals of this chapter. A second

(a)

(b)

Figure 4.1 Representative mineral crystals: (a) quartz; (b) tourmaline. (Photos courtesy of the Smithsonian Institute.) (For color version, see Plate 4.1, between pp. 248 and 249.)

goal of this chapter is to build connections between crystal chemistry (Chapters 2 and 3) and crystallography by explaining the relationships between chemical composition and coordination polyhedra and the form, symmetry and crystal faces that develop as crystals grow.

4.1.2 Motifs and nodes

When minerals begin to form, atoms or ions bond together, so that partial or complete coordination polyhedra develop (Chapter 2). Because the ions on the edges and corners of coordination polyhedra have unsatisfied electrostatic charges, they tend to bond to additional ions available in the environment as the mineral grows. Eventually, a small cluster of coordination polyhedra is formed that contains all the coordination polyhedra characteristic of the mineral and its chemical composition. In any mineral, we can recognize a small cluster of coordination polyhedra that contains the mineral's fundamental composition and unit of pattern or motif. As the mineral continues to grow, additional clusters of the same pattern of coordination polyhedra are added to form a mineral crystal with a three-dimensional geometric pattern – a long-range, three-dimensional crystal structure. Clusters of coordination polyhedra are added, one atom or ion at a time, as (1) the crystal nucleates, (2) it becomes a microscopic crystal, and, if growth continues, (3) it becomes a macroscopic crystal. Growth continues in this manner until the environmental conditions that promote growth change and growth ceases.

Long-range, geometric arrangements of atoms and/or ions in crystals are produced when a fundamental array of atoms, a unit of pattern or motif, is repeated in three dimensions to produce the crystal structure. A **motif** is the smallest unit of pattern that, when repeated by a set of symmetry operations, will generate the long-range pattern characteristic of the crystal. In minerals, the motif is composed of one or more coordination polyhedra. In wallpaper, it is a basic set of design elements that are repeated to produce a two-dimensional pattern, whereas in a brick wall the fundamental motif is that of a single brick that is repeated in space to form the three-dimensional structure. The repetition of these fundamental units of pattern by a set of rules called **symmetry operations** can produce a two- or three-dimensional pattern with long-range order. When several different motifs could be repeated by a similar set of symmetry operations, we may wish to emphasize the general rules by which different motifs may be repeated to produce a particular type of long-range order. In such cases it is useful to represent motifs by using a point. A point used to represent any motif is called a **node**. The pattern or array of atoms about every node must be the same throughout the pattern the nodes represent.

4.2 SYMMETRY OPERATIONS

4.2.1 Simple symmetry operations

Symmetry operations may be simple or compound. **Simple symmetry operations** produce repetition of a unit of pattern or motif using a single type of operation. Compound symmetry operations produce repetition of a unit of pattern or motif using a combination of two types of symmetry operation. Simple symmetry operations include (1) translation, by specific distances in specified directions, (2) rotation, about a specified set of axes, (3) reflection, across a mirror plane, and (4) inversion, through a point called a center. These operations are discussed below and provide useful insights into the geometry of crystal structures and the three-dimensional properties of such crystals.

Translation

The symmetry operation called **translation** involves the periodic repetition of nodes or motifs by systematic linear displacement. One-dimensional translation of basic design elements generates a row of similar elements (Figure 4.2a). The translation is defined by the **unit translation vector (t)**, a specific length and direction of systematic displacement by which the pattern is repeated. Motifs other than commas could be translated by the same unit translation vector to produce a one-dimensional pattern. In minerals, the motifs are clusters of atoms or coordination polyhedra that are repeated by translation.

Two-dimensional translations are defined by **two unit translation vectors (t_a and t_b or t_1 and t_2, respectively)**. The translation in one direction is represented by the length and direction of t_a or t_1; translation in the second direction is represented by the length and direction of t_b or t_2. The pattern generated depends on the length of the two unit translation vectors and the angles between their directions. The result of any two-dimensional translation is a **plane lattice** or **plane mesh**. A plane lattice is a two-dimensional array of motifs or nodes in which every node has an environment similar to every other node in the array (Figure 4.2a,b).

Three-dimensional translations are defined by **three unit translation vectors (t_a, t_b and t_c**

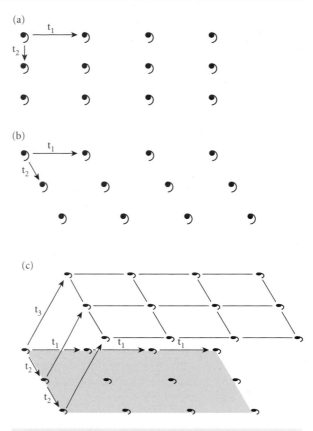

Figure 4.2 (a) Two-dimensional translation at right angles (t_1 and t_2) to generate a two-dimensional mesh of motifs or nodes. (b) Two-dimensional translation (t_1 and t_2) not at right angles to generate a two-dimensional mesh or lattice. (c) Three-dimensional translation (t_1, t_2 and t_3) to generate a three-dimensional space lattice. (From Klein and Hurlbut, 1985; with permission of John Wiley & Sons.)

or t_1, t_2 and t_3, respectively). The translation in one direction is represented by the length and direction of t_a or t_1, the translation in the second direction is represented by t_b or t_2 and the translation in the third direction is represented by t_c or t_3. The result of any three-dimensional translation is a **space lattice**. A space lattice is a three-dimensional array of motifs or nodes in which every node has an environment similar to every other node in the array. Since crystalline substances such as minerals have long-range, three-dimensional order and since they may be thought of as motifs repeated in three dimensions, the resulting array of motifs is a **crystal lattice**. Figure 4.2c illustrates a space lattice produced

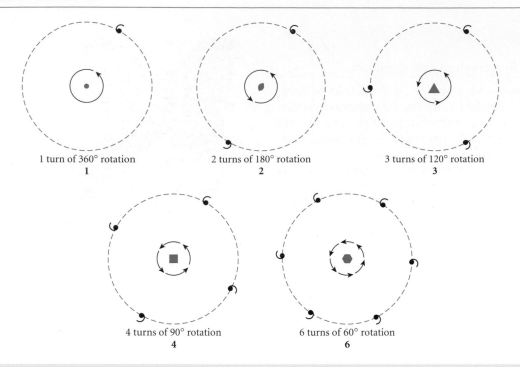

Figure 4.3 Examples of the major types of rotational symmetry (n = 1, 2, 3, 4 or 6) that occur in minerals. (From Klein and Hurlbut, 1985; with permission of John Wiley & Sons.)

by a three-dimensional translation of nodes or motifs.

Rotation

Motifs can also be repeated by non-translational symmetry operations. Many patterns can be repeated by rotation (n). **Rotation (n)** is a symmetry operation that involves the rotation of a pattern about an imaginary line or axis, called an **axis of rotation (A)**, in such a way that every component of the pattern is perfectly repeated one or more times during a complete 360° rotation. The symbol "n" denotes the number of repetitions that occur during a complete rotation. Figure 4.3 uses comma motifs to depict the major types of rotational symmetry (n) that occur in minerals and other inorganic crystals. The axis of rotation for each motif is perpendicular to the page. Table 4.1 summarizes the major types of rotational symmetry.

Reflection

Reflection is as familiar to us as our own reflections in a mirror or that of a tree in a

Table 4.1 Five common axes of rotational symmetry in minerals.

Type	Symbolic notation	Description
One-fold axis of rotation	(1 or A_1)	Any axis of rotation about which the motif is repeated only once during a 360° rotation (Figure 4.3, (1))
Two-fold axis of rotation	(2 or A_2)	Motifs repeated every 180° or twice during a 360° rotation (Figure 4.3 (2))
Three-fold axis of rotation	(3 or A_3)	Motifs repeated every 120° or three times during a complete rotation (Figure 4.3 (3))
Four-fold axis of rotation	(4 or A_4)	Motifs repeated every 90° or four times during a complete rotation (Figure 4.3 (4))
Six-fold axis of rotation	(6 or A_6)	Motifs repeated every 60° or six times during a complete rotation (Figure 4.3 (5))

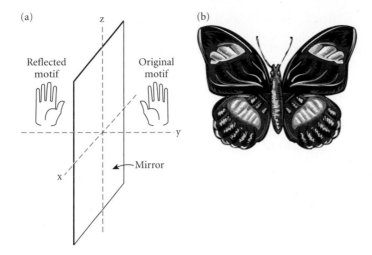

Figure 4.4 Two- and three-dimensional motifs that illustrate the concept of reflection across a plane of mirror symmetry (m). (a) Mirror image of a hand. (From Klein and Hurlbut, 1985; with permission of John Wiley & Sons.) (b) Bilateral symmetry of a butterfly; the two halves are nearly, but not quite, perfect mirror images of each other. (Image from butterflywebsite.com.)

still body of water. It is also the basis for the concept of bilateral symmetry that characterizes many organisms (Figure 4.4). Yet it is a symmetry operation that is somewhat more difficult for most people to visualize than rotation. **Reflection (m)** is a symmetry operation in which every component of a pattern is repeated by reflection across a plane called a **mirror plane** (m). Reflection occurs when each component is repeated by equidistant projection perpendicular to the mirror plane. Reflection retains all the components of the original motif but changes its "handedness"; the new motifs produced by reflection across a mirror plane are mirror images of each other (Figure 4.4). Symmetry operations that change the handedness of motifs are called **enantiomorphic operations**.

One test for the existence of a mirror plane of symmetry is that all components of the motifs on one side of the plane are repeated at equal distances on the other side of the plane along projection lines perpendicular to the plane. If this is not true, the plane is not a plane of mirror symmetry.

Inversion

Inversion is perhaps the most difficult of the simple symmetry operations to visualize. **Inversion** involves the repetition of motifs by

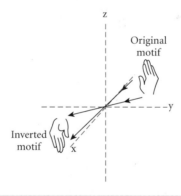

Figure 4.5 Inversion through a center of symmetry (i) illustrated by a hand repeated by inversion through a center (inversion point). (From Klein and Hurlbut, 1985; with permission of John Wiley & Sons.)

inversion through a point called a **center of inversion (i)**. Inversion occurs when every component of a pattern is repeated by equidistant projection through a common point or center of inversion. The two hands in Figure 4.5 illustrate the enantiomorphic symmetry operation called inversion and show the center through which inversion occurs. In some symbolic notations centers are symbolized by (c) rather than (i).

One test for the existence of a center of symmetry is that all the components of a pattern are repeated along lines that pass through a common center and are repeated at equal distances from that center. If this is not the case, the motif does not possess a center of symmetry.

4.2.2 Compound symmetry operations

Three other symmetry operations exist but, unlike those discussed so far, they are **compound symmetry operations** that combine two simple symmetry operations. **Glide reflection** (g) is a symmetry operation that combines translation (t) or (c) parallel to a mirror plane (m) with reflection across the mirror plane to produce a glide plane (Figure 4.6).

Rotoinversion ($\bar{n}$) is an operation that combines rotation about an axis with inversion through a center to produce an axis of rotoinversion. Figure 4.7b illustrates an axis of four-fold rotoinversion ($\bar{4}$) in which the motif is repeated after 90° rotation followed by inversion through a center so that it is repeated four times by rotoinversion during a 360° rotation. Axes of two-fold rotoinversion ($\bar{2}$) are unique symmetry operations, whereas axes of three-fold rotoinversion ($\bar{3}$) are equivalent to a three-fold axis of rotation and

a center of inversion (3i), and axes of six-fold rotoinversion ($\bar{6}$) are equivalent to a three-fold axis of rotation perpendicular to a mirror plane (3/m). **Screw rotation** (n_a) is a symmetry operation that combines translation parallel

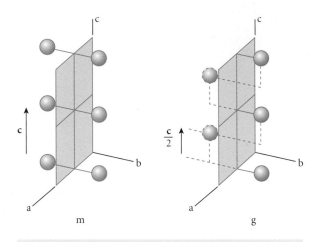

Figure 4.6 Mirror plane (m) with the translation vector (c) on the left, contrasted with a glide plane (g) with the translation vector (c/2) combined with mirror reflection on the right. (From Wenk and Bulakh, 2004; with permission of Oxford University Press.)

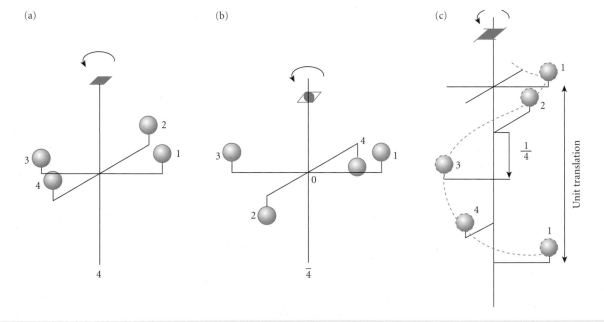

Figure 4.7 (a) An axis of four-fold rotation (4). This contrasts with (b) an axis of four-fold rotoinversion ($\bar{4}$) that combines rotation with inversion every 90°, and (c) a four-fold screw axis (4_1) that combines translation with 90° rotations every one-fourth translation. (From Wenk and Bulakh, 2004; with permission of Oxford University Press.)

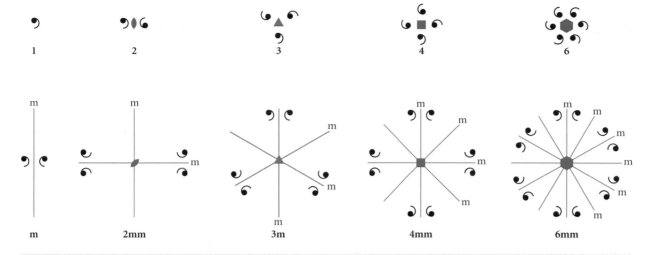

Figure 4.8 The ten plane point groups defined by rotational and reflection symmetry. (From Klein and Hurlbut, 1985; with permission of John Wiley & Sons.)

to an axis with rotation about the axis (Figure 4.7c).

Readers interested in more detailed treatments of the various types of compound symmetry operations should refer to Klein and Dutrow (2007), Wenk and Bulakh (2004) or Nesse (2000).

4.3 TWO-DIMENSIONAL MOTIFS AND LATTICES (MESHES)

The symmetry of three-dimensional crystals can be quite complex. Understanding symmetry in two dimensions provides an excellent basis for understanding the higher levels of complexity that characterize three-dimensional symmetries. It also provides a basis for learning to visualize planes of constituents within three-dimensional crystals. Being able to visualize and reference lattice planes is of the utmost importance in describing cleavage and crystal faces and in the identification of minerals by X-ray diffraction methods.

4.3.1 Plane point groups

Any fundamental unit of two-dimensional pattern, or motif, can be repeated by various symmetry operations to produce a larger two-dimensional pattern. All two-dimensional motifs that are consistent with the generation of long-range two-dimensional arrays can be assigned to one of **ten plane point groups**

based on their unique plane point group symmetry (Figure 4.8). Using the symbolic language discussed in the previous section on symmetry, the ten plane point groups are **1, 2, 3, 4, 6, m, 2mm, 3m, 4mm and 6mm**. The numbers refer to axes of rotation that are perpendicular to the plane (or page); the m refers to mirror planes perpendicular to the page. The first m refers to a set of mirror planes that is repeated by the rotational symmetry and the second m to a set of mirror planes that bisects the first set. Note that the total number of mirror planes is the same as the number associated with its rotational axis (e.g., 3m has three mirror planes and 6mm has six mirror planes).

4.3.2 Plane lattices and unit meshes

Any motif can be represented by a point called a node. Points or nodes can be translated in one direction by a unit translation vector t_a or t_1 to produce a line of nodes or motifs. Nodes can also be translated in two directions t_a and t_b or t_1 and t_2 to produce a two-dimensional array of points called a plane mesh or **plane net**. Simple translation of nodes in two directions produces five basic types of two-dimensional patterns (Figure 4.9). The smallest units of such meshes, which contain at least one node and the unit translation vectors, are called **unit meshes (unit nets)** and contain all the information necessary to produce the

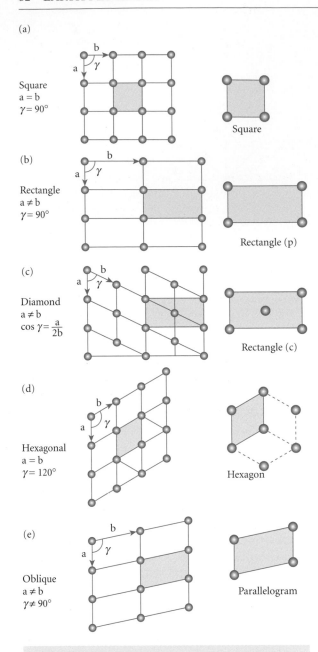

(a)

Square
a = b
γ = 90°

Square

(b)

Rectangle
a ≠ b
γ = 90°

Rectangle (p)

(c)

Diamond
a ≠ b
$\cos \gamma = \dfrac{a}{2b}$

Rectangle (c)

(d)

Hexagonal
a = b
γ = 120°

Hexagon

(e)

Oblique
a ≠ b
γ ≠ 90°

Parallelogram

Figure 4.9 The five principal types of meshes or nets and their unit meshes (shaded gray): (a) square, (b) primitive rectangle, (c) diamond or centered rectangle, (d) hexagonal, (e) oblique. (From Nesse, 2000, with permission of Oxford University Press.)

larger two-dimensional pattern. The unit meshes contain only translation symmetry information. The five basic types of unit mesh are classified on the basics of the unit translation vector lengths (equal or unequal), the angles between them (90°, 60° and 120° or none of these) and whether they have nodes

only at the corners (primitive = p) or have an additional node in the center (c) of the mesh.

Square unit meshes (Figure 4.9a) are primitive and have equal unit translation vectors at 90° angles to each other (p, $t_a = t_b$, γ = 90°). **Primitive rectangular unit meshes** (Figure 4.9b) differ in that, although the unit translation vectors intersect at right angles, they are of unequal lengths (p, $t_a \neq t_b$, γ = 90°). Diamond unit meshes have equal unit translation vectors that intersect at angles other than 60°, 90° or 120°. Diamond lattices can be produced and represented by **primitive diamond unit meshes** (p, $t_a = t_b$, γ ≠ 60°, 90° or 120°). They can also be produced by the translation of **centered rectangular unit meshes** (Figure 4.9c) in which the two unit mesh sides are unequal, the angle between them is 90° and there is a second node in the center of the mesh (c, $t_a \neq t_b$, γ = 90°). In a centered rectangular mesh there is a total content of two nodes = two motifs. If one looks closely, one may see evidence for glide reflection in the centered rectangular mesh and/or the larger diamond lattice. The **hexagonal unit mesh** (Figure 4.9d) is a special form of the primitive diamond mesh because, although the unit translation vectors are equal, the angles between them are 60° and 120° (p, $t_a = t_b$, γ = 120°). Three such unit meshes combine to produce a larger pattern with hexagonal symmetry. **Oblique unit meshes** (Figure 4.9e) are primitive and are characterized by unequal unit translation vectors that intersect at angles that are not 90°, 60° or 120° (p, $t_a \neq t_b$, γ ≠ 90°, 60° or 120°) and produce the least regular, least symmetrical two-dimensional lattices. The arrays of nodes on planes within minerals always correspond to one of these basic patterns.

4.3.3   Plane lattice groups

When the ten plane point groups are combined with the five unit meshes in all ways that are compatible, a total of **17 plane lattice groups** are recognized on the basis of the total symmetry of their plane lattices. Note that these symmetries involve translation-free symmetry operations including rotation and reflection, translation and compound symmetry operations such as glide reflection. Table 4.2 summarizes the 17 plane lattice groups and their symmetries. Primitive lattices are

Table 4.2 The 17 plane lattice groups and the unique combination of point group and unit mesh that characterizes each.

Lattice	Point group	Plane group
Oblique (P)	1	P1
	2	P2
Rectangular (P and C)	m	Pm
		Pg
		Cm
	2mm	P2mm
		P2mg
		P2gg
		C2mm
Square (P)	4	P4
	4mm	P4mm
		P4gm
Hexagonal (P) (rhombohedral)	3	P3
	3m	P3m1
		P3lm
Hexagonal (P) (hexagonal)	6	P6
	6mm	P6mm

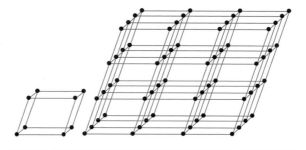

Figure 4.10 A primitive unit cell and a long-range space point lattice that results from its repetition by symmetry operations in three dimensions.

denoted by "P" and centered lattices by "C". Axes of rotation for the entire pattern perpendicular to the plane are noted by 1, 2, 3, 4 and 6. Mirror planes perpendicular to the plane are denoted by "m"; glide planes perpendicular to the plane are denoted by "g".

The details of plane lattice groups are well documented, but beyond the scope of this text.

4.4 THREE-DIMENSIONAL MOTIFS AND LATTICES

Minerals are three-dimensional Earth materials with three-dimensional crystal lattices. The fundamental units of pattern in any three-dimensional lattice are **three-dimensional motifs** that can be classified according to their translation-free symmetries. These three-dimensional equivalents of the two-dimensional plane point groups are called **space point groups**.

Space point groups can be represented by nodes. These nodes can be translated to produce three-dimensional patterns of points called **space lattices**. Space lattices are the three-dimensional equivalents of plane nets or meshes. By analogy with unit meshes or nets, we can recognize the smallest three-dimensional units, called unit cells, which contain all the information necessary to produce the three-dimensional space lattices. In this section, we will briefly describe the space point groups, after which we will detail Bravais lattices, unit cells and their relationship to the six major crystal systems to which minerals belong.

4.4.1 Space point groups

In minerals, the fundamental motifs are parts of clusters of three-dimensional coordination polyhedra sufficient to establish the composition of the mineral. When these are repeated in three dimensions during mineral growth, they produce the long-range order characteristic of crystalline substances (Figure 4.10). Like all fundamental units of pattern, these three-dimensional motifs can be classified on the basis of their translation-free symmetries.

Only 32 different three-dimensional motif symmetries exist. These define 32 space point groups, each with unique space point group symmetry. In minerals, the 32 **crystal classes** – to one of which all minerals belong – correspond to the 32 space point group symmetries of the mineral's motif. That the crystal classes were originally defined on the basis of the external symmetry of mineral crystals is another example of the fact that the external symmetry of minerals reflects the internal symmetry of their constituents. The 32 crystal classes belong to six (or seven) crystal systems, each with its own characteristic symmetry.

Table 4.3   The six crystal systems and 32 crystal classes, with their characteristic symmetry and crystal forms.

System	Crystal class	Class symmetry	Total symmetry
Isometric	Hexoctahedral	$4/m\,\overline{3}\,2/m$	$3A_4$, $4\overline{A_3}$, $6A_2$, $9m$
	Hextetrahedral	$\overline{4}3\,m$	$3\overline{A_4}$, $4A_3$, $6m$
	Gyroidal	432	$3A_4$, $4A_3$, $6A_2$
	Diploidal	$2/m\overline{3}$	$3A_2$, $3m$, $4\overline{A_3}$
	Tetaroidal	23	$3A_2$, $4A_3$
Tetragonal	Ditetragonal–dipyramidal	$4/m2/m2/m$	i, $1A_4$, $4A_2$, $5m$
	Tetragonal–scalenohedral	$\overline{4}2\,m$	$1\overline{A_4}$, $2A_2$, $2m$
	Ditetragonal–pyramidal	$4mm$	$1A_4$, $4m$
	Tetragonal–trapezohedral	422	$1A_4$, $4A_2$
	Tetragonal–dipyramidal	$4/m$	i, $1A_4$, $1m$
	Tetragonal–disphenoidal	$\overline{4}$	$\overline{A_4}$
	Tetragonal–pyramidal	4	$1A_4$
Hexagonal (hexagonal)	Dihexagonal–dipyramidal	$6/m2/m2/m$	i, $1A_6$, $6A_2$, $7m$
	Ditrigonal–dipyramidal	$\overline{6}m2$	$1A_6$, $3A_2$, $3m$
	Dihexagonal–pyramidal	$6mm$	$1A_6$, $6m$
	Hexagonal–trapezohedral	622	$1A_6$, $6A_2$
	Hexagonal–dipyramidal	$6/m$	i, $1A_6$, $1m$
	Trigonal–dipyramidal	$\overline{6}$	$1\overline{A_6}$
	Hexagonal–pyramidal	6	$1A_6$
Hexagonal (rhombohedral or trigonal)	Hexagonal–scalenohedral	$\overline{3}2/m$	$1\overline{A_3}$, $3A_2$, $3m$
	Ditrigonal–pyramidal	$3m$	$1A_3$, $3m$
	Trigonal–trapezohedral	32	$1A_3$, $3A_2$
	Rhombohedral	$\overline{3}$	$1\overline{A_3}$
	Trigonal–pyramidal	3	$1A_3$
Orthorhombic	Rhombic–dipyramidal	$2/m2/m2/m$	i, $3A_2$, $3m$
	Rhombic–pyramidal	$mm2$	$1A_2$, $2m$
	Rhombic–disphenoidal	222	$3A_2$
Monoclinic	Prismatic	$2/m$	i, $1A_2$, $1m$
	Sphenoidal	2	$1A_2$
	Domatic	m	$1m$
Triclinic	Pinacoidal	$\overline{1}$	i
	Pedial	1	None

Table 4.3 summarizes the six (or seven) crystal systems, the symmetries of the 32 space point groups or classes and their names, which are based on general crystal forms. It is important to remember that a crystal cannot possess more symmetry than that of the motifs of which it is composed, but it can possess less, depending on how the motifs are arranged and how the crystal grew.

4.4.2   Bravais lattices, unit cells and crystal systems

As noted earlier, any motif can be represented by a point called a node. Nodes, and the motifs they represent, can also be translated in three directions (t_a, t_b and t_c) to produce three-dimensional space point lattices and unit cells (see Figure 4.10).

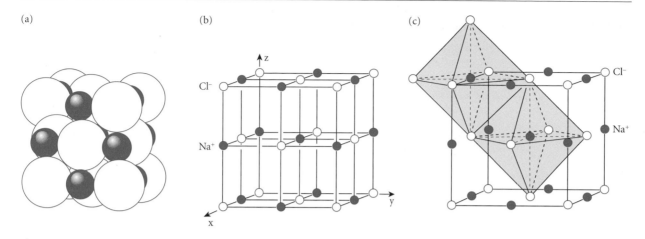

Figure 4.11 Relationship between (a) atomic packing, (b) a unit cell, and (c) octahedral coordination polyhedra in halite (NaCl). (From Wenk and Bulakh, 2004; with permission of Cambridge University Press.)

Unit cells are the three-dimensional analogs of unit meshes. A **unit cell** is a parallelepiped whose edge lengths and volume are defined by the three unit translation vectors (t_a, t_b and t_c). The unit cell is the smallest unit that contains all the information necessary to reproduce the mineral by three-dimensional symmetry operations. Unit cells may be primitive (P), in which case they have nodes only at their corners and a total content of one node (= one motif). Non-primitive cells are multiple because they contain extra nodes in one or more faces (A, B, C or F) or in their centers (I) and possess a total unit cell content of more than one node or motif.

Unit cells bear a systematic relationship to the coordination polyhedra and packing of atoms that characterize mineral structures, as illustrated by Figure 4.11.

Bravais (1850) recognized that only 14 basic types of three-dimensional translational point lattices exist; these are known as the **14 Bravais space point lattices** and define 14 basic types of unit cells. The 14 Bravais lattices are distinguished on the basis of (1) the magnitudes of the three unit translation vectors t_a, t_b and t_c or more simply a, b and c; (2) the angles (alpha, beta and gamma) between them, where ($\alpha = b \wedge c$; $\beta = c \wedge a$; $\gamma = a \wedge b$); and (3) whether they are primitive lattices or some type of multiple lattice. Figure 4.12 illustrates the 14 Bravais space point lattices.

The translational symmetry of every mineral can be represented by one of the 14 basic types of unit cells. Each unit cell contains one or more nodes that represent motifs and contains all the information necessary to characterize chemical composition. Each unit cell also contains the rules according to which motifs are repeated by translation; the repeat distances, given by $t_a = a$, $t_b = b$, $t_c = c$, and directions, given by angles α, β and γ (where $b \wedge c = \alpha$; $c \wedge a = \beta$; $a \wedge b = \gamma$). The 14 Bravais lattices can be grouped into **six crystal systems** on the basis of the relative dimensions of the unit cell edges (a, b and c) and the angles between them (α, β and γ). These six (or seven) systems in which all minerals crystallize include the **isometric** (cubic), **tetragonal, orthorhombic, monoclinic, triclinic, hexagonal** (hexagonal division or system) and hexagonal (**trigonal** division or system). Table 4.4 summarizes the characteristics of the Bravais lattices in major crystal systems.

4.5   CRYSTAL SYSTEMS

Imagine yourself in the minerals section of a museum. Large crystals are partially or completely bounded by planar crystal faces that are produced when minerals grow. Many other mineral specimens are partially or completely bounded by flat, planar cleavage faces produced when minerals break along planes of relatively low total bond strength. The

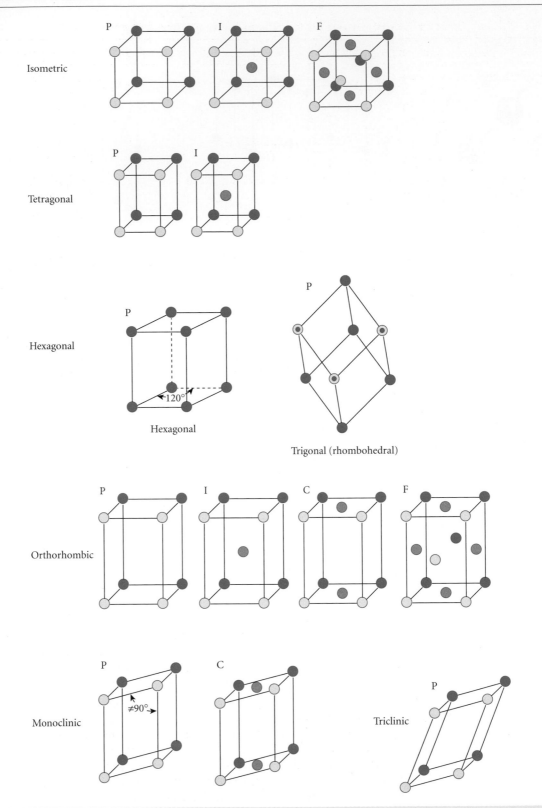

Figure 4.12   The 14 Bravais lattices and the six (or seven) crystal systems they represent. (Courtesy of Steve Dutch.)

Table 4.4 Major characteristics of Bravais lattice cells in the major crystal systems.

Crystal system	Unit cell edge lengths	Unit cell edge intersection angles	Bravais lattice types
Isometric (cubic)	(a = b = c) Preferred format for edges of equal length is ($a_1 = a_2 = a_3$)	$\alpha = \beta = \gamma = 90°$	Primitive (P) Body centered (I) Face centered (F)
Tetragonal	($a_1 = a_2 \neq a_3$) or a = b ≠ c	$\alpha = \beta = \gamma = 90°$	Primitive (P) Body centered (I)
Hexagonal (hexagonal)	($a_1 = a_2 \neq c$) (a = b ≠ c)	($\alpha = \beta = 90° \neq \gamma = 120°$)	Primitive (P)
Hexagonal (trigonal or rhombohedral)	($a_1 = a_2 = a_3$)	$\alpha = \beta = \gamma \neq 90°$	Primitive (P)
Orthorhombic	a ≠ b ≠ c	($\alpha = \beta = \gamma = 90°$)	Primitive (P) Body centered (I) End centered (A,B,C) Face centered (F)
Monoclinic	a ≠ b ≠ c	($\alpha = \gamma = 90° \neq \beta$)	Primitive (P) End centered (C)
Triclinic	a ≠ b ≠ c	(α, β and $\gamma \neq 90°$)	Primitive (P)

shapes of the crystals, the number and orientation of the crystal faces and the nature of the cleavage depend on the crystal structure of the mineral. That is, they depend on the basic motif and the symmetry operations that produce the three-dimensional crystal lattice. The nature of the crystal forms and cleavage surfaces depends on the crystal system and crystal class in which the mineral crystallized.

4.5.1 Crystallographic axes

To identify, describe and distinguish between planes in minerals, including cleavage planes, crystal faces and X-ray diffraction planes, a comprehensive terminology has been developed that relates each set of planes to the three **crystallographic axes** (Figure 4.13). For all but the rhombohedral division of the hexagonal system, the three crystallographic axes, designated a, b and c, are generally chosen to correspond to the three unit cell translation vectors (t_a, t_b, and t_c). With the exception noted, the three crystallographic axes have lengths and angular relationships that correspond to those of the three sets of unit cell edges (Table 4.4). The rules for labeling the three crystallographic axes are specific to each system; some systems have

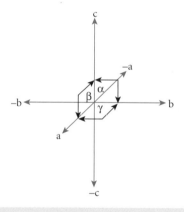

Figure 4.13 Conventional labeling of crystallographic axes, illustrating the positive and negative ends of the three crystallographic axes and the angles between the axes for crystals in the orthorhombic system.

multiple sets of rules. The details are beyond the scope of this text.

When referencing crystallographic planes to the crystallographic axes, a **standard set of orientation rules** is used (Table 4.5). To indicate their similarity, crystallographic axes with the same length are labeled a_1, a_2 and/or a_3 instead of a, b and c. In the isometric, tetragonal and orthorhombic systems (Figure 4.14), the b-axis (or a_2-axis) is

Table 4.5 The relationships of crystallographic axes and the rules for orienting crystals in each of the crystal systems. Note that the trigonal system (division) is listed independently of the hexagonal system (division) in this table.

Crystal system	Verbal description	Symbolic description
Isometric (cubic)	Three mutually perpendicular axes (a_1, a_2, a_3) of equal length that intersect at right angles	($a_1 = a_2 = a_3$) ($\alpha = \beta = \gamma = 90°$)
Tetragonal	Three mutually perpendicular axes; axes (a_1, a_2) are of equal length; the c axis may be longer or shorter	($a_1 = a_2 \neq c$) ($\alpha = \beta = \gamma = 90°$)
Orthorhombic	Three mutually perpendicular axes of different lengths (a, b, c); two axial length ratios have been used to identify the axes: c > b > a (older) or b > a > c (newer)	($a \neq b \neq c$) ($\alpha = \beta = \gamma = 90°$)
Monoclinic	Three unequal axes lengths (a, b, c) only two of which are perpendicular. The angle (β) between a and c is not 90°. The a-axis is inclined towards the observer. The b-axis is horizontal and the c-axis is vertical	($a \neq b \neq c$) ($\alpha = \gamma = 90°$; $\beta \neq 90°$)
Triclinic	Three unequal axes, none of which are generally perpendicular. The c axis is vertical and parallel to the prominent zone of crystal faces	($a \neq b \neq c$) ($\alpha \neq \beta \neq \gamma \neq 90°$)
Hexagonal	Four crystallographic axes; three equal horizontal axes (a_1, a_2, a_3) intersecting at 120°. One longer or shorter axis (c) perpendicular to the other three. a_1 oriented to front left of observer; a_2 to right; a_3 to back left; c vertical. Six-fold axis of rotation or rotoinversion	($a_1 = a_2 = a_3 \neq c$) ($\alpha = \beta = 90°$; $\gamma = 120°$)
Trigonal (or rhombohedral)	Axes and angles are similar to the hexagonal system; crystal symmetry is different with the c-axis a three-fold axis of rotation or rotoinversion	($a_1 = a_2 = a_3 \neq c$) ($\alpha = 120°$; $\beta = \gamma = 90°$)

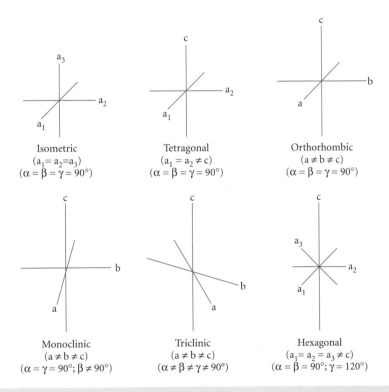

Figure 4.14 Crystallographic axes (positive ends labeled) and intersection angles for the major crystal systems: isometric, tetragonal, orthorhombic, monoclinic, triclinic and hexagonal systems.

oriented from left to right, with the right end designated as the positive end of the axis (b or a_2) and the left end designated as the negative end of the axis ($\bar{b}$ or $\bar{a}_2$). The a-axis (or a_1-axis) is oriented so that it trends from back to front toward the observer. The end of the a-axis toward the observer is designated as the positive end of the axis (a) and the end of the a-axis away from the observer is designated as the negative end ($\bar{a}$). The c-axis is oriented vertically with the top end designated as the positive end of the axis (c) and the bottom end designated as the negative end ($\bar{c}$). There are small exceptions to these rules in the hexagonal, monoclinic and triclinic systems that result from the fact that unit cell edges are not perpendicular to one another and not all crystallographic axes intersect at right angles. Even in the hexagonal and monoclinic systems, the b-axis is oriented horizontally with the positive end to the right, and in all systems the c-axis is vertical with the positive end toward the top and the negative end toward the bottom. The orientations, lengths and intersection angles between crystallographic axes in each of the major crystal systems are illustrated in Figure 4.14. The characteristics of the crystallographic axes in each system and the standard rules for orienting them are summarized in Table 4.5.

4.5.2 Crystal forms

Each of the crystal systems has an associated set of common crystal forms. **Crystal forms** consist of a three-dimensional set of one or more crystal faces that possess similar relationships to the crystallographic axes. Some natural crystals possess only one crystal form; others possess multiple or combined crystal forms. Crystal forms can be subdivided into two major groups: closed forms and open forms.

Closed crystal forms have the potential to completely enclose a mineral specimen and therefore to exist alone in perfectly formed (euhedral) crystals. Common closed forms include all the forms in the isometric system and many forms in the tetragonal, hexagonal, trigonal and orthorhombic systems. The pyritohedron (Figure 4.15) is a typical closed form, common in the mineral pyrite. Each closed form possesses a different shape that is related to the number and shape of faces in

Figure 4.15 A pyritohedron, a closed form in which all faces have the same general relationship to the crystallographic axes.

the form and their angular relationships to the crystallographic axes. Figure 4.16 illustrates common dipyramid closed forms in the trigonal, tetragonal and hexagonal systems.

Open crystal forms (Figure 4.17) do not have the potential to completely enclose a mineral specimen and so must occur in combination with other open or closed crystal forms. Common open forms include: (1) **pedions**, which consist of a single face, (2) **pinacoids**, a pair of parallel faces, (3) **prisms**, three or more faces parallel to an axis, (4) **pyramids**, three or more faces that intersect an axis, (5) **domes**, a pair of faces symmetrical about a mirror plane, and (6) **sphenoids**, a pair of faces symmetrical about an axis of rotation. Figure 4.17b illustrates the kinds of prisms that occur in the trigonal, tetragonal and hexagonal crystal systems.

The most common crystal forms in each system are discussed later in this chapter, after we have presented the language used to describe them. More detailed discussions are available in Klein and Dutrow (2007) and Nesse (2000).

4.6 INDEXING CRYSTALLOGRAPHIC PLANES

4.6.1 Axial ratios

Whatever their respective lengths, the proportional lengths or **axial ratios** of the three crystallographic axes (a:b:c) can be calculated. The standard method for expressing axial ratios is to express their lengths relative to the length of the b-axis (or a_2-axis) which is taken to be unity so that the ratio is expressed as

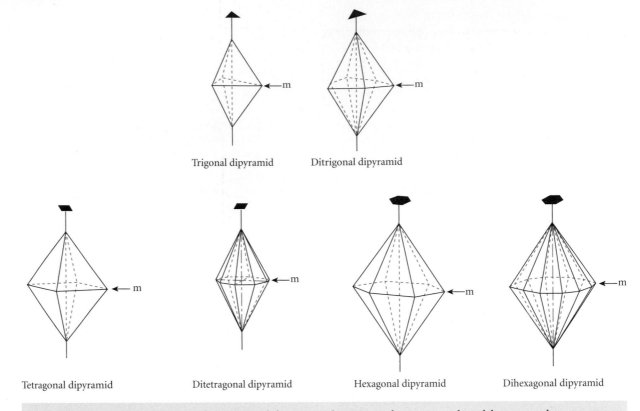

Trigonal dipyramid   Ditrigonal dipyramid

Tetragonal dipyramid   Ditetragonal dipyramid   Hexagonal dipyramid   Dihexagonal dipyramid

Figure 4.16   Different types of dipyramid forms in the trigonal, tetragonal and hexagonal systems. (From Klein and Hurlbut, 1985; with permission of John Wiley & Sons.)

a:b:c; b = 1. This is accomplished by dividing the lengths of all three axes by the length of the b-axis (a/b : b/b : c/b). An example from the monoclinic system, the pseudo-orthorhombic mineral staurolite, will illustrate how axial ratios are calculated. In staurolite, the unit cell edges have average dimensions expressed in angstrom (Å) units of: a = 7.87Å, b = 16.58Å and c = 5.64Å. The axial ratios are calculated from a/b : b/b : c/b = 7.87/16.58Å : 16.58/16.58Å : 5.64/16.58Å. The average axial ratios of staurolite are 0.47 : 1.00 : 0.34.

Axial ratios are essential to understanding how crystallographic planes and crystal forms are described by reference to the crystallographic axes as discussed in the section that follows.

4.6.2   Crystal planes and crystallographic axes

Crystalline substances such as minerals have characteristic planar features that include: (1) crystal faces that develop during growth, (2) cleavage surfaces that develop during breakage, and (3) crystal lattice planes that reflect X-rays and other types of electromagnetic radiation. All these types of planes possess a number of shared properties.

Each type of plane is a representative of large sets of parallel lattice planes. As a mineral with a particular crystal form grows freely it may be bounded by a sequence of planar faces. When it stops growing, it is bounded by crystal faces that are parallel to many other lattice planes that bounded the mineral as it grew over time. When a mineral cleaves, it breaks along a specific set of parallel planes of relative weakness, but these cleavage planes are parallel to large numbers of planes of weakness or potential cleavage surfaces in the mineral structure along which the mineral did not happen to rupture. When X-rays are reflected from a reflecting plane, they are reflected simultaneously from all the planes in the crystal that are parallel to one

(a)

(b)

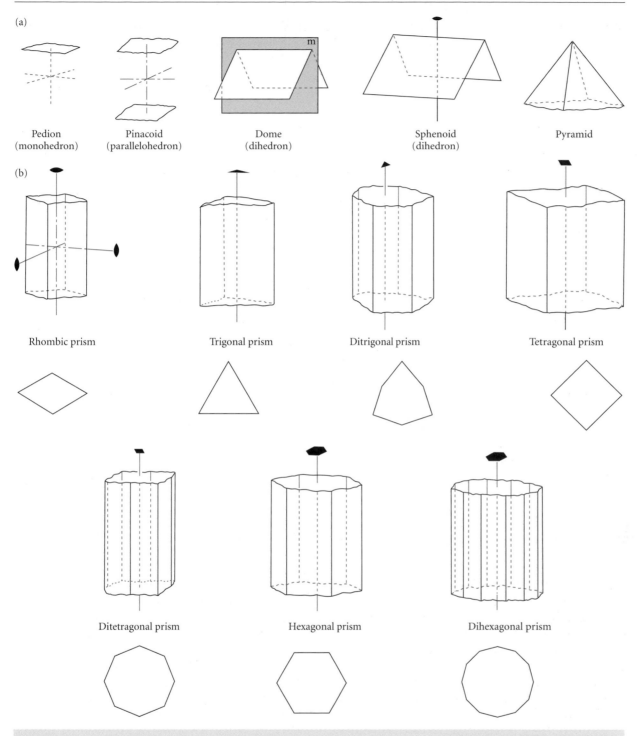

Figure 4.17 (a) Common open forms: pedions, pinacoids, domes, sphenoids and pyramids. (b) Different types of prisms that characterize the trigonal, tetragonal and hexagonal systems. The illustrated prisms are bounded by pinacoids at the top and bottom. (From Hurlbut and Klein, 1985; with permission of John Wiley & Sons.)

another to produce a "reflection peak" that is characteristic of the mineral and can be used to identify it. In addition, any set of parallel planes in a crystal is characterized by a particular molecular content; all the parallel planes in the set possess the same molecular units, spacing and arrangement. A molecular image of one of these planes is sufficient to depict the molecular content of all the planes that are parallel to it. Third, all the planes in a set of parallel planes have the same general spatial relationship to the three crystallographic axes. This means that they can be collectively identified in terms of their spatial relationship to the three crystallographic axes. This is true for crystal faces, for cleavage surfaces, for X-ray reflecting planes or for any set of crystallographic planes that we wish to identify. A universally utilized language has evolved that uses the relationship between the planar features in minerals and the crystallographic axes to identify different sets of planes. A discussion of this language and its use follows.

Figure 4.18 depicts several representative crystal planes with different relationships to the three crystallographic axes. Some crystal planes, or sets of parallel planes, intersect one crystallographic axis and are parallel to the other two (Figure 4.18a, b). Alternatively, a set of crystal planes may intersect two crystallographic axes and be parallel to the third (Figure 4.18c, d). Still other sets of planes

intersect all three crystallographic axes (Figure 4.18e, f). No other possibilities exist in Euclidean space; sets of planes in crystalline substances must intersect one, two or three axes and be parallel to those they do not intersect.

Of course, some sets of planes or their projections intersect the positive ends of crystallographic axes (Figure 4.18b, c, e). Others, with different orientations with respect to the axes, intersect the negative ends of crystallographic axes (Figure 4.18a). Still others, with yet different orientations, intersect the positive ends of one or more axes and the negative ends of the other crystallographic axes (Figure 4.18d, f). Given the myriad possibilities, a simple language is needed that allows one to visualize and communicate to others the relationship and orientation of any set of crystal planes to the crystallographic axes. The language for identifying and describing crystallographic planes involves the use of symbols called Miller indices, which has been employed since the 1830s and is explained in the following sections.

4.6.3   Unit faces and planes

In any crystal, the three crystallographic axes have a characteristic axial ratio, typically grounded in the cell edge lengths of the unit cell. No matter how large the mineral becomes during growth, even if it experiences preferred growth in a particular direction or inhibited growth in another, the axial ratio remains constant and corresponds to the axial ratio implied by the properties of the unit cell.

In the growth of any mineral one can imagine the development of a crystal face, one of many potential crystal faces, that intersects the positive ends of all three axes at lengths that correspond to the axial ratio of the mineral (Figure 4.19). For crystals with a center (i), such faces would intersect each axis at a distance from the center of the crystal that corresponds to the axial ratio. For the monoclinic (pseudo-orthorhombic) mineral staurolite discussed in the previous section, such a face could cut the a-axis at 0.47 mm, the b-axis at 1.00 mm and the c-axis at 0.34 mm from the center, or, for a larger crystal, it could cut the three axes at 0.47, 1.00 and 0.34 cm from the center. Any face or plane that intersects all three axes at

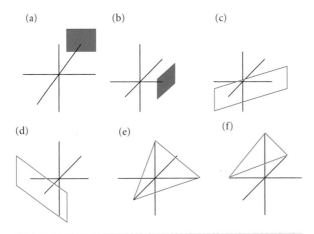

Figure 4.18 Representative crystal faces that cut one, two or three crystallographic axes. See text for further discussion of parts (a) to (f).

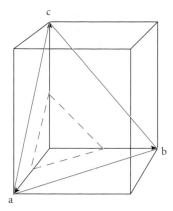

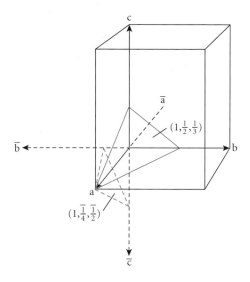

Figure 4.19 Unit face (outlined in solid blue) in an orthorhombic crystal with three unequal unit cell edges and crystallographic axes that intersect at right angles. All parallel faces (e.g., outlined in dotted blue) will have the same general relationship to the crystallographic axes and the same atomic content and properties.

Figure 4.20 Faces with different Weiss parameters on an orthorhombic crystal.

distances from the center that correspond to the axial ratio of the mineral is a **unit face** or **unit plane**. It is part of a set of parallel planes all of which are unit planes because they intersect the three crystallographic axes at lengths that correspond to the axial ratios.

4.6.4   Weiss parameters

Weiss parameters provide a method for describing the relationships between sets of crystal faces or planes and the crystallographic axes. They are always expressed in the sequence a:b:c, where a represents the relationship of the planes to the a-axis (or a_1-axis), b represents the relationship between the planes and the b-axis (or a_2-axis) and c depicts the relationship between the planes and the c-axis (or a_3-axis). A unit face or plane that cuts all three crystallographic axes at ratios corresponding to their axial ratios has the Weiss parameters (1:1:1). Mathematically, if we divide the actual lengths at which the face or plane intercepts the three axes by the corresponding axial lengths, the three intercepts have the resulting ratio 1:1:1, or unity, which is why such planes are called unit planes. Again, using the pseudo-orthorhombic mineral staurolite as an

example, if we divide the actual intercept distances by the axial lengths, the resulting ratios are 0.47/0.47 : 1.00/1.00 : 0.34/0.34 = 1:1:1. Even if we utilize the magnitudes of the dimensions, the three resulting numbers have the same dimensional magnitude, and so their ratio reduces to 1:1:1.

As discussed in the section on unit faces and planes, many planes intersect all three crystallographic axes. Unit faces or planes (1:1:1), such as those in Figure 4.19, intersect all three axes at lengths that correspond to their axial ratios. Other sets of planes, however, intersect one or more axes at lengths that do not correspond to their axial ratios. A face or plane that intersects all three crystallographic axes at different lengths relative to their axial ratios is called a **general face**. The Weiss parameters of such a face or plane will be three rational numbers that describe the fact that each axis is intersected at a different proportion of its axial ratio. There are many sets of general planes. For example, a general plane with the Weiss parameters (1:1/2:1/3), shown in Figure 4.20, would be a plane that intersects the c-axis at one-third the corresponding length of the c-axis and the b-axis at one-half the corresponding length of the b-axis. Since many general faces are possible, for example (1:1/2:1/3) or (1/2:1:1/4), it is possible to write a general notation for all the faces that intersect the three crystallographic

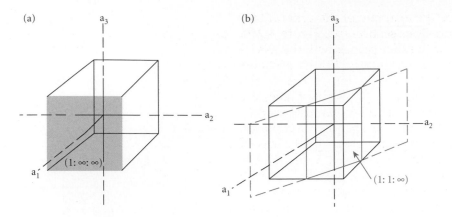

Figure 4.21 (a) The darkened front crystal face possesses the Weiss parameters: $(1:\infty:\infty)$. (b) The face outlined in blue possesses the Weiss parameters $(1:1:\infty)$.

axes as $(h:k:l)$ where h, k and l are intercepts with a, b and c.

Many crystal planes intersect the negative ends of one or more crystallographic axes. The location and/or orientation of these planes are not the same as those of planes that intersect the positive ends of the same axes. Planes that intersect the negative ends of one or more crystallographic axes are indicated by placing a bar over their Weiss parameters. For example, the dashed plane in Figure 4.20 has the Weiss parameters $(1:\bar{1}/4:\bar{1}/2)$.

Mineral planes may be parallel to one or two crystallographic axes. How do we determine the Weiss parameters for such faces and planes? The Weiss parameters of any face or plane that is parallel to a crystallographic axis are infinity (∞) because the plane never intersects the axis in question. If a set of planes is parallel to two crystallographic axes and intersects the third, it is assumed to intersect that axis at unity. Planes that are parallel to the a- and b-axes and intersect the c-axis have the Weiss parameters $(\infty:\infty:1)$. Planes parallel to the b- and c-axes that intersect the a-axis have the Weiss parameters $(1:\infty:\infty)$. Planes that cut the b-axis and are parallel to the a- and c-axes (Figure 4.21a) have the Weiss parameters $(\infty:1:\infty)$. Each set of planes, with its unique relationship to the crystallographic axes possesses its own unique Weiss parameters. If a set of planes intersects two axes and is parallel to the third, only one of the Weiss parameters will be infinity. The other two will be one if, and only if, the two axes are intersected at lengths corresponding to their axial

ratios. Therefore, the Weiss parameters $(1:\infty:1)$ represent planes that parallel the b-axis and intersect the a- and c-axes at unit lengths. Similarly the Weiss parameters $(1:1:\infty)$ are those of planes that cut the a- and b-axes at unit lengths and are parallel to the c-axis (Figure 4.21b).

Having begun to master the concepts of how Weiss parameters can be used to represent different sets of crystal planes with different sets of relationships to the crystallographic axes, students are generally thrilled to find that crystal planes are commonly referenced, not by Weiss parameters, but instead by Miller indices.

4.6.5 Miller indices

The **Miller indices** of any face or set of planes are the **reciprocals of its Weiss parameters**. They are calculated by inverting the Weiss parameters and multiplying by the lowest common denominator. Because of this reciprocal relationship, large Weiss parameters become small Miller indices. For planes parallel to a crystallographic axis, the Miller index is zero. This is because when the large Weiss parameter infinity (∞) is inverted it becomes the Miller index $1/\infty \rightarrow 0$.

The Miller index of any face or set of planes is, with a few rather esoteric exceptions, expressed as three integers **hkl** in a set of parentheses **(hkl)** that represent the reciprocal intercepts of the face or planes with the three crystallographic axes (a, b and c) respectively.

We can use the example of the general face cited in the previous section (see Figure 4.20), where a set of parallel planes cuts the a-axis at unity, cuts the b-axis at half unity and cuts the c-axis at one-third unity. The Weiss parameters of such a set of parallel planes are $1:1/2:1/3$. If we invert these parameters they become 1/1, 2/l and 3/1. The lowest common denominator is one. Multiplying by the lowest common denominator yields 1, 2 and 3. The Miller indices of such a face are (123). These reciprocal indices should be read as representing all planes that intersect the a-axis at unity (1) and the b-axis at one-half unity (reciprocal is 2), and then intercept the c-axis at one-third unity (reciprocal is 3) relative to their respective axial ratios. Every parallel plane in this set of planes has the same Miller indices.

As is the case with Weiss parameters, the Miller indices of planes that intersect the negative ends of one or more crystallographic axes are denoted by the use of a bar placed over the indices in question. We can use the example from the previous section in which a set of planes intersect the positive end of the a-axis at unity, the negative end of the b-axis at twice unity and the negative end of the c-axis at three times unity. If the Weiss parameters of each plane in the set are 1, $\bar{2}/3$ and $\bar{1}/2$, inversion yields 1/1, $\bar{3}/2$ and $\bar{2}/1$. Multiplication by two, the lowest common denominator, yields 2/1, $\bar{6}/2$ and $\bar{4}/1$ so that the Miller indices are $(2\bar{3}\bar{4})$. These indices can be read as indicating that the planes intersect the positive end of the a-axis and the negative ends of the b- and c- crystallographic axes with the a-intercept at unity and the b-intercept at two-thirds unity and the c-intercept at one-half unity relative to their respective axial ratios.

A simpler example is the cubic crystal shown in Figure 4.22. Each face of the cube intersects one crystallographic axis and is parallel to the other two. The axis intersected is indicated by the Miller index "1" and the axes to which it is parallel are indicated by the Miller index "0". Therefore the six faces of the cube have the Miller indices (100), $(\bar{1}00)$, (010), $(0\bar{1}0)$, (001) and $(00\bar{1})$.

Miller indices are a symbolic language that allows us to represent the relationship of any crystal or cleavage face or crystallographic plane with respect to the crystallographic axes.

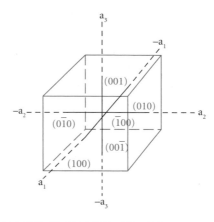

Figure 4.22   Miller indices of various crystal faces on a cube depend on their relationship to the crystallographic axes.

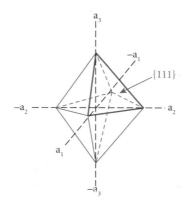

Figure 4.23   An isometric octahedron outlined in blue possesses eight faces; the form face {111} is outlined in bold blue.

4.6.6 Form indices

Every face in a form has the same general relationship to the crystallographic axes and therefore the same general Miller index, yet every face in a form has a different specific relationship to the crystallographic axes and therefore has a different Miller index. These statements can be clarified by using an example. Figure 4.23 shows the common eight-faced isometric form called the octahedron.

Each face in the octahedron has the same general relationship to the three crystallographic axes in that each intersects the three crystallographic axes at unity. The Miller indices of each face are some form of (111). However, only the top, right front face intersects the positive ends of all three axes. The bottom, left back face intersects the negative ends of all three axes, and the other six faces intersect some combination of positive and negative ends of the three crystallographic axes. None of the faces are parallel to one another; each belongs to a different set of parallel planes within the crystal. The Miller indices of these eight faces and the set of planes to which each belongs are (111), ($\bar{1}$11), (1$\bar{1}$1), (11$\bar{1}$), ($\bar{1}\bar{1}$1), ($\bar{1}$1$\bar{1}$), (1$\bar{1}\bar{1}$) and ($\bar{1}\bar{1}\bar{1}$). Their unique Miller indices allow us to distinguish between the eight faces and the sets of planes to which they belong. But they are all parts of the same form because they all have the same general relationship to the crystallographic axes. It is cumbersome to have to recite the indices of every face within it. To represent the general relationship of the form to the crystallographic axes, the indices of a single face, called the form face, are chosen and placed in brackets to indicate that they refer to the form indices. The rule for choosing the **form face** is generally to select the top face if there is one, or the top right face if there is one, or the top right front face if there is one. In the case of the octahedron, the top right front face is the face that intersects the positive ends of the a_1-axis (front), the a_2-axis (right) and the a_3-axis (top) and has the Miller indices (111). The form indices for all octahedral crystals are the Miller indices of the form face placed between curly brackets, {111}. Similarly the form indices for the cube (see Figure 4.22), in which the faces intersect one axis and are parallel to the other two are {001}, the Miller indices of the top face, whereas the form indices for the dodecahedron, in which each face intersects two axes

at unity and is parallel to the third is {011}, the indices for the top, right face.

Many other forms exist. Every crystal form has a form index, which is the Miller index of the form face placed in brackets. Each form consists of one or (generally) more faces and each face possesses a Miller index different from that of every other face in the form. Every crystal system has a characteristic suite of forms that reflect the unique characteristics of the crystal lattice of the system, especially the relative lengths of the three crystallographic axes that directly or indirectly reflect the lengths of the unit cell edges. The forms characteristic of each class (space point group) in each crystal system are beyond the scope of this text. A brief review of some common forms in each crystal system follows.

4.6.7 Common crystal forms in crystal systems

Isometric (cubic) system forms

All forms in the isometric system are closed forms. Common crystal forms in the isometric system include the cube, octahedron, dodecahedron, tetrahedron and pyritohedron (Figure 4.24), all closed forms. These forms often occur in combination with each other. Common isometric minerals, their crystal forms and form indices are summarized in Table 4.6.

These form indices are also used to describe cleavage in isometric minerals such as halite and galena, which possess cubic cleavage {001} with three orientations of cleavage at right angles; fluorite, which possesses octahedral cleavage {111} with four orientations; and sphalerite, which possesses dodecahedral cleavage {011} with six orientations of cleavage.

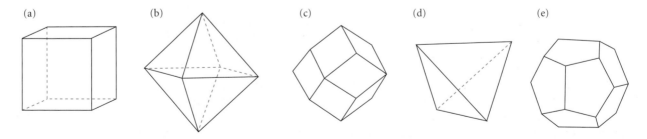

Figure 4.24 Five common forms in the isometric system: (a) cube, (b) octahedron, (c) dodecahedron, (d) tetrahedron, (e) pyritohedron.

Table 4.6 Common isometric crystal forms, form indices, form descriptions and minerals.

Crystal form	Form indices	Form description	Minerals that commonly exhibit crystal form
Cube	{001}	Six square faces	Halite, galena, pyrite, fluorite, cuprite, perovskite, analcite
Octahedron	{111}	Eight triangular faces	Spinel, magnetite, chromite, cuprite, galena, diamond, gold, perovskite
Dodecahedron	{011}	12 diamond-shaped faces	Garnet, sphalerite, sodalite, cuprite
Tetrahedron	{1$\bar{1}$1}	Four triangular faces	Tetrahedrite, sphalerite
Pyritohedron	{h0l}	12 pentagonal faces	Pyrite

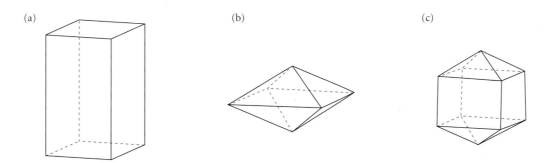

(a)   (b)   (c)

Figure 4.25 Common crystal forms in the tetragonal crystal system: (a) tetragonal prism in combination with a pinacoid, (b) tetragonal dipyramid, (c) tetragonal dipyramid in combination with a tetragonal prism.

Table 4.7 Common tetragonal crystal forms, form indices, form descriptions and minerals.

Crystal forms	Form indices	Form description	Minerals that commonly exhibit crystal form
Tetragonal dipyramid	{111} {hh1} {011} {0kl} and variations	Eight triangular faces; top four separated from bottom four by mirror plane	Zircon, rutile, cassiterite, scheelite, wulfenite, vesuvianite, scapolite
Tetragonal prism	{010} {110} and variations	Four rectangular faces parallel to c-axis	Zircon, scheelite, vesuvianite, rutile, malachite, azurite, cassiterite, scapolite
Tetragonal disphenoid	{0kl}	Four triangular faces; alternating pairs symmetrical about c-axis	Chalcopyrite
Basal pinacoid	{001}	Pair of faces perpendicular to c-axis	Vesuvianite, wulfenite

Tetragonal system forms

Tetragonal crystals can possess either closed or open forms, often in combination. Common closed crystal forms in the tetragonal crystal system include different eight-sided dipyramids. Common open forms include four-sided prisms and pyramids, as well as pinacoids and pedions. Typical crystal forms and associated minerals in the tetragonal crystal system are shown in Figure 4.25 and Table 4.7.

Hexagonal system (hexagonal division) forms

Common crystal forms in the hexagonal system include 6–12-sided prisms, dipyramids and pyramids. Pinacoids and pedions are also common. Some selected examples of common crystal forms and minerals in the hexagonal system are illustrated in Figure 4.26 and Table 4.8.

Trigonal system (hexagonal system, trigonal division) forms

Common crystal forms in the trigonal system include the six-sided rhombohedron, the 12-sided scalenohedron, six-sided trigonal dipyramids and three-sided trigonal pyramids.

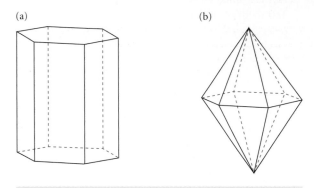

Figure 4.26 Common crystal forms in the hexagonal crystal system (hexagonal division): (a) hexagonal prism with pinacoids, (b) hexagonal dipyramid.

Pinacoids and pedions are also common. Many forms common in the hexagonal division also occur in trigonal crystals, but not vice versa. Common crystal forms and representative minerals in the trigonal crystal system are summarized in Figure 4.27 and Table 4.9.

Orthorhombic crystal system

Common crystal forms in the orthorhombic system include four-sided rhombic prisms, dipyramids and pyramids. Pinacoids are the dominant form and pedions are also common. Common crystal forms and associated minerals in the orthorhombic crystal system are indicated in Figure 4.28 and Table 4.10.

Monoclinic crystal system

Because of its lower symmetry, the only crystal forms in the monoclinic crystal system are four-sided prisms, two-sided domes, sphenoids and pinacoids, and pedions. More

Table 4.8   Common hexagonal crystal forms and associated hexagonal and trigonal minerals.

Crystal forms	Common form indices	Form description	Minerals that commonly exhibit crystal form
Hexagonal dipyramid	$\{11\overline{2}1\}$ and variations	12 triangular faces inclined to c-axis; top six separated from bottom six by mirror plane	Apatite, zincite
Hexagonal prism	$\{11\overline{2}0\}$ and variations	Six rectangular faces parallel to c-axis	Apatite, beryl, quartz, nepheline, corundum, tourmaline
Basal pinacoid	$\{0001\}$	Pair of faces perpendicular to c-axis	Apatite, beryl, corundum

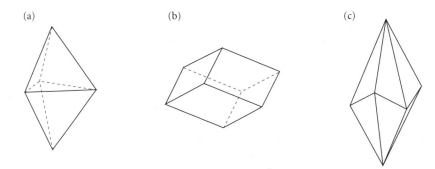

Figure 4.27   Common crystal forms in the trigonal system: (a) trigonal dipyramid, (b) rhombohedron, (c) scalenohedron.

Table 4.9 Common trigonal crystal forms and associated minerals.

Crystal forms	Form indices	Form description	Minerals that commonly exhibit crystal form
Rhombohedron	$\{10\bar{1}1\}$ $\{h0\bar{h}l\}$	Six parallelogram faces inclined to c-axis	Dolomite, calcite, siderite, rhodochrosite, quartz, tourmaline; chabazite
Trigonal Scalenohedron	$\{hk\bar{i}l\}$ and variations	12 scalene triangle faces inclined to c-axis	Calcite
Trigonal prism	$\{hk\bar{i}0\}$ and variations	Three rectangular faces parallel to c-axis	Tourmaline, calcite, quartz
Trigonal dipyramid	$\{hk\bar{i}l\}$ and variations	Six triangular faces; top three separated from bottom three by a mirror plane	Tourmaline

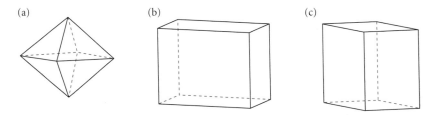

Figure 4.28 Common crystal forms in the orthorhombic crystal system: (a) rhombic dipyramid, (b) front, side and basal pinacoids, (c) rhombic prism with a pinacoid.

Table 4.10 Common crystal forms, form indices and minerals in the orthorhombic system.

Crystal forms	Form indices	Form description	Minerals that commonly exhibit crystal form
Rhombic dipyramids	$\{111\}$ $\{hkl\}$ and variations	Eight triangular faces; top four separated from bottom four by a mirror plane	Topaz, aragonite, witherite, olivine
Rhombic prisms; first, second and third order	$\{011\}$ $\{0kl\}$ $\{101\}$ $\{h0l\}$ $\{011\}$ $\{0kl\}$	Four rectangular faces parallel to a single crystallographic axis	Stibnite, aragonite, barite, celestite, topaz, enstatite, andalusite, cordierite, epidote, olivine
Pinacoids; front, side and basal	$\{001\}$ $\{010\}$ $\{001\}$	Two parallel faces perpendicular to a-, b- or c-axis	Barite, celestite, olivine, andalusite, topaz, hemimorphite

complex forms cannot exist in systems with low symmetry in which crystallographic axes do not all intersect at right angles. Common crystal forms and associated minerals in the monoclinic crystal system are indicated in Figure 4.29 and Table 4.11.

Triclinic crystal system

The only crystal forms in the triclinic system, with its extremely low symmetry, are pinacoids and pedions. Common forms and minerals in the triclinic system are illustrated in Figure 4.30 and listed in Table 4.12.

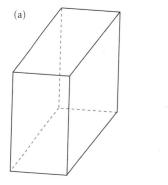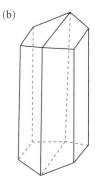

Figure 4.29 Monoclinic crystal forms: (a) front, side and basal pinacoids, (b) two monoclinic prisms and a side pinacoid.

Table 4.11 Common crystal forms, form indices, and minerals in the monoclinic system.

Crystal form	Form indices	Form description	Minerals that commonly exhibit crystal form
Monoclinic prisms; first, third and fourth order	{011} {0kl} {110} {hk0} {hkl}	Four rectangular faces	Gypsum, staurolite, clinopyroxenes, amphiboles, orthoclase, sanidine, sphene (titanite), borax
Pinacoids; front, side and basal	{001} {010} {001}	Pair of rectangular faces perpendicular to a-, b- or c-axis	Gypsum, staurolite, sphene (titanite), epidote, micas, clinopyroxenes, amphiboles

Table 4.12 Common crystal forms, form indices and minerals in the triclinic system.

Crystal forms	Form indices	Form description	Minerals that commonly exhibit crystal form
Pinacoids	{001}{010}{001} {0k1} {hk1} and variations	Two parallel faces	Kyanite, plagioclase, microcline, amblygonite, rhodonite, wollastonite
Pedions	{hk1}	Single face	Similar

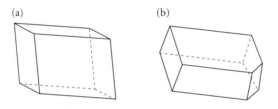

(a)   (b)

Figure 4.30 Triclinic crystal forms: (a) front, side and basal pinacoids, (b) various pinacoids and a pedion to the lower right.

4.7 TWINNED CRYSTALS

Many crystals are **twinned crystals** that contain two or more parts called twins. **Twins** have the following characteristics: (1) they possess different crystallographic orientations, (2) they share a common surface or plane, and (3) they are related by a symmetry operation such as reflection, rotation or inversion (Figure 4.31). Because twins are related by a symmetry operation, twinned crystals are not random intergrowths.

The symmetry operation that relates twins in twinned crystals is called a twin law. A **twin law** describes the symmetry operation that produces the twins and the plane (hkl) or axis involved in the operation. For example, swallowtail twins in gypsum (Figure 4.31a) are related by reflection across a plane (001),

which is not a mirror plane in single gypsum crystals. Carlsbad twins in potassium feldspar (Figure 4.31f) are related by a two-fold axis of rotation that is parallel to the c-axis (001), which is not a rotational axis in single potassium feldspar crystals.

The surfaces along which twins are joined are called **composition surfaces**. If the surfaces are planar, they are called **composition planes**, which may or may not be equivalent to twin planes. Other composition surfaces are irregular. Twins joined along composition planes are called **contact twins** and do not appear to penetrate one another. Good examples of contact twins are shown in Figure 4.31a and b. Twins joined along irregular composition surfaces are usually related by rotation and are called **penetration twins** because they appear to penetrate one another. Good examples of penetration twins are shown in Figure 4.31c–f.

Twinned crystals that contain only two twins are called **simple twins**, whereas **multiple twins** are twinned crystals that contain more than two twins. If multiple twins are repeated across multiple parallel composition planes, the twins are called **polysynthetic twins**. Polysynthetic albite twins (Figure 4.31b) are repeated by the albite twin law, reflection across (010), and are very common in plagioclase. They produce small ridges and troughs on the cleavage surfaces of

(a)

(010)

(100)

(110)

(0$\bar{1}$1)

(b)

(001)

(010)

(110)

(10$\bar{1}$)

(c)

{110}

(100)

(d)

{001}

(201)

(e)

(001)

(031)

(110)

(10$\bar{1}$)

(f)

{001}

(001)

(010)

(110)

(20$\bar{1}$)

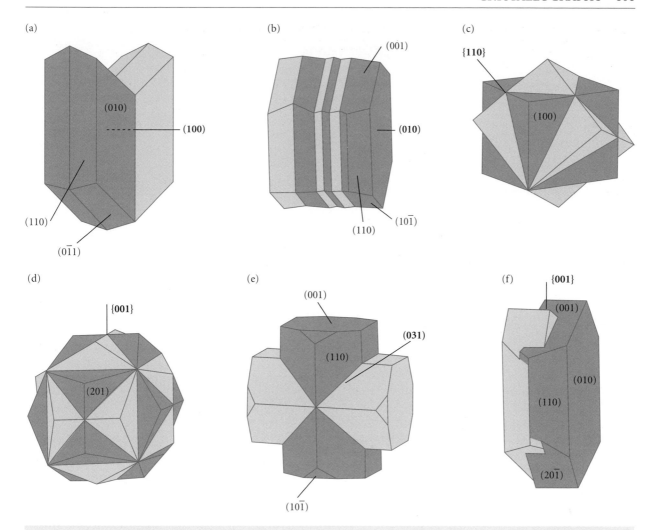

Figure 4.31 Examples of twinned crystals: (a) swallowtail twins in gypsum; (b) polysynthetic albite twins in plagioclase; (c) penetration twins in Galena; (d) penetration twins in pyrite; (e) penetration twins in staurolite; (f) Carlsbad twins in potassium feldspar. (From Wenk and Bulakh, 2004; with permission of Cambridge University Press.)

plagioclase, which the eye detects as striations – a key to hand-specimen identification of plagioclase.

Most twins are **growth twins** that form during mineral crystallization. Less commonly, twins result from displacive mineral transformations or from deformation. Deformation twins are called **mechanical twins**. The common mineral calcite typically develops mechanical twins (102) during deformation, and their development can play a significant role in the deformation of marbles and other metamorphic rocks, especially at low temperatures.

4.8 CRYSTAL DEFECTS

Ideally, crystals are perfectly formed with no defects in their lattice structures (Figure 4.32a). However, nearly all crystals contain small-scale impurities or imperfections that cause mineral composition and/or structure to vary from the ideal. These local-scale inhomogeneities are called **crystal defects**. Crystal defects have some profound effects on the properties of crystalline material that belie their small scale (Box 4.1). A convenient way to classify crystal defects is in terms of their dimensions.

Box 4.1 Frenkel and Schottky defects

Frenkel defects (Figure B4.1a) are formed when the ions in question move to an interstitial site, leaving unoccupied structural sites or holes behind. Frenkel defects combine omission and interstitial defects. Because the ion has simply moved to another location, the overall charge balance of the crystal is maintained, but local lattice distortions occur in the vicinities of both the holes and the extra ions. **Schottky defects** (Figure B4.1b) either are formed when the ions migrate out of the crystal structure or were never there. Schottky defects create a charge imbalance in the crystal lattice. Such charge imbalances may be balanced by the creation of a second hole in the crystal structure; for example, an anion omission may be created to balance a cation omission. They may also be balanced by the substitution of ions of appropriate charge difference elsewhere in the structure. The highly magnetic mineral **pyrrhotite** ($Fe_{1-x}S$) provides a good example. When a ferrous iron (Fe^{+2}) ion is omitted from a cation site in the crystal structure, leaving a charge deficit of 2, two ferric iron (Fe^{+3}) ions can substitute for ferrous iron (Fe^{+2}) ions to increase the charge by 2 and produce an electrically neutral lattice (Figure B4.1b). The formula for pyrrhotite reflects the fact that there are fewer iron (total Fe^{+2} and Fe^{+3}) ions than sulfur (S^{-2}) ions in the crystal structure due to the existence of a substantial number of such Schottky omission defects.

Point defects can occur on still smaller scales. In some cases electrons are missing from a quantum level, which produces an electron hole in the crystal structure. In others, an electron substitutes for an anion in the crystal structure. As with other point defects, the existence of electron holes plays an important role in the properties of the crystalline materials in which they occur. In most minerals, as temperature increases, the number of omission defects tends to increase. This allows minerals to deform more readily in a plastic manner at higher temperatures.

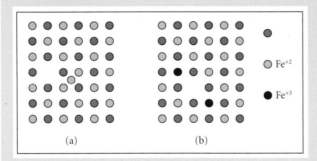

$\bigcirc$ Fe^{+2}

$\bullet$ Fe^{+3}

Figure B4.1 (a) Frenkel defect, with a vacancy due to an ion displaced to the interstitial site. (b) Shottky defect in pyrrhotite ($Fe_{1-x}S$) where a vacancy (absent Fe^{+2}) is balanced by the substitution of Fe^{+3} for Fe^{+2} in two lattice sites.

4.8.1 Point defects

Point defects involve individual atoms and therefore do not have longer range extent; they are considered to be **zero-dimensional defects**. Many types of point defects exist, and they are important in explaining the properties of minerals as well as other materials such as steel, cement and glass products:

1 **Substitution defects** (Figure 4.32b) occur when anomalous ions of inappropriate size and/or charge substitute for ions of appropriate size and/or charge in a structural site. These anomalous ions tend to distort the crystal lattice locally and to be somewhat randomly distributed within the crystal lattice.

2 **Interstitial defects** (Figure 4.32c) occur when anomalous ions occupy the spaces between structural sites. Such "extra" ions are trapped in the interstices between the "normal" locations of ions in the crystal lattice.

3 **Omission defects** (Figure 4.32d) occur when structural sites that should contain ions are unoccupied. In such cases, ions that should occur within the ideal crystal structure are omitted from the crystal lattice leaving a "hole" in the ideal crystal structure.

4.8.2 Line defects

Line defects are called **dislocations**. Like lines, they possess extent in one direction and are

therefore **one-dimensional defects**. Dislocations commonly result from shearing stresses
produced in crystals during deformation
that cause atomic planes to shift position,

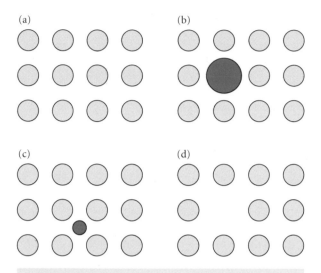

Figure 4.32   (a) Perfect crystal lattice; (b)
substitution defect; (c) interstitial defect; (d)
omission defect.

producing distortions in the crystal lattice
that can be represented by a line called a
dislocation line. Two major types of dislocations are recognized: **edge dislocations**
(Figure 4.33a) and **screw dislocation** (Figure
4.33b).

Dislocations are extremely important in
the plastic deformation of crystalline materials that leads to changes in rock shape and
volume without macroscopic fracturing.
Dislocations permit rocks to flow plastically
at very slow rates over long periods of time.
Detailed discussions are available in many
books on mineralogy (e.g., Wenk and
Bulakh, 2004) and structural geology (e.g.,
Davis and Reynolds, 1996; van der Pluijm
and Marshak, 2003). Figure 4.34 shows how
an edge dislocation can migrate through a
crystal by breaking a single bond at a time.
The result of dislocation migration is a change
in the shape of the crystal that has been
accomplished without rupture. Changes in
shape during deformation that do not involve
rupture are called plastic deformation, and
dislocations are critical to its occurrence
(Chapter 16).

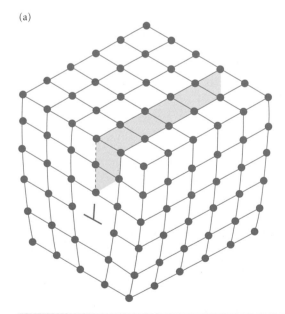

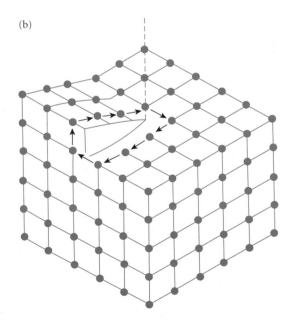

Figure 4.33   (a) Edge dislocation with an extra half plane of atoms; this is a line defect because the
base of the half plane can be represented by a dislocation line (⊥). (b) Screw dislocation, where a
plane of atoms has been rotated relative to the adjacent plane. (After Klein and Hurlbut, 1985; with
permission of John Wiley & Sons.)

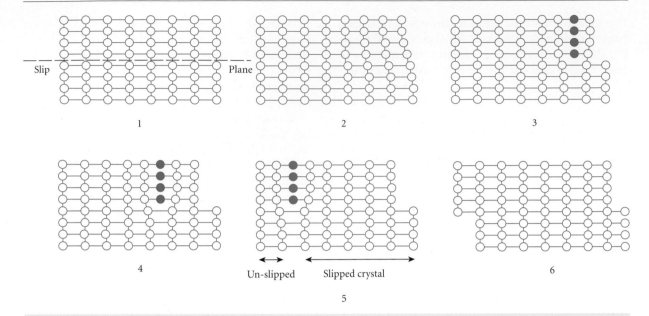

Figure 4.34 Two-dimensional depiction of how an edge dislocation created by slip due to shear can migrate through a crystal by breaking one bond at a time, so that no fractures develop as the crystal changes shape during deformation (steps 1 to 6). (After Hobbs et al., 1976.)

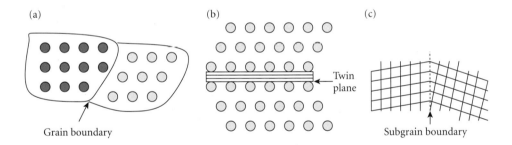

Figure 4.35 Three types of planar defect (shown in two dimensions): (a) intergranular grain boundary between two different crystals; (b) intragranular mechanical twin boundary resulting from mechanical slip; (c) intragranular subgrain boundary within a crystal, separated by a wall of dislocations. Imagine each extending in a second dimension perpendicular to the page and note how (b) and (c) accommodate changes in crystal shape.

4.8.3   Planar defects

Planar defects (Figure 4.35) extend in two dimensions and are therefore called **two-dimensional defects**. They are places within a crystal where the crystal structure changes across a distinct planar boundary. Examples include: (1) the boundaries between exsolution lamellae, for example between albite and potassium feldspar in perthites; (2) the subgrain boundaries between twins in twinned crystals; (3) the subgrain boundaries within crystals between out-of-phase crystal structures generated during ordering transformations; (4) grain boundaries between different crystals; and (5) extra atomic planes or missing planes called stacking faults.

Point defects, line defects and planar defects are critically important in the study of deformed rocks, particularly in the elastic and plastic deformation processes discussed in Chapter 16 (Box 4.2).

4.9 POLYMORPHS AND PSEUDOMORPHS

4.9.1 Polymorphs

As noted in the opening chapter, different minerals can have the same chemical

Box 4.2 Defects and plastic deformation in crystals

You may recall from earlier courses in which folds, faults and metamorphic foliations were discussed that when stresses are applied to rocks, they experience changes in shape and/or volume. These changes in shape and/or volume that occur in response to stress are called strains. They are analogous to the strains that occur in bones and muscles when they change shape in response to stress. Non-elastic strains are subdivided into those in which rocks break along fractures such as faults or joints and those in which shape changes are accomplished without fracturing. Irreversible strains that involve fracturing are called rupture; those that do not are called plastic strains and accommodate plastic deformation. Rupture is favored by rapid strain rates (think how fast the bone changes shape as it fractures), low confining pressures and low temperatures. On the other hand, plastic strain is favored by very low strain rates, high confining pressures and high temperatures (Figure B4.2a). Under such conditions, deep below the surface, rocks respond very slowly to stress in a manner more like Playdough® or modeling clay than like the rigid rocks we see at Earth's surface. How can rocks undergo significant strain without rupturing? A major key lies in the large number of defects that the minerals in rocks contain.

Plastic deformation at high temperatures and low strain rates largely results from two significant types of diffusion creep (Figure B4.2a) that are dependent on the existence of omission defects in minerals: (1) Coble (grain boundary diffusion) creep, and (2) Herring–Nabarro (volume diffusion) creep. Elevated temperatures are associated with elevated molecular vibration in an expanded crystal lattice. Such vibrations lower bond strength and increase the number of omission defects (also called holes or vacancies) in the crystal structure. As holes are created, adjacent atoms can migrate into the vacancy by breaking only one weak bond a time. The movement of the ions in one direction causes the holes or vacancies to migrate in the opposite direction (Figure B4.2b).

Under conditions of differential stress, ions tend to be forced toward the direction of least compressive stress (σ_3), which tends to lengthen the crystal in that direction. Simultaneously, holes tend to migrate toward the direction of maximum compressive stress (σ_1) until they reach the surface of the crystal where they disappear, causing the crystal to shorten in this direction (Figure B4.2b). In Coble creep, the vacancies and ions migrate near grain boundaries to achieve the strain, whereas in Herring–Nabarro creep, the vacancies and ions migrate through the interior of the crystals. Since thousands of omission defects are created over long periods of time even in small crystals, the long-term summative effects of plastic strain as each crystal changes shape by diffusion creep can be very large indeed.

At higher strain rates related to higher differential stresses, dislocation creep processes become dominant (Figure B4.1). In these environments edge dislocations and screw dislocations migrate through the crystal structure, once again breaking only one bond at a time, while producing plastic changes in shape. Because such dislocations result from strain, large numbers are produced in response to stress, and their migration accommodates large amount of plastic strain. Imagining the summative plastic changes in shape that can be accomplished by the migration of thousands of diffusing vacancies and/or migrating dislocations in a small crystal or 10^{20} dislocations migrating through the many crystals in a large mass of rock offers insight into the power of crystal defects to accommodate plastic deformation on scales that range from microcrystals to regionally metamorphosed mountain ranges.

Continued

Box 4.2   *Continued*

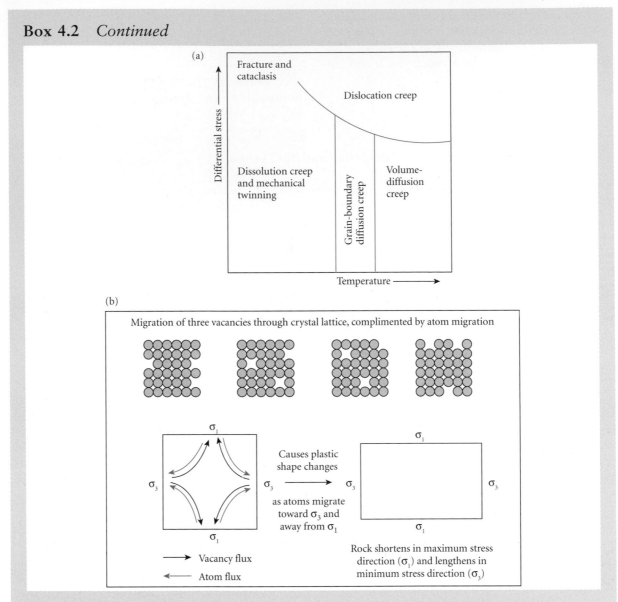

Figure B4.2   (a) Deformation map showing the significant role of omission defects and dislocations in the high temperature plastic deformation of crystals. (From Davis and Reynolds, 1996; with permission of John Wiley & Sons.) (b) Diagrams showing how the existence of omission defects permits adjacent ions to move into their former locations, effectively causing the omission or hole to migrate in one direction as the ions migrate in the other. The flux of atoms (blue arrows) toward regions of least compressive stress (σ_3) and and of vacancies (black arrows) toward areas of maximum compressive stress (σ_1) cause crystals to change shape.

composition, but different crystal structures. This ability for a specific chemical composition to occur in multiple crystal structures is called **polymorphism**. The resulting minerals with the same chemical compositions but different crystal structures are called **polymorphs**. In most cases, the crystal structure or form taken by the mineral is strongly influenced by the environment in which it forms. Polymorphs therefore record important information concerning the environments that produced them. Many polymorphs belong to very common or economically significant mineral groups, such as the examples summarized in Table 4.13.

The polymorphs of carbon can be used to illustrate how environmental conditions

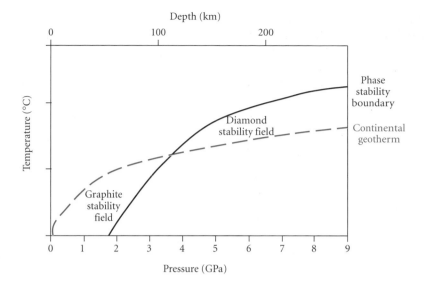

Figure 4.36   Phase stability diagram showing the conditions under which graphite, the low pressure polymorph of carbon, and diamond, the high pressure polymorph of carbon, are stable beneath continental lithosphere.

Table 4.13   Important rock-forming mineral polymorphs.

Chemical composition	Common polymorphs
Calcium carbonate ($CaCO_3$)	Calcite and aragonite
Carbon (C)	Diamond and graphite
Silica (SiO_2)	α-quartz, β-quartz, tridymite, cristobalite, coesite, stishovite
Aluminum silicate ($AlAlOSiO_4$)	Andalusite, kyanite, sillimanite
Potassium aluminum silicate ($KAlSi_3O_8$)	Orthoclase, microcline, sanidine
Iron sulfide (FeS_2)	Pyrite, marcasite

during growth determine which crystal structure a chemical compound possesses. Figure 4.36 is a phase stability diagram for systems composed of pure carbon. This phase stability diagram clearly indicates that diamond is the high pressure polymorph of carbon, whereas graphite is the low pressure polymorph. If we add **geotherms**, lines showing the average temperature of Earth at any depth, to this diagram, we can infer that diamonds are the stable polymorph of carbon at pressures of more than 3.5 GPa, corresponding to depths of more than 100 km below the surface of old continental shields, whereas graphite is the stable polymorph of carbon at all shallower depths. Inferences must be tempered by the fact that Earth's interior is not pure carbon and temperature distributions with depth are not constant, but it is widely believed that most natural diamonds originate at high pressures far below Earth's surface. If graphite is the stable polymorph of carbon at low pressures, why do diamonds occur in deposits at Earth's surface where pressures are low? Obviously, as diamonds rise toward Earth's surface into regions of substantially lower pressure, something keeps the carbon atoms from rearranging into the graphite structure. What keeps the transformation from unstable diamond to stable graphite from occurring?

Reconstructive transformations

Reconstructive transformations involve the conversion of one polymorph into another through bond breakage so that a significant change in structure occurs. Such transformations require large amounts of energy, and this requirement tends to slow or inhibit their occurrence. In the transformation of diamond to graphite, a large amount of energy is required to break the strong bonds that hold carbon atoms together in the isometric diamond structure, so that they can rearrange into the more open, hexagonal structure of graphite. This inhibits the transformation of diamonds into graphite as diamonds find themselves in lower pressure and lower temperature environments near Earth's surface. Minerals that exist under conditions where they are not stable, such as diamond near Earth's surface, are said to be **metastable**. All polymorphs that require reconstructive transformations have the potential to exist outside their normal stability ranges as metastable minerals. This allows them to preserve important information about the conditions under which they, and the rocks in which they occur, were formed and, in the case of diamonds, to grace the necks and fingers of people all over the world.

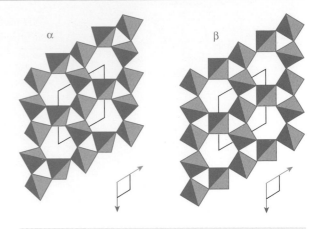

Figure 4.37 The closely similar structures of α- and β-quartz. (Courtesy of Bill Hames.)

Displacive transformations

Some polymorphs are characterized by structures that, while different, are similar enough that the conversion of one into the other requires only a rotation of the constituent atoms into slightly different arrangements without breaking any bonds. Transformations between polymorphs that do not require bonds to be broken and involve only small rotations of atoms into the new structural arrangement are called **displacive transformations** and tend to occur very rapidly under the conditions predicted by laboratory experiments and theory. Polymorphs involved in displacive transformations rarely occur as metastable minerals far outside their normal stability ranges and so may preserve less information about the conditions under which they and the rocks in which they occur originally formed.

Alpha quartz (low quartz) is generally stable at lower temperatures than beta quartz (high quartz). Although α- and β-quartz have different structures, the structures are so similar (Figure 4.37) that the conversion of one to the other is a displacive transformation. It is not at all unusual, especially in volcanic rocks, to see quartz crystals with the external crystal form of β-quartz but the internal structure of α-quartz. These quartz crystals are interpreted to have crystallized at the elevated temperatures at which β-quartz is stable and to have been displacively transformed into the α-quartz structure as they cooled, while retaining their original external crystal forms.

Other transformations between silica polymorphs are reconstructive. For example, the transformations between the high pressure minerals stishovite and coesite and between coesite and quartz are reconstructive. Therefore, both stishovite and coesite can be expected to exist as metastable phases at much lower pressures than those under which they are formed. Their preservation in rocks at low pressures allows them to be used to infer high pressure conditions, such as meteorite impacts, long after such conditions have ceased to exist.

Order–disorder transformations

Many polymorphs differ from one another in terms of the degree of regularity in the distribution of certain ions within their respective crystal structures. Their structures can range from perfectly ordered to a random distribution of ions within structural sites (Figure 4.38). The potassium feldspar minerals ($KAlSi_3O_8$) provide many examples of such variation in regularity or order in the distribution of aluminum ions within the structure. In the feldspar structure, one in every four tetrahedral sites is occupied by aluminum (Al^{+3}), whereas the other three are occupied by silicon (Si^{+4}). In the potassium feldspar **high sanidine**, the distribution of aluminum cations is completely random; the probability of finding an aluminum cation in any one of the four sites is equal. Crystal structures with such random distributions of cations are highly disordered

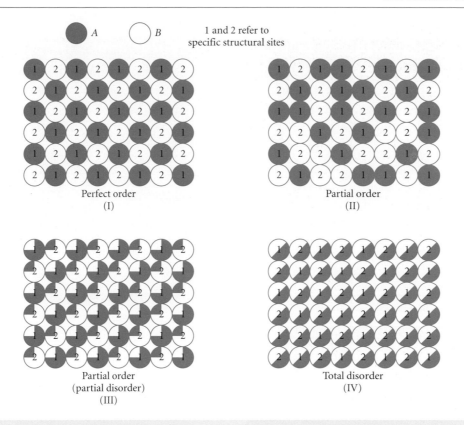

Figure 4.38 Variations in the order of minerals. (From Klein and Hurlbut, 1985; with permission of John Wiley & Sons.)

and are favored by high temperatures and low pressures of formation. On the other hand, in the potassium feldspar **low microcline**, the distribution of aluminum cations is highly ordered, with the aluminum distributed regularly in every fourth tetrahedral site. The probability of finding an aluminum cation in these sites is 100%, and the probability of finding one in the other three sites is zero. Crystal structures with such regular distributions of cations possess very low disorder, and their formation is favored by low temperatures and high pressures of formation. Intermediate degrees of order exist within the potassium feldspar group. Sanidine, with its high degree of disorder, crystallizes in the monoclinic system and is common in volcanic rocks formed at high temperatures and low near surface pressures, whereas microcline, with its low degree of disorder, crystallizes in the triclinic system and is common in rocks formed at higher pressures, and in some cases lower temperatures, below the surface.

4.9.2 Pseudomorphs

Minerals that take the crystal form of another, pre-existing mineral are called **pseudomorphs** and are said to be pseudomorphic after the earlier mineral (Figure 4.39). Pseudomorphs can be produced in many ways. All require that the original crystal possessed a significant number of crystal faces (was euhedral or subhedral) at the time it formed. Some pseudomorphs are produced by **replacement** in which the atoms in a pre-existing mineral are replaced by the atoms of a new mineral that retains the external crystal form of the original crystal. A common example is the replacement of pyrite (FeS_2) crystals by goethite (FeOOH) to produce goethite pseudomorphs after pyrite. Another common example is quartz (SiO_2) pseudomorphs after fluorite (CaF_2). Some pseudomorphs are **casts** produced by dissolution of the old mineral followed by precipitation of the pseudomorph to fill the cavity left behind. Other pseudomorphs

(a)

(b)

(c)

Figure 4.39 (a) Hematite replacing pyrite; (b) chalcedony encrusting aragonite; (c) quartz cast filling an aragonite solution cavity. (Photos courtesy of Stan Celestian, Maricopa Community College.) (For color version, see Plate 4.39, between pp. 248 and 249.)

are produced by the **loss of a constituent** from the original crystals. For example, the dissolution of carbonate ion from crystals of the copper carbonate mineral azurite $[Cu_3(CO_3)_2(OH)_2]$ can produce native copper (Cu) pseudomorphs after azurite. Still other pseudomorphs are produced when the new mineral forms a thin layer or crust over the original crystal. Still others form by inversion as when β quartz crystals are transformed into α quartz. The **encrustation** of the original mineral by the new mineral allows the new mineral to mimic the crystal form of the original mineral.

The properties of minerals and other crystalline materials are strongly influenced by their crystal structures and chemical compositions. These properties and the minerals that possess them are the subjects of Chapter 5.

Chapter 5

Mineral properties and rock-forming minerals

5.1   MINERAL FORMATION

Minerals are ephemeral; they have limited life spans. They represent atoms that have bonded together to form crystalline solids whenever and wherever environmental conditions permit. Ice is a good example. It forms whenever temperature and pressure conditions permit hydrogen and oxygen atoms to bond together to form crystals with a hexagonal structure. When temperatures increase or pressures decrease sufficiently, ice ceases to exist because the atoms separate into the partially bonded arrays that characterize liquid water. Ice, like all minerals, is ephemeral; more ephemeral under Earth's surface conditions than are most minerals. Another good example of the impermanence of minerals is the transitions between the polymorphs of carbon. At relatively low pressures, carbon atoms combine to form graphite. If the pressure increases sufficiently, the carbon atoms are rearranged so that graphite is transformed into diamond.

Earth Materials, 1st edition. By K. Hefferan and J. O'Brien. Published 2010 by Blackwell Publishing Ltd.

Minerals form via natural environmental processes that cause atoms to bond together to form solids. These include:

1   **Precipitation from solution.** Solutions from which minerals precipitate include:
  - **Surface water** in springs, rivers, lakes and oceans.
  - **Groundwater** in soils and underground aquifers.
  - **Hydrothermal solutions**, which are warm, aqueous solutions that have been heated at depth and/or by proximity to a body of magma.
2   **Sublimation from a gas.** Sublimation occurs where volcanic gases are vented at Earth's surface or where gas phases separate from solution in the subsurface.
3   **Crystallization from a melt or other liquid:**
  - **Lava flows** at the surface which form volcanic minerals and rocks.
  - **Magma** bodies in the subsurface, which form plutonic minerals and rocks.
4   **Solid state growth.** In solid state growth, new mineral crystals grow from the constituents of pre-existing minerals. This is

especially common during the formation of metamorphic minerals and rocks

5 **Solid–liquid or solid–gas reactions.** In such reactions, atoms are exchanged between the solid minerals and the liquid or gas phase with which they are in contact, producing a new mineral. These solid–liquid or solid–gas reactions are common in mineral-forming processes that range from weathering through vein formation to metamorphism.

This incomplete list of the ways in which minerals form illustrates some of the important processes that are constantly altering or destroying pre-existing arrays of atoms to form new minerals in ways that depend on the conditions and processes in the environment in which they form.

5.2 CRYSTAL HABITS

5.2.1 Habits of individual crystals

Minerals start small. Each mineral crystal begins with the bonding of a few atoms into a three-dimensional geometric pattern. Initial growth leads to the formation of small "seed crystals" called **nuclei**. If the appropriate atoms are available and the environmental conditions are suitable for growth, the nuclei will continue to attract appropriate atoms or ions and grow into larger mineral crystals. When it stops growing, the mineral can be bounded by crystal faces that reflect its internal crystal structure. Since minerals frequently occur as well-formed crystals and since crystal habits reflect the crystal structure of the mineral in question, being able to recognize crystal habits is very useful in mineral identification.

Single crystals can be described using a variety of terminology. The simplest terminology is based on the relative proportions of the crystals in three mutually perpendicular directions (a, b and c) where $a \geq b \geq c$. Table 5.1 and Figure 5.1 summarize the terminology used for individual crystal habits and illustrate examples of each.

5.2.2 Habits of crystal aggregates

When environmental conditions are suitable for the nucleation and growth of a single

Table 5.1 Crystal habits of individual crystals.

Crystal habit	Colloquial description	Crystal dimensions
Equant	Equal dimensions; shape may approach that of cube or sphere	$a = b = c$
Tabular	Tablet or diskette-like	$a = b > c$; c is thin
Platy	Sheet-like	$a \approx b \gg c$; c is very thin
Prismatic or columnar	Pillar-like or column-like; slender to stubby	$a > b = c$; a is long
Bladed	Blade- or knife-like	$a > b > c$; a is long, c is thin
Acicular	Needle-like; slightly thicker than filiform	$a \ggg b = c$; b and c are very thin
Capillary or filiform	Hair-like	$a \ggg b = c$; b and c are extremely thin

mineral crystal, they tend to be suitable for the formation of multiple crystals of the same mineral. The production and growth of multiple crystals closely adjacent to one another produces an assemblage of similar crystals called a **crystal aggregate**. In samples where crystal aggregates occur, at least two sets of crystal habits exist: one set describes the habit of individual crystals (Table 5.1) and the second set describes the habit of the aggregate. Examples of crystal aggregate habits are illustrated in Figures 5.2–5.4 and summarized in Table 5.2.

Geodes are crystal aggregates produced by the partial or complete filling of a subspherical cavity as crystallization proceeds from the walls of the cavity inward. Precipitation of layers of microscopic crystals produces most of the bands and precipitation of larger crystals produces the drusy, divergent or reticulated crystals that line the centers of many geodes (Figures 5.3 and 5.4). If the geode material is more resistant to weathering than the host rock, the geodes weather out as subspherical crystal aggregates whose concentric layering is revealed when they are sawed in two. Another intriguing type of concentrically layered crystal aggregate is called a **concretion.** Concretions grow outward from a

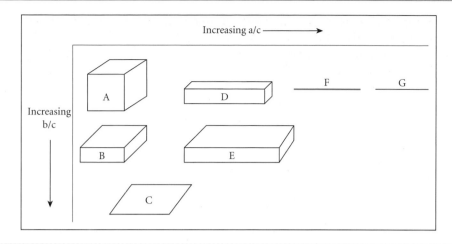

Figure 5.1 Individual crystal habits: A, equant; B, tabular; C, platy; D, prismatic or columnar; E, bladed; F, acicular; G, capillary or filiform.

Table 5.2 Summary of the habits of crystal aggregates.

Aggregate habit	Colloquial description
Fibrous	Parallel arrangement of acicular or filiform crystals
Radiating	Acicular–filiform crystals radiating outward from a central point
Divergent	Prismatic crystals diverging from a common area
Reticulated	Lattice-work of tabular to bladed crystals
Roseiform	Petal-like arrangement of tabular or bladed crystals
Drusy	Surface lined with very small (<2–3 mm), goosebump-like crystals
Dendritic, arborescent	Tree-like, branching network of crystals
Lamellar/foliated	Sub-parallel layers of minerals
Massive	Aggregate of very small crystals with a fine-grained appearance
Granular	Subequant, macroscopic crystal aggregate with a granular appearance
Banded	Parallel layers of same mineral with different color; as in agate
Concentric	Spherical to sub-spherical layers about a common center
Botryoidal/colloform	Rounded, mound-like aggregates; kidney-like
Oolitic	Spherical, concentrically layered, sand-sized (<2 mm) grain aggregates
Pisolitic	Spherical, concentrically layered, gravel-sized aggregates
Amygdaloidal	Spherical to ellipsoidal gas vesicles infilled with secondary minerals

central nucleus within a rock body and incorporate pre-existing mineral material as they do so. Thus concretionary aggregates consist of sub-spherical bodies that include both newly precipitated crystals and pre-existing material.

Individual and aggregate crystal habit minerals also possess a variety of other macroscopic and microscopic properties. In the following section we will consider the other macroscopic physical properties used in the identification of minerals. The microscopic properties used to identify minerals are discussed in Chapter 6.

5.3 MACROSCOPIC MINERAL PROPERTIES

Every mineral species possesses a unique set of properties that results from its unique combination of chemical composition, chemical bonding mechanisms and crystal structure.

Figure 5.2   Crystal aggregate habits, clockwise from upper left: fibrous serpentine; radiating pectolite; divergent stibnite; drusy calcite; dendritic pyrolusite; roseiform selenite gypsum. (Photo by John O'Brien.) (For color version, see Plate 5.2, between pp. 248 and 249.)

Figure 5.3   Crystal aggregate habits, clockwise from upper left: colloform (botryoidal) cassiterite; reticulate gypsum in geode; massive kaolinite; granular olivine. (Photo by John O'Brien.) (For color version, see Plate 5.3, between pp. 248 and 249.)

Figure 5.4   Crystal aggregate habits, clockwise from upper left: pisolitic calcite; concentrically banded chalcedony in geode; amydaloidal calcite and thomsonite filling vesicles in basalt; lamellar to foliated kyanite. (Photo by John O'Brien.) (For color version, see Plate 5.4, between pp. 248 and 249.)

Minerals are identified on the basis of their properties. In the field, correct mineral identification generally depends on the accurate identification of macroscopic properties and knowledge of the macroscopic properties that distinguish each mineral species.

5.3.1   Static and mechanical properties

Several common mineral properties depend on the fundamental static properties of minerals and/or how minerals respond mechanically to stress.

Hardness

Hardness is the resistance of a smooth mineral surface to scratching or abrasion by a sharp tool. Several scales are used to express the hardness of minerals, with the first being developed by Frederic Mohs in 1824 (Klein and Hurlbut, 1985). Mohs recognized that some minerals could scratch other minerals,

indicating different hardnesses. Mohs did not quantify these relationships but created a relative hardness scale that bears his name. Using ten readily available minerals, he assigned them integral values from 1 for the softest mineral, talc, to 10 for the hardest mineral, diamond (Table 5.3). Minerals with higher Mohs hardness numbers (H) can scratch minerals with lower Mohs hardness numbers. This is a relative hardness scale rather than a linear scale. Diamond (H = 10) is not twice as hard as apatite (H = 5) nor ten times harder than talc (H = 1).

Simple scratch tests can be used to determine the relative Mohs hardness of any other mineral, where the flat surface of one mineral is scratched with the sharp point of the other mineral while applying firm pressure. With a set of indenting tools or points composed of the minerals on the **Mohs scale**, the relative Mohs hardness of any mineral can be quickly ascertained. For example a mineral such as hornblende that can scratch apatite but is scratched by orthoclase possesses a Mohs

Table 5.3 Comparison between the Mohs and Knoop hardness scales; the former is a relative hardness scale, whereas the latter is an absolute hardness scale.

Mineral	Mohs hardness	Knoop hardness
Talc	1	1
Gypsum	2	32
Calcite	3	135
Fluorite	4	163
Apatite	5	430
Orthoclase	6	560
Quartz	7	820
Topaz	8	1340
Corundum	9	1800
Diamond	10	7000

hardness of between 5 and 6. Implements commonly used for approximate hardness determinations include a fingernail (H = 2.5), a penny (H = 3+), a steel nail or knife blade (H = 5–5.5) and a glass plate (H = 5.5). Minerals with hardness ≤3 are considered to be soft minerals. Minerals with a hardness of 3–5.5 are considered to have intermediate hardness. Hard minerals have a hardness greater than 5.5. In many situations, knowledge of the approximate hardness is sufficient to allow a mineral to be identified. For example, fluorite is commonly purple, translucent and vitreous, as is the variety of quartz called amethyst. But quartz is a hard mineral (H = 7) whereas fluorite is an intermediate mineral (H = 4), so a scratch test using a nail, knife blade or glass plate quickly distinguishes the two minerals.

More quantitative measures of the resistance of a mineral surface to abrasion exist. The **Knoop hardness scale** is an example of indentation hardness and uses an absolute hardness scale related to the stress required to indent a polished surface. Knoop hardness is calculated from data collected by using a diamond tool to indent a polished mineral surface. The tool is held against the specimen under a controlled force for a controlled period of time and the size of the indentation is then measured microscopically. The Knoop hardness is then calculated from the ratio between the force and the time over which it was applied to the size of the indentation. The larger the indentation, the softer the mineral.

The indentation is 32 times smaller for gypsum than it is for talc, so gypsum is 32 times more resistant to indentation than is talc. Table 5.3 compares the relative hardness of the minerals on the Mohs hardness scale with the absolute hardness on the Knoop hardness scale. The indentation in talc produced by the Knoop hardness test is 7000 times larger than that for diamond.

What accounts for the variation in hardness between minerals? When a mineral is scratched or abraded, a fine powder is produced. This implies that many of the bonds that held the atoms together in the mineral have been broken. Hardness depends first and foremost on bond strength and second on how densely concentrated the bonds are in the crystal structure. The stronger and more numerous the bonds are, the harder the mineral. Diamond is characterized by closely spaced, very strong bonds that result in the hardest known mineral. Talc, on the other hand, possesses some areas with a small number of very weak bonds, which helps to explain why it is "soft enough for a baby's skin" and used in cosmetics.

Density, specific gravity and weight

Density is the mass per unit volume of a material. It is expressed in units of mass divided by units of volume. These units are expressed in kilograms per cubic meter (kg/m^3) or, more commonly in the case of minerals, grams per cubic centimeter (g/cm^3). The density of a mineral is proportional to the number of atoms per unit volume, which is called the **packing index**, and to their **atomic mass number**. Minerals with very high density, such as native gold (Au), tend to have crystal structures with very closely spaced atoms (many atoms per volume) that have very high atomic mass numbers (Au = 197). Minerals with very low density, such as ice (H_2O), tend to have crystal structures with widely spaced atoms (few atoms per volume) and low atomic mass numbers (H = 1, O = 16).

Specific gravity (SG) is a dimensionless quantity. It is the ratio between the density of a material and the density of pure water at standard temperature and pressure (temperature = 3.9°C, pressure = 1 atmosphere). Since the density of pure water at standard temperature and pressure (STP) is 1.0 g/cm³, the specific gravity of a mineral will have the same numerical value as its density, but will be expressed without dimensions. An example will illustrate this point. The lead sulfide (PbS) mineral galena is quite dense because it contains very massive lead atoms; it has an average density of 7.5 g/cm³. It has a specific gravity of 7.5. Mathematically, the number is obtained from the ratio between the density of the mineral (7.5 g/cm³) and the density of water at STP (1.0 g/cm³):

$$SG = \frac{\text{density of mineral}}{\text{density of water}} = \frac{7.5 \text{ g/cm}^3}{1.0 \text{ g/cm}^3} = 7.5$$

The **weight** of a mineral or of any other material is simply its total mass accelerated by gravity. For example, when an object is placed on a scale, its mass is accelerated downward by gravity to produce a downward force on the scale, which is the object's weight. The total mass of the object, in grams (g) or kilograms (kg), is the total mass of all of the atoms it contains. Total mass therefore is proportional to the density of the material multiplied by its volume:

$$\text{Total mass (m)} = \text{density} \times \text{volume}$$
$$= \text{g/cm}^3 \times \text{cm}^3$$
$$= \text{grams or kilograms}$$

The downward acceleration that creates weight is the acceleration due to gravity (g) which varies from place to place but averages about 9.8 m/s² on Earth's surface. Weight (w) then is simply the total mass (m) of an object multiplied by the acceleration due to gravity (g). Mathematically, w = mg. Weight is a downward force and is expressed in force units called Newtons (N), which have the dimensions of kg m/s², since

$$w = mg = kg \cdot m/s^2 = \text{Newtons (N)}$$

The concept of weight and its relationship to density, specific gravity and volume can be illustrated using native gold as an example. Native gold, because of its closely packed, massive atoms has a density of 20 g/cm³. Its specific gravity is 20 because the ratio of its density to the density of pure water at STP is:

$$\text{Specific gravity (SG)} = \frac{\text{density of sample}}{\text{density of water}}$$
$$= (20 \text{ g/cm}^3)/(1.0 \text{ g/cm}^3)$$
$$= 20$$

A cubic centimeter (cm³) of gold has a small total mass of 20 g, since

$$\text{Mass (m)} = \text{density} \times \text{volume}$$
$$= 20 \text{ g/cm}^3 \times 1 \text{ cm}^3$$
$$= 20 \text{ g}$$

How much will this amount of gold weigh at Earth's surface? Since weight (w) = mg,

$$\text{Weight (w)} = mg = 20 \text{ g} \times 9.8 \text{ m/s}^2$$
$$= 196 \text{ g} \cdot \text{m/s}^2 = 0.196 \text{ kg} \cdot \text{m/s}^2$$
$$= 0.196 \text{ N}$$

A larger volume of gold, though its density and specific gravity is constant, will weigh more because it has more atoms and more mass.

Tenacity

Tenacity is defined as the manner in which minerals respond to short-term stresses at normal surface temperatures and pressures. Thus it is a subset of rheology, the way materials generally respond to stress, which is discussed in detail in Chapter 16.

Elastic minerals, including members of the mica group, can be bent when stressed but return to their original shape when the stress is released. Specimens that can be bent without breaking when stressed but do not return to their original shape possess a tenacity called **flexible**.

Some minerals respond to stress by deforming elastically by a small percentage and then undergo significant plastic strain in which they change shape and/or volume without visibly fracturing. Minerals such as native metals that are so plastic that they can be hammered into thin sheets are **malleable**, whereas those that can be drawn into a thin wire are **ductile**.

Brittle materials respond to stress by deforming elastically by a small percentage before fracturing after little or no plastic

Figure 5.5   From left: euhedral tourmaline, completely enclosed by crystal faces; subhedral quartz, partially enclosed by crystal faces; anhedral apatite, lacking crystal faces. (Photo by John O'Brien.) (For color version, see Plate 5.5, between pp. 248 and 249.)

strain. Most minerals at normal surface conditions fracture when stressed; their tenacity is therefore brittle. A few minerals, with a closely spaced set of cleavage planes, can be cut into thin shavings, and these otherwise brittle minerals have a tenacity called **sectile**. Since most minerals are brittle, it is the unusual tenacities such as elastic, flexible and malleable that are especially useful in mineral identification.

Growth surfaces and breakage surfaces

All mineral specimens possess external surfaces that enclose the mineral specimen and separate it from its surroundings. **Growth surfaces** form when the mineral ceases to grow. **Breakage surfaces** result when the specimen is broken from its host rock. Let us first consider growth surfaces.

Crystal faces
Crystal faces are relatively flat, geometric surfaces generated by mineral growth. As crystals grow, atoms are added in the geometric patterns that characterize the mineral in question. When the mineral stops growing, it is bounded by growth surfaces. If crystal faces are present, they represent an external expression of the mineral's internal crystal structure (Chapter 4). The beautiful crystals housed in many museums are testimony to the ability of mineral growth surfaces to express their

geometric internal crystal structures (Figure 5.5). Mineral crystals completely enclosed by crystal faces are said to be **euhedral**. Euhedral crystals develop under conditions that allow them to be completely enveloped by relatively flat, geometric crystal faces. Mineral crystals that are only partially enclosed by crystal faces are said to be **subhedral**. Subhedral crystals form under conditions that allow them to be only partially bounded by relatively flat crystal. Mineral crystals that possess no crystal faces are said to be **anhedral** and record growth conditions that did not permit the development of crystal faces. What factors determine whether a mineral will be bounded by crystal faces? Many crystals nucleate (begin to grow) on a pre-existing surface. The shape of that side of the mineral will reflect the shape of the pre-existing surface, rather than the crystal structure of the mineral. When mineral growth fills a restricted space, the mineral surfaces will reflect the shape of the space, rather than the internal geometry of the mineral's crystal structure. But when a mineral nucleates in a fluid and is free to grow in every direction, it will be completely enveloped by crystal faces.

Most mineral specimens are broken from the rock in which they occurred. Breaking the mineral produces breakage surfaces, which include cleavage, fracture and parting surfaces, as discussed below.

Cleavage surfaces

Some minerals invariably break along flat, planar, light-reflecting surfaces. These flat breakage surfaces are called **cleavage surfaces** and are related to the presence of weak planes in the crystal structure, along which the mineral cleaves preferentially. Fresh cleavage surfaces can be recognized by a combination of features that may include: (1) flat, planar surfaces, (2) strong reflection of light from these surfaces, and (3) the repetition of flat surfaces as sets of parallel surfaces. These flat, parallel, light-reflecting cleavage surfaces may occur on opposite sides of the mineral specimen or they may occur as smaller parallel surfaces on one side of the specimen analogous to parallel steps on a porch. In the latter instance, the parallelism of the step-like surfaces can be demonstrated by rotating the specimen in the light and noting that all the steps reflect light to the eye at the same moment in a manner analogous to the flash of light from the flat surface of a mirror. Since all the parallel cleavage surfaces on one or both sides of a mineral crystal have the same orientation, they together constitute one set, orientation or direction of cleavage. In minerals that possess more than one set or orientation of cleavage, the cleavage surfaces intersect each other at specific angles. When describing cleavage, one should carefully note the number of sets or directions of cleavage and the approximant intersection angles between them. It is often sufficient simply to note whether or not the cleavage sets intersect at right angles.

What causes minerals to possess cleavage? Since cleavage results from the existence of repeated sets of weakly bonded planes in the crystal structure, a set of cleavages reflects the existence of a set or direction of weak planes in the crystal structure. The number of sets or directions of cleavage equals the number of sets or directions of weakly bonded planes in the mineral's structure. The angles of intersection between sets of cleavage correspond to the angles between weakly bonded planes in the crystal structure. Minerals that do not possess parallel planes of weakness in their crystal structure do not possess cleavage.

Cleavage produces smaller mineral fragments of a predictable and repeatable shape that reflect the relative orientation of planes of weakness in the crystal structure. Minerals can have one, two, three, four or six sets of cleavage (Figure 5.6). Like all crystallographic planes, cleavages can be symbolically represented by their Miller indices, as discussed in Chapter 4.

Minerals, such as those in the mica group, that possess only one set of planes of low total bond strength in their crystal structure tend to break repeatedly along those planes and so possess one direction of cleavage. Feldspar minerals such as orthoclase possess two sets of cleavage that intersect at right (90°) angles, whereas amphiboles such as hornblende possess two sets of cleavage that intersect at angles of 57° and 123°. Still other minerals possess three sets or orientations of cleavage. In halite and galena, the three sets of weak planes intersect at 90° angles, producing cubic cleavage, whereas in calcite and dolomite they do not intersect at right angles and instead produce rhombic cleavage. Fluorite possesses four sets of cleavage (octahedral cleavage). The largest number of cleavage planes is six (dodecahedral cleavage), as exemplified by the mineral sphalerite (see Figure 5.6).

It is important to remember that mineral cleavage is a property of single crystals. In crystal aggregates, the characteristic cleavage of the mineral can often be obscured. The three varieties of the hydrated calcium sulfate mineral gypsum provide an instructive example. Gypsum possesses one set of weak planes in its internal crystal structure and therefore possesses one direction of cleavage. The variety of gypsum called selenite consists of large crystals. A single large crystal of pure selenite possesses one excellent direction of cleavage that is generally easy to discern. However, when selenite occurs in aggregates of multiple crystals with different orientations, the aggregate may appear to have more than one set of cleavage orientations, unless one is careful to observe only the cleavage of each single crystal. A second variety of gypsum, called satinspar, is composed of a fibrous aggregate with numerous very thin, needle-like crystals that grow parallel to one another during its formation. Because satin spar is an aggregate of thin crystals, the cleavage cannot be discerned macroscopically. A third common variety of gypsum is alabaster. Alabaster is a massive form of gypsum that is an aggregate of numerous very small, randomly oriented gypsum crystals that have

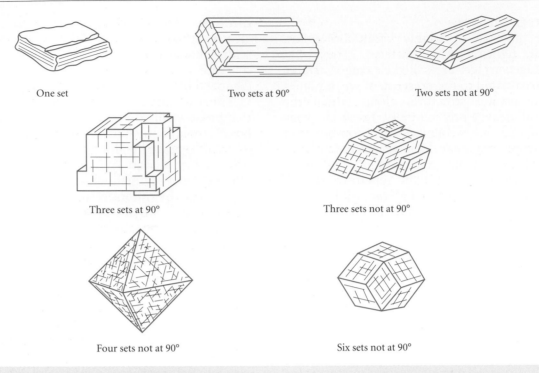

One set

Two sets at 90°

Two sets not at 90°

Three sets at 90°

Three sets not at 90°

Four sets not at 90°

Six sets not at 90°

Figure 5.6 Types of cleavage exhibited by minerals.

been compacted together. Because the crystals are so small and their cleavage planes are randomly oriented, alabaster does not display the cleavage observed in larger single crystals like selenite.

Fracture surfaces

When broken, many mineral specimens display non-flat, less reflective, non-parallel surfaces called **fracture surfaces**. Four types of fracture are described below: conchoidal, irregular, splintery and hackly (Figure 5.7). Fracture surfaces may completely enclose a mineral specimen, or they may occur in combination with cleavage surfaces. Note that two sets of flat, planar surfaces cannot enclose a volume; therefore, broken mineral specimens with fewer than three sets or orientations of cleavage surfaces will always possess fracture surfaces (and/or crystal faces) as well. For example, orthoclase has two sets of cleavage planes at right angles, and the remaining breakage surfaces exhibit irregular fracture.

Conchoidal fracture is marked by smooth, curved surfaces and occurs in both minerals and rocks. Common minerals such as quartz, chalcedony, opal and garnet are often recog-

Figure 5.7 Fracture types, clockwise from upper left: conchoidal fracture in opal; splintery fracture in selenite gypsum; uneven or irregular fracture in alabaster gypsum; hackly fracture in native copper. (Photo by John O'Brien.) (For color version, see Plate 5.7, between pp. 248 and 249.)

nized because of their lack of cleavage and the presence of well-defined conchoidal fracture. Conchoidal fracture develops in materials that have similar total bond strengths in all directions and is characteristic of glass, such as the volcanic rock obsidian, and finely crystalline rocks, such as chert and flint. Such materials obviously lack the long-range order necessary for planes of weakness and cleavage. When smooth, curved conchoidal fracture surfaces intersect, they have the potential to produce very sharp edges or points. Hard Earth materials with conchoidal fracture were prized for their use as stone scrapers, blades and arrowheads by pre-industrial societies.

Perhaps the most common type of fracture observed in mineral specimens is **uneven or irregular fracture**. As the names suggest, these fracture surfaces are uneven or irregular in a rather non-descript way. Some irregular fractures are produced when single mineral crystals break in directions that do not possess preferred planes of weakness in the crystal structure. In many instances, however, uneven or irregular fracture is produced in fine-grained, randomly oriented mineral aggregates such as alabaster gypsum or granular olivine specimens.

A third type of fracture is characteristic of fibrous aggregates. Many minerals grow to produce aggregates of numerous, thin, parallel, hair- or needle-like crystals. When these fibrous aggregates break, they tend to separate between the fibers, which gives rise to **splintery fracture**. The fibrous variety of gypsum called selenite is an excellent example of a mineral that possesses splintery fracture, as are the fibrous serpentine and amphibole minerals known as asbestos.

Some minerals, especially the malleable native metals, break in such a way as to produce ragged, sharp edges. These breakage surfaces are characteristic of **hackly fracture**.

Parting surfaces

Some minerals break along flat surfaces that are related stress or twinning. Surfaces produced when minerals break along planes of weakness produced by twinning and/or by stress after the mineral formed are called **parting surfaces**. These are not cleavage surfaces because they are not controlled by planes of weakness inherent in the mineral's essential crystal structure. Instead, the planes of weakness are produced when the mineral's structure is changed by twinning (Chapter 4) or in response to stress after its initial formation. For example, garnet does not possess cleavage, but stressed garnets commonly break along flat surfaces. Stressed corundum crystals also display well-defined parting surfaces not related to cleavage.

Striations

Some minerals, such as plagioclase, pyrite and calcite, exhibit parallel sets of linear features called striations that appear as engraved ridges and/or grooves on mineral surfaces. In the case of pyrite, the striations are due to the intersection of pyritohedron crystal faces on the face of cubic crystals (Hurlbut and Sharp, 1998). In the minerals calcite and plagioclase, striations develop due to crystallographic twinning, as described in Chapter 4. The presence of striations is the primary means by which plagioclase feldspar is distinguished from potassium feldspar. Because feldspars are the most abundant mineral group in Earth's crust and their proportions are central to the classification of igneous rocks (Chapter 7), this distinction is of great significance.

Taste

Certain minerals display other properties that can be useful in their identification. Some minerals possess a characteristic taste. The salty taste characteristic of halite, the bitter taste of sylvite and the sweet, alkaline taste characteristic of borax are good examples.

Feel

Very soft minerals, such as talc, graphite and molybdenite, possess a characteristic greasy feel. The greasy feel results from weak van der Waals bonds that allow the minerals to be broken into soft, dust-like fragments that "lubricate" the surface when the specimen is rubbed with the fingers.

Smell

Native sulfur (S) and many sulfide minerals such as marcasite (FeS_2) and sphalerite (ZnS) produce a sulfur smell when crushed or

Figure 5.8 Diaphaneity, from left: first pair, opaque, metallic galena and pyrite, which do not transmit light; second pair, plagioclase and potassium feldspar, which are somewhat translucent; third pair, rose quartz and aragonite, which are moderately translucent; fourth pair, calcite and fluorite, which are transparent and transmit the underlying image. (Photo by John O'Brien.) (For color version, see Plate 5.8, between pp. 248 and 249.)

powdered. Arsenic-bearing minerals such as arsenopyrite (FeAsS) and realgar (AsS) possess a garlicky smell.

Other static and mechanical properties

Some carbonate minerals, such as calcite, aragonite, witherite and rhodochrosite, possess the property of effervescence. When a drop of dilute hydrochloric acid ($HCl_{(a)}$) is placed on the specimen, it fizzes or effervesces by releasing carbon dioxide (CO_2) gas. Other carbonate minerals, such as dolomite, effervesce only when the specimen is powdered or the hydrochloric acid is heated to promote the chemical reaction that releases carbon dioxide gas.

Other mineral properties, such as melting and crystallization temperature, solubility and thermal conductivity, are discussed in relationship to chemical bonding (Chapter 2) and in the chapters on magma generation and igneous rocks (Chapter 8), weathering (Chapter 12) and ore deposits (Chapter 19).

5.3.2 Optical and electromagnetic properties

Many mineral properties result from interactions with light. Light is that part of the electromagnetic spectrum that can be sensed by the normal human eye. It is a very small portion of the total electromagnetic spectrum (Chapter 6), restricted to electromagnetic radiation with wavelengths between 700 nm (violet) and 300 nm (red). Light may be:

- Reflected by the mineral, either from its surface or internally.
- Transmitted through the mineral, in whole or in part.
- Absorbed by the mineral, either wholly or selectively.
- Dispersed, scattered or reradiated by the mineral.

A fuller analysis of the potentially complex ways in which minerals interact with light and with other wavelengths of the electromagnetic spectrum is given in the next chapter, which deals with optical microscopy.

Diaphaneity (opacity)

Diaphaneity or opacity depends on the amount of light transmitted by a mineral specimen (Figure 5.8). Mineral specimens that do not transmit any light are **opaque**. When light strikes these minerals, no light is transmitted. Instead the light is reflected, absorbed or both, even by a thin specimen of the mineral. Minerals that are truly opaque generally possess a dark gray to black streak.

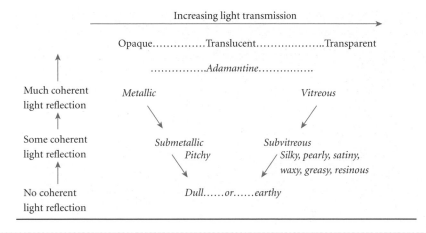

Figure 5.9   General relationships between light transmission (diaphaneity), light reflection/reradiation and luster. Light transmission increases from left to right and reflectivity increases from bottom to top.

Even a finely powdered sample of the mineral does not transmit light. Mineral specimens that transmit a significant portion of the incident light are **transparent**. Such specimens transmit enough light that an image can be transmitted through them. In this way they are analogous to a glass window that is transparent enough to clearly transmit an image. Whether or not a mineral specimen is transparent may depend on its thickness; thin specimens tend to be more transparent than thick ones because they scatter and absorb less light. Minerals that are macroscopically transparent often possess a white streak. Mineral specimens that transmit some light, but not enough to transmit an image, are **translucent**. Translucent minerals range from barely translucent to quite translucent or almost transparent. Diaphaneity is closely related to luster, as explained below.

Most mineral specimens are translucent. Translucent minerals commonly possess either white streaks or colored streaks that are neither gray nor black. One way to ascertain the diaphaneity or opacity of a specimen is to hold the specimen up to the light and peer through the thinnest edge you can find. This way you can determine whether no light, some light or enough light to transmit an image is passing through the mineral. But another way is to use the streak and luster of the mineral to test your hypothesis about diaphaneity. This will be clearer when you have finished reading this section.

Luster

Luster is the appearance of a mineral surface in reflected light. Because it involves appearance, it is a rather subjective property. The luster we perceive when looking at a macroscopic specimen is principally the result of the amount of light reflected from its surface, the scattering of light from the surface and internally, and the amount of light absorbed by the mineral. Figure 5.9 relates the different lusters possessed by minerals to these three major variables.

Luster is broadly subdivided into metallic and non-metallic lusters (Figure 5.9). Many minerals have a luster that is similar to the bright luster of metals, such as freshly polished silver, brass or chrome. We normally see shiny, opaque materials as having the luster associated with metals, thus the term metallic luster (Figure 5.10). This **metallic luster** is characteristic of minerals that reflect and/or reradiate large amounts of minimally scattered, coherent light and absorb the rest. They are therefore shiny but opaque (or nearly so). As a result, they tend to have a gray to black streak. Galena, pyrite and magnetite are excellent examples of minerals with metallic luster that are also opaque and possess a gray or black streak. Some metallic minerals, such as chromite, magnetite and bornite, possess a **submetallic luster**. This usually results from a larger amount of scattering of the light or from the mineral being not quite opaque.

Figure 5.10 Mineral lusters, from left: metallic stibnite; submetallic chromite; dull hematite. (Photo by John O'Brien.) (For color version, see Plate 5.10, between pp. 248 and 249.)

Some opaque mineral specimens with a gray to black streak reflect little or no coherent light and therefore appear to have a **dull or earthy luster**. Such specimens are often finely granular aggregates, so that they scatter light and do not reflect coherent light in a particular direction. Other nearly opaque minerals, such as rutile, reflect extremely bright coherent light, which gives them an unusually resplendent quality called an **adamantine luster**.

Most minerals transmit some light and therefore possess **non-metallic lusters** (Figure 5.11). These minerals commonly possess white streaks or colored streaks other than gray to black. Many of these minerals reflect light from a transparent or translucent background in a manner analogous to a mirror or a pane of glass. These minerals have the bright luster of shiny glass, which is formally called **vitreous** from the Latin word for glass. Quartz, calcite and fluorite are common minerals that possess a vitreous luster. Some transparent to translucent minerals, such as sphalerite, reflect very bright light that produces the resplendent surfaces that characterize adamantine luster. Many gem mineral specimens, such as diamond, emerald, ruby and sapphire, possess an adamantine luster which helps to give them their visual appeal. Some specimens of minerals that transmit light do not reflect coherent light and therefore have a **dull or earthy luster**. As with other dull or earthy specimens, these specimens are commonly fine-grained aggregates of multiple crystals that effectively scatter the available reflected light.

Several different types of **subvitreous luster** exist. These are generally produced by partial scattering of light waves by the mineral's surfaces or by internal scattering of light. All of these scattering effects produce lusters that are somewhat subdued when compared to those of vitreous and adamantine specimens. **Silky luster** is characteristic of minerals with a fibrous habit that consist of parallel fibrous aggregates. The parallel fibers reflect light in a manner reminiscent of silk. **Waxy** and **resinous lusters** are quite similar, having the subdued luster typical of floor wax and of tree resin or amber, respectively. **Greasy, pearly** and **satiny** are also somewhat subdued lusters that are similar to those of grease, pearls and satin, respectively. In practice, it is not generally necessary to distinguish between the various subvitreous lusters. In most mineral identification, it is only necessary to recognize that the specimen has a somewhat subdued, but not dull, luster and that it is translucent.

Streak

Streak is the color of the mineral powder and is typically obtained by scratching a mineral specimen on an unglazed porcelain plate called a streak plate. Since most streak plates

Figure 5.11 Non-metallic mineral lusters, clockwise from upper left: vitreous plagioclase; subvitreous garnet; resinous amber; waxy opal; silky serpentine; pearly soapstone; dull, earthy bauxite. (Photo by John O'Brien.) (For color version, see Plate 5.11, between pp. 248 and 249.)

have a hardness of about 6.5, only minerals that are somewhat softer than 6 will leave a powder when scratched on the plate. Harder minerals will not leave a powder on the plate but can be powdered by other means.

The streak plate is very useful for identifying minerals that are softer than the streak plate, especially those that produce a characteristically colored powder. The color of the streak is largely the product of which wavelengths of light are transmitted by the mineral powder. As a result, a mineral's streak is a fairly constant, often diagnostic property. Minerals that transmit a mixture of all wavelengths of light have a white streak. Such minerals are very common, so their streak is not a particularly useful property for identification. Opaque minerals absorb all wavelengths of light, producing a gray to black streak. Some minerals selectively absorb certain wavelengths of light while transmitting other wavelengths. These minerals possess a characteristic colored streak that is neither white nor gray to black. For example,

hematite possesses a diagnostic dark brick-red streak. This is because hematite absorbs nearly all wavelengths of light except for a narrow band in the red part of the spectrum. It is nearly opaque, with its characteristic brick-red streak sometimes accompanied by a dark gray streak as well.

Similarly, azurite possesses a distinctive blue streak because it transmits blue wavelengths while absorbing the red and yellow parts of the spectrum, and the closely related mineral malachite possesses a green streak because it transmits blue and yellow (blue + yellow = green) wavelengths while absorbing the red end of the visible spectrum. The determination of a characteristically colored streak is frequently the test required to clinch the macroscopic identification of mineral specimens.

Color

The color of a macroscopic mineral specimen is the result of a complex interplay among the reflection, absorption, transmission, refraction, scattering and dispersion of light as it interacts with the mineral's chemical and structural components. This interplay is strongly influenced by chemical impurities and structural irregularities called defects (Chapter 4) that are common in all naturally occurring solids. Minerals characterized by a relatively constant shade of color are said to be **idiochromatic**, which means they are "self-colored". Idiochromatic minerals possess essentially the same color, independent of any impurities and/or defects that occur. Color is a diagnostic property of idiochromatic minerals and can be used as a criterion for their identification. Excellent examples of idiochromatic minerals include azurite, which is always blue, sulfur, which is always yellow, and galena, which is always gray.

Other minerals, however, are characterized by colors that vary from one specimen to another or even within the same specimen. These minerals are said to be **allochromatic**, which means "foreign colored". Their color is strongly influenced by impurities and/or defects so that different specimens possess different colors that depend on the different impurities and/or defects they possess. A good example of an allochromatic mineral is quartz, which occurs in several different varieties that include colorless rock crystal, white milky quartz, pink-colored rose quartz, honey-brown to dark brown smoky quartz, yellow citrine, blue to green aventurine and purple amethyst. Color is not a diagnostic property of quartz or of any other allochromatic mineral. Many other minerals are allochromatic. For example, calcite can be colorless, white, gray, blue, green, yellow, pink or red. In the case of calcite, however, the different colors are not generally given varietal names.

If all naturally occurring solids have chemical impurities and/or defects in their crystal structures, why are some mineral species idiochromatic whereas others are allochromatic? Generally, if a mineral possesses a strong color in its pure state, it will be idiochromatic. The small amounts of chemical impurities and/or crystal defects permissible in the definition of a mineral are insufficient to affect its color significantly. If a mineral is colorless in its pure state, small amounts of impurities or defects will cause the selective absorption of light and the mineral will transmit or reflect selected wavelengths that give it color. Such minerals have a tendency to be allochromatic.

Craftspeople have long used the knowledge of impurities and color to dope glass with the appropriate impurities to impart color to this otherwise colorless material. Stained glass, with its gorgeous array of colors, often in artistic designs, bears witness to the concept of how impurities in colorless materials produce allochromatic effects. For a more detailed analysis of color, using the principles of crystal field theory, the reader is referred to Wenk and Bulach (2004) and Zoltai and Stoudt (1984).

Play of colors

Many minerals exhibit colors that change as the angle at which the light strikes them changes, as when the mineral is rotated in incident light. Such changes in color as the angle of incident light changes are collectively referred to as a **play of colors**. Several specific types of play of colors have been recognized. **Asterism** (Figure 5.12a) is caused by inclusions oriented according to the host mineral's crystal structure, which produces a six-sided, star-like pattern as light is scattered by the inclusions at specific incident angles. A closely related phenomenon is **chatoyancy** (Figure 5.12b), characteristic of certain fibrous minerals, in which a band of light moves perpendicular to the fibers, especially when the fibers are curved. Many minerals display a play of colors called **iridescence** in which the scattering of light from zones of contrasting composition within the mineral produces changes in color as the mineral is rotated. An excellent example of iridescence is the **labradorescence** that occurs in some specimens of plagioclase. The iridescence exhibited by many opals is referred to as **opalescence**.

Luminescence

Many minerals emit light or luminesce when subjected to an external source of energy. **Luminescence** is best observed in darkness, where such minerals appear to glow in the dark. Many minerals luminesce when subjected to some type of short wavelength radiation, such as gamma rays, X-rays or

(a)

(b)

Figure 5.12 (a) Asterism in the "Star of Bombay" sapphire. (Smithsonian Institution, with permission.) (b) Chatoyancy in tiger eye. (Photo by Kevin Hefferan.) (For color version, see Plate 5.12, between pp. 248 and 249.)

ultraviolet waves. Luminescence is produced when electrons are temporarily energized by an energy source. In their excited state, they release heat and visible light as they return to their original energy state or ground state. Specific types of luminescence are summarized in Table 5.4.

Magnetism

All minerals display some degree of **magnetism**; that is, a response to an external magnetic field. Three major types of magnetism are displayed by minerals, as indicated in Table 5.5.

Ferromagnetic and ferrimagnetic minerals have the ability to become magnetized parallel to an external magnetic field and are responsible for rock magnetization. The most magnetic ferrimagnetic minerals include magnetite ($FeFe_2O_4$) and pyrrhotite ($Fe_{1-x}S$). These two minerals are strongly attracted by small magnets. Magnetite that has been strongly magnetized by an external field becomes a magnet in its own right, as exemplified by the strongly magnetic variety of magnetite called **lodestone**. Prisms of lodestone cut parallel to the magnetization provided the first compass needles used by mariners about a thousand years ago.

Table 5.4 Four major types of luminescence exhibited by minerals.

Luminescence	Description
Fluorescence	Fluorescence occurs when materials are subjected to short wavelength radiation such as gamma rays, X-rays or ultraviolet waves. The color produced during fluorescence depends on the wavelength of visible light emitted
Phosphorescence	Material emits visible light after it is no longer subjected to the incident radiation. This is used in glow-in-the-dark toys, which continue to emit light after removal of the light source
Thermoluminescence	Materials emit visible light when heated to 50–475°C
Triboluminescence	Materials emit visible light in response to stress induced by rubbing or crushing the specimen

Table 5.5 Different varieties of magnetism exhibited by minerals.

Type of magnetism	Description
Dimagnetic minerals	Not attracted to even very powerful magnets; in fact they are slightly repelled by them
Paramagnetic minerals	Weakly attracted to strong magnets, become magnetized in an external magnetic field, but lose their magnetization when the external field is removed
Ferromagnetic– ferrimagnetic minerals	Strongly attracted to even weak magnets, and can retain magnetization for long periods of time

Electrical properties

Minerals exhibit a variety of electrical properties. **Pyroelectricity** is a phenomenon in which an increase in temperature induces an electric current that flows from one end of the crystal to the other. **Piezoelectricity** is similar but is produced by a pressure or stress applied to one end of the mineral. Both of these properties are characteristic of anisotropic minerals that lack a center of symmetry; the fact that one end of the crystal is different from the other allows an electric potential to be created across the crystal.

Using a combination of macroscopic properties, most minerals can be tentatively identified with reasonable accuracy. Tables summarizing the macroscopic properties used to identify significant rock-forming and/ or economic minerals are available as downloadable files from the website that supports this text.

5.4   SILICATE MINERALS

The widespread occurrence of silicate minerals makes them by far the most important group of rock-forming minerals; one with which every Earth scientist needs to be familiar. The unifying characteristic of silicate minerals is the presence of silica tetrahedra $(SiO^4)^{-4}$ composed of silica (Si^{+4}) in tetrahedral coordination with oxygen (O^{-2}). Silicate minerals are subdivided according to the degree to which silica tetrahedra are linked and the pattern in which they are linked through shared oxygen ions in the crystal structure (Chapter 2). Because oxygen and silicon are the two most abundant elements in Earth's crust and mantle, silicate minerals are abundant and widespread, comprising more than 92% of the roughly 3500 minerals discovered to date.

5.4.1   Nesosilicates (orthosilicates)

Nesosilicates, also known as **orthosilicates**, are silicate minerals characterized by isolated silica tetrahedra that are not linked through shared oxygen ions to other silica tetrahedra in the structure (Figure 5.13). The ratio of silicon (Si^{+4}) ions to oxygen (O^{-2}) ions in the tetrahedral sites of such minerals is $1:4$. This ratio is reflected in the formulas of nesosilicate minerals, which always contain an $(SiO_4)^{-4}$ component, and implies isolated tetrahedra. For example the **olivine group**, the most abundant mineral group in the upper mantle, has the formula $(Mg,Fe)_2SiO_4$. The formula indicates that it is a nesosilicate (SiO_4) in which the silica tetrahedra are isolated and linked by oxygen to different polyhedral elements. In this case, these are six-fold, octahedral elements that contain magnesium (Mg^{+2}) and/or iron (Fe^{+2}) cations, which electrically neutralize the silica tetrahedral component. As discussed in Chapter 3, olivine consists of two end member minerals in a complete solid solution series in which iron and magnesium substitute for one another. Forsterite is the magnesium-rich end member while fayalite is the iron-rich end member.

A second important mineral group, the **garnet group**, is widespread and abundant in metamorphic rocks; it also occurs in igneous rocks and the mantle. Garnet group minerals have variable compositions due to multiple substitutions and have the general formula $A_3B_2(SiO_4)_3$. The A in the formula signifies cations in a structural site in which oxygen ions are in coordination with a variety of +2 cations that may include Fe^{+2}, Ca^{+2}, Mg^{+2} and Mn^{+2}, among others. The B in the formula signifies a structural site for cations, an octahedral site in which oxygen atoms are in six-

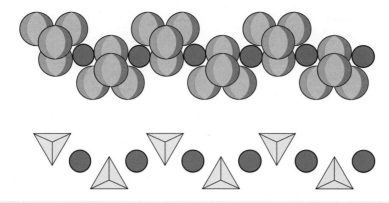

Figure 5.13 Basic nesosilicate (orthosilicate) structure with isolated tetrahedra linked to other polyhedral elements in the structure. (Courtesy of Steve Dutch.)

Table 5.6 Composition and occurrence of the major varieties of garnet.

Variety	Chemical formula	Common occurrence
Almandine	$Fe_3Al_2(SiO_4)_3$	Abundant in pelitic metamorphic rocks including schists, gneisses and granulites; occurs in some aluminum-rich pegmatites
Andradite	$Ca_3Fe_2(SiO_4)_3$	In regionally metamorphosed carbonate rocks and skarns
Grossularite	$Ca_3Al_2(SiO_4)_3$	In regionally metamorphosed carbonate rocks and skarns
Pyrope	$Mg_3Al_2(SiO_4)_3$	In ultrabasic rocks, including mantle peridotites and kimberlites
Spessartine	$Mn_3Al_2(SiO_4)_3$	Scarcer mineral, in skarns
Uvarovite	$Ca_3Cr_2(SiO_4)_3$	Scarce mineral, in chromium-enriched ultrabasic rocks

fold coordination with small +3 cations that may include Al^{+3}, Fe^{+3} or Cr^{+3}, among others. The general formula for garnet could be written as $(Fe,Ca,Mg,Mn)_3(Al,Fe,Cr)_2(SiO_4)_3$, but $A_3B_2(SiO_4)_3$ (also written $X_3Y_2(SiO_4)_3$) is preferred, not only for its simplicity but because real garnets have specific compositions that depend on which cations were available in the environment in which they formed. Both ways of writing the garnet formula, however, clearly show that garnets are nesosilicates with three principal polyhedral or structural sites into which cations of the appropriate radius ratio will fit. Names given to the principal end member varieties of garnet and the occurrences of garnets dominated by specific end member components are summarized in Table 5.6.

A third important nesosilicate group is the **aluminum silicate group**. This group comprises three polymorphs that are common in metamorphic rocks, especially in pelitic assemblages produced by the metamorphism of shales and mudrocks. The three polymorphs of aluminum silicate $(AlAlOSiO_4)$ are the low pressure polymorph **andalusite**, the high pressure polymorph **kyanite** and the high temperature polymorph **sillimanite**. The metamorphic conditions that produce each of these important minerals are detailed in Chapter 18.

Several other nesosilicates, including chloritoid, staurolite, topaz, titanite (sphene) and zircon, are significant rock-forming or economically important minerals.

5.4.2 Sorosilicates (disilicates)

Sorosilicate, also called **disilicate**, minerals possess the minimal amount of linkage possible between linked silica tetrahedra. In sorosilicates, pairs of silica tetrahedra are linked together through a single shared oxygen (O^{-2}) ion to form a basic unit that resembles a "bow tie" (Figure 5.14). The ratio of silicon (Si^{+4}) ions to oxygen (O^{-2}) ions in the linked pair of

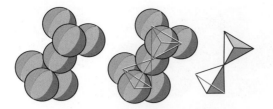

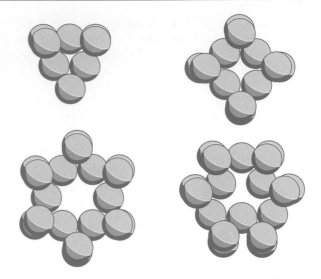

Figure 5.14 Basic unit of sorosilicate structure; pairs of silica tetrahedra are linked through a shared oxygen ion. (Courtesy of Steve Dutch.)

Figure 5.15 Triangular, square and hexagonal ring structures in cyclosilicates and the hexagonal net characteristic of beryl. (Courtesy of Steve Dutch.)

silica tetrahedra is 2:7, which is reflected in the formulas of sorosilicate minerals that always contain an $(Si_2O_7)^{-6}$ element, representing the linked pairing. For example, the mineral **hemimorphite** has the formula $Zn_4(Si_2O_7)(OH)_2 \cdot H_2O$. The formula indicates that hemimorphite is a sorosilicate (Si_2O_7) with paired silica tetrahedra. The Zn indicates that zinc occupies a second coordination site, the (OH) indicates the presence of a hydroxyl (OH^{-1}) anion in addition to oxygen (O^{-2}) anions and the H_2O signifies the presence of neutral, but dipolar, water molecules within the structure. The formula is written in a way that provides information on the chemical composition, the number of structural sites and the common elements that occupy each structural site in the mineral.

The most important group of sorosilicate minerals is the **epidote group**. The epidote group consists of five species; several of the monoclinic varieties display extensive substitution solid solution. These include epidote, clinozoisite and allanite. The rarer manganese-rich mineral piemontite and orthorhombic zoisite are the other members of the group. Epidote is an especially important metamorphic mineral in the greenschist and epidote–amphibolite facies discussed in Chapter 18. Other common and/or economically important sorosilicates minerals include lawsonite and vesuvianite (idocrase), which also occur largely in metamorphic rocks.

5.4.3 Cyclosilicates

When multiple silica tetrahedra link together through shared oxygen (O^{-2}) ions, they may form rings or chains of linked silica tetrahedra (Figure 5.14). In both cases, each silica tetra-

hedron is linked to two additional silica tetrahedra through the sharing of two of its oxygen ions. The ratio of silicon (Si^{+4}) ions to oxygen (O^{-2}) ions is 1:3 in both cases. In **cyclosilicates**, each silica tetrahedra is linked to two adjacent silica tetrahedra through a shared oxygen (O^{-2}) ion to form a structural element in the shape of a ring.

Three basic ring structures exist (Figure 5.15). In some minerals three silica tetrahedra link together through shared oxygen ions to form a triangular ring element in the shape of a triangle. This ring element has the formula (Si_3O_9), as in the rare cyclosilicate mineral **benitoite**, whose formula is $BaTiSi_3O_9$. This formula shows that, in addition to the triangular rings (Si_3O_9), benitoite has two other polyhedral elements: a large cation structural site that is occupied mostly by barium (Ba^{+2}) cations and a small cation structural site that is occupied mostly by titanium (Ti^{+4}) cations.

In other cyclosilicates, four silica tetrahedra join through shared oxygen ions to form a ring element in the shape of a square. This ring element has the formula Si_4O_{12}, as in the scarce cyclosilicate mineral **axinite**, whose chemical formula can be written as $(Ca,Fe,Mn)_3Al_2(BO_3)(Si_4O_{12})(OH)$. The formula shows that axinite possesses a square ring (Si_4O_{12}) structure, a structural site for divalent cations of moderate size (Ca,Fe,Mn) and a third structural site for smaller cations that is commonly occupied by aluminum (Al^{+3}) cation. The borate $(BO_3)^{-1}$ and hydroxyl $(OH)^{-1}$ anions occupy spaces within the open rings of the structure.

In the most common types of cyclosilicate, six silica tetrahedra link through shared oxygen ions to form six-sided, hexagonal ring elements. These ring components have the formula Si_6O_{18}, as in **beryl**, whose chemical formula can be written as $Be_3Al_2Si_6O_{18}$. As in the previous examples of cyclosilicates, beryl has three principal polyhedral structural sites: (1) a tetrahedral site occupied by Si^{+4} linked in the form of hexagonal rings (Si_6O_{18}), (2) a second tetrahedral site occupied by small beryllium (Be^{+2}) cations, and (3) an octahedral site occupied by slightly larger aluminum (Al^{+3}) cations. Tourmaline and cordierite are other common cyclosilicates with hexagonal ring structures.

5.4.4 Inosilicates

Inosilicate is the formal term for silicates in which silica tetrahedra are linked together through shared oxygen ions into one-dimensional chains of long-range extent. Because these chain structures extend from one side of the mineral crystal to the other, chain silicate structures are classified as **one-dimensional structures**. Many different types of inosilicate or chain structures exist, the most common of which are single-chain and double-chain inosilicate structures.

Single-chain inosilicates: pyroxenes and pyroxenoids

In **single-chain inosilicates**, each silica tetrahedron is linked to two adjacent silica tetrahedra through shared oxygen ions so that, as in the cyclosilicates, the Si/O ratio is 1 : 3. There are many different ways in which tetrahedra can be linked into single chains. In the **pyroxene group**, the most abundant group of single-chain inosilicate minerals, the tetrahedra alternate on opposite sides of the axis of the chain (Figure 5.16). One can visualize this structure as consisting of a basic unit of two doubly linked silica tetrahedra (Si_2O_6), on

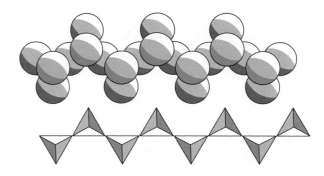

Figure 5.16 Single-chain inosilicate structure in pyroxene. (Courtesy of Steve Dutch.)

opposite sides of the axis, being repeated infinitely along the axis of the chain. This is reflected in the general formula for the minerals of the pyroxene group, which is $XY(Si_2O_6)$. The (Si_2O_6) infers single-chain inosilicates and the X and Y represent two other coordination sites. In real pyroxenes, the X represents a transitional octahedral–cubic structural site for cations. When it contains larger cations such as Ca^{+2} and Na^{+1}, it is somewhat distorted with many of the properties of an eight-fold site; when it contains only smaller cations, such as Fe^{+2}, Mg^{+2} and Mn^{+2}, it behaves more like a six-fold site. The Y represents a normal octahedral structural site for smaller cations, such as Fe^{+3}, Al^{+3} and Ti^{+4}, as well as Fe^{+2}, Mg^{+2} and Mn^{+2}. The general formula for pyroxene group minerals can be written as $(Ca^{+2},Na^{+1},Fe^{+2},Mg^{+2},Mn^{+2})$ $(Fe^{+3},Al^{+3},Ti^{+4},Fe^{+2},Mg^{+2},Mn^{+2})(Si_2O_6)$, but there are advantages in the simpler form $XY(Si_2O_6)$. As indicated by the assortment of elements in the X and Y coordination sites, widespread substitution occurs among elements of similar charge and atomic radii. Which elements are actually added to the site also depends on what is available in the environment in which the mineral crystallizes.

In **pyroxenoids**, the tetrahedra are distributed about the chain axis in a different fashion so that the repeat distance between tetrahedra occupying similar positions is larger. For example, in wollastonite ($Ca_3Si_3O_9$) the repeat distance is every third tetrahedron, whereas in rhodonite ($Mn_5Si_5O_{15}$) it is every fifth tetrahedron, as indicated by the manner in which their formulas are written.

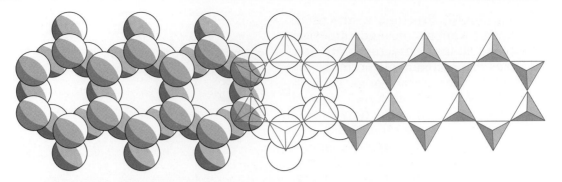

Figure 5.17   Double-chain silicate structure in amphiboles. (Courtesy of Steve Dutch.)

Double-chain inosilicates: amphibole group

In the **double-chain inosilicates**, two single chains are linked together through additional shared oxygen ions to form a double chain, commonly with an Si/O ratio of 4:11 (1:2.75). In the **amphibole group**, the most abundant group of double-chain inosilicate minerals, the basic structural unit consists of eight linked silica tetrahedra, four on each side of the axis of a double chain, which are repeated along the chain axis to form a long-range double chain (Figure 5.17). The basic unit has the formula Si_8O_{22}. But many of the oxygen atoms in the double chain are not linked to other silica tetrahedra and thus have unsatisfied charges (–1) that must be satisfied by bonding to other cations in other coordination polyhedra. This is reflected in the simplified general formula for minerals in the amphibole group, which is $X_2Y_5(Si_8O_{22})$ $(OH)_2$. Once again, the (Si_8O_{22}) signifies a double-chain inosilicate from the amphibole group, and the X and the Y denote the occurrence of two additional types of structural sites or coordination polyhedra. The $(OH)_2$ signifies the presence of hydroxyl (OH^{-1}) anion in addition to oxygen (O^{-2}) anion in the structure and conveys the important information that members of the amphibole group are hydrous silicates. The X structural site commonly contains larger cations, such as Fe^{+3}, Ca^{+2} and Na^{+1} in addition to Fe^{+2}, Mg^{+2} and Mn^{+2}. The Y structural sites are octahedral, six-fold sites that contain smaller cations, such as Fe^{+2}, Mg^{+2}, Mn^{+2} and Al^{+3}. As with the pyroxenes and many other silicates, formulas may be written in different ways to emphasize the structure of the mineral, its chemistry or both.

Minerals with triple- and quadruple-chain elements exist but are not common. Like other inosilicates, these chain silicates link silica tetrahedra into one-dimensional structures that possess long-range extent only in the direction parallel to the chain axis (Figure 5.17).

5.4.5   Phyllosilicates

When multiple chains of silica tetrahedra are linked through shared oxygen ions in a direction at a large angle to the chain axis, the chains combine to form a sheet of linked silica tetrahedra with long-range extent in two directions. Such two-dimensional sheet structures are characteristic of **phyllosilicate** minerals (Figure 5.18). Minerals in this group, often called sheet silicates, commonly have an Si/O ratio of 2:5 or 4:10. Their chemical formulas are often complicated by the fact that aluminum (Al^{+3}) cation substitutes in limited, and often variable, amounts for silicon (Si^{+4}) cation in the "silica" tetrahedra to form a limited number of aluminum tetrahedra.

Because one out of every four oxygen ions is not linked to another silica tetrahedron, one out of every four oxygen ions must bond to other cations in order to be electrically neutralized. Typically these cations occur in octahedral sites that alternate in some way with the tetrahedral layers to which they are linked. The two most common octahedral sites are those that contain magnesium (Mg^{+2}) bonded to oxygen (O^{-2}) and hydroxyl (OH^{-1}) ions and

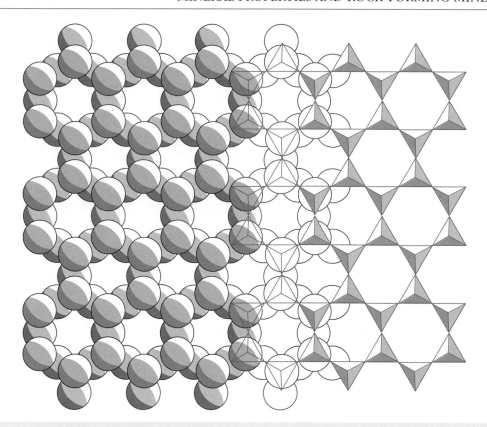

Figure 5.18 Two-dimensional sheet structure typical of phyllosilicate minerals. (Courtesy of Steve Dutch.)

those that contain aluminum (Al^{+3}), also bonded to oxygen and hydroxyl ions. The magnesium octahedral sites are called **brucite sites (b)** after the mineral brucite [$Mg(OH)_2$], and the aluminum sites are called **gibbsite sites (g)** after the mineral gibbsite [$Al(OH)_3$]. Iron frequently substitutes partially for the magnesium in brucite octahedral sites.

Common rock-forming phyllosilicate mineral groups include (1) serpentines, (2) talc, (3) chlorites, (4) micas, and (5) clays. Other moderately common phyllosilicate minerals include apophyllite, prehnite and stilpnomelane.

Serpentine, talc and chlorite group minerals

Serpentines, talc and chlorite provide an excellent basis for discussing the major phyllosilicate structures. **Serpentine** minerals (Figure 5.19a) are composed of alternating tetrahedral (t) layers (Si_2O_5)$^{-2}$ and octahedral layers (o) that give rise to a basic two-layer (t-o) structure. In serpentines, the octahedral layers are typically brucite layers (b), and the chemical formula $(Mg,Fe)_3Si_2O_5(OH)_4$ reflects the two-layer (t-o or t-b) structure typical of serpentine minerals. On the other hand, **talc** (Figure 5.19b) is composed of one octahedral (o) or brucite layer (b) sandwiched between two tetrahedral (t) layers, giving rise to a basic three-layer (t-o-t or t-b-t) structure that is reflected in its mineral formula $Mg_3(Si_4O_{10})(OH)_2$. Pyrophyllite [$(Al_2(Si_4O_{10})(OH)_2$] possesses a similar three-layer structure, but with a gibbsite layer (g) sandwiched between the two tetrahedral layers (t-g-t). **Chlorite group** minerals possess an extra brucite layer so that the basic structure is four layers (t-b-t-b), as reflected in its formula $(Mg,Fe)_3(OH)_6 \cdot (Mg,Fe,Al)_3(Si,Al)_4O_{10}(OH)_2$, where the portion before the stop represents the extra brucite layer. A more detailed discussion of phyllosilicate structures, especially clay minerals, occurs in Chapter 12, where they are discussed in conjunction with weathering and soils. Serpentines, chlorite group minerals and talc, common in metamorphic

(a)

Tetrahedral net

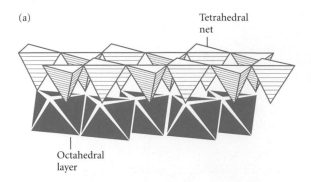

Octahedral layer

(b)

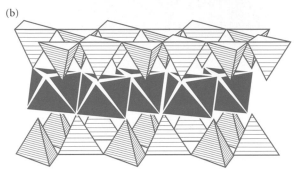

Figure 5.19 (a) Two-layer (t-o or t-b) structure of serpentine. (b) Three-layer (t-o-t or t-b-t) structure of talc. Both are viewed perpendicular to the sheets, with repeated stacking of the basic structural units to produce a three-dimensional crystal lattice. (From Wenk and Bulakh, 2004; with permission of Oxford University Press.)

rocks, will be discussed further in Chapters 15–18. These minerals are all relatively soft and possess one set of excellent cleavage because the fundamental, sheet-like structural units are held together by planes of weak bonds.

Mica group

Widespread and nearly ubiquitous in their occurrence are the members of the **mica group**. These minerals are significant rock-forming minerals in both igneous and metamorphic rocks and are not uncommon in sedimentary rocks. Micas are three-layer (t-o-t) phyllosilicates with two tetrahedral layers in which one out of every four silica tetrahedra has an aluminum (Al^{+3}) cation substituting for the silicon (Si^{+4}) cation as

reflected in the ($AlSi_3O_{10}$) component of their formulas. **Biotite** [$K(Mg,Fe)_3(AlSi_3O_{10})(OH)_2$] and **phlogopite** [$K(Mg)_3(AlSi_3O_{10})(OH)_2$] possess brucite octahedra sandwiched between the two tetrahedral layers (t-b-t), and potassium ions are weakly held in interlayer spaces between the basic structural units. **Muscovite** [$(KAl_3AlSi_3O_{10}(OH)_2)$] and **lepidolite** [$(K(Li, Al)_3AlSi_3O_{10}(OH,F)_2)$] possess gibbsite-related octahedra sandwiched between two tetrahedral layers (t-g-t), with potassium ions in interlayer spaces between the basic structural units (Chapter 12). Micas are characterized by the elasticity of their sheets, their perfect single set of cleavage and relatively low hardness.

Clay group

Clay group minerals commonly occur as microscopic crystals (<4 µm) in soils, sedimentary rocks and low temperature hydrothermal and metamorphic rocks, where they form by the low temperature alteration of aluminum silicate minerals, such as feldspars, micas and amphiboles. The clay minerals can be subdivided into two structural groups. The two-layer (t-o or t-g) **kandite group** clays include minerals, such as **kaolinite** [$(Al_2Si_2O_5)$], that possess serpentine-type structures. The three-layer (t-o-t) clays possess talc-type structures and belong to the **illite group** (t-g-t), with the approximate formula $KAl_3AlSi_3O_{10}(OH)_2$, and the **smectite group** (t-b-t), which includes minerals such as **montmorillonite** with the approximate formula $(Ca,Na)(Mg,Fe,Al)_3 AlSi_3O_{10}(OH)_2 \cdot nH_2O$. Smectites are called expansive clays due to their ability to absorb and release large amounts of water held in interlayer sites and thus to change volume significantly. As discussed in Chapter 12, these expansive clays are hazardous for use in foundation construction due to their shrink and swell properties. Clay minerals are widely used in industry as absorbents and fillers (Chapter 19). They are discussed in detail in Chapter 12 in conjunction with weathering, soil formation and environmental hazards.

5.4.6 Tectosilicates

Tectosilicate minerals are composed principally of silica tetrahedra linked through all

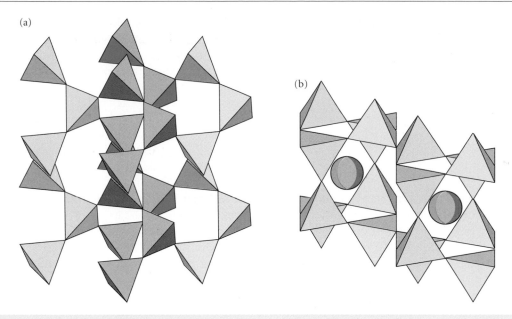

Figure 5.20   Three-dimensional framework structures typical of tectosilicates: (a) quartz; (b) potassium feldspar, with the tricolored atoms representing potassium. (Courtesy of Steve Dutch.)

their oxygen anions to adjacent silica tetrahedra to form three-dimensional framework structures (Figure 5.20). These silicates, often called **framework silicates**, have an Si/O ratio of 1:2, unless aluminum substitutes for some of the silicon ions in the tetrahcdral sitcs, in which case the (Si + Al)/O ratio is 1:2. This is because all four oxygen (O^{-2}) anions are shared with adjoining silica tetrahedra so that each silica tetrahedron possesses one-half of each of the four oxygen anions, making the Si/O ratio $1:4 \times 0.5 = 1:2$. Tectosilicates constitute an extremely important group of rock-forming minerals, comprising nearly 75% of the minerals in Earth's crust, and occur widely in igneous, metamorphic and sedimentary rocks.

The two most abundant groups of tectosilicate minerals are the pure SiO_2 **silica group** and the aluminum silicate **feldspar group**; the latter is the most abundant group of minerals in Earth's crust. Other important tectosilicate mineral groups include (1) the silica-poor, aluminum-rich **feldspathoid group**, (2) the aluminum-rich, hydrated **zeolite group**, and (3) the scapolite group. The first two groups are discussed in the sections that follow.

Silica group

The classic examples of tectosilicates are the **polymorphs of silica,** all of which have the chemical formula SiO_2. This implies that in their pure states they consist only of silica tetrahedra linked together through shared oxygen anions to form a three-dimensional framework structure. The silica polymorphs include the high pressure polymorphs **coesite** and **stishovite,** the high temperature/low pressure polymorphs **tridymite** and **cristobalite,** and the quartz polymorphs **alpha quartz (α-quartz)** and **beta quartz (β-quartz).**

The phase stability diagram for the silica group was discussed in Chapter 3. Coesite is the stable polymorph of silica at pressures of more than 20 kbar (2 GPa) which corresponds to burial depths of more than 60 km. Stishovite is a very high pressure polymorph of silica, with silicon and oxygen in six-fold coordination, which is stable only at pressures >75 kbar (7.5 GPa) or at depths that exceed ~250 km. These high pressure polymorphs, coesite and stishovite, occur in association with meteorite impact and thermonuclear bomb sites, and stishovite is likely an important constituent of the deep mantle. Tridymite and cristobalite are the stable polymorphs of silica under high temperature/low pressure conditions. Tridymite and cristobalite are especially common in silica-rich volcanic rocks produced under high temperature/low pressure conditions. The phase stability diagram (see Figure 3.6) illustrates that quartz is the stable polymorph of silica over a broad range of temperature–pressure conditions common in Earth's crust. This wide stability range and the abundance of silicon and oxygen help to explain why quartz is such an abundant rock-forming constituent of common igneous, sedimentary and metamorphic rocks. The diagram also illustrates that α-quartz (low quartz) is generally the stable form of quartz at normal near surface temperatures and pressures. Quartz is an important economic mineral, widely used in the manufacture of glass and optical fibers and as a source of silicon for the manufacture of microprocessors.

Opal ($SiO_2 \cdot nH_2O$) is an amorphous, hydrated form of silica. Electron microscopy has revealed that it is composed of minute (1500–8000 angstroms) spheres with fairly regular packing (Wenk and Bulach, 2004). This arrangement of spheres acts as a diffraction grating for light, which produces the opalescence characteristic of gem opals such as fire opals. Opal is also an important constituent of microscopic shells of organisms such as diatoms and radiolaria that accumulate to form siliceous sediments on ocean floors, as discussed in Chapter 14, which deals with biochemical sedimentary rocks.

The major properties and occurrences of the silica polymorphs and opal are summarized in Table 5.7.

Macroscopic quartz is easily recognized by its combination of hardness (H = 7), vitreous–greasy luster, lack of cleavage, conchoidal fracture and/or hexagonal prismatic crystals. Quartz is a classic example of an allochromatic mineral. Because quartz is colorless and transparent in its pure state, small amounts of impurities or defects may cause its color to change significantly. Because macrocrystalline quartz is so abundant, each of the major color varieties has its own name. The major varieties of macrocrystalline quartz, the cause of their colors and their common occurrences are summarized in Table 5.8.

Table 5.7 Macroscopic properties and common occurrences of the major silica minerals.

Silica mineral	Crystallography	Hardness	Specific gravity	Common occurrences
Alpha quartz	Hexagonal–trigonal; prismatic	7.0	2.6	Stable at relatively low temperatures and pressures; widespread in igneous, metamorphic and sedimentary rocks
Beta quartz	Hexagonal	7.0	2.5	Stable at elevated temperatures and relatively low pressures; occurs primarily in volcanic rocks
Cristobalite	Tetragonal (pseudoisometic)	6.0–7.0	2.3	Stable at high temperatures and low pressures; occurs in silica-rich volcanic rocks
Tridymite	Monoclinic (pseudohexagonal)	7.0	2.2	Stable at relatively high temperatures and low pressures; occurs in silica-rich volcanic rocks
Coesite	Monoclinic (pseudohexagonal)	7.0–8.0	2.9	Stable at high pressures; occurs in meteorite impactites, kimberlites and ultra high pressure metamorphic rocks produced at great depths
Stishovite	Tetragonal	8.0	4.3	Stable at very high pressures; occurs in meteorite impactites and is theorized to be an important constituent of the deep mantle
Opal	Amorphous; conchoidal fracture	5.5–6.0	2.1	Stable at low pressures and fairly low temperatures; forms around hot springs, in soils and in ocean basins, especially as accumulations of diatoms and radiolaria

Table 5.8   Colors and common occurrences of the major varieties of macroscopic quartz.

Variety	Colors	Cause of color	Common occurrences
Amethyst	Purple–violet	Fe^{+3} impurities	In open fractures and cavities
Aventurine	Green	Chrome mica inclusions	
Citrine	Yellow to yellow-brown	Fe^{+3} impurities	In open fractures and cavities
Milky quartz	White	Water bubble inclusions	Widespread in veins
Rock crystal	Clear, colorless		In open cavities and veins
Rose quartz	Pink	Inclusions of dumortierite nannocrystals	In pegmatites and veins
Smoky quartz	Brown–black	Al^{+3} substitution for Si^{+4}	Widespread in silicic igneous rocks and pegmatites

Several varieties of **microcrystalline to cryptocrystalline quartz** are quite common, especially as sedimentary rocks. Members of the **chert group** are composed of an aggregate of microcrystalline quartz crystals that are fairly equant, if not all the same size. As an aggregate of mineral crystals, chert qualifies as a rock, and the term is frequently used in that way. Members of the chert group include chert (white–medium gray), flint (dark gray–black), jasper (red–yellow) and prase (green). Chert group members are hard (H = 7), lack visible crystals and are smooth, with dull lusters and excellent conchoidal fracture. Members of the **chalcedony group** are characterized by aggregates of sheaves of radiated microscopic silica crystals that are often water bearing. The properties of chalcedony are similar to those of chert, but chalcedony is often slightly more transparent than chert and possesses a waxy or resinous luster, rather than a dull one. Major varieties of chalcedony include chalcedony (gray), carnelian (red), sard (yellow-brown) and chrysoprase (green). Banded varieties are called agate (concentric bands) and onyx (non-concentric bands). Chert and chalcedony are difficult to distinguish, not the least because they are often intergrown within a single rock, yielding specimens with intermediate characteristics. Opal, the amorphous form of silica, is often grouped with chert and chalcedony because of its similar properties, including smooth surface texture and conchoidal fracture. With its higher water content and unique internal structure, opal tends to be a little softer than chert and chalcedony and to have a decidedly waxy luster. Like chert and chalcedony, it also occurs in a wide variety of colors.

Feldspar group

Because aluminum (Al^{+3}) cations can substitute to some degree for silicon (Si^{+4}) cations in the tetrahedral site, many tectosilicate minerals have somewhat more complex formulas than the members of the silica group. This is largely because whenever an aluminum (Al^{+3}) cation replaces a silicon (Si^{+4}) cation in the tetrahedral site a charge deficiency of -1 is created in the structure. This charge deficiency must be balanced by the addition of sufficient cations to create a neutrally charged mineral. This is well illustrated in the most abundant group of minerals in Earth's crust, the feldspar group. The **potassium feldspar subgroup** includes several polymorphs including **orthoclase**, **microcline** and **sanidine**, whose simplified formula can be written as $KAlSi_3O_8$. The formula conveys the information that potassium feldspars are tectosilicates in which one out of every four silica tetrahedra contains an aluminum (Al^{+3}) cation instead of a silicon (Si^{+4}) cation so that the (Al + Si)/O ratio is 1:2 or more precisely 4:8. The substitution of the Al^{+3} for the Si^{+4} creates a charge deficiency of -1 which is balanced by the incorporation of potassium (K^{+1}) cation (tricolored atoms in Figure 5.20b) into the crystal structure to produce an electrically neutral mineral.

The polymorphs of potassium feldspar share several macroscopic properties (Table 5.9). They tend to be hard (H = 6.0–6.5) and possess similar specific gravity (2.5–2.6), colors that range from white to gray to green to pink to red, and two sets of excellent cleavage at right angles. Potash feldspars are commonly perthitic as the result of exsolved albite

Table 5.9   Macroscopic properties and occurrences of the major potassium feldspars.

Mineral	Crystallography	Other distinguishing properties*	Common occurrences
Microcline	Triclinic; stubby crystals of low symmetry	The variety amazonite is bright green; but also white–gray and pink–red	Felsic plutonic igneous rocks and pegmatites; metamorphic schists and gneisses; sedimentary arkoses
Orthoclase	Monoclinic; stubby prismatic with evident symmetry	White–gray–pink–red–green, but rarely bright green	Felsic plutonic igneous rocks; metamorphic schists and gneisses; sedimentary arkoses
Sanidine	Monoclinic; tabular with evident symmetry	More transparent than others; typically colorless to light gray	High temperature potassium feldspar, usually in felsic volcanic rocks

* Feldspars are best distinguished using the analytical techniques discussed in Chapter 6.

(Chapter 3). Unfortunately, potash feldspars possess overlapping macroscopic properties and are easier to distinguish in thin section using optical techniques (Chapter 6) than in hand specimens.

The other group of feldspar minerals is the **plagioclase group**, which has a general formula that can be written as $(Ca,Na)(AlSi)AlSi_2O_8$. As discussed in Chapter 3, this formula reflects the complete coupled ionic substitution series between plagioclase end members albite $(NaAlSi_3O_8)$ and anorthite $(CaAl_2Si_2O_8)$ that characterizes the plagioclase group. Recall also that the plagioclase phase diagram is discussed in Chapter 3 as an example of a two-component system with complete solid solution between end member components. Plagioclase compositions are typically expressed in terms of the proportion of the anorthite end member (%An), where the albite (%Ab) component is simply (100% – %An). Traditionally the plagioclase solid solution series is divided into six compositional ranges that are given specific names (Table 5.10), but it is more precise to simply express the plagioclase composition in terms of percent anorthite end member. The anorthite content and varieties of plagioclase are difficult to distinguish macroscopically but can be distinguished using the analytical techniques discussed in Chapter 6. Calcic plagioclases have more than 50% anorthite and include labradorite, bytownite and anorthite (Table 5.10). Sodic plagioclases contain less than 50% anorthite and include andesine, oligoclase and albite. All plagioclase varieties possess similar hardness (H = 6), specific gravity (2.6–2.8, increasing with %An), cleavage (two sets, one perfect, one good, near 90°), color (white–green–gray) and luster (vitreous–pearly), and crystallize in the triclinic system. Particularly useful in identifying plagioclase and distinguishing it from potash feldspars is its tendency to display parallel striations due to twinning on one set of cleavage surfaces.

Feldspathoid group

Like feldspars, feldspathoids are aluminum-bearing tectosilicates. However, the feldspathoids possess lower silica contents, higher aluminum contents and higher contents of alkali cations such as potassium, sodium and calcium, required to make them electrically neutral. Feldspathoids are relatively scarce minerals that occur primarily in silica-poor (silica undersaturated), alkali-rich peralkaline igneous rocks. They are excellent indicator minerals for such silica undersaturated rocks, which occupy fully half of the International Union of Geological Sciences (IUGS) standard classification chart for igneous rocks. The classification, occurrence and origin of alkaline igneous rocks are discussed in Chapters 7–10. The properties and occurrences of the major feldspathoid minerals are summarized in Table 5.11.

Zeolite group

Zeolite group minerals (Table 5.12) are hydrous silicates that form as secondary minerals at temperatures of 100–250°C as will be

Table 5.10 Major varieties of plagioclase by anorthite (An) and albite (Ab) content.

Plagioclase variety	Chemical formula	An content	Ab content
Albite	$(Ca_{0-0.1}Na_{0.9-1.0})(Al_{0-0.1}Si_{0.9-1.0})AlSi_2O_8$	An_{0-10}	Ab_{90-100}
Oligoclase	$(Ca_{0.1-0.3}Na_{0.7-0.9})(Al_{0.1-0.3}Si_{0.7-0.9})AlSi_2O_8$	An_{10-30}	Ab_{70-90}
Andesine	$(Ca_{0.3-0.5}Na_{0.5-0.7})(Al_{0.3-0.5}Si_{0.5-0.7})AlSi_2O_8$	An_{30-50}	Ab_{50-70}
Labradorite	$(Ca_{0.5-0.7}Na_{0.3-0.5})(Al_{0.5-0.7}Si_{0.3-0.5})AlSi_2O_8$	An_{50-70}	Ab_{30-50}
Bytownite	$(Ca_{0.7-0.9}Na_{0.1-0.3})(Al_{0.7-0.9}Si_{0.1-0.3})AlSi_2O_8$	An_{70-90}	Ab_{10-30}
Anorthite	$(Ca_{0.9-1.0}Na_{0.0-0.1})(Al_{0.9-1.0}Si_{0-0.1})AlSi_2O_8$	An_{90-100}	Ab_{0-10}

Table 5.11 Macroscopic properties and common occurrences of significant feldspathoid minerals.

Mineral/formula	Crystallography	Other distinguishing properties	Common occurrences
Cancrinite $Na_6Ca_2(AlSiO_4)_6(CO_3)_2 \cdot nH_2O$	Hexagonal; prismatic; rare	Yellow–rose–blue; greasy–waxy luster; granular–massive	By alteration of nepheline in silica-undersaturated, feldpathoidal igneous rocks
Lazurite $Na_3Ca(AlSiO_4)_3SO_4S_2$	Isometric; equant; rare	Deep azure blue–greenish blue; massive–granular	Scarce mineral in metamorphosed limestones/skarns
Leucite $KAlSi_2O_6$	Hexagonal; stubby prismatic; pseudo-isometric	White–gray; vitreous to dull luster	In silica-undersaturated, potassium-rich volcanic rocks
Nepheline $Na_3K(AlSiO_4)_4$	Hexagonal; prismatic; rare	White–gray; greasy luster; massive–granular habit	In silica-undersaturated, feldpathoidal igneous rocks
Scapolite* $(Na,Ca)_4(Al_{1-2}Si_{2-3}O_8)_3(CO_3,SO_4,Cl)$	Tetragonal; prismatic; square sections	Prismatic; square sections	In medium–high grade metamorphic carbonates/skarns and pelitic schists, gneisses
Sodalite $Na_4(AlSiO_4)_3Cl$	Isometric; equant; rare	Blue; also gray–green; massive–granular	In silica-undersaturated, feldpathoidal igneous rocks

* Scapolite is not a feldspathoid but is a closely related mineral.

detailed in Chapter 18. Zeolites form as secondary minerals, commonly as amygdules or cavity fillings in altered basalts and related rocks or as veins and alteration products in volcanic pyroclastic and glassy rocks. Upon heating, zeolites expel their water while their crystal structure remains intact. This allows such dehydrated zeolites to act as "sponges" and/or "molecular sieves" that can selectively absorb dissolved constituents such as hydrocarbons, heavy metals or other contaminants from water. Zeolite minerals and their synthetic counterparts are extensively utilized for the purposes of sewage treatment, water softening (by removal of Ca^{+2}) and water purification. They are also used in the catalysis of

high octane lead-free gasoline, the removal of radioactive isotopes from nuclear waste water and the scrubbing of pollutants from industrial stacks. Zeolite group minerals share several physical properties, including diaphaneity (translucent–transparent), luster (vitreous–pearly, silky in fibrous varieties), specific gravity (2.1–2.3) and streak (white). Their hardness is intermediate, ranging between 3.5 and 5.5.

In conclusion, silicate minerals, the predominant minerals in Earth's crust and upper mantle, consist of fundamental units called silica tetrahedra linked in various ways to produce the six major groups of silicate structures whose properties are summarized in

Table 5.12   Macroscopic properties of common zeolite minerals.

Mineral/formula	Crystallography	Other distinguishing properties
Analcime $NaAlSi_2O_6 \cdot 6H_2O$	Isometric; equant; trapezohedra	Colorless to white; no visible cleavage
Chabazite $(Ca,Na,K)_4Al_4Si_8O_{24} \cdot 12H_2O$	Trigonal; equant; pseudocubic rhombohedra	White–yellow; sometimes pink; poor cleavage
Clinoptilolite $(Ca,Na,K)_6Al_6Si_{30}O_{72} \cdot 20H_2O$	Monoclinic; platy–capillary; scaly–fibrous	White; one cleavage, usually not visible
Heulandite $(Ca,Na,K)_9Al_9Si_{27}O_{72} \cdot 24H_2O$	Monoclinic; prismatic–platy	Colorless–white–pale yellow; one perfect cleavage
Laumontite $Ca_4Al_8Si_{16}O_{48} \cdot 18H_2O$	Monoclinic; prismatic	White; two cleavages not at 90°
Natrolite $Na_2Al_2Si_3O_{10} \cdot 2H_2O$	Orthorhombic; prismatic–acicular; radiated–fibrous	Colorless–white–pale yellow; two perfect cleavages at 90°
Stilbite $(Ca,Na,K)_9Al_9Si_{27}O_{72} \cdot 28H_2O$	Monoclinic; tabular–platy; close radiated, sheaf-like groups	White–yellow–brown; one perfect cleavage set

Figure 2.21. Because of the abundance of oxygen and silicon in Earth's crust and mantle, these minerals are the major rock-forming minerals that Earth scientists have dealt with. Many are of economic importance as well. Knowledge of silicate minerals is essential to the practice of Earth science.

5.5   NON-SILICATE MINERALS

With the exception of native elements, non-silicate minerals are classified by their major anion or anionic group. Although non-silicate minerals are far less abundant in Earth's crust and mantle than are silicate minerals, many are of great economic value. This makes knowledge of non-silicate minerals extremely important to practicing Earth scientists.

5.5.1   Native elements

Minerals composed of a single element are called **native elements**. This group also includes several minerals that are composed of two or more closely related elements that possess very similar chemical characteristics. Fewer than 20 elements occur in their native state. Most native element minerals are quite rare; many have only been discovered with the advent of sophisticated instrumentation for examining Earth materials. Together they comprise less than 0.00002% of Earth's crust by weight (Wenk and Bulach, 2004). Native elements are divided into three sub-groups based on the chemical behavior of the elements: (1) metals, (2) semi-metals, and (3) non-metals.

Metals

The **native metals** are composed of metallic elements. All native metals crystallize in the isometric system, and almost all are characterized by cubic closest packing with face-centered cubic crystal lattices. As a result, substitution solid solution of metals of similar radii (e.g., gold and silver, iron and nickel) is common. Native metals are subdivided into three groups: the gold group, the platinum group and the iron group. These groups are distinguished on the basis of the chemical nature of the metallic elements and the properties they confer on their members.

The **gold group** metals include gold (Au), silver (Ag) and copper (Cu). Excellent metallic bonding causes these minerals to be soft, malleable and ductile and to be excellent thermal and electrical conductors. The metallic bonding of gold group metals also causes them to be opaque and to possess metallic luster, hackly fracture and low melting points. High atomic mass numbers and cubic closest packing produce minerals with high specific gravity. The members of the gold group, together with such elements as mercury, lead and palladium, are so similar in crystallographic and chemical properties that they readily substitute for one another.

The **platinum group** metals include several rare, but valuable, isostructural minerals that contain platinum (Pt), palladium (Pd), iridium (Ir) and osmium (Os). Solid solutions of these metals are common. Platinum group metals possess bonds that are less metallic than those of the gold group and so are generally harder and have higher melting points than gold group metals. Platinum group metals are excellent thermal and electrical conductors, are opaque and possess metallic luster and high specific gravity. They are important as catalysts for many chemical processes.

The **iron group** metals include minerals composed of iron and/or nickel. Native iron is rare, but two iron–nickel minerals, iron-rich kamacite and nickel-rich taenite, are important constituents of iron-rich meteorites.

Semi-metals

The **native semi-metals** are composed of semi-metallic elements such as arsenic (As), antimony (Sb) and bismuth (Bi). Because they possess metallic–covalent transitional bonds, they are brittle and much poorer thermal and electrical conductors than the native metals. This more directional bond type also results in lower symmetry, so that most semi-metals crystallize in the hexagonal system.

Non-metals

The **native non-metals** are composed of non-metallic elements, chiefly sulfur and carbon. Native sulfur forms by sublimation from gases at volcanic vents and by bacterial reduction of sulfate minerals occurring in the caprock of salt domes. Some atoms in sulfur are bound together by covalent bonds, while others are bound by van der Waals forces. This accounts for sulfur's translucency, brittle nature and low hardness.

In contrast, diamonds are the hardest minerals on Earth due to short, strong covalent bonds that bind atoms tightly in its cubic closest packing structure. Graphite, on the other hand, is very soft because some of its atoms are bound together by van der Waals forces. The loosely held electrons involved in these bonds also interact with light, which accounts for graphite's opacity and submetal-lic luster. Diamond is a high pressure polymorph formed at depth within the mantle, whereas graphite is a low pressure polymorph formed under near surface conditions. The formation and occurrence of diamonds in kimberlite pipes are discussed in Chapter 10.

Many native elements are of significant economic value. They are the principle sources for such elements as gold, platinum, iridium and osmium and for the carbon minerals graphite and diamond.

5.5.2   Halides

Halide minerals are characterized by large, highly electronegative, monovalent anions such as fluorine (F^{-1}), chlorine (Cl^{-1}), bromine (Br^{-1}) and iodine (I^{-1}). These elements, called the halogens, occur in row 17 (class VIIA) of the periodic table and are essential constituents of halide minerals in which they are ionically bonded with electropositive, metallic cations such as sodium (Na^{+1}), potassium (K^{+1}) and calcium (Ca^{+2}). The halides' ionic bonding and small ionic charge cause them to be brittle, translucent and quite soluble. They tend to possess low to moderate hardness and moderate to high melting temperatures, and to be poor conductors of heat and electricity. More than 80 halide minerals exist. However only three halides – halite (NaCl), fluorite (CaF_2) and sylvite (KCl) – are common. A fourth halide mineral, cryolite (Na_3AlF_6), was formerly essential to the refining of aluminum ores such as bauxite. Rarer halides typical of evaporite deposits are discussed further in Chapter 14, on biochemical sedimentary rocks.

5.5.3   Sulfides and related minerals

Sulfide minerals are composed of metallic and semi-metallic elements bonded with sulfide (S^{-2}) anions. A common example is pyrite (FeS_2). Closely related are the arsenides (As^{-2}), selenides (Se^{-2}) and tellurides (Te^{-2}), in which another element takes the place of sulfur. Nearly 500 sulfide and related minerals have been reported (Wenk and Bulach, 2004); only the common and/or economically important examples are discussed here. These minerals exhibit a variety of crystal structures that depend on ionic/atomic radii and on bond types which range from ionic–metallic through covalent–metallic. Most sulfides crystallize in the cubic, tetragonal or hexagonal systems, reflecting the high degree of symmetry of their crystal lattices.

The variety of bonding mechanisms in sulfides and related minerals leads to a great variety of characteristics, making generalizations difficult. Many sulfides with a significant component of metallic bonding are opaque with metallic luster, distinctive colors and characteristic streaks. Non-opaque minerals tend to possess extremely high refractive indices and transmit light only on thin edges. Most sulfides are relatively soft and are good electrical conductors, reflecting the metallic component of their bonds.

The sulfides and related minerals are economically significant as the major source of many metallic elements important to civilization, including copper (chalcopyrite, $CuFeS_2$; bornite, $CuFe_5S_4$; chalcocite, Cu_2S), lead (galena, PbS), zinc (sphalerite, ZnS), silver (argentite, AgS), nickel (nickeline, $NiAsS$), cobalt (cobaltite, $CoAsS$), molybdenum (molybdenum, MoS_2) and mercury (cinnabar, HgS). The description and origins of these economically important mineral deposits are detailed in Chapter 19.

5.5.4   Oxides

Oxide minerals contain metals or semi-metals ionically bonded with oxygen anions (O^{-2}) in a diverse range of fairly symmetrical, closely packed structures. As a result, most oxide minerals are relatively hard and dense, possess relatively high melting temperatures and crystallize in the isometric, tetragonal or hexagonal systems. Oxide minerals are subdivided into **simple oxides** that contain a single metal bonded with oxygen and **complex oxides** in which more than one metal is bonded with oxygen. Simple oxides can be further subdivided into: (1) the **cuprite group** (X_2O), (2) the **periclase group** (XO), (3) the **rutile group** (XO_2), and (4) the **corundum group** (X_2O_3), where X represents the single metal ion in the mineral. Complex oxides include the **ilmenite group** (XYO_2), **perovskite group** (XYO_3) and **spinel group** (XY_2O_4).

Oxides are widespread and abundant, and are often concentrated in economically valuable ore deposits. Notable examples include deposits of the iron oxide minerals hematite (Fe_2O_3) and magnetite ($FeFe_2O_4$), which are mined from iron-rich sedimentary and meta-sedimentary deposits, detailed in Chapter 14, and from magmatic and hydrothermal deposits, discussed in Chapter 19. Other oxide minerals yield economic deposits of manganese (pyrolusite, MnO_2), copper (cuprite, Cu_2O), titanium (ilmenite, $FeTiO_2$), tin (cassiterite, SnO_2), chromium (chromite, $FeCr_2O_4$) and uranium (uraninite, UO_2).

5.5.5   Hydroxides

The defining components of **hydroxide** minerals are metallic elements in combination with hydroxyl (OH^{-1}) ion. Oxygen is also an essential component of many hydroxide minerals. Hydroxide minerals tend to be softer than oxides and to possess somewhat lower specific gravity. Hydroxide minerals commonly develop by the weathering and alteration of other minerals under near surface conditions.

The most economically important hydroxide minerals are those that belong to the **bauxite** mineral group, which is the major source of aluminum ore. They include boehmite ($AlOOH$), diaspore ($AlOOH$) and gibbsite [$Al(OH)_3$]. Bauxite group minerals form during intense weathering of aluminum-bearing rocks in tropical environments (as detailed in Chapter 12). Another significant hydroxide mineral is the iron-bearing goethite [$FeO(OH)$], which is an important iron ore mineral in lateritic soils (Chapter 12) and in bog iron deposits (Chapter 14). **Limonite** is a term used for finely granular, sometimes amorphous, mixtures of iron hydroxide and hydrated iron hydroxide minerals with close affinities to goethite. Such granular masses are typically soft, with a characteristic rusty yellow to yellow-brown color. Other important hydroxide minerals include the manganese minerals brucite [$Mn(OH)_2$], manganite [$MnMnO_2(OH)_2$] and romanechite [$BaMnMn_9O_{20}\cdot3H_2O$].

5.5.6   Carbonates

All **carbonate** minerals contain carbonate [$(CO_3)^{-2}$] anions bonded with metallic or semi-metallic cations. Carbonate minerals are characterized by their relative softness and varying degrees of solubility in dilute HCl, which breaks the bonds in the carbonate radical releasing carbon dioxide gas (CO_2), causing the mineral to effervescence. Carbonate minerals occur widely in biochemical sedimentary rocks, where they are the major constituents of limestones and dolostones (Chapter 14). They also are major constituents of metamorphic rocks such as marble and skarns (Chapter 15). Carbonatites are rare igneous rocks containing carbonate minerals crystallized from carbonate magmas and lavas.

Approximately 70 carbonate minerals have been reported. The most common carbonate minerals belong to one of three groups. **Calcite group** carbonates constitute an isostructural group composed of small cations bonded to carbonate ions in trigonal (rhombohedral) crystal structures with rhombohedral cleavage. Calcite group minerals include calcite ($CaCO_3$), magnesite ($MgCO_3$), siderite ($FeCO_3$), rhodochrosite ($MnCO_3$) and smithsonite ($ZnCO_3$). **Aragonite group** carbonates constitute a second group composed of larger cations bonded to carbonate ions in orthorhombic structures. Aragonite group minerals include aragonite ($CaCO_3$), cerrusite ($PbCO_3$), strontianite ($SrCO_3$) and witherite ($BaCO_3$). The important rock-forming mineral dolomite [$CaMg(CO_3)_2$] bears some resemblance to the minerals in the calcite group. **Hydroxycarbonates,** such as the minor copper ore minerals azurite [$Cu_3(CO_3)_2(OH)_2$] and malachite [$Cu_2CO_3(OH)_2$], contain hydroxyl ion and/or water and commonly exhibit monoclinic structures. Carbonate minerals such as calcite and aragonite are economically important as the major raw materials in cement products, and with dolomite as significant sources of building stone.

5.5.7 Borates

The fundamental building blocks of **borate** minerals are $(BO_3)^{-3}$ triangles and, less commonly, $(BO_4)^{-5}$ tetrahedra, bonded with metal cations, most commonly sodium and/or calcium. The triangular and tetrahedral coordination polyhedra commonly link through shared oxygen ions into larger structural units such as rings and chains, in a manner similar to the linkage of silica tetrahedra in silicate

minerals. Most borate minerals contain appreciable hydroxyl (OH^{-1}) ions and/or water (H_2O) incorporated into their crystal structures. Borates are particularly abundant in enclosed lake basins of the Mojave Desert region in California, where their formation is attributed to evaporitic concentration of boron derived from volcanic hot springs and fumarole activity. More than 100 borate minerals have been described. Four significant examples are borax [$Na_2B_4O_5(OH)_4 \cdot 8H_2O$], colemanite [$CaB_3O_4(OH)_3 \cdot H_2O$], kernite [$Na_2B_4O_6(OH)_2 \cdot 3H_2O$] and ulexite [$NaCaB_5O_6(OH)_6 \cdot 5H_2O$].

5.5.8 Sulfates

Sulfates are minerals that contain $(SO_4)^{-2}$ combined with one or more metals or semi-metals. Major groups of sulfate minerals include **hydrated sulfates,** of which the most common example is gypsum ($CaSO_4 \cdot 2H_2O$), and **anhydrous sulfates,** of which the most common example is anhydrite ($CaSO_4$). Gypsum varieties include **selenite,** which is composed of large macroscopic crystals; **satinspar,** which consists of fibrous aggregates of parallel acicular–capillary crystals; and **alabaster,** which consists of masses of randomly oriented microsopic crystals. Gypsum, anhydrite and several other sulfate minerals are significant components of sedimentary evaporite deposits, which are discussed in Chapter 14.

Other sulfate minerals include alunite [$KAl_3(SO_4)_3(OH)_6$], anglesite ($PbSO_4$), barite ($BaSO_4$), celestite ($SrSO_4$) and several evaporite minerals that include epsomite ($MgSO_4 \cdot 7H_2O$) and polyhalite [$K_2Ca_2Mg(SO_4)_4 \cdot 2H_2O$].

5.5.9 Phosphates

Phosphate minerals contain $(PO_4)^{-3}$ anions bonded with metal cations. Although the phosphate group is large, only two minerals – apatite and monazite – are relatively common. **Apatite** [$Ca_5(PO4)_3(Cl,F,OH)$] forms in crystalline igneous rocks and in a marine sedimentary rock called phosphorites. In addition to its use as fertilizer, apatite is the main component of human teeth. **Monazite** [$(Ce,Y,La,Th)PO_4$], the other important phosphate mineral, is the principal source of rare Earth minerals such as thorium. Economic quantities of monazite are recovered from beach deposits along with other heavy, resistant minerals such as zircon, magnetite, rutile, ilmenite and garnet. Amblygonite ($LiAlFPO_4$) occurs in some pegmatites and is mined as a source of lithium used in refractory glass and metallic alloys. Turquoise [$CuAl_6(PO_4)_4(OH)_8 \cdot 4H_2O$] is a significant gemstone, widely used in jewelry.

5.5.10 Tungstates and molybdates

Tungstate group minerals are characterized by the tungstate [$(WO_4)^{-2}$] anion group bonded to metallic cations. Some 20 tungstate minerals have been identified. All are relatively rare, but scheelite ($CaWO_4$) and wolframite [$(Fe,Mn)WO_4$] are economically important sources of tungsten, whose high melting point and steel hardening properties make it an important element in the manufacture of hard, refractory steel alloys for use in high-speed cutting tools, incandescent lamps, electrical contacts, crucibles and spark plugs.

Molybdates are characterized by the molybdate [$(MoO_4)^{-2}$] anion group bonded to metallic cations. The only significant molybdate mineral is wulfenite ($PbMoO_4$), which is a source of molybdenum used with iron in the manufacture of high-quality steel alloys.

5.5.11 Other mineral groups

Other mineral groups, based on their principal anion group, are represented by relatively rare minerals and are not detailed in this text. These include **nitrates** [$(NO_3)^{-1}$] such as nitratite ($NaNO_3$) and niter (KNO_3), **chromates** [$(CrO_4)^{-2}$] such as crocoite ($PbCrO_4$), **vanadates** [$(VO_4)^{-3}$] such as vanadinite [$Pb_5(VO_4)_3Cl$] and carnotite [$K_2UO_2)_2(VO_4)_2 \cdot 3H_2O$], and **arsenates** [$(AsO^4)^{-3}$] such as erythrite [$Co_2(AsO_4)_2 \cdot 8H_2O$].

In this chapter, we have focused on the macroscopic properties of minerals. Chapter 6 details the microscopic techniques used to identify minerals and summarizes a few of the many advanced analytical methods used by scientists who seek a deeper understanding of Earth materials. As a result of space constraints, the macroscopic and microscopic properties of significant minerals, especially those used to identify them, are summarized in downloadable determinative mineralogy tables posted on the website that supports this text.

Chapter 6

Optical identification of minerals

In our efforts to understand Earth materials from as many different perspectives as possible, scientists have developed a variety of microscopic, imaging and analytical techniques that permit more precise and accurate characterization of such materials. Due to space constraints, only the optical methods are discussed here. A brief discussion of other imaging and analytical techniques is available in a downloadable file on the website that accompanies this text. Most of these techniques involve insights adapted from physics and chemistry to investigate the ways in which various forms of electromagnetic radiation interact with materials. It is these interactions that reveal the detailed structure and chemical compositions of both natural and synthetic materials. Insight into these interactions has opened the door for the production of synthetic materials that possess the desired structural and chemical properties for specific applications – which has marked the post-industrial revolution in materials science. Because most of these techniques involve the interaction of Earth materials with electromagnetic radiation, a brief review of the

major concepts of electromagnetic radiation is presented here.

Earth Materials, 1st edition. By K. Hefferan and J. O'Brien. Published 2010 by Blackwell Publishing Ltd.

6.1 ELECTROMAGNETIC RADIATION AND THE ELECTROMAGNETIC SPECTRUM

6.1.1 Electromagnetic radiation

Electromagnetic radiation, most familiar to us in the form of visible light energy, includes a much broader spectrum of energies, which are discussed below. Electromagnetic energy is generated by (1) the acceleration of electrically charged particles (e.g., electrons, protons or ions), or (2) changing magnetic fields. Once generated, electromagnetic radiation moves away from the place where it was produced carrying both energy and momentum. In a vacuum, such energy moves (propagates) at the speed of light ($c = 3.0 \times 10^8$ m/s); in matter it always propagates at less than the speed of light in a vacuum but still "lightening fast".

Electromagnetic radiation is characterized by both an **electrical component** and a **magnetic component**. The two components are mutually self-generating; the magnetic component can be generated by the electrical component and vice versa. Changes in one

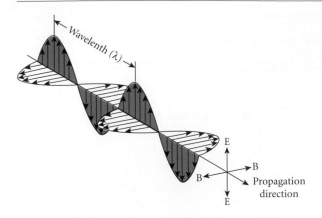

Figure 6.1 The electrical (E) and magnetic (B) components of electromagnetic waves vibrate perpendicular to each other and to the propagation direction.

component are accompanied by changes in the other.

Electromagnetic radiation possesses the properties of both particles and waves. In **particle models**, electromagnetic radiation is modeled as **photons**, which are discrete packets of energy that interact with atoms at the subatomic level. Photons have zero mass and carry an amount of energy characteristic of the type of electromagnetic radiation in question. Particle models work best for very small-scale (nanometer-scale) phenomena of short (nanosecond) duration.

For most larger scale phenomena, a wave model generally provides the most appropriate visualization. In **wave models**, electromagnetic radiation is modeled as a series of sinusoidal, transverse waves with features characteristic of the type of electromagnetic radiation in question. The electrical component (E) of the radiation vibrates in a plane (Y or E) perpendicular to the direction of propagation (X) while the magnetic component (B) of the radiation vibrates in a plane (Z or B) perpendicular to both (X and Y). These relationships are shown in Figure 6.1. For reasons of efficiency, in many of the discussions later in this chapter only the electrical vibration direction is considered.

Each wave series is characterized by **crests** and **troughs** that represent the maximum up and down or side to side (positive and negative) displacements from the propagation direction as the wave energy oscillates trans-

verse to it. This maximum displacement from the propagation direction is called the **wave amplitude (A)**. The **wavelength (λ)** of a single wave is the distance between successive wave crests or any other corresponding parts of successive wave forms. **Wave frequency (f)** is defined by the number of waves in the propagating wave series that pass a point in 1 s. The units of wave frequency are number per second (#/s) or cycles per second. This unit is called a **Hertz (Hz)**. For example, an electromagnetic wave with a frequency of 20,000 Hz would be a wave in which 20,000 crests passed a point during a single second of wave propagation. Wavelength (λ) and frequency (f) are related to velocity (υ) by the following equation:

$$\upsilon = \lambda f \qquad \text{(equation 6.1)}$$

For a wave propagating at a constant velocity, for example the speed of light in a vacuum (c), increasing wavelength is associated with decreasing frequency and vice versa.

6.1.2 The electromagnetic spectrum

The electromagnetic spectrum constitutes the full range of electromagnetic energy (Figure 6.2). Every part or **band** of the full spectrum is characterized by three properties: (1) its **energy (E)** expressed in electron volts (Ev), (2) its frequency (f) expressed in cycles per second or Hertz (Hz), and (3) its wavelength (λ). These three properties are closely related. Energy values are proportional to frequency which (equation 6.1) is inversely proportional to wavelength. Figure 6.2 illustrates the major, sometimes overlapping, parts of the electromagnetic spectrum and their characteristic energies, wavelengths and frequencies. Radio and TV waves have the longest wavelengths, smallest frequencies and least energy, whereas gamma rays and X-rays have the shortest wavelengths, highest frequencies and most energy. This brief introduction to the electromagnetic spectrum is necessary to understand optical crystallography and more advanced analytical techniques.

6.2 ESSENTIALS OF OPTICAL CRYSTALLOGRAPHY

Optical crystallography is the study of crystals based on how they interact with

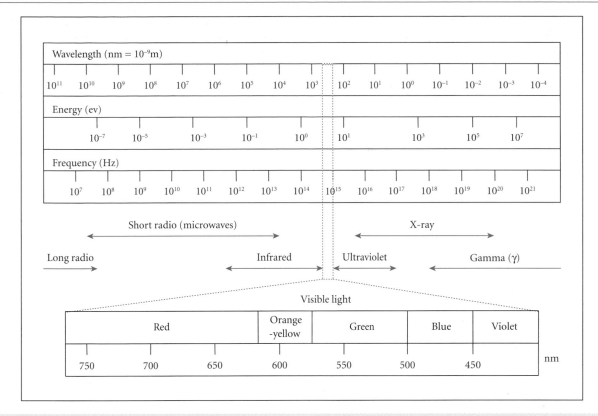

Figure 6.2   The continuous electromagnetic spectrum showing the wavelengths, frequencies and energies (in electron volts, ev) associated with each part or band of the spectrum. The narrow part of the spectrum visible to humans is shown along with the approximate wavelengths characteristic of each major color of the visible spectrum.

visible light incident upon their surfaces. It cannot be practiced well unless some basic concepts about light–crystal interactions are well understood.

6.2.1   Light and crystals

Whenever light is incident on the surface of a solid substance, a number of phenomena may occur. Light may be reflected, refracted, dispersed, transmitted, absorbed, scattered and/or reradiated by the substance. Such phenomena depend primarily on the substance's chemical composition and crystal structure. Because each mineral possesses a unique combination of chemical composition and crystal structure (including defects), the manner in which each mineral interacts with light is unique. Once these interactions are known, they can be used to identify minerals or any other crystalline substance. **Ordinary light**

(Figure 6.3) vibrates in all directions perpendicular to the direction of propagation (ray path). **Plane polarized light** (Figure 6.3) is constrained by a polarizing lens to vibrate in a single plane perpendicular to the direction of propagation.

Reflection, refractive index, refraction and dispersion

When light strikes the surface of a solid, part of the light is reflected from the surface, much like light from a mirror. The angle of **reflection** is the complement of the angle of incidence (i). Only incident light that is perpendicular to the boundary between two media is reflected back in the direction from which it came. All other light is reflected at the complementary angle (Figure 6.4). This is why light flashes from a mirror or the cleavage surfaces of a mineral to your eyes at a particular angle as the objects are rotated.

Light that is not reflected from the surface continues into the solid as refracted light. Refracted light changes both velocity and direction by an amount that depends on the solid and on the specific wavelength of visible light in question (Figure 6.4). The change in velocity is expressed by the **refractive index (n or RI)**, which is the ratio between the velocity of light in a vacuum (V_v) and the velocity of light in the material (V_m), and is given by:

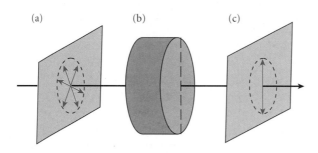

(a)   (b)   (c)

Figure 6.3 Ordinary light vibrates in all directions perpendicular to the direction of propagation. When ordinary light (a) is passed through a polarizing filter (b), it emerges as plane polarized light (c), constrained to vibrate in a single plane perpendicular to the propagation direction.

$$RI = V_v / V_m \qquad \text{(equation 6.2)}$$

The refractive index is inversely proportional to the velocity of light in the material. It increases as the velocity of light in the material decreases. Because light slows down as it interacts with the electric fields in atoms, the velocity of light in substances is always less than the velocity of light in a vacuum, and the refractive index of a substance is therefore always more than 1.0. For example, the refractive index of air, with its low proportion of atoms per volume, at standard temperature and pressure is 1.0008. Most minerals, with their much higher concentration of atoms per volume, possess refractive indices between 1.4 and 2.0, meaning that light slows down 20–50% as it passes from air into most crystals. The refractive index is critical in mineral identification.

The change in direction across the boundary between two media is called the **angle of refraction (r)**. This angle is related to the angle of incidence (i) and the refractive indices of the two materials (n_i and n_m) as given by Snell's law:

$$n_i \sin(i) = n_m \sin(r) \qquad \text{(equation 6.3)}$$

Snell's law predicts that incident light not perpendicular to the surface will always be refracted or bent toward the medium with the

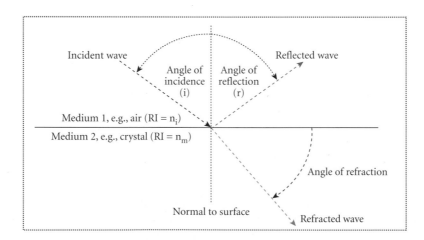

Figure 6.4 Incident light is reflected from a surface between two media, for instance air and a crystalline substance. It is also refracted as it moves into the crystalline substance with a new velocity, given by the refractive index (n), and in a new direction, given by the angle of refraction (r), as related by Snell's law.

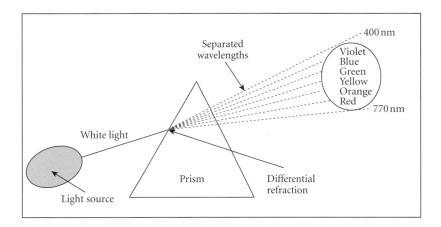

Figure 6.5 The dispersion of white light into different parts of the visible spectrum as it passes through a crystalline prism. Note that the amount of refraction increases as wavelength decreases, causing the separation of wavelengths into separate colors.

lowest velocity and highest refractive index. In Figure 6.4, the light is bent toward the mineral as it enters the crystalline material from the air, because the crystal has a higher refractive index and a lower velocity. The amount of refraction increases with the angle of incidence and with the change in refractive index (or velocity) across the boundary.

In a vacuum, all wavelengths of the magnetic spectrum are propagated at the velocity of light (c). In Earth materials, the velocity varies with wavelength. Higher energy, shorter wavelengths such as blue-violet slow less than longer wavelengths such as red and therefore possess a higher refractive index. As a result blue-violet wavelengths (~400 µm) are refracted more than red wavelengths (~700 µm). As white light, a mix of all wavelengths in the visible spectrum, is refracted through a crystal, the light is dispersed into separate wavelengths that travel different paths due to different amounts of refraction. This phenomenon, in which light is separated into its component wavelengths, is called **dispersion**. Good examples are the dispersion of light by raindrops to produce a rainbow or the dispersion of light by a prism (Figure 6.5).

Diaphaneity and color

Some materials, including many native metals, metallic sulfides and metallic oxides, are opaque. **Opaque** substances do not transmit light, even through very thin samples. Instead, light is absorbed as photons interact with electrons. When the energy of a photon is the energy required to excite an electron and cause it to "jump" to a higher quantum level (energy shell), the photon is absorbed as its energy is transferred to the electron. If all wavelengths (photons) of light are absorbed, no light is transmitted and the substance is opaque.

Translucent and transparent substances transmit some light. Many such substances transmit a distinct **color**. These colors result from the selective absorption and transmission of specific wavelengths (or particles) of light. For example, if the red wavelengths (red photons) are absorbed and the blue and yellow wavelengths are transmitted, the mineral will appear green, as in the mineral malachite. Many minerals possess the property of **pleochroism**, in which color varies with crystallographic orientation. Pleochroism occurs because these minerals selectively absorb and transmit light differently in different crystallographic directions.

Isotropic and anisotropic substances

Isotropic substances transmit light at the same velocity in every mutually perpendicular direction; they possess a single refractive index that is the same in every direction.

Minerals that crystallize in the isometric system (where $a_1 = a_2 = a_3$) and amorphous substances such as fluids and glass are isotropic substances. When plane polarized light enters an isotropic medium in which the distribution of atoms is statistically the same in every mutually perpendicular direction, it changes velocity but passes through the medium as a single light ray that vibrates in the same direction. Certain wavelengths may be selectively absorbed, but the light that is transmitted leaves the mineral as a single light ray vibrating in the same direction as when it passed through the polarizer.

All other crystals are **anisotropic**; they transmit light at different velocities in different directions. They possess a direction of maximum velocity, a direction of minimum velocity and many directions of intermediate velocity. As a result, they possess a minimum refractive index in one direction, a maximum refractive index in a second direction and intermediate refractive indices in between. Minerals that crystallize in the tetragonal, hexagonal, orthorhombic, monoclinic and triclinic systems (where $a \neq c$) are all anisotropic crystals.

Birefringence (B or δ) is the difference between the maximum refractive index (in the slowest direction) and the minimum refractive index (in the fastest direction) in a crystal. It is given by

$$B = RI_{max} - RI_{min} \qquad \text{(equation 6.4)}$$

Birefringence is zero in isotropic materials and has some finite value in all anisotropic crystals that increases with the difference between the maximum and minimum refractive indices. It can be observed and measured directly only in crystals oriented so that both the minimum and maximum refractive index directions are visible.

When plane polarized light enters an anisotropic mineral, the light generally changes significantly. In most orientations, **two light rays** are generated that vibrate perpendicular to each other but generally travel in different directions with different velocities. It is these two different light rays that are responsible for the **double refraction** seen in minerals such as calcite, where two images are transmitted separately along two different ray paths. As the two rays pass through the mineral, the

slow ray (traveling in the high refractive index direction) lags behind the **fast ray** (traveling in the low refractive index direction). The amount (measured in number of wavelengths) by which the slow ray lags behind the fast ray is called the **retardation (Δ)**. The retardation is proportional to the difference between the two refractive indices, the birefringence, and to the distance (d) traveled through the specimen, roughly its thickness. Retardation (Δ) is given by:

$$\Delta = d\,(RI_s - RI_f) = d\,(B) \qquad \text{(equation 6.5)}$$

where d is the distance or thickness, RI_s is the refractive index of the slow ray, RI_f is the refractive index of the fast ray and ($RI_s - RI_f$) is the birefringence. For thin sections where the thickness is relatively constant ($30\,\mu m$), the retardation is largely a function of birefringence.

6.2.2 The petrographic microscope

The light-transmitting microscope used in most optical investigations of rocks, minerals and other Earth materials is called a **petrographic microscope**. It is an essential tool in **petrography**, which involves the careful description of rock compositions, textures and small-scale structures. Figure 6.6 shows a typical petrographic microscope with its major components labeled.

The illuminator (light source)

Our discussion of the components of petrographic microscopes proceeds from the bottom up. All microscopes rest on a flat **base** that contains a **light source** or **illuminator**. The illuminator consists of a power source, an incandescent bulb and a blue filter. Light from the bulb is directed upward by lenses or mirrors and the blue filter passes light that approximates sunlight. Many models have a **rheostat** that permits the intensity of the light source to be adjusted.

The substage assembly

Between the illuminator and the stage, where the sample is placed, is the **substage assembly**. This assembly is designed to modify the light

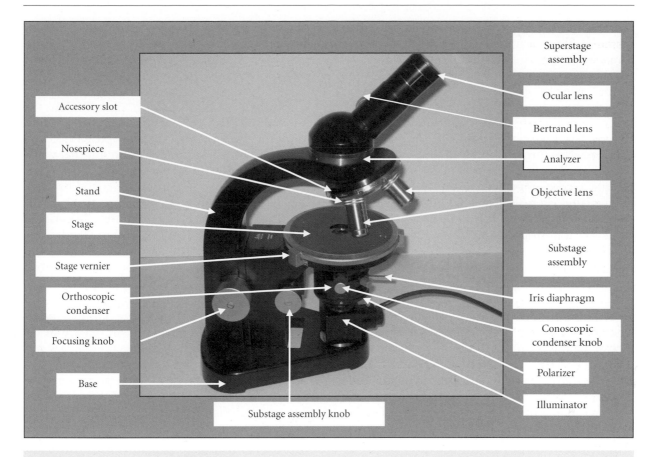

Figure 6.6 A standard petrographic microscope with the major components identified.

from the source that illuminates the sample from below.

The **polarizer** (also called the **lower polar** or **lower Nicol**) is a lens that causes random light from the illuminator to be polarized so that it is passed upward as light vibrating in a single plane. In most microscopes, the polarizer contains a polarizing film that only passes light vibrating in an east–west direction. All other components of light from the illuminator are absorbed. Once light leaves the polarizer on its way to the overlying stage, it is plane polarized light, constrained to vibrate in a single plane.

Above the polarizer are two condensing lenses. The first is a fixed condensing lens, the **orthoscopic condenser**, that focuses light on the area of the sample to be viewed. This lens provides what is called **orthoscopic illumination** and is the illumination under which most samples are initially viewed.

Above the orthoscopic condenser is an auxiliary condensing lens, called the **conoscopic lens** or **condenser**. It is auxiliary in the sense that it can be swung into the optical path of orthoscopic light when needed by means of an attached lever. The conoscopic lens produces a strongly convergent cone of light, **conoscopic illumination**, which is focused on only a tiny area of the sample on the stage. It is used for viewing interference figures in conoscopic mode, as discussed later in the chapter.

The final component of the substage assembly, which is placed between or below the two condensing lenses, is an **iris diaphragm**. The diaphragm can be opened or closed by means of an attached lever to vary the amount of light that reaches the sample. For most work, the diaphragm is kept in the full open position to maximize the amount of available light.

The microscope stage

The freely rotating, circular **stage** possesses a circular hole in the middle through which light from the substage assembly passes. Glass slides with samples are placed over the circular hole. The periphery of the stage is a **stage goniometer** subdivided into 360° for measuring rotation angles of the sample. Some models also possess a **stage vernier** for more precise angular measurements. Although the stage is ordinarily free to rotate, its position can be fixed by tightening a thumb screw on the stage margin. Additional screw holes on the stage permit instruments to be mounted on the microscope stage. These range from stage clips for holding sample slides in place to mechanical stages for counting the proportions of components as they are randomly identified.

The superstage assembly

Between the stage and the eyepiece at the top of the microscope is the **superstage assembly**. At the bottom of this assembly are three (or more) **objective lenses**. These lenses provide initial magnification of the sample image. One of these should always be in the optical path of the microscope. The **low power objective** typically provides 2.5–4× magnification, the **medium power objective** typically provides 8–10× magnification, and the high **power objective** provides 40–60× magnification. The objectives are mounted to a circular **nosepiece** that can be rotated to change the objective lenses. *Always change objectives by holding onto and rotating the nosepiece. Never change objectives by holding on to the objectives.* The latter will inevitably cause the objectives to become uncentered so that samples will not remain in the field of view as the stage is rotated. This will make proper identification impossible. Objective lenses may be recentered using two centering screws, but the process is time consuming for all but the very experienced.

Above the nosepiece on most microscopes is the **accessory slot,** used for the insertion of **accessory plates**, also called **compensator plates**. These plates change the quality of light from the objective lenses in ways that permit sophisticated mineral identifications. A very important component of the superstage assembly is the **analyzer** (also called the **upper polar** or upper Nicol). The analyzer is an auxiliary lens that can be swung into position by a lever. The analyzer lens is similar to the polarizer or lower polar except that its polarization direction is oriented perpendicular to that of the polarizer. Whereas the polarizer (lower polar) typically passes plane polarized light vibrating in an east–west plane, the analyzer (upper polar) only passes light vibrating in a north–south plane. That the upper polar can pass any light at all is testimony to the ability of anisotropic samples to change the vibration direction of the plane polarized light that enters them from the lower polar. When the analyzer is not inserted, the sample is said to be viewed in plane polarized light or **plane light mode**. When the analyzer is inserted, the sample is said to be viewed under **crossed polars mode** (crossed Nicols mode) because the vibration directions of the two polarizers are at right angles. Plane light mode and crossed polars mode are two of the major ways of viewing samples using petrographic microscopes. One can quickly switch back and forth between them by swinging the analyzer in and out of position.

Above the analyzer, commonly in the microscope tube, is the **Bertrand lens**. The Bertrand lens is yet another auxiliary lens that can be swung into the optic path by an attached lever. It is used only for conoscopic mode viewing using the high power objective with both the conoscopic condenser and the analyzer inserted into the optic path. It is in this mode that interference figures, magnified and focused on the eyepiece by the Bertrand lens, may be examined and interpreted by those doing advanced optical work. For most preliminary work, in plane light and under crossed polars, the Bertrand lens is not used.

The eyepiece or ocular head

In all major modes, the sample is directly viewed through the **eyepiece** or **ocular lens**, which fits into the top of the microscope tube. The ocular provides additional image magnification that ranges from 5× to 12× (most commonly 8–10×). The total magnification of the sample is given by the product of the objective magnification and the eyepiece magnification. A sample viewed using a 10× objective and a 10× eyepiece is magnified to 100× its true size. Most eyepieces contain **reticule markings** that include **cross-hairs**. These should ordinarily be oriented north–south and east–west, that is, parallel to the analyzer (upper polar) and to the polarizer (lower polar), respectively. The cross-hairs are used to center a feature of interest in the field of view. They also divide the field of view into quadrants for referencing the location of features relative to the center of the field of view. Some microscopes also contain **eyepiece micrometers**. These are scales that are used to measure distances between, and sizes of, objects in the sample. They must be calibrated with a stage micrometer for each objective and the correction factors recorded, after which the eyepiece micrometer can be used to measure distances and sizes accurately. An adjustable **focusing ring** on the circumference of the ocular can be used to focus the reticule markings. This should be done before starting work. **Stereographic microscopes** have two eyepieces, one of which has an adjustable focusing ring. **Trinocular microscopes** possess a third tube to which a camera or computer monitor may be attached and through which photomicrographs (microscopic images) and video images may be captured.

The focusing knobs and free working distance

The distance between the objective and the sample on the stage is called the **free working distance** (FWD) and can be changed by moving the stage up or down using knurled **focusing knobs** on the microscope arm. The larger knob is for coarse focusing; the smaller knob, for fine focusing. Moving the stage up decreases the free working distance; moving it down increases the free working distance. The specimen will be in focus at a particular free working distance that depends on the objective used and the eyesight of the investigator.

6.2.3   Modes of optical investigation

Grain mounts

One method for mineral identification involves preparing one or more **grain mounts**. This involves grinding the sample, for example with a mortar and pestle, into small grains that are passed through sieves to produce a sample with grains between 0.075 and 0.105 mm (Nesse, 2000). The finer material can be saved and used for other types of analyses, such as X-ray diffraction. Several dozen grains are randomly scattered onto a glass slide to achieve a variety of orientations, covered with a thin cover slip and then examined under a petrographic microscope. The grains may be mounted to the slide with a mounting medium or cement of known refractive index or may be immersed in an oil of known refractive index. In the **oil immersion method**, a sequence of samples is immersed in oils of different refractive indices. The refractive indices of the mineral are compared to that of the oils, until they have been matched. Identification is then made on the basis of the mineral's refractive indices and other properties. This method is generally time consuming but offers a very useful way to learn the theory and practice of optical mineralogy in an integrated manner.

Thin sections

Thin sections are thin slices of solid materials, typically rocks, mounted to a glass slide. They are the "bread-and-butter" of geological optical investigations. Rock thin sections permit multiple minerals to be magnified and identified and reveal textural relationships as well. Most "ground-up" work with rocks involves examining thin sections at some stage in the research process. Thin sections are prepared by first cutting a small piece or "chip" from a larger specimen, using a rock saw. One flat dimension of the chip must be smaller than the glass slide on which it is to be mounted. A flat side of the chip is smoothed and polished on a grinding wheel and that flat side of the chip is carefully mounted to a glass slide (Figure 6.7a). The traditional mounting medium was Canada balsam with a refractive index of 1.537, but this expensive medium has largely been replaced by epoxies with similar indices of refraction. The mounted chip is then cut on a special rock saw to produce a thin slice of even thickness (Figure 6.7b). The inverted sample is ground to a thickness of approximately 50 μm (0.05 mm) prior to a final polishing that produces a second smooth surface on a thin section of rock whose thickness should be 30 μm (0.03 mm). Any dyes that stain minerals for the purposes of easier identification are applied at this stage. A thin cover slip is mounted atop the thin section using epoxy (Figure 6.7c). The thin section is now ready for analysis. Preparing a good thin section

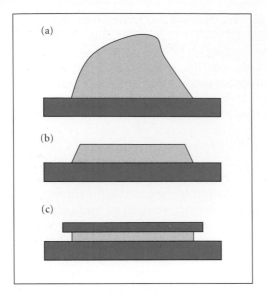

Figure 6.7   The essential steps in thin section preparation. (a) A chip with a flat surface is cut from the specimen, ground, polished and mounted to a slide. (b) The chip is cut and ground to a constant thickness. (c) The chip is polished to a thickness of 30 μm and covered with a thin cover slip.

with a constant thickness of 30 μm is a skill not mastered by everyone. Many commercial enterprises are available to prepare thin sections to order at a price.

Both grain mounts and thin sections can be examined in three principle modes using the petrographic microscope. These are, in order of increasing complexity: (1) plane polarized light (plane light) mode, (2) crossed polars (crossed Nicols) mode, and (3) conoscopic mode. Each reveals information about how the minerals in the sample transmit light and permit progressively more accurate mineral identification. The first two are especially useful modes for investigating rock textures and microstructures.

Plane polarized light mode

The initial investigation of most thin sections involves a combination of plane polarized light (plane light) mode and crossed polars (crossed Nicols) mode using the low and medium power objectives. Plane light investigations reveal how the sample transmits plane polarized light from the polarizer (lower

polar) when none of the auxiliary lenses, including the analyzer (upper polar), conoscopic lens and Bertrand lens, have been inserted into the optical path. What is observed through the eyepiece is plane polarized light that may have changed as it passed through the sample. Plane light mode is the preferred mode for observing (1) diaphaneity, (2) color, (3) pleochroism, (4) cleavage and fracture, (5) relief and Becke lines (relative refractive indices), and (6) some inclusions and alteration products. This combination of plane light properties is a starting point for distinguishing between different minerals. Plane light properties of some common minerals are summarized in Table 6.1.

Because they do not transmit light, **opaque minerals** can be recognized in plane light mode by the fact that they appear black in all orientations. Because they reflect light differently, opaque minerals can be distinguished from one another using a **reflecting light microscope** in which the specimen is illuminated from above by a strong light source. **Non-opaque minerals** transmit some light and so appear white or colored. The **transmitted light color** of minerals is best observed under plane light and depends on the wavelengths of visible light that are transmitted through the specimen. Common non-ferromagnesian rock-forming silicates such as quartz, feldspars and muscovite are colorless in plane light because all wavelengths are transmitted and peak transmission is in the middle of the visible spectrum. Most ferromagnesian rock-forming silicates, however, are colored to some degree because of the tendency of iron to absorb and transmit light selectively. Isotropic minerals transmit and absorb light equally in all directions and have constant colors. Anisotropic minerals tend to absorb light differently in different crystallographic directions and often display **pleochroism**, colors that change as the stage is rotated. Biotite shows pleochroism in shades of brown, yellow, red-brown and/or green, whereas hornblende is characterized by green to brown pleochroism in most orientations.

Cleavage and fracture are observed best in plane light mode (Figure 6.8). They are most clearly seen when the iris diaphragm is partially closed to reduce the light and increase contrast. **Cleavage** appears as sets of parallel fractures. Where the cleavage is perfect, the

Table 6.1 Plane light and crossed polars properties of some common rock-forming minerals.

Mineral	Mineral plane light properties			Crossed polars properties		
	Color and pleochoism	Relief	Cleavage	Birefringence	Extinction	Twinning and others
Quartz	Colorless; clear	Low	None	Low	Often undulatory	Clear; little alteration
Orthoclase	Colorless; "dusty"	Low	Two near 90°	Low		Carlsbad twins, if any
Microcline	Colorless; "dusty"	Low	Two near 90°	Low		Gridiron or scotch-plaid twins
Plagioclase	Colorless; "dusty"	Low	Two near 90°	Low	Albite twins alternate	Albite, Carlsbad, or pericline twins
Muscovite	Colorless	Moderate	One	Moderate to high	Almost parallel to cleavage	Birdseye extinction
Biotite	(Red) brown to yellow; some green	Moderate	One	Moderate to high	Almost parallel to cleavage	Birdseye extinction
Chlorite	Colorless to light green	Moderate	One	Low	Almost parallel to cleavage	Anomalous blue-violet interference colors
Hornblende	Brown to green	Moderate	Two not 90°	Low to moderate	Large angles	Some simple paired twins
Augite	Pale green	Moderate	Two near 90°	Moderate	Large angles	Some simple paired twins
Enstatite/ hypersthene	Pale pink or pale yellow	Moderate	Two near 90°	Low to barely moderate	Parallel or symmetrical in some orientations	
Olivine	Colorless	Moderate to high	None	Moderate to high		Curved fractures; alteration
Calcite/ dolomite	Colorless	Low (−) to moderate (+)	Three not 90°	Very high	Symmetrical to cleavage	Mottled extinction
Gypsum	Colorless	Low	One perfect	Low		
Garnet	Pale pink to pale red	High	None	None; isotropic	In all positions	Equant dodecaheral crystals
Staurolite	Colorless to yellow	High	Poor	Low to barely moderate	Parallel to long axis	

parallel fractures are easy to recognize, as are their intersection angles in some orientations. Where the cleavage is less perfect, the ruptures may only extend over small parts of the mineral, and careful observation under medium to high power is required to recognize their existence. It should be noted that a cleavage will not be visible when it is subparallel to the stage.

Another property that is clearly visible in plane light is relief. **Relief** results from the difference between the refractive index of a substance and the refractive index of the mounting medium or adjacent grains. It is seen under plane light as the sharpness or boldness of the grain boundary against the adjacent medium (Figure 6.8). The higher the relief is, the bolder the grain outline and

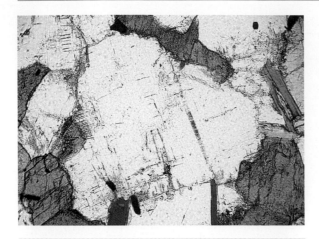

Figure 6.8 Plane polarized light photomicrograph of plutonic igneous rock (granitoid). Gray-green crystals with relatively high relief and two cleavages not at right angles are hornblende. Slightly dusty, colorless crystals with two cleavages at right angles and low relief are plagioclase. The light to darker brown crystals, with moderate relief and one cleavage are biotite. (Photo courtesy of Kurt Hollacher, Union College.) (For color version, see Plate 6.8, between pp. 248 and 249.)

the larger the difference in refractive index. When the refractive index difference between the mineral and its surroundings approaches zero, the relief becomes so low that the mineral boundaries are practically invisible. Among common rock-forming minerals, quartz and feldspars generally possess low relief because their refractive indices approach that of common mounting media (~1.537). On the other hand, minerals such as olivine, garnet and zircon possess much higher relief because their refractive indices are much higher than those of the mounting medium. Some minerals (e.g., calcite and dolomite) display relief that varies from high to low as the stage is rotated, producing a "twinkling" effect because in some orientations the refractive index is much higher than the mounting medium, whereas in others it is not. Table 6.1 also describes the relief of common rock-forming minerals.

Crystals may possess **positive relief**, in which case the crystal possesses a higher refractive index than the mounting medium, or **negative relief**, in which case its refractive index is lower. The sign of the relief can be determined by observing the behavior of **Becke lines**. Becke lines are bright lines produced by the refraction and dispersion of light as it crosses the boundary between the crystal and its surroundings. In such cases, light is refracted toward the slower medium with the higher refractive index, which generates a bright line near the boundary named after its discoverer. Becke lines are best observed by focusing on the grain boundary under medium to high power with the iris diaphragm partially closed down. As the working distance is very slowly increased, the Becke line will appear to move toward the medium with the highest refractive index. If the Becke line moves toward the mineral, the relief is positive. If it moves toward the adjacent medium, the relief is negative. When the refractive index of a mineral closely matches that of the mounting or immersion medium, its relief is extremely low and two Becke lines appear due to dispersion – a blue line that moves toward the medium of slightly higher refractive index and a yellow-orange line that moves away from it. This is very useful in determining the refractive indices by matching the refractive index of grain mounts to that of the mounting medium or oil in which it is immersed.

Inclusions and alteration products are best observed using a combination of plane light and crossed polars modes. Because they affect the appearance of minerals in plane light, they are introduced here. **Fluid inclusions** are commonly incorporated into many crystals as they grow, especially in crystals formed in the presence of meteoric, hydrothermal, magmatic or metasomatic fluids (Chapter 19). These inclusions tend to scatter light and make the mineral less transparent than it would otherwise be. **Solid inclusions** represent smaller mineral grains incorporated into larger crystals as they grew. **Alteration products** involve the partial or complete alteration of the earlier crystal into new materials. Generalizations are dangerous, but in many cases alteration causes altered minerals to be less transparent in plane light than are unaltered ones. For example, most quartz appears quite clear (hydrothermal vein quartz with many fluid inclusions is an exception) under plane light because quartz is generally quite resistant to alteration. One the other hand, feldspars, especially plagioclases, are more susceptible to alteration, which makes them appear "dusty" or "dirty" under plane light. When scanning colorless, low relief minerals in plane light, the combination of cleavage and alteration can allow one to distinguish feldspars from relatively unaltered quartz with its lack of cleavage. These preliminary ideas can be tested using crossed polars techniques.

To summarize, plane light investigations allow students of Earth materials to make preliminary distinctions between minerals based on combinations of diaphaneity, color, pleochroism, cleavage, fracture, relief, inclusions and alteration products (see Table 6.1). Many of these preliminary identifications can be further refined and often confirmed in crossed polars mode.

Crossed polars mode

The transition from plane light to crossed polars mode is accomplished by rotating the upper polar or analyzer into the optic path by means of a lever. Once this is done the light that reaches the eyepiece must pass through both the polarizer (lower polar), which polarizes light from the illuminator in an east–west direction, and the analyzer, which resolves components of light from the specimen that vibrate in a north–south direction. The two polarizers are oriented perpendicular to each other and the sample is being viewed under crossed polars. It is routine for an investigator to switch back and forth between plane light and crossed polars modes by alternately retracting and inserting the analyzer lens. This allows one to rapidly catalog both the plane light and crossed polars properties of the crystals being examined. Low and medium power objectives are suitable for most work. As Table 6.1 makes clear, this approach is often sufficient to distinguish between all the major minerals in a rock while also characterizing its textures and microstructures.

When plane polarized light enters an **isotropic medium** it slows down but passes through the medium as a single light ray that vibrates in the direction of polarization. Certain wavelengths may be selectively absorbed, but the light that is transmitted leaves the mineral as a single light ray vibrating in the same east–west direction as when it passed through the polarizer. Since there is only one polarized ray and no birefringence, no retardation occurs.

However, when plane polarized light enters an **anisotropic medium** in which the crystal structure varies with direction it is split into two light rays that vibrate perpendicular to each other but generally travel in different directions with different velocities. As the two rays pass through the mineral, the slow ray, traveling in the high refractive index direction, lags behind the fast ray, traveling in the low refractive index direction. This causes retardation to occur between the two waves. The amount of retardation is proportional to the difference between the two refractive indices, that is, the birefringence and to the distance traveled through the specimen, roughly the thickness. For thin sections where the thickness is relatively constant, the retardation is largely a function of birefringence.

When the fast ray and later the slow ray leave the mineral, they head toward the analyzer with whatever retardation they had when they left the crystal (no further retardation occurs in air, an isotropic medium). When they enter the analyzer, of necessity the two vibration directions are resolved so that only their north–south components are transmitted through the polarizer. Figure 6.9 summarizes this process.

For some wavelengths and retardations of visible light, the fast ray and the slow ray are canceled destructively by the analyzer and are not transmitted to the eyepiece. For other wavelengths and retardations of visible light, the two rays are added constructively, so that light passes through the analyzer and upward toward the eyepiece. These are seen as colors under crossed polars that, because they result from the interference of the two waves, are called **interference colors**. Several orders of interference colors can be produced, the number increasing with retardation, essentially with distance (specimen thickness) and birefringence. For a more detailed treatment of the theory behind these phenomena, the reader is referred to books by Nesse (2000, 2004). The relationships among interference colors, thickness and birefringence are summarized on a color chart such as that produced by Michel–Levy (Figure 6.10).

The orders of interference colors increase with retardation. If the thickness of the grain is closely known, as it is for well-prepared

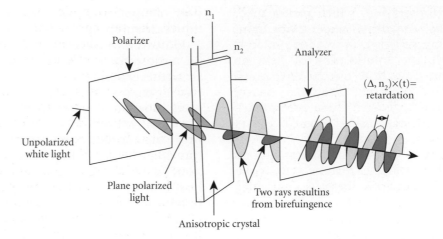

Figure 6.9   Plane polarized light from the polarizer is split into a fast ray and a slow ray by anisotropic (birefringent) crystals which are then resolved by the analyzer to produce a new polarized ray whose properties depend on the retardation of the two rays.

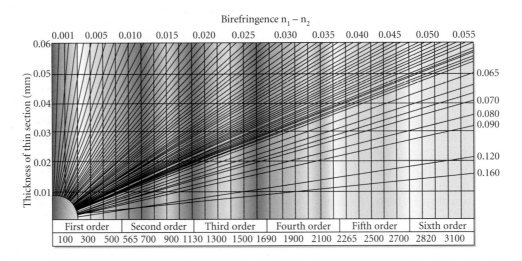

Figure 6.10   A modified version of the Michel–Levy color chart for interference colors viewed under crossed polars. The orders of color increase to the right as hue intensities decrease. (Courtesy of Olympus Microscopy.) (For color version, see Plate 6.10, between pp. 248 and 249.)

thin sections, the interference colors can be predicted from the birefringence or the birefringence can be estimated from the interference colors. Let us use the color chart by reading across the horizontal line that represents a thin section thickness of $30\,\mu m$ (0.03 mm). Lines that represent birefringence values radiate from the origin and birefringence values are labeled at the end of each line on the margins of the diagram. Isotropic minerals and glass that have zero birefringence produce no retardation and will appear dark

under crossed polars. Minerals with low birefringence values ($B = n_f - n_s < 0.10$) will show first order colors that range from grays and whites through pale yellow. Minerals with moderate birefringence ($B = 0.10$–0.30) will display second order colors, including brighter blues, yellows and purples. In practice, one must examine several grains to determine birefringence because the maximum birefringence and interference colors for such minerals will be observed only when both the fast and slow ray vibration directions are parallel

Figure 6.11 Photomicrograph of a thin section of gabbro viewed in crossed polars mode. The plagioclase is striped and has low birefringence (low first order white and gray interference colors). Brightly colored augite has moderate–high birefringence (well-developed second order interference colors in shades of blue, purple, red and orange). (Photo courtesy of Siim Sepp.) (For color version, see Plate 6.11, between pp. 248 and 249.)

to the stage. Minerals with higher birefringence (B > 0.30) may display third and fourth order colors. These can be recognized by their paler appearance and less intense hues. Among common rock-forming minerals, garnet is isotropic while quartz, feldspars and chlorite possess low birefringence. Orthopyroxenes, hornblende and most clinopyroxenes have moderate birefringence, and biotite, muscovite and olivine have relatively high birefringence. Figure 6.11 depicts several minerals under crossed polars.

There are specific directions in anisotropic minerals where only one ray travels through the mineral. These directions are called the **optic axes** of the mineral. For light waves traveling in the direction of the optic axes (and vibrating perpendicular to them), the refractive index is constant. Minerals in the tetragonal and hexagonal systems are **uniaxial minerals**. They have a single optic axis parallel to the c-axis. A ray traveling along the c-axis vibrates perpendicular to it and therefore encounters the same crystal structure ($a_1 = a_2$) with the same refractive index. Minerals in the orthorhombic, monoclinic and triclinic systems are **biaxial minerals**.

They have two optic axes – directions in which the ray encounters the same crystal structure with the same refractive indices. This topic is developed more fully in the section that follows.

Another important property observed under crossed polars is **extinction**. Whenever a mineral appears dark under crossed polars, no light is being transmitted to the eyepiece, all light from the illuminator has been extinguished and the mineral is said to be at extinction. Under crossed polars, **isotropic substances** remain in the extinction position through an entire 360° rotation of the stage. This occurs because east–west vibrating plane polarized light produced by the polarizer (lower polar) passes through the mineral with the same vibration direction and is completely extinguished by the analyzer (upper polar), which only passes light that has some north–south component of vibration. Since no light passes the analyzer, the mineral remains in extinction in all orientations. The quickest way to recognize a non-opaque isometric mineral or glass is this property, where all crystals remain in extinction as the stage is rotated.

There is one situation in which **anisotropic crystals** mimic the behavior of isotropic materials. When anisotropic crystals are viewed looking down an optic axis (with the optic axis perpendicular to the stage), only one ray is viewed moving up the optic axis, vibrating east–west, and the mineral remains at extinction through a complete 360° rotation because the analyzer cannot pass east–west vibrating light. On the other hand, in all other orientations, anisotropic minerals do not stay in the extinction position. Instead they go to extinction every 90° or four times during a 360° rotation of the stage. The extinction positions correspond to those crystal orientations where the vibration directions of the fast ray and the slow ray correspond to the orientation of the analyzer and the polarizer (upper and low polars) so that no light is transmitted to the eyepiece (Figure 6.12).

When anisotropic minerals are rotated to 45° from extinction, they display maximum brightness because they are in a position that maximizes the amount of light that passes through the analyzer as the two rays recombine.

Extinction angles are angles between extinction positions and mineral properties such as

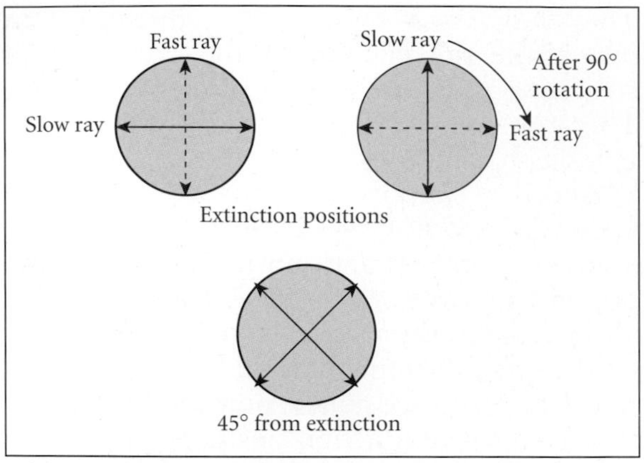

Figure 6.12 Diagram showing the correspondence between extinction positions and the vibration directions of the upper and lower polars (analyzer and polarizer lenses), which, in turn, are parallel to the cross-hairs when they are oriented N–S and E–W. At 45° from the extinction position, the vibration directions of the two rays are oriented NE–SW and NW–SE, respectively.

a cleavage, crystallographic axis, long axis or twin plane. To measure an extinction angle, the specimen is rotated until the property in question, for example, a prominent cleavage, is parallel to the north–south cross-hair. Its position, read from the stage goniometer, is recorded. The grain is then slowly rotated clockwise until it goes to maximum extinction. The new position is read from the stage goniometer, and the difference between the two angles recorded is the extinction angle.

Parallel extinction occurs when the extinction angle is zero; the mineral is at extinction when a cleavage, long dimension or other feature is parallel to the cross-hairs. Micas, such as biotite and muscovite, have one prominent cleavage with very small extinction angles (<5°); their extinction is commonly described as near-parallel to cleavage. Quartz crystals are commonly elongated parallel to the c-axis (the vibration direction of the slow ray); they show extinction parallel to the long dimension of such crystals. Micas such as muscovite and biotite and carbonates such as calcite and dolomite display an unusual extinction phenomenon, variously referred to as **mottled** or **birdseye extinction**. When in the maximum extinction position, many small points of light can be seen. Such extinction is

characteristic of these minerals (see Table 6.1). **Symmetrical extinction** occurs when a mineral goes to extinction when a feature such as a pair of cleavages or crystal faces is bisected into symmetrical halves by the cross-hair. Triclinic minerals with their low symmetry never display symmetrical extinction. **Uniform extinction** occurs when the entire crystal goes to extinction at one time. **Non-uniform** extinction occurs when different parts of a crystal go to extinction at different times. The major causes of non-uniform extinction are (1) twinning of crystals, (2) chemical zoning in crystals, and (3) strain distortion of crystals.

Twinned crystals are easily recognized under crossed polars. Because each twin component of the larger crystal possesses a different crystallographic orientation, each goes to extinction at a different stage of rotation; each has its own extinction angle. Simple **composition twins** are common in potassium feldspars such as orthoclase and sanadine and in ferromagnesian minerals such as hornblende and augite. The twinning can be observed only in thin sections cut across the twin plane, and many crystals of these minerals are not twinned. In those that are, as the stage is rotated, one twin goes to extinction while the other does not; as the stage is rotated further, their appearances are reversed as the second twin goes to extinction (Figure 6.13).

Microcline can generally be distinguished from other potassium feldspars such as sanidine and orthoclase by its complex twinning. The twins appear as spindle-shaped crystals that are intergrown at a high angle, giving rise to what is variously called **gridiron** or **scotch-plaid twinning** (Figure 6.14).

Plagioclase displays a variety of types of twining. The most prominent in many instances is **albite twinning**. Albite twins consist of crystals that possess one of two orientations that alternate with one another (Figure 6.13), providing the striations seen on some plagioclase cleavage surfaces. As the stage is rotated, one set of alternating crystals goes to extinction at one time; the second set goes to extinction at another time. The angles at which the two extinctions occur can be used to closely estimate the anorthite (An) content of the plagioclase. Albite twins sometimes occur with **Carlsbad twinning**, which consists of simple paired composition twins like those in sanidine or orthoclase. Spindle-

Figure 6.13 A sanidine crystal (clear, low birefringence) showing a simple composition (Carlsbad) twin crystal with the lower twin at extinction and the upper twin not at extinction. (Photo courtesy of Kurt Hollacher, Union College.) (For color version, see Plate 6.13, between pp. 248 and 249.)

Figure 6.15 Crossed polars photomicrograph of diorite (see Figure 6.8 for a plane light view) showing a plagioclase crystal in the center (low birefringence, "stripes" due to alternate extinction of albite twins with pericline twins at nearly right angles), hornblende (higher relief, moderate birefringence at lower left and right corners, two sets of cleavage not at right angles in the left-hand crystal) and biotite (thin, greenish crystal with moderately high birefringence and one cleavage orientation). (Photo courtesy of Kurt Hollacher, Union College.) (For color version, see Plate 6.15, between pp. 248 and 249.)

Figure 6.14 Crossed polars photomicrograph of a microcline crystal, displaying the gridiron twinning and low birefringence (first order gray and white colors) that make this mineral easy to recognize under crossed polars. (Photo courtesy of Kurt Hollacher, Union College.) (For color version, see Plate 6.14, between pp. 248 and 249.)

shaped **pericline twins** also occur, usually at an angle to the other twin types (Figure 6.15).

Minerals that are part of a solid solution series may change composition as they grow. In such cases, the crystals may display **chemical zoning** in which their composition changes from the early-formed inside of the crystal toward the later-formed margins of the crystal. Plagioclases and clinopyroxenes, especially in volcanic and shallow plutonic environments, are commonly zoned (Figure 6.16), as is garnet in many metamorphic rocks.

Non-uniform extinction can result from the deformation of crystals associated with strain. This distorts the crystal structure in a manner that causes different parts of the crystal to go to extinction at different times as the stage is rotated. Strained quartz (Figure 6.17) and micas crystal show **undulatory extinction** in which the extinction shadow sweeps across the crystal as the stage is rotated through several degrees.

Intergrowths between minerals can be observed most readily under crossed polars. **Perthitic intergrowths** caused by the exsolution of plagioclase from potash-rich feldspars (Chapter 3) as they cool occur as blebs, patches and stringers of sodic plagioclase in a potassium feldspar host. The perthitic intergrowths of plagioclase go to extinction at

Figure 6.16 Crossed polars photomicrograph of a volcanic rock showing a large plagioclase phenocryst (low birefringence, simple Carlsbad twins, narrower albite twins and chemical zoning shown by the shadows which indicate that the darker parts are near extinction and the lighter part farther from the extinction position). The phenocryst is set in a fine-ground mass of microscopic crystals and some cryptocrystalline or glassy material. (Photo courtesy of Kurt Hollacher, Union College.) (For color version, see Plate 6.16, between pp. 248 and 249.)

Figure 6.17 Crossed polars photomicrograph of a quartz crystal near the extinction position. Strain has produced non-uniform, undulatory extinction, so that the lower and upper parts of the crystal are at extinction while the central parts are not. The extinction shadow would sweep across the crystal toward the center as the stage is rotated. (Photo courtesy of Kurt Hollacher, Union College.) (For color version, see Plate 6.17, between pp. 248 and 249.)

different times than does the host potassium feldspar (Figure 6.18). Similar exsolution intergrowths between orthopyroxene and clinopyroxene are common and produce parallel intergrowths called **Schiller structures**.

Table 6.1 summarizes the optical characteristics of some common rock-forming minerals in plane light and crossed Nicols modes.

The more advanced conoscopic mode techniques discussed in the following section are not required for most of the preliminary work with thin sections, including preliminary rock identification and interpretation. What they do provide are multiple methods for testing one's hypothetical identifications and for making more specific and accurate identifications.

Conoscopic mode

A third important mode available to investigators of Earth materials using petrographic microscopes is **conoscopic mode**. In this mode the sample is viewed by (1) using the high

Figure 6.18 Crossed polars photomicrograph of perthite. Exsolved plagioclase blebs and stringers trend upper right to lower left across the potassium feldspar crystal. Note the low birefringence and first order colors characteristic of both feldspars. (Photo courtesy of Kurt Hollacher, Union College.) (For color version, see Plate 6.18, between pp. 248 and 249.)

power objective, with (2) the auxiliary condensing lens (conoscopic lens) in the substage assembly, and (3) the Bertrand lens in the superstage assembly rotated into the optical path. The conoscopic lens receives plane polarized light that has passed though the fixed condensing lens and focuses it as a cone of light (thus conoscopic lens) onto a very small portion of the sample. The light that passes through the sample is no longer orthoscopic light, but conoscopic light, which behaves in a rather different manner. One significant difference is that because a cone of light strikes the sample, most of the light strikes the sample at an angle less than 90°. Several new features are observed as conoscopic light interacts with the sample, the most significant of which are shadow-like figures called **interference figures**. Differential retardation of cones of light also produce rainbow-like colored lines, often of multiple orders, called **isochromatic lines** or **isochromes**. For thin sections, the number of orders of isochromatic lines is directly proportional to birefringence. The interpretation of interference figures and isochromes permits hypothetical identifications based on plane light and crossed polars characteristics to be tested and confirmed, modified or rejected. These interpretations commonly involve the use of accessory plates that are placed into the accessory slot just above the objective nosepiece.

Using conoscopic mode techniques, investigators can quickly determine whether an anisotropic material is uniaxial (tetragonal or hexagonal systems) or biaxial (orthorhombic, monoclinic or triclinic systems). With a bit more time and effort, one can determine which of the two rays is the fast ray and which is the slow ray. This allows one to determine whether the mineral is optically positive or optically negative. For biaxial minerals, the acute angle between the two optic axes (2V) can be closely estimated. The primary goals of conoscopic mode investigations are to improve the precision and accuracy of mineral identifications by determining (1) whether a crystal is uniaxial or biaxial, (2) whether it is optically positive or optically negative, and, for biaxial minerals, (3) the angle 2V between the two optic axes. Armed with knowledge of the plane light characteristics, the crossed polars characteris-

tics, the conoscopic mode characteristics and a good reference book (Nesse, 2004; Philpotts, 1989; Kerr, 1977; Heinrich, 1965), an experienced investigator can accurately identify most minerals. Determinative tables for the identification of rock-forming minerals optical techniques are available in downloadable form from the website that supports this text.

6.3 THE OPTICAL INDICATRIX, INTERFERENCE FIGURES AND OPTIC SIGN DETERMINATION

6.3.1 The optical indicatrix

An **optical indicatrix** is a three-dimensional geometric figure that is used to represent the refractive index in any direction in a crystal. The insights gained from understanding an optical indicatrix are absolutely essential for a full understanding of optical microscopy. The optical indicatrix is a sphere for isotropic materials and an ellipsoid for anisotropic materials. Radial distances from the center of the indicatrix to its margins are proportional to the refractive index of light that vibrates in that direction. Since ray paths are perpendicular to the direction of vibration, the ray paths associated with a particular refractive index are tangent to the surface of the indicatrix at the point where the radial distance representing the refractive index intersects it. A two-dimensional representation of these relationships is shown in Figure 6.19 in a central section through a typical indicatrix. Note that the semi-major axis of the ellipse corresponds to the higher refractive index and therefore to the vibration direction of the slow ray, and that the semi-minor axis of the ellipse corresponds to the lower refractive index and therefore to the vibration direction of the fast ray. Remember: light waves vibrate perpendicular to their propagation direction.

6.3.2 The isotropic indicatrix

The **isotropic indicatrix** is a sphere – an indicatrix of constant radius. This is because isotropic media have the same refractive index in every direction. The velocity of light moving through an isotropic medium is the same in every direction. Every central section through the isotropic indicatrix is a circular section

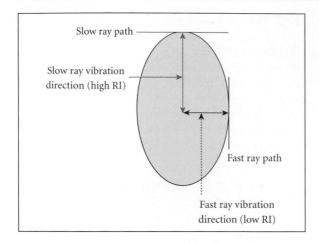

Figure 6.19   A central cross-section through a typical ellipsoidal indicatrix showing the relationship between the long axis (with higher RI) of the elliptical cross-section and the slow ray vibration direction and the short axis (with lower RI) of the elliptical cross-section and the fast ray vibration direction.

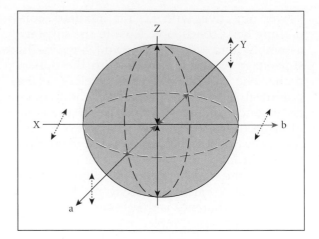

Figure 6.20   An isotropic indicatrix showing sample ray paths "a" (parallel to the Y-axis) and "b" (parallel to the X-axis) with vibration directions perpendicular to the ray paths. Two sample circular sections that contain the potential vibration directions for ordinary light traveling along "a" (YZ section) and "b" (XY section) are shown, as are the potential vibration directions (double arrows) for plane polarized light.

(Figure 6.20). No matter what the direction of the ray path, its velocity and inversely proportional refractive index are invariant. There is only one refractive index "omega" (n_ω) for any isotropic material.

When light enters an isotropic medium, no double refraction occurs; it does not split into two rays. This explains why (1) as plane polarized light enters an isotropic medium, it continues to vibrate in the same plane as it passes, and (2) all light that leaves the isotropic medium is extinguished by the analyzer so that isotropic materials remain at extinction under crossed polars in all positions of the stage.

6.3.3   The uniaxial indicatrix

A **uniaxial indicatrix** is an ellipsoid of revolution about the crystal's c-axis that represents the refractive index for waves vibrating in any direction within the crystal. Radial distances from the center of the ellipsoid to its surface are proportional to the refractive index for waves vibrating in that direction and being propagated at right angles to it. The semi-axis parallel to the c-axis has a length proportional to the refractive index (n) parallel to c and is called epsilon (n_ε). The second semi-axis, perpendicular to the c-axis (and thus in the a_1a_2 plane), has a length proportional to the refractive index (n) perpendicular to c and is called omega (n_ω). The maximum birefringence is given by the absolute value of the difference between the two refractive indices ($B = |n_\varepsilon - n_\omega|$).

A uniaxial indicatrix (ellipsoid of revolution) has a unique circular section that is perpendicular to the c-axis and therefore perpendicular to the epsilon refractive index (n_ε). In standard representations of the uniaxial indicatrix, the unique circular section is horizontal and has a radius equal to the omega refractive index (n_ω).

Elliptical principal sections through the center of the indicatrix that contain the c-axis are perpendicular to the unique circular section. In standard representations, principal sections are vertical and have one semi-axis parallel to the c-axis with a length proportional to the epsilon refractive index (n_ε), and a second semi-axis perpendicular to the c-axis with a length proportional to the omega refractive index (n_ω). If $n_\varepsilon > n_\omega$, the principal

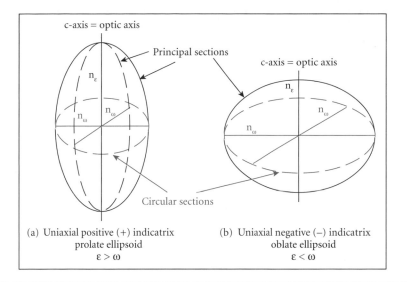

Figure 6.21 (a) Prolate uniaxial indicatrix with a vertical long axis parallel to the c-axis, the optic axis and the epsilon (n_ε) refractive index. These are perpendicular to the two horizontal short axes of the ellipsoid, which are parallel to the omega (n_ω) refractive index. (b) Oblate uniaxial ellipsoid with a vertical short axis parallel to the c-axis, the optic axis and the epsilon (n_ε) refractive index. These are perpendicular to the two equal long axes of the ellipsoid, which are parallel to the omega (n_ω) refractive index.

sections are prolate (extended) ellipses, and the indicatrix is a prolate ellipsoid with one long axis parallel to the c-axis and two equal short axes in the circular section (Figure 6.21a). If $n_\varepsilon < n_\omega$, the principal sections are oblate (flattened) ellipses, and the indicatrix is an oblate ellipsoid with one short axis parallel to the c-axis and two equal long axes in the circular section (Figure 6.21b).

All other sections through the uniaxial indicatrix that are neither parallel nor perpendicular to the c-axis and the epsilon refractive index are called random sections. Random sections have one semi-axis equal to the omega refractive index (n_ω) and a second axis equal to a refractive index called epsilon prime (ε') that possesses a refractive index between that of omega and epsilon (Figure 6.22).

Ordinary and extraordinary rays and the optic sign

A uniaxial indicatrix, with one optic axis, is appropriate for crystals in the tetragonal and hexagonal systems. When light enters such crystals where $a_1 = a_2 \neq c$, in any direction not parallel to the optic axis (c-axis), it doubly

refracts into two rays that move in different directions. Each wave vibrates in a plane perpendicular to its ray path and the two waves vibrate in mutually perpendicular planes.

One ray, called the **ordinary ray (o-ray = omega ray)**, always moves parallel to the c-axis, which is parallel to the crystal's optic axis. It vibrates in directions perpendicular to the c-axis, therefore in the a_1a_2 plane in which atomic configurations are uniform in every direction. As a result, the ordinary wave has a constant velocity and is associated with a constant refractive index (n) called **omega (ω or n_ω)**. Omega is proportional to the radius of the circular section perpendicular to the c-axis and to the optic axis in the uniaxial indicatrix.

The second wave, called the **extraordinary ray (e-ray = epsilon)**, moves in another direction that may be anywhere from perpendicular to nearly parallel to the ordinary wave. Its velocity varies with direction because atomic configurations vary with e-ray direction. The velocity of an extraordinary wave that propagates perpendicular to the ordinary wave is associated with a refractive index called **epsilon (ε or n_ε)**, which is perpendicular to omega (n_ω). The epsilon vibration direction

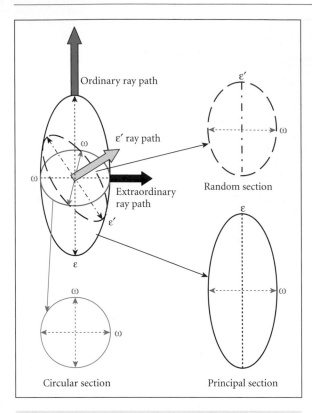

Figure 6.22 Uniaxial positive indicatrix and three major types of section. The unique circular section is perpendicular to the optic axis and ordinary ray path (blue arrow); no birefringence is displayed when such sections are parallel to the stage. When random sections are parallel to the stage, minerals display an intermediate apparent birefringence |ε′ − ω|. Principal sections are viewed when the optic axis and extraordinary ray path (black arrow) are parallel to the stage; the maximum birefringence |ε − ω| is observed in this orientation only.

split into two rays. Instead, it travels through the crystal as an ordinary wave with a constant refractive index equal to the radius of the circular section for light vibrating perpendicular to the optic axis. Crystals viewed looking down the optic axis (c-axis) with the circular section parallel to the stage appear isotropic because only the ordinary wave (omega = ω) exists and birefringence is zero.

On the other hand, **principal sections** (Figure 6.22) are viewed when the optic axis, which is parallel to the extraordinary wave vibration direction, is parallel to the stage. In such orientations, light passing through the crystal refracts into two rays. The ordinary ray (black arrow) moves parallel to the optic axis and vibrates perpendicular to it in the principal section with the refractive index omega (ω). At the same time, the extraordinary wave (blue arrow) moves perpendicular to the optic axis and vibrates parallel to it within the principal section with the refractive index epsilon (ε). Since both the minimum and maximum refractive indexes lie in the plane of principal sections parallel to the stage, the maximum birefringence between the two rays, |ε − ω|, is observed in this orientation only.

When any other **random sections** are parallel to the stage, two rays pass through the crystal. The ordinary ray (black arrow) moves along the optic axis and vibrates in the circular section with the refractive index omega (ω). However, the extraordinary ray (light blue arrow) moves perpendicular to the random section. It vibrates at right angles to the ordinary ray in the random section and has the refractive index epsilon prime (ε′). These sections display an intermediate apparent birefringence |ε′ − ω|.

The **optic sign** of a uniaxial mineral is determined by the relative velocities and associated refractive indices of the ordinary ray and the extraordinary ray. A crystal is **uniaxial positive** (+) if epsilon is more than omega (ε > ω); that is, if epsilon is the slow ray associated with the larger refractive index and omega is the fast ray associated with the smaller refractive index. In such cases the uniaxial indicatrix is a prolate ellipsoid (Figure 6.23). The long axis is labeled Z and is the optic axis parallel to the c-axis. The two short axes are radii of the circular section and so are equal in length and labeled X and Y.

is parallel to the c-axis and is contained in all principal sections through the indicatrix.

Intermediate velocities of epsilon that are associated with other e-ray directions possess intermediate refractive indices referred to as **epsilon prime (ε′ or n$_{ε′}$)** and are associated with random sections through the indicatrix.

Figure 6.22 depicts the three major types of central sections through a uniaxial indicatrix. The unique **circular section** is perpendicular to the optic axis, the c-crystallographic axis and ordinary ray path b (black arrow). When light enters a uniaxial crystal traveling parallel to the optic axis, it does not

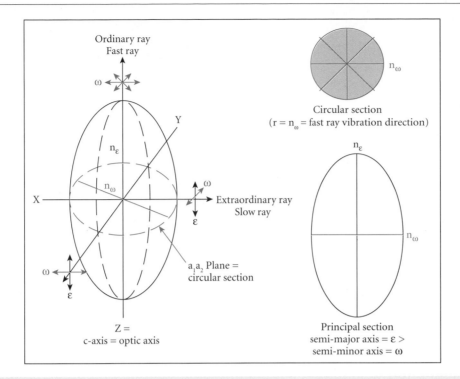

Figure 6.23 Uniaxial positive indicatrix. The crystal c-axis is vertical and parallel to the epsilon vibration direction. The a_1 and a_2 crystallographic axes are horizontal and lie in the unique circular section parallel to the omega (ω) vibration direction. The ordinary ray, traveling parallel to the c-axis and vibrating perpendicular to it, is the fast ray with the lower refractive index (n_ω). The extraordinary ray, traveling perpendicular to the c-axis (e.g., along X or Y) and vibrating parallel to it, is the slow ray with the higher refractive index (n_ε). One of an infinite number of possible prolate elliptical principal sections that contains the vibration directions for both the fast ray (ω) and the slow ray (ε) is shown. Random sections (with ε' and ω) are not shown.

A crystal is **uniaxial negative** (−) if epsilon is less than omega ($\varepsilon < \omega$); that is, if epsilon is the fast ray associated with the lower refractive index and omega is the slow ray associated with the higher refractive index. In such cases the uniaxial indicatrix is an oblate ellipsoid (Figure 6.24). The short axis is labeled Z and is the optic axis parallel to the c-axis. The two long axes are radii of the circular section and so are equal in length and labeled X and Y.

6.3.4 Uniaxial interference figures

Interference figures result from complex interference phenomena that occur between the fast and slow rays as they pass through the sample in conoscopic light and are processed by the analyzer. Detailed treatments of their origin are available in advanced treatments of optical mineralogy (Nesse, 2004; Kerr, 1977). To obtain an interference figure, it is necessary to follow several steps very carefully: (1) the crystal must be carefully focused under the high power objective, being sure not to touch the thin section; (2) the analyzer must be inserted so that the crystal is being viewed under crossed polars; and (3) both the conoscopic lens and the Bertrand lens must be swung into the optical path and focused. The type of interference figure observed will depend on the orientation of the crystal and its indicatrix with respect to the rotating stage.

Uniaxial optic axis figure

The **centered uniaxial optic axis (OA) figure** is observed when the optic axis is perpendicular to the stage so that the circular section (n_ω)

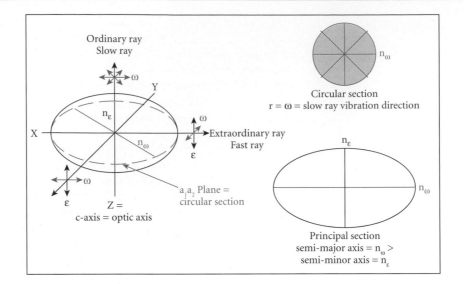

Figure 6.24 Uniaxial negative indicatrix. The crystal c-axis is vertical and parallel to the epsilon (ε) vibration direction. The a_1 and a_2 crystallographic axes are horizontal and lie in the unique circular section parallel to the omega (ω) vibration direction. Note that the ordinary ray, traveling parallel to the c-axis and vibrating perpendicular to it, is the slow ray with the higher refractive index (n_ω). The extraordinary ray, traveling perpendicular to the c-axis (e.g., along X or Y) and vibrating parallel to it, is the fast ray with the lower refractive index (n_ε). One of the many elliptical principal sections that contain the vibration directions for both the fast ray (ω) and the slow ray (ε) is shown. Random sections (with ε′ and ω) are not shown.

is parallel to the stage. To obtain a centered optic axis figure, one should select from a mineral population a crystal that shows minimum birefringence, that is, appears to be nearly isotropic so that it remains dark under crossed polars as the stage is rotated. A centered optic axis figure (Figure 6.25) appears as a cross that consists of two **isogyres**, shadow-like features that are tapered toward the center of the field of view. The center of the area where the two isogyres cross is called the **melatope** and represents the position of the vertical optic axis; the c-axis along which the ordinary ray is traveling. The isogyres are produced by the extinction of conoscopic light as it passes through the analyzer. One isogyre represents light vibrating parallel to the polarizer that does not pass through the analyzer; the other represents light vibrating in the preferred direction of the analyzer that does not pass through the polarizer. These extinctions produce the "uniaxial cross", as the centered optic axis figure is sometimes known (Figure 6.25). Any crystal that dis-

plays a uniaxial cross in conoscopic mode is a uniaxial mineral that crystallized in either the tetragonal or the hexagonal system.

In minerals with reasonably high birefringence ($n_s - n_f > 0.10$), colored lines will appear in concentric circles around the melatope. The order of these lines increases with distance from the melatope, and their hue intensities decrease in a manner consistent with the Michel–Levy color chart (see Figure 6.10). These lines of constant color are called isochromatic lines or isochromes (Figure 6.25). They represent cones of equal retardation produced when conoscopic light passes through the sample. When conoscopic light passes through the sample, light in the center of the cone travels the shortest distance and experiences minimum retardation, whereas light near the margins of the cone travels the longest distance and experiences maximum retardation. Retardation increases outward from the optic axis and is observed as isochromes of increasing order outward from the melatope.

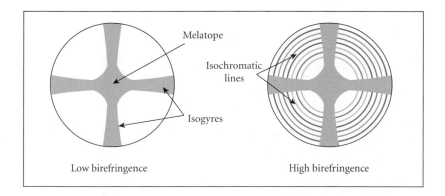

Figure 6.25 Centered optic axis figures for a uniaxial mineral display two isogyres in the form of a cross, the center of which is the trace of the optic axis or melatope. Minerals with low birefringence show first order grays and whites in the quadrants between the isogyres. Minerals with higher birefringence display isochromatic lines concentric about the melatope, whose numbers increase with retardation; that is with birefringence for a sample of constant thickness. (For color version, see Plate 6.25, between pp. 248 and 249.)

Because conoscopic light strikes the sample at an angle that decreases with distance from the center of the cone, not all the light travels up the optic axis, even when the latter is perpendicular to the stage. Therefore, both an ordinary ray and an extraordinary wave are generated by refraction. To determine the optic sign of a uniaxial mineral, one must determine the relative velocities (inverse to refractive indices) of the ordinary and extraordinary rays. If the extraordinary ray is the slow ray so that the epsilon refractive index is more than omega ($\varepsilon > \omega$), the mineral is positive. If the extraordinary ray is the fast ray so that the epsilon refractive index is less than omega ($\varepsilon < \omega$), the mineral is negative. A uniaxial centered optic axis figure (uniaxial cross) is ideal for this determination.

The vibration directions of the ordinary and extraordinary rays are always perpendicular to each other. However, their directions change with orientation. In (+) minerals the ordinary wave always vibrates tangent to the isochromes, whereas the extraordinary ray always vibrates perpendicular to the isochromes. At 45° from the isogyres, the extraordinary ray always bisects the quadrants so that it is oriented from NE to SW in the NE–SW quadrants and from NW to SE in the NW–SE quadrants. In (−) minerals, these relationships are reversed (Figure 6.26).

For minerals of low birefringence, a **gypsum accessory plate** (or 1λ plate) is generally used to make sign determinations. The gypsum plate has a uniform retardation of 550 μm; that is, one order of interference on the Michel–Levy color chart. The gypsum plate is cut parallel to a principal section. Because gypsum is anisotropic, it has a slow ray direction that is marked on the plate and is usually perpendicular to the plate long axis (Figure 6.26), which is oriented from NE to SW when the plate is inserted into the optical path. The vibration direction of the fast ray is perpendicular to that of the slow ray and is generally parallel to the gypsum plate long axis, which is oriented NW–SE. To use the gypsum plate, insert it into the accessory slot. When the slow ray direction of the gypsum plate is parallel to the slow ray direction of the crystal, all colors will increase by one full order, the retardation of the gypsum plate. First order grays and whites plus the retardation of the gypsum plate will add to produce shades of second order blues. When the slow ray direction of the gypsum plate is parallel to the fast ray direction of the crystal, the colors will subtract (partially cancel). First order grays and whites will increase by less than the retardation of the gypsum plate to produce first order yellows (Figure 6.26).

In **uniaxial positive minerals**, the extraordinary ray is the slow ray, its vibration direction epsilon is associated with the highest refractive index, and $\varepsilon > \omega$. If the extraordinary ray associated with the epsilon vibration

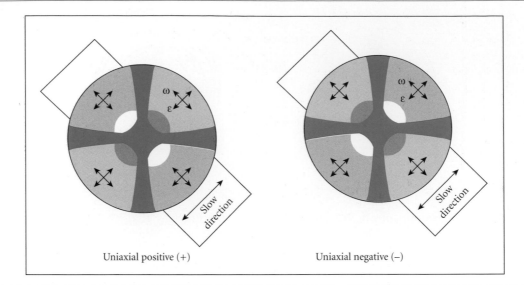

Uniaxial positive (+)   Uniaxial negative (−)

Figure 6.26 Application of the gypsum plate in optic sign determinations of uniaxial minerals. When the gypsum plate is inserted, addition (blues) occurs in the quadrants where the slow vibration direction is parallel to that of the gypsum plate and subtraction (yellows; shown in gray and white in figure) occurs where the slow vibration direction is perpendicular to that of the gypsum plate. Blue in the NE–SW quadrants means that the extraordinary ray is the slow ray associated with a higher refractive index so that $\varepsilon > \omega$ and the mineral is uniaxial positive. Yellow in the NE–SW quadrants indicates that the extraordinary ray is the fast ray associated with a lower refractive index (epsilon) so that $\varepsilon < \omega$ and the mineral is uniaxial negative. (Courtesy of Gregory Finn.) (For color version, see Plate 6.26, between pp. 248 and 249.)

direction is the slow ray, when the gypsum plate is inserted, addition will occur in the NE–SW quadrants where the slow ray vibration direction of the mineral is parallel to the slow ray vibration direction of the gypsum plate, and first order grays and white will turn blue when the plate is inserted. Subtraction will occur in the NW-SE quadrants where the slow ray vibration direction of the mineral is perpendicular to the slow ray vibration direction of the gypsum plate, and first order grays and whites will change to yellows. The dark gray–black isogyres will change to a reddish hue (Figure 6.26).

In **uniaxial negative minerals**, the extraordinary ray is the fast ray, its vibration direction epsilon is associated with the lowest refractive index, and $\varepsilon < \omega$. If the extraordinary ray associated with the epsilon vibration direction is the fast ray, when the gypsum plate is inserted, subtraction will occur in the NE–SW quadrants where the fast ray vibration direction of the mineral is parallel to the slow ray vibration direction of the gypsum

plate, and first order grays and white will turn yellow when the plate is inserted. Addition will occur in the NW–SE quadrants where the slow ray vibration direction of the mineral is perpendicular to the slow ray vibration direction of the gypsum plate, and first order grays and whites will change to blue. The dark gray–black isogyres will change to a reddish hue (Figure 6.26).

If the gypsum plate is used to determine the sign of high birefringent minerals, care must be taken to examine the color changes in those portions of each quadrant immediately adjacent to the melatope where the colors were first order grays or whites prior to the insertion of the gypsum plate. Then the rules about blue from addition and yellow from subtraction hold true. Blue in the NE–SW quadrants indicates a uniaxial positive mineral, and yellow in the NE–SW quadrants signifies a uniaxial negative mineral.

The behavior of isochromatic lines associated with an optic axis figure can be used to determine optic sign. In this case, a **quartz**

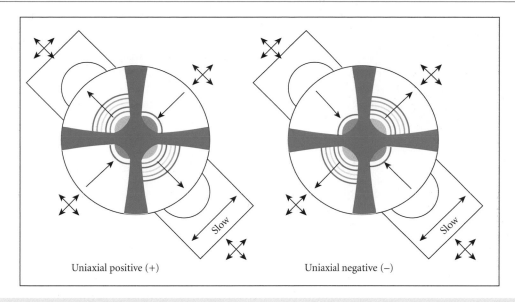

Uniaxial positive (+)   Uniaxial negative (−)

Figure 6.27 Application of the quartz wedge in optic sign determination for minerals with high birefringence and well-developed isochromatic lines. If isochromes move toward the melatope in the NE–SW quadrants and away from the melatope in the NW–SE quadrants, the mineral is positive. If isochromes move away from the melatope in the NE–SW quadrants and toward the melatope in the NW–SE quadrants, the mineral is negative. (Courtesy of Gregory Finn.) (For color version, see Plate 6.27, between pp. 248 and 249.)

wedge compensation plate is slowly inserted into the accessory slot and the movement of the isogyres is observed. As with the gypsum plate, the slow direction of the quartz wedge is NE–SW and the fast direction is NW–SE. Because the wedge is tapered, inserting it slowly increases the retardation and therefore the amount of addition or subtraction that occurs. If addition occurs, the isogyres appear to move inward toward the melatope. This occurs because first order isochromes become higher order isochromes during insertion so that higher order isochromes move closer to the melatope. If subtraction occurs, the isogyres appear to move outward away from the melatope. To summarize, addition in the NE–SW quadrants (isochromes move in) and subtraction in the NW–SE quadrants (isochromes move out) indicates that the mineral is uniaxial positive (Figure 6.27). Subtraction in the NE–SW quadrants (isochromes move out) and addition in the NW–SE quadrants (isochromes move in) indicates that the mineral is uniaxial negative (Figure 6.27).

When the optic axis is inclined more steeply than 60–70° with respect to the stage so that a random section is parallel to the stage, an **off-centered optic axis figure** may be observed. As the stage is rotated, the trace of the optic axis, the melatope, will move in the field of view. As long as the melatope remains in the field of view throughout rotation, quadrants can be recognized. In such cases, the quadrants of addition and subtraction can be determined using an accessory plate and the optic sign can be determined. When the angle of inclination of the optic axis with respect to the stage is less than 60°, the optic axis figure is so off-centered that the melatope leaves the field of view as isogyres sweep through it every 90°. In such cases, it is best to look for another crystal with lower birefringence in order to make an accurate determination of optic sign.

Optic normal or flash figure

When the optic axis lies within the plane of the stage so that a principal section is being viewed, both the omega and epsilon vibration directions are being observed (see Figure

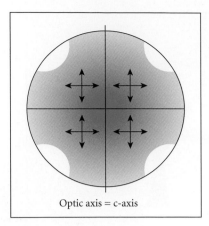

Optic axis = c-axis

Figure 6.28 The orientation of the uniaxial indicatrix that produces uniaxial flash figures and the resulting flash figure observed every 90° during rotation when the optic axis is parallel to one of the polarizing lenses. (Courtesy of Gregory Flinn.)

6.22). In this orientation the crystal will show maximum birefringence under crossed polars. In conoscopic mode, a **uniaxial flash figure** or **optic normal figure** (Figure 6.28) will be observed each time the optic axis is parallel to one of the two polarizing lenses. As the stage is rotated, an interference figure appears every 90°, in the form of a broad cross that nearly fills the field of view. With a few degrees of additional rotation, it breaks up into two curved isogyres that leave the field of view in the quadrants that contain the optic axes. It appears, and then disappears, in a "flash" – now you see it, now you don't – as the stage is rotated through a few degrees. The flash figure appears when both the extraordinary ray vibration direction (parallel to the optic axis) and the ordinary ray vibration direction (perpendicular to the optic axis) are parallel to the cross-hairs. The flash figure is not easy to use to determine the optic sign, but it does alert the viewer to the fact that a principal section is essentially parallel to the stage. In such orientations, both the omega and epsilon vibration directions lie in the plane of the stage. In grain mount investigations with immersion oils, these orientations permit epsilon to be measured and birefringence ($|\varepsilon - \omega|$) to be accurately determined.

6.3.5 The biaxial indicatrix

As their name implies, **biaxial crystals** possess two optic axes. You may recall that all minerals that crystallize in the orthorhombic, monoclinic and triclinic systems where $a \neq b \neq c$ (except incidentally) are biaxial. The optical features of biaxial minerals can be illustrated using a biaxial indicatrix.

A **biaxial indicatrix** (Figure 6.29) is a **triaxial ellipsoid** that represents the refractive indices of a biaxial crystal in every direction. It possesses three mutually perpendicular axes that can be labeled X, Y and Z. **X** is the short axis of the ellipsoid parallel to the vibration direction of the fast ray and associated with the **minimum refractive index called alpha (n_α or α)**. The long axis is the **Z axis** parallel to the vibration direction of the slow ray and is associated with the **maximum refractive index** called gamma (n_γ or γ). The intermediate axis of the ellipsoid is associated with the vibration direction of an intermediate ray and has an **intermediate refractive index called beta (n_β or β)**. For every biaxial mineral, alpha is less than beta which is less than gamma ($\alpha < \beta < \gamma$). The elliptical XY plane is called the **optic plane** (Figure 6.29) because it contains the two optic axes that characterize biaxial minerals. It also contains the vibration directions of the fast ray and slow ray and therefore the maximum (gamma) and minimum (alpha) refractive indices. Biaxial minerals exhibit maximum birefringence ($|\alpha - \gamma|$) when the optic plane is parallel to the stage.

All triaxial ellipsoids such as the biaxial indicatrix contain two circular sections (Figure 6.30). Each of these circular sections has a radius equal to the beta (β) refractive index. Light rays moving perpendicular to the circular sections have vibration directions that lie in the circular sections and that possess the same refractive index in every direction. The directions perpendicular to the two circular sections are the two optic axes. The angle between the two optic axes, measured in the optic plane, is called **optic angle (2V)**. The angle bisected by the X-axis is called 2Vx and the angle bisected by the Z-axis is called 2Vz. If just one angle 2V is given, it is the acute angle between the two optic axes (2Vz in Figure 6.30). As with light moving parallel to the optic axis and perpendicular to the circu-

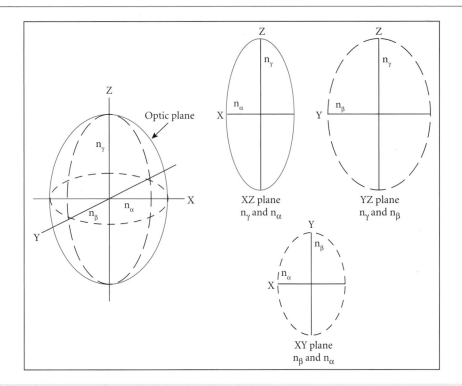

Figure 6.29   The general relationships between the three axes (X, Y and Z) of the biaxial indicatrix and refractive indices (α, β and γ) associated with each axis. The XY plane is called the optic plane because it contains the two optic axes. The three principle planes, each of which contains two semi-axes of the ellipsoid and two refractive indices, are also shown. (Courtesy of Gregory Flinn.)

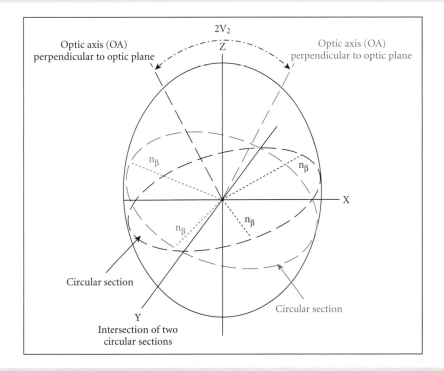

Figure 6.30   Diagram that depicts two circular sections (with refractive index β) perpendicular to which are two optic axes separated by an acute angle (2V) that is contained in the XZ (optic) plane. (Courtesy of Gregory Flinn.)

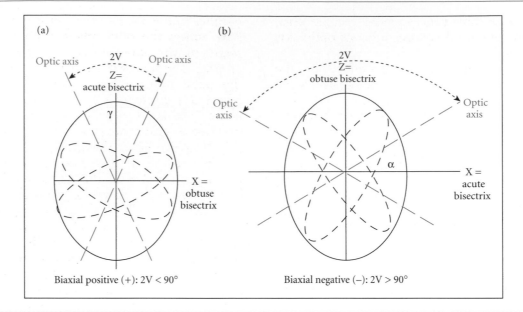

Figure 6.31   Simplified versions of the positive and negative biaxial indicatrix. (a) A positive indicatrix in which the acute bisectrix (Bxa) is Z and the slow ray vibration direction is gamma (γ). (b) A negative indicatrix in which the acute bisectrix (Bxa) is X and the fast ray vibration direction is alpha (α).

lar section in uniaxial minerals, light moving along either of the two optic axes in biaxial minerals does not doubly refract. The apparent birefringence in these two directions is zero. The circular sections are also called **optic normal sections** because they are perpendicular to the two optic axes. Note that intersection of the two circular sections is a line parallel to the Y-axis of the indicatrix. This line is an **optic normal** line parallel to the intermediate ray vibration direction with a refractive index of beta (β). Biaxial minerals exhibit minimum (zero) birefringence when an optic normal (circular) section is parallel to the stage and an optic axis is perpendicular to it.

A line in the optic plane that bisects the acute angle between the optic axes (optic angle or 2V) is called an **acute bisectrix (Bxa)**. Depending on mineral optics, the acute bisectrix may bisect 2Vz or 2Vx. The complementary line that bisects the obtuse angle between the optic axes is the **obtuse bisectrix (Bxo)**. In Figure 6.31a, the acute bisectrix (Bxa) bisects 2Vz and the obtuse bisectrix (Bxo) bisects 2Vx.

Light that enters biaxial minerals in any direction not parallel to one of the optic axes

refracts into two extraordinary rays. Unless the rays are vibrating parallel to one of the axes of the indicatrix, their refractive indices will be given by alpha prime ($n_{\alpha'}$ or α′) and gamma prime (n_{γ} or γ′). Birefringence of such grains will vary from near minimum when the optic axis is almost perpendicular to the stage to near maximum when the circular sections are nearly perpendicular to the stage and the optic section is nearly parallel to it.

The optic sign of biaxial minerals may be either positive or negative (or rarely neutral). Minerals are **biaxial positive (+)** when Z is the acute bisectrix, so that gamma (γ) is the acute bisectrix, so that the slow ray vibration direction (perpendicular to the slow ray path) is the acute bisectrix. The acute bisectrix is 2Vz and the obtuse bisectrix is 2Vx (Figure 6.31). Minerals are **biaxial negative (−)** when these relationships are reversed (Figure 6.31) so that X is the acute bisectrix associated with the alpha (α) refractive index, so that the fast ray vibration direction (perpendicular to the fast ray path) is the acute bisectrix. The acute bisectrix is 2Vx and the obtuse bisectrix is 2Vz. Biaxial minerals are **optically neutral** when their 2V equals 90° so that the acute bisectrix and the obtuse bisectrix are undefined. Note that if 2V were to become zero, the mineral would possess only one optic axis and be uniaxial.

The **optic orientation** is defined by the relationship of the optical indicatrix to the crystallographic axes (Nesse, 2004). For **orthorhombic minerals**, the three mutually perpendicular crystallographic axes (a, b and c) correspond in some way to the axes of the indicatrix X, Y and Z (α, β and γ refractive indices). Which crystallographic axis corresponds to which axis of the indicatrix depends on how the crystallographic axes were originally defined. In the **monoclinic system**, the b crystallographic axis (which is perpendicular to the ac plane) corresponds to one of the axes of the indicatrix (X, Y or Z), but because the a and c crystallographic axes are inclined with respect to each other, they do not correspond to axes of the indicatrix, except by chance. In the triclinic system, where the crystallographic axes do not intersect at right angles, there is no direct correspondence between the crystallographic axes and the axes of the indicatrix, except by chance.

6.3.6 Biaxial interference figures

As with uniaxial interference figures, **biaxial interference figures** are the products of complex extinction of light by the analyzer (upper polarizing lens) when crystals are viewed in conoscopic mode. To obtain any interference figure, it is necessary to follow several steps very carefully: (1) the crystal must be carefully focused under the high power objective, being sure not to touch the sample; (2) the analyzer must be inserted so that the crystal is being viewed under crossed polars; and (3) both the conoscopic lens and the Bertrand lens must be swung into the optical path. The interference figure will change as the stage and the orientations of the fast and slow rays are rotated. The type of interference figure observed will depend on the orientation of the crystal and its indicatrix with respect to the stage.

Acute bisectrix figures

An **acute bisectrix figure (Bxa figure)** is observed when the acute bisectrix is perpendicular to the stage. If the acute bisectrix is perpendicular to the stage, the optic normal parallel to the Y-axis of the bisectrix and the Bxo within the optic plane both lie in the plane of the stage at right angles to one another. As the stage is rotated, the trace of the optic normal (with refractive index β) remains perpendicular to the trace of the optic plane. The acute bisectrix figure consists of two isogyres that change as the stage is rotated (Figure 6.32). When the optic plane is parallel to the east–west cross-hair and the optic normal is parallel to the north–south cross-hair or vice versa, a crude cross centered on the acute bisectrix is observed (Figure 6.32, left). The **thicker isogyre** of the cross is **parallel to the optic normal** and the **thinner isogyre** is **parallel to the trace of the optic plane**. The two melatopes that represent the traces of the two optic axes occur in the optic plane on either side of the Bxa where the thinner isogyre is the thinnest (Figure 6.32, left). Every 90° rotation a cross will form; four times during a complete rotation. The orientations of the thick and thin isogyres will alternate between north–south and east–west every 90° as the orientations of the optic normal and the optic plane rotate through 90°. As the stage is rotated from these positions, the cross begins to break up into two separate curved isogyres (Figure 6.32, center) that move relatively slowly away from each other through 45° of rotation and then begin to converge again to reform the cross after 90°. For a centered Bxa figure, for minerals with an optic angle (2V) of less than 55–60° (<55°), both isogyres remain in the field of view throughout rotation. The maximum separation of the melatopes increases from zero for a 2V of zero (think uniaxial cross where no melatope separation occurs because there is only one optic axis) to progressively larger separations in the 45° position. For optic angles larger than 60°, the two melatopes leave the field of view before the 45° position is reached. In the 45° position (Figure 6.32, right) (1) the two melatopes are at the points of maximum curvature on the two isogyres; (2) the trace of the optic plane is a line through the two melatopes; (3) the Bxa lies on this line, halfway between the two melatopes; (4) the Bxo lies along this line, outside the field of view, on the concave side of the isogyres (other side of the melatopes); and (5) the optic normal is along a line perpendicular to the trace of the optic plane halfway between the two isogyres.

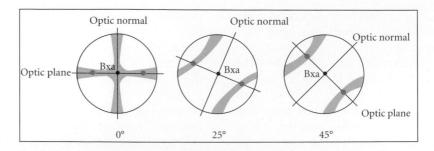

Figure 6.32 An acute bisectrix figure for a biaxial mineral with a moderate 2V. When the optic normal, parallel to the thick isogyre, and the optic plane, parallel to the thin isogyre that contains the two melatopes (blue dots), are parallel to the cross-hairs, a crude cross is formed. As the stage is rotated, the optic normal and optic plane rotate and the cross splits into two curved isogyres with the melatopes located at the points of maximum curvature. At 45° of rotation, the two isogyres reach maximum separation. When the Bxa figure is centered, the Bxa (black dot) remains in the center of the field of view during rotation. (Courtesy of Gregory Finn.)

Recall that crystals are biaxial positive when Z is the acute bisectrix, so that gamma (γ) is the acute bisectrix, and the slow ray vibration direction is the acute bisectrix. In such cases the acute bisectrix is 2Vz and the XY plane is parallel to the stage. Between the melatopes the optic plane trace (parallel to the X vibration direction) is the fast ray vibration direction associated with the minimum refractive index alpha (α), and the optic normal trace parallel to Y is the fast ray vibration direction associated with the intermediate refractive index beta (β). On the concave side of the melatopes, the optic plane trace is the slow ray vibration direction associated with a larger refractive index gamma prime (γ'), and the optic normal trace parallel to Y is the fast slow ray vibration direction associated with the intermediate refractive index β. Crystals are biaxial negative when X is the acute bisectrix associated with the α refractive index, so that the fast ray vibration direction is the acute bisectrix. In such cases the acute bisectrix is 2Vx and the YZ plane is parallel to the stage. Between the melatopes, the optic plane trace (parallel to the X vibration direction) is the slow ray vibration direction associated with the maximum refractive index γ, and the optic normal trace is the fast ray vibration direction associated with the refractive index β. On the concave side of the melatopes, the optic plane trace is the fast ray vibration direction associated with a smaller refractive index alpha prime (α'), and the

optic normal trace parallel to Y is the slow ray vibration direction associated with the intermediate refractive index β.

If a gypsum plate is inserted into the accessory slot, the quadrants of addition and subtraction will allow the optic sign of the mineral to be determined. As shown in Figure 6.32, this is most easily accomplished in the 45° (or 135°) position. The slow ray of the gypsum plate is oriented NE–SW and its fast ray is oriented NW–SE. If the slow ray vibration direction of the mineral is oriented parallel to that of the gypsum plate, addition will occur and first order white-gray colors will be increased to second order blue colors by the 550 μm retardation of the gypsum plate. If the slow ray vibration direction of the mineral is oriented perpendicular to that of the gypsum plate, subtraction will occur and first order gray-whites will become shades of yellow.

For **biaxial positive minerals**, in the 135° position, first order white-grays will be converted to blues in the NE–SW quadrants by addition. This occurs on the concave side of the isogyres in the 135° position because the optic plane that contains the slow ray vibration direction (γ) is parallel to the slow ray direction of the gypsum plate in the 135° position. In the 135° position, first order white-grays will be converted to yellows in the NW–SE quadrants by subtraction between the isogyres because the optic normal contains the fast ray vibration direction (β) that is perpendicular to the fast ray of the gypsum

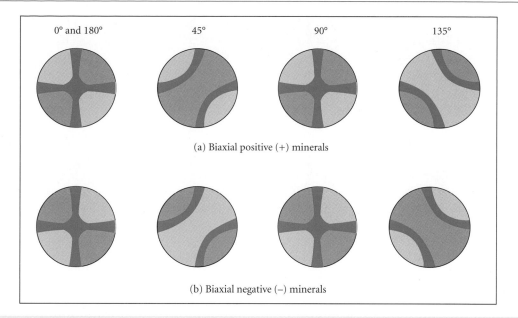

0° and 180°   45°   90°   135°

(a) Biaxial positive (+) minerals

(b) Biaxial negative (−) minerals

Figure 6.33   (a) The appearance of a Bxa interference figure for a biaxial (+) mineral of low–moderate birefringence viewed in various orientations with the gypsum plate inserted. When the gypsum plate is inserted, positive minerals show white-gray changing to blue in the NE–SW quadrants and to yellow in the NW–SE quadrants (shown in pale blue in figure). (b) A similar sequence that contrasts the appearance of a biaxial (−) mineral of similar birefringence where changes are to yellow in the NE–SW quadrants and to blue in the NW–SE quadrants. (For color version, see Plate 6.33, between pp. 248 and 249.)

plate. Figure 6.33a summarizes these relationships for several positions of the stage during rotation where 180° simply returns the situation to that seen in the 0° position. In the initial position, the optic plane parallel to X (α) is E–W and the optic normal parallel to Y (β) is N–S. They have been rotated clockwise in increments of 45° between images.

For **biaxial negative minerals**, the same types of arguments can be made to demonstrate that first order white-grays will be converted to yellows in the NE–SW quadrants where the slow ray vibration direction is perpendicular to the slow ray vibration direction of the gypsum plate so that subtraction occurs, and to blues in the NW–SE quadrants where the slow ray vibration direction is parallel to the slow ray vibration direction of the gypsum plate so that subtraction occurs. These relationships are summarized in Figure 6.33b. Their similarity to those that relate quadrants and colors for uniaxial minerals is striking.

Along acute bisectrix, two rays vibrate perpendicular to each other. At 45° addition

occurs between isogyres because slow rays parallel slow rays and fast parallel fast. At 135°, subtraction occurs between isogyres because slow parallel fast and fast parallel slow! This is the key.

As is the case with uniaxial minerals, biaxial minerals with moderate to high birefringence display one or more orders of isochromatic lines. Because biaxial minerals have two optic axes and two melatopes, the pattern of isochromes is somewhat different. As with uniaxial minerals, isochromatic lines are concentrically arranged about the melatope, but because there are two melatopes, the pattern roughly resembles a figure of eight (Figure 6.34a).

For high birefringent minerals, such as the one illustrated by Figure 6.34b, optic sign determinations are most easily obtained by inserting the **quartz wedge** when the mineral is in the 45° or 135° position with the optic normal NW–SE or NE–SW, respectively. As with uniaxial minerals, positive minerals are characterized by addition in the NE–SW

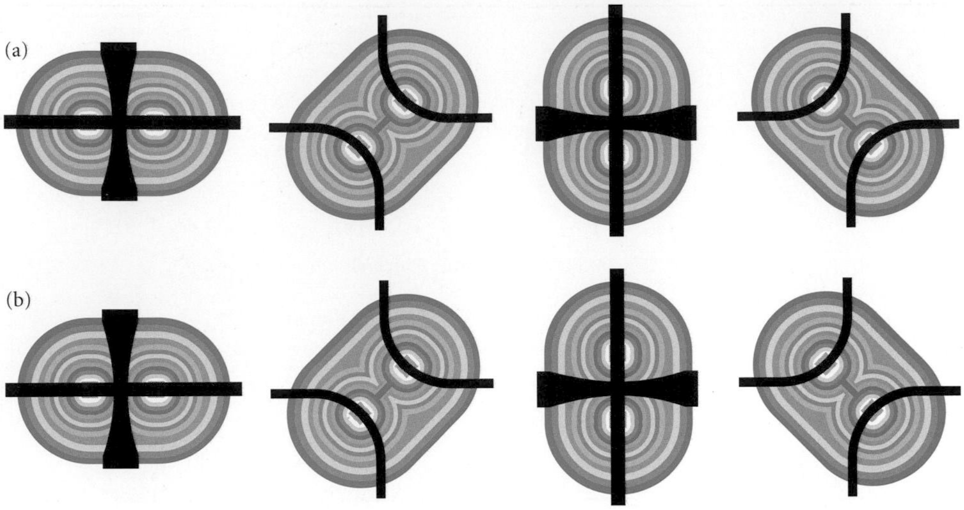

Figure 6.34 The appearance of a Bxa interference figure for a biaxial (+) mineral of moderate–high birefringence viewed in various orientations. When a quartz wedge is inserted into the accessory slot, positive minerals display isochromes that move in toward the melatopes in the NE–SW quadrants and outward in the NW–SE quadrants. (b) A similar sequence which contrasts the appearance of a biaxial (−) mineral of similar birefringence where changes are outward in the NE–SW quadrants and inward the NW–SE quadrants. (Courtesy of Steve Dutch.) (For color version, see Plate 6.34, between pp. 248 and 249.)

quadrants, which causes isochromes to move inward toward the melatopes, and by subtraction in the NW–SE quadrants, which causes isochromes to move outward away from the melatopes. For negative minerals, the movements are reversed; isochromes move outward in the NE–SW quadrants and inward in the NW–SE quadrants.

Optic axis figure

A **centered optic axis figure** is an interference figure seen when one of the optic axes is perpendicular to the stage. In such orientations, a circular section is parallel to the stage. In this orientation, only one refractive index (β) is parallel to the stage, so the mineral displays no birefringence. To obtain a centered optic axis figure, one should look for a biaxial mineral that displays minimal birefringence and remains dark as the stage is rotated and examine it under conoscopic mode.

The **centered biaxial optic axis figure** consists of a single isogyre whose curvature is inversely proportional to the crystal's 2V (Figure 6.35). The melatope is located on the thinnest part of the isogyre at its point of maximum curvature. On rotation of the stage, the melatope will remain centered in the field of view as the isogyre rotates around it. The isogyre will be straight when it is parallel to one of the cross-hairs, and it becomes concave to the NE (shown in Figure 6.34), SE, SW and NW in the 45°, 135°, 225° and 315° positions during a complete rotation. Because 2V is the optic angle between the two optic axes, when 2V is less than 25° a second isogyre may appear in the field of view. One can imagine that if the 2V became zero, there would be only one optic axis and the two isogyres would "merge" into a uniaxial cross.

In addition to being useful for estimating the 2V of a crystal, biaxial optic axis figures can be used to determine its optic sign. The stage is rotated until the isogyre is in a position of maximum curvature, in which orientation it will appear to largely "enclose" a quadrant. The opposite quadrant will not be visible, so that only three quadrants are defined. The procedure is similar to that used for interpreting Bxa figures, as is the explanation. Figure 6.36 illustrates both. For a biaxial positive mineral, the optic normal (β) is parallel to the tangent to the isogyre through the

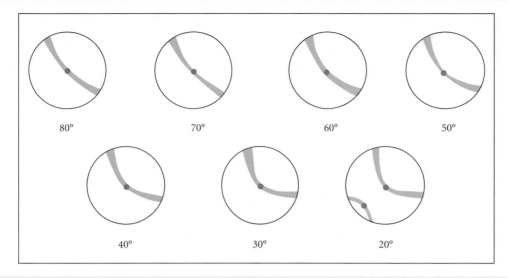

Figure 6.35 Centered optic axis figures in a position of maximum curvature, with the melatope (blue dot). The amount of curvature in such positions is inversely proportional to the crystal's optic angle (2V), with no curvature corresponding to a 2V of 90°.

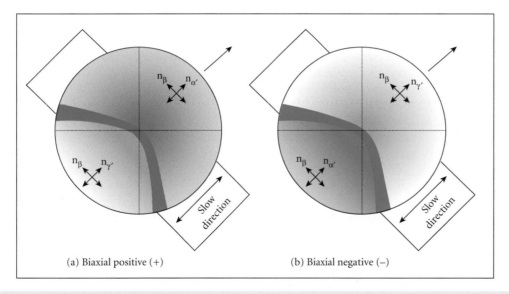

Figure 6.36 Gypsum plate determination of the optic sign using a biaxial centered optic axis figure in a position of maximum curvature of the isogyre. (a) If the concave side of the isogyre in the SW (illustrated) or NE quadrant turns blue, and the convex side turns yellow, the mineral is biaxial positive (+). (b) If the concave side of the isogyre in the SW (illustrated) or NE quadrant turns yellow, and the convex side turns blue, the mineral is biaxial negative (−). (Courtesy of Gregory Flinn.) (For color version, see Plate 6.36, between pp. 248 and 249.)

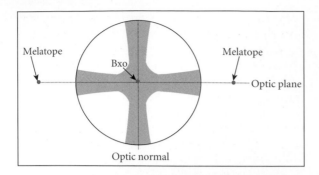

Figure 6.37 Bxo figure and the orientation of the optic normal and optic plane when it is observed. (Courtesy of Gregory Flinn.)

melatope. The acute bisectrix is Z and is located along the optic plane, which passes through the melatope on the convex side of the isogyre. The obtuse bisectrix lies on the optic plane on the concave side of the isogyre. On the convex side, the fast ray (α') is perpendicular to the optic normal (β) slow ray. On the concave side of the isogyre, the slow ray (γ) is perpendicular to the optic normal (β) fast ray. When the gypsum plate is inserted, its slow ray (NE–SW) is parallel to the crystal slow ray (γ) on the concave side so that addition occurs and white-grays change to blues (Figure 6.36a). On the convex side (toward the Bxa), the slow ray of the gypsum plate (NE–SW) is perpendicular to the crystal slow ray (β) so that subtraction occurs. Arguments similar to these can be made for all other orientations in which addition and subtraction occur. The end result is that biaxial positive minerals show addition (white-gray changes to blue) in the NE–SW quadrants and subtraction (white-gray changes to yellow) in the NW–SE quadrants for all orientations of the centered optic axis figures in maximum curvature orientations. For biaxial negative minerals, the quadrants of addition and subtraction and therefore of color changes are reversed. Biaxial negative minerals show subtraction (white-gray changes to blue) in the NE or SW quadrants and addition (white-gray changes to yellow) in the NW–SE quadrants for all orientations of the centered optic axis figures in maximum curvature orientations (Figure 6.36b).

For minerals with high birefringence and many isochromatic lines around the mela-tope, centered optic axis figures can be analyzed using a quartz wedge to determine optic sign. The arguments are similar to those used previously. When the isogyre is in a position of maximum curvature (so that it "encloses" a quadrant), the quartz wedge is slowly inserted. If the mineral possesses a positive optic sign, the isochromes will move in toward the melatope in the NE–SE quadrants and outward in the NW–SE quadrants. If the mineral has a negative optic sign, the movement of the isochromes will be reversed to outward in the NE–SW quadrants and inward in the NW–SE quadrants.

Obtuse bisectrix figure

An **obtuse bisectrix figure (Bxo figure)** is observed when the Bxo is essentially perpendicular to the stage (Figure 6.37). In this orientation, the Bxa and optic normal lie in the plane of the stage. As the stage is rotated, a cross may form that superficially resembles that of an acute bisectrix figure. However, there are two important differences. First, the isogyre that is parallel to the optic plane is thicker because it is farther away from the melatopes or optic axes in this orientation, so that the cross fills a larger proportion of the field of view. Second, as the stage is rotated, the cross breaks up quickly into separate isogyres that leave the field of view well before the stage has been rotated 45°. For minerals with large 2V (>70°), the distinction between an acute bisectrix figure and an obtuse bisectrix figure is difficult to make. Although the obtuse bisectrix figure can be used to determine sign, many workers prefer to find another crystal with an orientation appropriate for viewing an acute bisectrix or optic axis figure. If one does use an obtuse bisectrix figure to determine optic sign, the general rules for Bxa and optic axis figures are reversed. For positive minerals, subtraction (yellow, isochromes out) occurs in the NE–SE quadrants and addition (blue, isochromes in) occurs in the NW–SE quadrants. For negative minerals, addition (blue, isochromes in) occurs in the NE–SE quadrants and subtraction (yellow, isochromes out) occurs in the NW–SE quadrants. It is for this reason, and the difficulty in distinguishing between Bxo and Bxa figures, that many workers choose to look for clear examples of Bxa and/or optic axis figures when making sign determinations.

Biaxial minerals that appear isotropic and display no birefringence ($|\beta - \beta = 0|$) are minerals where one of the optic axes is perpendicular to the stage with one circular section parallel to the stage so that only one refractive index (β) is viewed. These crystals should yield approximately centered optic axis figures, with four well-defined quadrants. In grain mount studies, immersion crystals within a population of the same mineral that exhibit relatively low birefringence for the species are good candidates for crystals that have their Bxa perpendicular to the stage and the optic axes at a high angle to the stage. These often yield nearly centered acute bisectrix figures that are valuable for determining both the biaxial nature of the crystals and their optic sign. Crystals with very high birefringence are generally less useful for optic sign determinations.

Biaxial flash (optic normal) figure

When the y-axis (optic normal) is perpendicular to the stage, so that the two circular sections are also perpendicular to the stage, a biaxial flash figure is observed. As with the uniaxial flash figure, as the stage is rotated an interference figure appears every 90° in the form of a broad cross that nearly fills the field of view. With a few degrees of additional rotation, it leaves the field of view. It appears and then disappears in a "flash", as the stage is rotated through a few degrees. The flash figure is not easy to use to determine optic sign, but it does alert the viewer to the fact that the optic plane is essentially parallel to the stage. In such orientations, Z (γ) and X (α) and the two optic axes are parallel to the stage. In grain mount investigations with immersion oils, these orientations permit γ and α to be measured and birefringence ($|\gamma - \alpha|$) to be determined.

Those who study Earth materials have at their disposal a Pandora's box of ingenious instruments and methods that permit accurate and precise inferences concerning the previously hidden nature of Earth materials. For geoscientists, optical crystallography methods are supplemented, and often superseded, by a variety of chemical analytical techniques such as X-ray diffraction, mass spectroscopy, cathodoluminescence, electron microprobe analysis, X-ray fluorescence, optical emission spectroscopy, optical absorption spectroscopy, scanning electron microscopy, atomic force microscopy and a variety of other specialized methods that together reveal the chemical composition and structure of Earth materials. Due to space limitations, advanced analytical techniques are briefly reviewed in a downloadable file on the website that supports this book.

Chapter 7

Classification of igneous rocks

7.1   MAGMA AND IGNEOUS ROCKS

Igneous rocks are derived from magma. **Magma** is molten rock material generated by partial melting of Earth's mantle and crust. Magma contains liquids, crystals, gases and rock fragments in varying proportions depending upon temperature, pressure and chemistry conditions. At temperatures over 1200°C, above the crystallization temperature of most minerals, magma tends to be enriched in liquids and dissolved gases.

Gas solubility in liquids is related to pressure. At high pressures in the lower crust and mantle, gases are readily dissolved in liquid magma. As magma rises towards Earth's surface, decompression causes gases to segregate from the melt as a separate phase. Opening a carbonated beverage container produces a similar reaction. When the bottle is capped, carbon dioxide gases are dissolved in the liquid under pressure. The removal of the bottle cap induces a pressure reduction such that gases segregate (**exsolve**) from the

solution. This process, called exsolution, causes gas bubbles to develop, expand and rise towards the bottle top as a separate phase. In the same way, the liquid, solid and gas components within magma are largely dependent upon temperature and pressure conditions as well as the chemistry of the magma itself. Magma under high pressure conditions has a high volatile (dissolved gas) saturation point whereas magma under low pressure has a low volatile saturation point. We will look at these aspects further in this chapter when we consider non-crystalline rock textures.

With cooling, magma becomes progressively more enriched in solid material at the expense of liquid melt. Magma that solidifies within Earth produces **intrusive** or **plutonic** rocks. Intrusive rocks develop from magma that cools slowly within Earth producing large crystals visible to the eye. **Plutons** are magma chambers of various sizes, shapes and depths that store magma within Earth. The term "chamber" conjures images of a large room in which liquid magma is swishing around; this in fact is rarely the case. Because most magma is probably stored in rock fractures and pore spaces, plutons may contain

Earth Materials, 1st edition. By K. Hefferan and J. O'Brien. Published 2010 by Blackwell Publishing Ltd.

relatively small amounts of melt at any given time. We will explore this topic further in Chapter 8 as it is currently a very active research issue.

Magma may rise towards Earth's surface due to its low density relative to rock. Magmatic gases provide additional force propelling magma upward in explosive, volcanic eruptions. In a sense, the dissolved gases within magma are analogous to jet fuel blasting a rocket from its launch pad. Magma that rises and erupts onto the surface of Earth is called **lava**. Lava reaches Earth's surface via vents generating flows and associated volcanic debris. **Volcanic** or **extrusive** igneous rocks form by solidification of lava and volcanic debris on Earth's surface, producing rocks with small crystals and/or non-crystalline particles of various sizes.

7.1.1   How do we classify igneous rocks?

Igneous rocks are classified according to composition and texture. Composition is determined by magma chemistry. Texture refers to the size, shape, arrangement and degree of crystallinity of a rock's constituents. Together, these two sets of rock characteristics provide a means to classify rocks and to determine environmental conditions of rock formation. Nearly all magmas are **silicate magmas**, enriched in the elements silicon and oxygen which bond together to form the silica tetrahedron. Rare examples of non-silicate magmas do exist. For example, carbonatite magmas, as their name suggests, are rich in carbonate minerals such as calcite. Silicate magmas contain anywhere from ~40% to over 75% silica (SiO_2). As silica is generally the dominant chemical component, magma and igneous rocks are classified as **ultrabasic**, **basic**, **intermediate** and **acidic** based upon percent SiO_2 (Table 7.1). Acidic rocks are also referred to as **silicic**, based on their high SiO_2 content. In fact the terms acidic and basic are

Table 7.1 Rock description based on weight percent silica.

Rock group	Weight percent silica (SiO_2)
Ultrabasic	<45%
Basic	45–52%
Intermediate	52–66%
Acidic (silicic)	>66%

somewhat misleading as they have no reference to pH. We will look in more detail at rock classification by chemical composition in Section 7.3.

Magma chemistry determines the percentage of dark-colored or light-colored minerals as described in Table 7.2. Dark-colored minerals are generally enriched in the elements iron and magnesium and are referred to as ferromagnesian or **mafic** minerals. Light-colored **felsic** minerals are depleted in ferromagnesian elements and are generally enriched in elements such as silicon, oxygen, potassium and sodium. Rock color terms used to describe the relative percent of dark-colored minerals are listed in Table 7.2. A more detailed discussion of color classification will be provided in Section 7.5.

Figure 7.1 illustrates a simplified means by which some common crystalline (Figure 7.1a) and non-crystalline (Figure 7.1b) igneous rocks are identified based upon their texture, major minerals, percent SiO_2 and color:

- **Peridotite** is a very dark-colored (ultramafic) rock, depleted in SiO_2 (ultrabasic) and commonly enriched in the minerals pyroxene, olivine, amphibole and plagioclase. Ultramafic plutonic rocks occur in Earth's mantle. Ultramafic volcanic rocks are rare and will be discussed in Chapter 10.
- **Basalt** and **gabbro** are dark-colored (mafic), SiO_2-poor (basic) rocks rich in plagioclase, pyroxene and olivine. Basalt is a very common volcanic rock – encompassing the upper few kilometers of ocean

Table 7.2 Generalized description of rock color based on mineral composition.

Classification	Description
Ultramafic	Dark or greenish rocks rich in olivine; may also contain pyroxene or amphibole
Mafic	Dark-colored rocks containing pyroxene, amphibole ± olivine ± biotite
Intermediate	Grayish to salt and pepper-colored rocks rich in plagioclase, amphibole ± biotite ± quartz
Felsic	Light-colored or red rocks rich in potassium feldspar, quartz ± biotite or muscovite

(a)

Peridotite   Gabbro   Diorite   Granodiorite   Granite

Basalt   Andesite   Dacite   Rhyolite

Ultramafic	Mafic	Intermediate	Felsic		Crystalline textures
Peridotite	Gabbro	Diorite	Granodiorite	Granite	Coarse grains (phaneritic)
Komatiite (not shown)	Basalt	Andesite	Dacite	Rhyolite	Fine grains (aphanitic)

Dark-colored ferromagnesian minerals

Hornblende   Biotite

Pyroxene

Olivine

Feldspars

(Plagioclase)   (Alkali)

Quartz

100
80
60
40
20
0%

45 50 52 55 60 65 70 %SiO₂

(b)

Texture	Rock name
Vesicular	**Pumice** Light-colored, lightweight rock rich in gas holes (vesicles)
Vesicular	**Scoria** Dark-colored, lightweight rock rich in gas holes (vesicles)
Glassy	**Obsidian** Black to reddish rock with glassy luster and conchoidal (scalloped breakage) fracture
Pyroclastic	**Volcanic tuff** Rock composed of fine-grained ash- to sand-sized volcanic rock fragments
Pyroclastic	**Volcanic breccia** Rock composed of coarse-grained gravel and larger sized volcanic rock fragments

crust – that forms from rapid cooling. Gabbro crystallizes more slowly at depth in the lower crust of ocean basins.

- **Andesite** and **diorite** are gray-colored (intermediate) to salt and pepper-colored rocks rich in hornblende, pyroxene and plagioclase. Andesite and diorite contain more than half to almost two-thirds SiO_2. Andesite is a common volcanic rock around the Pacific Ring of Fire. Andesite volcanoes overlie diorite plutons.
- **Dacite** and **granodiorite** are light-colored (felsic) rocks, containing approximately two-thirds SiO_2, rich in plagioclase, alkali feldspar and quartz and also containing small amounts of hornblende and biotite. Dacite is a volcanic rock that, like andesite, occurs around the Pacific rim. Granodiorite is a plutonic rock that underlies andesite–dacite volcanoes.
- **Rhyolite** and **granite** are light-colored (felsic) rocks containing more than two-thirds SiO_2 (silicic or acidic) and rich in quartz, alkali feldspar with small percentages of plagioclase and biotite. Rhyolite is a volcanic rock that usually erupts in thick, continental crust. Granite plutons also occur in continental crust.
- Non-crystalline rocks (Figure 7.1b), those characterized by the absence of crystals, include frothy, vesicular rocks such as **pumice** (light colored) and **scoria** (dark colored). Other non-crystalline rocks include those with glassy textures such as **obsidian** or those enriched in rock fragments. Fragmental, also known as pyroclastic, volcanic rocks include **tuff** (volcanic ash to gravel size) and **breccia** (larger than gravel size).

Whenever possible, we classify rocks based upon texture as well as the minerals present.

Figure 7.1 (a) Crystalline igneous rocks are classified based upon their crystal size (texture) as well as the major minerals olivine, pyroxene, hornblende, biotite, feldspars and quartz. (b) Igneous non-crystalline rocks are classified based primarily on their texture – size and nature of rock particles, presence of vesicles or glass – as well as color. (Photos by Kevin Hefferan.) (For color version, see Plate 7.1, between pp. 248 and 249.)

Volcanic rocks – such as basalt, andesite, dacite and rhyolite – tend to have fine grains that are too small to be identified with the eye. In many cases, very fine grain size necessitates the use of color as a means to classify volcanic rocks. Rock color is used as a last resort because color is inherently unreliable.

Plutonic rocks – such as gabbro, diorite, granodiorite and granite – tend to have large crystals such that we can easily identify the minerals simply by looking at the rock with our eyes. As plutonic rocks have large crystals, we can identify these rocks based on the relative proportion of minerals (Figure 7.1a). A more detailed classification system based upon texture and composition will be presented later in this chapter.

7.2 IGNEOUS TEXTURES

Igneous textures may be broadly categorized on the degree of **crystallinity**. Given appropriate time, temperature and pressure conditions, silica tetrahedron structures within cooling magma link together to produce crystals. In some instances, extremely rapid cooling or the sudden loss of gas may result in solidification without the development of crystals, creating a glassy solid. The degree of crystallization can be classified into **holocrystalline**, **hypocrystalline** and **holohyaline** textures; these terms, simply stated, mean wholly crystalline, partially crystalline/partially glass and wholly glassy textures, respectively. In the following section, we will discuss crystalline and non-crystalline textures. This textural information will help us understand the conditions under which rocks form and how we use that information in classifying rocks. Let us now consider key aspects of crystalline texture that relate to rock classification.

7.2.1 Crystalline forms

Textural terms in crystalline rocks are based primarily on the completeness of crystal form as well as crystal size. **Euhedral** minerals contain complete crystal faces that are not impinged upon by other crystals (Figure 7.2). Euhedral crystals typically develop as early mineral phases in the crystallization of magma. Under such conditions the crystals have abundant free space for growth, enhancing the likelihood that perfectly formed crystal

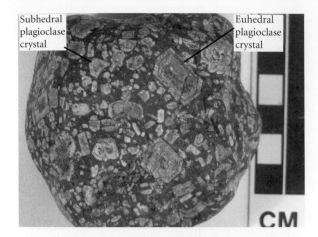

Subhedral plagioclase crystal

Euhedral plagioclase crystal

CM

Figure 7.2 Basalt porphyry containing euhedral and subhedral plagioclase crystals. Some euhedral crystals display zoning suggesting that the crystals continued to react with the melt during crystallization. Zoning is discussed further in the text. (Photo by Kevin Hefferan.) (For color version, see Plate 7.2, between pp. 248 and 249.)

faces develop. Later in the magma crystallization sequence, subhedral or anhedral crystals develop in the remaining void spaces between earlier formed crystals.

Subhedral crystal faces contain partially complete crystal forms in which at least one of the crystal faces is impinged upon by adjacent rock material (Figure 7.2). In subhedral textures, crystal growth may be aborted due to:

- Contact against previously formed minerals.
- Nucleation on pre-existing surfaces such as early formed crystals or the margins of the magma chamber.
- Resorption in which pre-existing euhedral crystals are partially remelted.
- Other secondary alteration processes that destroy pre-existing euhedral faces.

Anhedral crystals lack any observable crystal faces. As crystallization progresses in magma, the space available for the development of euhedral and subhedral crystals diminishes. As a result, anhedral crystal forms are determined by the shape of the existing space, rather than by mineral crystallography. The

remaining voids between existing crystal forms are referred to as **interstitial** space.

7.2.2 Crystalline textures

Crystalline textures provide critical information as to whether the rock solidified in a plutonic or volcanic setting. Magma that cools within Earth produces plutonic rocks with crystals easily visible to the eye. Plutons within Earth retain heat and allow for slow cooling, which enhances crystal growth. Igneous textures typical of plutonic rocks include coarse-grained pegmatitic, phaneritic and phaneritic–porphyritic textures. Lava that cools on Earth's surface loses heat quickly. As a result, many crystals are invisible to the eye and produce fine-grained aphanitic or aphanitic–porphyritic crystalline textures. These textures and the geological factors that influence them are described below.

Pegmatitic textures

Pegmatitic texture is characterized by large crystals averaging more than 30 mm in diameter. Pegmatites display large, early formed euhedral crystals surrounded by later formed subhedral crystals. In naming rocks with pegmatitic texture, the textural term (e.g., pegmatite) must be included in the rock name. Therefore, a rock with a pegmatitic texture and the composition of granite or granodiorite is a granite pegmatite or granodiorite pegmatite. Pegmatitic textures develop most commonly in granitic plutons (Figure 7.3) with high volatile contents. Gabbroic plutons rarely display pegmatitic textures, partly due to the lower volatile gas content. Because of the large, well-developed crystal forms, pegmatites are the source of many gemstones such as aquamarine and tourmaline. High volatile content also produces valuable ore deposits of metals such as tin, gold and silver in pegmatite deposits.

Phaneritic textures

Phaneritic texture implies crystal diameters ranging from 1 to 30 mm. Rocks with a phaneritic texture contain crystals visible to the naked eye (Figure 7.4). Early formed crystals are euhedral; later formed crystals are subhe-

Figure 7.3 Granite pegmatite with quartz, hornblende and muscovite. (Photo by Kevin Hefferan.) (For color version, see Plate 7.3, between pp. 248 and 249.)

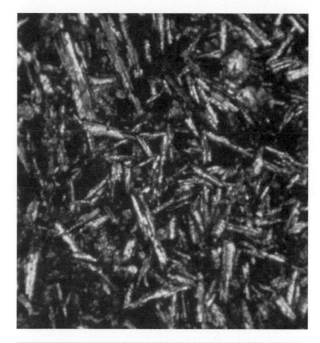

Figure 7.5 Aphanitic basalt with microcrystalline texture as viewed under a petrographic microscope. Note the euhedral to subhedral acicular plagioclase crystals embedded within a cryptocrystalline groundmass. Field of view is 2 cm. (Photo by Kevin Hefferan.) (For color version, see Plate 7.5, between pp. 248 and 249.)

Figure 7.4 Coarse-grained phaneritic granite with early formed, euhedral potassium feldspar. Subhedral quartz and hornblende represent later, void-filling minerals. (Photo by Kevin Hefferan.) (For color version, see Plate 7.4, between pp. 248 and 249.)

dral to anhedral. Phaneritic textures may be subdivided into fine (1–3 mm in diameter), medium (3–10 mm) or coarse (10–30 mm) grained. Fine-grained phaneritic textures commonly develop in shallow plutonic structures such as dikes and sills. Coarse-grained textures are generally associated with larger or deeper intrusions. We will discuss plutonic structures further in Chapter 8. Rock names

such as granite, diorite and gabbro imply a phaneritic texture so that we do not refer to "phaneritic granite" but simply "granite".

Aphanitic textures

Aphanitic textures contain small crystals less than 1 mm in diameter that are not generally discernible to the eye. With the use of a microscope or other analytical means, geologists can determine the composition and relative sequence of crystallization in the same manner as with phaneritic textures. Aphanitic textures are associated with volcanic rocks and subvolcanic rocks that cool quickly on or near Earth's surface. Aphanitic textures may be subdivided into microcrystalline and cryptocrystalline varieties. **Microcrystalline** textures contain **microlite** crystals large enough to be identified with a petrographic microscope (Figure 7.5). Aphanitic textures in which the crystal size is

Figure 7.6 Porphyritic–phaneritic texture with subhedral potassium feldspar phenocrysts in a groundmass of quartz and hornblende. As phenocrysts encompass >50% of the rock this is a granite porphyry. If phenocrysts encompass <50% of the rock volume, the term porphyritic granite would be used. (Photo by Kevin Hefferan.) (For color version, see Plate 7.6, between pp. 248 and 249.)

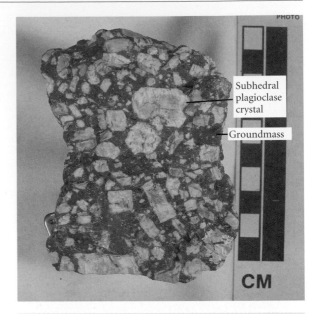

Figure 7.7 Andesite porphyry displaying aphanitic porphyritic texture. Note the subhedral plagioclase phenocrysts (white crystals) and the fine-grained aphanitic groundmass (gray). (Photo by Kevin Hefferan.) (For color version, see Plate 7.7, between pp. 248 and 249.)

too fine to be identified even with a petrographic microscope are termed **cryptocrystalline** (MacKenzie et al., 1984). Rock names such as rhyolite, andesite and basalt imply an aphanitic texture so that we do not refer to "aphanitic basalt" but simply basalt.

Porphyritic textures

Rocks with **porphyritic** textures consist of two distinctly different size crystals. Large crystals are referred to as **phenocrysts**; finer grained material constitutes the **groundmass**. In **porphyritic–phaneritic** textures, all crystals are visible to the eye, but the phenocrysts are distinctly larger than the groundmass crystals (Figure 7.6). In rocks with **porphyritic–aphanitic** textures, the larger phenocrysts are embedded in an aphanitic groundmass composed largely of microcrystalline, cryptocrystalline or glassy material (Figure 7.7).

In your introductory courses it may have been emphasized that the cooling rate of magma or lava determines crystal size. Slow cooling of magma deep within Earth produces coarse-grained pegmatitic or phaneritic textures. Rapid cooling of magma at shallow depths or as lava on Earth's surface generates fine-grained aphanitic textures. Rocks with

two distinctly different size crystals (porphyritic) are commonly explained by a two-stage cooling process. In two-stage cooling processes, the larger phenocrysts form slowly at depth, while the finer grained groundmass crystals cool rapidly as magma approaches Earth's surface. These straightforward explanations for the development of crystalline textures provide simple models that do not always hold true.

In the following section, we will discuss in greater detail the geological factors and conditions that determine crystalline textures.

7.2.3 The origin of crystalline textures

In addition to cooling rate and depth, other important factors influencing igneous textures include crystal nucleation rates, crystal growth rates, rate of magma undercooling, ion availability, chemical diffusion rates, viscosity, chemical composition and volatile content of the magma. Each of these factors will be discussed below.

Let us first consider magma crystallization using a simplified model that focuses on the

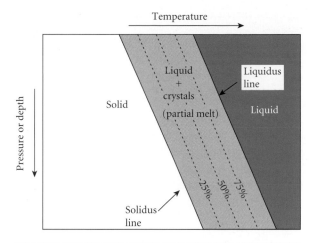

Figure 7.8 Pressure–temperature relations depicting solid, solid plus liquid and liquid fields separated by a solidus line and a liquidus line. Dashed lines indicate percent melt. (Courtesy of Stephen Nelson.)

solid and liquid states, ignoring the vapor state. At sufficiently high temperatures, an ideal magma exists in the liquid state. As temperatures decrease, the magma begins to crystallize so that liquid and crystals coexist. At sufficiently low temperatures, all the liquid is crystallized and the system becomes completely solid. On a phase stability diagram, three fields are recognized, separated by lines that represent the initial crystallization temperature and the final crystallization temperature (Figure 7.8). The high temperature "all liquid" field is separated from the lower temperature "liquid plus crystal" field by the **liquidus line**. The liquidus line represents the initial crystallization temperature below which crystals and liquids coexist. Below the liquidus temperature, crystals nucleate and continue to grow at **subliquidus** temperatures. Eventually, when the magma has cooled sufficiently, all remaining liquid crystallizes and the system crosses the solidus line at which point the magma completely solidifies. The **solidus line** separates a higher temperature field containing liquids and solid crystals from a lower temperature field in which only solid crystalline material exists. Temperatures below the solidus line are referred to as **subsolidus** temperatures.

The exact temperatures at which the ideal system changes from all melt to melt plus crystals (liquidus line), and from melt plus crystals to all solid (solidus line), depends on a variety of factors that include pressure and chemical composition. This means that for a given magma, we can not define a single temperature at which melting or solidification occurs. Instead, these changes occur along a temperature–pressure range as represented by the liquidus and solidus lines in Figure 7.8. Real magmatic systems are even more complicated, but this conceptual model approximates their behavior closely enough to be of value. We will now use this model to explain crystal nucleation.

Crystal nucleation

Below the liquidus line, crystal size depends primarily on the interplay between the crystal nucleation rate and the crystal growth rate (Swanson, 1977). Crystal nucleation involves the formation of new crystals, called nuclei or "seed" crystals, large enough to persist and grow into larger crystals. The **crystal nucleation rate**, the number of new seed crystals that develop per volume per time unit, is commonly expressed as nuclei per cubic centimeter per second (nuclei/cm^3/s). Crystal nuclei grow at rates that depend on the following:

1   The rate of undercooling.
2   Availability of the necessary ions.
3   The ease with which these ions migrate to the crystal growth site.

In real magmas, the rate at which crystal nuclei form varies for each mineral. One might expect nucleation rates to peak at the liquidus temperature because the liquidus temperature represents conditions in which the maximum amount of liquid magma is available to crystallize. In reality, nucleation rates peak at temperatures well below the liquidus temperature, for reasons discussed below.

Subliquidus crystal nucleation represents undercooling of the magma. **Undercooling** occurs when liquids are cooled to temperatures below the liquidus line. One reason why nucleation requires undercooling is that crystal formation requires the development of bonds between ions which produces the **heat of formation**. This heat increases the local temperature and remelts seed crystals. Only when the magma has been significantly undercooled below the liquidus temperature can incipient crystals nucleate, persist and grow (Brandeis et al., 1984). Therefore, nucleation rates are low near the liquidus temperature and increase at temperatures below the liquidus (undercooling). Nucleation rates decrease to approach zero at very large degrees of undercooling because increased magma viscosity impedes the assembly of nuclei. Very large degrees of undercooling imply that the magma is approaching the solidus line where all liquid has crystallized.

Crystal growth rate

The **crystal growth rate** is a measure of the increase in crystal radius over time. Growth rates are commonly expressed in centimeters per second (cm/s). Nucleation rates and growth rates are two independent processes which together determine the number and size of crystals. Crystal growth rates are governed by the rate of undercooling as well as the availability of elements and magma viscosity. Magmas that experience small levels of undercooling stay at temperatures just below the liquidus temperature for a long time. At such temperatures, the nucleation rate is low and the crystal growth rate is high, producing a small number of large crystals (Swanson, 1977). Prolonged undercooling conditions generate phaneritic, porphyritic or pegmatitic textures with euhedral to subhedral crystals (Figure 7.9).

On the other hand, magmas that experience large undercooling at temperatures well below the liquidus temperatures experience higher nucleation rates and lower growth rates (Swanson, 1977). As a result, a large number of small seed crystals produce aphanitic textures, some of which may display skeletal or dendritic textures. Extremely rapid temperature decreases prevent the growth of seed crystals. Failure to nucleate seed crystals produces non-crystalline glass such as obsidian that may contain spherulites (Figure 7.9).

Ion availability

Ion availability refers to the availability of ions that can fill specific ionic sites in a crystal lattice structure. If appropriate ions are readily available and can migrate to the lattice site, crystal growth is enhanced. The ion availabil-

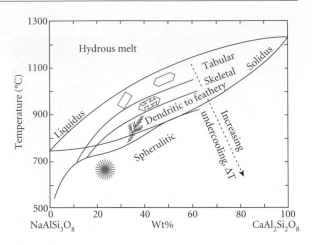

Figure 7.9 Plagioclase phase diagram illustrating crystalline textures that can form in response to rates of undercooling. Laboratory data (Lofgren, 1980, 1983) suggest that tabular, euhedral crystals form with cooling of a few °C per hour. Temperature changes of hundreds of °C per hour produce dendritic to spherulitic textures in lab experiments. (After Best, 2003; with permission of Wiley-Blackwell Publishers.)

ity factor helps explain why minerals that require relatively rare trace elements tend to be very small. For example, coarse-grained, crystalline, igneous rocks may contain apatite and zircon. Apatite is composed of minor elements such as phosphorous. Zircon consists of the trace element zirconium. Apatite and zircon crystals tend to be small, even within granite pegmatites, as their mineral growth rates are impeded by the limited availability of the minor and trace elements phosphorous and zirconium.

Viscosity

Diffusion, the rate at which elements migrate through magma, depends primarily on the viscosity of the melt. **Viscosity** (η), defined as the resistance of a fluid to shear stress, is measured by the ratio of shear stress (τ) to shear strain rate ($\dot{\varepsilon}$) given by $\tau = \eta\dot{\varepsilon}$. Shear stress is a force applied parallel to the surface of an object. Shear strain rate is a time-dependent change in the shape of an object in response to shear stress. We can rewrite the above formula as $\eta = \tau/\dot{\varepsilon}$. In this modified formula, viscosity is proportional to shear stress and inversely proportional to shear strain rate. Low viscosity materials, such as water, flow readily. High viscosity materials, such as honey, flow more slowly. For a given shear stress, the higher the viscosity, the lower the strain rate.

Chemical diffusion rate

Viscosity is a measure of resistance to fluid flow: the higher the resistance to flow, the higher the viscosity and the lower the strain (flow) rates. Therefore, viscosity affects the ability of ions to diffuse through the magma to crystal growth sites. Low magma viscosity increases the rate of diffusion and increases crystal growth rate. Conversely, high magma viscosity reduces the diffusion rate and decreases crystal growth rate. Viscosity is largely determined by the SiO_2 content, the development of silica tetrahedra, temperature and the dissolved gases in the magma. As most magmas contain about half to over two-thirds SiO_2, magma viscosity is strongly influenced by SiO_2 content, which determines the degree of molecular linkage in magma.

Viscosity and chemical composition

Viscosity is strongly related to the degree to which molecules are bonded or linked together in the fluid: an increase in molecular linkages results in higher magma viscosity.

Molecular linkage is determined by the relative abundance of network formers and network modifiers. **Network formers** are elements that tend to increase molecular linkage, thereby increasing viscosity. Network formers in silicate melts include silicon, oxygen and aluminum. Molecular linkage of the elements silicon and oxygen, creating the silica tetrahedron (SiO_4) structure, exerts primary control on magma viscosity. Silicon and oxygen behave as network formers that bond together through shared oxygen atoms to create polymerized chains of linked silica tetrahedra. The creation of these silica tetrahedron networks results in higher viscosity magmas. Under most conditions, magmas rich in SiO_2 have higher viscosities than SiO_2-poor magmas.

Elements such as magnesium and iron are referred to as network modifiers. **Network modifiers** are elements that decrease molecular linkage, thereby reducing viscosity. In silicate magmas, network modifiers inhibit the linkage of silica tetrahedron polymerization by filling sites otherwise occupied by silicon and oxygen. Therefore, magmas enriched in iron and magnesium and depleted in silicon and oxygen tend to have lower viscosities. Decreased linkage of silica tetrahedra results in a lower viscosity (Figure 7.10a) that characterizes basic (45–52% SiO_2) and ultrabasic (<45% SiO_2) melts (Spera, 2000).

Viscosity, molecular bonding and heat

The degree of molecular linkage is strongly influenced by both composition and temperature. Temperature is inversely proportional to viscosity and molecular bonding. As the temperature increases, molecular vibrations increase, bonds break and the amount of molecular linkage decreases thereby reducing viscosity. For example, honey is a high viscosity fluid at room temperature due to its relatively large number of organic molecular bonds. If honey is stored at cool temperatures, it tends to crystallize, developing organic molecular bonds resulting in greater viscosity. However, if you heat the crystallized honey in a microwave, the molecular bonds vibrate and break and viscosity is drastically reduced. This example illustrates the inverse relationship of temperature and viscosity: as temperature increases, magma viscosity decreases.

Low magma viscosity also promotes high crystal growth rates. Crystal nucleation is an exothermic process that releases energy in the form of heat. Low viscosity magma aids crystal growth by increasing the rate at which the heat of formation can be diffused away from the surface of the growing crystal. This allows the crystal to be cool enough for additional growth to occur. Crystal growth rates decline as the temperature and diffusion rates decrease, causing viscosity to increase.

The role of gases

Magma contains up to 7 wt % volatile gases, largely consisting of H_2O, CO_2 and SO_2. Minor gases include N, H, S, F, Ar, CO and Cl. Gases play a key role in magma viscosity by reducing molecular bonding. Gases dissolved in magma under pressure are extremely important network modifiers.

(a)

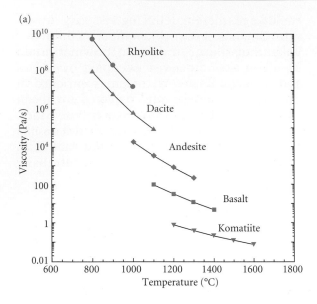

(b)

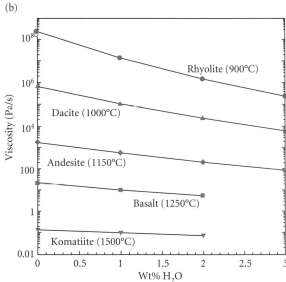

Figure 7.10 (a) Graph depicting viscosity versus temperature for five different lavas as determined through experimentation. Note that low-SiO$_2$, ultrabasic lavas (e.g., komatiite) have the lowest viscosity and, as SiO$_2$ content increases, viscosity increases. The graph also indicates that viscosity is inversely proportional to temperature as discussed in the text. (b) Graph plot of viscosity versus weight percent water illustrating that silicic (dacite and rhyolite) magmas experience the greatest decrease in viscosity with an increase in water content. (From Spera, 2000; courtesy of Frank J. Spera.)

Dissolved water vapor is a particularly potent network modifier (Figure 7.10b); an increase in water vapor pressure (P$_{H2O}$) decreases magma viscosity, particularly for silicic magmas. High dissolved water vapor content is an effective network modifier because it partially bonds with the corners of silica tetrahedra: the higher the magma SiO$_2$ content, the greater the effect. As water vapor occupies tetrahedra bonding sites, the water vapor effectively inhibits adjacent silica compounds from bonding together. This reduces the molecular linkage in the magma and reduces viscosity. A reduction in magma viscosity increases diffusion rates and crystal growth rates. The result is that high P$_{H2O}$ produces a low nucleation rate and a high growth rate generating small numbers of large crystals.

Silicic magmas (>66% SiO$_2$) possess a large capacity to dissolve water vapor and other volatiles. The unusually high (3–7%) P$_{H2O}$ in silicic magmas enhances both the diffusion rates and crystal growth rates, producing relatively few crystals that grow to a very large size (Wallace and Anderson, 2000). As a result, granite pegmatites and granodiorite pegmatites are relatively common. On the other hand, basic magmas (<52% SiO$_2$) generally contain <1% dissolved water vapor. P$_{H2O}$ is only a little higher in basic magma under pressure at depth than in the shallow subsurface. As a result, P$_{H2O}$ has a very limited effect on crystal size in gabbroic plutonic rocks. Phaneritic textures in gabbro develop by slow undercooling at depth while aphanitic textures in basalt develop by rapid undercooling on the surface. Very large plutonic crystals are uncommon in basic and ultrabasic rocks (Wallace and Anderson, 2000).

Variations in P$_{H2O}$ can result in variations in crystalline textures. For example, earlier formed granitic veins commonly display pegmatitic textures while later formed, cross-cutting veins may contain fine-grained (aplitic) textures. As silicic magma rises toward the surface, the confining pressure in the magma decreases enhancing exsolution within the magma. Exsolution allows gases such as water vapor to bubble out from the magma, drastically lowering magma P$_{H2O}$. P$_{H2O}$ reduction causes silica tetrahedra to link rapidly, which accelerates magma viscosity. High viscosity inhibits crystal growth and helps to explain the very small crystal size typical of fine-grained granitic dikes (aplites) or aphanitic rocks such as rhyolites. As will be discussed in the next section, this also explains the abundance of non-crystalline, glassy textures in silicic rocks as opposed to basic rocks.

7.2.4 Non-crystalline textures

Glassy, vesicular and pyroclastic are examples of non-crystalline igneous textures (see Figure 7.1b). Rocks may consist entirely of non-crystalline (holohyaline) components or contain a mixture of crystalline and non-crystalline (hypocrystalline) igneous textures. Each of the major types of non-crystalline textures is discussed below.

Glassy textures

A **glass**, such as the rock obsidian, is an amorphous solid. Amorphous solids possess a disordered form, thereby lacking an ordered crystalline structure. Many glasses contain small amounts of very small microlites and/or cryptocrystalline material. Glassy textures develop in lava that solidifies without experiencing significant crystallization. The lack of crystal structure in glasses is similar to the lack of long-range order characteristic of melts – glasses are essentially supercooled liquids. Glassy textures form by the near instantaneous solidification of melts preserving their disordered structure. Near instantaneous melt solidification results from two major mechanisms: quenching and rapid gas loss.

Quenching occurs when melts of any composition come into contact with liquid water or air. Water rapidly absorbs heat from the melt, causing it to solidify before crystals have time to nucleate and grow. Most basic (low SiO_2) glasses quench when volcanoes erupt on the ocean floor or as massive flood basalts. Thin glassy zones also occur on lava flow tops that have been quenched by contact with the atmosphere.

The second glass-forming mechanism, limited to silicic melts, occurs by **rapid loss of dissolved gas** from solution which rapidly lowers P_{H2O}. The rapid loss of dissolved water vapor allows silica tetrahedra to link together and causes melt viscosity to increase so rapidly that crystal nucleation and crystal growth are severely inhibited. The result is a glass, the product of the nearly instantaneous solidification of magma by loss of dissolved gas, rather than by extremely rapid cooling. This second model explains why glassy rocks, such as obsidian, are far more common in silicic rocks than in basic rocks. Unlike silicic magmas, basic magmas contain neither enough dissolved water nor sufficient silica tetrahedra to solidify rapidly due to loss of dissolved gases.

As noted earlier, many glasses do contain small microlites and cryptocrystalline minerals. Microlites and cryptocrystalline grains represent incipient crystal nucleation in a nearly solid magma of extremely high viscosity. Larger crystals may form by partial crystallization of the magma prior to its rapid solidification by quenching or gas loss. This can produce porphyritic glassy rocks called vitrophyres. **Vitrophyres** contain recognizable phenocrysts in a glassy groundmass and are said to have a **vitrophyric texture**. Over long periods of time, glasses may crystallize in the solid state by growing on pre-existing microlitic or cryptocrystalline nuclei in a process called **devitrification**. Growth commonly occurs outward from existing crystal nuclei to produce rounded masses of radiating crystals called **spherulites**. **Snowflake obsidian** is an excellent example in which cristobalite seed crystals grow into white snowflake forms within black glassy obsidian (Figure 7.11). **Perlites**, glassy SiO_2-rich volcanic rocks with

Figure 7.11 Snowflake obsidian displaying cristobalite seed crystals as well as conchoidal fracture. (Photo by Kevin Hefferan.)

higher water contents than obsidian, display a **perlitic texture** characterized by a cloudy appearance and curved or subspherical cooling cracks called perlitic cracks. Perlite is widely used as an aerating medium in potting soils.

Vesicular textures

Vesicular textures contain spherical to ellipsoidal void spaces called **vesicles**, which are analogous to holes in a household sponge. Vesicular textures develop due to exsolution and entrapment of gas bubbles in lava as it cools and solidifies. A model for the development of vesicular textures is illustrated in Figure 7.12. Plutons contain magmas at relatively high confining pressures such that gases are dissolved and the magma is undersaturated in volatile content. As volatile gases are of low density and tend to be buoyant, volatiles ascend within the pluton and can saturate magma in the upper part of the pluton. In conditions where magma rises toward the surface, the confining pressure decreases, and the ability of the magma to

retain dissolved gases decreases. As a result, magma becomes supersaturated with volatiles so that it can no longer hold all the gas in solution. At a depth referred to as the **level of exsolution**, volatiles exsolve from the liquid as a separate phase. Above the level of exsolution, volatiles nucleate as small bubbles in a process called vesiculation. As the magma continues to rise, gas solubility continues to decrease and the number and size of the gas bubbles increase. The increasing volume of gas bubbles decreases the density of the magma, resulting in rapid volume expansion, accelerating magma towards the surface. A **fragmentation surface** is encountered when bubbles constitute 70–80% of the magma volume. At the fragmentation surface, magma changes from a liquid with suspended gas bubbles to a buoyant gaseous mixture containing liquid blobs. At that level, explosive magma rapidly propels upward and lava erupts violently on Earth's surface (Chapter 9). Since exsolution is time dependent, as in bubbles released from champagne, gases often continue to exsolve even after lavas are extruded. It is the entrapment of such gas

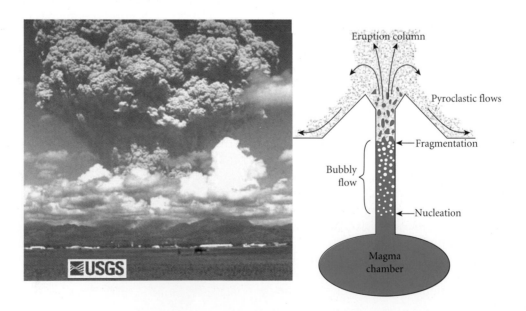

Figure 7.12   At depth within the magma pluton, gases are dissolved within the pressurized magma. As the magma rises towards the top of the pluton, pressure decreases and gases exsolve or separate from the magma. The separation of gases results in the generation of gas bubble growth, which produces a vesicular texture. (After Sparks, 1978; illustration courtesy of Ed Llewellin; photo courtesy of the US Geological Survey.)

bubbles in shallow plutons and erupted lava that produces frothy volcanic glass typical of vesicular textures (Sparks, 1978; Wilson, 1980; Cashman et al., 2000).

Vesicular rocks, defined as containing >30% vesicles by volume, include pumice and scoria. White- to gray-colored pumice solidifies as a frothy glass from silicic lava. Pumice is widely used as an abrasive soap. Scoria is a vesicular rock characterized by brownish red or black colors due to an abundance of iron and is used as a decorative stone. Scoria is derived from basic to intermediate lava and may be partially crystalline (hypocrystalline). Highly vesicular rocks are characterized by very low specific gravity and may actually contain such a large volume of vesicles that they float in water. Rocks that contain smaller amounts (5–30%) of vesicles are named using a modifier such as vesicular basalt or vesicular andesite, while those rocks with just a few vesicles (<5%) are given names such as vesicle-bearing basalt and andesite.

Hot fluids that flow through vesicular rocks may later precipitate secondary minerals in the void spaces of vesicles, producing **amyg-** **dules.** Common secondary minerals that infill pre-existing vesicles include quartz, calcite, epidote, zeolites and metals. Figure 7.13 depicts an amygdaloidal basalt in which quartz and epidote have precipitated in vesicles. Secondary fluid flow through rock can produce significant ore deposits of copper and other metals that precipitate in the void spaces. Amygdaloidal ore deposits are particularly important in rift basins such as the Keweenaw copper belt of North America (Chapter 19).

Pyroclastic textures

Volcanic eruptions eject broken rock particles of varying sizes, known as **pyroclasts** (which means fiery fragment). Pyroclasts may be ejected into the atmosphere as airborne **tephra** or transported along Earth's surface as pyroclastic flows. Following accumulation, these particles are cemented or welded together to produce volcanic rocks with fragmental or **pyroclastic textures.** Pyroclasts are classified according to their composition, size and shape (Figure 7.14; Table 7.3). Pyroclasts consist of several different types of materials:

- **Lithic** pyroclasts contain fragments such as basalt, andesite or other rocks.
- **Vitric** pyroclasts are composed of glassy fragments, most commonly pumice or scoria shards.
- **Crystal** pyroclasts contain minerals.

Figure 7.13 Amygdaloidal basalt in which vesicles have been infilled with quartz and epidote. This basalt has been altered by hydrothermal solutions which have changed the existing chemistry of the basalt and precipitated new (secondary) minerals in the vesicles. (Photo by Kevin Hefferan.) (For color version, see Plate 7.13, between pp. 248 and 249.)

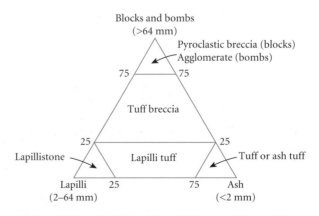

Figure 7.14 IUGS classification of pyroclastic debris based on diameter clast size. (After Fisher, 1961, 1966.)

Table 7.3 Common fragmental (pyroclastic) textures as classified by size and degree of roundness (Schmid, 1981, Fisher and Schmincke, 1984).

Clast diameter (mm)	Pyroclast name	Rock name
>64	Angular block	Breccia or tuff breccia
	Rounded bomb	Agglomerate
2–64	Lapilli	Lapillistone or lapilli tuff
0.0625–2	Coarse ash	Tuff or ash tuff
<0.0625	Fine ash or dust	Tuff or ash tuff

Figure 7.15 Angular volcanic blocks from Kilauea, Hawaii. (Photo by Kevin Hefferan.) (For color version, see Plate 7.15, between pp. 248 and 249.)

Pyroclasts are further divided by average grain size diameters:

- Pyroclasts, greater than 64 mm in diameter, are called **blocks** if angular (Figure 7.15) and **bombs** if rounded. Angular blocks lithify as breccia and rounded blocks form agglomerates.
- Gravel-sized pyroclasts (2–64 mm diameter) are called **lapilli**. Rocks that consist largely of lapilli are called **lapillistones**.
- **Ash** consists of sand-sized and finer sized pyroclasts (<2 mm diameter) which can be subdivided into coarse ash (0.0625–2 mm) and fine ash (<0.0625 mm) or **dust**. A rock composed of solidified volcanic ash is called **tuff**. Tuffs that contain significant amounts of gravel-size lapilli are called **lapilli tuffs**.

Large clasts provide information about distance from source vents and degree of transport. **Breccias** are deposited proximal to the volcano vent and subjected to minimal transport such that the angular block edges are not abraded. **Agglomerates** are composed of volcanic bombs abraded and rounded during transport.

Tuffs and lapilli tuffs are created by pyroclastic flows. Pyroclastic flows consist primarily of hot gases, lithic clasts, crystals, and pumice shards and fragments. Relatively small or single event eruptions produce unwelded tuffs. **Unwelded tuffs** display random shard orientations and spherical to ellipsoidal pumice vesicles. Due to the relative absence of compaction, heating and plastic deformation, unwelded tuffs tend to be soft, low density rocks with light colors and ashy lusters.

Increasing volumes of pyroclastic debris produce more compaction and the generation of **partially welded tuffs**. **Welding** results as fragments become progressively fused together as porosity decreases during compaction. Compaction induces plastic flow deformation in which hot shards are flattened and rotated from a relatively random to a more parallel orientation. Partially welded tuffs display parallelism among shards, elongated and partially flattened pumice vesicles, and draping of glassy pyroclasts around rigid fragments. Partially welded tuffs are harder, darker and higher density rocks compared to unwelded tuffs (Figure 7.16).

Multiple, hot pyroclastic flows emplaced over short periods of time may form a single unit in which the deposits cool and compact together. While the exteriors of such units cool rapidly, the interior portions can remain hot for substantial periods of time. The variable cooling rate allows the shards and pumice fragments in the interior portion to remain plastic as the unit undergoes compaction. Plastic fragments tend to bend around or drape over rigid crystal and lithic fragments.

Figure 7.16 Partially welded ash flow tuff deposit. (Photo by Kevin Hefferan.) (For color version, see Plate 7.16, between pp. 248 and 249.)

Figure 7.17 Densely welded tuff deposit. (Photo by Kevin Hefferan.) (For color version, see Plate 7.17, between pp. 248 and 249.)

The flattening of pumice fragments produces a foliated (layered) texture.

Large volcanic eruptions marked by the deposition of numerous, thick pyroclastic layers results in intense welding. Intense welding produces hard, **densely welded tuffs** with dark colors and glassy lusters that may resemble obsidian. In densely welded tuffs, shards show marked parallelism and flattening (Figure 7.17). Pumice fragments are extremely elongated, with completely closed vesicles and pronounced draping of vitric pyroclasts.

As discussed above, rock texture is a keystone component in the classification of igneous rocks. In the following section, we will discuss chemical composition in igneous rocks. The final section of this chapter presents more detailed classification schemes based on textural and composition factors.

7.3 CHEMICAL COMPOSITION OF IGNEOUS ROCKS

We have already introduced the fact that SiO_2 is fundamentally important in magma classification and affects igneous textures. In this section, we will offer a more detailed discussion of the importance of magma chemistry. Magma chemistry is described in terms of major, minor and trace elements.

7.3.1 Major elements

Major elements have concentrations greater than 1 wt % in Earth's crust. Note in Figure 7.18 that silicon and oxygen – the key components in silicate magma – represent the bulk of Earth's crust by weight (75%) and volume (94.7%). The remaining 90 naturally occurring elements constitute only 25% by weight and 5.3% by volume of Earth's crust. In addition to silicon and oxygen, the other six major elements in Earth's crust include aluminum, iron, calcium, sodium, potassium, and magnesium.

The eight major elements constitute the bulk of Earth's crust, encompassing 99.2% by weight and nearly 100% by volume. As silicon and oxygen bond together to form the compound SiO_2, it should be no surprise that most crustal minerals (~92%) are silicate minerals. SiO_2 concentration is the primary compositional means by which igneous rocks are segregated. Aluminum and iron are the third and fourth most abundant elements by weight in Earth's crust. Each of these plays a role in igneous rock classifications.

Note that oxygen is the only anion (negatively-charged ion) listed among the eight most common crustal elements. Therefore, the other major cation (positively-charged ion) elements tend to bond with the anion oxygen. The eight major crustal elements –

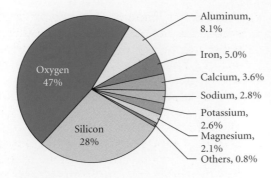

Element	Volume %	Ionic radius (angstroms)	Ionic charge
Oxygen	93.8	1.4	−2
Silicon	0.9	0.4	+4
Aluminum	0.5	0.5	+3
Iron	0.4	0.7	+2, +3
Calcium	1.0	1.0	+2
Sodium	1.3	1.0	+1
Potassium	1.8	1.4	+1
Magnesium	0.3	0.7	+2
Total	100.0		

Figure 7.18   The eight most abundant elements in Earth's crust by weight percent.

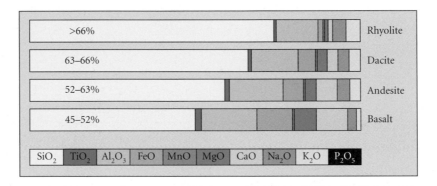

Figure 7.19   Oxide compounds in common volcanic rocks. (Courtesy of the US Geological Survey.)

seven cations and one anion – form seven major oxide compounds in igneous rocks: SiO_2, Al_2O_3, FeO, MgO, CaO, Na_2O and K_2O. In chemical analyses, geologists commonly express chemical components of igneous rocks as oxide compounds. For example, the chemical components of potassium feldspar ($KAlSi_3O_8$) can be expressed in terms of three different oxide compounds: $1/2(K_2O + Al_2O_3 + 6SiO_2)$. Therefore, in classifying igneous rocks (Section 7.6) several key questions relating to chemical composition and rock identification are:

- What is the percent SiO_2?
- What is the percent alumina Al_2O_3?
- What percent of the minerals are ferromagnesian bearing?

Figure 7.19 illustrates how the relative abundance of major oxide compounds serves to distinguish four common volcanic rocks. As might be expected, the SiO_2 abundance is the primary determining factor in classifying these rocks. Note that rhyolite contains >66% SiO_2, dacite 63–66% SiO_2, andesite 52–63% SiO_2 and basalt 45–52% SiO_2.

The second most abundant oxide compound after SiO_2 is the aluminum oxide compound. Aluminum oxide is primarily stored in the feldspar mineral group. The feldspar mineral group contains calcium, sodium and potassium oxides. Ferromagnesian minerals such as the pyroxene, amphibole and olivine minerals and biotite account for the iron oxide and magnesium oxide concentrations. Figure 7.19 also displays minor oxide compounds such as titanium oxide, manganese oxide and phosphate. Note the chemical trends depicted: as SiO_2 concentration increases, FeO, MgO, CaO and TiO_2 decrease while Na_2O and K_2O increase. These major element variations are critically important in the determination of magma and mineral chemistry as well as in rock nomenclature schemes.

7.3.2 Minor and trace elements

In addition to major elements, rocks also contain minor elements and trace elements. **Minor elements** consist of those elements that commonly occur in concentrations of 0.1–1.0% by weight. Chromium, manganese, phosphorous, hydrogen and titanium are among the most common minor elements in igneous rocks; in fact these elements may occur in concentrations normally associated with major elements (>1 wt %). Element concentrations vary somewhat among rock types. Minor element concentrations are related in part to SiO_2 concentrations. Igneous rocks with SiO_2 concentrations between 45 and 52% are relatively enriched in minor elements such as Cr, Ni and Cu. Igneous rocks with SiO_2 concentrations in excess of 66% tend to be enriched in minor elements such as Li, Be and Ba.

Trace elements occur in crustal rock in concentrations of <0.1% by weight and are typically measured in parts per million (<1000 ppm). Many minor and trace elements are economically important in part due to their scarcity. Although their concentrations are infinitesimally small, trace elements can provide information on the genesis and history of igneous rocks. Trace elements may be regarded as the DNA of magma. Trace elements are used as tectonic environmental indicators, essentially providing information on the origin and evolution of magma. How is this accomplished? Trace elements bear chemical characteristics that are said to be "compatible" versus "incompatible" or "mobile" versus "immobile".

7.3.3 Element compatibility

Compatibility is a measure of the ease with which an element fits into a crystal structure, and is analogous to the description of a compatible or incompatible person. A compatible person might be described as someone who gets along well with other people, fits into social structures, tends to form long-lasting bonds and does not leave when conditions

change. An incompatible person is someone who does not get along with other people, tends not to fit into social structures, moves around and quickly departs from relationships when conditions get hot.

The same descriptions aptly describe compatible and incompatible elements. **Compatible elements** tend to form long-lasting bonds and incorporate into crystal structures. Compatible elements are **immobile** in that they do not readily migrate from the crystal structure. **Incompatible elements** do not fit easily into crystal structures and their bonds are easily broken. Incompatible elements are **mobile** in that they tend to migrate from crystal structures into the melt when the rock is subjected to partial melting. Under conditions that lead to partial melting of a parent rock, compatible elements tend to remain within the solid crystalline structure and do not migrate with the melt. The solid **residual rock**, or **restite** remaining after melt removal, has a different chemical composition than the original parent rock. The restite is enriched in compatible elements and depleted in incompatible elements. Conversely, magmas derived from partial melting are enriched in incompatible elements and depleted in compatible elements.

Thus, the relative abundance of compatible and incompatible trace elements provides important information related to the melting history of igneous rocks. With small degrees of partial melting, incompatible elements such as K, Rb, Sr and Ba are depleted in the restite and enriched in magmas that migrate upward towards Earth's surface. Rocks derived from partial melts are enriched in incompatible elements and depleted in compatible elements. The restite rock, from which the partial melts have been removed, tends to be enriched in compatible elements such as Fe, Mn, Zn, Ti, V, Cr, Co, Ni and Cu, and depleted in incompatible elements. For example, igneous rocks generated by partial melts, such as ocean ridge basalts, are incompatible element enriched compared to the restite mantle. Detailed studies of trace element concentrations are used by petrologists to infer melting histories and to indicate tectonic environments for the genesis of rocks at ocean ridges, ocean islands or volcanic arcs (Box 7.1). We will return to this discussion in Chapter 10 when we address igneous associations.

Box 7.1 Trace elements

Trace elements include rare Earth elements, high field strength elements and large ion lithophile elements. **Rare Earth elements (REE)** have atomic numbers ranging from 57 to 71. In Earth's crust, REE with odd atomic numbers are more abundant than REE with even atomic numbers. This odd–even imbalance makes graphing REE abundances difficult. To correct this imbalance, geologists calculate the REE of Earth's crustal rocks and divide them by REE abundances in chrondritic meteorite samples. Why use chrondritic meteorites as a comparison rock? Chrondritic meteorites are thought to have formed at the same time as Earth (4.55 billion years ago). They are further thought to contain the same relative proportions of elemental abundances. However, unlike meteorites, Earth's crustal rocks have experienced partial melting. The partial melting has altered the REE abundances from their original (primordial) concentration, due to the mobility of incompatible elements. Thus, dividing the REE concentration of an Earth crustal rock sample by the REE concentration in a chondritic meteorite is thought to be akin to comparing a sample rock to Earth's original REE concentration. This process standardizes the differences between even and odd atomic numbered REE concentrations and makes data presentation graphs easier to plot and comprehend. The net result typically produces a relatively flat REE concentration pattern. This process is referred to as a chondrite-normalized pattern (Hess, 1989). REE are divided into light and heavy rare Earth elements. **Light rare Earth elements (LREE)** such as lanthanum (La), cerium (Ce), praseodymium (Pr), neodymium (Nd) and samarium (Sm) are situated on the left end of the periodic table. **Heavy rare Earth elements (HREE)**, situated at the right end of the periodic table, include europium (Eu), gadolinium (Gd), terbium (Tb), dysprosium (Dy), holmium (Ho), erbium (Er), thulium (Tm), ytterbium (Yb) and lutetium (Lu). Although all REE are incompatible, LREE elements are more incompatible than HREE and for this reason we distinguish between the two sets of REE.

High field strength (HFS) elements are characterized as having a relatively high ionic charge (+3 or +4) for a given radius. These elements are categorized as "immobile", meaning that they tend not to be transported away under limited partial melting. HFS elements have an ionic radius : valence charge ratio of less than 0.2 and include Ti, V, Zr, Hf, Nb, and Ta. As immobile elements, HFS elements and HREE tend to remain with the parent or restite rock. As these elements are retained with the original parent source material, these elements have "long memories" and are useful in tracing mantle-related processes.

Trace elements containing an ionic radius : valence charge ratio of greater than 0.2 are referred to as **large ion lithophile (LIL)** elements. These elements include Cs, Ba, Rb, Sr, Pb, K, Na and Eu. LIL elements tend to be mobile in partial melts and depleted in the restite. LIL elements are useful in determining the role of hydrous fluid interaction and the parental source of the partial melt. For example, a lithospheric magma source is indicated if an igneous rock is enriched in Sr and Nd (LREE); an asthenospheric magma source is suggested if Sr and Nd are depleted in an igneous rock.

Not all incompatible elements are incompatible to the same degree. Rb, Ba and Th are among the most incompatible elements. Sm and Hf are moderately incompatible. Sr is more compatible than Rb; Sm is more compatible than Nd and Lu is more compatible than Hf. Furthermore, an element's relative compatibility is dependent upon the mineral chemistry within the magma. Eu is incompatible, except in the presence of plagioclase. In an oxidizing environment more of the Eu will be in the Eu^{+3} state (versus Eu^{+2}) and the radius of Eu^{+3} closely matches the radius of Ca ions in feldspar. As a result, Eu^{+3} is compatible in Ca-plagioclase. Likewise, chromium is compatible in the presence of pyroxene minerals, garnet or chromite. Thus, the relative abundance of the REE is related to the mineral chemistry of the magma. What is the significance of trace elements? As will be discussed in Chapter 10, trace elements may be used to determine the magmatic source, tectonic environment and age of igneous rocks. In some cases, these studies utilize different proportions of isotopes. For example, Faure (1977) calculated the mean Sr^{87}/Sr^{86} ratios of rocks to distinguish between different magmatic environments such as ocean floor (<0.7028), ocean islands (0.7039) and island arcs (0.7044).

Continued

Box 7.1 *Continued*

Discrimination diagrams (Chapter 8) are particularly useful for ancient orogenic belts where tectonic environments are uncertain. Cann (1970) and Pearce and Cann (1973) pioneered tectonic discrimination diagrams that display trace element and minor element concentrations. Using minor and trace element concentrations of Rb, Sr, Y and Nb, Cann (1970) distinguished between mid-ocean ridge basalt and ocean island basalt. Pearce and Cann (1973) discriminated mid-ocean ridge basalt, ocean island basalt, volcanic arc basalt and continental basalt by plotting concentrations of Ti, Zr, Y and Sr. Discrimination diagrams have also been developed on the basis of variations in minor and major element concentrations. Mullen (1983) used MnO_2, TiO_2 and P_2O_5 to identify five basalt environments. Pearce et al. (1975, 1977) distinguished between oceanic and continental basalts through variations in major element concentrations. This approach is particularly important for complexly faulted regions where the rocks have been displaced from their original site of development. In complex fault zones, such as convergent margin settings, the point of origin and adjacent rock contacts have long since been disrupted by faulting. Thus, geochemical tracers provide detailed evidence regarding the conditions in which the rock originally formed. However, one must also be cognizant of the fact that these rocks have not existed in a closed system. Hydrothermal alteration and metamorphism can profoundly alter the original rock chemistry. The final rock chemistry may represent a complex developmental history. The critical point of this discussion is that trace elements, which represent an infinitesimally small component of magma, play a profound role in allowing us to determine the genesis, history and age of igneous rocks.

7.4   MINERAL COMPOSITION OF IGNEOUS ROCKS

Let us focus on minerals that compose the bulk of igneous rocks. Minerals may be regarded as primary or secondary, depending upon the time of their development. **Primary minerals** are those minerals that crystallize directly from magma at elevated temperatures. **Secondary minerals** form later in response to chemical changes that affect an existing rock. Secondary minerals replace primary minerals or infill voids (amygdules) through alterations by hot solutions, chemical reactions with country rock, or other secondary alteration processes. For the purpose of this chapter we will discuss only primary minerals, which are subdivided into major minerals and accessory minerals.

Major minerals

The seven oxide compounds combine to form eight common **major or essential mineral** groups. Major mineral groups are those constituents that commonly occur in abundances of greater than 5%. These include the quartz, potassium feldspar, plagioclase feldspar, feldspathoid, mica, amphibole, pyroxene and olivine mineral groups (Table 7.4). Common feldspar minerals include the plagioclase minerals (anorthite to albite) and potassium feldspar minerals (microcline, orthoclase and sanidine). Feldspathoid minerals, common in SiO_2-poor rocks, include leucite, nepheline and sodalite. Mica minerals include muscovite, biotite, phlogopite and lepidolite. Amphibole group minerals include hornblende, riebeckite and richterite. Pyroxene group minerals include augite, diopside, pigeonite, aegerine, hypersthene, enstatite and bronzite. Olivine minerals range from forsterite to fayalite. The major minerals are not of great economic importance; their value rests in the fact that they are the most common minerals in igneous rocks.

Accessory minerals

Less common minerals are considered accessory minerals, which occur in concentrations of less than ~5%. Accessory minerals consist largely of oxide, sulfide and less commonly silicate minerals. Accessory minerals include magnetite, hematite, ilmenite, spinel, sphene, rutile, fluorite, chromite, zircon, corundum, apatite, cassiterite, pyrite, chalcopyrite, molybdenite, pentlandite, pyrrhotite, uvaro-

Table 7.4 Common oxide compounds and major minerals in igneous rocks.

Oxide compound	Major igneous minerals or mineral groups
MgO	Olivine, pyroxene, amphibole, biotite
FeO	Olivine, pyroxene, amphibole, biotite
Al_2O_3	Plagioclase feldspar, potassium feldspar, micas
CaO	Plagioclase
Na_2O	Plagioclase feldspar, feldspathoids (nepheline, sodalite, cancrinite)
K_2O	Potassium feldspar, micas, feldspathoids
SiO_2	Quartz group minerals

Table 7.5 Two color index rock schemes based on percent of dark-colored minerals (DCM).

Dark minerals (Hyndman, 1985)	IUGS color index rock modifier terms
<40% DCM: felsic	<35% DCM: leucocratic
40–70% DCM: intermediate	35–65% DCM: mesocratic
70–90% DCM: mafic	>65% DCM: melanocratic
>90% DCM: ultramafic	

vite garnet, pyrope garnet, melilite, monazite, epidote, allanite, tourmaline, topaz, columbite and uraninite.

While accessory minerals are volumetrically small, some of these critical minerals are of great value in our society (Box 7.2). Accessory minerals are valued as metallic ore deposits, gems, abrasives and for many other useful applications.

7.5 MINERAL TERMINOLOGY

Whenever possible, it is best to name and describe rocks based upon their major mineral content. Mineral content yields important clues regarding rock chemistry, particularly with respect to major oxide compounds such as silica and aluminum oxide. Mineral composition and rock texture also help us understand the conditions in which the rock formed. Different sets of terminology have been developed to describe mineral components in igneous rocks.

7.5.1 The color index

The color index (CI), also called the mafic index, *is the proportion of mafic minerals in the total population of felsic and mafic minerals* given as:

$$CI = [\% \text{ mafic minerals}/(\% \text{ felsic minerals} + \% \text{ mafic minerals})] \times 100$$

The color index classification is based on the proportion of light- and dark-colored crystalline minerals and does not consider dark-colored, non-crystalline solids such as glass. In all cases, color should be used with caution. For example, rhyolites with high glass contents can be quite dark.

Igneous rocks can be subdivided into three or four groups (Table 7.5) based upon the percentage of dark-colored minerals (DCM). The four group classification consists of felsic (<40% DCM), intermediate (40–70% DCM), mafic (70–90% DCM) and ultramafic (>90% DCM) rocks. The International Union of Geological Sciences (IUGS) developed a color index scheme using rock name modifiers (Streckeisen, 1976). These modifiers consist of the terms leucocratic, mesocratic and melanocratic. Leucocratic rocks contain 0–35% DCM, mesocratic rocks contain 35–65% DCM and melanocratic rocks contain 65–100% DCM (Table 7.5).

7.5.2 Determination of mineral composition in igneous rocks

For rocks in which minerals are visible to either the naked eye or with a petrographic

Box 7.2   Elements and life

Look back over the chemical formulas for the major and accessory minerals. What do all these minerals have in common? The major and accessory minerals are enriched in the following major elements: Si, O, Al, Ca, Na, K, Fe and Mg as well as minor elements such as Ti, Mn, Cr, Co, Zn, Cu and P. Now compare these elements with the ingredients on your typical cereal box. These cereal ingredients may include Ca, Na, Zn, K, Fe, Cu, Mg and P – elements essential for the vitality of living organisms. Calcium, phosphorous and magnesium are critical to bone strength and growth. Calcium also stimulates hormonal secretions, activates enzymes, regulates heart rhythms and enhances the transmission of nerve impulses. Zinc minimizes macular degeneration. Magnesium lowers blood pressure and strengthens the heart. Sodium regulates blood pressure, but too much sodium may result in high blood pressure. Potassium is essential for structural proteins and reduces blood pressure and the risk of hypertension. Iron enhances dissolved oxygen levels in the blood. How important is oxygen? Oxygen not only directly promotes brain and heart function but also constitutes ~60% of the total body weight in the form of water, proteins, nucleic acids, carbohydrates and fat. Chromium allows the body to metabolize glucose. Other essential elements include nickel, molybdenum, carbon and iodine.

How do these elements become part of our foods? Well, crops are grown in fertile soil enriched in these essential elements. Crops absorb these elements, which have been dissolved and leached from the soil. Soil develops by the breakdown of rocks including igneous rocks, again generally involving water and exposure to oxygen. Igneous rocks, in particular, are known to provide the source material for among the most fertile soils on Earth. This fact has both positive and negative aspects. The positive aspect is that fertile farmland forms in regions where volcanic activity has occurred. For example, premium coffee is grown in Kona, Hawaii on volcanic soils enriched in nutrients. The negative side is that people tend to farm land in the shadow of dangerous volcanoes. In the event of an eruption, populations are at great risk.

Igneous rocks are also the source of most of Earth's metal deposits used in manufacturing, construction, medicine and high technology industries. Metals are widely used in the construction of steel for automobiles, skyscrapers and ships. Metals are also used extensively for their excellent conductive properties for electrical components such as batteries, wiring, cellphones and computers. Magma is enriched in metallic elements that include platinum, gold, copper, nickel, molybdenum, zinc, manganese, cobalt, titanium and chromium. Many of the sedimentary metal deposits, such as placer deposits, also are originally derived from the reworking of an igneous source. Elements within igneous rocks are fundamentally important to our health and the vitality of our global society.

microscope, one can determine the mineral percentages using modal composition. For very fine-grained or glassy rocks in which minerals are either too small to identify or not present, scientists use a system referred to as normative classification. Both of these techniques will be briefly discussed below. Normative classifications can be used for any igneous rock, whereas modal classifications are commonly used only for rocks in which the minerals can be identified.

Modal composition

The most straightforward approach to determining rock mineralogy involves visually identifying the minerals and determining their percentages by volume. The composition of a rock determined by actual mineral identification is referred to as its **modal composition** or **mode**. In coarse-grained phaneritic and pegmatitic rocks, a modal composition can be estimated by visual inspection of a rock, perhaps with the aid of a hand lens. For finer grained crystalline rocks, modal composition is determined using a petrographic microscope.

A much more accurate mode can be calculated for any coarse-grained rock by doing a **point count analysis** with a petrographic microscope. Accurate point count analysis requires moving a thin section incrementally on a grid system such that at least 400 mineral points are tabulated for each thin section. For each individual point, the scientist identifies the mineral and keeps a running total of all the major mineral points using a counter. A counter is similar to the device used by a baseball umpire to count balls, strikes and outs. However, in this case, the "umpire" is counting quartz, plagioclase and other modal minerals. The mineralogical point count results are then tabulated and expressed in volume percent. Modal analysis is a direct measuring technique in which constituent mineral grains are identified and the volume percentages determined. The modal composition approach works well for crystalline rocks in which minerals are readily identifiable but fails for very fine-grained or glassy igneous rocks in which individual minerals can not be identified or do not exist.

Normative composition

Normative mineralogy (Cross et al., 1902; Kelsey, 1965) is an indirect scheme using data derived from chemical analysis of a rock sample. The first norm classification was devised by Cross, Iddings, Pirsson and Washington (Cross et al., 1902), and is referred to as the **CIPW norm** classification in their honor. Normative classification systems are commonly used in aphanitic or glassy volcanic rocks, in which a rock's modal mineral composition can not be determined.

In normative calculations, specific rules are used to convert rock bulk chemical composition into a hypothetical suite of minerals, referred to as normative minerals, thought likely to occur. A norm is a process that takes a rock bulk chemical composition and creates a hypothetical set of minerals using that bulk chemical dataset. This method involves assumptions. First, one hopes that the hypothetical set of minerals chosen closely resembles the actual minerals in the rock or potential minerals that would have formed. Second, normative mineralogy is also calculated on the assumption that the magma crystallized near Earth's surface at low pressure, anhydrous conditions that approximate those under which many volcanic rocks form. Thus, standard CIPW normative calculations do not consider the role of high pressure or hydrous mineral phases such as micas and amphiboles. Given that hydrous igneous minerals include the mica and amphibole

mineral groups, this is a major omission. Special non-CIPW normative calculations exist to calculate normative minerals for rocks thought to be enriched in hydrous or other high pressure minerals.

Normative values are calculated using a multistep procedure utilizing whole rock chemical analysis of a rock sample. The chemical composition data are expressed in oxide weight percents (for example, FeO, Al_2O_3, etc.). These oxide weight percents are determined by powdering a rock and deriving geochemical results from an analytical spectrometer. Standard CIPW normative minerals include common anhydrous igneous minerals such as quartz, orthoclase, albite, anorthite, nepheline, magnetite, ilmenite, apatite, corundum, diopside, enstatite, hypersthene and olivine. For more detailed discussions beyond the scope of this text, the reader is referred to excellent textbooks by Winter (2009) and Best (2003).

7.5.3 Descriptive terminology based upon chemical composition

Numerous schemes have been developed to subdivide igneous rocks into distinctive groups based on major, minor and trace element abundances. In many cases these schemes have provided vital clues concerning the origin and significance of igneous rocks. Remember that the major elements of greatest abundance in Earth's crust include oxygen, silicon and aluminum which bond together to form silica and aluminum oxide. The most useful terminology for the chemical composition of igneous rocks naturally involves these three elements.

Abundance of silica

As silicon and oxygen are the primary chemical constituents in magma, the percentage of SiO_2 in rocks is an important means by which we classify magma and rocks. Two approaches have been developed as discussed below.

Acidic versus basic classification
On the basis of weight percent SiO_2, four major igneous rock types are defined: ultrabasic, basic, intermediate and acidic (Table 7.6).

Generally, the color index terms ultramafic, mafic, intermediate and felsic are consid-

Table 7.6 Two sets of terminology based on weight percent silica and color index (percent dark-colored minerals, DCM).

Weight percent silica	Percent dark-colored crystalline minerals
<45% SiO_2: ultrabasic	>90% DCM: ultramafic
45–52% SiO_2: basic	70–90% DCM: mafic
52–66% SiO_2: intermediate	40–70% DCM: intermediate
>66% SiO_2: acidic (silicic)	<40% DCM: felsic

Table 7.7 Silica saturation and key mineral indicators.

Silica saturation	Key mineral indicators
SiO_2 oversaturated	Quartz ± feldspars and/or magnesium orthopyroxene
SiO_2 saturated	Feldspars and/or magnesium orthopyroxene only
SiO_2 undersaturated	Forsterite olivine, nepheline, leucite and other feldspathoids ± feldspars and/or orthopyroxene minerals. Excludes quartz

ered loosely equivalent to the SiO_2 content terms ultrabasic, basic, intermediate and acidic, respectively. However, this approach is incorrect. Strictly speaking, the terms ultramafic and mafic refer to rocks containing >90% and >70% dark-colored minerals, respectively. The basic/acidic terminology refers to the percent SiO_2. The basic/acidic terms were derived from the mistaken idea that SiO_2 was an acidic compound and that elements such as iron and magnesium behaved as bases in magmas. Despite the fact that these terms are outdated, they are still used in the literature and therefore discussed in this text. A second igneous rock classification relates to SiO_2 saturation.

Silica saturation classification
Silica saturation is a concept that grew out of CIPW normative calculations. In such calculations, all oxides reported in the chemical analysis are used to make minerals according to a set of more than 20 steps that usually involves combining other oxides (e.g., Al_2O_3) with SiO_2. Three normative minerals provide useful examples of the concept of SiO_2 saturation:

- Normative orthoclase ($KAlSi_3O_8$) is produced by combining ½K_2O + ½Al_2O_3 + 3SiO_2.
- Normative albite ($NaAlSi_3O_8$) is created by combining ½Na_2O + ½Al_2O_3 + 3SiO_2.

- Normative enstatite ($Mg_2Si_2O_6$) is formed by combining 2MgO + 2SiO_2.

The concept of SiO_2 saturation (Table 7.7) is based on the presence or absence of three mineral groups: quartz, feldspars and feldspathoids.

Silica oversaturation

Silica oversaturation implies that all available cation oxides have been used to make normative minerals and additional SiO_2 remains available to generate normative quartz. The excess SiO_2 is indicated by the presence of "free quartz". All **quartz normative rocks are therefore oversaturated with SiO_2**. Most rocks with significant modal quartz are likely to also be quartz normative and therefore oversaturated with SiO_2. Granitic rocks tend to be oversaturated with SiO_2.

Silica saturation

Theoretically, a CIPW normative calculation may have exactly enough SiO_2 to consume all the other oxides. **SiO_2-saturated** rocks contain normative feldspars and/or orthopyroxene (enstatite or hypersthene) minerals, but lack either quartz – an indicator of SiO_2 oversaturation – or magnesium olivine or feldspathoids – indicators of SiO_2 undersaturation as discussed below.

Table 7.8 Description of igneous rocks based on aluminum oxide concentration.

Aluminum abundance	Al_2O_3 vs CaO, Na_2O, K_2O	Common minerals
Peraluminous	$Al_2O_3 > CaO + Na_2O + K_2O$	Muscovite, corundum, topaz, garnet, tourmaline, cordierite, andalusite, biotite
Metaluminous	$Na_2O + K_2O < Al_2O_3 < CaO + Na_2O + K_2O$	Hornblende, epidote, melilite, biotite, pyroxene
Subaluminous	$Al_2O_3 = Na_2O + K_2O$	Olivine, orthopyroxene, clinopyroxene
Peralkaline	$Al_2O_3 < Na_2O + K_2O$	Aegerine, riebeckite, arfvedsonite, aenigmatite, astrophyllite, columbite, pyrochlore

Silica undersaturation

Silica undersaturation occurs when, during CIPW normative calculations, SiO_2 is depleted before all the other oxides have been used to form normative minerals. In this case there may be insufficient SiO_2 to make quartz, feldspars or orthopyroxenes. SiO_2-undersaturated rocks commonly contain feldspathoid or magnesium olivine (forsterite) minerals that cannot coexist with quartz.

As we shall see, IUGS rock classifications based on modal mineralogies strongly reflect the concept of SiO_2 saturation because the primary mineral indicators include quartz (SiO_2 oversaturated), feldspars (SiO_2 saturated) and feldspathoids (SiO_2 undersaturated). It is also worth noting that most ultrabasic rocks are also SiO_2 undersaturated whereas most acidic rocks are SiO_2 oversaturated.

Relative abundance of aluminum oxide

Aluminum oxide is the second most abundant compound in Earth's crust. Igneous rocks are classified based upon the relative proportions of Al_2O_3 to CaO, Na_2O and K_2O. The relative proportion of these oxides yields the following descriptive terms: peraluminous, metaluminous, subaluminous and peralkaline (Table 7.8). The relative abundance of Al_2O_3 as compared to $CaO + Na_2O + K_2O$ largely determines the mineral assemblages that develop in igneous rocks (Shand, 1951; Hyndman, 1985). This classification is particularly useful for the discrimination of granitic rocks.

- **Peraluminous** rocks are characterized by minerals with unusually high Al_2O_3 contents.
- **Peralkaline** rocks contain normative or modal minerals with unusually high K_2O and/or Na_2O contents.

- **Metaluminous** rocks contain mafic minerals with average aluminum contents.
- **Subaluminous** rocks contain mafic minerals with low aluminum concentrations.

7.6 IGNEOUS ROCK CLASSIFICATION

Early in this chapter, we presented a simplified igneous rock classification for crystalline rocks using mineral components and texture. In reality, many different igneous rock classifications exist incorporating hundreds of possible rock names. The most widely used rock classifications identify igneous rocks based on texture and (1) modal minerals identified in the rock, (2) theoretical normative minerals calculated from chemical composition data from laboratory analyses, or (3) chemical composition of the rock based on laboratory analytical methods. One attempt to standardize rock nomenclature was initiated in the 1960s by Albert Streckeisen on behalf of the IUGS. The IUGS recommended a classification system for both plutonic and volcanic rocks using essential mineral groups as endpoints in triangular- and diamond-shaped diagrams (Streckeisen, 1976; LeBas and Streckeisen, 1991; LeMaitre, 2002). While the IUGS classification is generally accepted, it is not comprehensive to all igneous rocks and pre-existing rock nomenclature remains in use. The following discussion provides a summary of the IUGS classification system, pointing out the benefits of a unified classification approach as well as the drawbacks.

7.6.1 IUGS classification of mineral groups, QAPF

The IUGS classification is based upon the modal concentrations of five essential mineral groups abbreviated by the letters Q, A, P, F and M, respectively:

- **Q** = quartz, tridymite, cristobalite.
- **A** = alkali feldspar, including orthoclase, microcline, sanidine, perthite, anorthoclase and albite plagioclase with up to 5 mole % anorthite (An_0–An_5). Mole percent is calculated by taking the weight percent of a mineral and dividing by the mineral's molecular weight.
- **P** = plagioclase (An_5–An_{100}) and scapolite (altered plagioclase).
- **F** = feldspathoids, also known as foids. The term foid is derived from being *feld*-spath*oid* rich. Feldspathoids include the minerals nepheline, sodalite, cancrinite, leucite, analcite, nosean, hauyne and kalsilite. In naming a rock, we use the major feldspathoid mineral as either an adjective or as part of the noun. For example, instead of naming a leucite-rich syenite a "foid-bearing syenite", the rock would be called a "leucite-bearing syenite" or a nepheline syenite.
- **M** = mafic and related minerals, including olivine, pyroxene, amphiboles, micas, melilite, opaque minerals, garnet, epidote, calcite, allanite, zircon, apatite, sphene and titanite (Streckeisen, 1976).

The QAPF modal classification applies for igneous rocks with >10% felsic minerals and <90% mafic mineral (M) content by volume. The QAPF plot is a diamond-shaped diagram (Figure 7.20). The upper QAP portion of the diagram distinguishes SiO_2-oversaturated rocks containing quartz. The lower FAP portion of the diagram consists of SiO_2-undersaturated rocks containing feldspathoids minerals but not quartz. Line AP represents the line of SiO_2 saturation. The QAPF mineral constituents are normalized so that the sum total of these four mineral constituents equals 100% as follows:

$$\%Q = [Q/(Q + A + P + F)] \times 100$$
$$\%A = [A/(Q + A + P + F)] \times 100$$
$$\%P = [P/(Q + A + P + F)] \times 100$$
$$\%F = [F/(Q + A + P + F)] \times 100$$

The QAPF diagram was originally designed for the classification of plutonic igneous rocks.

Thereafter, it was adapted for volcanic crystalline rocks.

7.6.2   IUGS plutonic rock classification

The IUGS classification diagram for plutonic igneous rocks with >10% felsic minerals (Q, A, P or F) and <90% mafic minerals is presented in Figure 7.20. Thirty-five different plutonic rocks are recognized within the plutonic QAPF classification.

Careful observation illustrates some problems with the IUGS system. Near the plagioclase (P) corner, quartz diorite and quartz gabbro occupy the same region; diorite, gabbro and anorthosite also coexist in another location. Similarly, feldspathoid-bearing gabbro, diorite and anorthosite rock names also coexist in the same foid regions. How do we discriminate among these rocks? For anorthosite, the answer is straightforward as anorthosites contain more than 90% plagioclase. Distinguishing between gabbro and diorite as well as quartz gabbro and quartz diorite is a bit more complex.

As indicated in Figure 7.20, gabbro and diorite each contain <5% quartz; while quartz gabbro and quartz diorite each contain 5–20% quartz. For both of these sets of rocks, three different factors distinguish quartz gabbro and gabbro from quartz diorite and diorite:

1. Gabbros/quartz gabbros contain more than 35% mafic minerals whereas diorites/quartz diorites contain less than 35% mafic minerals. This approach is based on visual identification of hand samples and is the method used in the IUGS system.
2. Gabbros/quartz gabbros are more calcic with plagioclase anorthite contents >50. Diorites/quartz diorites are more sodic with plagioclase anorthite contents <50. Determination of anorthite content requires microscopic analysis of thin sections in a laboratory.
3. Gabbros/quartz gabbros contain 45–52% SiO_2 and diorites/quartz diorites contain 52–66% SiO_2. Determination of the SiO_2 content requires detailed geochemical analysis of powdered rock specimens in a laboratory.

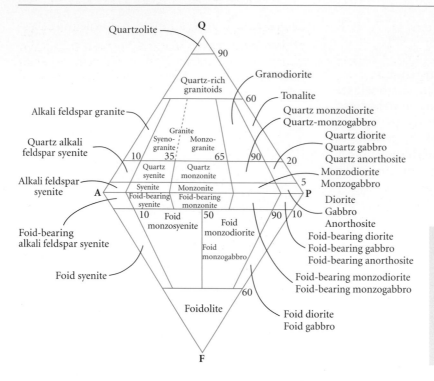

Figure 7.20 QAPF diagram for plutonic igneous rocks with >10% felsic minerals and <90% mafic minerals. Classification based upon the IUGS classification. (After Streckeisen, 1976; LeMaitre, 2002.)

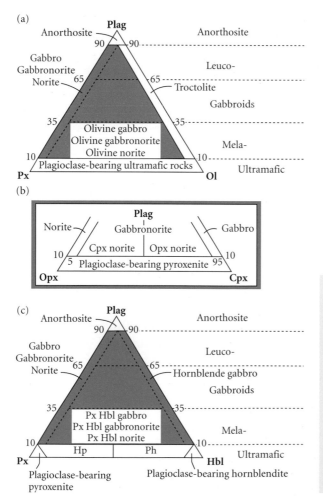

Figure 7.21 Modal classification of gabbroic rocks based on proportions of plagioclase (Plag), olivine (Ol), orthopyroxene (Opx), clinopyroxene (Cpx) and hornblende (Hbl). Note the terms leuco- and mela- may be used as adjectives to denote light colored and dark colored, respectively. (a) Gabbroic rocks enriched in plagioclase, pyroxene and olivine. (b) Gabbroic rocks enriched in plagioclase, orthopyroxene and clinopyroxene. (c) Gabbroic rocks enriched in plagioclase, pyroxene and hornblende. (After Streckeisen, 1976; LeMaitre, 2002.)

7.6.3 IUGS gabbroic rock classification

Gabbros may be further segregated based upon the modal mineral proportions of plagioclase (Plag), olivine (Ol), pyroxene (Px) and hornblende (Hb) as illustrated in Figure 7.21. Note that gabbros and norites coexist in the same region within this triangle. The distinction between these rocks is based upon whether the pyroxene minerals are orthopyroxenes (norite) or clinopyroxenes (gabbro). Orthopyroxenes crystallize in the orthorhombic system and include the minerals enstatite and hypersthene. Clinopyroxenes crystallize

in the monoclinic system and include the minerals augite and pigeonite. In these triangular diagrams, mineral abundances are recalculated so that the sum of the three mineral proportions equals 100%.

7.6.4 IUGS ultramafic rock classification

As stated previously, the QAPF classification scheme is applied when felsic minerals compose >10% of the rock (Streckeisen, 1973; LeBas and Streckeisen, 1991; LeMaitre, 2002). A different set of triangular rock discrimination plots (Figure 7.22) is

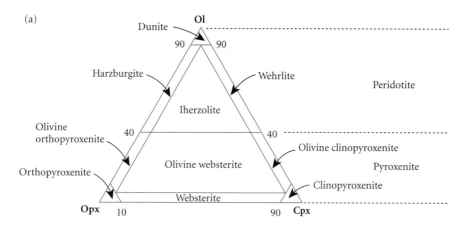

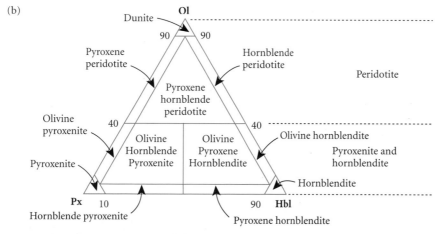

Figure 7.22 Modal classification of ultramafic plutonic rocks based on proportions of olivine (Ol), orthopyroxene (Opx), clinopyroxene (Cpx), pyroxene (Px) and hornblende (Hbl). (a) Plots of olivine, orthopyroxene and clinopyroxene. (b) Plots of olivine, pyroxene and hornblende for plutonic rocks that contain more than 10% essential hornblende. Note also that the terms pyroxenite and hornblendite may be used as generalized rock terms with preceding adjectives (e.g., olivine pyroxenite) for rocks enriched in either pyroxene or hornblende but containing less than 40% olivine. A more specific use for these rock names is shown at the lower two corners of (b) in which pyroxenites are defined as containing more than 90% pyroxene and hornblendites as containing more than 90% hornblende. (After Streckeisen, 1973; LeMaitre, 2002.)

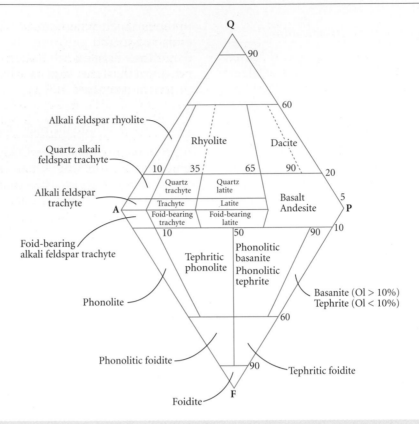

Figure 7.23 QAPF diagram for volcanic igneous rocks with >10% felsic minerals and <90% mafic minerals. The classification is based upon the IUGS classification. (After Streckeisen, 1976; LeMaitre, 2002.)

used for ultramafic plutonic rocks containing >90% dark-colored minerals. The ultramafic family includes peridotites (>40% olivine), pyroxenites (pyroxene-rich rock with <40% olivine) and hornblendites (hornblende-rich rock with <40% olivine). The minerals are recalculated so that the sum of the three mineral proportions of each triangular diagram equals 100%.

7.6.5 IUGS volcanic rock classification

The IUGS rock classification system has also been applied to QAPF volcanic rocks. Figure 7.23 presents a volcanic rock classification based upon the relative abundances of mineral groups: quartz (Q), alkali feldspars (A), plagioclase (P) and feldspathoids (F). The IUGS system does not include separate diagrams for mafic or ultramafic volcanic rocks.

7.6.6 IUGS classification drawbacks

Together the QAPF and the mafic/ultramafic classification diagrams adequately discriminate most plutonic igneous rocks based on modal mineralogy. The IUGS modal classification can be utilized for QAPF volcanic rocks for which mineral identification is possible. The IUGS classification system does have its drawbacks. The IUGS system focuses on rocks whose primary minerals are in the QAPF or M categories, and does not emphasize less common rocks such as carbonatites. Also, while the IUGS system is designed to enforce a uniformity in rock naming procedure, it cannot undo the wide variety of rock names already entrenched in the literature. The IUGS system also does not consider textural classifications other than phaneritic and aphanitic. The system does not address por-

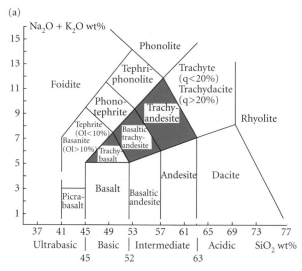

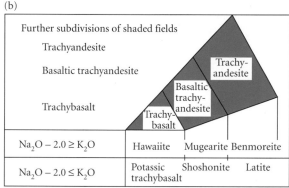

Figure 7.24   (a) Alkali oxide versus silica classification diagram for volcanic rocks.
(b) Rocks falling in the shaded areas may be further subdivided as shown below. This classification utilizes the IUGS rock terms. (After LeBas et al., 1986.)

phyritic or pegmatitic crystalline textures. Furthermore, it completely ignores noncrystalline textures. The IUGS system is a modal system, which requires that three or four key minerals can be identified. Applying this system to volcanic rocks is problematic in that accurate identification of minerals in fine-grained rocks is not always possible. As a result, the IUGS modal classification is not always useful in the case of fine-grained

igneous rocks wherein mineral identification is not possible.

Other classification schemes that utilize chemical analyses can be used to distinguish and name volcanic rock types. Chemical composition of very fine-grained rocks is determined in a geochemistry laboratory, typically by crushing and dissolving the rock specimen and using a spectrometer. One of the most common classification schemes utilizes alkali–SiO_2 plots for the classification of very fine-grained igneous rocks. The chemical parameters used to classify fine-grained igneous rocks are SiO_2 and $Na_2O + K_2O$.

7.6.7   Total alkali–silica classification

Several alkali–SiO_2 classification schemes have been developed for volcanic rocks (LeMaitre, 1984; LeBas et al., 1986). These classification schemes are useful for volcanic rocks wherein the mineral mode cannot be determined due to the presence of glass or cryptocrystalline texture. Where these conditions exist, chemical analyses present the only available option. The **total alkali to silica (TAS)** classifications are suitable for use with extremely fine-grained or glassy volcanic rocks that are relatively unaltered and contain less than 2% H_2O and 0.5% CO_2. In a TAS system, total alkalis ($Na_2O + K_2O$) are plotted on the ordinate (vertical y-axis) against SiO_2, which is plotted on the abscissa (horizontal x-axis). These rock fields are defined by total alkalis ($Na_2O + K_2O$) that range from 0 to 16% and percent silica that ranges from 35 to 77%. The TAS classification identifies 16 different volcanic rock fields (Figure 7.24).

Igneous rock classification is based upon texture and composition. Despite the rather straightforward approach to identifying rocks, hundreds of igneous rock names have been proposed over the past 200 years. In succeeding chapters we will look further at why such a great diversity of igneous rocks occur on Earth as well as the plutonic and volcanic settings in which we find igneous rocks.

Chapter 8

Magma and intrusive structures

8.1 ROCK MELTING

The great variety of magmas and lavas that solidify to produce igneous rocks in Earth's crust are initially formed by a process called **anatexis**, which refers to the partial melting of a source rock. Anatexis produces: (1) a liquid melt fraction enriched in lower temperature constituents, and (2) a residual rock component enriched in higher temperature, refractory elements. The type of magma produced by partial melting and subsequent processes depends upon factors such as:

- The composition, temperature and depth of the source rock.
- The percent partial melting of the source rock.
- The source rock's previous melting history.
- Diversification processes that change the composition of the magma after it leaves the source region.

The roles of each of these factors will be discussed in this chapter. Rock melting within

Earth Materials, 1st edition. By K. Hefferan and
J. O'Brien. Published 2010 by Blackwell Publishing Ltd.

Earth's interior is a complex process that has not been observed firsthand. However, rock melting has been experimentally modeled in the laboratory since the early 20th century, providing insight into both magma genesis and the crystallization processes that produce most igneous rocks. In the simplest approach, rock melting within Earth's interior is modeled as two idealized end member processes: equilibrium melting and fractional melting.

8.1.1 Equilibrium melting

Equilibrium melting occurs in a closed system where chemicals are neither added nor removed from the plutonic environment. Equilibrium melting requires that the melt remains in contact with the residual rock throughout the melting process. As a result, the overall composition of the system remains the same while the composition of melts and solids evolves as follows:

1 The melt becomes enriched in low melting temperature constituents. This is most pronounced with small degrees of partial melting (<10%) because the lowest

temperature constituents preferentially enter the melt first.

2 Increased degrees of partial melting (e.g., >30%) dilute the enrichment of low temperature constituents in the melt; the melt becomes progressively less enriched in low melting temperature as higher temperature constituents enter the melt.

3 The solid refractory residue is enriched in high melting temperature constituents.

4 With continued partial melting, the solid residue becomes progressively more enriched in the refractory constituents.

Real magmatic systems commonly display incomplete equilibrium assemblages due to incomplete chemical reactions between the crystals and melts. Incomplete chemical reactions may be due to large crystal size, high magma viscosity and/or low ion migration rates. The net result is the generation of zoned crystals (see Figure 7.2).

8.1.2 Fractional (disequilibrium) melting

Fractional melting implies that solids and melt separate into isolated fractions that do not continue to react together during the melting process. Because melts are separated from the refractory crystals, liquids and crystals do not remain in equilibrium. Fractional melting produces a melt that is more evolved than the parent source rock from which it was derived. For example, in fractional melting of rocks containing plagioclase and olivine, the early melts are highly enriched in low melting temperature constituents – such as sodium plagioclase and iron olivine – leaving behind a more refractory residual solid enriched in calcium plagioclase and magnesium olivine. Each succeeding melt will be less enriched in low melting temperature constituents than the initial melts. As a general rule, small degrees of partial melting of undepleted (previously unmelted) source rocks produce melts that are highly enriched in low temperature constituents. Larger degrees of melting of previously depleted source rocks produce melts that are significantly less enriched in low temperature constituents. Fractional melting is especially important in the generation of basalt by the partial melting of mantle peridotite.

While equilibrium melting and disequilibrium melting represent two idealized end member models, they provide useful starting points for the more complicated melting and diversification processes that affect magmas within Earth. These complications include: the exchange of elements between the system and an external source, melt interactions with surrounding mantle and crust rocks, the incorporation of exotic rock fragments and crystals, and multiple magma injections that can mix magmas of different composition. These are but a few possible complexities that may significantly alter magma chemistry within plutonic systems.

8.2 FACTORS IN ANATEXIS AND INITIAL MELT COMPOSITION

Why do solid rocks melt within Earth? Anatexis is initiated by some combination of increasing temperature, decreasing pressure or increasing volatile content, especially water vapor pressure (P_{H2O}). Let us briefly discuss these melting factors.

8.2.1 Increasing temperature

Increasing temperature is the most obvious cause for partial melting. Temperature increases with depth within Earth and can be represented by a sloping line called the **geothermal gradient**. The geothermal gradient is not uniform vertically or laterally within Earth. Earth's average geothermal gradient is ~25°C/km for the upper 10 km, but decreases with depth. The geothermal gradient also varies based upon rock age and tectonic setting, ranging from 5–10°C/km in old continental lithosphere to 30–50°C/km at hotspots, ocean spreading ridges and volcanic arcs. Elevated geothermal gradient sites constitute locations within Earth where mantle peridotite may melt as a result of increasing temperature. These higher temperature regions may be related to hotspots, magmatic intrusions or, less commonly, to localized frictional heating or high concentrations of radioactive elements.

8.2.2 Decreasing pressure

Decompression melting, also known as **adiabatic melting**, results from a decrease in pressure. Pressure is related to rock depth whereby 10 km depth corresponds approxi-

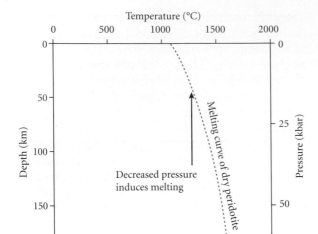

Figure 8.1 A pressure decrease may result in melting as the upwelling mantle rock intersects the dry melting curve for peridotite.

mately to 3.3 kbars. In volatile-poor systems, for example those with low water vapor contents, melting temperatures are proportional to pressure. Assuming all other factors are constant, the higher the pressure, the higher the melting temperature. Conversely, as pressure decreases, melting temperature decreases (Figure 8.1). As a result of lithosphere thinning in extensional environments, the underlying mantle rises upward, effectively decreasing the lithostatic stress. This decompression reduces rock melting temperatures. Decompression melting of mantle peridotite is the primary means by which basaltic magmas are generated at ocean spreading ridges and continental rifts. Decompression also plays an important role in hotspot regions where warm rocks become less dense, rise and undergo decompression melting.

8.2.3   Volatile-induced melting

Elevated volatile content under pressure, particularly P_{H2O}, may significantly lower melting temperatures (Figure 8.2). In addition to H_2O, other volatiles include compounds such as CO, CO_2, OH, SO_2, H_2S, NH_3, K_2O, Na_2O, HCl and HF, as well as the elements H, F, Cl, S, He and Ar. Volatiles act as a flux. A **flux** is an agent that reduces the melting tempera-

ture of a substance. Water vapor dissolved in magma under pressure tends to weaken Si–O bonds in silicate minerals. As silica bonds are progressively weakened, progressively lower temperatures are necessary to melt solid rock. As a result, an increase in P_{H2O} can induce rock melting at a constant temperature and pressure. The addition of relatively small amounts of water, even less than 1%, under high pressures can significantly lower rock melting temperatures. Elevated P_{H2O} also allows rocks to melt over a wider temperature range, strongly counteracting the effects of lithostatic pressure. It also tends to decrease the amount of FeO and MgO incorporated into partial melts. As a result, magmas produced by P_{H2O}-induced partial melting ("wet" melting) tend to be less mafic than the source rock.

The addition of volatile compounds such as H_2O, CO_2, K_2O and Na_2O is the major cause for partial melting in subduction zones. As subducted lithosphere penetrates to depths of 80–150 km, temperature–pressure conditions cause hydrous minerals such as amphibole, mica and serpentine to become unstable. Dehydration of these minerals releases water vapor and other volatiles, raises volatile pressures, and induces melting in the overlying mantle wedge. The addition of volatile compounds may depress melting temperatures in excess of several hundred °C at subduction zones (Figure 8.2).

Partial melting changes the parent rock composition by enriching the melt in low temperature components and enriching the refractory residue in high temperature components. But how can we account for such a wide variety of igneous rock types?

8.2.4   Partial melting and melt composition

Anatexis is one means by which diverse melts are generated. Anatexis generates a new magma derived from partial melting and leaves behind an unmelted, residual rock. Both the melt and the residual rock have compositions that differ from that of the source rock. Anatectic melts tend to be more enriched in low melting temperature constituents than the parent rock or the refractory residual rock. Partial melting removes incompatible elements, including large ion lithophile elements (Chapter 7), as well as SiO_2, K_2O and

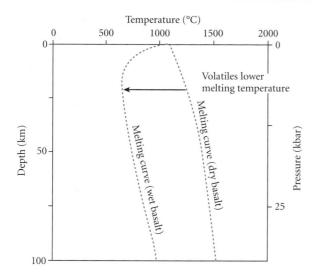

Figure 8.2 The addition of volatiles reduces the melting temperature of basalt by hundreds of °C. (After Green and Ringwood, 1967.)

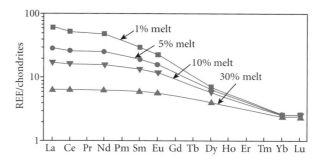

Figure 8.3 Incompatible light rare Earth elements (LREE) are progressively enriched with ever smaller degrees of partial melting. As greater degrees of partial melting occur, the rare Earth element pattern more closely approximates that of the parental material such as chondrites. Chondrites are thought to be chemically equivalent to the original Earth composition. (Courtesy of Stephen Nelson.)

Na_2O from the parent rock and concentrates these low melting constituents in the daughter magma. As a result of their concentration in the melt, the residual rock is depleted in incompatible elements and low melting temperature constituents. Rocks crystallizing from these melts tend to contain more silicic minerals, such as quartz, potassium feldspar and sodium plagioclase, than their parental source rock. The refractory residual rock tends to be enriched in high temperature constituents such as MgO, CaO and compatible elements such as high field strength (HFS) elements (Chapter 7). Common refractory minerals include olivine, pyroxene and calcium plagioclase. Despite these chemical changes, the anatectic melts and the residual rock remain genetically and chemically related to the original source rock. Thus, partial melting events generate magmas with greater incompatible elements and other low temperature constituent concentrations and refractory residues with progressively higher concentrations of compatible elements and other high temperature constituents.

The degree of partial melting plays a critical role in determining the degree of segregation between low melting temperature and high melting temperature constituents. As the degree of partial melting increases, the degree of enrichment of incompatible elements decreases. Recall from Chapter 7 that light rare earth elements (LREE) are generally more incompatible with solid minerals than are heavy rare earth elements (HREE). This means that LREE (La to Sm) are preferentially incorporated into melts; the smaller the degree of partial melting, the higher the LREE enrichment in the melt. Figure 8.3 illustrates how LREE enrichment may be used to infer the degree of partial melting of a garnet-bearing peridotite because garnet has a strong preference for HREE. At 1% partial melting, LREE are strongly enriched relative to HREE (Eu to Lu) and the degree of LREE enrichment decreases progressively as the degree of partial melting increases progressively to 30%.

Following partial melting, the residual mantle rock is depleted in incompatible elements, such as LREE and LILS (K → Cs), that are most incompatible compared to the original Earth composition. Melts derived from mantle melting are enriched in incompatible elements compared to original Earth concentrations (Figure 8.4). In this way, we can infer the nature and history of the source region by looking at the pattern of rare Earth elements and large ion lithophile (LIL) elements (discussed in Chapter 7) in the resulting melts and the rocks that form from them.

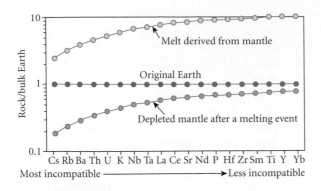

Figure 8.4 Melts derived from the mantle are enriched in incompatible elements; the residual mantle is, however, depleted in incompatible elements relative to primitive Earth concentrations. (Courtesy of Stephen Nelson.)

Multiple partial melting episodes of source rocks can produce the wide range of igneous rocks observed in Earth's crust. Because the low melting temperature constituents are concentrated in the melt, partial melts are characterized by a relatively high SiO_2 content and LIL content. In a very simplified model, with many local exceptions, partial melting of different rocks results in different products:

- Partial melting of ultrabasic peridotite generates a basic (basalt/gabbro) melt.
- Partial melting of basic rock produces an intermediate magma (andesite/diorite).
- Partial melting of intermediate rock yields a silicic melt (rhyolite/granite).

In this way, it is possible to imagine how a basic oceanic crust and a silicic–intermediate continental crust might have been extracted by multiple cycles of partial melting. These cycles are dominated by decompression melting of a refractory mantle, ocean ridge magmatism and flux melting of ocean lithosphere at subduction zones generating arc and continental crustal material. In addition to anatectic processes, the wide range of igneous rocks in Earth's crust is also due to magma diversification processes that occur after the melt separates from the source region, as discussed below.

8.3   MAGMA DIVERSIFICATION: DIFFERENTIATION, MIXING AND ASSIMILATION PROCESSES

In addition to partial melting of different source rocks, complex magma diversification processes are also responsible for producing Earth's wide array of magmas and igneous rocks. Diversification processes involve changes in bulk magma chemistry after its initial generation. Diversification processes include a number of closed-system differentiation processes as well as open-system processes related to magma–country rock interaction or interaction with other magmas. Closed-system and open-system diversification processes will be discussed in the following section.

8.3.1   Closed-system diversification

Differentiation includes all closed-system diversification processes in which the original melt evolves into one or more melts with a different composition, without material being exchanged with an external source. Closed-system differentiation processes include several **fractional crystallization** processes whereby early formed crystals are segregated from the remaining melt. Bowen (1928) was an early proponent of fractional crystallization. Bowen suggested that fractionation may be accomplished "through the relative movement of crystals and liquid". Bowen is regarded as the "father of modern petrology", so let us consider a few of his many important contributions.

Bowen's reaction series

In the late 19th century, geologists such as Reginald Daly began investigating the origin, composition and crystallization processes involved in generating igneous rocks. Norman Levi Bowen, a Canadian protégé of Daly, began working as an igneous petrologist at the Carnegie Institute of Washington in 1910. Petrologists are geologists who study the origin of rocks. Bowen studied rocks from a shallow igneous intrusion named the Palisades Sill, which is situated in New Jersey, just west of New York City. The Palisades Sill is a ~200 million-year-old intrusion that consists largely of basic rocks including basalt, diabase (coarse-grained basalt) and gabbro. Bowen noted that the mineralogy of the sill varied vertically. Certain levels within the sill were enriched in olivine, while other levels were enriched in orthopyroxene, clinopyroxene, plagioclase, and even in potassium feldspar and quartz. Through field observations and laboratory rock melting experiments, Bowen theorized that a crystal–liquid fractionation process must have been involved that allowed early formed crystals to be sequentially segregated from the remaining melt, causing the composition of the melt to change progressively as layers of different composition formed. Bowen melted igneous rock samples from the Palisades Sill in his Carnegie laboratory and developed a model based on the crystallization sequence of major igneous minerals. This model is referred to as **Bowen's reaction series** (Figure 8.5). Bowen recognized eight major mineral groups that constitute the bulk of most igneous rocks. These mineral groups are: olivine, pyroxene, plagioclase, amphibole, biotite, muscovite, potassium feldspar and quartz.

On the basis of his field and laboratory work, Bowen (1928) proposed a reaction series for cooling basic magma. Bowen's reaction series consists of a discontinuous and continuous crystallization series. The **discontinuous reaction series** involves distinctly different mineral groups, namely olivine, pyroxene, amphibole and biotite. In the discontinuous reaction series, high temperature refractory minerals such as Mg-rich olivine crystallize first, removing MgO from the remaining melt. Recall from Chapter 3 that as a melt cools to the peritectic, forsteritic olivine crystals react with the melt to produce the orthopyroxene mineral enstatite. This is a discontinuous melt–solid reaction that produces a new mineral with a different structure and composition. Such discontinuous reactions are possible if crystals are allowed to react (equilibrium conditions) with the melt. At still lower temperatures, discontinuous reactions convert pyroxene to amphibole and amphibole to biotite. These refractory minerals form a variety of basic rocks layers such as basalt, diabase and gabbro with various proportions of olivine, orthopyroxene, clinopyroxene, amphibole and biotite.

The **continuous reaction series** involves plagioclase which is characterized by a complete solid solution series within a single mineral group. The solid solution series exists between higher crystallization temperature CaO-rich plagioclase and progressively lower temperature, more Na_2O-rich plagioclase. In equilibrium conditions, early formed CaO-rich plagioclase crystals react with the melt to produce increasingly Na_2O-rich plagioclase crystals. Each of the minerals in the discontinuous and continuous reaction series forms in equilibrium conditions and follows the phase relationships detailed in Chapter 3.

When discontinuous and continuous reactions do not proceed to completion under equilibrium conditions, chemical variations can be preserved within crystals recording zoned crystals and reaction rims. **Zoned crystals** display a systematic pattern of chemical variations from the periphery of the crystal towards its center, recording an incomplete continuous chemical reaction between the crystal and the surrounding melt. For example, incomplete solid solution reactions involving plagioclase produce zoned crystals in which a CaO-rich plagioclase core may be rimmed by Na_2O plagioclase (see Figure 7.2). **Reaction rims** display a new mineral along the crystal periphery that surrounds a partially resorbed core composed of a different mineral. For example, hornblende rims commonly develop on augite cores. Reaction rims indicate incomplete discontinuous chemical reactions between crystals and melts.

Bowen's pioneering work on crystal–liquid reactions provided the foundation for under-

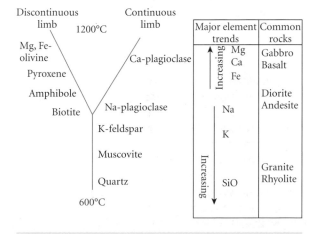

Figure 8.5 Bowen's reaction series.

standing mineral crystallization processes and magma evolution. From an original basic melt, Bowen inferred that a wide variety of rocks could crystallize through a sequential crystal–liquid separation process. Bowen outlined a fractional crystallization process whereby, as magma temperatures cooled below 1200°C, a distinct silicate mineral crystallization sequence ensued beginning with MgO-rich olivine, followed by CaO plagioclase and FeO olivine generating ultrabasic to basic rocks such as peridotite, basalt and gabbro. With the removal of MgO, FeO and CaO, the remaining melt is progressively enriched in SiO_2, K_2O, Na_2O and H_2O. With decreasing temperatures, increasingly more Na_2O-rich plagioclase continues to crystallize along with hydrous minerals such as hornblende and biotite producing intermediate rocks such as andesite and diorite. After most of the magma has crystallized, a SiO_2-, K_2O- and Na_2O-rich magma remains, from which the minerals muscovite, potassium feldspar and quartz crystallize at low temperatures. These three minerals are not part of a reaction series because they do not react with the magma, continuously or discontinuously, to produce minerals with a different composition. These three minerals form the silicic rocks such as granite and rhyolite.

Bowen suggested that fractional crystallization of basic magma could explain the widespread occurrence of silicic rocks such as the immense granite and granodiorite bodies exposed on continents. It is now recognized that fractional crystallization in itself cannot generate the vast volumes of silicic magma observed on Earth. Other processes must also be at play. Nevertheless, Bowen's reaction series remains a fundamental teaching tool useful in the visualization of magma crystallization sequences, melt–crystal interactions and the common association of major minerals in rock assemblages.

Fractional crystallization processes

Fractional crystallization involves the effective separation of crystals and melt from an originally homogeneous magma. Models for fractional crystallization involve marginal accretion, gravitational separation, convective flow and filter pressing. Each of these will be briefly discussed below.

Marginal accretion models

The interior of a magma chamber is well insulated, retains heat and cools relatively slowly. The periphery of the magma chamber in contact with surrounding, cool country rock loses heat by conduction, resulting in relatively rapid crystallization. In addition, crystals in the wall rocks provide nucleation sites for crystal growth. Crystallization along the walls of the magma chamber in which crystals preferentially form and adhere to the edges results in **marginal accretion**. As additional layers of crystal accrete to the walls, earlier formed crystals are no longer in contact with the magma and have been effectively segregated from it.

Marginal accretion may be subdivided based on the location (Figure 8.6). Heat rises so that the upper margin of the magma chamber may cool relatively quickly. As a result, **roof accretion** results from early crystallization of minerals along the ceiling or roof due to preferential heat loss. **Sidewall accretion** develops as the magma chamber walls release heat to the relatively cold country rock, generating crystals that adhere to the side margins of the magma chamber. **Floor accretion** occurs as crystals form along the base of the magma chamber. Each of these

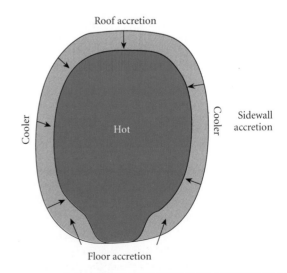

Figure 8.6 Marginal accretion due to preferential cooling of the perimeter of the magma chamber. (Courtesy of Stephen Nelson.)

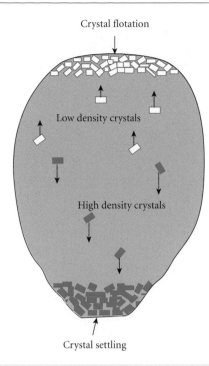

Crystal flotation

Low density crystals

High density crystals

Crystal settling

Figure 8.7 Gravitational separation involving crystal settling of high density crystals and crystal flotation of low density crystals in a magma chamber. (Courtesy of Stephen Nelson.)

marginal accretion processes may effectively separate early formed crystals from the remaining melt, enriching the melt in lower temperature constituents. Because they lose heat rapidly to the wall rocks, pluton margins adjacent to cool country rock commonly display finer grained **chilled margins** compared to the coarser grained crystals of the pluton interior.

Gravitational separation models
Gravitational separation includes fractionation processes that occur when crystals develop with significantly different densities than the surrounding magma (Figure 8.7). Gravitational separation includes crystal settling and crystal flotation processes.

Crystal settling occurs when higher density, ferromagnesian crystals settle to the base of a magma chamber relative to the lower density liquid magma. Crystal settling may result in discrete layers of crystal mush such that banded or layered cumulus crystals may pre-

cipitate via magmatic "sedimentation" on the pluton floor (Irvine et al., 1998). **Crystal flotation** can occur if early formed crystals, such as plagioclase, are less dense than the magma. As a result crystals may float towards the roof of a magma chamber, effectively segregating them from the remaining melt (Bottinga and Weill, 1970).

Both crystal settling and crystal flotation imply that individual crystals migrate through a low viscosity magma that has a different density than the crystals. However, in most cases, the differences in density between crystals and magma are vanishingly small and the viscosity of most magmas impedes crystal migration. However, gravitational migration may be possible for large clusters of early formed crystals with higher settling or flotation velocities in very high temperature, low viscosity magma. Crystal migration is far less likely to occur in lower temperature, higher viscosity magmas.

Other fractional crystallization processes
Convective flow segregation occurs whereby liquids and crystals are segregated due to factors such as velocity, density or temperature. Variations in convective flow velocity within low viscosity magma can result in the sorting of crystals and liquids. The largest or densest crystals will accumulate in regions of high velocity while smaller or lower density crystals accumulate in flow paths of lower velocity. Higher density crystals may also migrate by gravitational flow towards the base of the magma chamber. In some cases, large masses of unconsolidated or partially consolidated crystals break off roofs and side-walls and flow via density surge currents to the chamber floor. They may then settle at different velocities to produce graded bedding and other stratification features (Huppert and Sparks, 1984; Irvine et al., 1998).

In Bowen's fractional crystallization model, he alluded to variations in pluton heat loss as a contributing factor to crystal–liquid separation. The periphery of the pluton loses heat more readily, resulting in crystallization; the pluton interior retains heat and remains predominantly liquid. Thus, from an originally homogeneous magma, differential heat loss may generate convective flow and separate crystals from melt. Bowen (1928) recognized the role of temperature variations inducing

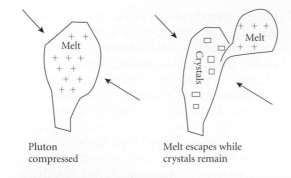

Pluton
compressed

Melt escapes while
crystals remain

Figure 8.8 Filter pressing process whereby a magma chamber containing a mix of crystals and liquids is compressed, squeezing out the more mobile liquid into a new chamber and leaving behind a crystal residue in the original chamber.

convective circulation in a magma chamber. Such convective currents can produce depositional structures usually associated with sedimentary processes such as graded bedding and cross-bedding. Bowen also envisioned the separation of crystals from liquid during the later stages of magma chamber cooling. In this case, cooling can be accompanied by deformation whereby crystals are compacted and rotated while liquids experience expulsion (Bowen, 1928). This variation on crystal–liquid fractionation involves a set of processes under the umbrella of **filter pressing** (Figure 8.8).

Crystal settling, crystal flotation, marginal (sidewall) accretion, convective flow and filter pressing are currently recognized as key liquid–crystal fractionation processes (Sleep, 1988; Irvine et al., 1998; Sisson and Bacon, 1999). Evidence for convective flow and crystal settling is perhaps best displayed in the Tertiary age Skaergaard Intrusion of Greenland. The Skaergaard Intrusion contains sedimentary-like features such as cumulate layering, graded bedding, cross-bedding and slump structures.

In addition to crystal–liquid fractionation, other processes such as liquid fractionation and open-system processes are now recognized as important magma diversification mechanisms, particularly at convergent plate boundaries and over hotspots.

Liquid fractionation

In addition to liquid–crystal fractionation, closed-system differentiation may occur due to liquid fractionation. In liquid fractionation, one parent magma fractionates to produce two or more distinctly different daughter magmas with different compositions. Liquid fractionation processes include differential diffusion and liquid immiscibility. **Differential diffusion** involves the preferential diffusion of select ions within the magma in response to compositional, thermal or density gradients as well as water content. Differential diffusion may play a key role in the transport and concentration of metallic ore deposits in plutonic systems. **Liquid immiscibility**, also called **liquid–liquid fractionation**, occurs when magma separates into two or more distinct immiscible liquid phases.

Bowen (1928) and Daly (1933), early proponents of liquid immiscibility, suggested that a magma may segregate into two separate magmas of different composition upon cooling. The separation of fat from chicken soup as it cools is an excellent analog for liquid immiscibility. The fat is soluble or miscible in soup at high temperatures, but becomes less so as the soup cools. Low density fat separates from the cooling soup as a separate phase and rises. Liquid immiscibility may occur to a limited extent with segregation of granitic liquids from basaltic magmas, as well as the segregation of sulfide-rich liquids from silicate magmas, and carbonate liquids from alkaline magmas. In laboratory studies, CO_2 exsolves as a separate fluid phase from alkalic magmas, generating a carbonate-rich fluid which crystallizes as a carbonatite. Liquid immiscibility has also been demonstrated in basaltic rocks from the Deccan traps (De, 1974). Philpotts (1979) also documented liquid immiscibility in two basaltic glasses from the Triassic–Jurassic Rattlesnake Hill Basalt of Connecticut where two different melts separated to form immiscible droplets in each other that solidified as glasses, resembling to some degree the fat droplets in the chicken soup analogy.

8.3.2   Open-system diversification

Open-system processes such as assimilation, magma mixing and magma mingling involve chemical changes due to interaction of magma with the surrounding country rock or with other magmas entering the system.

Assimilation

Assimilation occurs whereby the surrounding wall rock (country rock) is intruded by and reacts chemically with the magma. Magma composition evolves when elements from the wall rock are incorporated into the magma and/or when elements from the magma are transferred into the wall rocks (as during contact metamorphism). As is generally the case, the greater the surface area between the magma and the wall rock, the more extensively chemical components are exchanged between the two. Forceful injection of magma commonly fractures the surrounding wall rock by a process called **stoping**. Country rock fragments that fall (**stope**) into the magma are called **xenoliths** (Figure 8.9). Fracturing and fragmentation of country rock increases the potential for chemical reactions between wall rocks and magma by increasing the contact surface area. Many xenoliths display reaction rims, strongly suggesting that chemical reactions occur with the encasing magma.

Xenocrysts are foreign crystals not generated by crystallization of the surrounding magma (Figure 8.10). Chemical reactions between xenocrysts and magma may significantly alter the magma's chemical composition (Bailey et al., 1924; Daly, 1933).

Assimilation is fundamentally important in thick crustal regions such as continental and convergent margin settings. In the Thorr region of Donegal, Ireland, limestone and schist xenoliths stoped into a granitic magma. The xenoliths are several hundred meters in diameter and may have been larger prior to partial digestion by the surrounding hot magma. Clearly chemical reactions with xenoliths of these dimensions may significantly alter magma chemistry.

Magma mingling and mixing

In active plutonic environments, magma injection is unlikely to be a single event. Rather, intermittent pulses of magma are injected into the magma chamber from either the same source or different sources over long periods of time. **Magma replenishment**, involving multiple magma injections over time, produces complex relationships and constitutes an important diversification mechanism. **Magma mingling** occurs when two or more dissimilar magmas coexist, displaying contact relations but retaining their distinctive individual magma characteristics. Magma mingling implies that the magmas are interjected but do not combine thoroughly. In the case

Figure 8.9 Mafic gneiss xenolith entrained within a Proterozoic granite west of Golden, Colorado. Note the black reaction rim around the periphery of the xenolith.

Figure 8.10 Olivine xenocrysts in vesicular basalt. (For color version, see Plate 8.10, between pp. 248 and 249.)

where mafic and felsic magmas mingle, the rocks may display a "marble cake" appearance. The inability of two or more magmas to homogenize thoroughly may be due to significant differences in temperature, density or viscosity or to insufficient convection. D'Lemos (1992) cites magma mingling in the development of cylindrical pipes. These pipes formed as granitic melts intruded into diorite magma in Guernsey, Channel Islands. The pipes are thought to have developed from a less dense granitic magma that buoyantly injected itself into denser, overlying diorite magma. **Magma mixing** implies thorough mixing so that the individual magma components are no longer recognizable. For example, felsic and mafic magma can mix to produce an intermediate magma.

Magma mixing and mingling were discounted in the early 20th century when fractional crystallization hypotheses held sway. However, they are now recognized as significant means by which intermediate magmas may be generated by processes unrelated to fractionation. For example, Coombs and Gardner (2004) document pyroxene rims on olivine phenocrysts, and olivine phenocrysts occurring within an andesitic groundmass in rocks formed in a convergent plate tectonic setting. Such disequilibrium assemblages suggest that mixing of magma pulses of different compositions resulted in unusual mineral assemblages and reactions. Magma mixing is particularly important at continental rifts and convergent plate boundaries. In continental rift settings, hot basaltic magma generated by partial melting of the mantle rises through overlying intermediate to silicic crust. The hot basaltic magma partially melts the overlying crust generating a second, more silicic, magma source. These two magma regions may mingle or thoroughly mix. Magma mixing at convergent margins may be more complex, with various components. Dehydration of the subducted slab releases fluids increasing the volatile content. The addition of volatiles results in flux melting of the overlying lithosphere wedge above the subducting plate. Melting of the mantle wedge plays a critical role in altering magma chemistry at convergent plate boundaries. In addition to anatexis of the overlying mantle wedge, minor components can also be derived by partially melting the downgoing lithospheric slab, or from ocean basin or forearc basin material incorporated into the magma.

8.4   MAGMA SERIES

Despite compositional variations resulting from diversification processes, rocks derived from a given parent magma can display distinct geochemical signatures indicating common origin. As an analogy, consider the role of DNA in assessing lineage. Despite the genetic changes that offspring experience through successive generations, a genetic DNA link to a common ancestor may still be established over hundreds to thousands of years. Successive changes to magma composition through time bear a similar link to a common ancestor over hundreds of millions of years, so long as significant metamorphism has not affected rock chemistry.

Different processes have been introduced to explain the wide variety of magma and igneous rock compositions. Variations in magma and rock compositions may be due to:

- Widely varying chemistry of the initial magma, related to source rock composition, melting history and degree of melting.
- Alterations to the chemical composition of the magma through differentiation.
- Interactions between magmas such as magma mingling and mixing.
- Assimilation reactions with the surrounding country rock.
- Post-crystallization alteration processes (metamorphism).

The net result is that magma chemistry can evolve over time resulting in a series of genetically related magmas and rocks of different compositions. These compositional trends may be broadly grouped into "magma series". A **magma series** consists of genetically related magmas that have changed composition from a common original parental magma (Gill, 1981).

Igneous rock series include: (1) alkaline (or alkalic), and (2) sub-alkaline assemblages. Alkaline rocks have high Na_2O and K_2O concentrations compared to SiO_2 and are generally undersaturated with respect to SiO_2. The sub-alkaline series can be subdivided into calc-alkaline and tholeiite suites. Calc-alkaline rocks have moderate to high Na$_2$O, K$_2$O and SiO$_2$ concentrations. Tholeiites have low Na$_2$O and K$_2$O but high FeO and CaO concentrations. Major, minor and trace element concentrations provide critical information to understanding the origin and evolution of igneous rocks and rock series. However, assessing the data for large numbers of rock samples can be overwhelming. One approach to solving this dilemma involves using chemical data to construct variation diagrams that depict data graphically in a way that reveals patterns or trends.

8.4.1   Tracking magma series using variation diagrams

Variation diagrams provide a means to concisely and clearly display rock chemistry variations between samples collected in an area with an eye to establishing their origins and relationships. Bivariate diagrams consist of two variables plotted as an x-y coordinate system. Trivariate diagrams contain three variables and are plotted in a triangular field. The variables plotted may consist of major, minor or trace element proportions. Major and minor elements are generally plotted as weight percent oxides as discussed in Chapter 7. Linear or curvilinear data trends in which data points fall near a best fit line indicate a significant degree of relationship between rock samples. These relationships can usually be explained by progressive changes in magma composition related to anatexis of a common source and/or diversification processes. Random data suggest that the rocks are derived from different sources or have suffered severe secondary alteration associated with metamorphism such that their original chemical identity has been overprinted beyond recognition.

Bivariate variation diagrams: Harker diagrams

Harker (1909) devised a Cartesian (x, y) coordinate graph that can display data from hundreds of rock samples. **Harker diagrams** are bivariate diagrams in which the vertical ordinate (y-axis) represents weight percents of major or minor oxide compounds such as FeO, MgO, CaO, Na$_2$O, K$_2$O or TiO$_2$. The horizontal abscissa (x-axis) represents weight percent SiO$_2$ content (Figure 8.11). The objec-

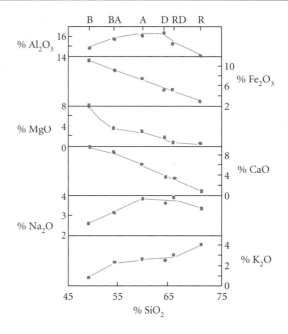

Figure 8.11   Harker diagram illustrating the major element variation of rocks derived by a fractional crystallization of a parent basalt magma with 49% SiO$_2$. Note the following relationships: (1) Al$_2$O$_3$ increases with the crystallization of ferromagnesian minerals (olivine, pyroxene, amphibole), and then decreases with the removal of plagioclase and micas; (2) MgO and Fe$_2$O$_3$ decrease with the crystallization of ferromagnesian minerals; (3) CaO decreases with the crystallization of calcium plagioclase; (4) NaO initially increases due to the removal of MgO, Fe$_2$O$_3$ and CaO, and then decreases as sodium plagioclase crystallizes in andesitic and more silicic rocks; (5) SiO$_2$ and K$_2$O increase as MgO, Fe$_2$O$_3$, CaO, Na$_2$O and Al$_2$O$_3$ are progressively removed by crystallization. B, basalt; BA, basaltic andesite; A, andesite; D, dacite; RD, rhyodacite; R, rhyolite. (After Ragland, 1989, with permission of Oxford University Press, Inc.)

tive is to concisely display variations in major or minor oxide concentrations with respect to changes in SiO$_2$. Harker diagrams permit overall trends of major element variation to be deduced. If Harker diagrams display smooth, curvilinear trends for all rock data points, then the following three inferences can be made:

1 The rocks are genetically related.
2 Major element variations reflect a liquid line of descent from a common source in which diversification caused major elements to either increase or decrease progressively with respect to variations in weight percent SiO_2.
3 The parent magma from which the rocks are derived has a composition near that of the sample with the least SiO_2, that is, of basaltic composition.

Figure 8.11 illustrates a stacked sequence of Harker diagrams for rocks derived from a basaltic magma (SiO_2 = 49%). Note that as predicted by Bowen's reaction series, MgO, Fe_2O_3 and CaO decrease with increasing SiO_2. These trends may be attributed to the early crystallization of olivine, pyroxene and calcium plagioclase to form basalt. Al_2O_3 initially increases but then decreases as plagioclase continues to crystallize, along with amphibole and biotite to form andesite. Na_2O and K_2O both increase with increasing SiO_2 until Na_2O begins to decrease as sodium plagioclase begins to crystallize, forming dacitic rocks. K_2O increases linearly with increasing SiO_2 content until the remaining K_2O-rich melt crystallizes, forming rhyolite.

Spider diagrams

In **spider diagrams**, trace element concentrations on the abscissa are plotted relative to a standard reference for each element on the ordinate. The standard reference varies depending upon the rock type analyzed and the tectonic relationship that the researchers wish to demonstrate. The standard reference may include chondrite meteorites, primitive mantle, normal mid-ocean ridge basalt (N-MORB), enriched mid-ocean ridge basalt (E-MORB) or an ocean island basalt (OIB). Flat trends indicate a close chemical relationship to the standard reference. Spider diagrams are useful in demonstrating the enrichment or depletion of compatible versus incompatible elements, LREE versus HREE variation, or other data that may indicate diversification trends or magmatic source. The relative abundance of specific elements may also yield information regarding the fractionation of particular minerals. For example, depletion of europium indicates plagioclase fractionation from the melt, which effectively removes europium from the magma. On the other hand, europium enrichment would indicate assimilation of plagioclase components into the melt.

Figure 8.12 displays data comparing the Ricardo basalts to N-MORB, E-MORB, OIB and primitive mantle compositions. Note that the trace element concentrations of the Ricardo basalts are most similar to OIB. Note also that LREE and the largest LIL are enriched relative to HREE in a way highly dissimilar to N-MORB and somewhat dissimilar to E-MORB. Ricardo basalts are between ten and 100 times more enriched in LIL and rare Earth elements compared to the primitive mantle. It is likely that the parent melts for the Ricardo basalts were produced by small degrees of partial melting of a relatively undepleted mantle source associated with hotspot melting.

Ternary variation diagrams

In studying Greenland's mafic Skaergaard Intrusion, Wager and Deer (1939) identified the calc-alkaline and tholeiitic series based on changes in iron concentration with progressive differentiation. These progressive changes in chemical composition were plotted on a triangular AFM (A, alkali; F, iron; M, magnesium) diagram. Tholeiitic suites display significant iron enrichment with increased fractionation while the calc-alkaline suite does not display meaningful iron enrichment, as illustrated in the AFM diagram (Figure 8.13).

When AFM diagrams are combined with Harker diagrams, it can be demonstrated that calc-alkaline rocks can be generated by diversification of basaltic or basaltic andesite parent magma. Whether this is always the case is still a matter of contention.

8.4.2 Major magma series

Calc-alkaline magmas

Upon fractionation, **calc-alkaline magmas** record a progressive decrease in iron and magnesium with increasing SiO_2 and alkali concentrations (Figure 8.13). Available FeO and MgO are progressively removed from the melt through crystallization of olivine, pyroxenes,

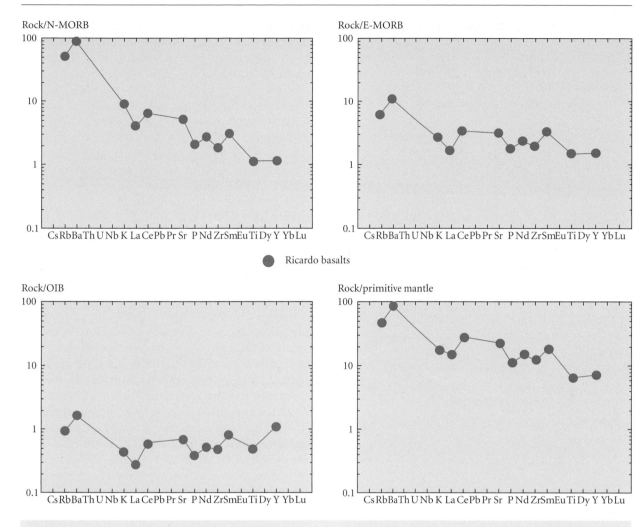

Figure 8.12 Spider diagrams of Ricardo basalts collected in the vicinity of the Garlock Fault in southeastern California (Anderson and Jessey, 2005). The Ricardo basalts display trace element similarities to ocean island basalts (OIB); E-MORB, enriched mid-ocean ridge basalt; N-MORB, normal mid-ocean ridge basalt. (Courtesy of David Jessey.)

amphiboles, biotite and iron oxide minerals such as magnetite. The removal of ferromagnesian minerals enriches the melt in Na_2O, K_2O and SiO_2. Thus, with increasing fractionation from magma, the calc-alkaline series exhibits increases in SiO_2, Na_2O and K_2O and decreases in CaO, MgO and total iron (Bowen, 1928; Miyashiro, 1974; Grove and Kinzler, 1986). The calc-alkaline series consists largely of andesite, dacite and rhyolite as well as high alumina (16–20% Al_2O_3) basalt (Irvine and Baragar, 1971). Calc-alkaline magmas dominate convergent margin environments, such as the modern circum-Pacific region, wherein batholiths and composite

volcanoes develop above subduction zones. The rocks in convergent margin volcanoes are known colloquially as the BADR series for the basalts, andesites, dacites and rhyolites that occur in them.

Tholeiitic magmas

With increasing fractionation, **tholeiitic magmas** experience iron enrichment at low to moderate SiO_2 concentrations (Figure 8.13). Early crystallization of forsterite olivine and calcium plagioclase depletes the melt of MgO and CaO, causing the remaining melt to become enriched in FeO, as well as SiO_2 and

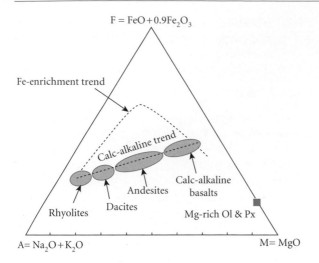

Figure 8.13 AFM diagram illustrating the calc-alkaline trend, with progressive iron depletion and alkali and SiO_2 enrichment, in which fractionation of a basaltic parent results in progressively more SiO_2- and alkali-rich rocks such as andesites, dacites and rhyolites. The tholeiite trend, displaying iron enrichment with fractionation is also illustrated.

alkali elements (Miyashiro, 1974; Grove and Kinzler, 1986). Thereafter, FeO-rich minerals, such as fayalitic olivine, and Fe-rich pyroxenes crystallize, depleting the melt in FeO while continuing to enrich it in SiO_2 and alkali elements. Tholeiitic magmas generate large volumes of basalt with subtle compositional variations (Irvine and Baragar, 1971). Extreme fractionation of tholeiitic magmas can produce small amounts of more silicic magmas and rocks. Tholeiitic magmas and basalts dominate extensional environments such as ocean ridges and continental rifts. Tholeiites also occur over hotspots in intraplate settings and in the frontal portions of immature (relatively young) volcanic arcs characterized by a thin volcanic arc crust (Miyashiro, 1974).

Alkaline magmas

Alkaline magmas and rocks are highly enriched in Na_2O and/or K_2O and are less common than either calc-alkaline or tholeiitic

magmas or rocks. Alkaline rocks never contain equilibrium quartz. Alkaline magmas contain high concentrations of the alkali (lithium, sodium, potassium, rubidium, cesium and francium) elements. Most alkaline rocks contain feldspathoids, alkali feldspars and/or sodic amphibole or pyroxene minerals (Sorensen, 1974). Alkaline rocks are characterized by extremely diverse chemical compositions with SiO_2 contents ranging from 0 to 65%, with magnesium numbers (MgO/MgO + FeO) ranging from 0 to 0.88. SiO_2-undersaturated alkaline rocks contain feldspathoids and/or olivine. Alkaline feldspathoid minerals include nepheline, leucite, cancrinite, sodalite and analcite (a zeolite mineral closely associated with the feldspathoids group). SiO_2-saturated alkaline rocks are enriched in alkali feldspars that include sanidine, anorthoclase, perthitic potassium feldspar, microcline, orthoclase and albite. Amphibole minerals that crystallize from alkaline magmas include riebeckite and richterite. Pyroxene minerals that occur in alkaline rocks include aegerine, aegerine-augite and spodumene.

Alkaline rocks encompass a wide variety of compositions that include volcanic rocks such as alkali basalts, hawaiite, benmoreite, mugearite, trachyte, nephelinite, phonolite, lamprophyre, carbonatite, komatiite, kimberlite and others. Plutonic alkaline rocks include syenite, nepheline syenite, monzosyenite, monzogabbro, nephelinite and others (Chapter 7). Alkaline rocks occur in a number of environments, which include stable cratonic interiors, continental rifts (Bailey, 1974) and ocean islands away from spreading centers (Sorensen, 1974) and subduction zones (Burke et al., 2003). Many alkaline rocks display geochemical signatures indicative of a mantle source.

Both alkaline and tholeiitic basaltic magmas occur at hotspots, ocean islands and rift environments and are produced by partial melting of mantle rocks. What is the relationship between these two magma series? Alkaline magmas and tholeiitic magmas may, in some cases, be generated from the same parental source. Alkaline magmas most likely form by limited melting (<10%) of a parent source under high pressure conditions, especially if the source rocks are relatively undepleted or even enriched in alkalis. Tholeiitic magmas are most likely generated by more extensive melting at lower temperature and pressure conditions of a previously depleted source rock. In addition, limited partial melting of tholeiitic source rock may yield alkali magmas, particularly under high pressure conditions in the lower crust/upper mantle (Frey et al., 1979, 1990; Naumann and Geist, 1999).

Bimodal suites

Bimodal magma suites are characterized by the voluminous occurrence of silicic and basic (e.g., rhyolite–dacite and basalt) rocks with few intermediate (e.g., andesite) rocks. Bimodal volcanism is associated with continental rifts. The basic component is derived by partial melting of the mantle. The silicic component is derived by partial melting of the continental crust heated by rising basic magmas. Bimodal volcanic activity is widespread within the Basin and Range province of the southwestern United States.

The different tectonic environments in which the major magma series commonly occur are summarized in Figure 8.14. Note that:

- Calc-alkaline magmas occur almost exclusively at convergent margins.
- Tholeiitic magmas are generated primarily at ocean spreading centers, but also occur in continental rifts, backarc basins, ocean islands and hotspots.

- Alkaline magmas occur primarily at hotspots, ocean islands and rift environments.
- Bimodal volcanism occurs at continental rifts.

So far in this chapter we have focused upon the origin and diversification of magma. For more detailed discussions beyond the scope of this textbook, the reader is referred to excellent petrology textbooks by Winter (2009), Raymond (2007), McBirney (2007), Best (2003), Blatt and Tracy (1996), Philpotts (1990), Ragland (1989) and Hyndman (1985) among others. In the following discussion, we will discuss plutonic rock bodies, created by magma cooling at depth within Earth.

8.5 INTRUSIVE IGNEOUS STRUCTURES

Plutonic structures form when magma intrudes pre-existing country rock within Earth. Many different types of plutonic structures exist. The critical distinctions in classifying these structures include the size and shape of the intrusion, as well as orientation. Intrusive contacts may be classified as concordant or discordant, depending upon the orientation of the contacts with respect to the surrounding country rock layers. The contacts of **concordant** intrusions are oriented parallel to the pre-existing layering in the

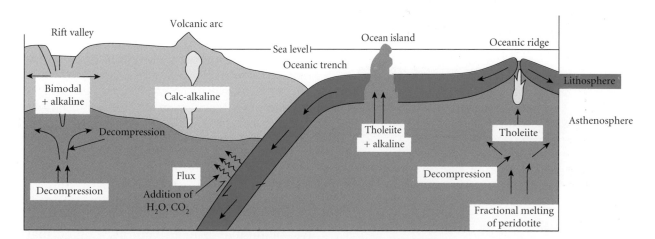

Figure 8.14 Generalized cross-section illustrating the relationships between anatexis, magma series and tectonic environments. (Courtesy of Stephen Nelson.)

surrounding country rock. The pre-existing geological layering may be bedding, metamorphic foliation, volcanic layers or other structural forms. In the event that no geological layering is discernible in the country rock, concordant intrusions are parallel to Earth's surface. The boundaries of **discordant** intrusions are oriented at a sharp angle to the country rock layering and cut across it. If no geological layering is present the boundaries are inclined at an angle to the surface of Earth. Intrusive structures discussed below include plutons (batholiths and stocks), sills, laccoliths, lopoliths and dikes (Figure 8.15).

8.5.1   Batholiths and stocks

Batholiths are defined as plutons of more or less irregular shape with surface exposures $\geq 100 \, km^2$. **Stocks** are plutons with surface exposures $\leq 100 \, km^2$. The $100 \, km^2$ surface area distinction between stocks and batholiths is an arbitrary number. The surface area measured may bear little relationship to the overall volume of the pluton. An immense pluton largely buried at depth is considered a stock if rock exposures on Earth's surface total $99 \, km^2$ or less – irrespective of the actual volume of the pluton body. Conversely, a relatively small but well-exposed pluton is classified as a batholith if surface exposures are slightly more than $100 \, km^2$.

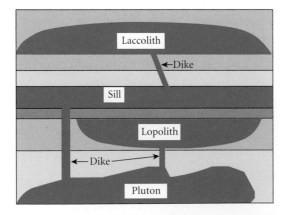

Figure 8.15 Simplified cross-section depicting plutonic structures intruding horizontal country rock layers. Note that dikes serve as feeder conduits to shallower structures.

Batholiths and stocks can form wherever magmatism occurs. As discussed in Section 8.1, magmatism occurs primarily at convergent and divergent plate boundaries and hotspots. Divergent margin and hotspot stocks and batholiths are largely gabbroic in composition, although these are not generally exposed for direct field analysis. Convergent margin batholiths are generally intermediate to silicic in composition and consist of rocks such as diorite, granodiorite, quartz diorite or granite. Convergent margin batholiths are well exposed in many localities where erosion has exposed rocks formed in magmatic arcs and orogenic belts, such as in the Cordilleran (Rocky Mountain) system of North America. Convergent margin batholiths may be enriched in valuable metals such as molybdenum, tin, gold and silver. Large silicic batholiths also form over hotspots in continental interiors as will be discussed further in Chapter 10.

Formerly, batholiths were considered to represent vast chambers of molten rock that existed at depth and cooled over thousands to a few million years (Buddington, 1959). However, seismic refraction studies have failed to identify large molten bodies, even at fast-spreading ocean ridge systems (Detrick et al., 1990). Modern models suggest that batholiths may consist of <10% melt at any time and can develop over tens of millions of years as incremental, composite assemblage of many separate plutons (Pitcher, 1979; Coleman et al., 2004; Glazner et al., 2004). Major batholiths do not represent a single intrusion but rather a diachronous set of individual plutons that occur within a spatial domain, particularly along convergent or divergent margins. For example, the Coastal Batholith in the Andes Mountains convergent margin is a composite batholith consisting of over 700 separate plutons (Pitcher, 1979). Magmatic intrusions at convergent margins play a pivotal role in the growth of continents through geological time. Batholith emplacement along convergent plate boundaries is the major mechanism by which continental crust has grown since the Archean, over 2.5 billion years ago.

Stocks and batholiths can display concentric or tabular zoning patterns. Zoning may develop due to fractionation and/or contamination of the magma with the surrounding country rock (Holder, 1979). Normal zoning within a pluton is characterized by a basic rim and a silicic core. Vance (1961) suggested that normal zoning develops as cooling and crystallization progress sequentially from the pluton margins towards the inner core. Zoning may result in the segregation of mafic minerals such as pyroxene and amphibole from more felsic minerals such as quartz and feldspars. In some batholiths, distinct couplets form whereby a pinstripe pattern develops with alternating mafic and felsic mineral layers (Figure 8.16). The cause for couplet development is uncertain but may involve multiple dike-like intrusions and/or fractionation processes.

8.5.2   Concordant plutonic structures

Concordant plutonic structures, whose boundaries parallel country rock layers, include sills, laccoliths and lopoliths. These concordant structures form when magma is injected into pre-existing planes of weakness between rock layers, prying them apart.

Sills

A **sill** is a tabular, concordant pluton that parallels country rock (see Figure 8.15). Tabular plutons have shapes that resemble those of table tops. They have two mutually

Figure 8.16   Anorthosite pluton in Montana displays rhythmic banding. (Photo by Kevin Hefferan.)

perpendicular long dimensions and a short dimension perpendicular to the other two. Sills develop through the injection of magma along a plane (e.g., bedding, foliations) of weakness parallel to the layering in the country rock. Sills commonly form in low viscosity magmas and are generally <50 m in thickness. However, sill thickness varies from thin, tabular, concordant plutons to much larger plutons such as the 300 m thick Palisades Sill in New Jersey.

Laccoliths

A **laccolith** (Gilbert, 1877, in Daly, 1933) is a blister-like concordant pluton characterized by a flat floor and domed roof (see Figure 8.15). Laccoliths are similar to sills in that the magma is injected parallel to the country rock layering. Laccoliths may begin as flat-topped sill structures. The intrusion of additional magma produces sufficient force to bulge the sill roof upward creating a convex structure. Laccoliths were first described in the Henry Mountains of Utah by G. K. Gilbert in the 1870s. Prominent laccoliths include the Prospect Intrusion in New South Wales, Australia and the Uwekahuna laccolith, Hawaii. The magma that produces laccoliths, sills and other concordant structures generally rises from deeper sources via discordant feeder dike systems.

Lopoliths

Lopoliths are dish-shaped to funnel-shaped concordant plutons that resemble a champagne glass in cross-section view (see Figure 8.15). According to Daly (1933), F. F. Grout originally defined lopoliths in 1918 based on studies of the Proterozoic Duluth Gabbro Complex of Minnesota. Other examples include the Skaergaard Intrusion of Greenland, the Sudbury Impact Complex in Ontario, the Bushveld Complex in South Africa, the Koillismaa Complex of Finland and the Muskox Intrusion in the Northwest Territories of Canada. Lopoliths extend from tens to hundreds of kilometers in length with thicknesses of up to several kilometers. The Bushveld Complex has a length of 550 km and is up to 8 km thick. Many large lopoliths, such as the Muskox Lopolith, consist of concentrically zoned mafic and ultramafic rocks. The champagne-glass lopolith structure can develop due to one or more of the following mechanisms: (1) meteorite impact and associated crustal melting as has been suggested for the Sudbury Complex, (2) normal faulting and crustal melting associated with rifting as in the Koillismaa Complex, or (3) a sill-like structure that receives upwelling magma from a conical feeder tube, which would correspond to the stem in the champagne glass. Lopoliths, like many layered mafic–ultramafic intrusions, may be highly enriched in valuable metals such as nickel, copper, chromium, platinum and palladium.

Veins

Veins form as hot fluids flow through fractures and then cool and crystallize. Veins may be concordant or discordant to the surrounding country rock. In a sense veins are small scale (millimeters to centimeters in diameter) sills or dikes. Veins can occur in great abundance resulting in **vein swarms**. Vein swarms may display random or preferred orientations. Preferred orientations of vein sets may be parallel (Figure 8.17), sub-parallel, orthogonal, en echelon or sigmoidal. Orthogonal vein sets occur at right angles to one another.

Figure 8.17 Swarm of parallel felsic veins intruding volcaniclastic rocks near Golden, Colorado. These veins, concordant to the country rock, can be considered centimeter-scale sills. Note the pen on the left for scale. (Photo by Kevin Hefferan.)

En echelon veins are parallel to sub-parallel but offset. Sigmoidal veins are "S-shaped" en echelon sets.

8.5.3 Discordant plutonic structures

Discordant structures, those whose contacts cut across pre-existing wall rock layers, include cylindrical necks, diatremes and dikes. Many of these discordant structures serve as conduits that transport magma from deeper plutons to shallow intrusions or even surface eruptions.

Necks and diatremes

Necks are cylindrical plutonic dikes exposed at the surface by subsequent erosion. They represent ancient conduit pipes that funneled magma upward to a volcano that has long since been removed by erosion. Solidification of magma within the cylindrical pipe produces hard, crystalline rock within the volcano. The poorly consolidated overlying volcanic edifice and surrounding country rock erode at a faster rate over time. The hard, crystalline rocks of the cylindrical neck are erosion resistant. As a result, the cylindrical pipe is preserved as a vertical rock column. Notable examples include Ship Rock, New Mexico, which has a central neck with radiating dike ridges, and Devil's Tower of Wyoming (Figure 8.18) which preserves radial columnar jointing. Necks are sometimes called "volcanic" necks because they were the central vent of a volcano. This is actually a misnomer

Figure 8.18 Devil's Tower neck, Wyoming. (Photo by Kevin Hefferan.)

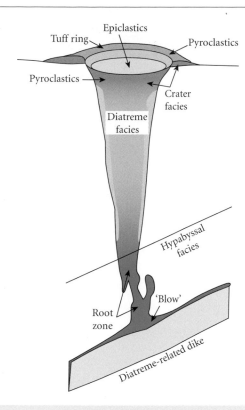

Figure 8.20   A granitic dike intruding Proterozoic metavolcanic rocks west of Golden, Colorado. Two locally derived metavolcanic xenoliths occur within the dike. (Photo by Kevin Hefferan.)

Figure 8.19   Idealized cross-section of a funnel-shaped, kimberlite diatreme with a basal root zone and sill feeders. (Courtesy of R. H. Mitchell.)

because necks represent plutonic cylindrical feeder dikes.

Diatremes are carrot-shaped, cylindrical pipes that can extend to depths of several Kms (Figure 8.19). Diatremes develop via explosive intrusions that originate deep within the mantle. The explosiveness is due to the high volatile content which propels magma and xenolith fragments upward towards the surface. As will be discussed in Chapter 10, diatremes are extensively studied due to their association with ultramafic diamond-bearing kimberlites. Volcanic structures such as tuff cones may have developed on Earth's surface above diatreme pipes. However, these surficial deposits are long since eroded, exposing deeper diatreme pipes.

Dikes

Dikes are tabular intrusions that cross-cut country rock layers. Because dike rocks are typically more resistant to weathering and erosion than surrounding country rock, dike exposures commonly form topographically elevated ridges. Dikes occur as individual structures or in dike sets or swarms. Figure 8.20 illustrates a single dike that changes its orientation relative to the vertical country rock layers which it cross-cuts.

While most dikes exposed on land measure up to tens of meters in thickness and from hundreds of meters to a few kilometers in length, exceptions exist. The 2.5 Ga Great Zimbabwe Dike of South Africa is ~700 km long and 8 km wide. The Great Dike is unusual for its length, ultrabasic composition and rich ore deposits of platinum group metals and chromite.

Dike swarms, radial dikes and concentric dikes

Dike swarms consist of multiple dikes that can occur either in parallel, sub-parallel, radiating, concentric or random orientations. **Radial dikes** are typically produced when vertical forces related to rising magma produce fractures that radiate outward from the central vent. The radial fractures are analogous to those produced in window glass when it is struck by an object. Radial dikes form when magma from the central vent is injected outward into the radial fractures. Figure 8.21 illustrates a topographic map (1 : 24,000 scale)

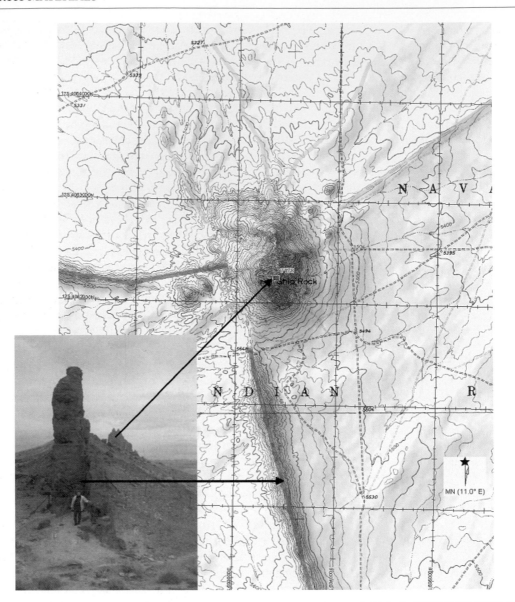

Figure 8.21   A radial dike swarm emanates outward from Ship Rock neck, which formed as a cylindrical plutonic pipe that fed a now eroded volcanic structure. (Courtesy of Neil Heywood.)

of Ship Rock, New Mexico, illustrating numerous mafic dikes that radiate outward from the Ship Rock neck. The inset image depicts the southern dikes (foreground) with Ship Rock neck (background).

Ring dikes and cone sheets constitute networks of interrelated, discordant intrusions associated with concentric fracture sets (Figure 8.22). The concentric fracture sets commonly develop following the collapse of a volcanic dome, in subsiding structures such as calderas or due to meteorite impacts. **Ring dikes** are nearly vertical in cross-section and circular in map view. Ring dikes occur in Tertiary volcanic suites on the Island of Mull, Scotland.

Cone sheet dikes are circular in map view and resemble ring dikes, except that cone sheet dikes converge at depth in cross-section view. Cone sheets possess an ice cream cone-like structure. Pressures associated with the intrusion of magma produce fractures, which propagate upwards and outwards from a central point. Magma is injected into these concentrically nested fractures, generating the cone sheet structures. Schirnick et al. (1999) report ~500 cone sheets, radial dikes and hypabyssal plutonic structures within the 20 km wide Tertiary age Tejeda Caldera complex on the Canary Islands.

En echelon and parallel dikes

En echelon dike sets consist of parallel, offset dikes that form in response to shear. Parallel dike sets form perpendicular to extension and are particularly common in rift environments. Parallel dikes commonly develop within sheeted dike complexes, discussed below.

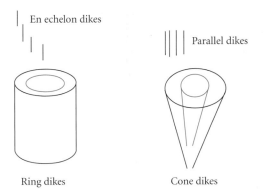

Figure 8.22 Three-dimensional diagram of en echelon, parallel, ring dikes and cone dike structures.

Sheeted dikes

Steeply inclined **sheeted dikes**, composed of gabbro, diabase and basalt, form layer 2B of the ocean floor at ocean spreading ridges (Figure 8.23). The sheeted dikes underlie volcanic pillow basalts (layer 2A) and overly plutonic gabbroic rock. Sheeted dikes form by the cooling and contraction of magma as it is injected into extensional fractures in oceanic rift valleys. Repeated injection into extension fractures occurs over millions of years as the ocean lithosphere diverges from the rift axis. As a result new magma injections intrude fractures parallel to previous dikes producing a very large number of parallel dikes called sheeted dikes. Hot magma injections into pre-existing dikes result in chilled margins along the periphery of the newly forming dike. Thus, sheeted dikes commonly contain finer grained chilled margins due to more rapid heat loss and crystallization, where younger dikes are in contact with older dikes. Sheeted dike formation is discussed in more detail in the discussion of oceanic ridge volcanism and ophiolite sequences in Chapter 10.

As will be discussed further in Chapter 10, magmas are generated primarily at divergent margins, convergent margins and hotspots. Magma is stored in the fractures and pore

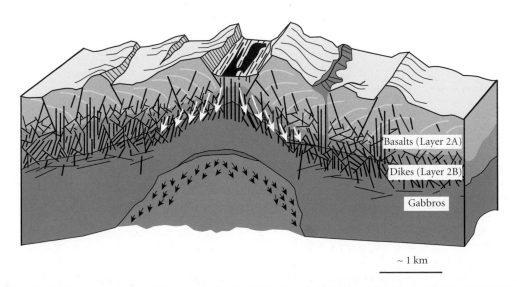

Figure 8.23 Block diagram of a mid-ocean spreading ridge (Karson, 2002). Black arrows refer to flow within gabbroic plutons at depth; white arrows indicate extension, faulting and rotation within sheeted dike complex. (Courtesy of Jeffrey A. Karson.)

spaces of deep plutons. Forces associated with faulting, buoyancy or changes in the temperature–pressure or fluid regimes propel magma upward to shallow plutonic structures such as sills, dikes and laccoliths. These processes are largely responsible for the distribution and concentration of economic metal deposits in plutonic igneous bodies. Magma that reaches Earth's surface generates a variety of volcanic processes and structures which we will address in the following chapter.

Chapter 9

Volcanic features and landforms

9.1 VOLCANOES, CRATERS AND VENTS

A **volcano** is a naturally occurring landform produced where lava erupts onto Earth's surface. Volcanic activity vividly displays the dynamic nature of our hot, turbulent planet, and profoundly impacts Earth in many ways. Volcanic eruptions generate new land area, valuable mineral deposits and arable soil. Volcanic eruptions can also kill tens of thousands of people and dramatically alter Earth's climate, causing major extinctions. Whether or not you ever climb a volcano or witness lava erupting first hand, volcanism is truly one of Earth's most impressive phenomena. In this chapter we will describe common volcanic landforms and their occurrence. We will also address why eruptive styles vary so much for different types of volcanoes and how volcanoes impact Earth's inhabitants.

Where does volcanic activity occur? While volcanic activity occurs on numerous planetary bodies (e.g., Venus and Jupiter's moon Io), we will focus upon volcanism on Earth. Earth's volcanoes are preferentially located where magmas are generated and rise to the surface at convergent and divergent boundaries, and over hotspots.

9.1.1 Crater and fissure vents

Lava commonly erupts from a **central vent** located in a crater near the summit of the volcano. A **crater** is a nearly circular depression produced by the ejection of rock during volcanic eruptions. From an analysis of 5564 volcanic eruptions, most of which were above sea level, Simkin et al. (1981) suggest that nearly half of all volcanic eruptions occur from central vents. Magma that solidifies in central vents forms cylindrical pipes which may be exposed by erosion as necks, as explained in the previous chapter.

Vents also occur as long, linear fractures called **fissure** vents from which lava is erupted onto the surface. Fissures are elongated cracks in Earth's crust and commonly develop at:

- Ocean ridge systems such as the Mid-Atlantic and East Pacific Rise ridge systems

Earth Materials, 1st edition. By K. Hefferan and J. O'Brien. Published 2010 by Blackwell Publishing Ltd.

Figure 9.1 "Curtain of fire" fissure eruptions on Mauna Loa produce an elongated spatter rampart ridge upon solidification of the lava. (Photo courtesy of D. A. Clague, taken March 1984, with permission of the US Geological Survey.) (For color version, see Plate 9.1, between pp. 248 and 249.)

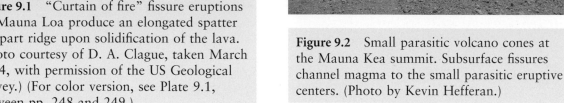

Figure 9.2 Small parasitic volcano cones at the Mauna Kea summit. Subsurface fissures channel magma to the small parasitic eruptive centers. (Photo by Kevin Hefferan.)

where the fissures are produced by horizontal extension.

- Continental rifts such as the East African Rift System or the Basin and Range of the southwestern USA, where the fissures are also produced by horizontal extension.
- Hotspots such as at Iceland (both a hotspot and an ocean ridge) and Hawaii (Figure 9.1). Hawaii's Mauna Loa and Kilauea contain a series of radial fissures emanating outward from a central vent. These form in response to the stretching of the crust during uplift.

Magma solidifying in such fissures produces discordant plutonic features such as the parallel dikes, radiating dikes and sheeted dikes discussed in the previous chapter.

9.1.2   Flank eruptions

Lava may also erupt from the sides or the base of a volcano producing **flank eruptions**. Flank eruptions may involve fissures or isolated cylindrical vents. In Hawaii, linear fissures provide conduits for lava to produce many, smaller flank eruptions. Low discharge flank eruptions produce smaller **parasitic volcanoes** along the summit, side or base of larger volcanoes. These are often aligned over the fissures that fed them. Mauna Kea, an immense volcano located on the big island of Hawaii,

contains ~100 parasitic volcanoes (Figure 9.2). We will discuss these different types of volcanoes further in Section 9.2.

Central vents, fissures and flank eruptions may occur alone or in combination in a given volcano, and individual eruptions may evolve from one emission style to another. For example, the 1983 eruption at Kilauea began as a fissure eruption and evolved into a central vent eruption over several months. On the other hand, SP Crater from the San Francisco volcanic field of Arizona (USA) provides a striking example of a flank eruption of basaltic lava following the development of a central vent crater (Figure 9.3).

9.1.3   Calderas

Calderas are large, generally circular to oval depressions caused by subsidence of Earth's surface. Subsidence is generally initiated by the rapid loss of magma stored in the shallow subsurface. Magma loss may be due to (1) subsurface withdrawal of magma from a shallow chamber as magma migrates to another location, or (2) cataclysmic eruptions that empty the shallow magma chamber. In either case, subsurface magma withdrawal undermines support of the overlying rocks, which progressively subside creating a collapsed basin or caldera. In the subsurface beneath the caldera, plutonic structures such

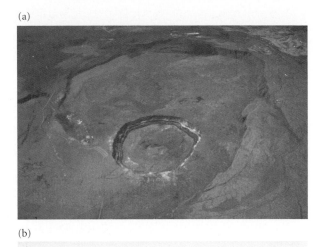

(a)

(b)

Figure 9.3 A beautiful summit crater is clearly visible at the top of SP Volcano in the San Francisco volcanic field, located north of Flagstaff, Arizona. In a more recent volcanic eruption, basaltic lava flow was released from a fissure at the base of the volcanic cone. (Photo courtesy of the US Geological Survey.)

Figure 9.4 (a) View to the northeast of Kilauea Caldera. The smaller (1 km diameter) Halema'uma'u Crater occurs within the center of the larger Kilauea Caldera. (Photo courtesy of the US Geological Survey.) (b) View to the south illustrating the caldera's western collapse wall in the foreground right and the Halema'uma'u Crater in the background in this 2006 photo. Beginning in March 2008, the Halema'uma'u Crater began erupting ash and steam in an explosive display not witnessed since 1924. (Photo by Kevin Hefferan.)

as ring dikes can occur by intrusion of magma into the fractures that formed as near surface rocks collapsed (Chapter 8). The Kilauea Caldera (Figure 9.4) provides an outstanding example of a nested "pit" caldera that formed by the removal of low viscosity basaltic magma from shallow, subsurface chambers. A series of flank eruptions, lava lake flows and lowering of the magma level in the chamber resulted in surface subsidence and collapse on different scales, producing a larger caldera in which a smaller pit caldera is nested.

While Hawaii's Kilauea Caldera represents a magma withdrawal caldera, most great calderas are generated by massive explosive volcanic eruptions such as Yellowstone, Mt Toba, Lake Taupo and Mt Mazama. Mt Mazama in the Cascade Range (USA) erupted in a cataclysmic event ~5677 BC ejecting an estimated 1000 km^3 of rock material (Zdanowicz et al., 1999). The huge caldera depression occupying the former site of Mt Mazama is referred to as Crater Lake. The Crater Lake Caldera averages 8 km in diameter and 1.6 km in depth.

Explosive caldera eruptions also occur in the ocean, despite the high hydrostatic pressure environment. Calderas have been observed in submarine volcanoes associated with subduction zones. McCauley Caldera, located near the Kermadec volcanic arc of New Zealand, is 6–10 km across and over 1–km deep.

9.2   CLASSIFICATION OF VOLCANOES

Volcanoes are classified based on their eruptive history as well as their chemical and physical characteristics. Eruptive history refers to the previous record of volcanic eruptions and likelihood for recurrence. **Active volcanoes** have erupted within the past several hundred years and are likely to do so again. **Dormant volcanoes** are potentially active, but have not erupted within thousands of years; thus they are "sleeping". **Extinct volcanoes** have not erupted for tens of thousands of years and volcanic activity is not likely to recur.

Volcanoes are classified based on their size, shape and explosiveness. The size of the volcanic landform is directly related to the frequency and duration of eruptions, which is directly related to the volume of magma available. The shape of the volcano and its explosiveness are related to the viscosity, composition, temperature and volatile content of the magma. As discussed in Chapter 7, viscosity increases with increasing silica and decreasing temperature so that lower temperature, silica-rich magmas are more viscous. Baker et al. (2004) suggest using food analogies when discussing magma viscosity. They state that at 25°C and 1 bar pressure (a typical kitchen environment):

- Smooth peanut butter has a viscosity of ~108 Pa/s, nearly equal to that of rhyolite melt with 2% dissolved water vapor by weight at 800°C.
- Ketchup has a viscosity of ~102 Pa/s, nearly equal to that of anhydrous (volatile-poor) tholeiitic basalt at 1200°C.

Viscosity plays a critical role in the development of volcanic structures, the aerial distribution and thickness of lava – and the explosiveness of eruptions. Highly viscous, silicic lavas do not flow readily. As a result viscous lava accumulates locally, producing thick deposits around the vent. If lava erupts from a central crater, a volcanic dome develops. As discussed in Section 7.2, viscous lavas rich in exsolved gas bubbles produce violent eruptions by entrapping gases and permitting gas pressure to build to the point of explosivity. High viscosity, volatile-rich lava eruptions can be devastatingly destructive as discussed later in this chapter. Low viscosity basic lava flows great distances from the source vent, producing thin, often aerially extensive flows. In some instances, flood basalts flow hundreds of kilometers from the vent and cover thousands of square kilometers. As a result, basaltic lava flows commonly generate extensive plateaus and broad, gently sloping, convex-shaped (shield) volcanoes.

Let us now investigate the major volcanic landforms you might encounter along the islands and mountains of the Pacific rim, eastern Indian Ocean, Iceland, Mediterranean Sea, East Africa or western North America. These locations provide stunning examples of Earth's vitality and gargantuan capacity for sudden, catastrophic, change.

9.2.1   Flood basalts

Flood basalts, also known as **large igneous provinces** (LIP), are the product of massive outpourings of low viscosity basaltic lava that envelop hundreds of thousands of square kilometers. Flood basalts form above mantle hotspots in the marine setting as **oceanic flood plateaus** or on land as **continental flood plateaus**. Oceanic flood basalts include the Ontong Plateau deposits in the western Pacific Ocean basin and the Kerguelen Plateau in the Indian Ocean. Spectacular examples of continental flood basalts include India's Deccan Traps, the Siberian flood basalts, East Africa's Karoo flood basalts and North America's Columbia River/Snake River Plain flood basalts. The largest flood basalt of all – the Central Atlantic Magmatic Province (CAMP) – includes both marine and continental deposits distributed throughout the Atlantic Ocean margins recording the break-up of the Pangea supercontinent. We will discuss these examples in greater detail in Chapter 10.

Flood basalts are somewhat unique among volcanic features in that no modern examples are known, which prevents the direct observation of eruptive style and their impact upon Earth's climate and inhabitants. The Columbia River/Snake River Plain flood basalts are the youngest flood basalt deposits on Earth, having erupted within the past 17 Ma (Figure 9.5).

Flood basalts typically release lava through multiple fissures. Multiple, long fissures and low lava viscosity together with large volumes

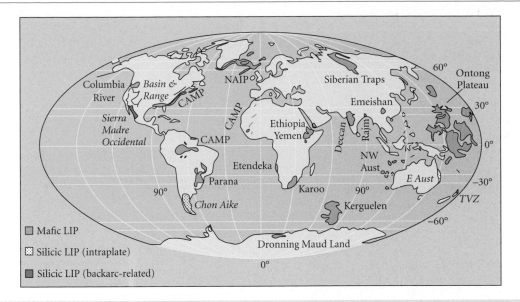

Figure 9.5 Flood basalts occur in both continental and oceanic environments in which low viscosity basaltic magma travels up to thousands of kilometers from the vent source (Bryan et al., 2002). CAMP, Central Atlantic Magmatic Province; LIP, large igneous provinces; NAIP, North Atlantic Igneous Province; TVZ, Taupo Volcanic Zone. (After Coffin and Eldholm, 1994; courtesy of M. F. Coffin.)

of magma promote widespread flooding as opposed to localized accumulation around central vents. Where fluid lavas flow into basins, deep lava lakes may form to produce unusually thick accumulations. Flood basalts may erupt repeatedly over millions of years to produce stacked sequences of flood deposits, hundreds to thousands of meters thick. The tops of each individual flow may show vesiculation as gas bubbles migrate toward the top of flows. Thick flows display columnar jointing, formed as the lava cools and contracts. The joint sets create polygonal columns less than 1 m in diameter but meters to hundreds of meters in height where lava lakes once existed. The relatively flat surfaces produced as fluid lavas fill in low areas generate immense **lava plateaus**. Figure 9.6 illustrates columnar joints exposed at Giant's Causeway in Ireland, an ancient lava lake that formed from a fissure eruption in the Irish Sea 60 million years ago.

9.2.2 Ocean ridge fissure eruptions

Ocean ridges consist of a 65,000 km long global network of submarine rift mountains characterized by horizontal extension and basaltic volcanism (Chapter 1). The crests of

ocean spreading ridges are generally 2–3 km below sea level. Iceland, the Galapagos Islands and the Azores Islands represent rare examples of ocean ridges exposed above sea level due to unusually large volumes of erupted lava; these ocean islands represent regions where hotspots locally underlie the spreading ridge system.

Ocean ridge eruptions are generally not explosive due to low volatile content within the magma, low viscosity and the relatively high hydrostatic pressures associated with water depth that inhibit exsolution. Tensional stresses associated with rifting result in eruptions from linear fissures sub-parallel to the ridge axis. As basaltic magma rises through ocean ridge fissure vents, interaction with cold seawater produces a variety of remarkable features which include pillow basalts, black smokers and mineralized mound deposits.

As hot (1100–1300°C) basaltic magma rises upward and reacts with cold seawater, spheroidal **pillow lavas** develop (Figure 9.7). The outer shell of the pillow lava is quenched instantaneously producing a glassy rind. The interior of the pillow cools more slowly. After the outer glassy rind forms, radial cooling

(a)

(b)

Figure 9.6 (a) Twelve meter high "pipe organ" basalt columns, Giants Causeway, Northern Ireland. (b) Most columns are six-sided (hexagonal) but four-, five- and eight-sided columns also exist. Note also the radial cooling fractures in each of the columns. Pen for scale. (Photos by Kevin Hefferan.)

Figure 9.7 Pillow lavas on the sea floor, East Pacific Rise, ~2500 m depth. The photo was taken from the Alvin submersible, operated by the Woods Hole Oceanic Institution, and is about 3 m across. (Photo courtesy of Jeffrey A. Karson.)

joints develop within the interior of the pillow. Repeated fracturing of the lava flow and pillow expansion provides conduits from which interior lava escapes to produce additional pillows. These propagate downslope and accumulate one atop the other. A series of pillows flow downslope and breach, generating cascading pillow trails. Individual pillows are generally less than 1 m in diameter and occur in pods. The cumulative thickness of pillow basalt pods ranges from meters to hundreds of meters. Pillow lavas are widespread along ocean ridge systems, but also form in continental lakes and volcanic arc settings or anywhere subaqueous basalt eruptions occur.

In addition to pillow basalts, ocean ridges are sites of hydrothermal (hot water) activity as heated seawater flows through the hot, juvenile crust produced by sea floor spreading. Hydrothermal activity produces **black smoker** vents that emit plumes of dark, hot (350–400°C), metal- and sulfide-rich solutions that precipitate black, sulfide-rich mounds which can grow to chimney-like tower structures (Figure 9.8). Black smoker towers and sulfide mounds were first discovered along the axis of the East Pacific Rise (Francheteau et al., 1979). The black color is due to the suspended, fine-grained sulfides. Black smoker chimneys observed along the East Pacific Rise and Mid-Atlantic Ridge grow at the rate of up to 6 m per year, with the largest chimneys ~50 m high and up to 12 m in diameter. Black smoker towers and associated mounds are enriched in metals

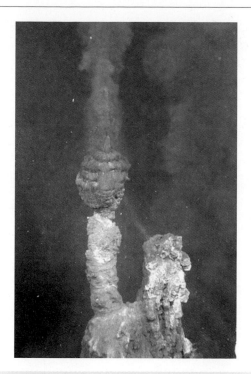

Figure 9.8 Black smokers from the Kermadec arc. (Photo courtesy of the New Zealand American Submarine Ring of Fire 2007 Exploration, NOAA Vents Program, Institute of Geological and Nuclear Sciences and NOAA.)

such as cobalt, nickel, copper, zinc and silver. This mineralization process is the source for economically viable, volcanogenic, massive sulfide deposits found in modern and ancient plate margins (Chapter 19).

Ocean ridges also contain **white smokers** that emit lower temperature (100–300°C) solutions that precipitate light-colored minerals such as calcite, gypsum, barite and quartz. In December 2000, calcite towers 20–30 m high were viewed by geologists in the Alvin submersible along a previously unknown volcanic vent field, located 14.5 km west of the Mid-Atlantic Ridge. These calcite towers, dubbed "Lost City Vent Field", have been formed by hydrothermal reactions between ocean mantle harzburgites and cold seawater over the past 30,000 years. As a byproduct of hydrothermal activity, harzburgite alters to serpentinite, and calcium carbonate, silica and sulfate minerals precipitate generating white-colored rock columns (Fruh-Green et al., 2003).

Amazingly, many life forms such as crabs, clams, tube worms, mussels and long-necked barnacles dwell in the vicinity of ocean ridge vent systems. These organic communities do not depend on sunlight and photosynthesis. Instead, chemosynthetic bacteria fix energy from the hydrothermal vents into organic molecules that provide nutrients for marine life forms. The diversity of life forms was first observed by researchers aboard the Alvin submersible along the East Pacific Rise from 1977 to 1979. The ocean ridge environment is now considered as one of the possible environments in which life first developed on Earth.

9.2.3 Shield volcanoes

Shield volcanoes are broad, sloping edifices that cover hundreds to thousands of square kilometers with shapes that resemble the defensive shields of ancient warriors. Of the conical volcanic landforms, shield volcanoes encompass the greatest volume. Shield volcanoes are produced by hot, low viscosity, basaltic lava that flows great distances from the vent. The slopes of shield volcanoes are not steep, generally 2–10°. Shield volcanoes occur over hotspots that emit large volumes of basaltic magma from central vents, fissure rifts and flank eruptions.

Shield volcano growth produced the Emperor Seamount and Hawaiian Island hotspot chain of the central Pacific Ocean. The Big Island of Hawaii is a composite of five shield volcanoes: Kilauea, Mauna Loa, Mauna Kea, Hualalai and Kohala (Figure 9.9a, inset). Since 1983, Kilauea's Pu'u O'o Volcano on the east fissure rift zone has been the primary source of basaltic lava flowing into the ocean along the southeast coast of Hawaii. Mauna Loa (Figure 9.9b) and Mauna Kea are immense shield volcanoes over 4 km in elevation. Mauna Kea is 4.2 km above sea level. From the base of the Pacific Ocean floor, Mauna Kea's ~11 km relief exceeds that of Mount Everest.

The Hawaiian eruptions that produce shield volcanoes also generate fiery basaltic **lava fountains** several hundred feet high (Figure 9.10). Airborne blobs of liquid lava emitted by fountains are referred to as **spatter**. Spatter that solidifies at the base of the lava fountains is referred to as **welded spatter** or

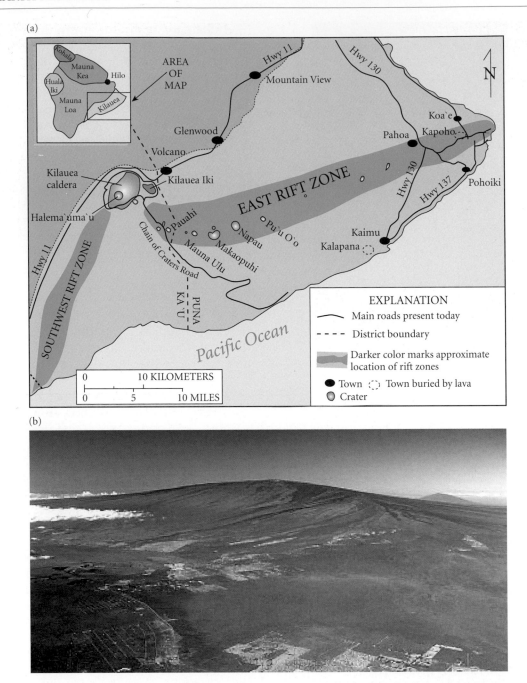

Figure 9.9 (a) The active Kilauea volcanic zone; the inset map illustrates the five shield volcanoes on Big Island of Hawaii. (b) The Mauna Loa shield volcano. (Photo courtesy of the US Geological Survey.)

agglutinate deposits. Spatter cones (Figure 9.11), typically less than 20 m in height, form where welded spatter accumulates around a central vent. Lava fountains that erupt via fissures (see Figure 9.1) produce linear ridges called **spatter ramparts** on either side of the fissure (Figure 9.12). These features are well displayed on the shield volcanoes of Hawaii, where they can be produced in a matter of days or weeks.

Lava propelled high into the air may also cool, solidify and fall to the ground as vesicular clots of scoria or as glassy fragments, especially if the lava comes in contact with water.

Figure 9.10 Pu'u 'O'o lava fountain on the Kilauea Volcano. (Photo courtesy of C. Heliker, taken September 19, 1984, with permission of the US Geological Survey.) (For color version, see Plate 9.10, between pp. 248 and 249.)

Figure 9.11 Five-meter-tall spatter cone constructed by a lava fountain eruption on the Pu'u 'O'o spatter and cinder cone complex, Hawaii. (Photo courtesy of T. N. Mattox, taken March 3, 1992, with permission of the US Geological Survey.) (For color version, see Plate 9.11, between pp. 248 and 249.)

Figure 9.12 Approximately 5 m high, 100 m long spatter rampart in Kilauea. (Photo by Kevin Hefferan.)

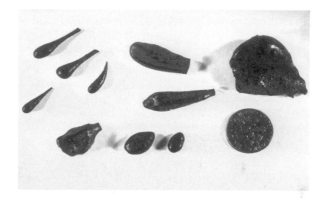

Figure 9.13 Pele's tears collected downwind from Kilauea Volcano, Hawaii. Note the US dime for scale in the lower right. (Photo courtesy of J. D. Griggs, taken November 1984, with permission of the US Geological Survey.)

When lava droplets are quenched in flight they form black, glassy, streamlined particles called **Pele's tears** (Figure 9.13), in honor of the Hawaiian goddess of fire. As small lava droplets are propelled through the air, some are partially stretched into golden, acicular strands called **Pele's hair** (Figure 9.14). Newly formed deposits display Pele's tears attached to the hair follicles like a lead weight attached to a fishing line. Pele's hair particles have diameters of ~0.5 mm, lengths up to 2 m and can remain airborne over distances of tens of kilometers.

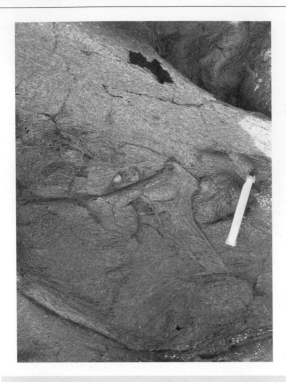

Figure 9.14 Pele's blonde hair created by windblown streamlining of silica glass. The metal scale is 10 cm in length. (Photo courtesy of Neil Heywood.) (For color version, see Plate 9.14, between pp. 248 and 249.)

Figure 9.15 Thurston lava tube, Kilauea Volcano. (Photo by Kevin Hefferan.)

Figure 9.16 Lava tube roof collapse allows a skylight view of hardened lava "stalactite" drips hanging downward into the partially evacuated tube. (Photo courtesy of C. Heliker, taken September 8, 1992, with permission of the US Geological Survey.) (For color version, see Plate 9.16, between pp. 248 and 249.)

Lava tubes

Low viscosity lava can erupt as massive sheet flows that inundate a landscape, or as channelized flows within lava levees or lava tubes. **Lava levees** are elevated lateral banks that contain the lava flow within a stream-like channel. **Lava tubes** (Figure 9.15) are shallow subterranean tunnels channeling lava beneath thin, solidified basaltic roofs. In some tubes, skylights (Figure 9.16) develop as a portion of the tunnel roof collapses, allowing the cautious observer a window to peer into the flowing lava tube. When the lava source is depleted, continued flow of lava can evacuate the tube. An evacuated lava tube may preserve stalactites hanging from the ceiling, lava drips and flow lines occurring on the walls, and gas vesicles around the tube periphery. Lava tubes range from less than 1 m to tens of meters in diameter and may extend several tens of kilometers in length. Lava tubes several kilometers long occur in the basaltic pahoehoe flows of Hawaii. Lava tubes also occur in volcanoes of the Cascades (USA), Italy, Japan, the Canary Islands, Tenerife Island (Spain) and New South Wales, Australia.

Pahoehoe lava

Pahoehoe lava consists of low viscosity, "runny" basaltic lava which produces thin flows with a billowing, rippled and/or ropey surface (Figure 9.17). Rapid cooling produces a glassy rind that becomes convoluted due to the continued lava flow and/or expansion of

Figure 9.17 Pahoehoe lava developing in Kilauea, Hawaii. (Photo courtesy of the US Geological Survey.) (For color version, see Plate 9.17, between pp. 248 and 249.)

Figure 9.18 An advancing lobe of glowing aa lava covers earlier pahoehoe lavas on the coastal plain of the Kilauea Volcano, Hawaii. (Photo courtesy of the US Geological Survey.) (For color version, see Plate 9.18, between pp. 248 and 249.)

gas beneath the thin, solid, glassy rind. Entrapment of gas bubbles within the solid, glassy rind typically produces a vesicular zone just beneath the surface. Pahoehoe lava is well developed on shield volcanoes but can occur anywhere low viscosity lava flows down moderate slopes.

Aa lava

More viscous **aa** lava tends to produce thicker, slower moving lava flows with angular, jagged, fractured surfaces (Figure 9.18). The angular blocks that cover the surface of aa

lava flows are described as spiny, rubbly or clinkery. Typical aa flows are 2–8 m thick. When aa lavas flow into a subaqueous environment, water seeps into the fractures. The injection of cold water into hot fractures at shallow depths produces steam, yielding an explosive reaction and the ejection of airborne pyroclastic fragments. Explosive volcanic eruptions are limited to shallow depths. At greater depths, higher pressures impede volatile expansion and explosive eruptions (Francis, 1993).

Pahoehoe and aa lavas may be generated from the same volcanic source. Pahoehoe lavas can evolve into more viscous aa lavas by cooling, loss of volatiles or change in topographic slope. However, whereas pahoehoe flows are restricted to basaltic lava, higher viscosity aa lavas may be basaltic to andesitic in composition. High viscosity results from higher molecular linkage due to higher SiO_2 content, lower dissolved gas content, lower temperatures and/or higher shear stresses acting on the flow surface (shear stresses increase with topographic slope angle). Spectacular transitional pahoehoe to aa basaltic flows occur in Hawaii where lava flows over steep crater walls and experiences increased shear stresses. At the base of the crater, aa lavas revert back to pahoehoe as they flow along the relatively flat crater floor. At Pisgah Crater in California, pahoehoe flows change to aa due to the effects of cooling, dissolved gas loss and increased slope gradients. As is usually the case, a number of factors (temperature, viscosity, volatile content, slope) play critical roles in determining lava flow characteristics.

9.2.4 Pyroclastic cone volcanoes

Whereas shield volcanoes encompass hundreds of square kilometers and attain heights of over 4 km above sea level, pyroclastic cones are modest in scale and can develop within a period of years. Pyroclastic cones are relatively small, usually encompassing areas of less than 20 km² and with heights typically of less than 500 m.

Pyroclastic cones are steep-sided (~30–35°) conical features composed of tephra. **Tephra** consists of volcanic rock fragments of various sizes and compositions emitted during explosive eruptions. Common ash- to bomb-sized rock fragments include basalt and andesite, with lesser amounts of dacite and rhyolite. Pyroclastic cones develop from tephra emitted from central vents. Pyroclastic cones occur in a wide variety of settings including continental rifts, convergent plate boundaries, divergent plate boundaries and over hotspots. The Mauna Kea shield volcano in Hawaii contains over 100 small satellite pyroclastic cones (see Figure 9.2). Pyroclastic cones include:

- **Scoria cones** composed predominantly of vesicular basaltic material.
- **Cinder cones** consisting of ash, lapilli and bomb-sized particles of various compositions that accumulate as circular to oval-shaped conical volcanoes (Box 9.1).

The Late Miocene to Holocene San Francisco volcanic field in Arizona, which occupies an area of approximately 5000 km², provides an excellent example of a **volcanic field**. The San Francisco field includes approximately 600 small cinder cones, the most famous of which is Sunset Crater. Sunset Crater is among the youngest volcanic features in the conterminous United States, having erupted approximately 1000 years ago (Smiley, 1958). San Francisco Peaks, the largest volcanic feature in the volcanic field, is a composite volcano that bears a striking resemblance to post-1980 Mt St Helens in having an amphitheatre-shaped crater in its northeastern flank (Figure 9.19).

Box 9.1   Parícutin

You probably live in an area entirely devoid of volcanic activity. Could you imagine steam rising from a newly formed crack in your yard, which over a period of months and years develops into a volcano? This is exactly what occurred in Mexico, shocking local farmers and geologists throughout the world. The Parícutin cinder cone volcano unexpectedly formed in a cultivated cornfield in February of 1943. Within a few years a level farmland tract catapulted into a volcanic cone rising over 400 m above the surrounding terrain and covering approximately 15 km². The fiery lava and ash eruption (Figure B9.1) continued to erupt until 1952. Parícutin is a monogenetic volcano; that is, lava and ash erupted from a singular eruptive event over a limited time frame (1943–1952). The 9-year eruption of the Parícutin Volcano provided an active example of the eruptive style responsible for the development of cinder cone volcanoes throughout the world. Having observed a modern-day cinder cone develop, geologists use Parícutin as a model in the analysis of older cones.

For more information, refer to the wonderful 1993 book entitled *Parícutin: a Volcano Born in a Mexican Cornfield* edited by Jim Luhr and Tom Simkin of the Smithsonian Institution.

Figure B9.1   A continuous fountain of viscous bombs erupting from Parícutin Volcano, Michoacan, Mexico. (Photo courtesy of L. Storm, taken October 21, 1943, with permission of the US Geological Survey. Published as plate 32A in *US Geological Survey Bulletin* 965D, 1956.)

9.2.5 Composite volcanoes

Composite volcanoes are majestic cone-shaped mountains encompassing tens to hundreds of square kilometers in area with slopes ranging from 10° to 30°. Composite volcanoes consist of alternating layers of pyroclastic debris and lava flows that build volcanic cones. In a sense, these volcanoes are a composite of many different rock types, generating stratified layers (hence the alternative name **stratovolcanoes**). Though composite volcanoes are large and majestic, many exceeding 4 km in elevation, their volume pales in comparison to immense shield volcanoes (Figure 9.20).

Figure 9.19 Cinder cones of the San Francisco volcanic field in the foreground with the San Francisco Peaks composite volcano in the background. (Photo by Kevin Hefferan.)

Composite volcanoes contain a wide variety of igneous rocks whose composition ranges from basalt to rhyolite but they are generally dominated by andesite. Andesitic lava typically has higher viscosity and greater yield strength than basalt. As a result, andesitic lava flows are thicker, slower and more likely to build up around the vent source as compared to basaltic flow in shield volcanoes. Andesitic lava's high viscosity generates block lava flows, which are similar to, but coarser, than aa lava. **Block lava** consists of smooth-sided blocks up to several meters in diameter that tumble downslope (Figure 9.21). Block lavas lithify as volcanic breccia deposits (Francis, 1993). Block lavas occur in a wide variety of settings such as convergent margins and hotspots.

Composite volcanoes are widespread throughout the Pacific "ring of fire" and in the eastern Indian Ocean, where they occur in two major settings:

1 Continent–ocean convergent margins, where **continental margin volcanic arcs** develop on the overlying continental lithosphere. Partial melting of a thick overlying wedge containing continental lithosphere produces the calc-alkaline series that include andesitic, dacitic, basaltic and rhyolitic rocks. Continental margin arcs occur along the west coast of South America, Central America, the Pacific Northwest and Alaska in North America.

2 Ocean–ocean convergent margins, where **island volcanic arcs** develop above the

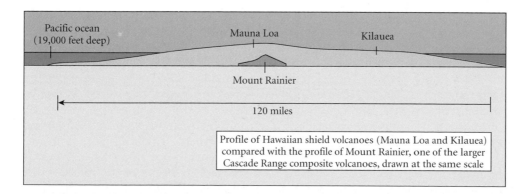

Figure 9.20 Relative scale of composite volcanoes versus shield volcanoes. (From Tilling et al., 1987; with permission of the US Geological Survey.)

Figure 9.22 Fin-shaped "whaleback" spine at Mt St Helens, April 28, 2006. (Photo courtesy of Dan Dzurisin, with permission of the US Geological Survey.)

Figure 9.21 Block lava composed of trachyandesite (an extrusive rock, intermediate in composition between trachyte and andesite) at Craters of the Moon National Monument, Idaho. (Photo courtesy of the US National Park Service.)

overlying ocean plate. Partial melting of a thin overlying wedge containing ocean lithosphere produces calc-alkaline basalts and andesites, which are the predominant volcanic rocks in these settings. However, tholeiitic basalts, boninites and other rocks also form in young island arc systems. Island arc volcanoes occur along the Aleutian Range of the northern Pacific Ocean, throughout the western Pacific Ocean, the eastern Indian Ocean region and in the Caribbean Sea. We will discuss both continental margin and island arc systems in greater detail in Chapter 10.

Because they erupt a wide variety of lava compositions, composite volcanoes consist of a variety of aphanitic, porphyritic aphanitic, pyroclastic and glassy rocks. The lava flows are dominated by andesitic rocks, but basal-

tic, dacitic and even rhyolitic rocks also occur. The pyroclastic rocks are dominated by andesitic, dacitic and rhyolitic tuffs and lapilli tuffs, with their higher viscosity and dissolved gas contents. Dacite and rhyolite magmas are even more viscous than andesitic magma, resulting in even slower rates of movement. The slow rate of magma movement can produce spines and domes in which viscous magma crystallizes within the composite volcano crater, essentially acting as a plug inhibiting the release of lava.

Spines

Spines develop when a solidified magma plug is pushed up through the conduit, forming a vertical column on Earth's surface. **Plugs** consist of magma that solidifies within the "throat" or conduit of the volcano. The term "plug" is entirely appropriate because it performs the same role as a stopper in a bathtub, preventing the release of fluid. While plugs may be referred to as volcanic plugs, the rocks of which they are composed actually formed as shallow (hypabyssal) plutonic features. When the hot plug is extruded onto the surface it becomes a spine. Following the 1902 eruption of Mt Pelée in the Caribbean, a 45 m high vertical spine was exposed. The Mt Pelée spine was not a long lasting feature, crumbling within a few years. In 2004, Mt St

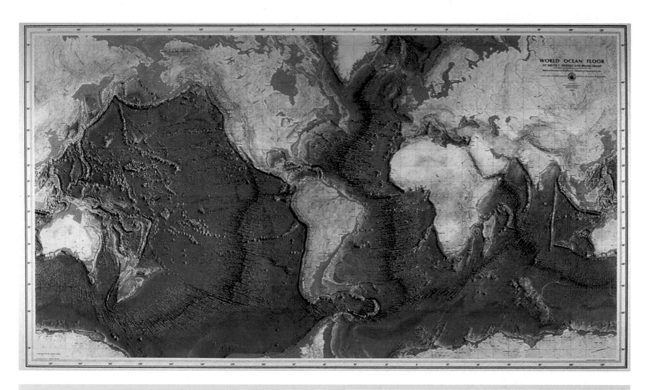

Plate 1.7 Map of the ocean floor showing the distribution of the oceanic ridge system. (Courtesy of Marie Tharp, with permission of Bruce C. Heezen and Marie Tharp, 1977; © Marie Tharp 1977/2003, by permission of Marie Tharp Maps, LLC, 8 Edward St, Sparkhill, NT 10976, USA.)

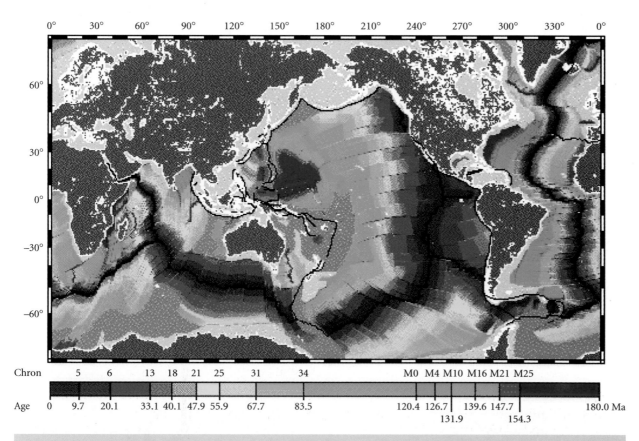

Plate 1.10 World map showing the age of oceanic crust; such maps confirmed the origin of oceanic crust by sea floor spreading. (From Muller et al., 1997; with permission of the American Geophysical Union.)

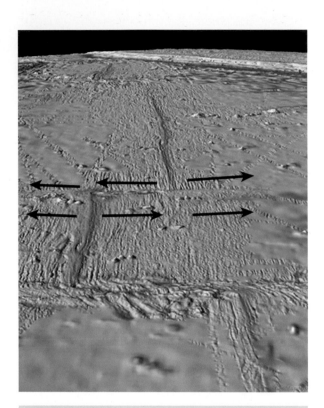

Plate 1.15 Transform faults offsetting ridge segments on the eastern Pacific Ocean floor off Central America. Arrows show the directions of sea floor spreading away from the ridge. Portions of the fracture zones between the ridge segments are transform plate boundaries; portions beyond the ridge segments on both sides are intraplate transform scars. (Courtesy of William Haxby, LDEO, Columbia University.)

(a)

(b)

Plate 4.1 Representative mineral crystals: (a) quartz; (b) tourmaline. (Photos courtesy of the Smithsonian Institute.)

(a)

(b)

(c)

Plate 4.39 (a) Hematite replacing pyrite; (b) chalcedony encrusting aragonite; (c) quartz cast filling an aragonite solution cavity. (Photos courtesy of Stan Celestian, Maricopa Community College.)

Plate 5.2 Crystal aggregate habits, clockwise from upper left: fibrous serpentine; radiating pectolite; divergent stibnite; drusy calcite; dendritic pyrolusite; roseiform selenite gypsum. (Photo by John O'Brien.)

Plate 5.3 Crystal aggregate habits, clockwise from upper left: colloform (botryoidal) cassiterite; reticulate gypsum in geode; massive kaolinite; granular olivine. (Photo by John O'Brien.)

Plate 5.4 Crystal aggregate habits, clockwise from upper left: pisolitic calcite; concentrically banded chalcedony in geode; amydaloidal calcite and thomsonite filling vesicles in basalt; lamellar to foliated kyanite. (Photo by John O'Brien.)

Plate 5.5 From left: euhedral tourmaline, completely enclosed by crystal faces; subhedral quartz, partially enclosed by crystal faces; anhedral apatite, lacking crystal faces. (Photo by John O'Brien.)

Plate 5.7 Fracture types, clockwise from upper left: conchoidal fracture in opal; splintery fracture in selenite gypsum; uneven or irregular fracture in alabaster gypsum; hackly fracture in native copper. (Photo by John O'Brien.)

Plate 5.8 Diaphaneity, from left: first pair, opaque, metallic galena and pyrite, which do not transmit light; second pair, plagioclase and potassium feldspar, which are somewhat translucent; third pair, rose quartz and aragonite, which are moderately translucent; fourth pair, calcite and fluorite, which are transparent and transmit the underlying image. (Photo by John O'Brien.)

Plate 5.10 Mineral lusters, from left: metallic stibnite; submetallic chromite; dull hematite. (Photo by John O'Brien.)

Plate 5.11 Non-metallic mineral lusters, clockwise from upper left: vitreous plagioclase; subvitreous garnet; resinous amber; waxy opal; silky serpentine; pearly soapstone; dull, earthy bauxite. (Photo by John O'Brien.)

(a)

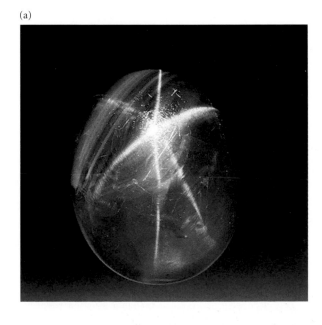

(b)

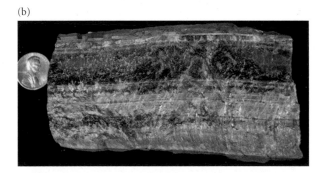

Plate 5.12 (a) Asterism in the "Star of Bombay" sapphire. (Smithsonian Institution, with permission.) (b) Chatoyancy in tiger eye. (Photo by Kevin Hefferan.)

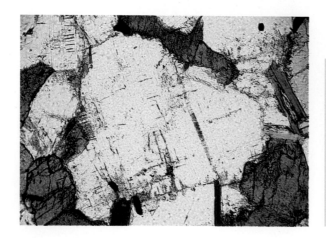

Plate 6.8 Plane polarized light photomicrograph of plutonic igneous rock (granitoid). Gray-green crystals with relatively high relief and two cleavages not at right angles are hornblende. Slightly dusty, colorless crystals with two cleavages at right angles and low relief are plagioclase. The light to darker brown crystals, with moderate relief and one cleavage are biotite. (Photo courtesy of Kurt Hollacher, Union College.)

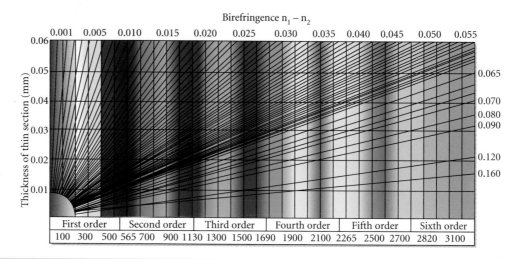

Plate 6.10 A modified version of the Michel–Levy color chart for interference colors viewed under crossed polars. The orders of color increase to the right as hue intensities decrease. (Courtesy of Olympus Microscopy.)

Plate 6.11 Photomicrograph of a thin section of gabbro viewed in crossed polars mode. The plagioclase is striped and has low birefringence (low first order white and gray interference colors). Brightly colored augite has moderate–high birefringence (well-developed second order interference colors in shades of blue, purple, red and orange). (Photo courtesy of Siim Sepp.)

Plate 6.13 A sanidine crystal (clear, low birefringence) showing a simple composition (Carlsbad) twin crystal with the lower twin at extinction and the upper twin not at extinction. (Photo courtesy of Kurt Hollacher, Union College.)

Plate 6.14 Crossed polars photomicrograph of a microcline crystal, displaying the gridiron twinning and low birefringence (first order gray and white colors) that make this mineral easy to recognize under crossed polars. (Photo courtesy of Kurt Hollacher, Union College.)

Plate 6.15 Crossed polars photomicrograph of diorite (see Figure 6.8 for a plane light view) showing a plagioclase crystal in the center (low birefringence, "stripes" due to alternate extinction of albite twins with pericline twins at nearly right angles), hornblende (higher relief, moderate birefringence at lower left and right corners, two sets of cleavage not at right angles in the left-hand crystal) and biotite (thin, greenish crystal with moderately high birefringence and one cleavage orientation). (Photo courtesy of Kurt Hollacher, Union College.)

Plate 6.16 Crossed polars photomicrograph of a volcanic rock showing a large plagioclase phenocryst (low birefringence, simple Carlsbad twins, narrower albite twins and chemical zoning shown by the shadows which indicate that the darker parts are near extinction and the lighter part farther from the extinction position). The phenocryst is set in a fine-ground mass of microscopic crystals and some cryptocrystalline or glassy material. (Photo courtesy of Kurt Hollacher, Union College.)

Plate 6.17 Crossed polars photomicrograph of a quartz crystal near the extinction position. Strain has produced non-uniform, undulatory extinction, so that the lower and upper parts of the crystal are at extinction while the central parts are not. The extinction shadow would sweep across the crystal toward the center as the stage is rotated. (Photo courtesy of Kurt Hollacher, Union College.)

Plate 6.18 Crossed polars photomicrograph of perthite. Exsolved plagioclase blebs and stringers trend upper right to lower left across the potassium feldspar crystal. Note the low birefringence and first order colors characteristic of both feldspars. (Photo courtesy of Kurt Hollacher, Union College.)

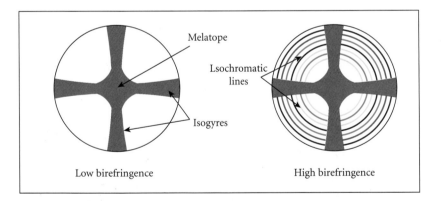

Plate 6.25 Centered optic axis figures for a uniaxial mineral display two isogyres in the form of a cross, the center of which is the trace of the optic axis or melatope. Minerals with low birefringence show first order grays and whites in the quadrants between the isogyres. Minerals with higher birefringence display isochromatic lines concentric about the melatope, whose numbers increase with retardation; that is with birefringence for a sample of constant thickness.

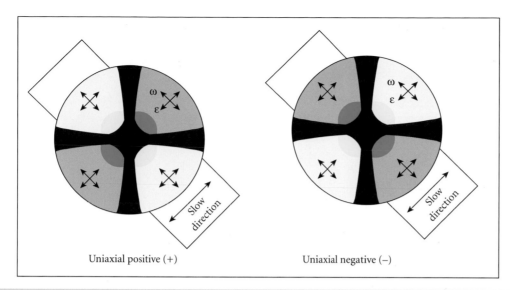

Uniaxial positive (+)   Uniaxial negative (−)

Plate 6.26 Application of the gypsum plate in optic sign determinations of uniaxial minerals. When the gypsum plate is inserted, addition (blues) occurs in the quadrants where the slow vibration direction is parallel to that of the gypsum plate and subtraction (yellows) occurs where the slow vibration direction is perpendicular to that of the gypsum plate. Blue in the NE–SW quadrants means that the extraordinary ray is the slow ray associated with a higher refractive index so that $\varepsilon > \omega$ and the mineral is uniaxial positive. Yellow in the NE–SW quadrants indicates that the extraordinary ray is the fast ray associated with a lower refractive index (epsilon) so that $\varepsilon < \omega$ and the mineral is uniaxial negative. (Courtesy of Gregory Finn.)

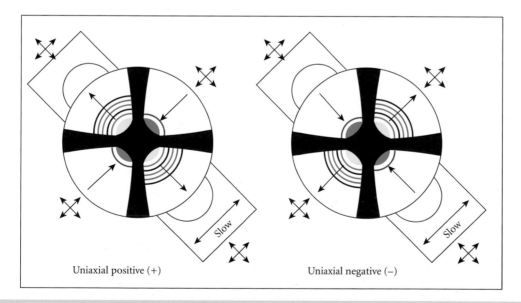

Uniaxial positive (+)   Uniaxial negative (−)

Plate 6.27 Application of the quartz wedge in optic sign determination for minerals with high birefringence and well-developed isochromatic lines. If isochromes move toward the melatope in the NE–SW quadrants and away from the melatope in the NW–SE quadrants, the mineral is positive. If isochromes move away from the melatope in the NE–SW quadrants and toward the melatope in the NW–SE quadrants, the mineral is negative. (Courtesy of Gregory Finn.)

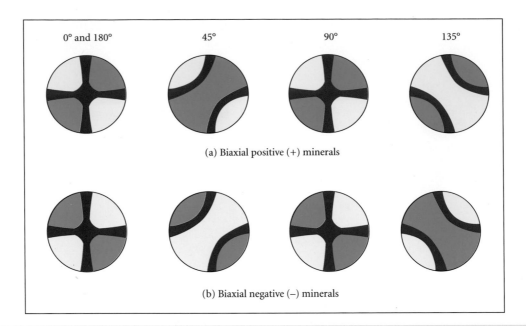

0° and 180°   45°   90°   135°

(a) Biaxial positive (+) minerals

(b) Biaxial negative (–) minerals

Plate 6.33 (a) The appearance of a Bxa interference figure for a biaxial (+) mineral of low–moderate birefringence viewed in various orientations with the gypsum plate inserted. When the gypsum plate is inserted, positive minerals show white-gray changing to blue in the NE–SW quadrants and to yellow in the NW–SE quadrants. (b) A similar sequence that contrasts the appearance of a biaxial (–) mineral of similar birefringence where changes are to yellow in the NE–SW quadrants and to blue in the NW–SE quadrants.

(a)

(b)

Plate 6.34 The appearance of a Bxa interference figure for a biaxial (+) mineral of moderate–high birefringence viewed in various orientations. When a quartz wedge is inserted into the accessory slot, positive minerals display isochromes that move in toward the melatopes in the NE–SW quadrants and outward in the NW–SE quadrants. (b) A similar sequence which contrasts the appearance of a biaxial (–) mineral of similar birefringence where changes are outward in the NE–SW quadrants and inward the NW–SE quadrants. (Courtesy of Steve Dutch.)

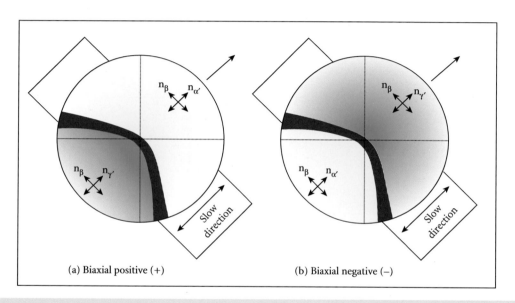

Plate 6.36 Gypsum plate determination of the optic sign using a biaxial centered optic axis figure in a position of maximum curvature of the isogyre. (a) If the concave side of the isogyre in the SW (illustrated) or NE quadrant turns blue, and the concave side turns yellow, the mineral is biaxial positive (+). (b) If the concave side of the isogyre in the SW (illustrated) or NE quadrant turns yellow, and the concave side turns blue, the mineral is biaxial positive (−). (Courtesy of Gregory Flinn.)

(a)

Peridotite   Gabbro   Diorite   Granodiorite   Granite

Basalt   Andesite   Dacite   Rhyolite

Ultramafic	Mafic	Intermediate	Felsic		Crystalline textures
Peridotite	Gabbro	Diorite	Granodiorite	Granite	Coarse grains (phaneritic)
Komatiite (not shown)	Basalt	Andesite	Dacite	Rhyolite	Fine grains (aphanitic)

Dark-colored ferromagnesian minerals

Hornblende

Biotite

Pyroxene

Olivine

Feldspars

(Plagioclase)   (Alkali)

Quartz

100
80
60
40
20
0%

45   50   52   55   60   65   70   %SiO$_2$

(b)

Texture	Rock name
Vesicular	**Pumice** Light-colored, lightweight rock rich in gas holes (vesicles)
Vesicular	**Scoria** Dark-colored, lightweight rock rich in gas holes (vesicles)
Glassy	**Obsidian** Black to reddish rock with glassy luster and conchoidal (scalloped breakage) fracture
Pyroclastic	**Volcanic tuff** Rock composed of fine-grained ash- to sand-sized volcanic rock fragments
Pyroclastic	**Volcanic breccia** Rock composed of coarse-grained gravel and larger sized volcanic rock fragments

Plate 7.1 (a) Crystalline igneous rocks are classified based upon their crystal size (texture) as well as the major minerals olivine, pyroxene, hornblende, biotite, feldspars and quartz. (b) Igneous non-crystalline rocks are classified based primarily on their texture – size and nature of rock particles, presence of vesicles or glass – as well as color. (Photos by Kevin Hefferan.)

Plate 7.2 Basalt porphyry containing euhedral and subhedral plagioclase crystals. Some euhedral crystals display zoning suggesting that the crystals continued to react with the melt during crystallization. Zoning is discussed further in the text. (Photo by Kevin Hefferan.)

Plate 7.3 Granite pegmatite with quartz, hornblende and muscovite. (Photo by Kevin Hefferan.)

Plate 7.4 Coarse-grained phaneritic granite with early formed, euhedral potassium feldspar. Subhedral quartz and hornblende represent later, void-filling minerals. (Photo by Kevin Hefferan.)

Plate 7.5 Aphanitic basalt with microcrystalline texture as viewed under a petrographic microscope. Note the euhedral to subhedral acicular plagioclase crystals embedded within a cryptocrystalline groundmass. Field of view is 2 cm. (Photo by Kevin Hefferan.)

Plate 7.6 Porphyritic–phaneritic texture with subhedral potassium feldspar phenocrysts in a groundmass of quartz and hornblende. As phenocrysts encompass >50% of the rock this is a granite porphyry. If phenocrysts encompass <50% of the rock volume, the term porphyritic granite would be used. (Photo by Kevin Hefferan.)

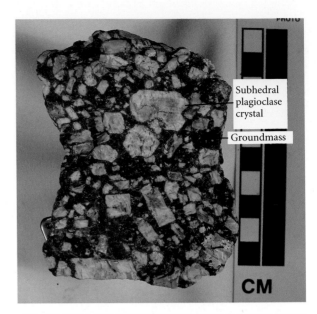

Plate 7.7 Andesite porphyry displaying aphanitic porphyritic texture. Note the subhedral plagioclase phenocrysts (white crystals) and the fine-grained aphanitic groundmass (gray). (Photo by Kevin Hefferan.)

Plate 7.13 Amygdaloidal basalt in which vesicles have been infilled with quartz and epidote. This basalt has been altered by hydrothermal solutions which have changed the existing chemistry of the basalt and precipitated new (secondary) minerals in the vesicles. (Photo by Kevin Hefferan.)

Plate 7.15 Angular volcanic blocks from Kilauea, Hawaii. (Photo by Kevin Hefferan.)

Plate 7.16 Partially welded ash flow tuff deposit. (Photo by Kevin Hefferan.)

Plate 7.17 Densely welded tuff deposit. (Photo by Kevin Hefferan.)

Plate 8.10 Olivine xenocrysts in vesicular basalt.

Plate 9.1 "Curtain of fire" fissure eruptions on Mauna Loa produce an elongated spatter rampart ridge upon solidification of the lava. (Photo courtesy of D. A. Clague, taken March 1984, with permission of the US Geological Survey.)

Plate 9.11 Five-meter-tall spatter cone constructed by a lava fountain eruption on the Pu'u 'O'o spatter and cinder cone complex, Hawaii. (Photo courtesy of T. N. Mattox, taken March 3, 1992, with permission of the US Geological Survey.)

Plate 9.10 Pu'u 'O'o lava fountain on the Kilauea Volcano. (Photo courtesy of C. Heliker, taken September 19, 1984, with permission of the US Geological Survey.)

Plate 9.16 Lava tube roof collapse allows a skylight view of hardened lava "stalactite" drips hanging downward into the partially evacuated tube. (Photo courtesy of C. Heliker, taken September 8, 1992, with permission of the US Geological Survey.)

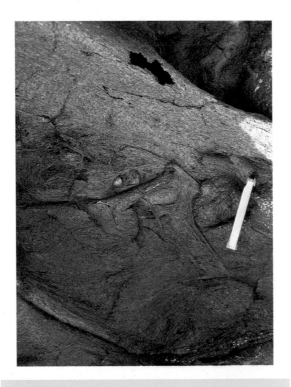

Plate 9.14 Pele's blonde hair created by windblown streamlining of silica glass. The metal scale is 10 cm in length. (Photo courtesy of Neil Heywood.)

Plate 9.17 Pahoehoe lava developing in Kilauea, Hawaii. (Photo courtesy of the US Geological Survey.)

Plate 9.18 An advancing lobe of glowing aa lava covers earlier pahoehoe lavas on the coastal plain of the Kilauea Volcano, Hawaii. (Photo courtesy of the US Geological Survey.)

(a)

(b)

Plate 9.35 (a) This fumarole at Kilauea Volcano is releasing sulfur gases which, upon cooling, sublimate (gas to solid) around the vent as sulfaterra. Sulfaterra is sulfur-rich soil and rock. (b) Sulfur crystals sublimate from gases released at Kilauea. Pen for scale. (Photo by Kevin Hefferan.)

(a)

(b)

Plate 11.2 (a) Thin laminations (above) and thicker beds (below) in the Jurassic Berlin Formation, Connecticut. (b) Cross-stratified beds (center) with initially inclined laminations, separated by a near horizontal erosion surfaces in the Jurassic Navajo Formation, Zion National Park, Utah. (Photos by John O'Brien.)

Plate 11.5 Transition from laminar flow (background) to turbulent flow (foreground) in a branch of the Delaware River, Pennsylvania. (Photo by John O'Brien.)

(a)

(b)

Plate 11.9 Current ripples. (a) Asymmetrical current ripples on a modern beach flat, where the current flowed left to right. (Photo by John O'Brien.) (b) Asymmetrical current ripples in Cretaceous sandstone, where the current flowed from upper left to lower right. (Photo courtesy of Duncan Heron.)

Plate 11.10 Slightly wavy-crested sand waves or subaqueous dunes, Rio Hondo, southern California; the current was from left to right. (Photo by John O'Brien.)

(b)

Plate 11.16b This single tabular set of planar cross-strata records the migration of a straight-crested bed form from left to right in the Sinian Sandstone, northwest China. (Photo courtesy of Marc Hendrix.)

(b)

Plate 11.17b Trough sets of festoon cross-strata in profile at a high angle to composite dune migration. (Courtesy of the US Geological Survey.)

Plate 11.18 Plane bed transition. (a) View of a plane bed under shallow unidirectional flow, with a lineation on the bed. (b) Diagram showing the horizontal strata and parting lineation developed in plane bed transition. (c) Horizontal strata in sandstone produced by plane bed flow regimes. (d) Parting lineation on a sandstone bedding surface. (a, b and d, courtesy of Marc Hendrix; c, photo courtesy of Duncan Heron.)

(a)

(b)

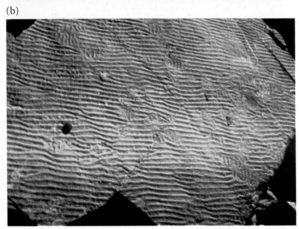

Plate 11.22 Oscillation ripple marks showing crest bifurcation, symmetry, short wavelengths and small ripple indices. (a) Modern tidal flat, Cape Cod Massachusetts. (Photo by John O'Brien.) (b) Preserved in sandstone. (Photo courtesy of Peter Adderley.)

(a)

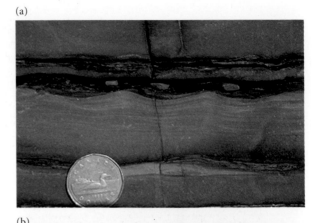

(b)

Plate 11.23 (a) Oscillation ripples in the Carboniferous Horton Group, Nova Scotia. (Photo courtesy of John Waldron.) (b) Oscillation combined flow ripples in the Jurassic Entrada Sandstone, Utah. (Photo courtesy of Tim Cope.)

Plate 11.28 Typical loess deposit; note the paucity of stratification in comparison to that displayed by wind-blown sands.

Plate 11.29 Wind ripples on back-beach sand dunes, Australia, with branching crests, strong asymmetry and large ripple indices; wind from left to right. (Photo by John O'Brien.)

Plate 11.31 Formation of aeolian cross-strata by dune migration. (a) Dune showing stoss-side erosion and lee-side deposition by sand flows. (b) Lobate sand flows on a dune avalanche face generate large-scale, steeply inclined cross-strata. (c) Large-scale, steeply inclined, tangential cross-strata in the Jurassic Navajo Sandstone, Utah, formed by aeolian dune migration with a left-to-right component. (Photos by John O'Brien.)

Plate 11.33 A striated bedrock surface in Cambrian dolostones, northwestern New Jersey. (Photo by John O'Brien.)

(a)

Plate 11.35 Glacial varves from the Pleistocene, Maine: coarser, lighter colored, resistant summer layers alternate with finer, darker colored, crumbly winter layers. (Photo courtesy of Thomas K. Weedle, with permission of the Maine Geological Survey.)

(b)

Plate 11.34 (a) Glacial till, Pleistocene, Ohio; note the polymictic composition, poor sorting and lack of stratification. (Photo courtesy of the US Geological Survey.) (b) Glacial till overlying a striated bedrock surface, Switzerland. (Photo courtesy of Michael Hambrey, with permission of www.glaciers-online.net.)

Plate 11.36 Large glacial dropstone and other ice-rafted debris from the Pleistocene, Lake Spokane, Washington. (Photo courtesy of the US Geological Survey.)

(a)

(b)

Plate 11.37 (a) Mud flow with a matrix strength sufficient to suspend boulders at the top in California. (b) Debris flow surge moving down a river valley, Mt Rainier National Park, Washington. (Photos courtesy of the US Geological Survey.)

(a)

(b)

Plate 11.38 (a) Debris flow deposit above an erosion surface, southern Utah. (Photo by Kevin Hefferan.) (b) Multiple debris flow deposits, Lone Pine, California. (Photo by John O'Brien.)

Plate 11.39 Turbidity current in a laboratory showing the head and main body, and the turbulent suspensions that developed over both.

(b)

(a)

(b)

Plate 11.41b Partial Bouma sequence with a massive, grade A unit over an erosional base, horizontal laminate B unit, convolute laminated C unit and laminated D unit, from the Tertiary, Simi Hills, California. (Photo by John O'Brien.)

Plate 11.42 Sole marks on bed bases. (a) Flute marks, Austen Glen Formation, Ordovician, New York; current from lower left to upper right. (b) Groove casts, Cretaceous, Wheeler Gorge, California. (Photos by John O'Brien.)

Plate 12.1 Differential weathering between durable cliff and pillar-forming sandstone and less durable mudrocks in Utah. (Photo by Kevin Hefferan.)

Plate 12.3 Bryce Canyon, Utah showing pillars and windows formed by differential erosion, controlled by near vertical and horizontal joint sets. (Photo by John O'Brien.)

Plate 12.4 Joints in anorthosite bedrock, Saranac Lake, New York. (Photo by Anita O'Brien.)

Plate 12.6 Biological weathering by tree root growth in Silurian dolostone, Door County, Wisconsin. (Photo by Kevin Hefferan.)

Plate 12.8 Spheroidal weathering of basaltic rock at Table Mountain, Golden, Colorado. (Photo by Kevin Hefferan.)

Plate 12.15 Australian rock art using hematite (red), limonite (brown), kaolinite (white) and manganese oxides/hydroxides (black), from Nourlangie Rock, Kakadu National Park, northern Australia. (Photo by John O'Brien.)

Plate 12.18 Layered soil produced by the disintegration and decomposition of rock materials near Earth's surface. From top to bottom, the O-, A-, E-, B- and C-horizons. (Courtesy of the Society of Soil Scientists of Southern New England.)

Alfisol

Andisol

Aridisol

Entisol

Gelisol

Histosol

Plate 12.20 Examples of the major soil orders in the USDA-NRSC soil taxonomy (see Table 12.5). (Courtesy of the US Department of Agriculture.)

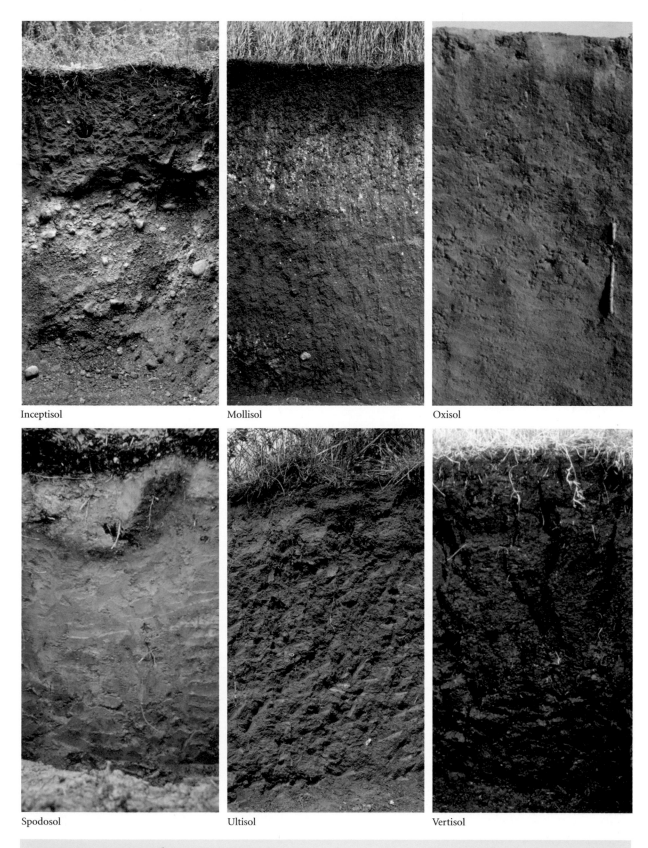

Inceptisol

Mollisol

Oxisol

Spodosol

Ultisol

Vertisol

Plate 12.20 *Continued*

(a)

(b)

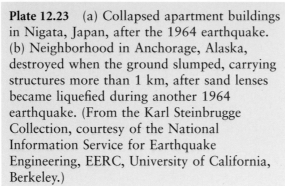

Plate 12.25   Italy's famed Leaning Tower of Pisa, which was constructed on compressible, clay-rich soil. As a result of differential consolidation, the tower requires additional reinforcement (here supplied by Maureen Crowe) to maintain its equilibrium. (Photo courtesy of Tony Crowe.)

Plate 12.23   (a) Collapsed apartment buildings in Nigata, Japan, after the 1964 earthquake. (b) Neighborhood in Anchorage, Alaska, destroyed when the ground slumped, carrying structures more than 1 km, after sand lenses became liquefied during another 1964 earthquake. (From the Karl Steinbrugge Collection, courtesy of the National Information Service for Earthquake Engineering, EERC, University of California, Berkeley.)

(a)

(b)

Plate 12.26   (a) Jurassic soil with plant root casts, buried by braided stream deposits, Connecticut. (b) Holocene soil with a well-defined O-horizon at the top, buried by pyroclastic deposits from the Tarawera Volcano, New Zealand. (Photos by John O'Brien.)

(b) (i)

(ii)

Plate B12.1b (i) Cave spring with an underground river emerging from a cave produced by subsurface dissolution, Virginia. (ii) Sinkhole, Winter Park, Florida produced by the sudden collapse of the surface into a subsurface cavity, May, 1981. (Courtesy of the USGS.)

(a)

Plate B12.3a Damage to the Van Norman (Lower San Fernando) Dam, 1971. (From the Karl Steinbrugge Collection, courtesy of the National Information Service for Earthquake Engineering, EERC, University of California, Berkeley.)

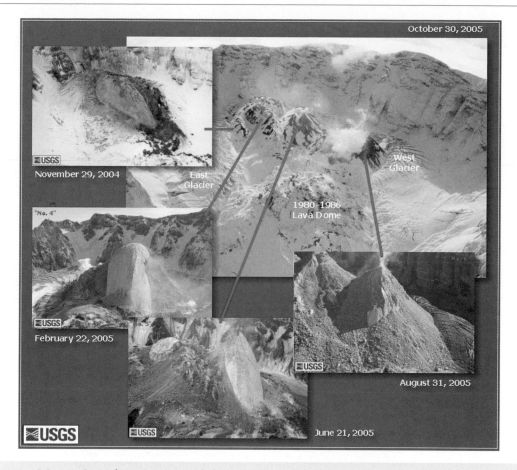

Figure 9.23   Mount St Helens' crater and dome with views indicating dome growth from 1980 to 2005. Dome growth has continued up to 2009. (Photo courtesy of John Pallister, taken October 30, 2005, with permission of the US Geological Survey.)

Helens developed a spine within its resurgent dome (Figure 9.22) which continues to grow as of this writing.

Domes

Domes are steep-sided, blister-like forms that occur within volcanic craters (Figure 9.23). Domes are inflated areas generated by the accumulation of viscous, often glassy, blocky lava of dacitic, rhyolitic or trachytic composition, on the surface or in the shallow subsurface. Domes also contain large, viscous bombs intermixed with tephra.

Dome growth requires monitoring because volcanic dome collapse generates deadly pyroclastic flows which, but for early evacuations, can result in massive loss of life. Recent catastrophic dome collapse eruptions occurred in 1980 at Mt St Helens (USA) (Box 9.2), 1991 at Mt Unzen (Japan) and 1997 at Soufriere Hills (Montserrat). The June 1991 dome collapse on Japan's Mt Unzen generated a pyroclastic flow (nuée ardente) that raged downslope. When the flow unexpectedly changed its direction, 38 geologists and photographers monitoring the event perished. Pyroclastic flows and lahars represent the two greatest hazards from composite volcanoes.

Pyroclastic deposits

Pyroclastic deposits form by a variety of processes that generate volcanic fragments. By far the most voluminous pyroclastic rocks produced in continental settings are those generated by highly explosive volcanic eruptions

Box 9.2   Mt St Helens eruption

Prior to 1980, Mt St Helens displayed a nearly perfect symmetrical cone form (Figure B9.2a). In the spring of 1980, a volcanic eruption was preceded by earthquake swarms, steam eruptions and the development of a cryptodome (a shallow dome beneath the surface) along the northern flank of the volcano. The Mt St Helens cryptodome grew at a rate of 1.8 m per day, resulting in a feature 90 m above the pre-existing mountain surface (Lipman and Mullineaux, 1981). On May 18, 1980, a 5.1 magnitude earthquake triggered motion on three large blocks, resulting in a landslide. The landslide effectively unroofed the magma, providing a pathway that allowed for a lateral blast eruption on the north side of the volcano. Immediately following the lateral blast, vertical ash plumes rose in a classic Plinian (explosive) eruption. This major eruption was followed by periodic eruptions throughout the summer of 1980. The tremendous eruption on May 18, 1980 drastically lowered the near perfect symmetrical peak of Mt St Helens by over 500 m, creating an amphitheatre-shaped crater depression on the northern slopes (Figure B9.2b). Following the explosive 1980 eruption, two periods of dacite dome growth ensued: from 1980 to 1986 and from 2004 through the publication of this text. Ongoing active dome development is closely monitored at Mt St Helens through geophysical measurements related to earth bulging (uplift and tilting), seismicity and phreatic eruptions. The cryptodome growth at Mt St Helens represents the slow rebuilding of a composite cone which one day may return to its perfectly symmetrical cone shape.

(a)

(b)

Figure B9.2   (a) Before the devastating May 18, 1980 eruption, Mt St Helens was considered to be one of the most symmetrical peaks in the Cascade Range. (Photo courtesy of Jim Nieland, taken before May 18, 1980, with permission of the US Forest Service.) (b) Post-eruption view from the north of the lateral blast caldera. (Photo courtesy of Lyn Topinka, taken May 19, 1982, with permission of the US Geological Survey.)

of silicic composition. Why are eruptions of silicic magma so explosive? As noted previously in Chapter 7, silicic magmas have the ability to dissolve large amounts of water vapor deep below Earth's surface under high confining pressures. As silicic magma rises towards the surface, the water vapor begins to separate (exsolve) as a separate phase from the magma as it crosses the level of exsolution. Exsolution of gas generates expanding gas bubbles and lowers the P_{H2O} of the magma. Decreased water vapor allows silica tetrahedra to rapidly link, dramatically increasing magma viscosity. This increase in magma viscosity severely inhibits the formation of crystals and creates a mass of hot, highly vesicular, low density, semi-solid, glassy material. As this viscous, glassy mass continues to ascend towards the surface, vesicles expand and increase in number to produce zones of frothy, glassy magma that are essentially hot pumice. Each expanding gas bubble is surrounded by glassy, semi-solid bubble walls that thin as the bubbles expand in much the same way that the walls of a balloon or a piece of bubble gum thin as they expand. If the confining pressure on this material is decreased slowly as the magma rises or suddenly, as when a vent opens to the surface, the material will cross the **level** (or **surface**) **of fragmentation**. Fragmentation occurs as the outward pressure of the expanding gas bubbles ruptures the thinned bubble walls generating glassy bubble wall fragments called **shards**. Once the glassy material has been shattered, a mixture of fragments and hot gases is created. This mixture, characterized by extremely low density and viscosity, accelerates rapidly upward towards the surface through any available opening. This triggers a chain reaction in which large volumes of frothy pumice fragments accelerate upward and are hurtled out of a vent to great heights producing cataclysmic explosive eruptions.

Explosive volcanic eruptions produce a **vertical plume** or **eruption column** that can be divided into three distinct parts (Sparks and Wilson, 1976; Sparks, 1986) (Figure 9.24):

- A lower **gas thrust region** that consists of material thrust from the vent by expanding gases at velocities ranging from 100 to 600 m/s.

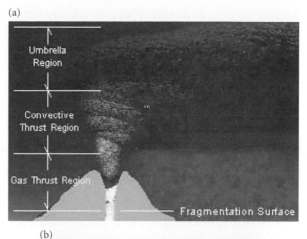

(a)

(b)

Figure 9.24 (a) Diagram depicting aspects of explosive eruptions, which include an exsolution surface above which bubbles separate from magma and expand, and a fragmentation surface above which the magma fragments to form an explosive gas thrust, convective thrust and an umbrella region of pyroclastic debris. (After Sparks 1978, 1986; courtesy of Stephen Sparks.) (b) In addition to the May 18 eruption, five more explosive eruptions of Mt St Helens occurred in 1980. This July 22 eruption sent pumice and ash up to 18 km into the air, and was visible in Seattle, Washington (160 km to the north). (Photo courtesy of Mike Doukas, with permission of the US Geological Survey.)

- An upper **convective thrust region** produced by the convective rise of heated atmospheric gases and fragments.
- An **umbrella region**, similar to the mushroom-shaped heat cloud produced by thermonuclear bomb explosions, begins to spread laterally as the result of temperature inversions in the atmosphere.

The vertical plume consists largely of expanding gases and glassy (vitric constituents) pumice fragments and shards created when the magma crosses the fragmentation surface. The vertical plume may also contain subordinate rock fragments (lithic constituents) ripped from vent walls and fragments of any crystals (crystal constituents) that formed in the magma below the level of exsolution. The buoyancy of the plume is largely dependent upon the degree of mixing and heating that occurs with the surrounding atmosphere (Sparks, 1986).

Two modes of eruptive column behavior – convective thrust and gravitational collapse – result in two kinds of deposits: pyroclastic fall (air fall) deposits and pyroclastic flow deposits.

1 **Convective thrust** occurs when the thrust region plume exhibits buoyancy. Buoyancy is due to low particle concentration within the hot plume as well as heating and mixing with surrounding atmospheric air. Convective thrust produces a vertical plume that equilibrates forming a mushroom-shaped region from which pyroclastic particles ultimately descend as air fall or pyroclastic fall deposits.
2 **Gravitational collapse** of the plume occurs due to negative buoyancy resulting from high particle concentrations within the dispersion and/or insufficient heating and mixing with the surrounding atmospheric air. Gravitational collapse of a plume due to higher density than the surrounding air is one means by which pyroclastic flows are initiated (Sparks and Wilson, 1976).

Let us consider air fall deposits and pyroclastic flow deposits in greater detail.

Air fall deposits

Pyroclastic fall (air fall) deposits are produced by airborne pyroclasts propelled upward in an eruption column (Figure 9.24b). Pyroclasts in the eruptive column (tephra) may stay aloft for considerable periods of time and be dispersed over great distances. The relatively small 1980 eruption of Mt St Helens resulted in thin ash fall tuff deposits 2000 km downwind from the volcano. The ejected ash eventually spreads outward and downwind and begins to accumulate as sorted, stratified air fall deposits on Earth's surface.

The densest air fall particles are deposited first, followed by increasingly finer grained particles (Wright et al., 1980). Because the largest pyroclasts tend to settle first and closest to the vent (bombs and breccia) and the smaller particles tend to settle more slowly and farthest from the vent, pyroclast size in air fall deposits generally decreases upward within the deposit and with distance from the source. Some proximal solidified air fall deposits grade upward from coarse breccias near the base through lapilli tuffs into fine tuffs near the top. Distal ash-sized particles are deposited and later compacted and cemented together as **ash fall tuffs** (Figure 9.25). Ash fall tuffs are generally lightweight,

Figure 9.25 Loosely welded tuff overlain by pahoehoe, Kilauea Crater, Hawaii, pre 2006. Although Hawaiian volcanoes generally erupt liquid basaltic lava, explosive ash eruptions do occur, as in 1924 and in 2007–2008. (Photo by Kevin Hefferan.)

porous, poorly cemented and rather soft rocks (Sparks and Wilson, 1976). The internal "blanket" layering, sorting and fining-upward graded bedding of air fall deposits allows them to be distinguished from pyroclastic flow deposits. Fining-upward means that the coarsest grains are deposited at the base and finer sized grains are deposited towards the top of a layer.

Pyroclastic flow deposits

Many pyroclastic rocks form by **pyroclastic flows**, consisting of turbulent mixtures of hot rock fragments and gases (Figure 9.26). Unlike air fall deposits, **pyroclastic flow deposits** generally lack stratification and are poorly sorted because chaotic mixtures of clast sizes are deposited rapidly. Some pyroclastic flows display inverse grading. Pyroclastic flows are commonly lobate in plan view (Figure 9.26) and lens like in cross-section.

Figure 9.26 This August 7, 1980 pyroclastic flow stretches from Mt St Helens' crater to the valley floor below. Pyroclastic flow velocities can exceed 100 km/h and reach temperatures of over 400°C. (Photo courtesy of Peter W. Lipman, with permission of the US Geological Survey.)

Three major types of pyroclastic flows are recognized:

1   **Pyroclastic surges** are very low density, extremely hot, gaseous flows containing ash- to lapilli-size particles that travel in excess of 100 km/h. Low densities allow pyroclastic surges to defy gravity and climb upwards from valleys enveloping higher slopes and ridges. In volcanically active areas, warning signs advise people to climb to higher elevations. In the case of pyroclastic surges, such upward mobility may be in vain. In 1995, pyroclastic surges inundated the Caribbean island of Montserrat turning the verdant, green island into an eerie monochromatic, gray landscape resembling a lunar surface (Figure 9.27). Surges are generated by plume collapse or directed blasts.
2   **Pumice flows** are low to moderate density, hot vesiculated flows. Siliceous pumice flows produce light-colored, vesicular **ignimbrites**. Andesitic to basaltic flows produce vesicular **scoria flows**. Massive explosive eruptions, such as the 1912 Katmai eruption in Alaska, produce immense ignimbrites. Thick ignimbrites contain features such as columnar joints which indicate relatively slow cooling. Ignimbrites are generated by the collapse of vertical ash plumes.
3   **Nuées ardentes** (French for "fiery clouds") are fluidized mixtures of hot, incandescent

Figure 9.27 Pyroclastic flows that buried Plymouth on the Island of Montserrat. (Photo courtesy of R. P. Hoblitt, taken January 8, 1997, with permission of the US Geological Survey.)

rock fragments and gases that flow along the surface as a glowing cloud of billowing pyroclastic debris. Searing gases generated by bubble fragmentation and the incorporation of atmospheric gases provide buoyant support for the mixture to behave as a fluid. Pyroclastic flows have temperatures that can approach 1000°C, can move at top speeds exceeding 150 km/h, over distances of as much as 100 km, and can form pyroclastic deposits up to 1 km thick. Pyroclastic flows can kill tens of thousands of people within minutes. High density, vesicle-poor **block and ash** pyroclastic flow deposits are generated by nuée ardentes (Wright et al., 1980; Francis, 1993). Nuées ardentes are produced by the collapse of lava domes.

Lahars

Lahars are volcanic mudflows up to tens of meters thick with the consistency of wet cement. They are dense mud slurries of ash- to bomb-sized blocks that are easily remobilized in the presence of water. Lahars are particularly dangerous in steep terrain, such as the slopes of composite volcanoes, and are capable of traveling tens of kilometers with velocities approaching 100 km/h; they can bury thousands of people in a matter of minutes (Box 9.3). Lahars can form at three different times:

- Synchronous with volcanism (syn-eruption lahars) and active pyroclastic flow.
- Soon after volcanism has ceased (early post-eruption lahars) where pyroclastic deposits are commonly remobilized. Factors that induce remobilization include ground shaking, slope instability, heavy precipitation and/or melting of ice and snow.
- On the slopes of dormant or inactive volcanoes (late post-eruption lahars) where old pyroclastic deposits are remobilized.

Syn-eruption lahars are generated by combining pyroclastic flows and/or air fall deposits with glacial meltwater or precipitation. The 1991 Mt Pinatubo (Philippines) eruption involved both syn-eruption and post-eruption lahars. Mt Pinatubo's eruption occurred during a monsoon so the precipitation mixed with the ash fall/flow deposits and other accumulated pyroclastic debris producing deadly syn-eruption lahars. Sadly, the lahars did not cease with the eruption activity. The immense pyroclastic deposits infilled stream channels and were easily remobilized with subsequent precipitation events as cool, post-eruption lahars. Fortunately, because two US military bases were in close proximity to the eruption, United States Geological Survey (USGS) volcanologists actively monitored Mt Pinatubo with the Philippine Institute of Volcanology and Seismology. Their combined efforts kept the human death toll to less than 500 people.

9.2.6 Rhyolite caldera complexes

Rhyolite caldera complexes are large volcanic features that lack the typical, highly elevated landform associated with most volcanoes. Rhyolite caldera complexes erupt with such unimaginable violence that the entire volcanic structure, rather than just the top of the volcano, collapses producing large caldera depressions. Because of the size of these explosive eruptions, rhyolite caldera complexes have been labeled "supervolcanoes". As low depression calderas are the signature aspect, rhyolite caldera complexes are also known as "inverse volcanoes". Following a major eruptive pulse, rhyolite caldera complexes may continue to emit smaller scale eruptions over long time intervals and experience regional uplift, referred to as "resurgence". Rhyolite caldera complexes occur above continental rifts (Long Valley Caldera, California and Valles Caldera, New Mexico, USA), hotspots (Yellowstone, USA) and subduction zones (Lake Taupo, New Zealand and Lake Toba, Indonesia).

Lake Taupo is located within a rhyolite caldera complex system created by a cataclysmic eruption 26,500 years ago in which 800 km^3 of tephra erupted. Subsequent major eruptions have occurred as recently as 181 AD. Lake Toba, on the island of Sumatra, Indonesia is the largest of Earth's rhyolite caldera complex supervolcanoes. Lake Toba is situated in a caldera which measures 100 by 30 km. The Toba complex has been volcanically active for the past 1.2 Ma, releasing ~3400 km^3 of magma. Approximately 74,000 years ago 2800 km^3 of rhyolitic debris erupted in an immense explosion covering an area greater than 20,000 km^2. Ash was deposited

Box 9.3 A tale of two cities: past and future?

Major metropolitan cities, towns and villages throughout the world thrive in the shadow of composite volcanoes. Composite volcanoes pose substantial hazards, especially from pyroclastic flows and lahars. Let us consider two regions built in similar geological environments. The Nevado del Ruiz eruption in Colombia was a particularly gruesome disaster – with a massive loss of life which was preventable. Nevado del Ruiz is a 5389 m high, snow-capped composite volcano in the Andes volcanic chain; 75 km downstream from the volcano, the village of Armero was built directly on top of an 1845 lahar that killed over 1000 people. However, fertile soils in the valleys surrounding Nevado del Ruiz enticed people to rebuild on these deposits.

Throughout 1985, minor earthquakes and steam explosions rocked Nevado del Ruiz, prompting a scientific study of the largest volcano in Colombia. The National Bureau of Geology and Mines produced a report and hazard assessment map on October 7, 1985 that indicated that lahars could severely impact Armero and surrounding regions downslope from the volcano. Fearing public unrest and costs, government officials discounted the report, ignoring its findings as well as the hazard assessment map (Figure B9.3a). On November 13, 1985 an eruption occurred that melted the glacial ice on top of Nevado del Ruiz. A combination of glacial meltwater and pyroclastic debris released by the eruption combined to produce multiple lahars that flowed down tributary valleys and joined

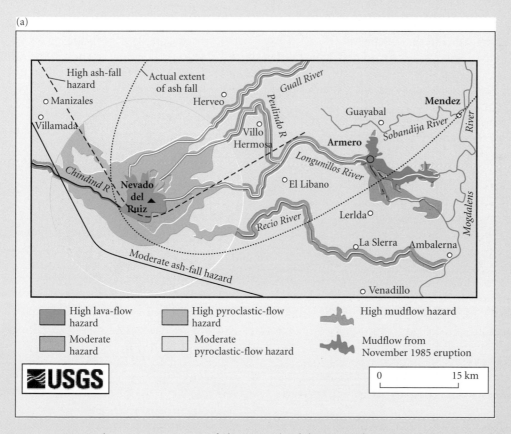

Figure B9.3 (a) Hazard assessment map of the Nevado del Ruiz Volcano in Colombia. (After Wright and Pierson, 1992; with permission of the US Geological Survey.) (b) Armero buried by a deadly lahar. (Photo courtesy of the US Geological Survey.) (c) Image of lahar flows and pyroclastic flow zones in the Seattle–Tacoma region. (Courtesy of the US Geological Survey.)

Continued

(b)

(c)

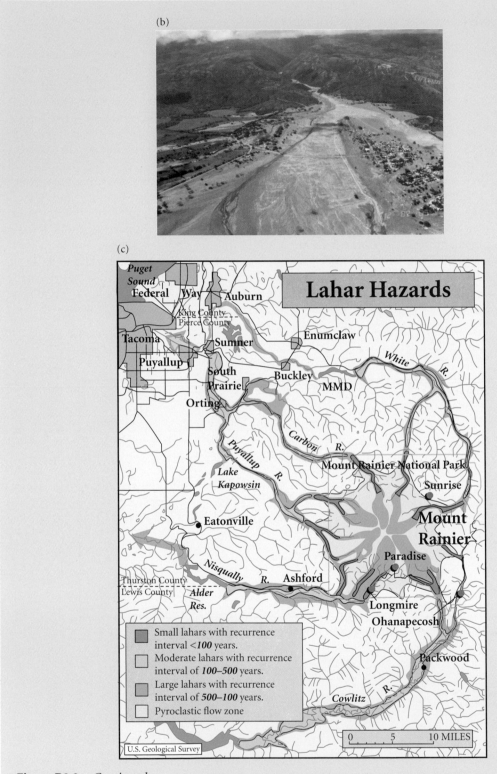

Figure B9.3   *Continued*

Box 9.3   *Continued*

to form massive lahars. Within minutes, one of the lahars raged down a valley burying approximately 23,000 people under approximately 10 m of debris (Figure B9.3b).

Several thousand kilometers to the north, the Cascade composite volcano chain occurs on the west coast of North America. The Cascade volcanoes are produced by the same tectonic processes as Nevado del Ruiz and contain similar characteristics: steep, majestic, glacier-covered volcanoes capable of producing immense pyroclastic flows and lahars. Approximately 5600 years ago, the Osceola volcanic mudflow (lahar) flowed northwest from Mt Rainier to the Pacific coast burying everything in its path. Land use has changed substantially since the Osceola roared down Rainier's slopes; the mudflow now lies under the rapidly growing Seattle–Tacoma metropolitan area. Note that the recurrence interval for lahars in the Seattle–Tacoma metropolitan region suggests that moderate-size lahars are expected to reach the area an average of once every 100–500 years (Figure B9.3c). One should consider the ramifications of urban development in the shadows of Mt Rainier and on a geologically young volcanic mudflow. A lahar warning system has been set up for Mt Rainier. If a lahar begins to flow down-valley on the mountain, pressure sensors will be activated that set off an alarm system at sites further down the valley, so that people will be warned to move to higher ground. If warning procedures are effective, massive loss of life should not occur.

Volcanic eruptions in marine settings may induce immense tsunamis that overwhelm populations scattered thousands of kilometers from the eruptive center. The Banda Aceh tsunami of December 26, 2004, which killed over 200,000 people, was caused by a displacement of the sea floor related to a major earthquake. However, volcanic eruptions, such as the 1883 Krakatoa eruption, produce displacements of the sea floor that are also capable of generating massive tsunamis. Krakatoa is located between the islands of Java and Sumatra in Indonesia. Prior to the 1883 eruption, the island of Krakatoa consisted of three contiguous stratovolcanoes – Perboewatan, Danan and Rakata – situated in an ancient caldera. Beginning in May 1883, ships in the eastern Indian Ocean noted ash plumes ejected to heights greater than 10 km. On August 26, a series of cataclysmic eruptions occurred that produced violent pyroclastic plumes up to 25 km high. These eruptions ultimately caused a caldera collapse, generating a catastrophic tsunami that resulted in the deaths of over 30,000 people. An additional 5000 people were killed by the hot pyroclastic flows and airborne debris. Krakatoa and Banda Aceh are both located along Indonesia's Java–Sumatra subduction zone: history continues to repeat itself.

up to 3100 km from the vent source (Rose and Chesner, 1987; Chesner and Rose, 1990).

Perhaps **Yellowstone** is the best known rhyolite caldera complex located in the interior of a lithospheric plate. Yellowstone is among the largest resurgent calderas on Earth at ~45 × 75 km in diameter. Currently, Yellowstone volcanic activity is limited to spectacular geysers and thermal springs, which we will discuss shortly. Yellowstone has experienced three major cataclysmic eruptions within the Quaternary Period (Figure 9.28); these explosive eruptions occurred 2 Ma, 1.2 Ma and 600,000 years ago. The Yellowstone eruptions have been among the largest explosive volcanic events in Earth's history.

Rhyolite caldera complexes explosively erupt very infrequently, even by geological standards. At the present time, we can enjoy the beautiful hydrothermal features within Yellowstone National Park (see next section) without any reasonable fear for our safety.

9.2.7   Phreatomagmatic and phreatic eruptions

Phreatomagmatic and phreatic eruptions are steam-driven explosions produced when magma or lava heats groundwater, converting it to steam. As a result of steam generation within rock fractures, violent explosions occur in which steam, hot water and/or pyroclastic debris are hurled into the air.

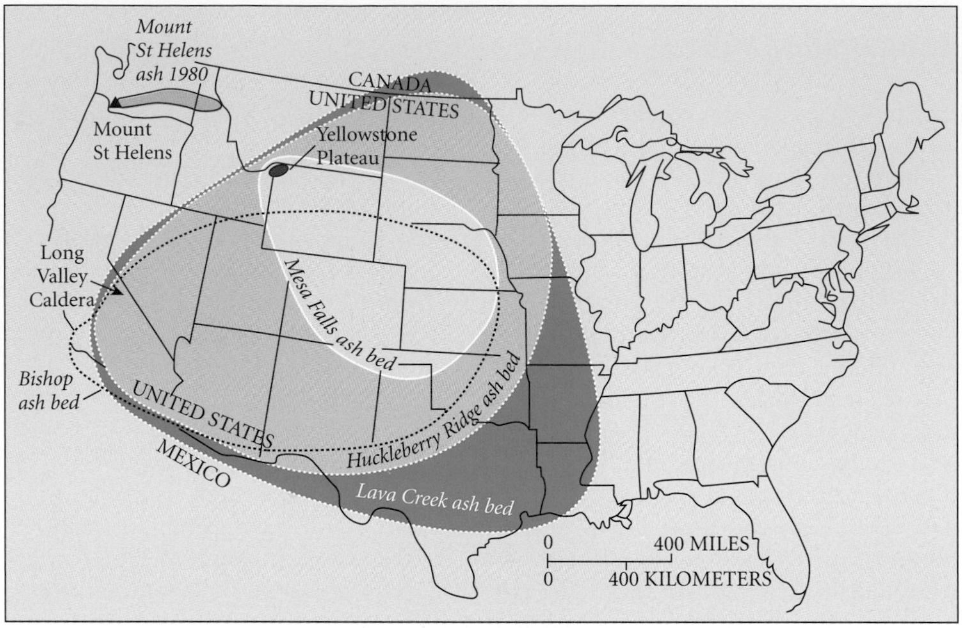

Figure 9.28 Yellowstone's Quaternary eruptions include three of the four largest North American volcanic eruptions, distributing ash throughout the region west of the Mississippi. The fourth is the Bishop tuff, which erupted from the Long Valley Caldera 700,000 years ago. (Courtesy of the US Geological Survey.)

Phreatomagmatic eruptions

Phreatomagmatic eruptions involve both magma and heated groundwater. Phreatomagmatic eruptions in Hawaii are attributed to magma levels dropping below the water table. As a result of the interaction of magma and water near the water table, liquid water vaporizes and serves as a propellant fuel for the explosive eruption of tephra (Figure 9.29).

Volcanic features produced by phreatomagmatic eruptions include tuff rings, tuff cones and maars. Tuff rings and tuff cones are both positive relief features. In contrast, maars are negative relief depressions.

1 **Tuff rings** are gently sloping, circular structures composed of stratified, glassy volcanic debris and scoria. Tuff rings develop due to the explosive eruption of basalt in a lake, beach or wetland environment. Essentially tuff rings are shallow water cinder cones in which pyroclastic material has been reworked by wave action, distributing volcanic debris around the vent. Hawaii's Diamond Head, the best known tuff ring (Figure 9.30), has a relief of ~100 m to over 230 m high and contains a central crater 1 km in diameter (Francis, 1993).

2 **Tuff cones** are (Figure 9.31) circular volcanic cones formed by the eruption of basalt in water. Tuff cones are associated with tuff rings; however, they tend to be smaller and steeper features. The smaller scale of tuff cones may be due to less explosive eruptions of shorter duration.

3 **Maars** are low relief volcanic craters that form by shallow explosive phreatomagmatic eruptions (Figure 9.32). Commonly, the volcanic crater fills with water to create either a freshwater or saline lake. As an example, Lake Becharof, Alaska's second largest freshwater lake, occupies Ukinrek Maars in Alaska. Ukinrek Maars, which is approximately 100 m deep and 300 m in diameter, formed due to explosive eruptions triggered by the injection of basaltic magma into water-saturated glacial till.

Zuni Salt Lake in New Mexico is a saline lake that occupies a 2 km wide crater over 120 m

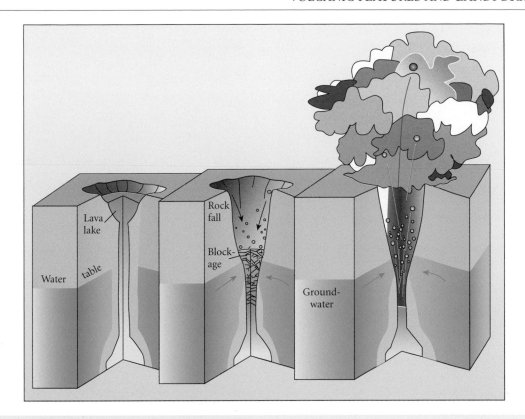

Figure 9.29 Phreatomagmatic volcanic activity is particularly common in places such as Hawaii where both magma and water are in large supply. (Courtesy of the US Geological Survey.)

Figure 9.30 Diamond Head tuff ring. (Photo courtesy of Steve Dutch.)

Figure 9.31 Aerial view of a 260 m high tuff cone inside Okmok Caldera on Umnak Island in the eastern Aleutians, Alaska. This cone formed ~2000 years ago during the hydromagmatic collapse of Okmok Volcano. (Photo courtesy of Christina Neal, with permission of the Alaska Volcano Observatory and US Geological Survey.)

Figure 9.32   View north to Ukinrek Maars, Alaska. This maar is about 300 m in diameter. The maar crater formed during a 10-day eruption in March and April of 1977. In the distance (center) are the Gas Rocks, an older volcanic center. (Photo courtesy of C. Nye, taken May 9, 1994, with permission of the Alaska Division of Geological and Geophysical Surveys.)

Figure 9.33   Hot spring pools precipitating opaline (silica) sinter deposits within Yellowstone National Park, Wyoming, USA. (Photo by Kevin Hefferan.)

Figure 9.34   Beehive geyser at Yellowstone National Park. (Photo by Kevin Hefferan.)

deep. Zuni Salt Lake maar contains a low rim of intermixed basalt fragments and shale, sandstone and limestone country rock. Tuff cones, tuff rings and maars are landforms produced by phreatomagmatic eruptions. Next we will address landforms produced by phreatic eruptions.

Phreatic eruptions

The eruption of heated water and steam without magma, characterizes **phreatic eruptions**. Phreatic eruptions produce hot springs, geysers and fumaroles (**solfataras**). These features form when groundwater percolates downward toward a high temperature magma reservoir. As the water is heated, it expands, becomes less dense and rises along fractures allowing for the release of steam and hot water. Phreatic eruptions produce siliceous sinter deposits, tufa limestone, sulfides and other minerals associated with hydrothermal deposits. Three types of phreatic eruptions occur:

1   **Hot springs** contain groundwater heated by proximity to magma (Figure 9.33).
2   **Geysers** are eruptive hot springs that eject fountains of heated water periodically (Figure 9.34). Geysers erupt as groundwater is heated to temperatures above 100°C. Because the boiling point of water is higher under pressure, superheated water rises, converts to steam and is ejected explosively as a geyser.
3   **Fumaroles** (Figure 9.35) emit mixtures of steam and other gases such as hydrogen sulfide. Sublimation of the sulfide gases to solid form produces economically viable sulfur deposits used in making agricultural fertilizers and other uses in the chemical industry.

(a)

(b)

Figure 9.35   (a) This fumarole at Kilauea Volcano is releasing sulfur gases which, upon cooling, sublimate (gas to solid) around the vent as sulfaterra. Sulfaterra is sulfur-rich soil and rock. (b) Sulfur crystals sublimate from gases released at Kilauea. Pen for scale. (Photos by Kevin Hefferan.) (For color version, see Plate 9.35, between pp. 248 and 249.)

Hot springs, geysers and fumaroles develop in areas of unusually high geothermal gradients and are important energy sources where hot rocks and/or magmas lie close to the surface.

Volcanic activity beneath glaciers is also significant in alpine settings, glaciated continents and islands. Sub-glacial volcanic eruptions can result in the release of glacial meltwater floods called **jökulhlaups** (Icelandic term pronounced yo-kul-hloips). Jökulhlaups are the sudden flood burst of glacial lake water or water contained within a glacier.

For example, Iceland's Vatnajokull ice cap overlies seven active volcanoes including Grímsvötn. On September 30, 1996, scientists detected seismic activity associated with a sub-glacial eruption of the Grímsvötn Caldera. On November 5, vertical fractures on the glacier released a large jökulhlaup.

So far in this chapter we have classified volcanic features largely on the basis of landform size, shape and physical characteristics of the volcanic rock material. These factors are related to: magma budget, silica content, magma viscosity and the presence/absence of water. Volcanoes are also classified by their explosiveness.

9.3   CLASSIFYING VOLCANIC ERUPTIONS

A useful way to classify volcanic activity is on the basis of eruptive style or process. This includes explosivity, volume of pyroclastic debris and height of the eruption column during eruptive episodes. Volcanoes vary from being "quiet" to cataclysmic in eruptive style, as summarized in Table 9.1.

The vast majority of quiescent eruptions – **Hawaiian, Icelandic, Strombolian** and **Surtseyan** – occur at ocean spreading ridges and hotspots where low viscosity basaltic lava is erupted onto Earth's surface. Gas bubbles migrate upward at a faster rate than the low viscosity, basaltic magma. As a result of quiescent degassing, basaltic eruptions are typically non-explosive. Quiescent eruptions produce lava flows, lava lakes and lava fountains punctuated by only occasional explosive bursts.

Explosive eruptions – **Vulcanian, Vesuvian, Plinian** and **Ultraplinian** – occur at convergent margin volcanoes and rhyolite caldera complexes. These explosive eruptions are characterized by massive ash clouds >10 km in height, glowing nuées ardentes and devastating pyroclastic flows and lahars. Explosive eruptions commonly consist of a series of short, violent bursts, generated by the shattering of solid rock plugs of andesite to rhyolite composition. The early, explosive bursts are considered a throat-clearing phase in which the magma conduit is evacuated of viscous plugs and constraining debris. Plug shattering is followed by an extensive period of eruptive phases ranging from moderate to cataclysmic in scale.

Table 9.1   Common classification terms used to describe volcano explosivity. (After Walker, 1973.)

Eruption name	Description	Examples
Quiescent eruptions		
Hawaiian	Eruptions begin as fissures, evolving to central vent flows and the generation of large shield volcanoes, fiery basaltic lava fountain eruptions, quiet lava flows and cinder cones	Kilauea, Hawaii
Icelandic	Persistent fissure eruption of low viscosity basaltic lava flows. Prolonged quiet eruptions may generate lava plateaus and flood basalts	Laki, Iceland, 1783 $12 \, km^3$ of lava
Surtseyan (phreatomagmatic)	Explosive, steam-blast eruptions with lava flows and pyroclastic debris. Surtseyan eruptions are named after the volcanic island of Surtsey, which rose above sea level on November 14, 1963. Within 2 months Surtsey, a newly created island south of Iceland was 1.3 km long and 174 m high (Decker and Decker, 2006). Surtsey continued to erupt until 1967	Surtsey, Iceland, 1963
Strombolian	Periodic bursts ("burps") of moderately explosive eruptions (<5 km high) with great concentrations of pyroclastic fragments and incandescent basaltic lava flows	Stromboli, Italy
Explosive eruptions		
Vulcanian	Explosive eruptions of basaltic to rhyolitic viscous lava and large volumes of volcanic ash plumes (<25 km high) and pyroclastic debris	Vulcano, Italy
Vesuvian	Violent eruptions of volcanic debris ejected, scattering ash over thousands of square kilometers	Mt Vesuvius, Italy
Plinian	Tephra eruptions emit immense ash clouds >11 km in height into the stratosphere	Krakatoa, 1883
Ultraplinian	Violent tephra eruptions of volumes $>1 \, km^3$ and ash cloud heights 25–55 km	Mt Taupo, New Zealand, 181 AD

The most explosive eruptions are termed Plinian or Ultraplinian eruptions. Plinian eruptions involve rapidly ascending andesitic to rhyolitic magma and exsolved gases moving upward at velocities of several hundred meters per second. Instead of quiescent degassing prior to magma eruption, the entrained volatile bubbles rise within the magma stream resulting in sustained jets of violently propelled lava. As a result, the gases are violently discharged, creating mushroom-shaped plumes of ash. Recent Plinian eruptions include: the 1815 Tambora eruption (Indonesia), the 1883 Krakatoa eruption (Indonesia), the 1912 Katmai eruption (Alaska), the 1980 Mt St Helens eruption (USA) and the 1991 eruption of Mt Pinatubo (Philippines). Ultraplinian eruptions include the 5677 BC Mazama eruption that created Crater Lake Caldera (Oregon), the 74,000-year-old Toba eruption in Sumatra, the Taupo (New Zealand) eruption and the 2 Ma Yellowstone Lava Creek eruption. Ultraplinian eruptions are climate changing, cataclysmic events.

9.4   VOLCANIC RESOURCES AND HAZARDS: GLOBAL IMPACTS

Volcanic activity has dramatically enhanced our planet on a global scale. Volcanic activity has produced most of Earth's water from volcanic degassing of Earth's interior, generated valuable metal deposits (Chapter 19), created vast tracts of arable land noted for their rich productive soils, and may be the source for Earth's early life forms at hydrothermal vents.

On the other hand, volcanic activity presents significant hazards to our world in the form of pyroclastic flows, lahars and tsunamis that can kill hundreds of thousands of people within a single day. On slightly longer time scales, Plinian and ultraplinian eruptions, as well as flood basalt eruptions, can dramatically alter Earth's climate. The April 5, 1815 eruption of Mt Tambora (Indonesia) killed 92,000 people and caused disastrous crop failures and summer snowfalls as far away as North America. Volcanic eruptions affect climate in three ways. Firstly, volcanic dust ejected into the atmosphere causes short-term cooling, on the order of days to months. The cooling is due to the blockage of solar radiation, preventing the Sun's warm infrared rays from penetrating to Earth's surface. The amount and duration of cooling depends upon the volume of dust suspended in the atmosphere. Secondly, volcanic eruptions release volatiles such as sulfur oxide and sulfur dioxide into the stratosphere. Sulfur oxides react with water to form sulfuric acid haze in the stratosphere. The reflective properties of the sulfuric acid haze result in decreased penetration of infrared rays on the order of 1–5 years, depending upon the amount and duration of haze in the atmosphere. Presumably, cataclysmic eruptions such as occurred at Toba and Yellowstone would produce a global cooling period of significantly longer than 5 years. Exactly how many years? That is a good question without a suitable answer. The third means by which climate change may occur is the release of greenhouse gases, such as carbon dioxide, during a volcanic eruption. These greenhouse gas emissions may absorb infrared heat radiation and retain that heat in the atmosphere, causing near surface warming.

Obviously these three climate alteration mechanisms are competing and the third warming mechanism is counterbalanced by the two cooling mechanisms. As carbon dioxide is quickly absorbed by plants and dissolved in the oceans, the warming effect is typically short term such that cooling mechanisms usually dominate. However, massive eruptions over tens of thousands or millions of years may dramatically increase greenhouse gas emissions so that warming dominates. Massive flood basalt eruptions could be responsible for extreme climate change. It is not a coincidence that nearly all major extinctions on Earth coincide with massive flood basalt eruptions.

Volcanism provides a spectacular example of our young Earth's vitality. Volcanic activity is concentrated at convergent or divergent plate boundaries or over hotspots. Magmatism at convergent margins is largely due to the partial melting of the lithospheric wedge overlying the subduction zone. Convergent margin volcanoes may be quite explosive due to a high magma viscosity and high volatile content. Convergent plate boundaries are characterized by the development of composite volcanoes. Andesite to rhyolite compositions are the most common crystalline volcanic rocks at composite volcanoes, although basalt may also be abundant. Pyroclastic rocks such as tuff, breccias and scoria are also commonly deposited.

Divergent plate boundaries are dominated by basaltic lava flows. Divergent plate margins occur as a continuous belt of mountain ridges along the ocean floor. Ocean ridges are the site of plate extension, wherein rift valleys develop. Magmatism at divergent margins is due to the partial melting of the upper mantle due to mantle uplift and decompression. Low viscosity basaltic lava tends to erupt in a quiescent fashion such that violent explosions are rare.

Hotspots may occur at plate boundaries, such as in Iceland, or within lithospheric plates, such as in Hawaii. Hotspots are locations on Earth where unusually large volumes of magma are supplied to Earth's surface over a long period of time. Hotspots may erupt a wide variety of lava compositions; however, basaltic lavas are the most widespread. Hotspots located within continental crust may produce basalt as well as rocks such as rhyolite. Hotspots may produce rhyolite caldera complexes, pyroclastic cones, flood basalts and other volcanic features.

Volcanic eruptions may hold the key to the development of life through chemosynthetic activity at mid-ocean ridges. Volcanism improves the quality of life through the generation of rich arable soils and valuable metal deposits. Yet there is a dark side: despite the media attention given to meteorite impacts, volcanic eruptions may be the culprit responsible for the sudden demise of life forms on Earth by serving as the catalyst for major extinctions. No humans have ever witnessed the fierce destructive nature of flood basalt or rhyolite caldera complex eruptions. The consequences to life from such massive eruptions are beyond human experiences on Earth.

Chapter 10

Igneous rock associations

10.1   IGNEOUS ROCK ASSOCIATIONS

The purpose of this chapter is to relate igneous rock associations to a petrotectonic framework, incorporating information presented in Chapters 7–9. **Petrotectonic associations** are suites of rocks that form in response to similar geological conditions. These associations most commonly develop at divergent plate boundaries, convergent plate boundaries and hotspots (Figure 10.1). Hotspots can occur at lithosphere plate boundaries (e.g., Iceland) or in intraplate settings (e.g., Hawaii). Fisher and Schmincke (1984) estimate the percent of magma generated at modern divergent, convergent and hotspot regions as 62, 26 and 12%, respectively.

While plate tectonic activity plays a critical role in the development of petrotectonic associations, it is not the sole determining factor. For example, the earliest onset of modern plate tectonics continues to be debated, with some researchers (Kusky et al., 2001; Parman et al., 2001) favoring Archean (>2.5 Ga) onset and others (Hamilton, 1998, 2003; Stern, 2005, 2008; Ernst, 2007) proposing Proterozoic initiation of deep subduction ~1 billion years ago (1 Ga). If the latter is true, then over 75% of Earth's magmatic history occurred under conditions that pre-date the onset of modern plate tectonic activity. In addition to questions regarding magmatism at Precambrian plate tectonic boundaries, Phanerozoic intraplate magmatism may or may not be influenced by lithospheric plate boundaries (Hawkesworth et al., 1993; Dalziel et al., 2000). So while the plate tectonic paradigm is very useful, it does not address all igneous rock assemblages produced throughout Earth's tumultuous history. In the following sections we will address major igneous petrotectonic associations, beginning with divergent plate boundaries.

10.2   DIVERGENT PLATE BOUNDARIES

Decompression of the asthenosphere in response to lithospheric extension results in partial melting of mantle peridotite at divergent margins. Basic (basaltic) melts rise and

Earth Materials, 1st edition. By K. Hefferan and J. O'Brien. Published 2010 by Blackwell Publishing Ltd.

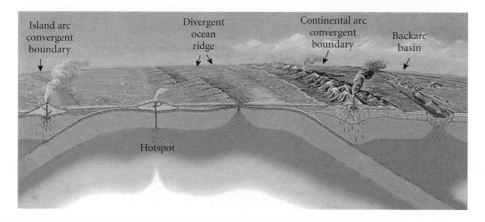

Figure 10.1 Major tectonic environments where igneous rocks occur. (Courtesy of the US Geological Survey and US National Park Service.)

solidify to produce oceanic crust, while refractory residues cool below a critical temperature to form the thickening mantle layer of ocean lithosphere. Ocean lithosphere is created primarily at spreading ridges such as the Mid-Atlantic Ridge, East Pacific Rise and Indian–Antarctic ridge systems. A small percentage of ocean lithosphere is generated in backarc basin spreading ridges (e.g., Marianas Trough) and ocean hotspots (e.g., Hawaii). In all cases, anatexis of ultramafic mantle is the primary magmatic source of ocean lithosphere.

Ocean lithosphere contains four distinct layers as indicated in Figure 10.2a. Layer 1 contains well-stratified marine pelagic sediments and sedimentary rocks that accumulate on the ocean floor. Layer 2 can be subdivided into two basaltic rock layers. An upper layer contains pillow basalts that develop when basic lavas flow onto the ocean floor, rapidly cool in the aqueous environment and solidify in spheroidal masses (Chapter 9). Beneath the pillow basalt pile, basic magma injects into extensional fractures producing steeply inclined diabase dikes as the magma cools and contracts. Repeated horizontal extension and magma intrusions generate thousands of dikes arranged parallel to one another in a sheeted dike complex (Chapter 8). Beneath the sheeted dike layer, basic magma cools slowly, allowing phaneritic crystals to nucleate and grow as layer 3. Layer 3 contains massive (isotropic) gabbro in the upper section, layered (cumulate) gabbro in a middle section, and increasing amounts of layered (cumulate)

peridotite towards the bottom of the section, marking the base of ocean crust. The Mohorovičić discontinuity (Moho) occurs at the contact between cumulate rocks in layer 3 and non-cumulate, metamorphosed rocks in layer 4, marking the rock boundary between the ocean crust and mantle. Layer 4 is composed of depleted mantle peridotite refractory residue (e.g., harzburgite, dunite). Layer 4 mantle peridotite is marked by high temperature, solid state strain fabric (metamorphosed) and represents the lowest layer of the oceanic lithosphere.

Layers 3 and 4 are generally unexposed on ocean floors because they are overlain by layers 1 and 2. In rare locations, these deep layers are exposed on the ocean floor in ultra slow (<1 cm/yr), or magma-starved, spreading ridges and transform zones where brittle faulting and uplift processes bring them to the surface. Slices of ocean lithosphere are also preserved in alpine orogenic belts as ophiolite sequences. Let us now consider petrotectonic assemblages that form at ocean ridge spreading centers.

10.2.1 Mid-ocean ridge basalts

At ocean spreading centers (Figure 10.2b–c), partial melting of lherzolite (peridotite) generates voluminous, geochemically distinct, **mid-ocean ridge basalts** (MORB) and gabbros containing minerals such as plagioclase, augite, hypersthene, pigeonite, diopside and olivine. MORB are the most abundant volcanic rocks on Earth. Typical major and

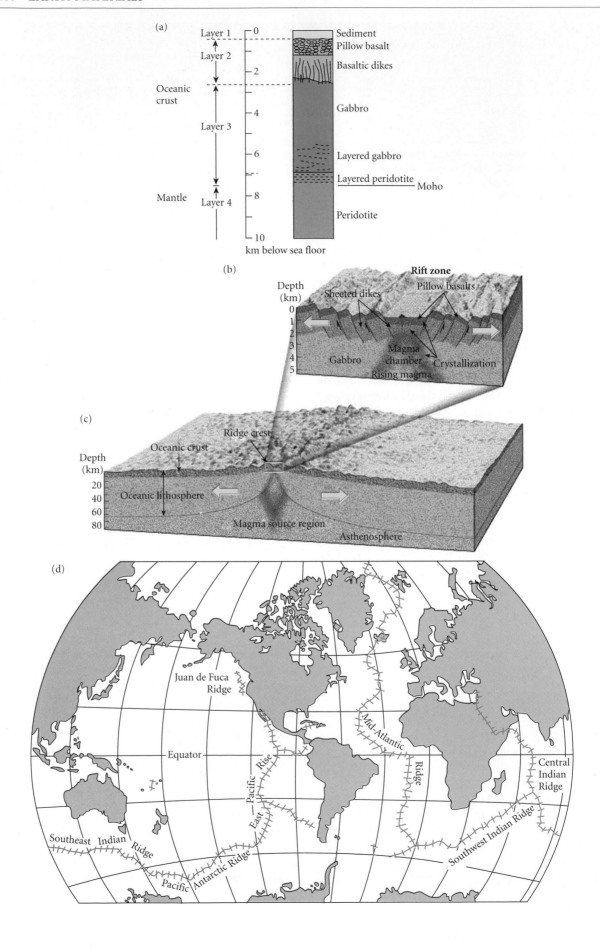

(a)

Layer 1

Oceanic crust

Layer 2

Layer 3

Mantle

Layer 4

0

2

4

6

8

10

km below sea floor

Sediment
Pillow basalt
Basaltic dikes

Gabbro

Layered gabbro
Layered peridotite — Moho

Peridotite

(b)

Rift zone

Depth (km)

Sheeted dikes   Pillow basalts

0
1
2
3
4
5

Gabbro

Magma chamber

Crystallization

Rising magma

(c)

Ridge crest

Oceanic crust

Depth (km)

Oceanic lithosphere

20
40
60
80

Magma source region

Asthenosphere

(d)

Juan de Fuca Ridge

Mid-Atlantic

Equator

East Pacific Rise

Ridge

Central Indian Ridge

Southeast Indian Ridge

Pacific Antarctic Ridge

Southwest Indian Ridge

Table 10.1 Trace element abundances for N-MORB and E-MORB in parts per million (ppm). (After Best, 2003; data from Sun and McDonough, 1989.)

	LIL					HFS		LREE					HREE				
	Cs	Rb	Ba	Th	U	Nb	Ta	La	Ce	Pr	Nd	Sm	Zr	Eu	Gd	Yb	Lu
N-MORB	0.007	0.56	6.3	0.12	0.47	2.33	0.132	2.5	7.5	1.32	7.3	2.63	74	1.02	3.68	3.05	0.455
E-MORB	0.063	5.04	57	0.6	0.18	8.3	0.47	6.3	15	2.05	9	2.6	73	0.91	2.97	2.37	0.354

E-MORB, enriched mid-ocean ridge basalt; HFS, high field strength; HREE, heavy rare Earth elements; LIL, large ion lithopile; LREE, light rare Earth elements; N-MORB, normal mid-ocean ridge basalt.

minor element concentrations are indicated in Table 10.1. MORB are low SiO_2 (45–52%), low potassium (<1% K_2O) tholeiites with high MgO (~7–10%), Al_2O_3 (15–16%) and compatible element concentrations (Ni and Cr ~100–500 ppm). MORB develop from partial melting of a depleted mantle source, as indicated by low $^{87}Sr/^{86}Sr$ ratios (0.702–0.704), low volatile and incompatible element concentrations, and high compatible element concentrations (Cann, 1971). "Depleted source" refers to mantle lherzolite that has undergone previous melt cycles that largely removed mobile incompatible elements (Chapter 7).

Mid-ocean ridge basalts can be subdivided into **normal MORB** (N-MORB) and **enriched MORB** (E-MORB) based upon minor and trace element abundances (Figure 10.3; Table 10.1). N-MORB are strongly depleted in highly incompatible elements such as large ion lithophile (LIL) elements (such as Cs, Rb and Ba), high field strength (HFS) elements (such as Nb and Ta) and light rare Earth elements (LREE, such as La, Ce, Pr, Nd and Sm). These geochemical characteristics imply that

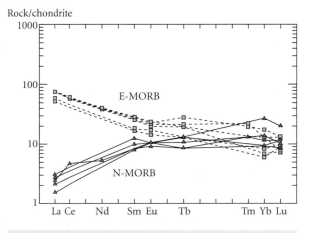

Figure 10.3 Chondrite-normalized rare Earth element patterns for enriched and normal mid-ocean ridge basalt (E-MORB, squares; N-MORB, triangles) samples collected from the Mid-Atlantic Ridge. (From Schilling et al., 1983; with permission of the *American Journal of Science*.)

Figure 10.2 (a) Idealized stratigraphy of ocean lithosphere and ophiolites. Note the petrological Moho between layers 3 and 4, separating the base of the crust from the upper mantle. (Courtesy of the Ocean Drilling Project.) (b) Block diagram of ocean ridge divergent margins. (c) Ocean ridges are primary sites for the generation of ocean lithosphere. (d) The global distribution of divergent margins.

N-MORB magma represent 20–30% partial melting of a well-mixed, depleted mantle source (Frey and Haskins, 1964; Gast, 1968).

Although the major element and heavy rare Earth elements (HREE, ranging from Eu to Lu) concentrations are comparable, E-MORB have higher incompatible element (LREE, HFS, LIL) concentrations relative to N-MORB. Specifically, E-MORB are defined by having chondrite normalized La/Sm ratios of >1. Lanthanum may occur in concentrations of 1–5 ppm in N-MORB but up to 100 ppm in E-MORB.

How can we account for chemical variations between N-MORB and E-MORB?

Several different hypotheses have been proposed. First, E-MORB may represent smaller degrees (~10–15%) of partial melting of residual mantle rock so that the incompatible elements are more highly concentrated in E-MORB magmas. Second, E-MORB could be tapping a deep mantle source that has not been previously melted. Third, E-MORB could represent magma enriched from magma mixing, assimilation or partial melts derived from subducted ocean lithosphere. For example, Eiler et al. (2000), based on a study of 28 basalt samples from the Atlantic, Pacific and Indian ridges, propose that E-MORB include a component of partially melted oceanic lithosphere that has been recycled into the upper mantle from ancient subduction zones.

While MORB is the dominant volcanic rock type at divergent margins, other rock types occur in varying proportions. Ocean ridges also produce high aluminum basalts, where the Al_2O_3 concentrations are >16%. Other rocks such as andesite, icelandite, ferrobasalt, trachyte, hawaiite, mugearite, trachybasalt, trachyandesite, dacite and rhyolite can occur as minor components at ocean ridges as well as in "leaky" transforms, continental rifts and ocean islands. The andesitic to rhyolitic volcanic rocks at ocean ridges have higher TiO_2 (>1.3%) concentrations compared to more common convergent margin varieties and are always subordinate to basalt (Gill, 1981).

Divergent margins (see Figure 10.2d) generate the bulk of ocean floor rocks, which represent ~70% of Earth's area. As a result, the ocean ridge basalt and underlying gabbro and peridotite are widespread in our relatively young ocean basins, all of which are less than 200 million years old. What happens to old ocean lithosphere? At least for the past 1 billion years, it has been subducted and recycled at convergent margins, as discussed below.

10.3 CONVERGENT PLATE BOUNDARIES

While divergent plate boundaries are dominated by MORB, chemically diverse igneous assemblages erupt in the convergent margins widely distributed in the Pacific Ocean, eastern Indian Ocean and the Caribbean and Scotia Seas (Figure 10.4).

Convergent margin magmatism may occur for thousands of kilometers parallel to the trench, and up to 500 km perpendicular to the trench in the direction of subduction (Gill, 1981). Plutonic rocks at convergent margins include diorite, granodiorite, quartz diorite, granite, gabbro, tonalite and rocks referred to as trondhjemite. This plutonic suite of rocks occurs in batholiths above subduction zones and provides magma to overlying volcanic arcs. The spectrum of possible volcanic rock types varies widely from youthful island arc environments – dominated by arc basalts and basaltic andesites – to mature continental arc systems – comprised largely of andesites, with lesser amounts of basalt, dacites, rhyodacites and rhyolites.

As opposed to relatively simple decompression melting of the mantle at divergent margins, convergent margin magmatism is affected by more variables, each of which can diversify magma composition. These variables include:

- Composition (continental versus oceanic) and thickness of the overlying converging plate: thinner ocean lithosphere in the overlying plate generally produces metaluminous, mafic to intermediate rocks. Thicker continental lithosphere overlying the subduction zones commonly yields peraluminous, potassic, intermediate to silicic rocks.
- Composition of rock material experiencing anatexis: Earth materials experiencing partial melting may include overlying ultrabasic mantle wedge, basic to silicic forearc basement, subducted basic–ultrabasic ocean lithosphere, and marine sedimentary material. The relative proportion of each of these components affects the composition of plutonic and volcanic rocks generated in the arc system.
- Flux melting whereby volatile-rich minerals such as micas, amphiboles, serpentine, talc, carbonates, clays and brucite release H_2O, CO_2 or other volatile vapors that lower the melting temperature of mantle peridotite and eclogite (high pressure metagabbro) overlying the subduction zone.
- Diversification processes such as fractionation, assimilation and magma mixing (Chapter 8) as well as metamorphic reac-

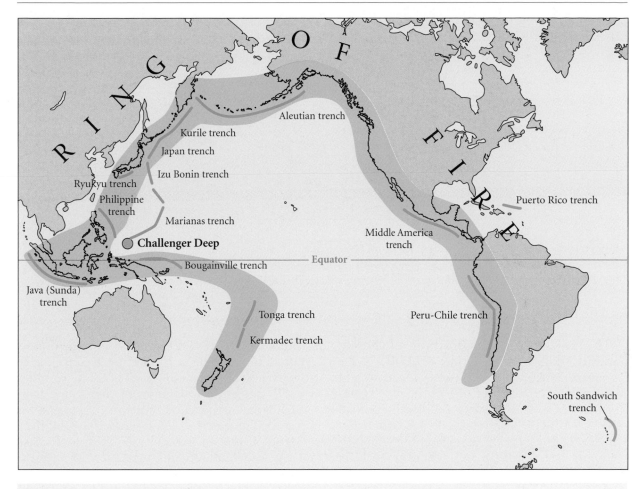

Figure 10.4   Earth's convergent margins. (Courtesy of the US Geological Survey.)

tions (Chapter 15) strongly alter magma composition generated in the overlying wedge of the arc system.

- Dip angle of the subduction zone wherein old, cold, dense lithosphere favors steep subduction and young, warm, buoyant lithosphere produces shallow subduction zones (Figure 10.5). Steeply inclined subduction zones allow for the melting of thick wedge-shaped mantle slabs in the overlying plate. Shallowly dipping subduction zones allow only thin wedge-shaped mantle slabs to intervene above the subducting plate, minimizing overlying mantle wedge input. The negative buoyancy of old, cold, dense ocean lithosphere is the key force driving deep lithosphere subduction in modern plate tectonics over the past 1 Ga. This negative buoyancy may not have been present in the hot, buoyant Archean ocean lithosphere such that deep subduction may not have been possible (Davies, 1992; Ernst, 2007; Stern, 2008). This may explain the origin of some unique Archean rock assemblages as well as the virtual absence of Archean blueschists discussed later in this text.

While magma composition is highly variable based on the factors described above, Phanerozoic convergent margins are dominated by the calc-alkaline suite of rocks whose chemistry is enriched in SiO_2, alkalis (Na_2O and K_2O), LIL, LREE and volatiles and is relatively depleted in FeO, MgO, HFS and HREE concentrations (Miyashiro, 1974; Hawkesworth et al., 1993; Pearce and Peate, 1995). The presence of hydrous minerals such as hornblende and biotite indicates that arc magmas contain >3% H_2O. Volatiles play an important role in subduction zone flux melting.

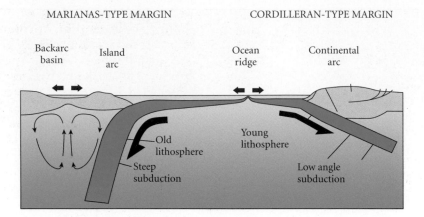

MARIANAS-TYPE MARGIN   CORDILLERAN-TYPE MARGIN

Figure 10.5 The steeply dipping Marianas-type island arc subduction model and the shallowly dipping Cordilleran continental arc subduction model. Note the thick, deep, mantle wedge overlying the Marianas-type margin and the thin mantle wedge in the Cordilleran model.

The calc-alkaline association of **basalt, andesite, dacite and rhyolite (BADR)** is the signature volcanic rock suite of convergent margins and constitutes one of the most voluminous rock assemblages on Earth, second only to MORB (Perfit et al., 1980; Grove and Kinzler, 1986). Harker diagram plots of major elements (see Figure 8.11) generally indicate a liquid line of descent from a common source, such that BADR rocks are derived from a common parent magma of basaltic composition. Andesite, named for South America's Andes Mountains, which overlie the Peru–Chile trench, is by far the most common calc-alkaline volcanic rock forming at convergent margins (see Figure 10.4). The more silicic (dacite, rhyolite) members of the BADR group represent more highly fractionated daughter products. We will discuss each of these below.

The major rock types in volcanic arc systems can be distinguished based upon major element concentrations such as SiO_2, K_2O and Al_2O_3 content. Basalts contain 45–52% SiO_2 and can be subdivided into a number of different varieties based upon major and minor element concentrations. Basalts common in convergent margins include aphanitic and aphanitic–porphyritic varieties of arc tholeiites (low K_2O) and calc-alkaline basalts (moderate K_2O). Plagioclase phenocrysts are common. The arc tholeiites differ from other tholeiitic basalts (MORB and ocean islands) in containing higher concentrations of Al_2O_3, typically in concentrations greater than 16 wt %. As a result, arc tholeiites are also referred to as high aluminum basalts. The calc-alkaline basalts differ from tholeiites in having higher alkali (notably K_2O) concentrations and not displaying iron enrichment typical of tholeiitic fractionation trends (Figure 10.6). As discussed in Chapters 7 and 9, magma viscosity and explosiveness are proportional to SiO_2 increases. As a result, the more siliceous volcanic rocks described below commonly produce pyroclastic tuff and breccia deposits in addition to aphanitic to aphanitic–porphyritic crystalline textures.

Andesites are volcanic rocks containing >52–63% SiO_2. Andesites can be subdivided based upon the range of SiO_2: basaltic andesites, common in youthful island arc systems, contain >52–57% SiO_2 while more silicic andesites, common in mature continental arc systems, contain >57–63% SiO_2 (see Figure 7.24). Andesites commonly occur as gray, porphyritic–aphanitic volcanic rocks with phenocrysts of plagioclase, hornblende, pyroxene or biotite. Plagioclase phenocrysts are most common and may display euhedral, zoned crystals. Hornblende phenocrysts are also common and may display reaction rims. Pyroxene (principally augite, hypersthene or pigeonite) and biotite phenocrysts less commonly occur. Quartz, potassium feldspar, olivine or feldspathoid phenocrysts are rare. Interestingly, the bulk composition of andesite (and its plutonic equivalent diorite) approximates that of terrestrial crust, suggesting that subduction zone processes have played a significant role in the development

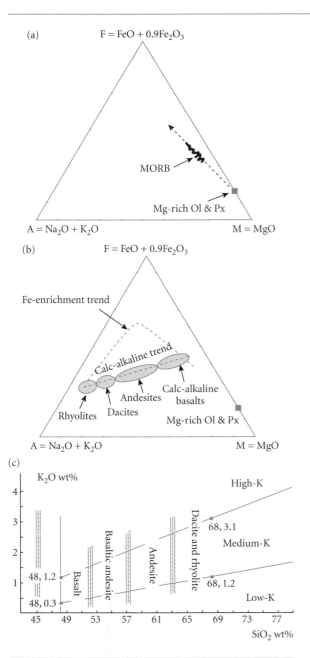

of continental crust (Hawkesworth and Kemp, 2006). The generation of voluminous andesite is favored by subduction angles greater than ~25°, anatexis of thick (greater than ~25 km), continental, hanging wall plates and partial melting of subducted slabs at depths of 70–200 km (Gill, 1981).

Dacites (Chapter 7) are quartz–phyric volcanic rocks, intermediate between andesite and rhyolite (Gill, 1981). While most dacites contain 63–68% SiO_2, the total alkali to silica (TAS) dacite classification extends to 77% SiO_2 (see Figure 7.24). Dacites are enriched in plagioclase and are the volcanic equivalent of granodiorites, in which alkali feldspars are subordinate to plagioclase. When present, phenocrysts are commonly subhedral to euhedral, zoned and generally consist of oligoclase to labradorite plagioclase or sanidine. Minor minerals commonly include biotite, hornblende, augite, hypersthene and enstatite.

Trachyandesites (also known as latites and **shoshonites**) are generally composed of ~66–69% SiO_2, although the lower TAS limit begins at 57% SiO_2 (see Figure 7.24). Trachyandesites commonly contain phenocrysts of andesine to oligoclase plagioclase feldspar amidst a groundmass of orthoclase and augite.

Rhyolites (>69% SiO_2) and **rhyodacites** (~68–73% SiO_2) are associated with explosive silicic eruptions producing fragmental, glassy and aphanitic to aphanitic–porphyritic textures. Rhyodacite is a rock term, not recognized by the IUGS system, for intermediate volcanic rocks that bridge the dacite/rhyolite boundary. These rocks can occur as glasses (obsidian or pumice), pyroclastic tuffs and breccias, or as aphanitic to aphanitic–porphyritic crystalline rocks. Common phenocrysts include alkali feldspar or quartz, with minor concentrations of hornblende and biotite.

In addition to variations in SiO_2, arc rocks display significant variation in K_2O concentrations, ranging from low (tholeiitic), medium (calc-alkaline) and high K_2O (calc-alkaline to shoshonite) rock suites (Gill, 1981). The progression from tholeiite to calc-alkaline to shoshonite (trachyandesite) reflects increasing K_2O and K_2O/Na_2O and decreasing iron enrichment (Jakes and White, 1972; Miyashiro, 1974). K_2O content in convergent margin volcanic suites broadly correlates with the thickness of the overlying slab in convergent margin systems (Figure 10.6c). Low

Figure 10.6 (a) Tholeiitic mid-ocean ridge basalts (MORB) display iron enrichment due to the early crystallization of magnesium-rich olivine and pyroxenes. (b) The calc-alkaline suite does not display significant iron enrichment but displays alkali enrichment with progressive crystallization. (c) Three volcanic rock suites are recognized on the basis of percent Si_2O and KO_2: low potassium assemblages consist of tholeiitic basalt; medium potassium assemblages contain calc-alkaline assemblages; high potassium suites consists of high potassium calc-alkaline rocks and trachyandesite rocks referred to as shoshonites (Gill, 1981; LeMaitre, 2002).

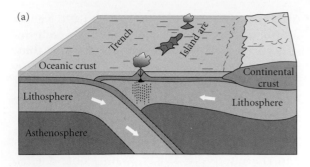

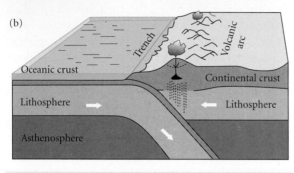

Figure 10.7   (a) Ocean–ocean convergence producing island arc volcanoes and backarc basins. (b) Ocean–continent convergence producing continental volcanic arcs. (Courtesy of the US Geological Survey.)

potassium tholeiites dominate with overlying slab thicknesses ranging from ~0 to 20 km; medium- to high potassium calc-alkaline andesites are associated with overlying slab thicknesses of ~20–40 km; high potassium shoshonites commonly develop where the overlying slab is >40 km thick (Gill, 1981).

Three major types of convergent margins occur: (1) ocean–ocean convergence generating youthful island arc volcanic complexes (Figure 10.7a), (2) ocean–continent convergence generating mature continental arc complexes (Figure 10.7b), and (3) continent–continent convergent margins marked by the cessation of subduction and consequent continental collision. Many of the same rock types can be found in all three environments; however, continental systems contain a greater proportion of silicic calc-alkaline rocks enriched in quartz and potassium feldspar; in contrast, island arcs contain a greater proportion of mafic to intermediate rocks as described below.

10.3.1   Island arcs

Ocean lithosphere is subducted beneath an overlying plate composed of oceanic lithosphere (Figure 10.7a) producing island arc chains in the eastern Indian Ocean, the Caribbean and Scotia Seas and the western Pacific Ocean, notably the Marianas Islands. Island arcs develop on the overlying ocean lithosphere plate, above the subduction zone. Island arc volcanoes are underlain by intermediate to mafic plutonic suites dominated by diorite, quartz diorite, granodiorite, tonalite and even gabbro. **Diorites** contain <5% quartz and **quartz diorites** contain 5–20% quartz (see Figure 7.20); both of these rocks are enriched in plagioclase and hornblende, with lesser amounts of pyroxene and biotite. Hornblende (and to a lesser degree pyroxene and biotite) imparts dark (mafic) colors, while the plagioclase tends to result in lighter (felsic) hues; together, these mineral suites tend to occur in approximately equal concentrations, producing a speckled light and dark coloration. Diorite, quartz diorite and granodiorite batholiths intrude beneath youthful volcanic arcs. **Granodiorites**, which represent the plutonic equivalent of dacites and rhyodacites, contain >20% quartz and more plagioclase than potassium feldspar. Island arc granodiorites are generally metaluminous, containing hornblende, biotite and minor amounts of muscovite. Island arc plutons can also consist of tonalites and trondhjemites, which are plutonic rocks enriched in plagioclase feldspar and quartz. **Tonalites**, first described from Monte Adamello near Tonale in the eastern Alps, contain calcium plagioclase and quartz with minor amounts of potassium feldspar, biotite and hornblende. **Trondhjemites**, also known as plagiogranites, are granodioritic rocks in which sodium plagioclase represents half to two-thirds of the total feldspar component.

In addition to the voluminous calc-alkaline rock suite dominated by andesites and basaltic andesites discussed earlier, young island arc systems also produce low potassium arc tholeiite basalts as well as relatively rare rocks named boninites and adakites. Low potassium arc tholeiites occur on the oceanward side of the volcanic arc, nearest the trench. Tholeiitic magmas commonly form at subduction zones where the overlying plate is relatively thin. Major element concentrations of the tholeiitic island arc basalts are very similar to MORB, as indicated by their relatively low K_2O concentrations and iron enrichment, suggesting a similar depleted mantle source – most likely by flux melting of the ocean lithosphere wedge overlying the subducted slab as well as the subducted slab itself. Island arc tholeiite basalts can be distinguished from MORB by greater concentrations of potassium and other LIL elements (such as Ba, Rb, Sr, Cs, Rb and U) and lower concentrations of HFS elements (such as Th, Hf, Ta, Ti, Zr, Nb and Y) (Perfit et al., 1980; Hawkesworth et al., 1993; Pearce and Peate, 1995). Tholeiitic island arc magmas commonly produce basalts, basaltic andesites and andesites in the volcanic arc and diorite, tonalite (plagiogranite) or lesser granodiorite plutons in the underlying magmatic arc.

Boninites, named for the Bonin Islands in the western Pacific Ocean, are high magnesium (MgO/MgO + total FeO > 0.7) intermediate volcanic rocks that contain a SiO_2-saturated (52–68% SiO_2) groundmass. These rare rocks contain phenocrysts of orthopyroxene, and notably lack plagioclase phenocrysts (Bloomer and Hawkins, 1987). Boninites are enriched in chromium (300–900 ppm), nickel (100–450 ppm), volatile elements and LREE as well as zirconium, barium and strontium. Boninites are depleted in HREE and HFS elements. These unusual rocks occur proximal to the trench and bear the geochemical signature of primitive mantle-derived magmas produced early in the subduction cycle (Hawkins et al., 1984; Bloomer and Hawkins, 1987; Pearce and Peate, 1995). Thus, boninites are a product of subduction-related melting in the forearc of youthful island arc systems. Van der Laan et al. (1989, in Wyman, 1999) suggest that boninites are produced by high temperature, low pressure remelting of previously subducted ocean lithosphere. Interestingly, boninites can be associated with rare ultrabasic komatiites that we will discuss later in this chapter.

Adakites are silica-saturated (>56% SiO_2) rocks with high Sr/Y and La/Yb ratios (LREE enriched relative to HREE) and low HFS (such as Nb and Ta) concentrations. Adakites, named for Adak Island of the Aleutian Island chain, have been thought to be derived by slab melting of eclogite and/or garnet amphibolite from the descending ocean lithosphere (Kay, 1978; Defant and Drummond, 1990; Stern and Killian, 1996; Reay and Parkinson, 1997). While it was initially believed that adakites only form where young (<25 Ma), thin, hot ocean lithosphere is subducted beneath island arc lithosphere, adakites are now known to form at continent–continent collision sites as a result of shallow slab subduction of continental lithosphere (Chung et al., 2003). Shallow slab subduction and lithosphere recycling at subduction zones may play a significant role in the development of adakites as well as their plutonic equivalents trondhjemites and tonalites. Research continues to determine possible relationships of adakite formation with Archean **tonalite, trondhjemite and granodiorite (TTG) associations** and the evolution of continental crust (Drummond and Defant, 1990; Castillo, 2006; Gomez-Tuena et al., 2007) (Box 10.1).

Insofar as the overlying arc lithosphere is relatively thin in immature island arc systems, young volcanic arcs are dominated by basalts and basaltic andesites with rare boninites and adakites. Prolonged subduction in island arc systems generates increasingly thicker arc lithosphere. As island arc lithosphere thickens, andesites and dacites predominate as the Si_2O and K_2O contents of all rocks increase with the development of continental-type arc lithosphere (Miyashiro, 1974). In nearly all convergent margins, the calc-alkaline association is generated by fractional crystallization of basaltic magma derived by partial melting of overlying mantle peridotite – fluxed by fluids released from the dehydrated subducted oceanic lithosphere slab. The continued removal of crystals from melt leads to continuous variation in the residual liquid (liquid line of descent) generating basalt, andesite, dacite and rhyolite. In addition to fractionation, open-system diversification processes (Chapter 8) such as assimilation and magma mixing alter magma chemistry (Grove and Kinzler, 1986) and alter the chemical composition of magmas generated in the mantle wedge overlying the subducted slab.

In addition to magmatism within the island arc complex, igneous activity can also occur behind the island arc in backarc basins.

Box 10.1   Tonalite, trondhjemite and granodiorite (TTG) association

Plutons containing TTG are found in subduction zone environments ranging from the Archean to the Recent. However, Archean (>2.5 Ga) subduction zone plutonic rocks consist dominantly of TTG. Trondhjemites were named by V. M. Goldschmidt in 1916 for holocrystalline, leucocratic Norwegian rocks enriched in sodium plagioclase and quartz and depleted in biotite and potassium feldspar (Barker, 1979). Trondhjemites are similar to tonalites but contain greater concentrations of sodium plagioclase (oligoclase to albite) and more variable potassium feldspar concentrations (Figure B10.1a). Tonalites and trondhjemites are also known as plagiogranite. **Charnockites**, an orthopyroxene-bearing suite of rocks of generally granitic composition, also occur with the TTG association.

TTG associations occur in Archean rocks such as the Pilbara Craton of Australia, and the Beartooth and Big Horn Mountains of Wyoming. In contrast, Proterozoic and younger convergent margin granitic rocks consist predominantly of granite and granodiorite, with TTG associations representing a very small component in ocean–ocean convergence. What conditions changed at ~2.5 Ga? Shallow-dipping subduction zones "pinch out" the overlying mantle wedge such that the wedge component plays a relatively minor role in magma genesis. Archean subduction involved higher geothermal gradients, shallow subduction and melting of downgoing ocean lithosphere, with minimal input from the overlying lithosphere wedge. Shallow subduction of oceanic lithosphere at unusually low angles (Figure B10.1b) has been proposed as a model for the growth of Archean continental crust through the generation of TTG plutonic rocks and their volcanic equivalents – adakites (Martin, 1986; Smithies et al., 2003).

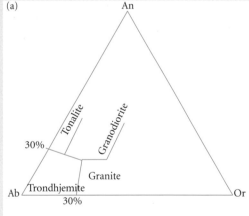

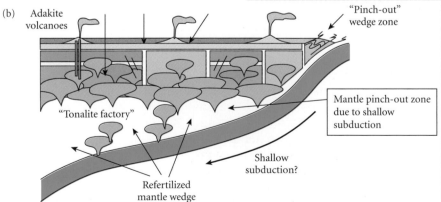

Figure B10.1   (a) Classification of some granitoid rocks enriched in plagioclase and >20% quartz. Trondhjemite is a light-colored tonalite containing sodium-rich oligoclase albite. Ab, albite (sodium plagioclase); An, anorthite (calcium plagioclase); Or, orthoclase (potassium feldspar). (b) Shallow subduction of the Archean ocean lithosphere may have produced tonalite plutons and adakite volcanic rocks. (Courtesy of the Geological Survey of Canada; with permission of the Natural Resources of Canada 2009.)

Backarc basins

Although compressional forces dominate island arc settings, lithospheric extension can occur in the overlying plate, behind the arc, resulting in the development of **backarc basins** (Figure 10.8). How does backarc extension occur? "Trench pull" forces move the volcanic arc towards the subduction zone resulting in the seaward movement of the trench and volcanic arc (Chase, 1978). Extension is manifested as normal faults and backarc spreading. The western and northern Pacific Ocean (Figure 10.8a) provide excellent examples of backarc basins, including the Sea of Japan, the Bering Sea, the Lau Basin–Havre Trough, Manus Basin and the Marianas Trough.

Backarc basins (Figure 10.8b) erupt a diverse suite of volcanic rocks including

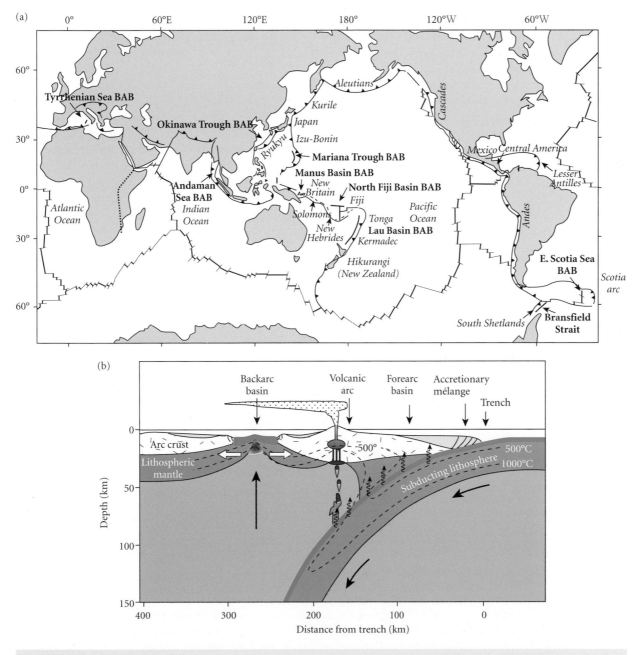

Figure 10.8   (a) Modern backarc basins (BAB) are concentrated in the western Pacific Ocean. (b) Backarc basins form by extension within the arc crust. (Courtesy of Wikipedia.)

basalt, basaltic andesite, andesite and dacite; however, tholeiitic and alkalic basalts commonly dominate. As suggested in our earlier discussion of island arc tholeiitic basalts, the relative proportions of andesitic versus basaltic magmas is related to the nature and thickness of the lithospheric wedge above the subduction zone (Gill, 1981). In some island arc settings (e.g., Kermadec, Marianas, Scotia and Vanuatu arcs), basalts dominate over andesites in the backarc and the volcanic arc regions. Fryer et al. (1981), on the basis of trace element chemistry, identified a distinctive group of rocks known as **backarc basin basalts** (BAB). BAB are tholeiitic, with geochemical similarities with both MORB and arc tholeiite trends. Relative to MORB, BAB display greater enrichment of H_2O, alkali elements and LIL elements. BAB are slightly depleted in titanium, yttrium and niobium and display flat rare Earth element patterns 5–20 times that of chondrites (Sinton et al., 2003). BAB may show relative enrichment in volatile elements, thorium and LREE, which suggests the involvement of subduction-related fluids in magma genesis (Pearce and Peate, 1995).

Why do backarc basins produce a wide array of rock types that range from near MORB to calc-alkaline compositions? Extension in the backarc (Figure 10.8b) results in partial melting of mantle peridotite, producing MORB-like magmas. However, these magmas interact to varying degrees with calc-alkaline sources. Calc-alkaline magma sources include the hydrated mantle wedge situated above the downgoing slab, recycled subducted lithospheric slab, and subducted marine sediment. Thus, BAB are produced by a combination of partial melting of lherzolite upper mantle wedge that has been fluxed by volatiles released by the subducted ocean lithosphere, as well as decompression melting of mantle peridotite at backarc spreading ridges (Fretzdorff et al., 2002). As indicated in the discussion above, distinction between different types of basalts generated in different tectonic environments is largely dependent upon geochemistry (Box 10.2).

10.3.2   Continental margin arcs

Mature convergent margins, involving the subduction of ocean lithosphere beneath thick continental lithosphere, occur along the eastern Pacific region extending from the Cascades southward to the Andes Mountains (Figure 10.8a). Ascending hydrous melts from the subducted ocean slab chemically react with the overlying wedge composed of mantle and thick continental lithosphere. These magmas produce continental arc plutons that are more silicic than island arc plutons as a result of the thick overlying continental lithosphere through which subduction zone fluids must penetrate. Extensive assimilation and magma mixing within the overlying continental slab result in K_2O and SiO_2 enrichment within plutons.

Box 10.2   Geochemical approaches to petrotectonic associations

Basalt, basalt, basalt! As indicated in the preceding discussion, basalt can be produced in a number of different tectonic environments. Petrotectonic studies utilize a number of different approaches to identifying sites of basalt genesis. One of these approaches utilizes geochemical indicators discussed in Chapter 7. Pearce and Cann (1973) combined geochemical variations of minor and trace elements to infer tectonic origin. The Pearce and Cann (1973) classification (Figure B10.2a) is widely used in the tectonic analysis of basalts and is particularly useful in accretionary terranes (Chapter 1) where the original source of basalt is ambiguous. The petrotectonic environments defined by these discrimination diagrams include: within plate basalts, island arc tholeiites, calc-alkaline basalts, mid-ocean ridge basalts, ocean island tholeiites and ocean island alkaline basalts. Shervais (1982) also developed discrimination diagrams using minor element concentrations such as vanadium and titanium (Figure B10.2b). These geochemical techniques designed to determine petrotectonic origin are extremely useful if used in conjunction with field studies – and if metamorphism has not chemically altered the rock.

Box 10.2 *Continued*

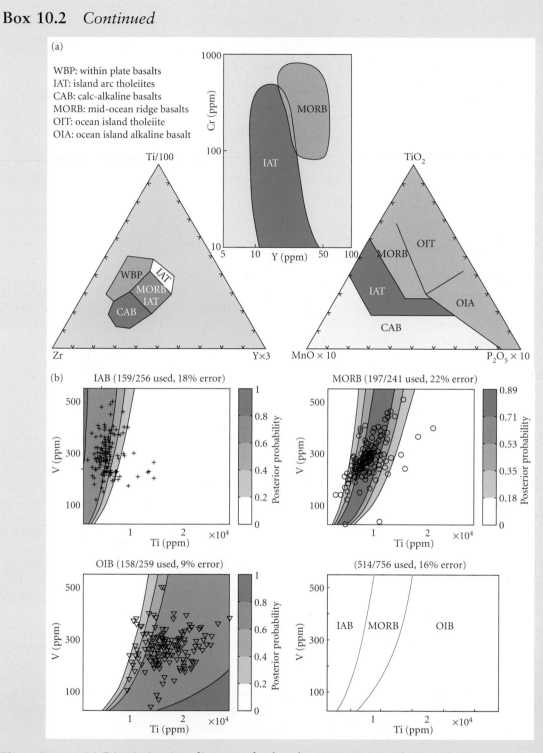

Figure B10.2 (a) Discrimination diagrams for basalt using minor and trace element concentrations such as Zr, Ti, Y, Cr, MnO and P₂O₅. (After Pearce and Cann, 1973; with permission of Elsevier Publishing.) (b) Discrimination diagrams using vanadium and titanium concentrations as discriminating elements. IAB, island arc basalts; OIB, ocean island basalts; MORB, mid-ocean ridge basalts. (After Shervais, 1982; with permission of Elsevier Publishers.)

Ocean–continent convergent margins produce voluminous granodiorite, diorite, granite and tonalite plutons. Large granodiorite plutons commonly dominate ocean–continent convergent plate boundaries. For example, the Sierra Nevada consists of 25–30 km thick upper crustal rocks containing hundreds to thousands of individual granodioritic, dioritic and tonalitic plutons. The Sierra Nevada composite batholith developed due to eastward-dipping subduction beneath western North America. Arc magmatism continued from 220 to 80 Ma (Fliedner et al., 2000; Ducea, 2001).

Magma from these intermediate–silicic plutons erupts onto Earth's surface producing composite volcanoes. Together with voluminous andesites, rocks such as dacites, rhyodacites, rhyolites and latites display aphanitic–porphyritic to pyroclastic textures in composite volcanic settings such as Mt St Helens and Crater Lake in the Cascade Range (USA).

In late stages of ocean–continent subduction, highly alkalic shoshonites can erupt over thick continental lithosphere. **Shoshonites** are dark-colored, potassium-rich trachyandesites, commonly containing olivine and augite phenocrysts with a groundmass of labradorite plagioclase, alkali feldspar, olivine, augite and leucite. Shoshonites occur in thickened lithosphere farthest from the trench region, in continent–continent collisions and in some backarc basins.

10.3.3 Continental collision zones

In long-lived convergent margins, the termination of ocean lithosphere subduction marks the transition from mature arc–continent convergence to a final collision of two continental blocks. The Alpine–Himalayan orogenic system is the modern classic model of continent–continent collision following tens of millions of years of ocean lithosphere subduction. Ancient examples include the complete subduction of ocean basins 250 million years ago, resulting in continent–continent collision creating the supercontinent Pangea.

In continent–continent collisions, the lower continental lithosphere does not subduct to great depths but essentially breaks off and underplates the overlying continental lithosphere plate producing a doubly thick lithosphere. Melting at the base of this lithospheric stack produces Al_2O_3-, K_2O- and SiO_2-rich igneous rocks such as rhyolites, rhyodacites and shoshonites and plutonic rocks of increasingly granitic composition. These magmas are the result of relatively flat subduction (<25°), thick continental lithosphere (>25 km), higher degrees of partial melting of continental lithosphere and/or arc basement, and the diminished role of ocean lithosphere subduction. Alkaline basalts also occur in continent–continent collisions as a result of upwelling mantle melts.

Rhyolites and rhyodacites are characterized by high viscosity, which retards lava flow, resulting in thick accumulations of limited aerial extent. Common minerals include quartz, potassium feldspar, biotite, plagioclase, anorthoclase and magnetite. Shoshonitic magmas are generated farthest from the trench, wherein melts assimilate K_2O and Na_2O as they rise through thick slabs of overlying continental lithosphere. Volcanic eruptions of rhyolite to shoshonite lavas can erupt explosively, generating voluminous pyroclastic tuffs and breccias, or produce lava flows that solidify to produce glassy and/or aphanitic–porphyritic textures.

The plutonic equivalent of rhyodacite and rhyolite are granitic rocks or granitoids (Box 10.3). The term "granitic" or "granitoid" is loosely used for silica-oversaturated plutonic rocks that contain essential potassium feldspars and quartz. Granitoids include the IUGS fields of quartzolite, quartz-rich granitoid, alkali feldspar granite, granite, quartz granitoid, granodiorite and tonalite (see Figure 7.20). Granitoid rocks that form at mature convergent margins tend to be peraluminous to metaluminous, containing hornblende, biotite and/or muscovite. Although granites of variable composition occur, S-type and I-type granites tend to predominate. I-type magmas form by partial melting of basic to intermediate igneous rocks in or above the subduction zone at ocean–ocean or ocean–continent convergent margins. Peraluminous, potassium-rich, S-type granites and granodiorites are particularly common at continent–continent collisions. The peraluminous sedimentary component is derived from phyllosilicate minerals in graywackes and mudstones of the continental crust and accretionary wedge. These sedimentary materials melt to

Box 10.3 Granite classification

Strictly speaking, the term "granite" is restricted to plutonic rocks containing 20–60% quartz and 35–90% alkali to plagioclase feldspars (see Figure 7.20). Thus, the two essential mineral groups in granite are quartz and feldspars. Other minor minerals include hornblende, biotite and muscovite. Accessory minerals include magnetite, rutile, tourmaline, sphene, apatite, molybdenite, gold, silver and cassiterite. Granite plutons are genetically associated with Precambrian cratons and convergent margins (Pitcher, 1982). Phanerozoic granitic plutonic belts are found along continent–ocean subduction zones or at continent–continent suture zones. Within orogenic settings, granites may be emplaced synchronous (syn-kinematic) with convergence, as late-stage collisional plutons or as post-kinematic intrusions.

Granites have been subdivided by a number of methods, one of which attempts to infer source rock origin. Chappell and White (1974) and others recognize four distinct types of granite (M, I, S, and A types) based upon the nature of the inferred parental source rock (Table B10.3). M-, I-, and S-type granites are orogenic granites associated with subduction, whereas A-type granites are anorogenic in origin.

- **M-type granites** (Pitcher, 1982) are derived from mantle-derived parental magmas, as indicated in the low Sr^{87}/Sr^{86} ratios (<0.704). M-type granites are associated with calc-alkaline tonalites, quartz diorites and gabbroic rocks. In addition to quartz and feldspars, hornblende, clinopyroxene, biotite and magnetite are among the major minerals. M-type granites develop in island arc settings. Copper and gold mineralizations are associated with M-type granites.

- **I-type granites** (Chappell and White, 1974) are generated by the melting of an igneous protolith from either the downgoing oceanic lithosphere or the overlying mantle wedge. I-type granites are enriched in Na_2O and CaO and contain lower Al_2O_3 concentrations. I-type granites have Sr^{87}/Sr^{86} ratios of less than 0.708, usually in the range 0.704–0.706, indicating magma derived from a mantle source. Because they are primarily derived from a mantle source, I-type granites may be enriched in mafic minerals such as hornblende, biotite, magnetite and sphene. Porphyry copper, tungsten and molybdenum deposits are associated with I-type granites. I-type granites are prevalent along the Mesozoic–Cenozoic Andes Mountains (Chappell and White, 1974; Beckinsale, 1979; Chappell and Stephens, 1988).

- **S-type granites** (Chappell and White, 1974) are produced by the melting of sedimentary crustal rocks in collision zones. S-type granites are depleted in Na_2O but enriched in Al_2O_3 (peraluminous). S-type granites have Sr^{87}/Sr^{86} ratios of >0.708, indicating that source rocks had experienced an earlier sedimentary cycle. S-type granites are also known as two-mica granites in that they commonly contain both muscovite and biotite, reflecting the peraluminous content of the sedimentary source rock rich in phyllosilicate minerals. Hornblende is conspicuous by its absence. Other minerals include monazite and aluminosilicate minerals such as garnet, sillimanite and cordierite. Tin deposits are associated with S-type granites (Chappell and White, 1974; Beckinsale, 1979).

Table B10.3 The major features of M-, I-, S- and A-type granitoids.

	M	I	S	A
SiO_2	54–73%	53–76%	65–79%	60–80%
Na_2O	Low, <3.2%	High, >3.2%	Low, <3.2%	>2.8%
K_2O/Na_2O	Very low	Low	High	High
Sr^{87}/Sr^{86}	<0.704	<0.706	>0.706	0.703–0.712

Continued

Box 10.3 *Continued*

- **A-type granites** (Loiselle and Wones, 1979) are anorogenic rocks produced by activities that do not involve the subduction and collision of lithospheric plates. A-type granites are enriched in alkaline elements with high K/Na and (K + Na)/Al ratios as well as high Fe/Mg, F, HFS elements (Zr, Nb, Ga, Y, Zn) and rare Earth element concentrations. A-type granites are depleted in Mg, Ca, Al, Cr, Ni and have lower water contents and high Ga/Al ratios (Collins et al., 1982; Whalen et al., 1987). Relative to I-, S- and M- type granites, A-type granites are more enriched in LIL elements and depleted in refractory elements (Creaser et al., 1991). A-type granites are peralkaline and commonly contain biotite, alkali pyroxenes, alkali amphiboles and magnetite (Collins et al., 1982). Associated with A-type granites are alkali-rich, relatively anhydrous rocks that can include alkali granite, syenite, alkali syenite and quartz syenite.

produce two-mica granites containing biotite and muscovite.

Convergent margins contain a diverse range of rock types. In addition to intermediate and silicic igneous rocks prominently discussed thus far in this chapter, basic and ultrabasic assemblages also occur at convergent margins due to magmatic processes and/or tectonic displacement. In some cases, these involve metamorphic processes that we will address in Chapters 15–18. Let us first consider tectonically emplaced Alpine orogenic complexes and then we will briefly discuss basic–ultrabasic zoned intrusions.

10.3.4 Alpine orogenic complexes

Alpine orogenic complexes are fault-bounded, deformed rock sequences that mark the site of present or former convergent margins. Unlike the intermediate to silicic igneous rocks that develop in situ (in place) as a result of subduction-induced magmatism, alpine orogenic complexes have been transported far from their site of origin by thrust faulting and shearing. Because such tectonism can be intense, these complexes are commonly dismembered into fault blocks and jumbled together in a haphazard fashion such that their original layering may be disrupted. Alpine orogenic bodies contain disrupted pelagic sediment layers, basalt, cumulate basic and ultrabasic layers as well as tectonized mantle slices of ocean lithosphere and calc-alkaline intrusive and volcanic assemblages. Alpine orogenic bodies are commonly associated with tectonic mélanges.

A tectonic **mélange** (from the French word for mixture) is an intensely sheared, heterogeneous rock assemblage embedded within a highly deformed mud matrix. Mélanges form at subduction zones where rocks and tectonic blocks are sliced from the downgoing oceanic lithosphere and often mixed with rocks formed in forearc settings. The diverse suite of rocks may include: (1) deformed and altered mid-ocean ridge, ocean island and ocean plateau basalts, (2) limestone, chert and other marine sedimentary rocks, and (3) slices of eclogite, peridotite and blueschist from subducted oceanic or forearc lithosphere. Eclogites and blueschists are high pressure metamorphic rocks characteristic of subduction zones (Chapter 18).

Ophiolites constitute one type of Alpine deposit in which the oceanic or backarc basin lithosphere or volcanic arc basement rocks are preserved in orogenic belts. The term ophiolite was first proposed by Steinmann (1905) for serpentinized rocks in the Alps. Over the next several decades, Steinmann recognized a suite of rocks that, thereafter, became known as the "Steinmann trinity". These three rock types consist of pelagic chert, serpentinite (hydrothermally altered peridotite) and spilites (altered pillow basalts). As the term is currently used, ophiolites are thought to represent coherent slices of oceanic lithosphere, volcanic arc basement or backarc basin lithosphere "obducted", or thrust, onto the edge of continents above subduction zones.

While it is intuitively obvious that ophiolites may originate by sea floor spreading at ocean ridges, petrologists recognize that many ophiolites represent oceanic fragments produced in either forearc or backarc settings. These ophiolites are referred to as **suprasubduction zone (SSZ) ophiolites**. SSZ ophiolites

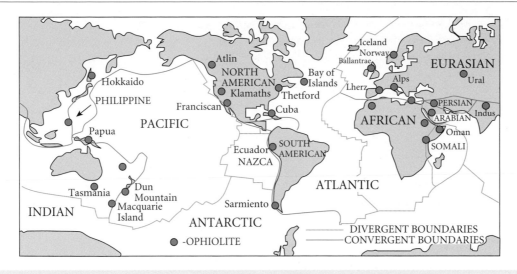

Figure 10.9   Ophiolite locations throughout the world. (Courtesy of William Church.)

develop due to extensional tectonics that result in backarc spreading or forearc spreading producing oceanic lithosphere. Researchers use immobile elements as petrogenetic indicators, such as chromium, to determine ophiolite sites of origin (Dick and Bullen, 1984). Both the origin of ophiolites and their means of emplacement on the edges of continents remain areas of intense research.

Complete ophiolite sequences display a stratigraphic sequence similar to that of ocean lithosphere (see Figure 10.2a). The stratigraphy of an idealized ophiolite sequence was defined by the first Penrose Conference in 1972 as follows:

- **Layer 1:** pelagic, marine sedimentary rock such as ribbon chert, thin shale beds and limestone derived from the lithification of siliceous ooze, clay and calcareous ooze sediments, respectively.
- **Layer 2A:** basic volcanic complex, which may contain pillow basalt.
- **Layer 2B:** basic sheeted dike complex.
- **Layer 3:** cumulate gabbroic complex with basal cumulate peridotites and pyroxenites.
- **Layer 4:** tectonized ultrabasic complex consisting of variably metamorphosed harzburgite and dunite. Podiform chromite deposits occur with dunite bodies. The tectonized ultrabasic complex overlies a metamorphic basal sole thrust.

While the idealized stratigraphic layering of ophiolites mimics ocean lithosphere layering, the complete four-layer stratigraphic sequence is rarely preserved. Tectonically disrupted ophiolites, missing one or more layers, are referred to as partial, or dismembered, ophiolites. Ophiolites occur throughout the world (Figure 10.9) and mark the former location of ocean lithosphere subduction. Excellent examples of ophiolites include the following localities: Oman, Troodos (Cyprus), Coast Range (California), Newfoundland and Morocco. Ophiolites are important in providing the following:

1 Valuable ore deposits containing Cu, Ag, Au, Zn, Ni, Co, Cr and other metals.
2 Evidence documenting oceanic lithosphere subduction dating from the Precambrian to the present. Well-documented ophiolites less than 1 Ga occur throughout the world. Archean examples, dating as far back as 3.8 Ga, are highly controversial and may (Furnes et al., 2007) or may not (Hamilton, 2007; Nutman and Friend, 2007) represent true ophiolites.

In addition to their occurrence in Alpine orogenic complexes and ophiolites, basic and ultrabasic magmas intrude convergent margin assemblages to form the concentrically zoned or layered plutons discussed below.

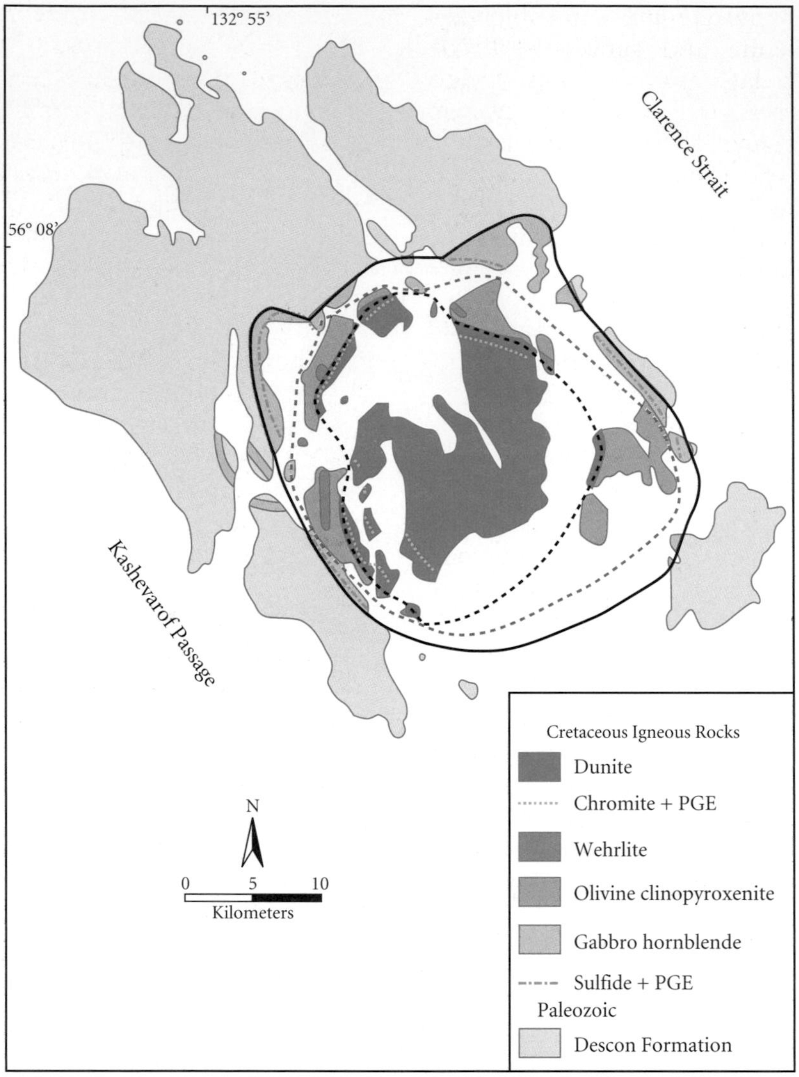

Figure 10.10   Zoned intrusion from the Blashke Islands Complex, southeast Alaska. PGE, platinum group elements. (After Kennedy and Walton, 1946.)

10.3.5   Alaska-type (zoned) intrusions

Alaska-type intrusions consist of concentrically layered (zoned) plutons formed in convergent margin settings. Alaska-type intrusions are commonly several kilometers in diameter, exhibit a dunite core and pyroxenite shell, and are surrounded by massive gabbro. Late granitic zones may also occur around the perimeter of the intrusive structure. In contrast to tectonically emplaced alpine suites, Alaska-type plutons form in situ by intrusion of magma into the surrounding country rock. Originally recognized in Alaska and in the Ural Mountains, these plutonic bodies have since been identified in many other localities throughout the world. Buddington and Chapin (1929) noted the concentric layers in the Blashke Island Complex of Alaska (Figure 10.10).

Irvine (1959) conducted extensive work on the Cretaceous age Duke Island Complex (Alaska). The Duke Island Complex consists of a 3 km diameter plutonic body at Judd Harbor and a 5 km diameter intrusion at Hall Cove. These two intrusions, which are likely to be continuous at depth, contain a dunite core surrounded by successive rings of peridotite, olivine pyroxenite, hornblende–magnetite pyroxenite and gabbro – all of which are cut by late granitic rocks. Irvine (1959) noted layers that exhibited graded beds within ultrabasic rocks at Duke Island, in which crystals are coarse grained at the base and fine upward within a layer. Duke Island remains an active exploration site with economic deposits of copper, nickel, platinum and palladium.

Alaska-type ultrabasic–basic plutons commonly occur as post-orogenic intrusions in volcanic arc or accretionary mélange terrains. A number of different processes have been suggested for their formation; these include fractionation of ultrabasic or basic parent magma from the upper mantle, magma mixing at convergent plate boundaries, or magmas from deep mantle plumes (Taylor, 1967; Tistl et al., 1994; Sha, 1995; Ishiwatari and Ichiyama, 2004). Alaska-type intrusions are economically important as sources of metals, particularly platinum group elements (PGE).

Thus far in this chapter we have focused upon petrotectonic assemblages from divergent and convergent plate boundaries. While divergent and convergent margins produce the bulk of Earth's magmatism, igneous rocks also develop within lithospheric plates without any direct link to plate boundary processes, as discussed below.

10.4 INTRAPLATE MAGMATISM

Intraplate magmatism refers to magma generation and igneous rock suites generated within lithospheric plates, rather than at plate boundaries. Intraplate magmatism may be initiated by hotspot activity, continental rifts or overthickened continental lithosphere. Intraplate magmatism produces a wide range of igneous rock types including:

- Tholeiitic to alkalic basalt and related gabbros of hotspots and LIP.
- Siliceous anorogenic granite and rhyolite.
- Silica-undersaturated rocks.
- Basic–ultrabasic suites including komatiites and kimberlites.
- Carbonatites.

Large igneous provinces (LIP), encompassing volumes $>10^6 \text{km}^3$ (Mahoney and Coffin, 1997), are the greatest manifestation of intraplate magmatism on Earth (Figure 10.11). Most LIP are basaltic in composition although silicic examples, known as **SLIP (silicic large igneous provinces)**, such as Yellowstone, also occur. The most widespread Phanerozoic intraplate magmatic features consist of massive tholeiitic flood basalts. These massive volcanic landforms occur as both **oceanic flood basalts** and **continental flood basalts**.

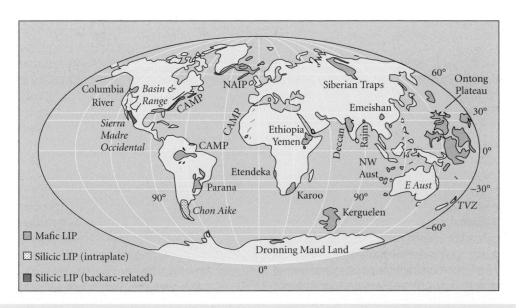

Figure 10.11 Earth's large igneous provinces (LIP). CAMP, Central Atlantic Magmatic Province; NAIP, North Atlantic Igneous Province; TVP, Taupo Volcanic Zone. (After Coffin and Eldholm, 1994; courtesy of M. F. Coffin.)

In the following section, we will first consider oceanic intraplate magmatism and then later discuss continental intraplate assemblages.

10.4.1   Oceanic intraplate magmatism

Ocean islands and ocean plateaus form above mantle hotspots that erupt anomalously high volumes of tholeiitic and alkalic basaltic lava onto the ocean floor. **Ocean islands** are volcanic landforms that rise upward above sea level. **Seamounts** are volcanically produced peaks below sea level. **Oceanic plateaus** are broad, flat-topped areas that result from massive outpourings of lava flowing laterally from source vents. Oceanic plateaus cover large areas of the ocean floor, ranging up to 10 million km^2.

Ocean island basalts

Ocean island basalts (OIB) are a geochemically distinct suite of rocks distinctly different from MORB (Figure 10.12). In contrast to MORB, OIB are more alkalic and are less depleted – and may in fact be somewhat enriched with respect to incompatible elements such as potassium, rubidium, uranium, thorium and LREE (Hofmann and White, 1982; Hofmann, 1997). The different geochemical signatures have been interpreted to represent different mantle source areas. While MORB were considered to represent partial

melts of an upper, depleted (previously melted) mantle, OIB were considered to perhaps represent partial melts from a deeper, undepleted mantle source. However, ocean island basalts display large variations in strontium, neodymium and lead and other isotopic ratios, suggesting the role of multiple sources and processes. Various hypotheses proposed for OIB chemistry include:

* Small degrees of melting of a primitive mantle source.
* Melting of a mantle source enriched in alkali elements.
* Incorporation of subducted oceanic crust in the source region.
* Entrainment of subducted sedimentary rocks in the source region (Hofmann and White, 1982; Hofmann, 1997; Kogiso et al., 1998; Sobolev et al., 2005).

The isotopic signatures of many OIB indicate that magmas were derived from non-primitive sources of variable mantle composition. For example, Rb/Sr and Nd/Sm ratios are lower than primitive mantle ratios while U/Pb, Th/Pb and U/Th ratios are higher than primitive mantle sources (Hofmann and White, 1982). In fact, most OIB display isotopic ratios indicative of an enriched mantle source, particularly their elevated incompatible element and NiO concentrations. Why is the mantle composition so variable? One explanation involves mantle enrichment due to the incorporation of a recycled oceanic lithosphere derived from ancient subduction zones (Hofmann and White, 1982; Hofmann, 1997; Turner et al., 2007). Hirschmann et al. (2003) and Kogiso et al. (2003) propose that undersaturated, nepheline normative OIB magmas may be derived by partial melting of a garnet pyroxenite, which itself is derived by the mixing of subducted MORB basalt and mantle peridotite. Thus, at least some OIB hotspots tap mantle melts enriched by ocean lithosphere subducted up to 2.5 Ga (Turner et al., 2007). Let us consider the best known OIB location – the Hawaiian Islands.

Hawaii

From the ocean floor up, the Hawaiian Islands constitute the highest mountains on Earth with a relief of ~10 km high; Mt Everest in contrast has an elevation of just over 9 km.

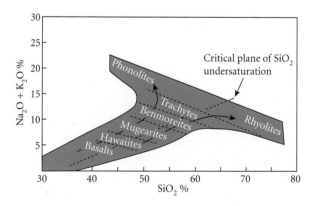

Figure 10.12   The fractionation sequence occurring at ocean islands produces a diverse suite of volcanic rocks with variable alkali and silica concentrations. (Courtesy of Stephen Nelson.)

The Hawaiian hotspot has been active for over 80 million years, generating a chain of seamounts and islands extending for a distance of 5600 km. The Hawaiian Islands are dominated by olivine tholeiites; tholeiitic basalts comprise ~99% of the exposed Hawaiian volcanic rocks with alkalic basalts contributing only a small fraction. Early eruptions of alkali basalts are followed by extensive tholeiitic basalts that generate massive shield volcanoes. Late-stage Hawaiian volcanism reverts to alkali basalts (hawaiites to benmoreites), perhaps indicating a lower degree of melting as the island moves away from the hotspot magma source and temperatures decline. The dominant tholeiitic iron-enrichment trend (Chapter 8) likely results from fractional crystallization of early formed, magnesium-rich olivine and pyroxene. As magmas become more enriched in iron, alkali-enriched hawaiite forms by fractionation of tholeiitic basalt. The addition of magnetite to the crystallizing assemblage causes the remaining magmas to become progressively less enriched in iron. Continued crystallization produces an iron-depletion and alkali-enrichment trend in which magmas evolve successively from hawaiite to mugearite and benmoreite (see Figure 7.24b). Depending upon the degree of silica saturation, the final minerals to crystallize may include feldspathoids in silica-undersaturated phonolite, potassium feldspar in silica-saturated trachyte or quartz in silica-oversaturated rhyolite (Figure 10.12).

Hawaiian basaltic magmas are thought to form by partial melting of a heterogeneous mantle plume source, perhaps composed of primitive garnet lherzolite, enriched by a subducted ocean lithosphere component. Using geochemical data such as Ni, Mg, Pb, O, Hf and Os isotopic ratios, researchers (Blichert-Toft et al., 1999; Sobolev et al., 2005) propose that the Hawaiian magma source is generated from a mantle enriched from the recycling of eclogite (subducted ocean lithosphere), which magmatically mixes with mantle peridotite. Using geochemical constraints such as high NiO concentrations, Sobolev et al. (2005) estimate that recycled (subducted) ocean crust accounts for up to 30% of the Hawaiian source material. These important studies suggest that the Hawaiian intraplate magmatism may be related to ancient subduction processes. While OIB volcanism at locations such as the Hawaiian Islands is impressive, humans have yet to witness the immense volcanism necessary to erupt gargantuan ocean plateaus.

Oceanic flood basalt plateaus

The Ontong–Java Plateau, located near the Solomon Islands in the western Pacific Ocean, is the largest oceanic flood basalt plateau on Earth. As summarized by Fitton and Godard (2004), the dates for the Ontong–Java eruptions are somewhat enigmatic. The Ontong–Java flood basalts erupted either in a single massive flood eruption (~122 Ma) or in a series of eruptions spread over 10 million years with the initial massive outpouring occurring ~122 Ma. The Ontong–Java Plateau encompasses a surface area of 2 million km^2 and a volume of 60 million km^3. On the basis of ten Ocean Drilling Project (ODP) rock core analyses, the Ontong–Java Plateau region is thought to consist largely of a relatively homogeneous low potassium tholeiite that erupted as massive sheet flows and pillow basalts, accompanied by minor volcaniclastic and vitric tuff deposits (Coffin and Eldholm, 1994; Fitton and Godard, 2004). On the basis of geochemical data (e.g., enrichment in incompatible elements Zr to Lu), Tejada et al. (2004) suggest that the Ontong–Java basalt magma was derived by 30% melting of a primitive, enriched, high magnesium (15–20 wt % MgO) mantle source.

The origin of oceanic intraplate magmatism continues to be an active area of research. At least in some cases, intraplate magmatism may be partially derived by deep convection cells involving ancient subducted ocean lithosphere. In other cases, intraplate magmatism may be driven by mantle plumes unrelated to plate boundary activities. As discussed in Chapter 1, hotspots tap magmas from different depths such as the lower crust, upper mantle or mantle–core boundary. Geochemical analyses of basalt and seismic tomography continue to provide insight into our attempts to understand these perplexing igneous processes. Our discussion of oceanic intraplate magmatism has centered on varieties of basalt, which dominate these settings. In sharp contrast, continental intraplate magmatism produces a wider range of igneous rock types, discussed in the following section.

10.4.2 Continental intraplate magmatism

Continental intraplate magmatism and volcanism produce:

- Continental flood basalts.
- Continental rift assemblages.
- Bimodal volcanism.
- Layered basic and ultrabasic intrusions.
- Ultrabasic suites that include komatiites and kimberlites.
- An unusual array of alkaline rocks and anorogenic granites.

Continental flood basalts

Examples of huge outpourings of continental flood basalts (CFB) include the Deccan traps of India, Karroo basalts of Africa, Siberian flood basalts of Russia and the Columbia River, Snake River plain and Keweenaw flood basalts of the United States. The three largest flood basalt events – the Permo-Triassic Siberian traps, the Triassic–Early Jurassic Central Atlantic Magmatic Province and the Cretaceous–Tertiary Deccan traps – correspond with the largest extinction events in Earth's history (Renne, 2002).

Although less common, silicic large igneous provinces (SLIP) also occur in association with continental break-up, intraplate magmatism and backarc basin magmatism. SLIP are silicic-dominated provinces containing rhyolite caldera complexes and ignimbrites. SLIP occur notably in the Whitsunday volcanic province of eastern Australia, the Chon Aike Province of South America and Yellowstone (Bryan et al., 2002). Below we briefly describe several well-known continental flood basalt provinces beginning with the CAMP.

Central Atlantic Magmatic Province (CAMP)
The CAMP formed during the Early Jurassic break-up of the Pangea supercontinent, which produced rift basins and flood basalts in North America, South America, Europe and Africa (see Figure 10.11). These once contiguous tholeiitic basalts are now widely dispersed across the Atlantic Ocean realm, encompassing a total area of more than 7 million km². The Ar^{40}/Ar^{39} ages indicate that the CAMP basalts erupted between 191 and 205 Ma, with a peak age of 200 Ma. CAMP rocks consist of tholeiitic to andesitic basalts, with rare alkaline and silicic rocks. CAMP tholeiites have low TiO_2 concentrations, negative mantle normalized niobium anomalies and moderate to strongly enriched rare Earth element patterns. These geochemical patterns indicate an anomalously hot mantle plume that resulted in the partial melting of the overlying lithosphere (Marzoli et al., 1999).

Siberian flood basalts
The **Siberian flood basalts** (see Figure 10.11) consist predominantly of tholeiitic basalt flows tens to a few hundreds of meters thick with minor trachyandesites, nephelinites, picrites, volcanic agglomerates and tuffs (Zolotukhin and Al'Mukhamedov, 1988; Fedorenko et al., 1996). The 251 Ma Siberian flood basalts were already recognized as one of the greatest known outpourings of lava when, in 2002, the western Siberian Basin flood basalt province was discovered which effectively doubled the aerial extent of the Siberian traps to approximately 3,900,000 km² (Reichow et al., 2002). It is analogous to burying half of the contiguous United States in lava. In the Maymecha-Kotuy region of Russia, Kamo et al. (2003) suggest that the entire 6.5 km thick basalt sequence erupted within ~1 million years, based upon U/Pb dates obtained from the base (251.7 ± 0.4 Ma) and top (251.1 ± 0.3 Ma) of the basalt sequence – truly mind boggling in scale.

Deccan traps
Over 1,000,000 km³ of flood basalt erupted in southwestern India between 65 and 69 Ma (Courtillot et al., 1988). The Deccan traps (see Figure 10.11) encompass an area of 500,000 km² in western India. Individual lava flows generally vary from 10 to 50 m in thickness with total flow thicknesses varying from less than 100 m to more than 2 km (Ghose, 1976; Sano et al., 2001). The flood basalts and related dike swarms are interpreted to result from rifting as the Indian Plate migrated over a mantle plume (Muller et al., 1993). The Deccan traps are dominated by tholeiitic basalts with minor amounts of alkalic basalts. Geochemical studies suggest that the Deccan basalts originated by fractional crystallization of shallow magma chambers (~100 kPa,

1150–1170°C). The basaltic magma experienced variable degrees of contamination as it ascended and assimilated granitic crustal rocks (Mahoney et al., 2000; Sano et al., 2001).

Columbia River flood basalts

Although relatively small compared to the flood basalt provinces listed above, the Columbia River flood basalts (see Figure 10.11) are among the most studied CFB on Earth. The Columbia River flows consist largely of quartz tholeiites and basaltic andesite, with 47–56 wt % silica (Swanson and Wright, 1980; Reidell, 1983). Columbia River basalts crop out in the US states of Washington, Oregon and Idaho, encompassing an area of approximately 163,700 ± 5000 km^2 (Tolan et al., 1989). The total volume of erupted lava has been estimated to be approximately 175,000 ± 31,000 km^3 (Tolan et al., 1989). The Columbia River basalt group has been subdivided into five formations: the Imnaha Basalt, Grande Ronde Basalt, and coeval Picture Gorge, Wanapum Basalt and Saddle Mountain Basalt. The Grande Ronde Basalt, which erupted 15.5–17 Ma, comprises approximately 87% of the total volume of the Columbia River basalt (Swanson and Wright, 1981). Over 300 individual lava flows erupted from northwest trending fractures between 6 and 17 Ma, making this the youngest continental flood basalt province on Earth. Individual flows traveled as much as 550 km, erupting in north–central Idaho and flowing to the Pacific Ocean (Hooper, 1982). Unlike most other flood basalt provinces, the Columbia River basalts lack early picritic basalt eruptions and interbedded silicic lavas and have less than 5% phenocrysts (Durand and Sen, 2004). The low concentrations of phenocrysts are thought to be related to either rapid ascent of magma (McDougall, 1976) or to a high water content of ~4.4%, which effectively lowered the melting temperature and inhibited the early development of large crystals (Lange, 2002).

Various hypotheses have been proposed for the origin of the Columbia River basalt flows. One set of hypotheses propose that the basalts crystallized from primary magma (Swanson and Wright, 1980, 1981). The relatively low Sr^{87}/Sr^{86} (0.7043–0.7049) ratios indicate a mantle source (McDougall, 1976). However,

the high SiO$_2$ concentrations, high total FeO (9.5–17.5 wt %) and low MgO concentrations (3–8 wt %) suggest that the parental magma was not primary (Hooper, 1982; Lange, 2002). A second set of hypotheses assert that the basalts are the product of diversification processes (McDougall, 1976; Reidel, 1983). Viable diversification models suggest that partial melting of pyroxenite (Reidell, 1983) or eclogite (Takahashi et al., 1998) parental rock was followed by the injection of separate magmatic pulses, subsequent magma mixing and assimilation of crustal rock. Trace element data suggest that the Columbia River basalts were not derived by fractionation of a single magmatic pulse. Thus, a unified model suggests that the Columbia River basalts were created by multiple pulses of heterogeneous mantle-derived magmas, contaminated by continental crust during magma ascent and magma mixing (Hooper, 1982; Riedel, 1983). The tectonic origin of the Columbia intraplate magmatism has been the subject of debate. Possible tectonic causes include: heating following subduction of the Juan de Fuca ridge, backarc spreading, the Yellowstone hotspot and continental rifting.

Continental rifts

Continental rifts produce a wide array of rocks that include alkalic basalt as well as alkaline and silicic rocks. Alkaline rocks include phonolite, trachyte and lamproite. Silicic rocks include rhyolite and rhyodacite, which occur in lava domes or as pyroclastic flow and ash fall deposits. Plutonic rocks vary from syenite and alkali granite to gabbroic rocks.

Continental rifting occurs in regions such as the East African rift basin, Lake Baikal (Russia), the Basin and Range and the Rio Grande rift system (USA). The East African rift system (Figure 10.13) erupts abundant alkali basalt as well as phonolite, trachyte, rhyolite and carbonatite lava. Ancient continental rifts include the Permian age (~250 Ma) Rhine Graben (Germany) and Triassic (~200 Ma) Oslo Graben (Norway) and rift basins of the Atlantic Ocean basin and the 1.1 Ga Keweenaw rift of the Lake Superior basin (USA). Continental rift zones can contain important hydrocarbon reservoirs because of

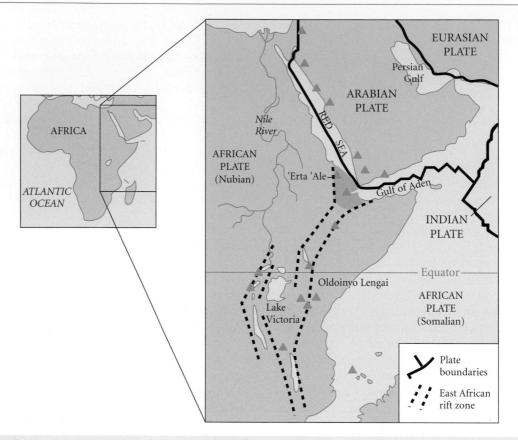

Figure 10.13   The East African rift system represents the third leg to the Gulf of Aden and Red Sea rift chain. (Courtesy of the US Geological Survey.)

the rapid deposition of organic-rich sediments. Volcanic flows and associated shallow intrusives can also provide valuable metallic ore deposits such as nickel and copper and platinum group elements.

What is the driving force behind the lithospheric extension that leads to the development of continental rifts? Various hypotheses have been proposed, which include (Figure 10.14):

- Upwelling of hot plumes generated by the return convective loop of downgoing oceanic lithosphere.
- Partial melting at great depths of over-thickened continental lithosphere following supercontinent assembly.
- Subduction of ocean spreading ridges resulting in shallow sub-lithospheric melting producing backarc basin type extension within the continental lithosphere.

All of these forces have the potential to generate continental rifts.

Bimodal volcanism

The widespread occurrence of basalt and rhyolite without significant andesite is referred to as **bimodal volcanism** (Section 8.4). Bimodal volcanism occurs at continental rifts and hotspots underlying continental lithosphere. Partial melting of the mantle generates basaltic magma. The rising basaltic magma partially melts continental crust, resulting in the dual occurrence of basalt and rhyolite. A classic example occurs in Yellowstone National Park in Wyoming (USA). Yellowstone's magmatic source is related to a mantle hotspot that has been active for at least 17 million years. The Yellowstone hotspot may have provided the source material for the immense Columbia River flood basalts in Idaho, Oregon and Washington as well as the

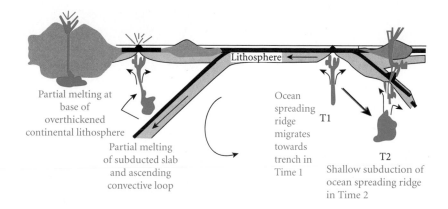

Partial melting at
base of
overthickened
continental lithosphere

Partial melting
of subducted slab
and ascending
convective loop

Lithosphere

Ocean
spreading
ridge
migrates
towards
trench in
Time 1

T1

T2

Shallow subduction of
ocean spreading ridge
in Time 2

Figure 10.14   Possible tectonic causes for continental rifts.

northern parts of California and Nevada. Most of the magma producing the Columbia River flood basalts erupted 15–17 million years ago. Since that time, as North America has migrated in a southwest direction, the position of the active hotspot has migrated ~800 km in a northeasterly direction to its present location at Yellowstone.

Christiansen (2001) recognized three immense rhyolitic lava deposits at Yellowstone's silicic large igneous province: the 2.1 Ma Huckleberry Ridge Tuff, the 1.3 Ma Mesa Falls Tuff and the 640,000-year-old Lava Creek Tuff. Together, these three tuff deposits constitute the Yellowstone Group. The Huckleberry Ridge eruption dispersed 2450 km^3 rhyolite deposits over an area of 15,500 km^2 and produced a caldera over 75 km long. The Mesa Falls eruption produced tuff deposits largely within the Huckleberry Ridge Caldera. While the Mesa Falls eruptive deposits were restricted to the pre-existing caldera, a new 16 km caldera developed along the northwest end of the Huckleberry Ridge Caldera. The youngest Lava Creek cycle of eruptive activity began around 1.2 Ma and continued for approximately 600,000 years. The Lava Creek eruption produced a large caldera and scattered rhyolitic deposits over an area of 7500 km^2.

Thus the Yellowstone Caldera is a composite caldera generated by three separate rhyolitic eruptive events. In the intervening time between each of these rhyolitic eruptions, basaltic lava also erupted (Figure 10.15). The

Figure 10.15   Dual columnar basalt flows are separated above and below by massive rhyolite ignimbrite deposits in the Yellowstone Caldera, USA. (Photo by Kevin Hefferan.)

basalt eruptions appear to be independent of the rhyolite eruptive cycles. The rhyolite and basalt eruptions represent two distinctly different magmatic sources. The rhyolitic magma is derived from the successive emplacement of granitic batholiths within the crust. The basaltic magma is generated by partial melting of the peridotite-rich upper mantle. The Yellowstone Caldera consists of two ring fracture zones within this composite caldera structure. Ring fractures are circular fracture sets generated by ground subsidence following the release of magma from a shallow pluton (Chapter 8).

Approximately 40 rhyolite eruptions have occurred in the past 640,000 years, since the last of the three cataclysmic Quaternary eruptions at Yellowstone. No lava has erupted in Yellowstone over the past 70,000 years. Two resurgent domes are currently being constructed within the Yellowstone Caldera and the ground surface is slowly being inflated, with uplift as much as 1 m since the 1920s. While the eruption of lava at Yellowstone is not anticipated in the next few thousand years, the area is presently experiencing uplift, perhaps the early warning signs of a new eruptive phase (Christiansen, 2001). Yellowstone has been the subject of a movie entitled *Supervolcano*, which is entirely appropriate: eruptions there were among the largest on Earth. The magma that erupted from Yellowstone 2.1 million years ago was approximately 6000 times greater than the volume released in the 1980 eruption of Mt St Helens. The smallest of Yellowstone's three Quaternary eruptive events released five times more debris than the massive 1815 Tambora (Indonesia) eruption. It is estimated that 25,000 km³ of magma are contained within the 7 km deep Yellowstone batholith. Should a portion of that magma erupt from the Yellowstone Caldera, North America would experience a devastating eruption unlike any other witnessed in human history.

Layered basic–ultrabasic intrusions

Layered basic–ultrabasic intrusions are anorogenic bodies injected into stable continental cratons at moderate depths. Layered intrusions include shallow tabular sills and dikes as well as funnel-shaped lopoliths. These intrusions commonly contain layers of rocks such as norite, gabbro, anorthosite, pyroxenite, dunite, troctolite, harzburgite and lherzolite. Minor silicic rocks such as granite can also occur. Common major minerals include olivine, orthopyroxene (enstatite, bronzite, hypersthene), clinopyroxene (augite, ferroaugite, pigeonite) and plagioclase.

Layered intrusions develop by differentiation of eclogite–peridotite parent magmas resulting in mineral segregation within a pluton. In addition to closed-system differentiation processes, open-system diversification processes (Chapter 8) such as multiple injec-

Figure 10.16   Close up of rhythmic layers within a channel structure in the Stillwater Complex. (Photo by Kevin Hefferan.)

tions, magma mixing or chemical diffusion can produce discrete layering in complex intrusive bodies. Layers generated by these processes occur on the scale of meters, centimeters or as microscopic cryptic lenses. Layers may occur as flat, planar structures or display features commonly associated with sedimentation such as cross-bedding, graded bedding, channeling (Figure 10.16) or slump structures. Cryptic (hidden) layering is revealed only by subtle changes in chemical composition.

As with Alaskan-type intrusions, layered basic–ultrabasc intrusions are highly valued for metal deposits, particularly platinum group elements (PGE) as well as chromium, nickel and cobalt. Metallic ore enrichment is likely due to a combination of factors that include original high concentrations of chromium, nickel, cobalt and PGE in magnesium-rich, refractory magmas as well as subsequent remobilization and concentration by halogen-rich (e.g., chlorine) fluids derived from the assimilation of crustal rock (Boudreau et al., 1997).

Three of the largest layered intrusions on Earth are the Stillwater Complex in Montana, the Bushveld Complex in South Africa and the Skaergaard Intrusion in Greenland. Other significant layered intrusions include the Muskox Intrusion of the Northwest Territories (Canada), the Keweenaw and Duluth Intrusion of Minnesota (USA) and the Great Dike of Zimbabwe. The 1.1 Ga **Duluth Complex**, formed during the Keweenaw rift event, is a major undeveloped PGE source. Plans are currently underway to begin mining PGE in the Duluth Complex within the next few years.

Stillwater Complex

The 2.7 Ga **Stillwater Complex** is a large, layered basic–ultrabasic igneous intrusion in the Beartooth Mountains of southwestern Montana. The Stillwater Complex is exposed along a northwesterly strike for a distance of 48 km, with observable thicknesses up to 6 km. The Stillwater Complex, which formed when basic magma intruded meta-sedimentary rocks, is the finest exposed layered intrusion in North America and contains economic deposits of platinum group metals as well as chromium, copper and nickel sulfides (McCallum et al., 1980, 1999; Premo et al., 1990).

The Stillwater Complex consists of three main units, which include a lowermost basal zone, an ultramafic zone and an upper banded zone. The basal zone consists of norite, harzburgite and bronzite-rich orthopyroxenite layers. The ultramafic zone consists of dunite, harzburgite, bronzite-rich orthopyroxenite and chromite-rich peridotite layers. The basal and ultramafic zones contain copper, chromium and nickel sulfide ore deposits. The upper banded zone consists largely of repetitive layers of alternating norite, gabbro, anorthosite and troctolite and is enriched in copper, nickel and PGE ore deposits (McCallum et al., 1980, 1999; Todd et al., 1982). Chlorine-rich magmatic fluids played a key role in leaching background metal deposits within the intrusion and concentrating these metals in discrete enriched layers called reefs within the banded zone (Boudreau et al., 1986, 1997; Meurer et al., 1999).

Bushveld Complex

South Africa's 2.06 Ga **Bushveld Complex**, a massive laccolith or domal structure, is the world's largest layered igneous intrusion. Extending over 400 km in length, up to 8 km thick and underlying an area of 60,000 km², this complex contains a layered sequence of basic and ultrabasic rocks, capped locally by granite.

The Bushveld Complex consists of four main zones (Daly, 1928; Vermaak, 1976; Cawthorn, 1999). These include, from top to bottom, the following:

1 Upper zone consisting of gabbro and norite.
2 Main zone containing gabbro and anorthosite.
3 Critical zone consisting of anorthosite, norite and pyroxenite.
4 Basal zone consisting of orthopyroxenite, harzburgite, dunite and peridotite. A chromite horizon occurs at the top of the basal series.

The Bushveld Complex hosts the largest reserves of vanadium, chromium and platinum group metals in the world. PGE are concentrated within what is referred to as the Merensky Reef within the critical zone. Anorogenic granitic rocks capping the complex contain tin, fluorine and molybdenum. The Bushveld Complex layering formed through differentiation processes accompanied by a series of magmatic injections, resulting in a massive laccolith or domal structure. As in the Stillwater Intrusion described above, chlorine-rich magmatic fluids are thought to have played a role in concentrating PGE in the Bushveld Complex (Boudreau et al., 1986).

Skaergaard Intrusion

Whereas most layered ultrabasic–basic intrusions are Precambrian in age, Greenland's 55 Ma **Skaergaard Intrusion** is the youngest of the great PGE-enriched intrusions. The Skaergaard lopolith intrusion crops out along Greenland's eastern shores and offers exceptionally good exposures of layering formed by differentiation and convective current structures. The Skaergaard Intrusion, with a volume of 500 km³, is heralded as the finest example on Earth of fractional crystallization, displaying layered sequences of euhedral to subhedral crystals as well as distinctive structures usually associated with sedimentary beds. These structures include cross-bedding, graded bedding and slump structures (Wager and Deer, 1939; Irvine, 1982; Irvine et al., 1998).

Zoned and layered ultrabasic–basic intrusive complexes provide rare but massive examples of magma diversification yielding segregated mineral zones and valuable metallic ore deposits.

(a)   (b)

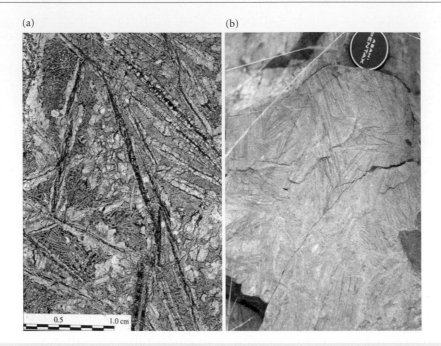

0.5   1.0 cm

Figure 10.17   Microphotograph (a) and field photograph (b) of spinifex texture komatiites. (Photos courtesy of Maarten de Wit.)

Other ultrabasic suites: intraplate volcanics and shallow intrusives

Komatiites

Komatiites are ultrabasic volcanic rocks found almost exclusively in Archean (>2.5 Ga) greenstone belts. Greenstone belts are metamorphosed assemblages of green-colored rocks that contain layers of ultrabasic and basic rocks overlain by silicic rocks and sediments (Chapter 18). Komatiites, named after the 3.5 Ga Komatii region of Barberton, South Africa, are high magnesium (>18% MgO), olivine-rich volcanic rocks, depleted in titanium and LREE. The high magnesium content and LREE depletion indicate a previously depleted mantle source (Sun and Nesbitt, 1978, in Walter, 1998). Komatiite flows, first recognized in the Barberton region of South Africa in 1969, commonly contain spinifex texture (Figure 10.17). **Spinifex texture** consists of needle-like, acicular olivine, pyroxene (augite and/or pigeonite) and chromite phenocrysts in a glassy groundmass (Viljoen and Viljoen, 1969; Arndt, 1994). Spinifex texture commonly occurs in the upper parts of komatiite flows or in the chilled margins of sills and dikes where rapid quenching produced skeletal, acicular crystals (Arndt and Nesbitt, 1982). In addition to spinifex texture, circular varioles, radiating spherulites and tree-like dendritic textures also occur. These textures are attributed to rapid undercooling (Chapter 8) or quenching of extremely hot lavas (Fowler et al., 2002).

Nearly all komatiites erupted during the Archean Eon when the early Earth was much hotter. Komatiites indicate elevated liquidus temperatures of 1575–1800°C (1 atmosphere pressure) in the Archean upper mantle (Green et al., 1975; Arndt, 1976; Wei et al., 1990; Herzberg, 1992, 1993, in de Wit, 1998). The virtual absence of Phanerozoic komatiites may be attributed to lower upper mantle temperatures which precludes the extensive mantle melting required to produce ultrabasic melts. The only known Phanerozoic (<544 Ma) komatiites occur on Gorgona Island, Colombia, where 88 Ma komatiites erupted as >1500°C ultrabasic lava flows. Gorgona Island, located 80 km west of Colombia in the Pacific Ocean, is composed largely of gabbro and peridotite (Echeverria, 1980; Aitken and Echeverria, 1984). Gorgona Island is also notable for the rare occurrence of ultrabasic pyroclastic tuffs which record explosive volcanism (Echeverria and Aitken, 1986).

Hypotheses for the origin of komatiites include:

- Melting in the hydrated mantle wedge above the subduction zones (Allegre, 1982; Grove et al., 1997; Parman et al., 2001).
- A deep mantle plume hotspot that led to large degrees of partial melting producing oceanic plateaus (Storey et al., 1991).
- Partial melting (10–30%) of a garnet peridotite at pressures of 8–10 GPa (Walter, 1998).

Komatiites, like layered gabbroic intrusions, are associated with valuable metallic ore deposits such as nickel, copper and platinum metals. For example, komatiite metallic ore deposits occur in the 2.7 Ga Yilgarn Craton of Western Australia, the 3.5 Ga South African Barberton region and the 2.7 Ga Canadian Shield. The nickel sulfide ore deposits are thought to have originated in ultrabasic lava tubes (Chapter 9) that concentrated high density metals in channel beds.

Kimberlites

Kimberlites are magnesium-rich, ultrabasic rocks that rapidly rise to Earth's surface via cylindrical diatremes (Chapter 8) from the crust and mantle. Diatremes vary greatly in surface area, ranging from a few square meters to square kilometers. Most diatremes taper downward, resembling an inverted cone in cross-section view. Kimberlite pipes occur with other plutonic structures such as dikes and sills. Kimberlites, which originate at temperatures of 1200–1400°C and depths exceeding 150 km, rise explosively through thick continental lithosphere.

Volatile, enriched, very low viscosity mantle melts rocket upward towards Earth's surface at velocities of ~15–72 km/h (Sparks et al., 2006). The magma is propelled upward by either the degassing of CO_2-enriched magma or by phreatomagmatic processes (Chapter 9). Phreatomagmatic processes require a water source to interact with the kimberlite magma. The high volatile content serves two primary purposes in that (1) it lowers the melting temperature preventing crystallization, and (2) it provides the propellant "jet fuel" to accelerate kimberlite magma to Earth's surface. Sparks et al. (2006) suggest

that kimberlite eruptions generate up to 10,000 m³ of pyroclastic debris over hours to months, producing Plinian ash plumes up to 35 km high (Chapter 9). Strangely, no ultrabasic lavas have been documented with kimberlite deposits. This is probably due to their low preservation potential and the extremely high volatile (up to 20%) content of kimberlite magma, which can produce 70% vesiculation in the erupting lava (Sparks et al., 2006).

Kimberlite eruptions form maar craters (Chapter 9) that largely fill with brecciated, pyroclastic debris (Dawson, 1980; Mitchell, 1986; Sparks et al., 2006). Due to the association of high temperature, pressure, volatile content and velocity, kimberlites commonly exhibit extensive hydrothermal alteration and are intensely fractured (Dawson, 1980). Altered olivine and phlogopite phenocrysts occur within a fine-grained groundmass of serpentine, calcite and olivine. Olivine constitutes the major mineral in the vast majority of kimberlites. However, in many samples olivine is completely replaced by serpentine, mica or clay minerals (Skinner, 1989). Kimberlites also contain the high pressure minerals pyrope garnet, jadeite pyroxene and diamond, which are stable at mantle depths >150 km.

Kimberlites were first discovered in the Kimberly region of South Africa where they are intimately associated with diamonds. Although best known from South Africa, kimberlites crop out in continental lithosphere throughout the world, commonly occurring with carbonatites and alkaline igneous rocks. Carbonatites – igneous rocks enriched in carbonate minerals such as calcite, dolomite or ankerite – are important CO_2 energy sources propelling kimberlites up from mantle depths. Kimberlites are also associated with reactivated shear zones and fracture zones (White et al., 1995; Vearncombe and Vearncombe, 2002). Kimberlites commonly intrude Early Proterozoic to Archean age cratons (2–4 Ga); intrusion ages vary and kimberlites as young as Tertiary age (~50 Ma) are known (Dawson, 1980).

Carbonatites, lamprophyre, lamproites and anorogenic granites

In addition to kimberlites, other rare and unusual rocks that occur in continental lithosphere include carbonatites, lamprophyres and lamproites. These SiO_2-undersaturated rocks typically occur in shallow (hypabyssal), volatile-rich dikes and may be associated with kimberlites.

Carbonatites are shallow intrusive to volcanic rocks that contain >20% CO_3 minerals such as natrolite, trona, sodic calcite, magnesite and ankerite as well as other minerals such as barite and fluorite. The origin of carbonatite was a contentious issue prior to the 1960 eruption of the Oldoinyo L'Engai Volcano in Tanzania. Oldoinyo L'Engai erupted unusually low viscosity pahoehoe carbonatite lava at temperatures of ~500°C. Carbonatites form in stocks, dikes and cylindrical structures primarily at continental rifts (Dawson, 1962).

Lamproites may be diamond-bearing such as the Argyll lamproite in Western Australia, which is a major producer. **Lamprophyres** are magnesium-rich, volatile-rich, porphyritic rocks containing mafic phenocrysts such as olivine, biotite, phlogopite, amphibole, clinopyroxene and melilite. Lamprophyres are associated with kimberlites and continental rift zones, but also occur as dikes intruding granodiorite plutons at convergent margin settings.

Lamproites are potassium-rich, peralkaline rocks containing minerals such as leucite, sanidine, phlogopite, richterite, diopside and olivine. Lamproites are enriched in barium (>5000 ppm), lanthanum (>200 ppm) and zirconium (>500 ppm). In contrast to lamprophyres and carbonatites, lamproites are relatively poor in CO_2 (<0.5 wt %). Lamproites occur in areas of thickened lithosphere that have experienced earlier plate convergence or rifting episodes.

Anorogenic (A-type) granites are silicic plutonic rocks that are not associated with convergent margin tectonism. A-type granite environments include stable cratons, continental rifts, ocean islands and inactive, postcollisional continental margins. Anorogenic granite, alkali granite and syenite were particularly common 1.1–1.4 Ga following the assembly of the mid-Proterozoic Columbia Supercontinent. These A-type, granitic intrusions are widespread in North America, extending from Mexico to the Lake Superior region. Significant volumes of anorogenic granites occur in Precambrian cratons throughout the world. These mid-Proterozoic granitoid rocks are remarkably similar in age, composition and appearance, displaying rapakivi texture. **Rapakivi** texture refers to sodium plagioclase overgrowths on pre-existing orthoclase crystals.

A number of models have been proposed for the origin of A-type granites. One model proposes the overthickening of continental lithosphere such that the upper mantle and base of the crust partially melt generating silicic magma that subsequently rises and cools at shallower depths to form anorogenic granite. Other "residual source models" propose that A-type granites, such as Pikes Peak Batholith in Colorado (USA), develop from the partial melting of residual silicic granulite rocks (Chapter 18) that had previously generated I-type granites (Barker et al., 1975; Collins et al., 1982). Alternative models suggest that A-type granites are derived by melting quartz diorite, tonalite or granodiorite parent rocks (Anderson, 1983).

In Chapters 7–10 we have presented a logical approach to the description, classification and origin of igneous rocks and landforms. We have also demonstrated the tectonic relations in an understandable framework. Hopefully we have been somewhat successful in helping you understand igneous processes. The scope of this text requires us to limit our discussion of important topics. For more detailed discussions beyond the scope of this textbook, the reader is referred to excellent petrology textbooks by Winter (2009), Raymond (2007), McBirney (2007), Best (2003), Blatt and Tracy (1996), Philpotts (1990), Ragland (1989) and Hyndman (1985) among others.

In succeeding chapters, we will investigate how igneous rocks are altered in two ways: (1) by weathering and erosion at Earth's surface, and (2) through the effects of high temperatures, pressures and hot fluids via metamorphic reactions. In all of these reactions, water plays a critical role in altering and mobilizing elements within Earth's crust.

Chapter 11

The sedimentary cycle: erosion, transportation, deposition and sedimentary structures

11.1 SEDIMENTS AND SEDIMENTARY ROCKS

Sedimentary materials including soils, sediments and sedimentary rocks, cover more than 80% of Earth's surface (Tucker, 2001). They contain most of the fluid resources such as groundwater, natural gas and petroleum on which modern societies depend. They harbor important deposits of coal, metallic ores and aggregate materials used in the production of products such as plaster, cement, concrete and asphalt. They provide the substrate on which structures such as houses, commercial buildings, highways and industrial complexes are built and the soils in which agricultural and forest products are produced.

Because sedimentary rocks record essential information about the processes that produced them, they are also vast repository of the history of Earth's surface over time. This repository includes the fossils of organisms that record the history of life on Earth from its emergence more than 3.5 billion years ago to the present time. It also includes rocks that preserve evidence of the specific environment and tectonic setting in which they accumulated, which provide most of the evidence on which maps depicting Earth's ancient surface (paleogeographic maps and paleotectonic maps) are based. These rocks also contain components that record major changes in Earth's climate and atmospheric composition through time. Much time, effort and creativity has gone into interpreting this repository, often with truly extraordinary results.

Sediments are rock materials, generally but not always unconsolidated, that form on or close to Earth's surface. This definition excludes (1) solid volcanic materials that accumulate directly from the solidification of lava flows, (2) solid plutonic igneous and metamorphic rocks that generally form well below Earth's surface, and (3) most high temperature hydrothermal deposits. Sediments are classified genetically, according to the processes involved in their formation. Most sediment can be genetically classified into three major components: (1) detrital sedi-

Earth Materials, 1st edition. By K. Hefferan and J. O'Brien. Published 2010 by Blackwell Publishing Ltd.

ment, (2) organic sediment, and (3) chemical sediment. The latter two are sometimes grouped together as biochemical sediment.

Detrital sediments (Chapter 13) are the solid products of weathering, the breakdown of rocks at or near Earth's surface. Detrital sediments include most of the gravel, sand and mud particles that accumulate on Earth's surface. **Organic sediments** (Chapter 14) are the solid products of organic synthesis or precipitation and include hard materials such as shells, bones and teeth and soft materials such as cellular materials composed of organic molecules. **Chemical sediments** (Chapter 14) are the solid products of inorganic precipitation such as mineral crystals precipitated from solution. These three types of sediment may occur in any proportions in a particular sediment or sedimentary rock.

Many other terms are used to describe major groups of sediment. **Clastic** is a general term for solid sedimentary particles regardless of origin. The term **epiclastic** is used for clastic sediment that is transported as solid particles across Earth's surface. Related terms include **bioclastic**, which is used for clastic particles of organic origin such as shell fragments transported by waves or currents, **siliciclastic**, which refers to clastic particles composed of silicate minerals, and **volcaniclastic**, which is used for clastic particles produced initially by volcanic processes. Yet another related term is **terrigenous**, which refers to clastic sediment derived from a land mass.

As long ago as the 16th century, observers noticed that some solid rock bodies possessed many of the same features that are present in largely unconsolidated sediments. By the 18th century, geologists had begun to speculate that these solid **sedimentary rocks** had been produced by the solidification of initially unconsolidated sediments. Eventually it was recognized that for each type of unconsolidated sediment there is an equivalent sedimentary rock that is produced by its solidification. The general name for the set of processes by which unconsolidated sediment is converted into more consolidated sedimentary rock is **lithification**, literally to "turn into rock" (from the Greek work *lithos* = rock).

11.2 THE SEDIMENTARY CYCLE

The **sedimentary cycle** (Figure 11.1) is a simple model of the processes responsible for the production of sediments and sedimentary rocks. Even simple versions provide an excellent overview of the interrelationships between sediments, sedimentary rocks and the processes that produce them. Simple models of the sedimentary cycle commonly illustrate five or more significant sets of processes that affect the production, dispersal and accumulation of sediments and their subsequent conversion into solid sedimentary rocks.

The first is the production of sedimentary materials in a **source area**, largely by weathering, at or near Earth's surface, perhaps accompanied by uplift. **Weathering** (Chapter 12) involves the physical, chemical and/or organic breakdown of rock material into smaller mineral and rock fragments and into dis-

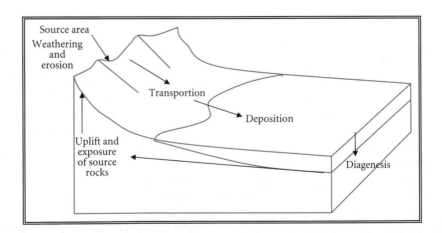

Figure 11.1 A simple model of the sedimentary cycle.

solved components. Two major types of weathering processes are recognized: **disintegration** processes, which break rock materials down into smaller pieces of the original material, without changing their composition, and **decomposition** processes, which change the composition of the original material, producing new materials.

Weathering, by producing smaller particles, and uplift, by increasing slope angles, enhance a second set of processes called **erosion** by which sedimentary materials are removed from a place on Earth's surface. Erosion is an instantaneous process.

Erosion initiates a period of **transportation** during which sedimentary material is moved or transported in solid or dissolved form across Earth's surface. Most transportation is by water, glaciers, mass flows or wind. Since the first three of these generally flow downhill under the influence of gravity, sedimentary processes tend to lower relatively high areas by sediment erosion and then disperse the sediment downhill to sites of accumulation in relatively low areas that tend to be filled in by sediment deposition.

Transportation ceases when sediments accumulate on Earth's surface by processes collectively referred to as **deposition**. It is conceptually convenient to think of erosion and deposition as opposites. Erosion involves the removal of rock material from a place on Earth's surface, whereas deposition involves the accumulation of sediment at a place on Earth's surface. Acting alone, erosion and deposition tend to decrease the relief of Earth's surface. So why, after 4.6 billion years, is Earth's surface not flat? Tectonic processes, such as folding, faulting and volcanism, and isostasy generate significant relief and prevent the development of a "flat Earth". Erosion, transportation and deposition normally occur multiple times during the history of a sedimentary particle as it is dispersed from its source area to its current resting place. A grain of sand in a back-beach dune field may well have originated in a source area high in the mountains, then been eroded, transported and deposited multiple times by glaciers, mass flows and water during its dispersal into and down a drainage system to the coastline. What might its journey be in the future?

Diagenesis (Chapters 13 and 14) encompasses a suite of low temperature processes that affect sediments after their accumulation, typically after burial. It includes the processes of lithification such as compaction and cementation. **Compaction** results from the expulsion of intergranular fluids caused by increases in confining pressure during progressively deeper burial. Compaction aids lithification by increasing the attractive forces between grains as they move into closer contact. **Cementation** occurs when subsurface fluids precipitate minerals in the spaces between grains that bind or cement grains to one another.

A single sedimentary cycle ideally ends with burial and diagenesis. If, after a significant period of burial, a sedimentary rock is eventually re-exposed at Earth's surface where a new cycle of weathering, erosion, transportation, deposition and diagenesis can occur, these constitute a second sedimentary cycle and the products are referred to as **polycyclic** sediments and sedimentary rocks.

11.3 STRATIFICATION AND SEDIMENTARY ENVIRONMENTS

One of the most striking features of sediments and sedimentary rocks is their layering, which is called **stratification**. All sediments are deposited in layers called **strata**. Each stratum records a period of net sediment accumulation on Earth's surface and records information about the conditions and processes that produced it. Layer thickness tends to increase with the rate and duration of deposition and is decreased by subsequent erosion. Thick strata (>1 cm thick) are called **beds**, whereas thin strata (<1 cm thick) are called **laminations** or **laminae** (Figure 11.2a). Beds are commonly preserved as compound layers, defined for the purpose of describing sedimentary rocks. In general, the boundaries between strata reflect changes in the type of material deposited through time or record intervals of erosion between periods of net deposition.

Most strata form when sediments come to rest on relatively flat surfaces. For this reason, most strata are horizontal or very nearly so at the time they form. This notion is summarized in the **principle of original horizontality**, which states that strata are nearly horizontal at the time they form. The major value of this principle is the recognition that very steeply inclined and/or folded strata must have been

(a)

(b)

Figure 11.2 (a) Thin laminations (above) and thicker beds (below) in the Jurassic Berlin Formation, Connecticut. (b) Cross-stratified beds (center) with initially inclined laminations, separated by a near horizontal erosion surfaces in the Jurassic Navajo Formation, Zion National Park, Utah. (Photos by John O'Brien.) (For color version, see Plate 11.2, between pp. 248 and 249.)

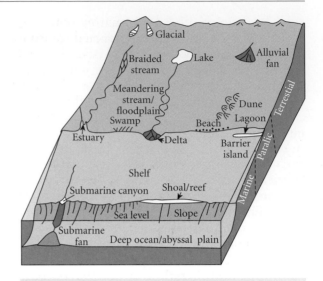

Figure 11.3 Major terrestrial, paralic and marine depositional environments.

subjected to forces that rotated them from the horizontal position after they accumulated. There is one very important exception to the principle of original horizontality. Cross strata (Figure 11.2b) form when sediments accumulate on steeper slopes of up to 35°. They are commonly truncated by originally subhorizontal erosional surfaces and are interlayered with nearly horizontal strata.

Because each stratum records the type of sediment that accumulated on Earth's surface and because the conditions under which each type of sediment accumulated can potentially be inferred, sedimentary strata represent a vast storehouse of information concerning the changes that have occurred on Earth's surface over time. Because sedimentary layers accumulate on Earth's surface, one atop the other over time, the relative ages of sedimentary strata can be inferred from the **principle of superposition**, first stated by Niels Stensen (Nicolas Steno) in the 17th century. In simple terms, this principle states that in any sequence of strata that have not been overturned, strata become younger from the bottom of the layered sequence toward the top. By inferring the surface conditions that produced each layer and knowing the sequence in which those conditions occurred, we can infer a long history of how surface conditions changed over time. With radiometric and other dating techniques, one can determine the absolute ages of layers and place their history into an absolute time context. With correlation techniques to relate the histories preserved in strata at many different locations, scientists have been able to develop a global history of Earth's surface, albeit one limited by the incomplete record preserved in sediments and our ignorance of how to extract that record completely.

Figure 11.3 illustrates several of the major surface environments in which sediments accumulate. These **depositional environments** may be broadly subdivided into terrestrial, transitional (paralic) and marine environments. Major **terrestrial** depositional environments include glacial, alluvial fan, braided stream, meandering stream and floodplain, lacustrine (lake), aeolian (dune) and paludal (swamp) environments. **Transitional (paralic)** environments typically occur in the terrestrial–marine transition and include coastal–deltaic environments such as deltas, estuaries, tidal flats, beaches and back-beach dunes and barrier island–lagoon systems. Important **marine** depositional environments include epicontinental epeiric seas, reefs, restricted seas, continental shelves, continental slopes, submarine fans and continental rises, forearc basins, trenches and deep ocean floors. The temporal and spatial distributions of such environments strongly depend on the existing tectonic setting and climatic conditions. Where applicable, the sediments deposited in response to processes that operate in these environments are discussed in the following sections of this chapter, in Chapter 13 on detrital sedimentary rocks and in Chapter 14 on organic and chemical sedimentary rocks. However, a detailed discussion of sedimentary environments is beyond the scope of this text. Interested readers should consult Boggs (2005), Prothero and Schwab (2004), Tucker (2001), Reading (1996) and Selley (1988) for further information.

11.4 SEDIMENT DISPERSAL, DEPOSITION AND PRIMARY SEDIMENTARY STRUCTURES

As noted in the previous section, sediments are eroded, transported and deposited by a variety of agents that flow across Earth's surface. Chief among these are water, wind, glaciers and mass flows. These flows are either fluid flows or plastic flows, and they flow in response to the stresses that act on them.

Fluid flows, such as water and wind, have no shear strength. In water on Earth's surface, the shear stress that initiates flow is typically the tangential force of gravity (g_t) acting parallel to the flow surface, which causes the fluid to flow downhill, as in rivers. But water can also be set in motion by stress provided by wind blowing across the surface, as in surface currents. In addition, both wind and water flow in response to differential heating, which generates parcels of fluid with different densities, which in turn move in response to gravity, as in thermohaline and atmospheric circulation. Whatever the cause, such fluid flows move with sufficient velocity to be important agents in the dispersal of sediments across Earth's surface.

In **plastic flows**, which possess shear strength, the major stress that initiates flow is the tangential force of gravity (g_t), which increases with increasing slope angle. For example, once critical threshold shear stresses are achieved, glaciers flow from areas of maximum surface elevation toward areas of minimum surface elevation, whether that is downhill, as in mountain glaciers, or from areas of maximum thickness toward areas of minimum thickness, as in many continental glaciers. Likewise, mass flows such as debris flows and turbidity currents are initiated once critical threshold values of g_t are reached, typically on relatively steep slopes.

Flows may also be distinguished as laminar or turbulent flows (Figure 11.4). In **laminar flow**, adjacent parcels of the flow move roughly parallel to one another in a well-organized pattern, with negligible mixing between them. In **turbulent flow**, adjacent parcels of the flow move in chaotic patterns and random mixing between parcels is common.

11.4.1 Water and water-deposited sediments

Water is the major agent by which sediments are eroded, transported and deposited on Earth's surface. Water is a Newtonian fluid. Like all fluids, Newtonian fluids lack shear strength, but they are also characterized by a constant resistance to shear stress, which is called **dynamic viscosity (μ)** and is given by:

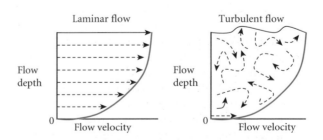

Figure 11.4 Laminar and turbulent flow profiles; flow lines are dashed.

$$\mu = \tau_o/(dU/dY)$$

where dU/dY is the change in velocity with depth in the flow and τ_o is the shear stress (e.g., g_t) acting on the flow's surface. Why is viscosity important? It is easier to understand the importance of dynamic viscosity if the equation is rearranged to:

$$dU/dY = \tau_o/\mu$$

It can then be seen that flow velocity is proportional to shear stress, such as increased tangential force of gravity with increased slope, and is inversely proportional to viscosity. The resistance to shear stress may be thought of as the internal resistance to flow. The viscosity of water increases with the degree of molecular linkage between water molecules, which is inversely proportional to temperature. Colder water possesses more molecular bonds between molecules, thus a higher viscosity or resistance to flow. In addition to determining flow velocity at a particular shear stress, viscosity, along with other variables, also plays an important role in determining whether flows are laminar or turbulent. These variables combine to produce several expressions for **Reynolds' number (Re)**, one of which is:

$$Re = UL\rho/\mu$$

where U is a velocity factor such a mean flow velocity, L is a length factor such as depth or particle diameter and ρ is a density factor such as fluid density.

Reynolds' number is essentially a ratio between the inertial forces and the viscous forces in a fluid medium. The Reynolds' number can be used to predict whether flow will be laminar or turbulent. Viscous forces tend to damp out turbulence and so favor laminar flow. Inertial forces tend to favor the development of fluid turbulence. For any situation, there is a critical Reynolds' number, below which flow is laminar and above which flow is turbulent. Figure 11.4 illustrates cross-sections of subcritical laminar and supercritical turbulent flow such as might occur in a stream or tidal channel. On smooth beds, a laminar flow sublayer may occur near the base of a turbulent flow; on rough beds

Figure 11.5 Transition from laminar flow (background) to turbulent flow (foreground) in a branch of the Delaware River, Pennsylvania. (Photo by John O'Brien.) (For color version, see Plate 11.5, between pp. 248 and 249.)

with a larger length factor, flow will be turbulent.

Figure 11.5 illustrates the transition from subcritical laminar to supercritical turbulent flow in a natural channel, probably caused by an increase in flow velocity, depth and/or bed roughness.

Entrainment, transportation and deposition

Both laminar and turbulent water have the ability to erode, transport and deposit sediment. The process by which epiclastic sediment transportation is initiated by erosion is called **entrainment**. Particles are said to be entrained by the flow when their movement is initiated. As might be expected, by analogy with using a garden hose to sweep particles of different sizes from one's driveway, there is a relationship between the flow velocity of the water and the size of the particles that can be entrained by the flow. But this relationship is not as straightforward as one might initially expect. Hjulstrom (1939) developed a diagram (Figure 11.6) that summarizes the critical velocity required for entrainment, transportation and deposition of epiclastic particles of various diameters. It should be noted that the diagram applies specifically to a 20°C, 1 m deep aqueous flow and that the details change

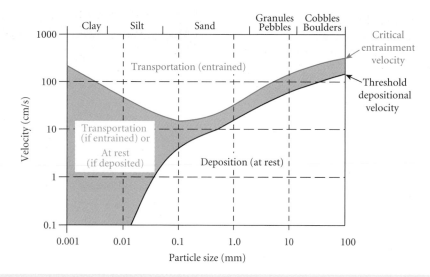

Figure 11.6   Hjulstrom's diagram showing velocity conditions for entrainment (erosion), transportation and deposition of sediment particles of different sizes.

when these conditions are changed. But, as we will see, generalizations from **Hjulstrom's diagram** offer powerful insights into sediment entrainment, transportation and deposition.

Hjulstrom's diagram contains two curves that divide the diagram into three fields. The upper curve is the erosion or entrainment curve that plots the **critical entrainment velocity** required to initiate particle movement against particle size. Perhaps surprisingly, the critical entrainment velocity curve (upper curved line in Figure 11.6) is smallest (~20 cm/s) for particles in the very fine sand range (~0.1 mm diameter). The critical entrainment velocity increases progressively, as expected, for larger particles up to >100 cm/s for pebble-size particles. This proportional relationship between critical entrainment velocity and particle size reflects the fact that progressively higher velocities and bed shear stresses are required to overcome the rest mass and initiate the movement of progressively more massive particles. However, for cohesive silt and clay particles finer than 0.0625 mm, the critical entrainment velocity increases with decreasing particle size! What is responsible for this seemingly counterintuitive relationship? The higher the amount of surface area/volume, the stronger the attractive forces, and the more cohesive the sediment will be. This **cohesiveness** is accelerated in the clay range because clays possess very

large surface area:volume ratios and have large amounts of unsatisfied electric charge on their surfaces. The inverse relationship between particle and critical entrainment velocity in silts and clays reflects the increasing cohesiveness of such particles as average particle size decreases. Because they tend to stick together, they behave as larger particles with increasingly large rest masses that are increasingly difficult to entrain. The top field on the diagram includes flow conditions under which all sizes shown on the diagram will have been eroded and entrained and so are being transported. Under such conditions, these particles will continue to be transported unless the velocity decreases substantially.

The second curve on Hjulstrom's diagram represents the **threshold depositional velocity**. It illustrates the conditions under which various particle sizes cease to be transported and are deposited. The threshold velocity of deposition (lower curved line in Figure 11.6) is lower than that required for entrainment for all particle sizes. This is because once entrained, a particle has inertial momentum due to its mass and velocity. The lower velocities for deposition of a particular size reflect the need for the particle to lose momentum before it can stop moving and be deposited; the higher velocities required for entrainment of any particle size reflect the need for the particle's inertial tendencies to be overcome before it starts moving. Note that the velocity at which a particle stops moving, its threshold velocity of deposition, is proportional to particle size for all particle sizes (Figure 11.6). The bottom right field on the diagram includes the conditions under which sizes shown on the diagram will be at rest. Under such conditions, these particles will continue to remain at rest unless the velocity increases substantially to a place above the entrainment curve.

The central field (shaded region in Figure 11.6) lies between the critical entrainment velocity and the threshold velocity for deposition. Why is this field labeled **transportation or at rest**? Where flow velocity is increasing over a bed where all particles are at rest, the critical entrainment velocity has not yet been reached; no entrainment occurs under the particle size–velocity relationships in this field. But where flow velocity is decreasing into this field from values above the critical entrainment velocity curve, particles with size–velocity relationships in this field will continue to move because they have not yet reached the lower threshold velocity of deposition. Once entrained, coarse sand and gravel particles are transported at velocities slightly lower than the velocities required for entrainment, yet higher than those required for **deposition**. However, they will be deposited if the flow velocity decreases by a rather small amount to velocities below the critical depositional velocity. For example, a 10 mm pebble will be entrained at 300 cm/s and will continue to be transported as long as the velocity does not drop below 200 cm/s, a decrease in flow velocity of 33%. However, a clay particle entrained at 100 cm/s will continue to be transported unless the velocity approaches zero. This explains the much larger field for the velocity conditions under which transportation occurs in fine sediments and why very calm waters are required for their deposition.

Once entrained, the **sediment load** of different particles sizes tends to be carried in different parts of the flow (Figure 11.7). The **bed load** is carried in continual or intermittent contact with the bed over which the water is flowing. It is subdivided into a traction load, which is in continual contact with the bottom, and a saltation load, which is in intermittent contact with the bottom. The **traction load** (Figure 11.7) contains the coarsest particles,

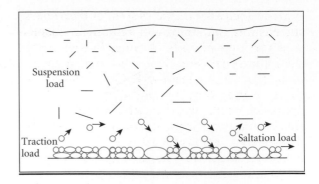

Figure 11.7 Sediment loads in an idealized aqueous medium.

those with sufficient mass that they are not lifted from the bottom by turbulence and lift forces. Particles in the traction load move down-current by (1) **rolling** along the bed, (2) **sliding** along the bed, and (3) **creep**, in which particle collisions set particles in motion. Where the traction load is more than one particle deep, the term **traction carpet** can be used. The **saltation load** contains smaller, less massive particles that can be lifted from the bottom by turbulence, lift forces and particle impacts, but which rapidly settle back to the bed before being lifted again. Particles in the saltation load inscribe roughly parabolic arcs as they "skip" or saltate across the bed in a down-current direction (Figure 11.7). The **suspension load** contains the smallest, least massive particles – those particles that once lifted from the bed have such small settling velocities that they remain suspended above the bed for long periods of time. A good example of suspended particles is the dust particles visible in a shaft of light shining through a window.

Particles are transferred between loads. As flow velocity increases, particles in the traction load may begin to saltate and particles in the saltation load may become suspended for longer periods of time. Conversely, as flow velocity decreases, particles in the suspension load may begin to saltate and particles in the saltation load may return to the traction carpet. Particles larger than coarse silt can be suspended at high velocities but generally fall back into the bed load as velocities decrease. As a result, particles larger than coarse silt are generally deposited from the traction load of aqueous flows. The very low critical depositional velocity for particles smaller than coarse silt ensures that they will remain in suspension for long periods of time. Particles finer than coarse silt are carried in and deposited from the suspension load. Only under very calm conditions, when flow velocities approach zero, are they able to settle from suspension and be deposited. Water also transports a **solution load** that consists of the dissolved solids produced by dissolution of rocks and minerals with which the water has been in contact. In the oceans, the solution load typically constitutes 3.5% of the total mass of seawater. In freshwater bodies it is considerably less, averaging about 0.015%.

Aqueous flow patterns can be very complex. However, two major types of aqueous flow are recognized in sedimentary environments: (1) unidirectional flows, and (2) oscillatory flows. If we understand how each interacts with unconsolidated sediments over which they flow, the sediments produced by each can be recognized in the sedimentary record.

Unidirectional flow

Unidirectional flows are characterized by flow in one direction over a period of time. Most stream flow, sheetwash, surface and deep ocean currents, longshore currents, flood tidal currents and ebb tidal currents can be modeled as unidirectional flows. For this reason, they have been carefully studied. Engineers involved in flood control, channel diversion, dam building, irrigation, pipelines and coastal erosion studies have a real need to understand the nature of unidirectional flow. Laboratory investigations and field studies have demonstrated conclusively that as unidirectional flows move over unconsolidated beds, those beds change in ways that are related to the flow characteristics. This work has been summarized in the **flow regime concept**, which offers insight into the origin of the primary sedimentary structures produced by unidirectional flows. A simplified version of the flow regime concept for initially flat beds composed of unconsolidated medium sand is presented in Figure 11.8. Flow regimes can be varied by changing the particle size of the bed, flow velocity, flow viscosity (e.g., temperature), flow depth and/or flow density. Figure 11.8 uses changes in flow regime to illustrate these concepts. Two flow regimes are recog-

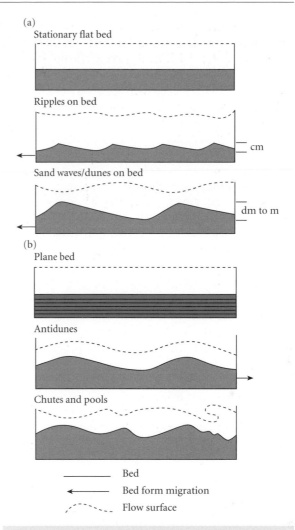

Figure 11.8   A simplified version of the flow regime concept: (a) lower flow regime bed forms; (b) upper flow regime bed forms. Flow is from right to left; flow regime increases downward.

nized, the lower flow regime, where the bed is out of phase with the flow surface (Figure 11.8a), and the upper flow regime, where the bed is generally in phase with the flow surface (Figure 11.8b).

At very low flow (e.g., very low U) in the **lower flow regime**, no movement occurs on the bed, because the critical entrainment velocity for medium sand has not been achieved. Once the critical entrainment velocity is achieved, the bed is quickly deformed into small, symmetrical **current ripples** that possess relatively gentle up-current or stoss-side slopes and steeper down-current or lee-

(a)

(b)

Figure 11.9 Current ripples. (a) Asymmetrical current ripples on a modern beach flat, where the current flowed left to right. (Photo by John O'Brien.) (b) Asymmetrical current ripples in Cretaceous sandstone, where the current flowed from upper left to lower right. (Photo courtesy of Duncan Heron.) (For color version, see Plate 11.9, between pp. 248 and 249.)

Figure 11.10 Slightly wavy-crested sand waves or subaqueous dunes, Rio Hondo, southern California; the current was from right to left. (Photo by John O'Brien.) (For color version, see Plate 11.10, between pp. 248 and 249.)

side slopes (Figure 11.9). Straight-crested ripples typically form first, followed by various curved-crested ripples at slightly higher flow regimes. Current ripples (1) possess relatively continuous, non-branching crests, (2) possess small ripple indices (RI = wavelength : height ratio), and (3) tend to migrate in a downflow direction over time. The asymmetry of current ripples permits them to be used to infer the direction of current flow, from the gentle stoss side toward the steep lee side, which is inclined in the flow direction. Current ripples record both lower flow regime conditions and flow directions.

As flow regime or velocity continues to increase within the lower flow regime, larger bed forms called subaqueous **sand waves** or **dunes** begin to develop on the bed. Initially these bed forms possess relatively straight crests transverse to flow, but at higher flow regimes their crests become more curved and their forms more complex. Sand waves are straight-crested two-dimensional bed forms where, ideally, every two-dimensional cross-section parallel to flow is the same; dunes are curved-crested three-dimensional bed forms where cross-sections parallel to flow vary. Smaller current ripples commonly develop on the stoss side of dunes and lee sides are inclined in the flow direction (Figure 11.10).

At still higher flow regimes that mark the transition to **upper flow regime** conditions, dunes begin to be sheared out into a flat or plane bed that is parallel to the nearly flat flow surface (see Figure 11.8b). This flow regime is called **plane bed** or the plane bed transition. Bed load sand moves along the plane bed at relatively high velocities. With further increases in flow regime, unconsolidated sand beds are deformed into bed forms called **antidunes** that are in phase with standing waves on the flow surface (Figure 11.11). These bed forms are called antidunes because they migrate with the standing wave in an up-flow direction by lee-side erosion and stoss-side deposition. With continued flow,

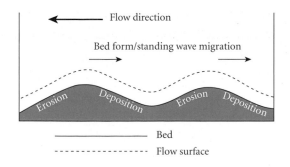

Figure 11.11 Antidunes in phase with standing waves on a flow surface. Antidunes migrate in the opposite direction to the flow direction by lee-side erosion and stoss-side deposition.

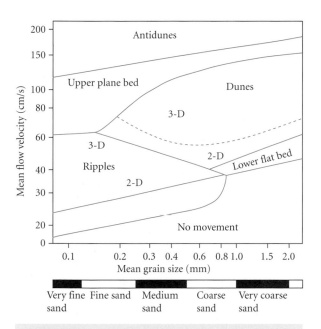

Figure 11.12 Flow regimes with respect to mean flow velocity and grain size. (After Harms et al., 1982.)

the standing wave collapses and antidunes are smeared out into an uneven surface characterized by shallow **chutes** and deeper **pools** (see Figure 11.8b).

As noted above, the sequence of bed forms with increasing flow regime depends on the grain size of the bed over which the flow moves. Figure 11.12 illustrates the bed-form sequence for other grain sizes in the traction

load. For coarse silt through fine sand, sand waves and dunes do not form. Instead the sequence of bed forms is (1) lower flow regime current ripples, (2) upper flow regime plane bed transition, and (3) antidunes. Current ripples do not form in beds composed of coarse sand or gravel. Instead the sequence of bed forms once traction movement is initiated is (1) lower flow regime flat bed, then (2) sand waves or dunes, followed by (3) upper flow regime plane bed, and (4) antidunes.

The importance of the bed forms associated with different flow regimes becomes more apparent when they are related to sediment deposition from traction. Net deposition requires that net bed aggradation occurs as sediment is transferred from the saltation and/or suspension loads to the traction load over time. Once this important concept is understood, it is much easier to understand the relationship between bed-form migration and the sedimentary structures that form during net deposition. These structures have the potential for long-term preservation and frequently record the flow regime conditions under which they formed.

Net aggradation of sediment on a flat or plane bed produces **horizontal stratification**. This occurs on upper flow regime plane beds for all sizes of sand and gravel, with minimum depositional velocities increasing with particle size as expected. It also occurs on lower flow regime flat beds for coarse sand and larger sediment (Figure 11.12).

Depending on the rate of bed-form migration and bed aggradation, sediment accumulation on a lower flow regime **current rippled bed** can generate (1) **current ripples**, (2) **horizontal laminations**, (3) **small-scale cross-lamination**, or (4) **climbing ripple laminations**. One key to understanding how such a variety of structures can form from a current rippled bed is to understand how current ripples migrate. During flow, entire trains of current ripples migrate down-current by erosion of sediment from their up-current or stoss sides and deposition on their down-current or lee sides (Figure 11.13). Very coarse silt to medium sand is entrained on the stoss side, migrates to the ripple crest and then slides down the steep lee side to form an inclined sand layer. Repetition of this process over time causes the entire ripple form to migrate down-current with the formation of small-

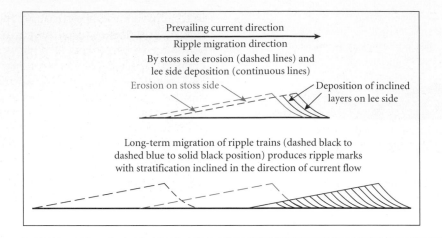

Figure 11.13   Ripple migration by stoss-side erosion and lee-side deposition forming internal cross-laminations.

scale internal cross-laminations that dip in a down-current direction (Figure 11.13).

As ripple trains migrate, entire ripple forms move along the bed including crests, but also troughs. Careful examination of Figure 11.13 will persuade one that the area of maximum sediment accumulation beneath the initial crest (dashed outline to left) has become an area of no sediment accumulation under a trough (dashed outline to right). What has happened is that erosion has progressively removed sand from the stoss side, permitting the trough to migrate down-current. The trough may be visualized as an erosion surface that removes all the sand deposited by the ripple that preceded it down-current. If no sediment is added to the bed during ripple migration, no net deposition occurs because troughs erode the current ripples that migrate across it. Now let us add bed aggradation or "fallout from above" to the mix. If sand is being added to the bed over time, the elevation of all parts of the ripple – including the ripple trough – will increase. As troughs migrate down-current, their elevation will increase slightly. As a result, troughs climb slightly and do not erode all sediment deposited during the migration of the previous ripple. At very slow climb rates that result from very slow bed aggradation relative to ripple migration rates, layers a few grains thick will be preserved with the passage of each ripple trough. The passage of multiple ripple troughs will generate a sequence of nearly horizontal laminations, each bounded by erosion surfaces, which cannot easily be distinguished from horizontal laminations generated on flat or plane beds. If, however, the aggradation and climb rates are slightly higher, but less than the slope of the stoss side, layers thick enough to preserve recognizable cross-laminations will be preserved (Figure 11.14a). The passage of multiple ripple trains will be preserved as multiple centimeter-thick sets of ripple-scale cross-laminations that dip in the direction of current flow, even though current ripple forms are not preserved. Each set is bounded by erosion surfaces with such gentle inclinations that they are nearly horizontal.

Under conditions of rapid fallout on the aggrading bed relative to ripple migration, troughs will climb at angles equal to or exceeding the stoss-side slope. In the former case, the entire ripple form is preserved as sequences of lee-side cross-laminations separated by erosion surfaces produced as troughs migrated up the stoss side at the stoss-side angle (Figures 11.14b and 11.15). Such features are called **type I climbing ripple laminations**. With extreme fallout from above, even the stoss side experiences net aggradation so that troughs climb at angles higher than the stoss-side slope angle. This produces **type II climbing ripple laminations** that preserve both stoss-side and lee-side laminations and entire ripple forms that climbed in a down-current direction (Figures 11.14c and 11.15b).

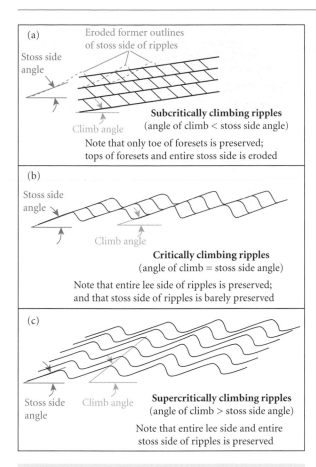

Figure 11.14 Progressive increases in trough climb rates due to bed aggradation by fallout from above. These generate: (a) cosets of small-scale ripple laminations; (b) type I climbing ripple laminations; (c) type II climbing ripple laminations. (Courtesy of Marc Hendrix.)

Figure 11.15 Climbing ripple laminations produced by down-current (left to right) migration of ripple trains under rapid fallout and bed aggradation in the Oligocene Sacate Formation, California. Type I with no stoss-side lamination preserved can be seen to the right of the blending stub and type II with stoss-side lamination preserved is above the load-casted sand layer. (Photo by John O'Brien.)

To summarize, as bed aggradation rates increase from zero to extreme fallout from above during the migration of ripple trains, the sequence of preserved structures evolves from (1) isolated ripple marks, through (2) horizontal laminations, (3) small-scale cross-laminations, (4) type I climbing ripple laminations, to (5) type II climbing ripple laminations. For very useful sets of visual aids, readers should refer to the many video images that relate bed-form migration and sedimentary structures (Rubin and Carter, 2006; http://walrus.wr.usgs.gov/seds/).

Because lower flow regime sand waves and dunes are larger features than ripples and possess longer lee-side faces, larger scale sets of cross-strata can form as they migrate down-current. Very small rates of bed aggradation generate very small trough climb rates, which can theoretically preserve thicknesses of a few grains from the preceding sand wave or dune as horizontal stratification. Slightly larger rates of aggradation and climb generate centimeter-thick sets of small-scale cross-laminations similar to those produced by current ripple migration at lower flow regimes. Their origin from dune or sand wave migration can be inferred if the sediments in such small-scale cross-laminations are coarse sand as ripples do not form in such coarse sediments. More commonly, bed aggradation rates permit widely spaced troughs to climb at angles that preserve decimeter-thick portions of longer wavelength sand waves and dunes as **medium-scale cross-stratification**. The migration of unusually large sand waves and dunes combined with rapid bed aggradation can produce meter-thick **large-scale cross-stratification**.

Laboratory experiments, supported by video images (see Rubin and Carter, 2006; http://walrus.wr.usgs.gov/seds/), have shown conclusively that the three-dimensional geometry of cross-stratification generated by ripple, sand wave and dune migration depends on the three-dimensional geometry of the migrating and evolving bed forms. Steady-state

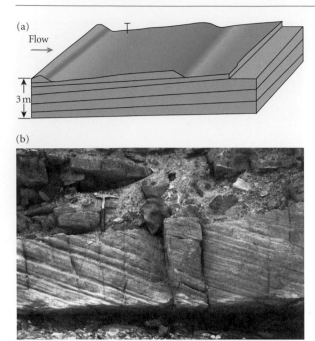

Figure 11.16 Tabular sets of planar cross-strata. (a) Formation by aggrading, straight-crested ripples or sand waves with stoss erosion at small climb angles. (Courtesy of the US Geological Survey.) (b) This single tabular set of planar cross-strata records the migration of a straight-crested bed form from left to right in the Sinian Sandstone, northwest China. (Photo courtesy of Marc Hendrix.) (For color version, see Plate 11.16b, between pp. 248 and 249.)

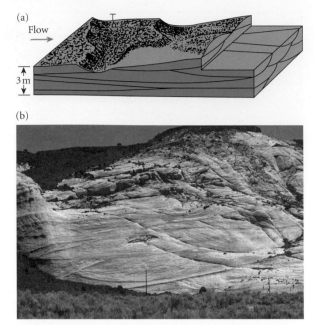

Figure 11.17 (a) Wedge sets and trough sets of festoon cross-strata formed by aggrading curved-crested dunes with stoss erosion at moderate climb angles. (b) Trough sets of festoon cross-strata in profile at a high angle to composite dune migration. (a, b, courtesy of the US Geological Survey.) (For color version, see Plate 11.17b, between pp. 248 and 249.)

migration of straight-crested ripples and sand waves can produce tabular sets of planar cross-strata (Figure 11.16). Note that the tabular–planar nature of these cross-strata is most apparent in sections sub-parallel to the current direction and that the strata appear to be horizontal in sections perpendicular to flow where their apparent dip is 0°. In plan view, the traces of cross-strata are fairly straight, reflecting fairly straight sand wave crests that yield cross-strata with consistent strike.

The migration of curved-crested ripples and dunes can produce wedge sets of planar cross-strata when viewed in sections parallel to flow (Figure 11.17). In sections at high angles to flow, cross-strata occur in trough sets of festoon cross-laminations. Trough sets result from the erosion of preceding dunes by offset "scoop-shaped" troughs during bed aggradation. Festoon cross-strata record the curvature of lee faces in cross-sections perpendicular to flow. Cross-strata traces are also curved in plan view, which reflects the curvature of dune crests. Flow direction in festoon cross-strata is roughly perpendicular to the place of maximum curvature in plan view (Figure 11.17). Although multiple sets of small- to medium-scale cross-strata can form by processes other than bed-form migration, most probably form by mechanisms similar to those outlined here, albeit in complicated ways for complex bed forms.

Plane or flat bed aggradation during a decreasing flow regime leads to the formation of **horizontal strata** (Figure 11.18b, c). Upper flow regime horizontal strata range from very fine to very coarse sand or gravel and commonly exhibit parallel orientation of elongate sand particles parallel to flow (Figure 11.18a). These are preserved on stratification surfaces

(a)

(b)

(c)

(d)

Figure 11.18   Plane bed transition. (a) View of a plane bed under shallow unidirectional flow, with a lineation on the bed. (b) Diagram showing the horizontal strata and parting lineation developed in plane bed transition. (c) Horizontal strata in sandstone produced by plane bed flow regimes. (d) Parting lineation on a sandstone bedding surface. (a, b and d, courtesy of Marc Hendrix; c, photo courtesy of Duncan Heron.) (For color version, see Plate 11.18, between pp. 248 and 249.)

as **parting lineations** parallel to flow direction, which document formation under upper flow regime plane bed transition conditions (Figure 11.18b, d). Aggradation of coarse sand and gravel on lower flat bed surfaces produces horizontal strata that lack parting lineation. Coarse sand and gravel deposits are dominated by sometimes crude horizontal stratification.

It is probable that the higher portions of the upper flow regime under which antidunes, chutes and pools form are characterized more by erosion than by deposition. However, very high flow regime flows do decelerate, and under such conditions deposition is likely to occur. What kinds of stratification are produced by deposition from antidunes and chutes and pools? Field and laboratory studies

show that antidunes begin as in phase stationary waves (Figure 11.19a) that migrate upcurrent due to stoss-side deposition from currents that decelerate and lee-side erosion from currents that accelerate over the bed form (Figure 11.19b). Eventually antidunes shear out (Figure 11.19c) to form chutes and pools (Figure 11.19d). Figure 11.19e shows the stratification that developed in a laboratory where repeated antidune formation occurred on an aggrading bed. The resulting strata occur as bundles of low angle crossstrata separated by curved erosion surfaces. Some cross-strata are convex upward due to partial preservation of antidunes and/or chutes; others are convex downward, perhaps due to deposition in pools. These cross-strata very closely resemble the hummocky and

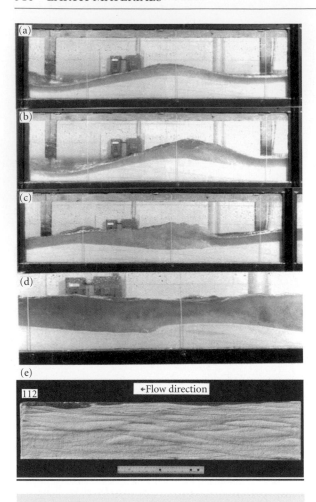

Figure 11.19 Upper flow regime and antidunes. (a–d) Antidunes in laboratory flume experiments: (a) in-phase antidunes (b) migrate up the flow to the right and (c) the standing wave breaks with the development of (d) chutes and pools. (e) Antidune cross-stratification generated in a laboratory. (Photos courtesy of John Bridge.)

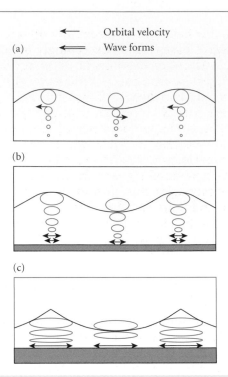

Figure 11.20 (a) Deep water waves showing roughly circular orbitals whose diameters decrease systematically to approach zero at the wave base (L/2). (b) Transitional waves (L/2 > depth > L/20) showing progressive flattening of the orbitals as the waves interact with the bed. (c) True shallow water waves (depth < L/20) with flattened elliptical orbitals that produce oscillatory flow on the bed.

swaley cross-strata produced by oscillatory flows, which are discussed in the next section.

Oscillatory flow

Oscillatory flows are characterized by back-and-forth flow in opposite directions over time and are generated by shallow water waves. Most wind-driven progressive waves that reach near shore environments are generated by storm winds blowing over deep water. Once generated, such waves are dispersed out of the storm area and can travel long distances (10,000 km) through the body of water in which they formed. As long as they are moving through water deeper than one-half the distance between wave crests, which is called the wave length (L), they are **deep water waves**. As deep water waves move through water, water molecules inscribe roughly circular orbits in which water moves in the direction of wave motion as the crest approaches, then downward and backward as the trough approaches, then upward and forward to approach its original position as the next crest approaches, so that there is one orbit for each wave period (Figure 11.20). Orbital diameters are equal to the wave height and gradually decrease with depth until at a depth of L/2 they approach zero. Below this depth, called the **wave base**, the waves have negligible heights and essentially do not exist. The wave base (L/2) is exceedingly important in near shore environments. In water deeper than the wave base, sediments are not entrained by waves because waves do not interact with the bottom. In shallower water, waves move large quantities of sediment.

The position of the wave base varies with wavelength. Really large storm-produced waves, generated by very strong winds, begin to interact with the bottom in much deeper water than do smaller waves with shorter wavelengths. It is useful to divide shallow water coastal areas into areas of (1) shallow water, above the normal wave base, where fine sediments are continually suspended by wave action; (2) deeper water, between the normal wave base and storm wave base, where sand is entrained only during storms and fines are deposited between storm events; and (3) still deeper water, below the storm wave base, where sand is not entrained and fines are continually deposited from suspension.

Oscillatory flows are produced by shallow water, wind-driven waves, generally in near shore environments. Waves moving through water whose depth is less than the wave base (<L/2) are called **transitional** and/or **shallow water waves**. Such waves interact with and entrain sediment from the bottom as their orbitals become progressively flattened. True shallow water waves (depth <L/20) have orbitals so severely flattened that the water moves forward as the crest approaches and backward as the trough approaches. Such back-and-forth movement of water is called oscillatory flow and causes entrained sediments to move back and forth as well. Those who have been to a beach might try to visualize the back-and-forth movement of water (backwash and surge) as wave trains approach the shoreline.

Flume experiments with oscillatory flow have demonstrated that a sequence of flow regimes and bed forms develops in response to increased oscillatory flow velocity on the bed. At very low velocity, no sand is entrained, so no traction load develops. As velocity is increased, initial entrainment velocities are reached and a traction load of very coarse silt and very fine sand is produced. Subsequent entrainment velocities on non-cohesive beds are proportional to particle size. Very small relief bed forms called rolling grain ripples

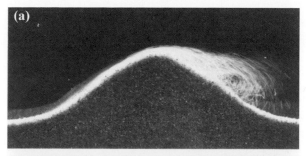

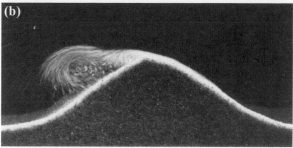

Figure 11.21 Oscillatory flow and sand movement (left in (a), then right in (b)) about a symmetrical oscillation (two-dimensional vortex) ripple in a flume experiment. (Photo courtesy of Sergey Voropayev.)

develop early, but these have not been observed in nature. These small asperities on the bed are quickly transformed into two-dimensional vortex ripples, better known as **oscillation ripples**. Oscillation ripples are straight-crested forms with small ripple indices (length:height ratios) produced by the back-and-forth movement of sand across the bed and across the ripple forms (Figure 11.21). Oscillation ripples differ from current ripples in that they possess (1) fairly straight to sinuous, often branching, somewhat pointed crests, (2) symmetrical profiles (Figure 11.22), and (3) internal cross-strata that dip away from the crest in opposite directions (Figure 11.23a).

Any flow process that causes one oscillatory flow component to be enhanced relative to the other so that components of both unidirectional and oscillatory flow occurs produces what is called **combined flow**. Most oscillatory ripples display evidence of combined flow with one set of cross-strata more pronounced than the other. They also possess bases that are less even than those of most current ripples produced by unidirectional

(a)

(b)

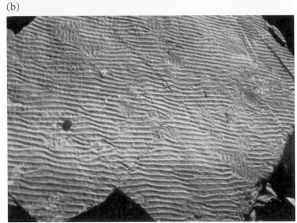

Figure 11.22 Oscillation ripple marks showing crest bifurcation, symmetry, short wavelengths and small ripple indices. (a) Modern tidal flat, Cape Cod Massachusetts. (Photo by John O'Brien.) (b) Preserved in sandstone. (Photo courtesy of Peter Adderley.) (For color version, see Plate 11.22, between pp. 248 and 249.)

(a)

(b)

Figure 11.23 (a) Oscillation ripples in the Carboniferous Horton Group, Nova Scotia. (Photo courtesy of John Waldron.) (b) Oscillation combined flow ripples in the Jurassic Entrada Sandstone, Utah. (Photo courtesy of Tim Cope.) (For color version, see Plate 11.23, between pp. 248 and 249.)

flow (Figure 11.23b). Two-dimensional vortex ripple heights increase with increasing oscillatory flow velocities that occur when larger waves with higher orbital velocities are transformed into shallow water waves.

Rippled sands deposited in the lower parts of the lower flow regime are commonly interstratified with muds deposited from suspension under calm flow conditions. This is especially common on tidal flats where combinations of bimodal, unidirectional tidal currents and oscillatory waves produce current,

oscillatory and combined flow rippled sands, and periods of slack flow cause muds to settle from suspension. Three major types of intimately interstratified sand and mud deposits, each with different sand : mud ratios, are recognized in such sequences. **Lenticular stratification** is produced in protected areas with low sand : mud ratios and contains isolated lenses of ripple-laminated sand deposited from traction encased in dominant muds deposited from suspension (Figure 11.24c). **Wavy stratification** is characterized by subequal amounts of relatively continuous ripple-laminated sand and mud layers with the rippled sand giving the mud layers a wavy appearance (Figure 11.24a). **Flaser stratification** is produced in

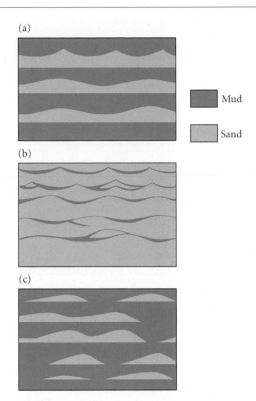

| | Mud |
| | Sand |

Figure 11.24 Interlayered ripple-laminated sandstone and mudstone from tidal flat sediments, Pennsylvanian, Kentucky: (a) wavy strata; (b) flaser strata with a mud-drape flaser preserved in ripple-laminated sand; (c) lenticular strata with isolated ripples preserved in mud. (From Friedman and Sanders, 1978; with permission of Gerald Friedman.)

less protected areas with high sand:mud ratios. Flaser are thin mud drapes deposited from suspension that partially preserve ripple forms in areas dominated by the deposition of ripple-laminated sand (Figure 11.24b).

What happens when oscillatory flow regimes increase, as when storm waves approach the coast? At still higher velocities, vortex ripples assume three-dimensional forms in which small, elevated hummocks alternate with depressions called swales. The sequence from rolling grain ripples to two-dimensional oscillation ripples and hummocks and swales is considered to represent the lower flow regime. At higher flow regimes, the hummocks and swales are replaced by a near flat plane bed.

Hummocky–swaley surfaces can be seen at low tide in very shallow water but also appear to be associated with storm waves in somewhat deeper water in the geological record. Deposition on hummocks produces **hummocky cross-stratification (HCS)**, and deposition in swales generates **swaley cross-stratification (SCS)**. The two sometimes occur together. HCS is characterized by gently inclined, convex up cross-strata, whereas SCS is characterized by gently inclined, convex down cross strata, both in sets with thicknesses less than 0.5 m (Figure 11.25). In older rocks, these types of cross-strata generally occur in relatively fine sandstones inter-stratified with marine fossil-bearing mudrocks that were deposited from suspension. This suggests that they form chiefly on bottoms below the normal wave base but above the storm wave base.

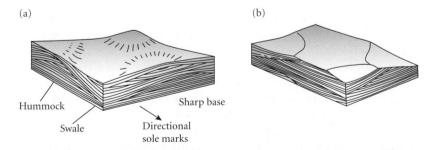

Figure 11.25 (a) Block diagram showing hummocky cross-stratification (HCS). (b) Block diagram showing swaley cross-stratification (SCS). (From Tucker, 2001; with permission of Wiley-Blackwell.)

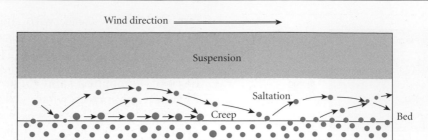

Figure 11.26 Major modes of sediment transport by winds: creep, saltation and suspension.

11.4.2 Wind (aeolian) and wind-deposited sediments

Winds are set in motion by a combination of differential heating, which creates air masses of different density, and the acceleration of gravity, which causes denser air masses to displace lighter ones. Because air possesses a much lower viscosity (resistance to flow) than water, it can attain much higher velocities and is generally characterized by turbulent flow. Because air has a much lower density than water, wind generally exerts smaller shear stresses on the bottom and therefore tends to entrain and transport smaller particles. In the range of normal wind speeds (<80 km/h), winds do not have the ability to entrain gravel.

The erosion, transportation and deposition of sediment by winds resemble those of unidirectional water flows in several regards (Figure 11.26). Larger grains are transported in a traction load by the process called **creep**, which is maintained by grain to grain collisions. Somewhat smaller grains are transported in intermittent suspension by saltation. During major wind storms the saltation load moves in part by **sheet flow** in which many saltating sand grains collide in mid-air, before reaching the bed, and stay suspended for longer than normal times. Large amounts of sand can be moved during such sandstorms. The finest grains are carried above the bed load in longer term suspension.

As wind velocity increases, a critical threshold velocity is reached where entrainment of very fine sand begins (Figure 11.27). Progressive increases in velocity entrain progressively coarser and finer (more cohesive) particles as their critical entrainment velocities are reached. As with water, grains finer than coarse silt are transported in the suspension load with fine silt and clay carried in long-term suspension. Coarse silt to medium sand is transported primarily by saltation, although it may be suspended at very high wind velocities. Fine to coarse sand and perhaps fine gravel are transported slowly by creep in the traction load, although sand is transferred from the creep load to the saltation load with increases in velocity. Coarser gravels are not generally entrained by wind. Even more effectively than water, wind sorts material by size during transportation. This sorting is retained during deposition so that aeolian sediments are generally very well sorted. In addition, sand grains tend to undergo significant rounding during aeolian transport because they are large enough to collide with considerable force, especially during saltation, and the low viscosity of air cushions such impacts less than those that occur in water. As a result, aeolian sands tend to be well rounded and aeolian lag gravels often show evidence of significant abrasion by "sand-blasting".

As wind velocity decreases below the critical velocity for deposition (Figure 11.27), any transported gravel and coarse sand is deposited first, from the creep load, followed by the deposition of progressively finer sand, then coarse silt, from the saltation load. As with water, very low wind velocities are required for fine silt and clay to settle from suspension. Think of the fine material suspended in a ray of sunlight in a calm room that slowly settles onto the top of your dresser or wardrobe over long periods of time. Where prevailing winds blow across a sediment source area over time, the suspension load is transported longer and farther than the saltation load, which is transported longer and farther than the traction load. Over time, repeated entrainment, transportation and deposition produces a down-

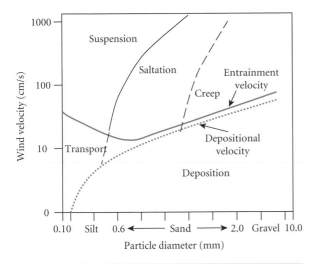

Figure 11.27 Velocity conditions for wind erosion, transportation by suspension, saltation and creep and deposition of various sediment sizes. (After Tucker, 2001; with permission of Wiley-Blackwell.)

Figure 11.28 Typical loess deposit; note the paucity of stratification in comparison to that displayed by wind-blown sands. (For color version, see Plate 11.28, between pp. 248 and 249.)

Figure 11.29 Wind ripples on back-beach sand dunes, Australia, with branching crests, strong asymmetry and large ripple indices; wind from left to right. (Photo by John O'Brien.) (For color version, see Plate 11.29, between pp. 248 and 249.)

wind sequence from (1) coarse lag gravels where the finer materials have been removed by wind erosion, a process called deflation; to (2) well-sorted sand deposits such as dune fields in which sand has been deposited from the bed load; and (3) well-sorted mud deposits formed as muds settle from suspension. The extensive, poorly stratified deposits of **loess** produced by winds that entrained, transported and deposited abundant silt from glacial deposits during the last ice age are an excellent example of the latter (Figure 11.28). Loess soils are very fertile and support significant grain production in the Great Plains of North America, Europe and Asia.

Wind blowing over non-cohesive, coarse silt and sand generates various bed forms that depend on flow velocities and the attendant shear stresses on the surface. At relatively low velocities, **wind ripples** develop in coarse silt and sand. Wind ripples are current ripples that superficially resemble aqueous unidirectional current ripples but differ from them in a number of ways. Wind ripples tend to (1) have lower heights relative to wavelengths and thus larger ripple indices (RI >12); (2) be strongly asymmetrical; and (3) unlike continuous-crested subaqueous current ripples, show branching crests (Figure 11.29). Nothing analogous to lower flow regime dunes develops on such surfaces. Instead, at higher veloci-

ties, roughly three times the critical entrainment velocity, the ripples disappear and the surface is transformed into a flat or plane bed.

However, much larger forms are commonly produced by winds in areas with sufficient sand supply. **Sand dunes** are very complex features with wavelengths of tens to hundreds of meters and heights of centimeters to meters that occur in a variety of forms determined by sand supply, vegetative cover and long-term wind conditions. Figure 11.30 shows the

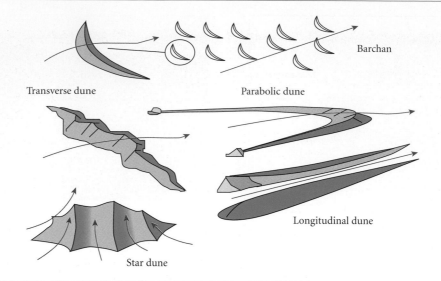

Figure 11.30 Major types of sand dunes: transverse, barchan, parabolic, star and longitudinal, and their relationship to wind directions. (Courtesy of Steve Dutch.)

major types of sand dune and their relationship to the prevailing wind directions.

Relatively straight-crested transverse dunes and curved-crested barchan and parabolic dunes have well-defined upwind (stoss) sides with rather gentle slopes and downwind (lee) sides with steep (20–35°) slopes. The upwind sides experience net erosion as sand is entrained and transported in migrating wind ripples toward the dune crest. Sand that accumulates at the crest becomes unstable and slides down the steep downwind side of the dune, called the avalanche face, in the form of lobate sand flows whose dip decreases toward the base of the avalanche face (Figure 11.31a, b). Repeated sand flows generate multiple, steeply inclined (20–35°) cross-strata formed by lee-side deposition on the avalanche face. As with subaqueous dunes, erosion of sand from the upwind (stoss) side and deposition on the downwind (lee) side causes wind-blown dunes to migrate down-flow over time. Net aggradation of the bed, essential to long-term preservation and migration of multiple dunes, produces multiple sets of cross-strata separated by erosional surfaces that record the passage of interdune troughs. Relatively straight-crested transverse dunes migrate to produce tabular sets of planar cross-strata, whereas the migration of curved-crested dunes generates wedge sets or trough sets of festoon cross-strata. Aeolian cross-

strata are more tangential to the underlying erosion surface than are most cross-strata generated by the migration of subaqueous dunes and so tend to be more strongly concave upward in profile (Figure 11.31c).

Longitudinal and star dunes tend to display much more complicated bidirectional and multidirectional cross-beds. **Dras** are larger features (2.0×10^2 to 2.0×10^4 m) composed of multiple dune forms whose migrations give rise to very complex cross-stratification. Aeolian sand deposits tend to be quartz rich, fine grained and well rounded, possess excellent sorting, and contain multiple sets of large-scale cross-stratification.

11.4.3 Glaciers and glacial sediments

Glaciers are significant agents of sediment erosion, transportation and deposition, especially during periods of global cooling that produce so-called ice-house conditions. At the present time, glaciers cover nearly 10% of Earth's surface, but as recently as 18,000 years ago they covered nearly 30% (Paterson, 1999). Glaciers form when net snow accumulation exceeds losses by ablation (melting and sublimation) over extensive periods of time. As the volume of snow increases, a combination of repeated thawing and freezing, along with increased burial pressure, slowly convert the snow into ice. As the ice volume continues

(a)

(b)

(c)

Figure 11.31 Formation of aeolian cross-strata by dune migration. (a) Dune showing stoss-side erosion and lee-side deposition by sand flows. (b) Lobate sand flows on a dune avalanche face generate large-scale, steeply inclined cross-strata. (c) Large-scale, steeply inclined, tangential cross-strata in the Jurassic Navajo Sandstone, Utah, formed by aeolian dune migration with a left-to-right component. (Photos by John O'Brien.) (For color version, see Plate 11.31, between pp. 248 and 249.)

to increase, the ice sheets thicken and begin to flow plastically in response to gravitational forces. Valley glaciers in mountainous areas flow down valleys. Larger continental glaciers, covering large areas (10^5–10^6 km^2) with thicknesses up to several kilometers, flow radially outward from their area of maximum thickness under the tangential force of gravity, in a manner analogous to honey or molasses flowing away from their areas of maximum thickness. As long as the volume of glacial ice expands, the margins of the glacier tend to advance. Only when the ablation exceeds accumulation, so that glacial ice volumes contract, do glacial margins tend to retreat from areas formerly covered with ice.

The details of glacial erosion, transportation and deposition depend in part on whether the glacier is a cold, dry glacier or a warm, wet glacier. **Cold, dry glaciers** occur in very cold regions where ice remains frozen most of the year. Such glaciers are dominated by processes that involve flowing ice. **Warm, wet glaciers** occur in regions that partially thaw for extended periods so that glacial ice is melted, especially along its upper, lateral and lower margins. Warm, wet glaciers are characterized by processes that involve a mix of glacial ice flow and unidirectional aqueous flow.

How are sediments entrained by glaciers? Because glacial ice is a plastic solid, flow rates tend to be quite slow (<1 m/day). Even at

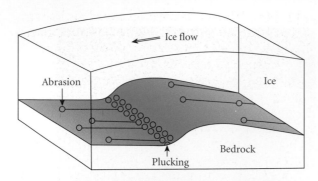

Figure 11.32 Erosion of bedrock by glacial plucking and glacial abrasion.

Figure 11.33 A striated bedrock surface in Cambrian dolostones, northwestern New Jersey. (Photo by John O'Brien.) (For color version, see Plate 11.33, between pp. 248 and 249.)

these slow speeds, glaciers tend to entrain any loose material with which they are in contact. **Glacial entrainment** occurs primarily by some combination of three processes: (1) erosion of unconsolidated sediment; (2) plucking; and (3) abrasion of bedrock. When a glacier flows over an area it entrains unconsolidated material in its path, eventually eroding the surface down to the underlying bedrock. Anyone familiar with the thin to non-existent soils that overlay bedrock in areas subject to recent glacial erosion is familiar with the results of this process. As the glacier continues to flow, additional sediment is entrained from the underlying bedrock by plucking and/or abrasion. **Plucking** occurs when a glacier encounters an obstruction in an uneven bedrock surface, often where it is flowing downhill across bedrock (Figure 11.32). Pressure melting occurs; water infiltrates into joints and other cracks and then freezes and expands. Repeated freezing and expansion enlarges cracks to the place where rock fragments are dislodged from the bedrock and incorporated into the glacier. Plucking requires that glaciers be warm enough for some melting to occur. **Abrasion** occurs when rock materials carried in contact with bedrock under the pressure of the overlying ice mechanically wear away the underlying bedrock.

Medium-sized particles moving in contact with bedrock produce flow parallel scratches called **striations** (Figure 11.33). Larger particles or clusters of particles produce larger **grooves**. Very small particles tend to **polish** the surface. Each of these processes removes material from the underlying bedrock, which is then entrained within the glacier, including

very small particles referred to as **rock flour**. Striated, polished bedrock surfaces (Figure 11.33) are a compelling indication of glacial erosion.

Continental glaciers entrain all sizes of sediment from huge areas, so that the entrained sedimentary particles that are eventually deposited possess a great variety of compositions and sizes. These materials are carried below, within and atop the glacier by slow, plastic flow toward the terminus of the glacier. There is little tendency for sediments to be sorted by size during either transportation or deposition by glacial ice. As a result, most glacial deposits are characterized by poorly sorted debris with a wide variety of compositions. How is glacial debris deposited? Many glaciers contain a debris-rich basal layer. Some sediment is deposited beneath the glacier, often behind bedrock obstacles, by a process called lodgment, which occurs when the frictional resistance between the debris-rich basal layer and the bedrock is higher than that between the basal layer and the glacier that slides over it. Warm glaciers deposit sediment by local subglacial melting, which releases material from the basal layer. Even more glacial sediment is deposited at glacial margins as glaciers stagnate and/or begin to melt so that their entire load can no longer be carried. No matter which set of depositional processes is involved, sediment deposited directly from glaciers is called **glacial till**.

(a)

(b)

Figure 11.34   (a) Glacial till, Pleistocene, Ohio; note the polymictic composition, poor sorting and lack of stratification. (Photo courtesy of the US Geological Survey.) (b) Glacial till overlying a striated bedrock surface, Switzerland. (Photo courtesy of Michael Hambrey, with permission of www. glaciers-online.net.) (For color version, see Plate 11.34, between pp. 248 and 249.)

Glacial tills and lithified glacial tills called **tillites** tend to be very poorly sorted, boulder-bearing and polymictic (contain a wide variety of fragment compositions) and lack the well-defined internal stratification produced by agents such as wind and water that sort sediment by size during transportation (Figure 11.34a). When tills cover grooved and/or striated bedrock surfaces (Fig. 11.34b), a glacial origin is, to coin a phrase, a "slam dunk".

Because ice eventually melts, especially in warm, wet glaciers, glacial sediments are intimately associated with water-laid sediments. In some cases, melt water simply modifies original glacial textures by removing many of the fines, leaving poorly sorted mixtures of gravel and/or sand. In other cases, unidirectional flow within the glacier, at the margins of the glacier or beyond the margins of the glacier deposits sediments whose characteristics are those of sediments deposited by unidirectional aqueous flow. These include sands and gravels deposited by (1) subglacial streams and preserved as eskers, (2) glacial margin streams preserved as kame terraces, and (3) braided stream channels in periglacial environments beyond the margins of the glacier and preserved as outwash plain deposits. Because of their enhanced porosity, these subglacial and periglacial water-laid sands and gravels act as important aquifers in many parts of the world, and some older ones are significant oil reservoirs.

In addition to outwash plains, lakes commonly develop in periglacial basins near glacial margins or behind ice dams. Fine-grained deposits of clay and silt or fine sand are commonly deposited in the calm bottom waters of such lakes. Where such lakes freeze over in winter and thaw in summer, alternating dark, organic-rich winter layers and lighter, coarser silt/sand summer layers are common. Each couplet represents a year of sediment deposition. Such annual layers of sediment are called **varves** (Figure 11.35). Like tree rings, which they resemble to some extent, varves yield clues to climatic variations (e.g., summer–winter length) over time.

Icebergs dislodged from the glacial margin float into adjacent lakes and marine settings carrying poorly sorted sediment. When icebergs melt, ice-rafted debris sinks to the bottom. Large clasts dropped from icebergs are called **dropstones**. Dropstones tend to disturb delicate laminations, and their presence in environments so calm that muds settled from suspension is an excellent indicator of contemporaneous glaciation (Figure 11.36).

A detailed treatment of glacial sedimentary processes and features is beyond the scope of

Figure 11.35 Glacial varves from the Pleistocene, Maine: coarser, lighter colored, resistant summer layers alternate with finer, darker colored, crumbly winter layers. (Photo courtesy of Thomas K. Weedle, with permission of the Maine Geological Survey.) (For color version, see Plate 11.35, between pp. 248 and 249.)

Figure 11.36 Large glacial dropstone and other ice-rafted debris from the Pleistocene, Lake Spokane, Washington. (Photo courtesy of the US Geological Survey.) (For color version, see Plate 11.36, between pp. 248 and 249.)

this text. For more detailed information, the reader is referred to Hambrey and Alean (2004), Evans and Benn (2004), Menzies (2002), Mickelson and Attig (1999), Paterson (1999) and Benn and Evans (1998).

11.4.4   Mass (sediment gravity) flows and their deposits

In the previous sections we have discussed the entrainment, transportation and deposition of sediment by fluid flows such as moving water and wind and plastic flows such as glaciers. Large amounts of sediment also are dispersed across Earth's surface by a fourth set of processes, variously referred to as **mass flows** and/or **sediment gravity flows**. These names reflect the fact that these flows involve relatively dense masses of sediment moving downslope under the tangential force of gravity by virtue of their possessing higher densities than the media that surround them. There are many types of subaerial and subaqueous mass flows. Because of their abundance as Earth materials, special attention will be paid to debris flows and turbidity currents and their deposits.

Significant factors in most classification and nomenclature schemes include (1) the internal cohesiveness of the rock materials, (2) the proportions of sediments and ambient fluids such as water, (3) the manner in which sediments are supported within the flow, and (4) maximum flow speed.

In **cohesive flows**, such as rock falls, rock slides and slumps, movement occurs due to a loss of cohesion between the rock mass and its substrate. The rock mass largely maintains internal cohesion as it moves downslope. In all other mass flows, internal cohesion is not maintained as flow constituents separate and change relative positions during flow. **Granular flows** (Pierson and Costa, 1987) are characterized by large proportions of sedimentary particles (≥80%) relative to water (≤20%), though they may contain significant air. Granular flows include rapidly moving debris avalanches, slower grain and earth flows and very slow creep. Grain-to-grain collisions and compressed air are significant in maintaining granular flows. **Slurry flows** (Pierson and Costa, 1987) are characterized by smaller amounts of sedimentary particles (~60–80%) and larger amounts of liquid (~20–40%) and are sometimes called liquefied flows. These include debris flows and mud flows with maximum velocities of 100 km/h, some high concentration turbidity currents and very slow flows such as those involved in solifluction. Liquids

THE SEDIMENTARY CYCLE   321

and suspended fines are important in maintaining slurry flows. **Hyperconcentrated flows** (Pierson and Costa, 1987) possess lower sediment (~20–60%) to water (~40–80%) ratios and include hyperconcentrated stream flows and many, but not all, subaqueous turbidity currents. Most hyperconcentrated flows form when slurry and/or granular flows mix with and incorporate a significant amount of ambient water, which then plays a significant role in maintaining flow.

Mass flows are initiated by slope failure. Mass flows begin as masses of sediment and/or fluid with some internal cohesiveness or shear strength. In order for them to lose cohesion and be set in motion, their shear strength must be overcome. This requires that the tangential force of gravity, a shear stress, be larger than the yield stress required to overcome the internal shear strength and to cause loss of sediment cohesion. This can be accomplished by increasing the slope so that the tangential force of gravity increases or decreasing the internal shear strength of the sediment mass. Triggering mechanisms that cause slopes to oversteepen include (1) deposition, especially if rapid, (2) tectonic steepening that results from surface deformation, and (3) erosional undercutting. Mechanisms that cause loss of internal cohesion include (1) shaking by earthquake vibrations or storm waves, (2) increases in pore fluids by infiltration, (3) increases in pore fluid pressure related to loading, and (4) liquefaction in which sediments are transformed into liquids, and which is related to the first three. Whatever the mechanism, once the tangential force of gravity exceeds the yield strength of the material, the mass will begin to flow downhill and will continue to do so until the tangential force of gravity decreases and/or the shear strength of the material increases, causing deposition to occur. Many mass flows that bury houses and cause loss of life are triggered by human activities including (1) oversteepening of slopes during construction projects, (2) loss of vegetative cover from construction or fire, (3) decrease in material strength from movement of Earth materials during construction, (4) increased pore fluids and/or pore fluid pressures due to water infiltration, and (5) loading from increased water use and building loads (Menzies, 2002).

Debris flows and **mud flows** are types of slurry flows (Figure 11.37). Most geologists distinguish between the two based on a poorly defined ratio between mud and larger particles, with debris flows containing significant concentrations of larger particles, commonly boulders (Prothero and Schwab, 2004). In the following discussion, we will use the term debris flow for both types of slurry flow. Lahars (Chapter 9) are debris flows with significant volumes of volcaniclastic sediment. Most debris flows form in subaerial environments, but they are well known from subaqueous environments as well. Because of their relatively low water content, debris flows have substantial yield strengths that result from frictional and/or electrical forces between grains. Most debris flows are generated in unconsolidated material on steep slopes after periods of heavy rainfall, and perhaps are associated with a loss of vegetative cover that otherwise would have helped to anchor the material in place. Some form when rainwater infiltration decreases frictional forces between grains, thus lowering the shear strength of the material to where its yield strength is less than the tangential force of gravity on steep slopes. This is especially true in soils with smectite clays that expand by absorbing water (see Chapter 12). Other debris flows form when granular flows slide into rain-swollen river systems and mix with the ambient fluid to become slurry flows. Many debris flows are characterized by multiple surges (Figure 11.37b) caused by additional debris flows that enter the dispersal system. Once formed, the sediment/water slurry flows as a well-mixed flow that retains surprising cohesive strength. Sediment concentration is high and the flow so strong and "viscous" that boulders can be supported in suspension (Figure 11.37a). In many debris flows, large boulders are concentrated near the top of the flow so that during transportation, average particle size increases upward within the flow.

Debris flows begin to slow down when they reach a gentler slope. Small amounts of fluid may be expelled as grains begin to settle. These processes cause the shear strength of the debris flow to exceed the tangential force of gravity and flow ceases. This usually happens quite rapidly so that the flow is essentially frozen in place. The entire mass is

(a)   (b)

Figure 11.37 (a) Mud flow with a matrix strength sufficient to suspend boulders at the top in California. (b) Debris flow surge moving down a river valley, Mt Rainier National Park, Washington. (Photos courtesy of the US Geological Survey.) (For color version, see Plate 11.37, between pp. 248 and 249.)

deposited at once rather than grain by grain, and this **frozen bed** retains many of the characteristics it possessed during transportation (Figure 11.38).

Most debris flow deposits display several characteristics helpful in their identification, including (1) poor sorting of material by size, (2) crude or no internal stratification, (3) common inverse grading, (4) boulders and cobbles dispersed in and supported by a finer grained matrix, sometimes as diamictitites, and (5) many different types of gravel particles derived from different parts of a drainage basin as in polymictic conglomerates.

Turbidity currents (Natland and Kuenen, 1951) form when masses of sediment begin to flow in subaqueous marine or lacustrine environments and mix with and incorporate ambient water. Turbidity current deposits are called **turbidites** and may possess any or all of the properties of slurry, hyperconcentrated and/or low concentration unidirectional flow deposits.

Turbidity currents are generated by subaqueous slope failures that occur when the tangential force of gravity exceeds the shear strength of unconsolidated sediment that is usually water saturated. Triggering mechanisms include those discussed earlier that involve increasing the slope by rapid deposition, tectonic steepening and/or erosional undercutting, and/or decreasing the internal shear strength of the sediment mass by vibration, increased pore fluid pressure and/or liquefaction. Marine turbidity currents and turbidites are commonly generated on (1) continental slopes associated with passive margins, (2) steep slopes on the margins of pull-apart basins, (3) slopes in trench arc systems, and (4) slopes associated with forearc basins adjacent to orogenic belts. But any slope will do, and some are apparently generated on slopes of only a few degrees.

Once the sediment mass begins to move downslope, it mixes with the ambient water. The rheology of the flow strongly depends on

(a)

(b)

Figure 11.38   (a) Debris flow deposit above an erosion surface, southern Utah. (Photo by Kevin Hefferan.) (b) Multiple debris flow deposits, Lone Pine, California. (Photo by John O'Brien.) (For color version, see Plate 11.38, between pp. 248 and 249.)

the amount of mixing that occurs. If little mixing occurs, the flow will possess the characteristics of a slurry (debris) flow. If more mixing occurs, it will be transformed into a hyperconcentrated flow. If still more mixing occurs, it will be transformed into a fluid flow with low sediment concentrations that resembles a unidirectional flow. Much has been learned from laboratory experiments in which liquefied sediment mixtures are released into a long tank called a flume. Figure 11.39 shows a lab-generated turbidity current.

A simplified model of turbidity currents can be used to illustrate how they evolve, recognizing that the details are both more complex than our discussion and not well understood. Repeated observations show that turbidity currents possess (1) a turbulent, multilobed region at the front of the flow, which is called the **head**, behind which is (2) a faster flowing **main body** that continually feeds sediment into the head and contains relatively high sediment concentrations, which grades back into (3) a tail that contains relatively low sediment concentrations (Figures 11.39 and 11.40). The turbulence developed in the head continually entrains water and/or sediment into the flow, and turbulent suspensions extend from the head back over the body of the flow (Figure 11.39).

Most erosion by turbidity currents occurs by scour concentrated in the turbulent head region. Repeated erosion by turbidity currents

Figure 11.39   Turbidity current in a laboratory showing the head and main body, and the turbulent suspensions that developed over both. (For color version, see Plate 11.39, between pp. 248 and 249.)

is responsible for the formation of most submarine canyons as well as valley systems on continental slopes, delta slopes, trench slopes and submarine fans. The sediment dispersion mechanism necessary to maintain flow in turbidity currents ranges from fluid turbulence in low concentration flows to cohesive matrix strength in very high concentration flows. Grain-to-grain collisions and pore pressure liquefaction may also be important. Deposition from turbidity currents is varied and complex. As one might expect, it tends to

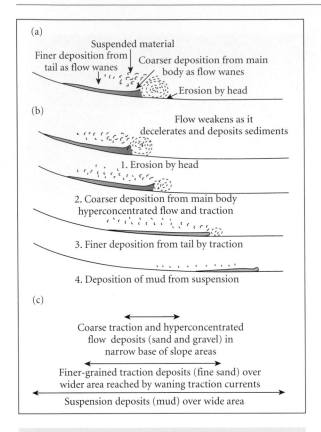

(a)

Suspended material
Finer deposition from
tail as flow wanes
Coarser deposition from main
body as flow wanes
Erosion by head

(b)

Flow weakens as it
decelerates and deposits sediments

1. Erosion by head

2. Coarser deposition from main body
hyperconcentrated flow and traction

3. Finer deposition from tail by traction

4. Deposition of mud from suspension

(c)

Coarse traction and hyperconcentrated
flow deposits (sand and gravel) in
narrow base of slope areas

Finer-grained traction deposits (fine sand) over
wider area reached by waning traction currents

Suspension deposits (mud) over wide area

Figure 11.40 (a) Model of a turbidity current with a head, main body and tail. Erosion occurs primarily in the turbulent head; coarse material is deposited from the main body, finer material from the tail and very fine material from suspension. (b) Sequence of deposition during the passage of a turbidity current. (c) Distribution of sediments across an idealized basin.

occur when flows reach gentler slopes and begin to decelerate.

Those portions of turbidity currents with high sediment concentrations are called **high concentration turbidity currents**. When they possess very elevated sediment concentrations (>60%), they behave as slurry (debris) flows. Such slurry flow turbidity current deposits are coarse-grained sandstones or conglomerates, display little internal stratification, may contain matrix-supported gravel clasts and may display inverse grading. At lower sediment, especially mud, concentrations (~20–60%), high concentration turbidity currents behave as hyperconcentrated flows. Unlike slurry flows, hyperconcentrated flows lack sufficient matrix strength to suspend particles.

Instead, particles are suspended primarily by fluid turbulence. The sediments in hyperconcentrated flows begin to separate by size, with larger particles moving closer to the bed and mud carried upward by fluid turbulence. Deposition occurs as individual particles accumulate on the bed, sometimes quite rapidly, as the flow decelerates. Continued deceleration generates graded beds in which grain sizes decreases upward. In hyperconcentrated flow deposits, gravel particles are in contact with one another, rather than matrix supported, and sandstones and conglomerates display a moderate degree of sorting. High concentration turbidity current slurry flow and hyperconcentrated flow deposits are the first sediments to be deposited from any turbidity current that contains them. As a result, such deposits are most abundant in **proximal turbidites** deposited closest to the sources of the turbidity currents at the base of prominent slopes and on the most active portions of submarine fan depositional systems.

Portions of turbidity currents with low sediment concentrations (<20%) are called **low concentration turbidity currents**. Such dilute turbidity currents behave in a manner analogous to that of unidirectional flows. The coarsest sediments, usually sand, are carried in a traction load, finer sand or coarse silt are carried in a saltation load, and finer silt and mud are transported in suspension. As such flows decelerate, sand is deposited from the traction carpet. Plane bed transition horizontally laminated sands are commonly deposited from the traction load of low concentrated turbidity currents. Further flow deceleration leads to the formation of current ripples in finer sand and/or coarse silt. As the flow wanes, silts and clays settle from suspension. The low concentration portions of turbidity currents may be deposited atop earlier high concentration turbidites. However, the lowest concentration turbidity currents, carrying only fine sand and suspended mud, frequently flow beyond the area where high concentration deposits form (Figure 11.40). As a result, the deposits of low concentration flows are most abundant in **distal turbidites** deposited farthest from the sources of turbidity currents. These deposits dominate turbidites some distance from the base of major slopes, on the less active and lower portions of submarine fans and on abyssal plains.

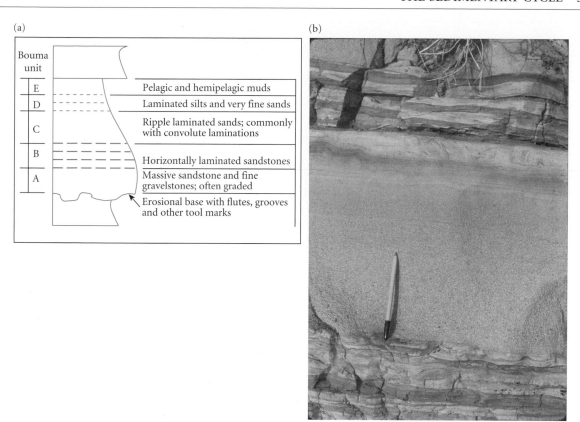

Figure 11.41   (a) Classic Bouma sequence showing an erosional base overlain by units A, B, C, D and E. (b) Partial Bouma sequence with a massive, grade A unit over an erosional base, horizontal laminated B unit, convolute laminated C unit and laminated D unit, from the Tertiary, Simi Hills, California. (Photo by John O'Brien.) (For color version, see Plate 11.41b, between pp. 248 and 249.)

A pivotal model for a classic turbidite, which contains most of the deposits discussed above, was proposed by Bouma (1962) and is now called a Bouma sequence. A complete **Bouma sequence** (Figure 11.41) consists of five major units, labeled A, B, C, D and E, over an erosional base. Complete Bouma sequences begin with an erosional base produced by scour as the sediment-laden head and/or main body of the turbidity current passed over the bottom, often a cohesive mud bottom. Above the erosional surface is unit A, a massive sandstone and/or conglomerate unit lacking clear evidence of internal stratification. Although this unit is sometimes ascribed to deposition from the upper flow regime, it is more likely the result of deposition from high concentration turbidity currents. If unit A is graded, moderately sorted and with any gravel clasts in contact, it is likely the product of rapid deposition from hyperconcentrated flows. Where the A unit is ungraded and/or contains matrix-supported clasts, it was likely produced from a slurry flow. Deposits from slurry flows were not included in Bouma's sequence but are common in models for high concentration turbidites (Lowe and Guy, 2000). Unit B consists of sandstones with horizontal laminations that record traction load deposition by low concentration turbidity currents in the plane bed transition. Unit C, with prominent current ripple cross-stratification in finer grained sandstones, records lower flow regime deposition by low concentration turbidity currents. Unit D, which consists of weakly laminated siltstones, records small amounts of suspension deposition, perhaps by multiple turbidity currents. Unit E consists of very fine mudstones and is not the direct product of the turbidity currents. Instead, it is the product of normal pelagic deposition where tiny mud particles settle to the calm bottom waters of the deep sea floor. Complete Bouma sequences are

(a)   (b)

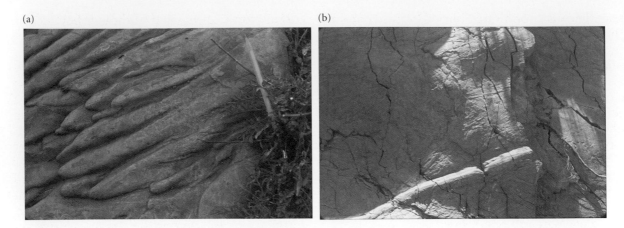

Figure 11.42   Sole marks on bed bases. (a) Flute marks, Austen Glen Formation, Ordovician, New York; current from lower left to upper right. (b) Groove casts, Cretaceous, Wheeler Gorge, California. (Photos by John O'Brien.) (For color version, see Plate 11.42, between pp. 248 and 249.)

Table 11.1   Properties of sediments deposited by major depositional agents.

Agent	Composition	Texture	Structures	Other
Water: unidirectional flow	Variable sandstone compositions	Well-sorted to moderately sorted sandstones; mudrocks; conglomerates	Asymmetrical current ripples; unidirectional cross-strata; horizontal strata common	Small ripple indices
Water: oscillatory flow	Variable sandstone compositions	Well-sorted to moderately sorted sandstones; mudrocks; conglomerates	Symmetrical oscillation ripples; with bidirectional cross-strata; hummocky and swaley cross-strata	Wavy, lenticular and/or flaser bedding common
Wind	Quartz-rich sandstones	Rounded, very well-sorted sandstones; mudrocks; negligible conglomerates	Aeolian ripples; large-scale cross-stratification; tangential bases	Terrestrial flora and fauna
Glacier	Polymictic conglomerates with many clast compositions	Very poorly sorted, boulder-bearing, conglomerates; many matrix supported	Little or no internal stratification	Striated, polished bedrock surfaces; dropstones
Mass flow: mud–debris flow	Often polymictic conglomerates with many clast compositions, but less variable than glacial deposits	Poorly sorted, boulder-bearing, conglomerates; many matrix supported	Crude, if any, internal stratification; some inverse grading	
Mass flow: turbidity current	Sandstone, mudrocks and fine conglomerates; variable sandstone compositions	Moderate to moderately poor sorting; mud matrix common in sandstones	Bouma sequences common; massive sandstone; laminated sandstone; cross-laminated sandstone; interlayered sandstone and mudrock	Sharp, erosional base; sole marks common

rare because not all turbidity currents contain high concentration portions and because it is rare for a single turbidity current to deposit all the units at the same geographic location (see Figure 11.40).

Where erosion occurs on cohesive mud bottoms, subsequent deposition fills the erosional features to produce sole marks on the sole or base of turbidite sand and/gravel beds (Figure 11.42). When turbulent eddies in the head impinge on firm mud bottoms, they produce erosional scours that are deep on the up-current side and become shallower down-current. When they are filled with sand, protuberances are produced on the sole of the turbidite bed that are large on the up-current side and become less pronounced in the down-current direction. Subsequent erosion of soft mudstones exposes these sole marks on the base of the turbidite bed. Such turbulent scour fill sole marks are called **flute marks** (Figure 11.42a). When stone or wood tools are dragged across cohesive mud bottoms, they produce grooves or gutters in the underlying mud that, when filled with sand, become **groove or gutter casts** on the sole of the turbidite (Figure. 11.42b). Many other sole marks are produced on the base of turbidite beds and, as might be expected, on the base of some sandstones produced by unidirec-tional flows. Discussion of these is beyond the scope of this text.

In this chapter we have focused on primary sedimentary structures that form at the time of sediment deposition and record significant information concerning the processes and conditions of deposition. Some **penecontemporaneous structures** that form shortly after sediment deposition and **secondary structures** that form well after deposition are discussed in Chapters 13 and 14. We have seen that each of the major agents of erosion, transportation and deposition possesses a unique set of properties that impart different sets of erosional and depositional characteristics to the sediments and sedimentary rocks they produce. The major contrasts in compositions, textures, structures and other properties of the sediments produced by each agent are summarized in Table 11.1.

Two major goals of this textbook are to help people understand how to interpret the record of Earth surface processes preserved in sedimentary rocks and how to apply the knowledge of sediments and sedimentary processes to the solution of resource needs and geohazard problems that involve such Earth materials. These will provide recurring themes in the chapters that follow.

Chapter 12

Weathering, sediment production and soils

12.1 WEATHERING

Weathering is the in place (in situ) breakdown of rock materials at or near Earth's surface. Weathering processes are fundamentally important in the generation of the soils, sediments and sedimentary rocks that cover more than 80% of Earth's surface. As noted in Chapter 11, most sediment originates as solid detrital particles and dissolved solids produced during weathering. These materials are subsequently eroded and dispersed by water, wind, glaciers and mass flows across Earth's surface to be deposited as detrital and biochemical sediments.

Weathering is also the dominant process in the production of soils upon which so many essential human activities depend. Soils and sediments provide critical wildlife habitats in both terrestrial and aquatic environments, serve as aquifers and aquitards critical for the storage and transmission of water, contain critical supplies of coal, petroleum, natural gas and ore deposits, and are widely used as raw materials in the construction of roads, dams, buildings and other structures. Soils are essential to agriculture and the production of forest products. Weathering processes determine soil texture, soil nutrient and water retention properties and therefore the types of crops that can be successfully grown in a given area. In short, we could not survive on Earth without the sediment and soil-producing processes involved in weathering.

Soils and sediments are directly impacted by surface contaminants. Knowledge of the factors that permit soils and sediment to transmit pollutant plumes away from the initial site of contamination is especially critical, as is knowledge of how to construct containment barriers or remediate near surface contamination. Soils present many challenges to engineers who must be able to evaluate their behavior before construction begins or risk the problems associated with ground failure.

In addition, because many types of soils are strongly dependent on climate and organic activity, ancient soils offer major clues to the emergence and evolution of life on continents,

Earth Materials, 1st edition. By K. Hefferan and J. O'Brien. Published 2010 by Blackwell Publishing Ltd.

major long-term changes in global and local climates, and the long-term evolution of Earth's atmosphere. For these reasons, this chapter is devoted to developing a deeper understanding of weathering, sediment production and soils.

Weathering involves an interactive set of physical, chemical and biological processes that result in the in situ breakdown of rock material at or near Earth's surface. Weathering may occur in the original source area where bedrock is exposed, or in rock materials that have been eroded, transported and deposited thousands of kilometers away from their original source area.

Weathering and erosion are two important, but different, sets of processes. The distinguishing factor between weathering and erosion is that weathering processes involve the breakdown of rock material in a particular location, whereas erosion processes involve the removal of rock material from a geographic location which initiates its transportation to another location. Weathering and erosion are intimately related because weathering generally breaks rock materials down into smaller detrital, organic or dissolved constituents whose small size (unless they are cohesive) makes them more easily removed by erosion and dispersed by transportation. As will be seen in the discussions that follow, different rock materials weather at different rates, a process called **differential weathering**. Less resistant rocks that break down more rapidly tend to be eroded more rapidly. More resistant rocks that weather more slowly tend to erode more slowly. Differential weathering, combined with differences in rock durability, tends to produce differential erosion as the products of weathering are removed at different rates. Figure 12.1 illustrates differential erosion between durable, cliff-forming sandstones and more easily eroded mudrocks in southern Utah.

Major weathering processes may be subdivided into disintegration and decomposition processes. **Disintegration** is the breakdown of larger, more coherent rock bodies into smaller fragments of the same composition. As discussed below, disintegration may involve both physical (mechanical) and biological processes. Disintegration generates an increased number of smaller rock or mineral fragments of the original material that is being

Figure 12.1   Differential weathering between durable cliff and pillar-forming sandstone and less durable mudrocks in Utah. (Photo by Kevin Hefferan.) (For color version, see Plate 12.1, between pp. 248 and 249.)

disintegrated. When such fragments are eventually transported and deposited as sediment they may allow us to recognize the kinds of rock material originally exposed in the source area and to infer their dispersal pathways (Chapter 13). This may allow us to pinpoint the source or **provenance** of a particular sediment as being from a particular place where such rocks are (or were) exposed.

Decomposition is any breakdown of rock materials that involves changes in chemical composition. Decomposition generally alters a rock's mineralogy so that minerals stable at higher temperatures or pressures are altered to minerals stable at the temperatures and pressures near Earth's surface. Decomposition involves both inorganic and organic chemical processes and is strongly dependent on the availability of water, which plays a significant role in decomposition processes.

In the sections that follow, we will examine how the types of weathering processes and the degree of weathering that occur at a particular place vary greatly depending upon factors such as climate, rock type, slope and time. Disintegration is more prevalent in cold and dry climates; decomposition processes dominate in warmer and wetter climates (Figure 12.2). Steep slopes favor short-term, incomplete decomposition as rock materials are removed rapidly by erosion before decompo-

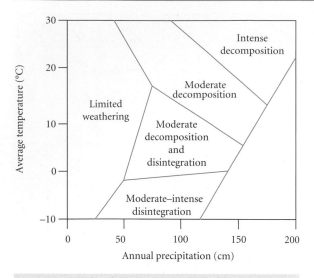

Figure 12.2 The relative roles of mechanical disintegration and chemical decomposition as a function of average annual temperature and rainfall.

Figure 12.3 Bryce Canyon, Utah showing pillars and windows formed by differential erosion, controlled by near vertical and horizontal joint sets. (Photo by John O'Brien.) (For color version, see Plate 12.3, between pp. 248 and 249.)

sition is complete, whereas gentle slopes favor longer term, more thorough weathering.

12.1.1 Disintegration

Disintegration includes any mechanical or organic weathering process that breaks rocks into smaller pieces of the same material. Several common disintegration processes are discussed in the paragraphs that follow.

Joint formation

Joints are fractures in rocks relative to which little or no tangential (fracture parallel) movement has taken place. Joints originate in response to stresses of various kinds and can have rather complicated histories. Most of the joints that act as conduits for fluids during weathering are produced by the decrease in confining pressure that occurs as formerly buried rocks approach the surface. Figure 12.3 illustrates joints in Eocene sandstones from Bryce Canyon, Utah. Stratification is not significantly offset parallel to the near vertical joints. Weathering is clearly accelerated along the joints, which in turn influence the differential weathering and erosion responsible for the development of spires (hoodoos) bounded by joint surfaces and capped by resistant rock layers. For more detailed information con-

cerning the relationships between joints and landscape evolution see Ritter et al. (2006), Huggett (2002) and Burbank and Anderson (2000).

Rocks can be buried to depths of many kilometers. The pressure exerted on buried rock objects at depth is referred to as **lithostatic pressure** or **confining pressure**. Lithostatic pressure is an isotropic confining stress (force per unit area) that results from the weight of the overlying rocks that push inward equally in all directions. As rocks move closer to Earth's surface, lithostatic pressure diminishes with decreasing burial depth. This decrease in lithostatic pressure or load is called **unloading** or **decompression**. Unloading results from either erosion or faulting that removes overlying rocks. As buried rocks experience unloading they tend to expand, and when they expand by more than 1 or 2%, they tend to fracture, resulting in joints. Joints may have many different orientations. They commonly occur along pre-existing weaknesses in rocks, many of which originated from earlier tectonic stresses that affected the rocks during burial. As decompression and expansion proceed, such fractures propagate and become more numerous. As a result, formerly intact rock is progressively fractured into smaller pieces. An interesting example occurs in quarries where workers excavate intact rock; within a few days, many fractures

Figure 12.4 Joints in anorthosite bedrock, Saranac Lake, New York. (Photo by Anita O'Brien.) (For color version, see Plate 12.4, between pp. 248 and 249.)

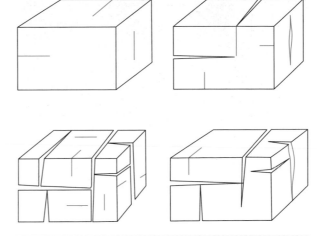

Figure 12.5 Sequential diagram (clockwise from top left) showing disintegration as visible fractures are enlarged and lengthened during frost action (shattering) and microfractures are enlarged to become visible fractures.

develop in the unexcavated rock on the quarry floor.

Sheet joints are rock fractures that open sub-parallel to Earth's surface (Figure 12.4) and develop to depths that may exceed 100 m. They tend to form under upwardly convex surfaces such as domes and ridges in homogeneous rocks like granites. In such rocks, maximum tensile stress is roughly perpendicular to the convex surfaces. The term **exfoliation** is used to describe sheet joints that resemble the curved surface of an onion. Other joints open at high angles to the surface (Figure 12.4).

Fracture propagation is strongly abetted in some climatic settings by a second set of processes called frost action.

Frost action

Frost action (shattering) occurs when pre-existing fractures and weak surfaces are enlarged by the expansion of water as it freezes (Figure 12.5). Water penetrates small fractures or boundaries between crystals. When it freezes, water expands by nearly 10%, creating tensile stresses that cause the crystals to separate or the small fractures to become enlarged. When the ice melts, the enlarged fracture can hold more water. When the water freezes again, the fracture grows again. Repeated freezing and thawing leads to

the formation of many new and enlarged, intersecting fractures that facilitate the progressive disintegration of the original rock into smaller pieces of the same material. **Frost wedging** occurs along fractures oriented steeply to Earth's surface. **Frost heaving** develops along surfaces parallel to Earth's surface when water freezes along bedding planes and/or sheet fractures, often lifting soil particles several centimeters above the surface.

Although joint formation and frost action dominate disintegration globally, several other processes are locally important. **Crystal growth** (other than ice crystals), may pry rock material apart as the crystal grows in a fracture or pore space. Rock expansion due to the growth of evaporate minerals such as gypsum is particularly important in desert environments and can be especially significant in the development of some arid climate soils, as discussed later in this chapter. **Slaking** occurs when minerals such as clays and micas expand when wetted. Significant mineral expansion generates stresses that cause adjacent minerals to disintegrate. Many workers have suggested that **thermal volume changes (insolation)** that result from daily or seasonal changes in rock temperature may cause significant amounts of disintegration. However, rocks have a low

Figure 12.6 Biological weathering by tree root growth in Silurian dolostone, Door County, Wisconsin. (Photo by Kevin Hefferan.) (For color version, see Plate 12.6, between pp. 248 and 249.)

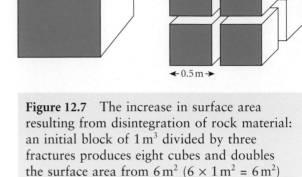

Figure 12.7 The increase in surface area resulting from disintegration of rock material: an initial block of $1\,m^3$ divided by three fractures produces eight cubes and doubles the surface area from $6\,m^2$ ($6 \times 1\,m^2 = 6\,m^2$) to $12\,m^2$ (8×6 faces $= 48 \times 0.25\,m^2 = 12\,m^2$).

heat capacity and are generally poor conductors of heat, so significant volume changes should only occur in response to thermal volume changes very close to the surface of rock bodies. Most solids, including rocks and wood and concrete structures such as highways, bridges and sidewalks, expand with increasing temperature and contract as temperatures decrease. Thermal expansion and contraction could induce stresses that would cause spalling of small particles of rock off the surface of a rock body. But the coefficient of thermal expansion (the rate at which volume changes with temperature change) in rocks and minerals is very small and likely insufficient to produce spalling.

Biological processes in disintegration

Root growth, in which rocks are pried apart and fractures enlarged as root systems expand during growth, is important in disintegration (Figure 12.6).

A host of **animal activities** in which organisms crack, drill, bore, burrow, mix and feed on rock material, causing it to be broken down into smaller pieces, are also significant in disintegration of surficial rock materials. But these activities pale compared to the activities of humans. Every time we move surficial material to farm or plant crops, to build houses, skyscrapers and shopping centers, to construct dams, canals, bridges, tunnels and highways, and to strip or otherwise mine subsurface resources, humans disintegrate huge volumes of surficial material. By changing the way such surface materials are organized, we create multiple geohazards and change the ways in which the materials undergo further weathering.

Disintegration and decomposition

Rock disintegration processes enhance chemical decomposition by increasing the surface area of the resulting rock fragments (Figure 12.7). This increase in surface area enlarges the surface over which chemical reactions can occur when rock surfaces are in contact with a solvent such as soil water or groundwater. This increase in chemically reactive surface area accelerates the rate at which decomposition reactions occur, so that a positive interaction between disintegration and decomposition results.

A closely related example of the interplay between disintegration and decomposition is shown by **spheroidal weathering** in which massive, well-jointed rocks such as granite, gabbro and basalt rocks weather into spheroidal forms (Figure 12.8). Spheroidal weathering begins with the formation of spaced rectangular joint sets that split the bedrock into multiple blocks or parallelepipeds, similar in shape to sugar cubes. This increases both

Figure 12.8 Spheroidal weathering of basaltic rock at Table Mountain, Golden, Colorado. (Photo by Kevin Hefferan.) (For color version, see Plate 12.8, between pp. 248 and 249.)

the surface area of the rock and access for the infiltration of groundwater. As with sugar cubes, dissolution proceeds most rapidly at the corners of the blocks where three chemically active faces intersect. Dissolution is slower along the edges of the blocks where two faces intersect and slower still in the centers of the individual faces. As the corners and edges are decomposed and eroded, blocks become rounded. Decomposition proceeds from the surface inward, so that more decomposed outer layers peel off before less decomposed inner layers, producing the forms shown in Figure 12.8.

12.1.2 Decomposition

Decomposition processes produce compositional changes in rocks and are induced through chemical and biological (biochemical) reactions between minerals and pore fluids such as water (aqueous solutions) and air (soil gases), organisms and organic compounds. Of these, the most significant agent in decomposition is downward percolating water. Every time it rains, water percolates through joints and between particles in the soil, causing additional increments of decomposition to occur. These change the composition of the original minerals and the composition of the pore fluids and often involve the formation of new minerals that can become significant components of soils over time. Biochemical decomposition

includes a complex set of interactive processes that depend upon:

- Bedrock composition and the ease with which different minerals decompose.
- Climatic factors such as temperature range, rainfall and groundwater levels.
- Organic factors such as vegetative cover, soil communities and bacterial activity.
- Soil water geochemical factors such as ionic concentrations, acidity–alkalinity and oxidation reduction potential.
- Topographic factors such as slope.

The following section deals with some of the major processes of decomposition summarized in Table 12.1.

Dissolution

Dissolution occurs when a mineral or other soil component is wholly or partially dissolved during chemical decomposition. The dissolution of the mineral clearly changes the chemical composition of both the original rock from which the mineral has been removed and the solution to which the dissolved solids have been added. Two common examples, the dissolution of halite (NaCl) and the dissolution of calcite ($CaCO_3$), are shown below. In the first example, halite reacts with polarized water molecules that separate sodium and chloride into ions that are dissolved in the water and surrounded by hydration sheaths.

$$1 \quad NaCl + H_2O \rightarrow Na^{+1}_{(aq)} + Cl^{-1}_{(aq)} + H_2O$$

In the second example, carbon dioxide gas dissolves in water to produce dissolved carbonic acid. During dissolution, the mildly acidic carbonic acid reacts with the calcite to produce dissolved calcium ions and dissolved bicarbonate ions in a process called **carbonation**.

$$2 \quad H_2O + CO_2 \rightarrow H_2CO_{3(aq)}$$

$$3 \quad H_2CO_{3(aq)} + CaCO_3 \\ \rightarrow Ca^{+2}_{(aq)} + 2(HCO_3)^{-1}_{(aq)}$$

In areas where soluble rocks such as limestones, dolostones or evaporites are common, long-term dissolution may lead to the formation of caves, sinkholes and other features characteristic of karst topography. Karst development, named after the Karst region in Slovenia, poses significant environmental issues related to surface collapse events and aquifer flow complexity (Box 12.1). Karst features are widespread in the United States (Figure 12.9) and throughout the world.

Ion exchange

Ion exchange occurs when ions are directly exchanged between a mineral and a solution. In the first example below, a hydrogen ion in aqueous solution is exchanged with a

Table 12.1 Major processes and products of decomposition.

Decomposition process	Examples	Decomposition products
Dissolution	$NaCl + H_2O \rightarrow Na^{+1}_{(aq)} + Cl^{-1}_{(aq)} + H_2O$ $H_2CO_{3(aq)} + CaCO_3 \rightarrow Ca^{+2}_{(aq)} + 2(HCO_3)^{-1}_{(aq)}$	Dissolved solids including Ca^{+2}, Mg^{+2}, Na^{+1}, K^{+1}, H_4SiO_4, CO_3^{-2}, SO_4^{-2}
Ion exchange	$KAlSi_3O_8 + H^{+1}_{(aq)} \rightarrow HAlSiO_3 + K^{+1}_{(aq)}$ $NaAlSi_3O_8 + H^{+1}_{(aq)} \rightarrow HAlSi_3O_8 + Na^{+1}_{(aq)}$	Dissolved solids including Na^{+1} and K^{+1}
Hydrolysis	$2KAlSi_3O_8 + 2H^{+1}_{(aq)} + 9H_2O$ $\rightarrow Al_2Si_2O_5(OH)_4 + H_4SiO_{4(aq)} + 2K^{+1}_{(aq)}$	Clay minerals such as kaolinite, illite, smectite
	$Mn_2SiO_4 + 4H_2O \rightarrow 2Mn(OH)_2 + H_4SiO_{4(aq)}$	Other hydroxides
Hydration	$CaSO_4 + 2H_2O \rightarrow CaSO_4 \cdot 2H_2O$ $Fe_2O_3 + H_2O \rightarrow 2FeO \cdot OH$	Hydrated and hydrous oxide minerals
Oxidation	$2Fe^{+2}_2SiO_4 + 4H_2O + O_2 \rightarrow 2Fe^{+3}_2O_3 + 2H_4SiO_{4(aq)}$ $4Fe^{+2}S^{-2}_2 + 15O_2 + 8H_2O$ $\leftrightarrow 2Fe^{+3}_2O_3 + 8S^{+6}O^{-2}_{4(aq)} + 16H^{+1}_{(aq)}$	Oxide mineral and dissolved solids, e.g., SO_4^{-2}, H_4SiO_4
Chelation		Dissolved metals contained in organic ring complexes (chelates)

Box 12.1   Karst development and its implications

The effects of dissolution are well illustrated by the formation of caves, sinkholes (dolines) and other karst features. Atmospheric and soil carbon dioxide, produced largely by organic respiration and bacterial decomposition, combine with water to produce carbonic acid, which progressively dissolves carbonate or evaporite minerals (Figure B12.1a). Karst features are particularly common in warm, humid climates where abundant water and biotic activity combined with high temperatures favor dissolution. As acidic groundwater flows along joints, faults and bedding planes, dissolution gradually enlarges them. This produces a positive feedback loop in which enlarged conduits permit more groundwater flow, which produces more dissolution. Dissolution occurs during relatively rapid flow in the vadose zone above the water table, in the zone near the fluctuating top of the water table and possibly during slower flow in the phreatic zone below the water table.

One result of such large-scale dissolution is the formation of networks of large cavities in the form of **caves** that frequently contain underground streams that enter the subsurface down dissolution features and emerge as cave springs (Figure B12.1b(i)). **Sinkholes (dolines)** are circular to ovoid depressions (Figure B12.1b(ii)) that form by (1) gradual loss of surface soluble rocks by accelerated dissolution; (2) gradual surface sinking of less soluble overburden (soil, sediment) by infiltration into cavities produced by dissolution of subsurface rock layers; or (3) sudden collapse of the surface by

Continued

Box 12.1   *Continued*

the collapse of rock or soil into an underlying cavity. The features and processes associated with karst development present many natural hazards that can be compounded by human activities. Natural karst hazards include severe flooding of sinkholes and karst valleys, continued dissolution and subsidence and/or sudden collapse of the surface during sinkhole formation. Many human activities accelerate subsidence and collapse. Diversion of water into underground systems can accelerate dissolution, leading to collapse of bedrock or soil into underground cavities. Overextraction of water from karst aquifers lowers the water table to the point where loss of water pressure triggers the collapse of surface materials into underlying cavities. In situations where collapse is imminent, additional surface loading during construction projects may trigger subsidence or collapse. Building houses, roads and commercial structures in karst regions presents real challenges. In addition, surface pollutants that rapidly enter groundwater in karst areas due to high recharge rates are quickly dispersed, thus polluting wells, springs and underground rivers over wide areas. Extreme care must be used when disposing of any hazardous materials in karst areas.

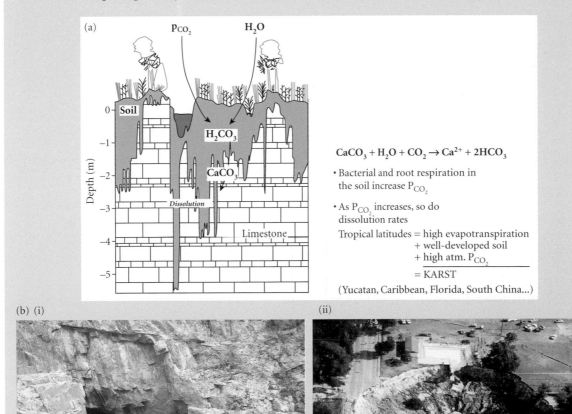

Figure B12.1   (a) Dissolution of carbonates by carbonated acidic groundwater. (b) (i) Cave spring with an underground river emerging from a cave produced by subsurface dissolution, Virginia. (ii) Sinkhole, Winter Park, Florida produced by the sudden collapse of the surface into a subsurface cavity, May, 1981. (a, b courtesy of the USGS.) (For color version, see Plate B12.1b, between pp. 248 and 249.)

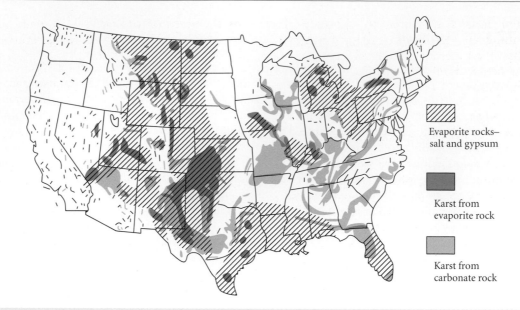

Evaporite rocks–
salt and gypsum

Karst from
evaporite rock

Karst from
carbonate rock

Figure 12.9 Distribution of karst dissolution features in the United States. (After Davies and LeGrand, 1972; with permission of the US Geological Survey.)

potassium ion in the **potash feldspar** ($KAlSi_3O_8$) with which the solution is in contact. Repetition of such exchanges enriches soil water in dissolved potassium, and the addition of hydrogen ions weakens the structure of the potash feldspar as it is progressively decomposed. Soil waters in warm, humid climates have a low pH, are rich in hydrogen ions and are acidic. Such solutions decompose feldspars faster than more alkaline solutions with higher pH and fewer hydrogen ions, so that feldspar decomposition is accelerated in areas with warm, wet climates.

$$1 \quad KAlSi_3O_8 + H^{+1}_{(aq)} \rightarrow HAlSiO_3 + K^{+1}_{(aq)}$$

The second example shows a similar ion exchange involving **albite** ($NaAlSi_3O_8$), the sodium plagioclase feldspar. Since feldspars are the most abundant mineral group in Earth's crust, ion exchange processes are an important part of chemical decomposition.

$$2 \quad NaAlSi_3O_8 + H^{+1}_{(aq)} \rightarrow HAlSi_3O_8 + Na^{+1}_{(aq)}$$

The third example involves a reversible ion exchange between the potassium clay mineral **illite** [$KAl_2AlSi_3O_{10}(OH)_4$] and dissolved hydrogen ion. Removal of potassium from illite has the potential to convert the

illite into the clay called kaolinite that commonly dominates the clay mineral assemblages in acidic soils.

$$3 \quad KAl_2AlSi_3O_{10}(OH)_4 + H^{+1}_{(aq)}$$
$$\leftrightarrow HAl_2AlSi_3O_{10}(OH)_4 + K^{+1}_{(aq)}$$

Hydrolysis

In weathering, **hydrolysis** is a chemical reaction between a mineral and water in which dissolved hydrogen ions and/or hydroxyl ions are added to form one or more new minerals. In most hydrolysis reactions, the original mineral is a silicate mineral and the new mineral is a hydroxide or clay mineral. Feldspars, the most abundant group of silicate minerals in Earth's crust, decompose by hydrolysis into clays. As a result, clay minerals are the most abundant group of new minerals produced during chemical decomposition. Because clay mineral crystals tend to be very small (<4 μm), clay minerals are the most abundant constituents of the mud fraction of detrital sediments in soils that are later dispersed by erosion and transportation and then deposited in surface environments.

In the first example below, potassium feldspar ($KAlSi_3O_8$) reacts with dissolved hydrogen ions and water to produce the clay mineral kaolinite $[Al_2Si_2O_5(OH)_4]$ plus dissolved orthosilicic acid (commonly referred to as dissolved silica) and dissolved potassium ions.

1 $2KAlSi_3O_8 + 2H^{+1}_{(aq)} + 9H_2O$
$\rightarrow Al_2Si_2O_5(OH)_4 + H_4SiO_{4(aq)} + 2K^{+1}_{(aq)}$

The second reaction is similar but involves the sodium plagioclase feldspar albite $(NaAlSi_3O_8)$ reacting with hydrogen ions and water to form kaolinite $[Al_2Si_2O_5(OH)_4]$ plus orthosilicic acid and dissolved sodium ions. In both cases, the reactions may continue over long periods of time so that feldspar crystals are progressively converted into clay minerals by hydrolysis.

2 $2NaAlSi_3O_8 + 2H^{+1}_{(aq)} + 9H_2O$
$\rightarrow Al_2Si_2O_5(OH)_4 + H_4SiO_{4(aq)} + 2Na^{+1}_{(aq)}$

The third example generally occurs during the weathering of manganese-bearing minerals such as ferromagnesian silicates in which manganese (Mn^{+2}) substitutes for ferrous iron (Fe^{+2}). In this reaction, manganese-bearing olivine (Mn_2SiO_4) reacts with water to produce the hydroxide mineral **pyrolusite** $[Mn(OH)_2]$ plus dissolved orthosilicic acid.

3 $Mn_2SiO_4 + 4H_2O$
$\rightarrow 2Mn(OH)_2 + H_4SiO_{4(aq)}$

Hydration and dehydration

Hydration involves the addition of water to a crystal structure during the reaction between a mineral and the aqueous solution. In the first example below, the anhydrous calcium sulfate mineral **anhydrite** is converted into the hydrated calcium sulfate mineral **gypsum** by the addition of water. The reversal of this reaction is called **dehydration** and involves the conversion of hydrated gypsum into anhydrous anhydrite with loss of water from the crystal structure.

1 $CaSO_4 + 2H_2O \rightarrow CaSO_4 \cdot 2H_2O$

The second example illustrates the conversion of the anhydrous iron oxide mineral **hematite** into the hydrous iron oxide mineral **goethite**

by the addition of water. This reaction is also reversible with goethite being converted to hematite by dehydration.

2 $Fe_2O_3 + H_2O \rightarrow 2(FeO \cdot OH)$

Oxidation

Oxidation is a chemical reaction in which one or more electrons are transferred from a cation in the mineral to an anion, increasing the valence of the cation. Oxygen is strongly electronegative (Chapter 2) and therefore tends to capture electrons. Because oxygen is abundant in many weathering environments, the majority of oxidation reactions involve the transfer of electrons from a cation to oxygen as they react chemically to produce an oxide mineral. The production of an oxide mineral is not required by oxidation reactions. Only the loss of an electron is required. But because oxygen is the most abundant electronegative element on Earth, it dominates oxidation reactions.

In the first example below, ferrous (Fe^{+2}) iron-bearing **olivine** $(Fe^{+2}_2SiO_4)$ combines with dissolved oxygen to form the ferric (Fe^{+3}) iron mineral hematite $(Fe^{+3}_2O_3)$ plus dissolved orthosilicic acid. The reaction involves the loss of an electron from the iron to the oxygen, that is, oxidation, which increases the valence state of the iron from +2 to +3.

1 $2Fe^{+2}_2SiO_4 + 4H_2O + O_2$
$\rightarrow 2Fe^{+3}_2O_3 + 2H_4SiO_{4(aq)}$

The second example is similar but involves the conversion of the manganese silicate mineral **rhodonite** $(Mn^{+2}_2SiO_3)$ by reaction with water and dissolved oxygen into the manganese oxide mineral **manganite** $(Mn^{+3}O_2)$ plus dissolved orthosilicic acid.

2 $2Mn^{+2}_2SiO_3 + 4H_2O + O_2$
$\rightarrow 2Mn^{+3}O_2 + 2H_4SiO_{4(aq)}$

These two oxidation equations are simple, but representative, examples of the many oxidation reactions involved in the decomposition of ferromagnesian silicate minerals.

The third example illustrates the oxidation of the iron sulfide mineral **pyrite** $(Fe^{+2}S_2)$ to the iron oxide mineral hematite $(Fe^{+3}_2O_3)$. The oxidation (electron loss) of the iron is accomplished when the ferrous (Fe^{+2}) sulfide mineral pyrite reacts with water and dissolved oxygen to produce the ferric (Fe^{+3}) oxide mineral hematite plus dissolved sulfate ions and dissolved hydrogen ions. In the reaction below, the sulfur in the sulfide (-2) pyrite gains electrons to become sulfur ($+6$) in the sulfate (SO_4^{-2}) anion.

$$3 \quad 4Fe^{+2}S^{-2}_2 + 15O_2 + 8H_2O$$
$$\leftrightarrow 2Fe^{+3}_2O_3 + 8S^{+6}O_4{}^{-2}{}_{(aq)} + 16H^{+1}{}_{(aq)}$$

Under reducing conditions (low oxidation–reduction potential), chemical reaction number three is easily reversible. When it is reversed, the iron ions gain electrons so that the valence state is reduced from ferric iron (Fe^{+3}) to ferrous iron (Fe^{+2}). This occurs as hematite combines with sulfate and hydrogen ions and is converted to pyrite with the release of oxygen and water. Such reactions that involve the loss of electrons are called reducing reactions and the general name for processes involving the loss of electrons is **reduction**. Oxidation–reduction reactions are further discussed in Chapter 13.

Organic decomposition and chelation

Organic activity plays a major role in chemical decomposition processes. **Chelates** are organic hydrocarbon ring complexes produced directly by lichen and indirectly by the decay of humus. These highly soluble organic molecules tend to bind metallic elements such as Ca^{+2}, Mg^{+2}, Al^{+3}, Fe^{+2}, Fe^{+3}, K^{+1} and Na^{+1}, in effect removing them from solution. This process is called **chelation**. Chelation often involves the exchange of hydrogen (H^{+1}) ions from the chelating agent to the solution and metal ions from the solution to the chelate. The increase in hydrogen ions decreases the pH of the solution, making it more acidic, while the decrease in metal ions in solution makes the metals in the remaining minerals more soluble. Both processes tend to increase the decomposition of metal-bearing minerals significantly. Chelates are also very important to plants as they provide nutrient metals to plants in a form that makes the metals readily available for absorption.

Organic activity also indirectly affects decomposition rates. Respiration produces carbon dioxide (CO_2) as a byproduct. When CO_2 combines with water (H_2O), carbonic acid (H_2CO_3) forms (equation 12.1).

$$CO_2 + H_2O = H_2CO_3 \quad \text{(equation 12.1)}$$
$$H_2CO_3 = (HCO_3)^{-1} + H^{+1} \quad \text{(equation 12.2)}$$
$$(HCO_3)^{-1} = (CO_3)^{-2} + H^{+1} \quad \text{(equation 12.3)}$$

When carbonic acid dissociates, a hydrogen ion is released during the formation of a bicarbonate $[(HCO_3)^{-1}]$ ion (equation 12.2). In certain situations, the bicarbonate ion dissociates into carbonate $[(CO_3)^{-1}]$ ion (equation 12.3) and an additional hydrogen (H^{+1}) ion is produced. The release of hydrogen ions lowers the pH of the soil water and generally increases its acidity. Thus significant increases in soil CO_2 content ultimately generate significant increases in dissolved hydrogen (H^{+1}) ions making soil waters significantly more acidic. This tends to speed chemical decomposition of most common rock-forming minerals.

12.2   DISSOLVED SOLIDS

As noted above, when water infiltrates into fractures and spaces between grains, it dissolves constituents from the rocks and minerals with which it is in contact. An essential byproduct of the decomposition reactions discussed in the previous section is an abundance of dissolved solids in soil and groundwater. Some of these solids are reprecipitated in soils and others are carried underground and precipitated as mineral cements during sedimentary rock diagenesis. A large proportion of the dissolved solids in groundwater are discharged by springs into surface waters (Box 12.2). Some dissolved solids reside in surface water for a considerable period of time, especially in lakes. However, most of the dissolved load in surface waters eventually flows into the oceans via surface runoff where it is joined by smaller amounts of dissolved solids discharged directly into the ocean by submarine springs. In this way dissolved solids are widely dispersed through groundwater, surface water and ocean water. Over time, these dissolved solids are removed from solution to form solid biochemical sediments that accumulate on Earth's surface (Chapter 14) and hydro-

Box 12.2   People and Earth materials: mineral water

Mineral water sales worldwide exceed 10^{13} liters per year and are rising rapidly. But just what is mineral water and where does it come from? Natural mineral water is obtained from springs or from wells drilled into the aquifer that supplies the spring. The "minerals" in mineral water are dissolved solids in concentrations that exceed 250 parts per million (250 ppm). Where do these dissolved solids come from? Natural rainwater has less than 10 ppm dissolved solids. Once it infiltrates into the ground, it finds itself in contact with rocks and minerals with varying degrees of solubility. As decomposition and related chemical reactions occur, natural waters leach constituents from the rocks and minerals and their total dissolved solid concentrations increase to where they become mineral waters. Mineral water taste depends largely on what "minerals" are dissolved, and an analysis of a mineral water permits one to infer the types of rocks and minerals that the groundwater was in contact with prior to being discharged as a spring. For example, hydrogen (H^{+1}) ions are sensed as sour, sodium (Na^{+1}) ions as salty and certain organic sulfate and chloride substances as bitter. In addition, natural mineral waters are carbonated to different degrees, which depend primarily on the amount of dissolved carbon dioxide gas they contain; thus the familiar choice between mineral water with and without gas. To capture the "spritz", naturally carbonated waters are usually recovered from a well and bottled under pressure so that the dissolved carbon dioxide is not lost. Many carbonated mineral waters are artificially carbonated. Given the commercial importance of the spring and mineral water industries, great care is taken to prevent the infiltration of pollutants into the groundwater that supplies these waters.

Mineral water is also important in the manufacture of beer. The English city of Burton-upon-Trent, home of the Bass Brewery, is famed for its hard, calcium-rich mineral waters, which are believed to be ideal for brewing fine ales. These waters, obtained from wells, owe their properties primarily to the dissolution of gypsum beds that underlie the valley of the Trent River. Breweries all over the world "burtonize" their own water sources by the addition of soluble gypsum in their quest for a better brew.

---

thermal metamorphic rocks on the sea floor (Chapter 18).

In the sections that follow, we focus on the detrital sediments and soils that are generated by weathering processes.

12.3   DETRITAL SEDIMENTS

The solid, inorganic components of residual soils are **detrital sediments**. These detrital sediments are either **resistates**, which are residual mineral and rock fragments of the original parent rock that have survived (resisted) decomposition, or **new minerals** generated by decomposition processes during the weathering process. New minerals are produced by processes such as hydrolysis (e.g., clay minerals), oxidation (e.g., hematite, goethite, pyrolusite), hydration (e.g., gypsum) and carbonation (e.g., calcite), as discussed in the previous section.

12.3.1   Resistates and chemical stability

The population of resistate rock and mineral fragments that occur in a particular source area depends on several factors. One important factor is the mineral composition of the source bedrock. One cannot find a resistate rock or mineral fragment in a residual soil that was not in the original bedrock. When such resistate fragments are eroded, transported and deposited, they carry with them information about the rock types and minerals that existed in the source area at the time they were produced (Chapter 13). These in turn may offer vital clues to the tectonic setting in which deposition likely occurred and help answer a variety of other questions.

Resistate rock and mineral fragments from the parent rock occur in residual soils only if they survive decomposition. Whether such rock and mineral fragments survive decomposition depends primarily on:

Table 12.2 Chemical stability of major minerals under average weathering conditions.

Mineral stability (lowest to highest)	Rate of decomposition	Goldlich's series (after Bowen's reaction series)
Halite	Fastest	
Calcite		
Olivine		
Pyroxenes (augite)		
Calcic plagioclase		
Amphiboles (hornblende)		
Sodic plagioclase (albite)		
Biotite		
Orthoclase		
Muscovite		
Clay minerals		
Zircon, rutile, tourmaline		
Quartz		
Aluminum oxides (gibbsite)		
Iron oxides (hematite)	Slowest	

Olivine → Pyroxene → Hornblende → Biotite

Anorthite → Labradorite → Andesine → Oligoclase/albite

Orthoclase

Muscovite

Quartz

1 The resistance of each mineral to decomposition, based upon its chemical stability and the geochemical environment in the soil.

2 The rate of decomposition, which depends primarily on climatic factors such as precipitation, temperature, vegetation and organic activity.

3 The duration of decomposition, which depends primarily on erosion rates, which in turn depend on relief (slope), vegetative cover and rainfall.

Minerals that strongly resist decomposition are said to be chemically stable; minerals easily decomposed are said to be chemically unstable. As discussed in Chapter 2, ionically bonded substances with weak bonds are especially susceptible to dissolution, whereas strongly bonded covalent minerals are much more resistant to dissolution.

Chemical stability of resistates

The **chemical stability** of any mineral – its resistance to decomposition – depends on the details of climate and soil geochemistry, but some useful generalizations can be made. Goldlich (1938) essentially inverted Bowen's reaction series (Chapter 8) and applied it to an entirely separate set of geological processes involving mineral stability during weathering.

Goldlich's rule states that the susceptibility of common igneous minerals is inversely proportional to their crystallization temperatures as summarized in Bowen's reaction series (Table 12.2). Minerals that crystallize at high temperatures, such as olivines, pyroxenes and calcium-rich plagioclases, are chemically unstable in the low temperature and low pressure environment of Earth's surface. As a result, they are far more susceptible to decomposition on Earth's surface than minerals that crystallize at lower temperatures, such as potassium feldspars, muscovite and quartz. As a result, these low temperature minerals tend to be preserved as resistate minerals.

Chemically unstable minerals, such as halite, calcite, olivine and pyroxenes, tend to become relatively depleted as they are decomposed and removed from the resistate population. In contrast, minerals stable at low temperatures and pressures, such as quartz, clays and iron oxides, tend to become relatively enriched in the resistate population. The relative chemical stability or susceptibility to chemical decomposition of common minerals is generally known and is shown by selected examples on the left side of Table 12.2. Among the rarer heavy minerals (specific gravity >2.8) that are least susceptible to decomposition are rutile, tourmaline and zircon, and these too are likely to survive decomposition. The tendency of chemically

Table 12.3 Factors that affect the survival of resistate fragments during decomposition.

	Decomposition rate		
	Slow ⟶		Rapid
Mineral resistance to decomposition	High	Moderate	Low
Rainfall	Low	Moderate	High
Temperature	Low	Moderate	High
Vegetation/organic activity	Sparse/low	Moderate	High
	Duration of decomposition		
	Short ⟶		Long
Erosion rate	Rapid	Intermediate	Slow
Relief/slope	High/steep	Intermediate/moderate	Low/gentle
Vegetative cover	Sparse	Moderate	Extensive
	Surviving detrital assemblage		
	Many unstable components survive	Metastable and resistant components survive	Only resistant components survive

stable minerals to survive decomposition means that they have the greatest potential to be dispersed from the source area and to occur in sediments deposited elsewhere on Earth's surface. The high chemical stability of quartz is one of the reasons why it is the most abundant mineral in most sandstones and gravelstones (conglomerates and breccias), even though feldspars are more common than quartz in primary source rocks. The high chemical stability of clay minerals helps to explain why they are the major constituents of mudrocks such as shales.

Rate and duration of decomposition

Rapid rates of decomposition accelerate depletion of chemically unstable minerals in the resistate population while simultaneously accelerating the enrichment of chemically stable constituents that resist decomposition. As noted earlier, heat and rainfall increase decomposition rates because heat and water catalyze decomposition reactions. Thus rates of decomposition are highest in areas with warm, humid climates (see Figure 12.2; Table 12.3).

The duration of weathering in the source area is extremely important in determining the survival rates of residual rock and mineral fragments and depends primarily on rates of erosion. If erosion rates are high, rock material will be removed and dispersed from the

source area before significant amounts of decomposition have occurred, so that a higher proportion of chemically unstable residual detritus will be dispersed into areas of deposition. If erosion rates are low, rock material will stay in the source area and will decompose for a longer period of time, which results in smaller proportions of only the most resistant residual detritus being delivered to areas of deposition (Table 12.3).

Erosion rates depend primarily on (1) relief, (2) vegetative cover, (3) precipitation, and (4) the type of erosion agents involved. Erosion rates are generally proportional to relief. The steeper the slope, the more rapidly detrital sedimentary materials are removed from it by mass flows and running water.

However, the proportional relationship between relief and erosion is affected by other factors such as vegetative cover. The root systems of vegetative cover tend to hold soils in place, thus retarding rates of erosion, while aiding decomposition by the production of more acidic soil water. You may be familiar with examples of wind blowing dust from vacant lots or from freshly tilled fields but not from adjacent lots covered with grass or other vegetation. Vegetation is also planted to reduce erosion rates in coastal sand dunes.

In addition, the type of erosion agent strongly affects erosion rates. Because glaciers generally erode all sizes of material down to the bedrock, when glaciers flow over an

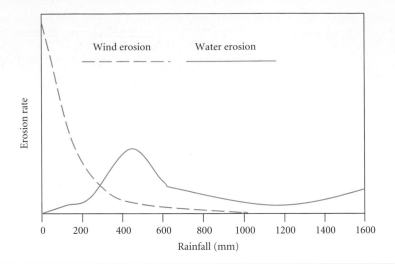

Figure 12.10   Generalized erosion rates for wind and water as a function of rainfall.

area, vegetated or not, erosion rates are initially very rapid indeed. However, once the soil has been removed down to bedrock, rates of glacial erosion decelerate rapidly. Mass wasting processes such as landslides and debris flows erode extremely rapidly. Erosion by water increases with flow velocity, runoff volume and slope. Wind erosion is most effective in dry areas with minimal vegetative cover and low rainfall (Figure 12.10).

One might expect erosion rates to increase systematically with rainfall, but, as shown in Figure 12.10, this is not always the case. Where rainfall is less than 350–400 mm/yr, erosion rate increases with rainfall as expected. This is largely due to the lack of vegetation and relatively impermeable soils that characterize desert regions. As a result of dry, hardpan soils, infiltration is minimal so that most precipitation remains on the surface as overland flow, thereby increasing erosion rates of poorly vegetated soils.

Above 350–400 mm/yr, erosion rates actually fall up to 890–1015 mm/yr. In this range, an increasingly thick vegetative cover of grasses and other plants tends to retard erosion rates. Of course one fire or land-clearing project can destroy the vegetative cover and substantially increase rates of erosion by wind, mass wasting and surface runoff. At still higher precipitation rates, erosion rates again increase, primarily because they promote higher rates of mass wasting, which not only

disperses sediments downslope, but also removes much of the vegetative cover in the process.

In addition to residual minerals, new minerals develop as a result of decomposition.

12.3.2   New minerals

The detrital fraction of residual soils generally contains a proportion of non-resistate or new minerals that are produced by decomposition of the original rock. These new minerals are commonly concentrated in the lower portions of mature soils. The most abundant group of new minerals produced by decomposition (hydrolysis) processes is clay minerals.

Clay minerals

Clay minerals constitute a large group of aluminum-bearing phyllosilicate minerals (Chapter 5). Most clay crystals are of small size (<4 μm). For this reason, the term clay has two distinct but overlapping meanings in the context of sedimentary rocks:

1   Clay is used as a compositional term for a group of phyllosilicate minerals with a specific set of chemical compositions and structures.
2   Clay is also used as a textural term for any very small particles (<0.004 mm diameter) that may or may not be clay mineralogically.

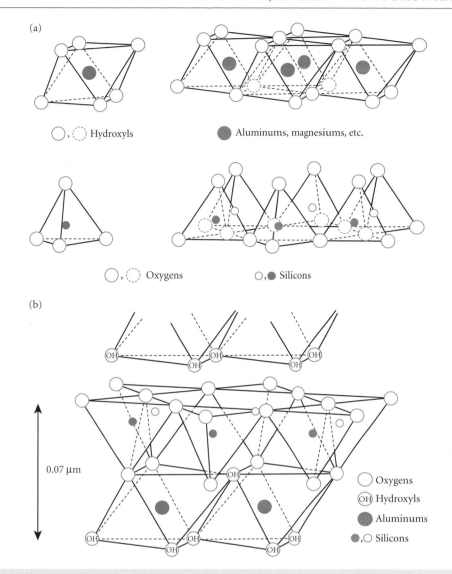

Figure 12.11 (a) The basic components of clay minerals: a single silica tetrahedral layer below and a single aluminum or magnesium octahedral layer above. (b) Coordination polyhedra model of a two-layer, 7 angstrom (0.07 μm) kandite clay mineral such as kaolinite. (After Grim, 1968; with permission of McGraw-Hill.)

The following discussion focuses on clay minerals defined by composition rather than texture. These phyllosilicate minerals (Chapter 5) are best understood in terms of their structures, which consist of two or more layers (sheets) connected by shared bonds, and in some cases separated by interlayer sites. Three major types of layers (sheets) occur (Figure 12.11a):

1 S-layers, which are silica-rich tetrahedral layers (Si_2O_5) with some Al_2O_5 in which silicon and/or aluminum are in tetrahedral coordination with oxygen; tetrahedral layers are also called T-layers.

2 G-layers, which are gibbsite octahedral layers [$Al_2(OH)_4$].

3 B-layers, which are brucite octahedral layers [$(Mg,Fe)_2(OH)_4$] in which aluminum, magnesium or iron are in octahedral coordination with oxygen and hydroxyl ions. Octahedral G- and B-layers are also called O-layers.

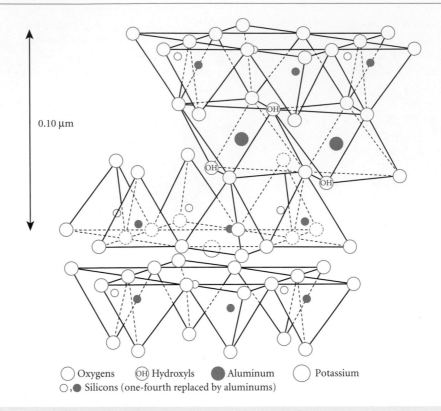

0.10 μm

○ Oxygens (OH) Hydroxyls ● Aluminum ○ Potassium
○, ● Silicons (one-fourth replaced by aluminums)

Figure 12.12   A three-layer illite model depicting an aluminum octahedral layer (G) sandwiched between two silica tetrahedral layers (S) in the form S-G-S, with interlayer potassium (K^{+1}). (After Grim, 1968; with permission of McGraw-Hill.)

Variations in the types and arrangements of these three types of layers permit four major groups of clay minerals to be distinguished, each with a different phyllosilicate structure. Hybrids of all four groups are common.

Kandites are two-layer (S-G or T-O) clays with a basic structure that consists of a single tetrahedral layer (S) bonded to a single octahedral layer (G). Pure kandites are composed of the repeated stacking of these basic structural units (S-G-S-G … S-G). Kandites are sometimes referred to as 7 angstrom clays because the two layers have an aggregate thickness of approximately 7Å (0.07μm), which is the repeat distance for the basic structural units. This repeat distance between layers is called d-spacing and is extremely helpful in distinguishing between clay minerals using methods such as X-ray diffraction. The best known example of a kandite clay mineral is **kaolinite**, also known as China clay, an important natural resource used as the raw material in the production of fine

ceramics and to produce glossy paper products. In its pure form, kaolinite consists of alternating tetrahedral sheets of Si_2O_5 and octahedral gibbsite sheets of $Al_2(OH)_4$ with an aggregate thickness of 7Å (Figure 12.11b). The chemical formula of pure kaolinite is written as $Al_2(Si_2O_5)(OH)_4$ or $Al_4(Si_4O_{10})$ $(OH)_8$, which clearly reflects its two-layer structure and composition. Some kaolinites possess H^+ ions between the S-G layer pairs, which changes their composition slightly.

Illites are three-layer (S-G-S or T-O-T) clays with a basic structure that consists of a single octahedral layer (G) sandwiched between two tetrahedral layers (S) (Figure 12.12). In addition, one-fourth of the silica tetrahedra [$(SiO_4)^{-4}$] in the tetrahedral layer are replaced by aluminum tetrahedra [$(AlO_4)^{-5}$]. This substitution creates a positive charge deficiency that necessitates the paired substitution of a similar number of positively-charged cations such as potassium (K^{+1}) and hydrogen (H^{+1}). This paired cation substitution occurs in the interlayer space (//) between adjacent S-G-S units. Pure illites are composed of the repeated stacking of structural units (S-G-S//S-G-S ... S-G-S) with interlayer K^+ ions. Illites are also referred to as 10 angstrom clays because the three layers have an aggregate thickness or repeat distance of approximately $10\,\text{Å}$ ($0.1\,\mu\text{m}$) A few water molecules can also occur in these interlayer sites. The chemical formula for illite can be written in several different ways. One common form is $(K,H)Al_2AlSi_3O_{10}(OH)_2 \cdot nH_2O$ and clearly reflects the two tetrahedral layers ($AlSi_3O_{10}$), the octahedral layer $[Al_2(OH)_4]$ and the interlayer cations and water. Clay mineral compositions are quite varied, and most illites depart from the standard composition noted above in that the silicon:aluminum (Si/Al) ratio in the tetrahedral site is a little larger than 3:1. This reduces the charge imbalance produced when aluminum (Al^{+3}) replaces silicon (Si^{+4}) in the tetrahedral site and thus reduces the amount of interlayer potassium (K^{+1}) required to electrically balance the crystal structure.

Smectites are three-layer (S-G-S, S-B-S or T-O-T), expandable lattice clays with a basic structure that consists of an octahedral layer (G and/or B) sandwiched between two tetrahedral layers. Interlayer sites are highly expandable and can incorporate many ions, including large amounts of water. Pure smectites are composed of the repetition of structural units such as (S-G-S//S-B-S//S-B-S ... SGS) separated by expandable interlayer sites in which water and such ions as calcium (Ca^{+2}), sodium (Na^{+1}) and hydrogen (H^{+1}) may be absorbed. Because smectites are expandable layer clays, their repeat distance may range from $10\,\text{Å}$ ($0.10\,\mu\text{m}$) to more than $21\,\text{Å}$ ($0.21\,\mu\text{m}$), depending on how much interlayer absorption has occurred. Such clays tend to expand by hydration when wet and to contract by dehydration when dry. Some sodium-rich montmorillonite can expand to ten times its normal thickness when wet (Prothero and Schwab, 2004). Cation absorption is balanced by paired substitutions of cations of lower charge in the octahedral and tetrahedral sites. A common smectite clay mineral is **montmorillonite** (Figure 12.13).

Montmorillonite's composition is quite variable; its chemical formula may be written as $(Ca,Na,H)(Al,Mg,Fe)_2(SiAl)_4O_{10}(OH)_2 \cdot$

nH_2O. The first set of parentheses contains the common interlayer cations; the second set contains the octahedral G- or B-layer cations; the third set contains the tetrahedral S-layer cations; and the nH_2O refers to the variable amounts of absorbed interlayer water. Soils that contain large amounts of montmorillonite are called **expansive soils**. Not only does their volume change as water is absorbed and released, but also their strength and plasticity. Such soils are often involved in landslides and ground subsidence events that cause severe property damage and loss of life. Expansive soils are discussed in more detail in the section of this chapter devoted to the engineering properties of soils.

Chlorites are four-layer (S-B-S+B) clay minerals. The basic chlorite structure consists of an octahedral brucite layer (B) sandwiched between two tetrahedral layers (S) with an additional octahedral brucite layer (B), which may have a somewhat different composition than the other brucite layer (Figure 12.14). Pure chlorites, of which many varieties exist, have the basic stacking pattern (S-B-S+B, S-B-S+B ... S-B-S+B). Chlorites are 14 angstrom clays because the basic structural unit (S-B-S+B) is $14\,\text{Å}$ ($0.14\,\mu\text{m}$) thick and their repeat stacking distance is $14\,\text{Å}$. A common chlorite group mineral is **clinochlore**, whose formula can be written as $(Fe,Mg)_3(Fe_3)AlSi_3O_{10}(OH)_8$. The first set of parentheses indicates the cation content of the octahedral site sandwiched between the tetrahedral sites in which aluminum (Al^{+3}) substitutes for every fourth silicon (Si^{+4}), and the second set of parentheses shows the content of the second octahedral site. Many chlorites contain much more magnesium (Mg^{+2}) in the octahedral sites. Chlorite is also a common metamorphic mineral and is therefore more abundant in metamorphic rocks than other clay minerals.

Many clay minerals are complex hybrids of the clays discussed above. Hybrid clays, which contain stacked layer sequences characteristic of more than one type of clay mineral, are called **mixed layer clays**. For example, a clay with the structure S-B-S+B//S-G-S//S-B-S//S-B-S+B would be a mixed chlorite–smectite clay mineral, whereas one with the structure S-G-S-G-S-G-S-S-G-S-SG would be a mixed kandite–illite clay mineral. The most common group of mixed layer clay minerals are illite–smectite clays.

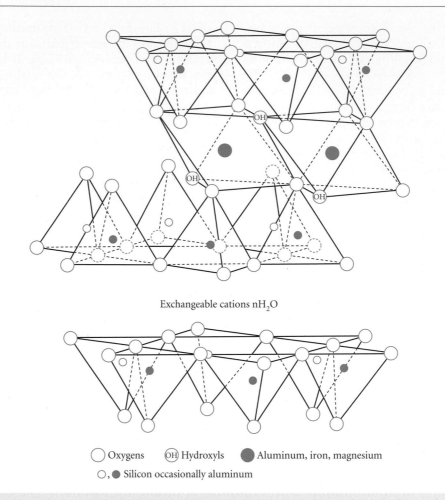

Exchangeable cations nH$_2$O

○ Oxygens   ⓞ Hydroxyls   ● Aluminum, iron, magnesium

○, ● Silicon occasionally aluminum

Figure 12.13 A three-layer smectite clay, montmorillonite, in a partially expanded state, with absorbed water molecules and cations occupying the space between the triple layers. (After Grim, 1968; with permission of McGraw-Hill.)

Degraded clay minerals lack the interlayer constituents that should occur. In most cases, such interlayer constituents have been removed from them by pore waters during decomposition or other types of chemical alteration. For example, illites that lack their full complement of interlayer potassium (K^{+1}) ions are degraded illites, and smectites that lack their full complement of interlayer calcium (Ca^{+2}) and/or sodium (Na^{+1}) ions are degraded smectites. Clay degradation occurs most commonly under acidic conditions that tend to leach cations from the clay structures. Such leaching processes are important in the development of soils and soil horizons, as discussed later in the chapter.

Insoluble iron and manganese oxides and hydroxides

Oxides and hydroxides of iron and manganese are significant constituents of many soils. In some soils, especially lateritic soils (oxisols) formed in warm, humid environments, oxides and hydroxides are especially abundant. These minerals are largely produced by the decomposition of ferromagnesian silicates and iron-bearing sulfides by processes that include dissolution and reprecipitation, oxidation, hydrolysis and/or hydration. The most common products are listed in Table 12.4.

Hematite is responsible for the reddish color, and limonite and goethite are responsi-

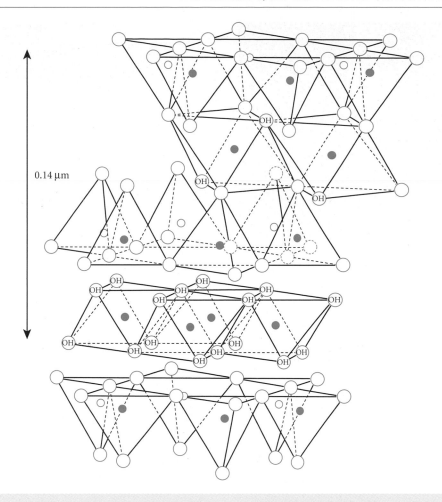

Figure 12.14 A four-layer, 14 angstrom (0.14 μm) structure typical of chlorites, with two tetrahedral layers that sandwich an octahedral layer in three-layer clays (S-B-S) and an octahedral brucite (B) layer below. Key as on Figure 12.13. (After Grim, 1968; with permission of McGraw-Hill.)

Table 12.4 Common iron and manganese oxides and hydroxides produced by weathering.

Mineral or mineraloid	Chemical composition
Hematite	Fe_2O_3
Goethite	$FeOOH$
Limonite*	$FeOOH \cdot nH_2O$
Pyrolusite	$Mn(OH)_2$
Manganite	MnO_2
Romanechite	$BaMnMn_8O_{16}(OH)_4$

* Mineraloid.

ble for the yellow-brown color of many soils and weathered rock surfaces Manganese minerals yield dark gray to black colors in soils and on weathered surfaces. Native popula-tions have long collected these soil constitu-ents, mixed them with water and used them as pigments for artistic conceptions (Figure 12.15).

Highly insoluble aluminum oxides and hydroxides

In highly acidic soils in warm, arid climates, decomposition may generate aluminum oxide and hydroxide minerals rather than alumi-num silicate minerals such as clay minerals. In such cases, the silica in clay minerals is dissolved, which leaves aluminum combined with hydroxyl ions, oxygen and some water to form a mineral suite characteristic of the major aluminum ore called **bauxite**. Most bauxite deposits consist of a variety of alumi-num-bearing minerals and mineraloids that

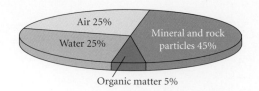

Figure 12.16   Proportions of the major components in average soil.

Figure 12.15   Australian rock art using hematite (red), limonite (brown), kaolinite (white) and manganese oxides/hydroxides (black), from Nourlangie Rock, Kakadu National Park, northern Australia. (Photo by John O'Brien.) (For color version, see Plate 12.15, between pp. 248 and 249.)

are commonly associated with the insoluble iron and manganese minerals in oxisols. Three common minerals in bauxite deposits are **gibbsite** [Al(OH)$_3$], **diaspore** [α-(AlOOH)] and **boehmite** [β-(AlOOH)]. Because the aluminum in bauxite is relatively weakly bonded to the oxygen and hydroxyl ions, it is more easily separated during refining than is the aluminum bonded to silicon in clay minerals, feldspars and other silicates. Without the widespread formation of bauxite minerals in deeply weathered, tropical soils, aluminum would be much more expensive than it is.

Other common soil minerals

The calcium carbonate (CaCO$_3$) minerals **calcite** and **aragonite** are abundant in some soils, especially in caliche soils called aridosols that develop in fairly arid climates, as discussed later in this chapter. The calcium sulfate minerals gypsum (CaSO$_4$·2H$_2$O) and **anhydrite** (CaSO$_4$) also occur in some aridosols.

Soils developed in anoxic areas with reducing environments, such as histosols, may contain sulfide minerals such as pyrite (FeS$_2$).

Under special circumstances, dozens of other minerals can be precipitated from soil

pore waters, but their discussion is beyond the scope of this chapter. Precipitation of these minerals and of the more common ones discussed previously may produce anything from small clumps of partially indurated material called soil **peds** to truly solid crusts called **durisols** or **petrosols**. For this reason, soils are not completely unconsolidated materials, as will be explored in the following section.

12.4   SOILS

Soils are largely unconsolidated surficial deposits produced by weathering processes and capable of supporting rooted plant life. Weathering of bedrock generates in situ **residual soils**. The detrital constituents of residual soils may then be eroded and dispersed by transportation to be deposited elsewhere on Earth's surface. Further weathering of such transported sediments generates **transported soils**. Most soils are relatively thin, typically up to 2 m thick according to the United State Department of Agriculture's Natural Resources Conservation Service (USDA-NRCS, 1999). Figure 12.16 summarizes the average composition of soils, following the 2 m definition, in terms of mineral, organic, water and air content. Humus is partially decayed, dead, organic matter. These percentages represent averages that do not portray the great variation in soil compositions, which depends on factors such as parent rock type, vegetative cover, seasonal moisture content, soil compaction and soil age.

Many textural classifications of soils exist. Figure 12.17 shows one commonly used textural classification based on the percentages of sand, silt and clay that are present.

The terminology is straightforward, if elaborate. **Clay soils** contain more clay than sand or silt, **sand soils** contain much more sand

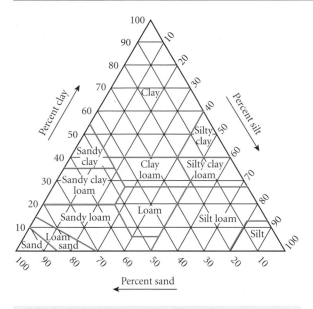

Figure 12.17 Textural classification of soils. (Courtesy of the US Department of Agriculture.)

Figure 12.18 Layered soil produced by the disintegration and decomposition of rock materials near Earth's surface. From top to bottom, the O-, A-, E-, B- and C-horizons. (Courtesy of the Society of Soil Scientists of Southern New England.) (For color version, see Plate 12.18, between pp. 248 and 249.)

than clay and silt, and **silt soils** contain much more silt than sand or clay. **Loam** is a term used for soils that contain subequal proportions of sand and silt and significant amounts of clays. Various modifiers are used for clay, sand, silt and loam soils that contain more accessory constituents than the defined limits of the four main soil types based on texture.

12.4.1   The importance of soils

Because human beings inhabit Earth's surface, soils are extremely important to many aspects of human affairs and have been and continue to be the focus of intensive study. Soils are important as **natural resources** that support a wide range of natural vegetation, as well as agricultural and forest products. Knowing how to match crops, irrigation and fertilizers to specific soils is an essential aspect of enhancing food production worldwide.

Soils are also important as **structure-bearing materials** that support our homes, offices, factories, power plants, highways, pipelines, reservoirs and aqueducts. When such structures fail, the consequences can be catastrophic. Understanding the load-bearing properties of soils and how their behavior might change over time as the result of human activities or natural processes is extremely important to engineers.

As discussed earlier in the chapter, water moving through unconsolidated surficial materials commonly reacts chemically with them. Where reactions involve the removal of contaminants and the purification of the water, soils act as **water filters** and as **contaminant sinks**. Groundwater flow through soils is also capable of **dispersing contaminants** over broad areas. Understanding the roles played by soils in such processes is of fundamental importance to geohydrologists and environmental scientists, for example, those interested in site protection and site remediation programs.

Soils, as the products of weathering, are also the **source of detrital sediments** and of many of the **dissolved solids** that occur in groundwater, surface water and the oceans. The many varieties of natural mineral and spring waters (see Box 12.2) attest to the variety of chemical reactions that can take place between soils and the groundwater that passes through them.

12.4.2   Soil horizons and soil profiles

Many soils are clearly layered (Figures 12.18 and 12.19). When soil layers are the product

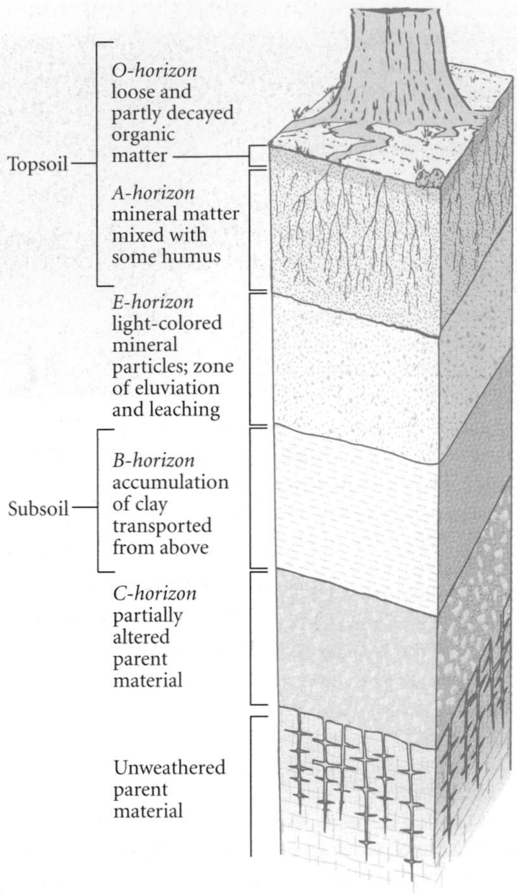

O-horizon
loose and
partly decayed
organic
matter

Topsoil

A-horizon
mineral matter
mixed with
some humus

E-horizon
light-colored
mineral
particles; zone
of eluviation
and leaching

B-horizon
accumulation
of clay
transported
from above

Subsoil

C-horizon
partially
altered
parent
material

Unweathered
parent
material

Figure 12.19 The ideal distribution of soil horizons in a fully developed soil with vegetative cover. (From Prothero and Schwab, 2004.)

of in place weathering processes they are called horizons. Each **horizon** is characterized by sets of properties produced by soil-forming processes that distinguish the horizon from the layers above and below. In many layered soils, multiple horizons occur in a vertical sequence of layers that constitutes a **soil profile**.

The classification of soil horizons is complex and has some tongue twisting jargon; please bear with us. The classification system used by USDA-NRCS (1999) divides horizons into **epipedons**, which constitute the top layer of soil profiles that have not been truncated by later erosion, and **subsurface horizons**. Eight different epipedons of diverse origins are recognized, ranging from those

produced by human activities to those produced without human intervention. Eighteen different subsurface horizons are defined, each the product of a different set of soil-forming processes. Although some of these soil horizons can be tentatively identified in the field, accurate identification of many requires extensive laboratory work that involves chemical, mineralogical and organic compound analysis.

To simplify the notion of soil horizons in a way that still carries meaning, many scientists utilize an older classification of soil horizons. One advantage of this system is that these horizons can commonly be identified in the field when a trench has been dug to expose a soil profile. This classification recognizes five major types of soil horizons, each with subdivisions, which combine to produce soil profiles. These soil horizons are generally distributed from the top down in soil profiles as (1) the O-horizon, (2) the A-horizon, (3) the E-horizon, (4) the B-horizon, and (5) the C-horizon which, in residual soils, is underlain by unaltered parent material (Figure 12.19). Each horizon may include multiple, recognizable sub-horizons.

The **O-horizon**, where present, is generally a dark brown to black epipedon that occupies the upper portion of the soil in which it occurs (Figures 12.18 and 12.19). It is characterized by being rich in organic (O) material, mostly humus, that represents the incompletely decomposed plant debris that has accumulated near the surface over time.

The **A-horizon** is dominated by mineral material, with or without a significant proportion of admixed organic matter. The A-horizon is also known as the **zone of leaching** because significant amounts of material are removed by dissolution, cation exchange or the physical removal of fine material by the downward percolation of aqueous solutions over time. This process of downward removal of solid particles and dissolved ions is known as eluviation so that the A-horizon is also known as the **zone of eluviation**.

The lower parts of some A-horizons contain a light-colored subdivision called the **E-horizon**. The light color is caused by substantial leaching of iron and/or aluminum from the lower part of the organic-poor A-horizons. E-horizons are chemically resistant and often quartz-rich zones. The E-horizon is the classic expression of the zone of eluviation from which materials have been removed by leaching or eluviation.

The **B-horizon**, also known as the **zone of accumulation**, is characterized by enrichment in some of the constituents leached from the A-horizon. The process by which materials are translocated downward to be added to the lower part of a soil is known as illuviation. For this reason, the B-horizon is also known as the **zone of illuviation**. In relatively warm, humid climates, reprecipitation of amorphous or crystalline iron oxides (e.g., hematite = Fe_2O_3) or hydroxides (e.g., limonite ≈ $FeOOH$) commonly gives the B-horizon a distinctly reddish or yellowish hue (see Figure 12.18). In humid climates, where chemical decomposition is thorough, clay minerals and bauxite are also concentrated in the B-horizons to form aluminum-rich horizons.

In dryer climates, calcium carbonate ($CaCO_3$) precipitates, producing B_k soil horizons. In many cases, mineral precipitation in the B-horizon binds soil particles together into hard, nodular zones or into completely indurated sub-horizons called **duricrusts**. The most common examples of duricrusts are the calcium carbonate **calcrete** or **petrocalcic** horizons common in caliche soils. Caliche soils form in arid and semi-arid climates with seasonal deficiencies in rainfall where evaporation of soil moisture initiates calcium carbonate precipitation. Calcrete horizons occur closer to the surface in progressively dryer climates and may occur at the surface of aridosols formed in warm, arid, desert climates. Similar hard sub-horizons of silica (**silcrete**) and gypsum (**petrogypsic**) occur less commonly in soils.

Other characteristic soil structures occur in the B-horizon interval, including:

1 **Peds,** which are partially cemented clods of soil particles of various sizes that give the soil a crumbly lump appearance.
2 **Cutans,** which are concentrations of illuviated material such as clays or iron oxides that occur as layers or that envelope less-altered cores.
3 **Glaebules,** which are prolate to equant hard lumps formed by mineral precipitation and include concretions and nodules of all sizes.

The **C-horizon**, also called the **soil mantle**, represents moderately to minimally weathered, slightly altered materials that are transitional to the underlying, unaltered parent material. Unlike the materials in the B-horizon, C-horizons are not significantly enriched in illuviated materials from above.

Where soils are developed over bedrock, the largely unweathered bedrock constitutes the so-called **R-horizon** or regolith horizon.

As will be seen from the discussion that follows, complete soil profiles form only in mature soils developed over 10^3–10^5 years. Ideal conditions include the presence of vegetation, sufficient precipitation and the absence of erosion or other disturbances. The major reasons for the incomplete development of soil horizons include (1) insufficient duration of weathering processes, (2) climatic conditions that inhibit the formation of one or more horizons, and (3) soils truncated by erosional processes.

12.4.3   Soil classifications

Given the tremendous variety of unconsolidated materials that exist on Earth's surface and the overarching importance of surficial materials in a great variety of human enterprises, it is not surprising that the description and classification of soils is exceedingly complex. Two rather different soil classification systems have evolved in the United States – one used by most agricultural soil scientists and a second utilized primarily by engineers. Many other systems have evolved internationally, leading to current efforts to develop a worldwide reference system. These are discussed in the sections that follow.

Agricultural classification of soils

Soil scientists in the United States have developed a soil classification system or taxonomy that attempts to organize the thousands of individual soil types that have been recognized. This has been accomplished by organizing all soils into a hierarchy of soil categories in much the same way as the Linnaean classification system in biology organizes all life forms into a hierarchical classification system involving kingdoms, phyla, classes, orders, families, genera and species:

Orders
    Suborders
        Great groups
            Subgroups
                Families
                    Series

The purpose is to make a complex system more comprehensible and to make it easier to discern relationships between different soils. The soil taxonomy is based on the properties of the soil profile, especially the following:

1 The number and types of soil horizons present.
2 Available nutrient chemicals.
3 The distribution of organic materials.
4 Soil color.
5 Seasonal soil moisture content.
6 Overall climate.

The USDA-NRCS (1999) has organized soils into 12 major **orders**, which are fairly easy to learn (Figure 12.20). Each order is subdivided into as many as seven **suborders** of which there are a total of 64. The suborders are subdivided into more than 300 **great groups**, which are subdivided into some 2400 **subgroups**, which are further subdivided into **families** and lastly into soil **series**. Some 19,000 different soil series have been mapped in the United States alone. Some soils can be identified on the basis of field investigations, but most require substantial laboratory work as well. Once a soil is placed into its appropriate soil series, a great deal of descriptive, relational and interpretive information, gathered over many decades, is available to the practicing soil scientist. The characteristics of the 12 major soil orders are briefly described and summarized in Table 12.5; images of examples of each major soil order are provided in Figure 12.20.

Unfortunately, soil classifications and classification criteria vary between different countries. An international effort is underway to facilitate communication between people using different classification systems. In 1998 a **world reference base (WRB) for soil sciences** was developed with the endorsement of the International Union of Soil Scientists (IUSS) and the Food and Agricultural Organization (FAO) of the United Nations. The WRB has provided a worldwide classification of soils to which soils classified by different systems can be compared. It divides soils into 25 orders and 98 groups based on the physical characteristics of the soil. Unlike the USDA-NRSC classification scheme, climate is not considered in the WRB. The long-range goal of these efforts is to promulgate adherence to a single standard worldwide soil classification scheme.

Engineering classification of soils

Engineers approach soil classification from a different set of perspectives than those of soil scientists. They are not especially focused on soil horizons and the suitability of soils for agriculture and forestry. Instead, engineers are concerned with the mechanical properties of all unconsolidated surficial deposits, regardless of origin. This is reflected in the descriptors and classification systems of soils utilized by engineers. Most geotechnical and engineering personnel in the United States use the **Unified Soil Classification System** (Tables 12.6 and 12.7). In this system, soils are given names and symbols according to their particle size distributions, notably the proportions of gravel, sand, silt and expansive clays, and to the content of non-expansive clays and organic materials in the soils. Engineering definitions of gravel, sand, silt and clay do not correspond exactly to those used by geologists, who employ the Wentworth–Udden grade scale (W-U scale) (Chapter 13), with which comparisons are given in parentheses, nor to that used by the USDA-NRCS discussed above.

In the system employed by the Unified Soil Classification System, the main groups are:

1 **Gravel (G)**, where particle diameters exceed 4.0 mm (as compared with 2.0 mm in the W-U scale).
2 **Sand (S)**, where particles range from 0.074 to 4.0 mm (as compared with 0.0625–2.0 mm in the W-U scale).
3 **Silt (M)**, where particles range from 0.004 to 0.074 mm (as compared with 0.004–0.0625 in the W-U scale).
4 **Clays (C)**, which are defined in the same way in both systems as particles smaller than 0.004 mm (4 μm). The USDA-NRCS classification system, however, defines clays as particles smaller than 2 μm.

Alfisol

Andisol

Aridisol

Entisol

Gelisol

Histosol

Figure 12.20 Examples of the major soil orders in the USDA-NRSC soil taxonomy (see Table 12.5). (Courtesy of the US Department of Agriculture.) (For color version, see Plate 12.20, between pp. 248 and 249.)

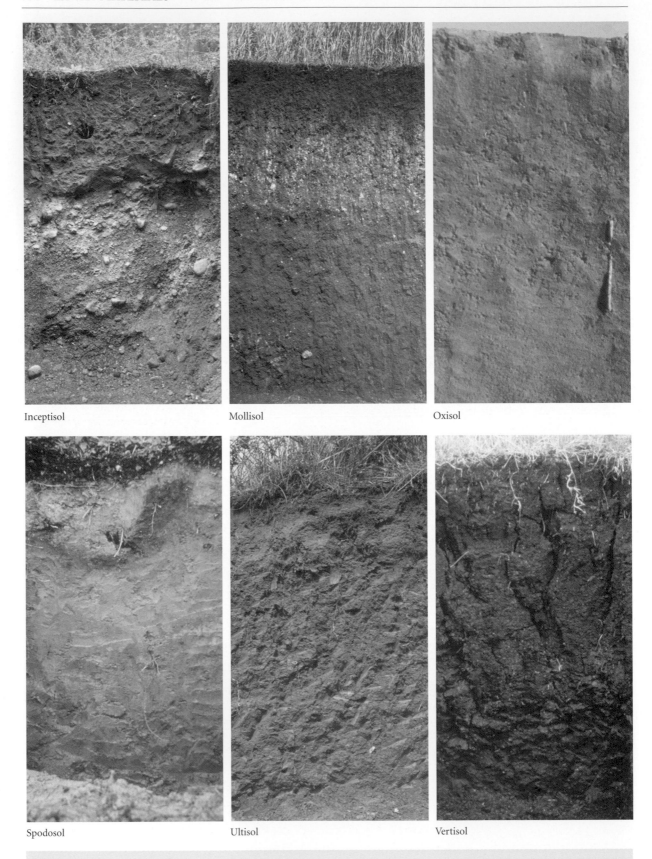

Inceptisol

Mollisol

Oxisol

Spodosol

Ultisol

Vertisol

Figure 12.20 *Continued*

Table 12.5   Soil orders in the USDA-NRSC soil taxonomy; their diagnostic features and environments of formation.

Order	Diagnostic features	Environment
Alfisols	Gray to brown A-horizon epipedon; B sub-horizons rich in clays with reasonably high concentrations of bases such as Ca, Na, Mg; reasonably high moisture content	Relatively humid areas with sparse forest or savannah cover; base and water content yield fertile soils
Andisols	Weak horizon development; rich in disordered clays and Al–humus complexes; high phosphorous retention; good moisture capacity and cation exchange capacity	Form in a wide range of non-arid climates; mostly on volcaniclastic materials; tend to be quite fertile
Aridosols	Sparse organic material in A-horizon epipedon; well-developed B-horizons, often rich in Ca-carbonates, even gypsum; low moisture content for long periods of time	Dominate in arid regions with sparse rainfall and vegetative cover; suitable for agriculture only if irrigated
Entisols	Lack significant soil horizon development; soils only because they have the capability to support rooted plants; often sand rich	Occur in any climate or setting; mostly on young surfaces; also in chemically inert parent materials or on slopes where erosion occurs
Gelisols	Permafrost soils and soil features; patterned ground, broken horizons and incorporation of organic matter in lower horizons produced by frost heaving and churning	In high latitude and/or high elevation areas where soils freeze for long periods
Histosols	Mostly very organic rich O-horizon; deeper horizons tend to be poorly developed, if at all	Mostly peat and muck from partially decomposed plant debris in swamps or bogs or water-saturated soils in areas of poor drainage
Inceptisols	Weak horizon development; less clay concentration in B-horizon than alfisols; carbonate and silica-rich B-horizons may occur; reasonably high moisture content	Form in a range of non-arid regions from subpolar to tropical; often with forest cover; less suitable for agriculture than alfisols
Mollisols	Very dark, thick, organic-rich O and A epipedon; high base content, especially calcium; clays with high cation exchange potential	Common under grasslands in semi-arid plains and steppes with seasonal moisture deficits; some under forest cover; great for grain production
Oxisols	Weak horizon development; extreme decomposition and base depletion; clays, mostly kaolinite, with low cation exchange capacity; bauxite under extreme conditions; quartz and iron oxides	Develop over long periods of time in tropical/subtropical settings with high rainfall and thick vegetative cover; generally infertile
Spodosols	Thick O-horizon; well-leached A-horizon with low Fe, Al, Ca; well-developed B-horizons with clays, reddish iron oxides or black humic material; good cation exchange	Dominate under coniferous forests; in areas with reasonable rainfall; generally suitable for agriculture
Ultisols	Well-leached A-horizon with some organics; clay-rich B-horizons with generally low base contents as Ca, Na and K are largely removed which distinguishes them from alfisols	Humid climates; low base content; soils unsuitable for sustained agriculture unless fertilized with Na and K
Vertisols	High expansive clay content; large changes in volume associated with wetting and drying; cracks when dry and other evidence of soil movement; may have horizons	Poor soil for structures given the volume changes and tendency for strength and plasticity to change during wetting and drying

Table 12.6 Unified Soil Classification System for coarse-grained soils.

Soil divisons	Soil characteristics	Soil group name	Soil group symbol
Clean gravel	<5% fines; continuous size variation over a range	Well-graded gravel	GW
Clean gravel	<5% fines; mostly one size or polymodal	Poorly graded gravel	GP
Dirty gravel	>12% fines; mostly silt	Silty gravel	GM
Dirty gravel	>12% fines; mostly clay	Clayey gravel	GC
Clean sand	<5% fines; continuous size variation over a range	Well-graded sand	SW
Clean sand	>5% fines; mostly one size or polymodal	Poorly graded sand	SP
Dirty sand	>12% fines; mostly silt	Silty sand	SM

Table 12.7 United Soil Classification System for fine-grained soils.

Soil divisions	Soil characteristics	Soil group name	Soil group symbol
Silt	Inorganic silts with very slight plasticity	Silt	ML
Silt	Inorganic silts, with mica giving soil more elasticity	Micaceous silt	MH
Silt	Organic silts with low plasticity	Organic silt	OL
Clay	Inorganic clays and silty clays of low–medium plasticity	Silty clay	CL
Clay	Inorganic clays of high plasticity	High plastic clay	CH
Clay	Organic clays with medium–high plasticity	Organic clay	OH

One need not memorize these numbers; simply remember that the size classification systems are similar, but different in detail. Look them up as required.

Coarse-grained soils (Table 12.6) are those that contain more than total 50% sand and gravel by weight. Gravels contain more gravel than sand, whereas sands contain more sand than gravel. Coarse-grained soils are further subdivided according to the percentage of fine-grained components (clays + silts). **Clean soils** contain less than 5% fines and **dirty soils** contain more than 12% fines. Coarse-grained soils are subdivided further on the precise percentages of fines and on whether they are mostly silt or clay. Each soil type is represented by an appropriate symbol. In addition, transitional names can be used. For example a well-graded gravelly soil with 5–12% fines of which the majority is silt could be called a GW-GM. Other soil characteristics are typically recorded as well by engineers, including (1) maximum particle size, (2) color, (3) layering, (4) compactness or compressibility, (5) moisture characteristics, (6) drainage conditions, (7) structure, (8) strength, and (9) plasticity.

Fine-grained soils (Table 12.7) contain more than 50% silt plus clay. Silts are defined as soils with more silt than clay, whereas clays contain more clay than silt. Fine-grained soils are further subdivided according to their mica content, organic content and their degree of plasticity. Highly organic soils are prone to compaction, dehydration and decomposition resulting in volume loss, which makes these soils unsuitable for construction. Soil plasticity is largely determined by the soil's ability to absorb water and therefore by their smectite (expandable lattice) clay content. As will be seen in the following section, plasticity is an extremely important measure of the mechanical properties of soils and allows one to predict how they will react in different circumstances. The "L" in the group symbols stands for loam, a soil that contains appreciable amounts of both silt and clay in the fine fraction.

The Unified Soil Classification System contains a separate class for soils that are especially rich in organic materials. These **organic soils** are mostly peats and mucks and are roughly equivalent to the gelisols in the USDA-NRCS classification.

12.4.4   Soil mechanics

Soil engineers are especially concerned with the mechanical properties of soils. The mechanical properties of soils are critical factors in the suitability of soils for use in the construction of roads, dams and buildings and in many other aspects of land utilization. Many geohazards are the direct result of ignoring or failing to understand the implications of the mechanical properties of soils. Let us consider some of these, including soil strength, soil sensitivity, shrink and swell potential and compressibility. Each is fundamental to understanding the engineering aspects of soils.

Soil strength

Soil strength is the amount of stress a soil can bear without failing by rupture or plastic flow. It is an expression of the ability of a soil to resist irreversible deformation such as inelastic changes in shape, volume and position. Strong soils are quite resistant to stress and, along with many kinds of bedrock, generally provide excellent substrates for buildings. Weak soils are subject to compression, collapse or flow when stressed and therefore provide poor substrates for structures. Problems for engineers arise because soil strength can change, especially in response to changes in water content (Box 12.3), so that formerly strong soils loose strength and become weak soils that fail by rupture, flow plastically or even flow like a liquid.

Soil sensitivity

The measure of a soil's tendency to change strength is expressed by **soil sensitivity**, a measure of the change in soil strength that results from changes in water content and various kinds of disturbances such as vibrations, excavations and loading that stress soils. Soil sensitivity in response to water content, easily determined in the lab, is commonly expressed by **Atterberg limits** that permit the subdivision of fine-grained soils into four classes on the basis of how they behave as their moisture content changes (Figure 12.21).

On the basis of their behavior, soils may be subdivided into four Atterberg classes:

(1) brittle solids, (2) semi-solid soils, (3) plastic soils, and (4) liquid soils. The boundary between brittle solids and semi-solid soils is called the **shrinkage limit (SL)**, which is the water content below which soils do not shrink as additional moisture is lost during drying (Figure 12.21). Above the shrinkage limit, semi-solid soils shrink and crack as they lose moisture and become progressively more stiff and brittle. Soils that remain brittle or semi-solid under all conditions of potential moisture content tend to be strong and provide excellent substrates for most construction projects so long as they are not loaded beyond their rupture strength. Solid bedrock is even better.

The **plastic limit (PL)** separates semi-solid soils from plastic soils and is the water content at which soil deformation changes from rupture to plastic flow (Figure 12.21). Plastic substances change shape and/or volume in response to stress or pressure but do not rupture visibly. Because they retain cohesive strength, they do not flow like a liquid. Plastic soils are moisture sensitive in that their strength decreases and they deform more easily as they become progressively less cohesive with increasing moisture content. This helps to explain the many slope failure incidents that occur following heavy rainfall and the concurrent infiltration of groundwater into soils. The **plasticity** of a soil is a measure of its cohesiveness, which is sensed as a sticky, cohesive feel to the touch. It generally increases with clay content (especially expandable smectites) and water content. Soils with low clay content tend to be relatively non-cohesive and therefore possess relatively low plastic limits. Soils with high clay content tend to be much more cohesive and to possess significantly higher plastic limits. Because plastic soils deform when loaded, they do not make good substrates for major construction projects.

The **liquid limit (LL)** separates plastic soils from liquid soils (Figure 12.21). It is the water content at which soils lose their shear strength and begin to flow. When a sufficient amount of moisture has been added to a soil, it may begin to behave as a liquid; that is, it will loose cohesive strength and begin to flow under its own weight. This can have disastrous consequences for the structures placed such soils. The liquid limit tends to be rela-

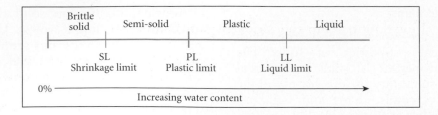

Figure 12.21   The major Atterberg classes of fine-grained soils and the limits that define them.

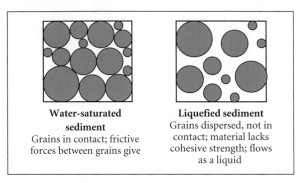

Figure 12.22   Liquefaction occurs when grains (gray) are separated; cohesive strength due to friction, present when grains are in contact is lost when grains are separated and dispersed in the liquid (white).

(a)

(b)

Figure 12.23   (a) Collapsed apartment buildings in Nigata, Japan, after the 1964 earthquake. (b) Neighborhood in Anchorage, Alaska, destroyed when the ground slumped, carrying structures more than 1 km, after sand lenses became liquefied during another 1964 earthquake. (From the Karl Steinbrugge Collection, courtesy of the National Information Service for Earthquake Engineering, EERC, University of California, Berkeley.) (For color version, see Plate 12.23, between pp. 248 and 249.)

tively low for non-cohesive soils such as sands and coarse silts and helps to explain the liquefaction of sands during an earthquake, as discussed in the succeeding paragraph. It tends to be higher for clay-rich soils, which are much more cohesive.

One important example of a sensitive soil is the tendency of water-saturated sands and coarse silts to lose their strength during an earthquake. The vibrations destroy the grain contact strength possessed when the sand grains are at rest. Vibrations cause the grains to separate. Individual grains become dispersed in the water that occupied the spaces between grains resulting in liquefaction as the soil is turned into quicksand-like material (Figure 12.22).

Liquefied sand has no strength and therefore cannot support a load (Figure 12.23a). Liquefaction is particularly prevalent in unconsolidated sand–silt soils disturbed by earthquake ground shaking. On slopes,

Box 12.3   Liquefaction and the Van Norman Dam

As the population of Los Angeles expanded rapidly during the early years of the 20th century, the need for larger supplies of water increased. A large aqueduct system was built to bring water to the city from central California. A reservoir was required to store the water flowing from the aqueduct system. Between 1912 and 1915, the Van Norman Dam (Lower San Fernando Dam) was built to create that reservoir. The construction methods involved the building of an earthen dam, some 640 m long, composed of silty sand over a clay-rich core. A concrete parapet was placed on top, giving the dam an overall height of more than 43 m. Reservoir capacity, expanded to meet growing needs in 1930, was more than 34 billion liters of water in a 2.5 km long reservoir with a maximum depth of 40 m. The dam was built in the hinterlands of the San Fernando Valley, but it was not long before a growing population of suburbanites built their homes in the valley below.

The winter of 1970–71 was unusually dry, even by the standards of southern California. In early February the reservoir was at half capacity but still contained approximately 18 billion liters of water. Low reservoir levels proved fortuitous. Early on the morning of February 9, a magnitude 6.7 earthquake occurred in the San Gabriel Mountains near Sylmar, California. The earthquake destroyed two hospitals, collapsed freeway overpasses, killed 65 people and injured some 2000. But it was almost much worse. The Sylmar earthquake lasted nearly a minute. The shaking of the water-saturated sediments of which the earthen dam was constructed induced liquefaction. A major portion of the dam structure, some 550 m long, began to slide into the reservoir. Some 610,000 m³ of dam embankment were being displaced as the water-saturated reservoir side of the embankment turned to liquid. The concrete parapet was carried with it (Figure B12.3a). The dam had lost 9 m of its original height. Only a 30 cm thick portion of the downstream side of the embankment remained above water level (Figure B12.3b). Even with the lower than normal water level in the reservoir, a major disaster was about to occur.

What happened next? The earthquake ended, ground shaking stopped and the earthen material regained its strength as the sand and silt grains regained contact and frictional resistance was restored. Of course the water in the reservoir was still exerting pressure on a flimsy 30 cm thick barrier. Fearing imminent disaster, officials evacuated 80,000 people from the valley below, while engineers drained the water from the reservoir over the next 3 days. Had the earthquake lasted a few more seconds, Sylmar might have a much larger place in the pantheon of historic earthquakes in California.

Los Angeles still needs water. A new dam was built close by to a much higher standard and survived the 6.7 Northridge earthquake in 1994 with only minor damage.

(a)

Figure B12.3   (a) Damage to the Van Norman (Lower San Fernando) Dam, 1971. (From the Karl Steinbrugge Collection, courtesy of the National Information Service for Earthquake Engineering, EERC, University of California, Berkeley.) (For color version, see Plate B12.3a, between pp. 248 and 249.)

Box 12.3   *Continued*

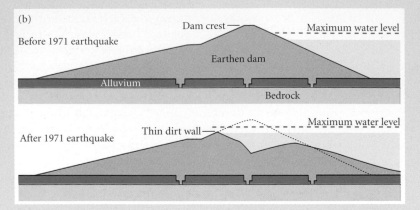

(b)

Before 1971 earthquake

Dam crest

Maximum water level

Earthen dam

Alluvium

Bedrock

After 1971 earthquake

Thin dirt wall

Maximum water level

Figure B12.3   (b) Cross-section of the Van Norman Dam before and after failure. (Courtesy of the US Geological Survey.)

liquefied layers can cause entire areas to flow downhill carrying structures with them (Figure 12.23b). Water-rich, liquefied **quick clays** can also lose their strength and flow, especially when loaded. Since placing structures on the surface increases the load pressures on the substrate, careful studies of soil sensitivity must be carried out prior to the initiation of many types of construction projects.

Shrink–swell potential

Other soil parameters have proven useful in soil evaluations by engineers. The **plasticity index (PI)** of a soil is the range of water contents over which the soil behaves as a plastic substance (see Figure 12.23). It is the difference between the liquid limit and the plastic limit:

$$PI = LL - PL$$

Where PI is less than 5%, small changes in water content can convert soil from a semi-solid to a liquid state. Sand- and silt-rich soils tend to possess low plastic indexes because they are not very cohesive. Clay-rich soils tend to have much higher plastic indices. Soils rich in expandable smectite clays tend to have the highest plastic indices of all (Figure 12.24). The presence of minerals that can absorb large quantities of water in the soil, such as

smectite clays, can produce soils that are potentially extremely unstable. One criterion that expresses this concept is the shrink–swell potential of a soil.

Shrink–swell potential expresses the tendency of soils to change volume when wetted. Soils that contain large amounts of **expansive clays** (smectites), with their tendency to absorb water when wetted, tend to have large shrink–swell potentials. Such soils often possess large degrees of sensitivity as well and are responsible for many ground failure episodes. Ground failures include severe ground subsidence, dam failures and major landslide events. As a general rule, such weak soils exhibiting high sensitivity, shrink–swell ratios and plasticity indexes make poor substrates for structures unless their shrink–swell potential can be significantly reduced.

The behavior of soils can often be predicted from laboratory experiments. Figure 12.24 shows a plot of plasticity index versus liquid limit that is used extensively in laboratory testing of fine-grained soils. These plots are called **Casagrande diagrams** after the person who introduced them to the field. The A-line, given by PL = 0.73 (LL – 20), separates fine-grained soils by plasticity. The vertical line, given by LL = 50%, separates such soils by their liquid limit. Four fields result on Casagrande diagrams. Silt-rich sediments (ML or OL) tend to possess low plasticity because of

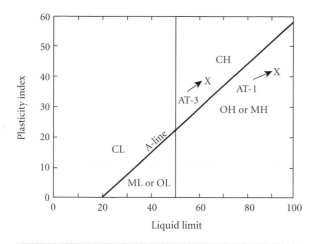

Figure 12.24   A Casagrande plot of soil sensitivity using the plasticity index (PI) and liquid limit (LL); sand- and silt-rich soils have little or no plasticity and low liquid limits. (Courtesy of the US Geological Survey.)

Figure 12.25   Italy's famed Leaning Tower of Pisa, which was constructed on compressible, clay-rich soil. As a result of differential consolidation, the tower requires additional reinforcement (here supplied by Maureen Crowe) to maintain its equilibrium. (Photo courtesy of Tony Crowe.) (For color version, see Plate 12.25, between pp. 248 and 249.)

their low clay contents and low liquid limits because of their minimal cohesiveness. The clay content of silty clays (CL) gives them more cohesiveness and therefore higher plasticity, while the silt content helps them retain a relatively low liquid limit. High plastic clays (CH) are rich in expansive smectite (montmorillonite) clays or somewhat less expansive illites. This results in very high plasticity and correspondingly high liquid limits. Organic clays (OH) and mica silts (MH) testify to the importance of mica in increasing the plasticity and cohesiveness of the soils in which it occurs. CH, OH and MH soils require elevated water contents to exceed the liquid limit.

Compressibility

Compressible soils undergo decreases in volume when loaded. **Compressibility** expresses the tendency of soils to consolidate and lose volume. Soils with variable compressibility tend to settle unevenly causing variations in the subsidence of surfaces on which structures have been placed (Figure 12.25). If differential compaction is significant, structural damage may result.

Porosity

Soils hold water both by absorption in the interlayer sites of expansive clays and by storing water in the spaces between individual soil particles. The capacity of a material to hold water in its intergranular spaces is called porosity. **Porosity (Ps)** is simply the volume percentage of void spaces, called pores, in a rock. It is given by the simple formula

$$Ps = \% \text{ pore space} = (\text{volume of pores}/\text{total rock volume}) \times 100$$

In general, porosity increases with particle size, although shape factors also play a role (Chapter 13). The primary importance of soil porosity is that it represents the total capacity of a rock to hold or store fluids such as groundwater in the void spaces between solid grains. The search for groundwater often begins in subsurface materials with substantial amounts of porosity.

Permeability

Permeability (K) expresses the rate of fluid flow through a material. Permeability can be

Table 12.8 Generalized hydraulic conductivity of sediments.

Sediment type	Hydraulic conductivity
Clays	$\sim 10^{-6}$ cm/s
Silts	$\sim 10^{-4}$ cm/s
Sands	$\sim 10^{-2}$ to 10^{-3} cm/s
Gravels	$\sim 10^{-1}$ to 10^{1} cm/s

determined by using Darcy's law. In 1856, Henry Darcy, a French engineer, conducted a study of groundwater flow through a porous medium. Darcy discovered that the rate of water flow through a bed was proportional to the difference in height (h) of the water between two points and inversely proportional to the flow path length (L); in essence, the change in height divided by the distance represents the **hydraulic head** or slope of the water table. Darcy developed a "law" describing the groundwater flow where flow rate (Q) equals the cross-sectional area (A) of flow multiplied by the hydraulic conductivity (K) and hydraulic gradient (h/L) of the sediment or rock. Hydraulic conductivity is a measure of permeability that varies depending upon the fluid's viscosity (μ) and density (ρ), the effective permeability (κ) and the acceleration of gravity (g), as given by $K = \kappa\rho g/\mu$. The hydraulic gradient (h/L) is often the slope of the water table. **Darcy's law** can be expressed as:

$$Q = A[K \times (h/L)]$$

This formula states that flow rates are proportional to hydraulic conductivity and therefore to effective permeability. Permeability increases with pore size and pore interconnectedness, both of which tend to increase with particle size and sorting. It decreases with increasing surface tension between water and the minerals with which it is in contact. As a result, sandy and gravelly soils with their larger pore spaces have higher permeability and hydraulic conductivity than clay-rich soils with their smaller pore spaces and higher surface tension, which retard fluid flow. Typical hydraulic conductivities for unconsolidated sediments are given in Table 12.8. Note the general increase in permeability with increasing particle size. Sands and gravels tend to be quite permeable and act as **aquifers**, which store and transmit water. On the other hand, clay-rich layers tend to be quite impermeable and act as **aquitards**, which retard fluid flow. Rock permeability is influenced by a variety of other factors that are discussed in more detail in Chapter 13.

Permeability is also fundamentally important in aquifer studies and in determining contaminant flow behavior in the subsurface. Soil permeability is of particular importance to hydogeologists because it determines the rate at which water is able to flow into and through porous storage rocks in aquifers. Septic systems require permeable soils that will disperse waste materials efficiently; if such soils should become clogged with waste so that their permeability is reduced, septic system backup may occur. Permeability is also of great importance to environmental geologists interested in the rates and directions of water-borne pollutant dispersal in the subsurface. Materials with low permeability impede the dispersal of water-borne pollutants and are used as confinement barriers for hazardous waste sites.

12.4.5 Buried soils and paleosols

Buried soils are former soils that have been buried beneath the surface by subsequent deposition. Buried soils occur beneath (1) glacial tills and outwash, (2) wind-blown sand, (3) river-deposited sand and mud (Figure 12.26a), (4) mass-wasting deposits produced by processes such as landslides, mudflows and debris flows, (5) marine sediments formed during transgression onto a land surface, and (6) volcanic materials produced by lava flows and pyroclastic eruptions (Figure 12.26b).

Buried soils are common along regional unconformities where long-term weathering in continental environments is followed by a period of deposition. Buried soils that have been uncovered and exposed at Earth's surface by subsequent erosion are called **exhumed soils**.

Paleosols are ancient soils that formed under conditions not related to the present climate. They have been the subject of increasing study over the last two decades as scientists have come to realize their potential as important aids in inferring aspects of Earth's

Figure 12.26   (a) Jurassic soil with plant root casts, buried by braided stream deposits, Connecticut. (b) Holocene soil with a well-defined O-horizon at the top, buried by pyroclastic deposits from the Tarawera Volcano, New Zealand. (Photos by John O'Brien.) (For color version, see Plate 12.26, between pp. 248 and 249.)

history (Retallack, 2001b). The study of ancient soils is called **paleopedology**.

Recognition of paleosols

One major problem in recognizing paleosols is that they may be severely truncated by erosion and/or extensively altered during burial. Significant erosion may remove enough of the upper soil horizons to make soil order recognition difficult or impossible. Alteration may do the same by causing:

1 Decomposition of organic matter which deprives the O- and some A-horizons of a defining characteristic.
2 Oxidation of ferrous iron (Fe^{+2}) to ferric iron (Fe^{+3}), producing an oxidized layer where one did not originally exist.
3 Dehydration of goethite/limonite to hematite to produce a reddish layer where one did not originally exist.

In addition, subsequent diagenetic processes may alter other geochemical and mineralogical trends that help pedologists to recognize soil types. Despite these problems, paleopedologists have been able to recognize good examples of all the major soil orders in older rocks as well as some extinct soil types. Given Earth's long history and substantial changes in climate, biota and composition of the atmosphere, one might expect ancient soils to be very different from modern ones. Yet most ancient soils are sufficiently similar to modern ones that the USDA-NRCS soil taxonomy can be used to classify them. As Retallack (2001b) states, very few of the soils found in the paleopedological record are extinct types.

Soils also yield clues to the evolving composition of Earth's atmosphere. Perhaps the most significant insight comes from the emergence of oxisols some 2.0 Ga. Prior to this time, red ferric (Fe^{+3}) oxide minerals such as hematite in soils are rare to absent. Instead, the dominant soils belong to an unnamed soil order with green to gray-white soil horizons in which the iron-bearing minerals contain reduced ferrous iron (Fe^{+2}). What caused soils to change in this way some 2.0 Ga? Noting the widespread expansion of red-colored sediments or "red beds" at this time, most geologists believe that 2.0 Ga marks the time when free oxygen first became abundant in Earth's atmosphere. This change in Earth's atmosphere from a reducing atmosphere in which methane CH_4 was a significant component to an oxidizing atmosphere in which O_2 was a major component is one of the most significant changes in Earth history, for it eventually allowed the evolution of organisms that use O_2 during respiration and decomposition.

Significant insights are being gained from the study of Archean (>2.5 Ga) soils concerning the emergence of life in terrestrial environments. Because living tissues selectively utilize carbon-12 (^{12}C) relative to carbon-13 (^{13}C), organic carbon possesses lower $^{13}C/^{12}C$ ratios than atmospheric carbon. The presence of organisms in a soil can be detected by such depressed $^{13}C/^{12}C$ ratios in carbonate minerals precipitated by soil waters, even when organic matter has not been preserved. Several Archean soils show such depressed $^{13}C/^{12}C$ ratios (Retallack, 2001b), which suggests that microbial populations inhabited terrestrial soils more than 2.5 Ga. Because the atmosphere was depleted in free oxygen (O_2) prior to 2.0 Ga, such microbial populations probably utilized atmospheric methane (CH_4) abundant in Earth's early atmosphere. Altinok (2006) sampled the oldest known unconformity (~3.5 Ga) by drilling and discovered a paleosol that contains possible organic matter, which, if confirmed, would push the emergence of terrestrial life back even further.

Using the ratios of carbon isotope ^{13}C and ^{12}C, scientists can monitor fluctuations in the amount of the greenhouse gas CO_2 in the atmosphere as well. As expected, the abundance of CO_2 in the atmosphere generally mirrors atmospheric temperatures. When CO_2 levels are high, temperatures tend to be high; when they are low, temperatures tend to be low. Using these methods, scientists have been able to document a dramatic trend of global cooling that began in the late Eocene period and has continued, with some fluctuations, to the present time. This long-term cooling corresponds to a long-term decrease in atmospheric CO_2. What might have caused this trend? Greg Retallack (2001a) thinks he

has the answer. He notes that grasses evolved following the greenhouse period that marked the early Tertiary. With the expansion of grasslands, mollisols formed for the first time in the Eocene and had spread rapidly by the Miocene to cover as much as 20% of the land surface. With their thick accumulations of slow-decaying organic material, mollisols sequester huge amounts of carbon, effectively removing CO_2 and another greenhouse gas, methane (CH_4), from the atmosphere. Mollisols also may have stimulated biological productivity in oceans and helped to increase the rate at which light is reflected from Earth's surface, accelerating the cooling trend. If Retallack is right, the rise of mollisols may well have been a major cause of the coeval long-term global cooling that eventually led to widespread glaciations during the Pleistocene.

Paleopedologists continue to push the envelope. They are speculating about soils as proxies for tectonic flexure associated with the formation of orogenic belts and foreland basins (Decelles, 2006); about very high $^{13}C/^{12}C$ ratios in soils that may indicate an Ordovician greenhouse atmosphere with 16–18 times more CO_2 than exists at present (Retallack, 2001b); and about the soil record of vascular plants containing lignin that begins in the Silurian and that may have caused severe reductions in atmospheric CO_2 content, which led to a long-term episode of global cooling that extended into the Permian. Weathering and soil development may be even more important than we have thought.

Chapter 13

Detrital sediments and sedimentary rocks

Sediments and sedimentary rocks cover more than 85% of Earth's surface. More than 75% of the sediments on Earth's surface are detrital sediments composed of particles that are produced by and/or survive weathering to be eroded, transported and eventually deposited on Earth's surface. The resource value of detrital sediments cannot be overstated. Detrital sediments provide substrates on which we grow our food and timber resources and conduct our construction projects, while providing vegetative cover for habitat. In addition, detrital sediments and sedimentary rocks store valuable fluid resources such as oil, natural gas and water. This chapter discusses detrital sediments and sedimentary rocks with a focus on (1) textures, (2) compositions, (3) classification, (4) provenance, (5) tectonic implications, (6) diagenesis, and (7) uses.

its constituents – both its solid particles and the void spaces between them. Most rocks are classified and named primarily on the basis of their texture and composition, both of which yield clues concerning their history and origin. Texture determines the ability of sedimentary materials to store and transmit fluids such as water, oil and natural gas. The susceptibility of sediments to mass wasting processes such as landslides and debris flows is strongly influenced by their texture. Texture also plays an important role in the diagenesis of detrital sediments. Understanding the textures of sedimentary rocks provides an essential basis for their classification and interpretation, for the prediction of their resource potential and for the evaluation of their potential as geohazards.

13.1   TEXTURES OF DETRITAL SEDIMENTS

The **texture** of a sediment or sedimentary rock refers to the size, shape and arrangement of

Earth Materials, 1st edition. By K. Hefferan and J. O'Brien. Published 2010 by Blackwell Publishing Ltd.

13.1.1   Particle size

Several different classification schemes are used to describe particle size in epiclastic rocks. The size scale used by most geologists, but not by most engineers (Chapter 12), is the **Wentworth–Udden grade scale** (Figure 13.1) created in 1918. The Wentworth–Udden

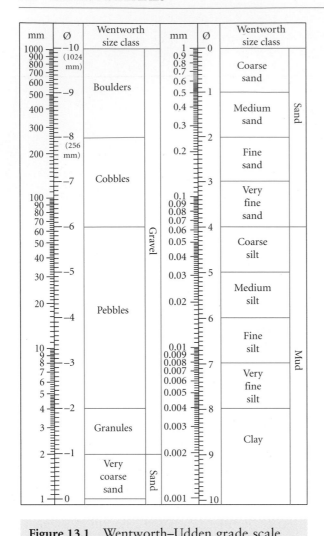

Figure 13.1 Wentworth–Udden grade scale with phi (Φ) equivalents. (After Lewis, 1984.)

grade scale defines gravel, sand and mud particles according to their mean diameter.

1. **Gravel** particles have diameters larger than 2.0 mm, about the size of a small grain of rice.
2. **Sand** particles have diameters between 2.0 and 0.0625 mm, the size range seen on sandpaper or between finely granulated sugar and small grains of rice.
3. **Mud** particles possess diameters smaller than 0.0625 mm, like processed flour or baby powder.
4. Mud particles can be further subdivided into **silt** particles, with diameters between 0.0625 and ~0.004 mm), and tiny **clay** particles, whose diameters are less than 0.004 mm (4 μm).

Each of these size classes can be subdivided further as shown in Figure 13.1. Gravel is subdivided into boulders (>256 mm), cobbles (64–256 mm), pebbles (4–64 mm) and granules (2–4 mm). Sand is subdivided into very coarse (1–2 mm), coarse (0.5–1.0 mm), medium (0.25–0.50 mm), fine (0.250–0.125 mm) and very fine (0.0625–0.1250 mm) sand fractions. Silt can be subdivided into coarse, medium, fine and very fine silt. All sizes of epiclastic grains can be described using the Wentworth–Udden grade scale.

Krumbein (1934) led a fundamental revolution in the mathematical analysis of populations of detrital particles, utilizing the fact that the Wentworth–Udden scale is a negative, base 2 logarithmic ($-\log_2$) scale. To facilitate mathematical and statistical analysis of detrital sediments, Krumbein created logarithmic phi scale (Φ scale) equivalents for the sizes in the Wentworth–Udden grade scale (Figure 13.1). A logarithm is just the power (x) to which a base (n) must be raised to produce a given number (a). Stated another way:

$$\text{If } n^x = a, \text{ then } \log_n a = x$$

A simple example may be helpful. The base 10 to the third power (10^3) equals 1000 and can be written in the form $n^x = a$ as $10^3 = 1000$. In this case, since $n^x = a$, then $\log_n a = x$, and we can write that the base 10 logarithm of 1000 is equal to three; that is, $\log_{10} 1000 = 3$. Since the Wentworth–Udden scale is a base 2 negative logarithmic scale, phi values are given by the formula $\Phi = -\log_2 d$, where d is the diameter in millimeters. If this is true then:

$$\text{If } 2^{-\Phi} = d, \text{ then } \log_2 d = -\Phi$$

Although all the boundary equivalents between millimeter scales and phi scales are given in Figure 13.1, some examples may be helpful.

1. The boundary between boulders and cobbles is 256 mm, which is equal to 2 mm raised to the eighth power, that is, $2^8 = 256$ mm, so that $\log_2 256 = 8 = -\Phi$; Φ is −8. Thus the phi scale equivalent of 256 mm is −8.
2. The boundary between pebbles and granules occurs at a phi value of −2 because

$2^2 = 4\,mm$, thus $\log_2 4 = 2 = -\Phi$; Φ is -2.

3 The boundary between gravel and sand occurs at $\Phi = -1$ because $2^1 = 2\,mm$, thus $\log_2 2 = 1 = -\Phi$; Φ is -1.

4 The very coarse and coarse sand boundary occurs at $\Phi = 0$ because $2^0 = 1\,mm$, thus $\log_2 1 = 0 = -\Phi$; Φ is -0.

5 Smaller particles have positive phi values. For example, the boundary between sand and silt occurs at 0.0625 mm. Since $0.0625\,mm = 2^{-4}\,mm = -\Phi = -4$, then Φ is 4. Lastly, the boundary between silt and clay occurs at a phi value of 8 because $0.004\,mm = 2^{-8}$, thus $\log_2 0.004 = -8 = -\Phi$, and Φ is 8.

In analyzing size distributions in sedimentary rocks, the phi scale is almost invariably used because differences between adjacent size classes are of the same whole number magnitude, whereas in the Wentworth–Udden grade scale they are not of the same value. For example, the boundaries between very coarse, coarse, medium and fine sand are 0, 1 and 2 phi, respectively, so that each size covers an interval equal to 1 phi unit. The same boundaries are at 1, 0.5 and 0.25 mm when expressed by the Wentworth–Udden scale, so that coarse sand covers an interval of 0.5 mm, medium sand an interval of 0.25 mm and fine sand an interval of 0.125 mm.

13.1.2 Textural classification of detrital sediments

The epiclastic constituents of a sediment or sedimentary rock can be subdivided into a:

1 **Coarse fraction** or **grains** that include(s) all gravel (G) particles and all sand (S) grains.
2 **Fine fraction** or **matrix** that includes all mud (M) particles.

Detrital sediments and sedimentary rocks can be classified and named according to the proportions of gravel (G), sand (S) and mud (M) in the detrital fraction. Many classifications exist; Folk et al. (1970) presented a widely used classification based on triangular, three-component diagrams (Figure 13.2a). The first criterion used in this classification is the

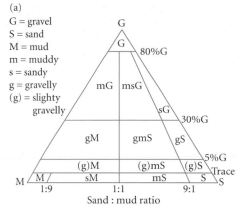

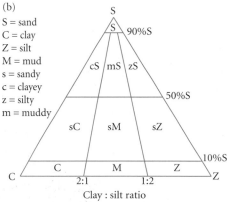

Figure 13.2 Three-component diagrams giving the textural names of detrital sediments. (a) GSM triangle for naming sands and gravels. (b) SCZ triangle for naming muds and mud-bearing sands. Lithified equivalents are gravelstones, sandstones and various mudrocks including claystones, mudstones and siltstones. (After Folk et al., 1970.)

percentage of gravel (%G), which decreases from the top apex of the triangle (100%G) to the base (0%G). The percentage of gravel is important because it reflects the competence of the transport/depositional medium to transport large gravel particles to the site of deposition and to keep smaller particles from coming to rest (Chapter 11). The second criterion used is the sand:mud ratio, which is important because it reflects the ability of the transport/depositional medium to keep mud in suspension and therefore to keep it from being deposited with the grain fraction. Different categories of detrital sediment defined on the proportions of gravel, sand and mud

(a)

(b)

Figure 13.3 Gravelstones. (a) A matrix-supported framework with gravel particles "floating" in and supported by finer particles; in this case a mud-supported matrix. (Photo courtesy of Mario Coniglio.) (b) A clast-supported framework with gravel particles in contact, from the Cretaceous, California. (Photo by John O'Brien.) (For color version, see Plate 13.3, opposite p. 408.)

are defined by the two diagrams in Figure 13.2 and are discussed below.

Gravels and gravelstones contain more than 30% gravel in their detrital fraction. They occupy the top four sectors (G, mG, msG and sG) on the GMS diagram (Figure 13.2a). Detrital materials that contain more than 80% gravel fragments or clasts (G) generally possess **clast-supported textures** because the gravel clasts are in contact and therefore support one another (Figure 13.3a). Smaller sand and/or mud particles, if present, simply fill the spaces between the clast-supported frameworks. Gravels and gravelstones that

contain much less than 80% gravel (G) commonly possess **sand-** or **mud-supported textures** because the gravel particles are generally not in contact and are supported by a sand and/or mud matrix (Figure 13.3b). If the matrix is mostly sand, they are sandy gravels or gravelstones (sG) with a sand-supported framework; if the matrix is mostly mud, they are muddy gravels or gravelstones (mG) with a mud-supported framework.

Sands and sandstones contain less than 30% gravel in their detrital fraction and contain more sand than mud (sand:mud ratio > 1:1). They occupy the six sectors in the bottom right portion of the GSM diagram (gS, gmS, (g)mS, (g)S, mS and S). Epiclastic sediments that contain 5–30% gravel in the epiclastic fraction are called gravelly sands (gS) or gravelly muddy sandstones (gmS).

If a sand or sandstone contains less than 5% gravel and has a sand:mud ratio greater than 9:1, it is a relatively pure sand or sandstone (S). Such relatively pure sandstones are also called **arenites**. If a sand possesses a sand:mud ratio less than 9:1, it is a muddy sand or muddy sandstone (mS). Such mud-rich sandstones are called **wackes**. If either of these has even a trace of gravel, it is a slightly gravelly sand or sandstone ((g)S or (g)mS).

Muds and mudrocks contain less than 30% gravel in their detrital fraction and contain more mud than sand (sand:mud ratio < 1:1). They occupy the four sectors in the bottom left portion of Folk's GSM diagram (Figure 13.2a). Those with more than 5% gravel are gravelly muds or gravelly mudrocks (gM). Those with less than 5% gravel are divided into sandy muds and sandy mudrocks (sM) with sand:mud ratios > 1:9 and relatively pure muds and mudrocks (M) with sand:mud ratios < 1:9. As with sands and sandstones, the prefix "gravel-bearing" may be used if any gravel occurs in the muds or mudrocks. Muds and mudrocks may be further subdivided on the basis of clay (C), silt (Z) and sand (S) ratios using Folk's SCZ triangle (Figure 13.2b).

13.1.3 Central measures

The **central measure** of a particle population is an attempt to represent the typical particle size in the population. Detrital grain populations can be described by three central meas-

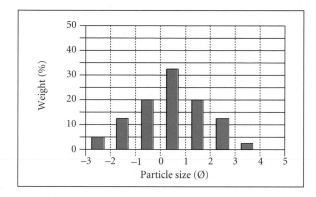

Figure 13.4 Typical histogram of weight percent sediment versus phi size classes; the mode is between 0Φ and 1Φ and the range is between -3Φ and $+4\Phi$.

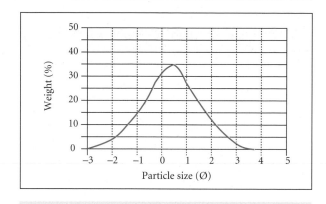

Figure 13.5 Typical frequency curve for weight percent versus phi size classes, for data similar to those used in Figure 13.4.

ures: (1) modal particle size, (2) median particle size, and (3) mean particle size. Ideally these central measures should be calculated from diameter measurements of all the particles in a population. In reality, there are too many grains of sand on a beach or of mud on a lake bottom to make this feasible. Instead, a representative or random sample of the population is collected and analyzed. For finer sediments it is impractical even to measure sample particles. Instead, the sediments are sieved and/or passed through a settling column and the weight percentage in each phi size fraction is determined, after which the results are graphed in terms of weight percent versus phi size classes (Figures 13.4 and 13.5).

Three types of graphs are typically used to portray the size data from which central measures may be determined:

1 A **histogram** in which the weight percent in each phi size class is plotted in the form of a vertical bar graph (Figure 13.4).
2 A **frequency curve** in which a best-fit line is plotted for similar data (Figure 13.5).
3 A **cumulative curve** (Figure 13.6) in which the cumulative weight percent of all size fractions coarser than and including the phi size fraction under consideration are plotted against phi size classes.

The **mode** is the most abundant particle size and is most easily determined when size data are plotted on a histogram (Figure 13.4) or a

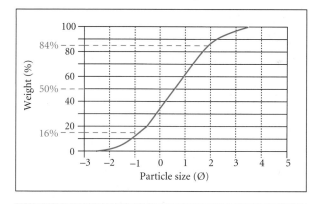

Figure 13.6 Typical cumulative frequency curve for cumulative weight percent versus phi size classes; the 16th percentile (Φ_{16}), 50th percentile (Φ_{50}) and 84th percentile (Φ_{84}), for data similar to those used in Figure 13.4.

frequency curve (Figure 13.5) in which weight percents are plotted against phi size classes. It is usually expressed in terms of the midpoint of the size class in which the largest weight percent occurs (e.g., 0.5Φ in Figures 13.4 and 13.5). Grain size distributions in which two size classes are more abundant than those adjacent to them contain two modes and are said to have **bimodal** size distributions. The more abundant mode is the **primary mode** and the less abundant is the **secondary mode**.

The **median** size is the particle size such that half the population is larger and half the population is smaller. Because measuring individual particles is impractical, the median is most easily estimated from a cumulative frequency curve (Figure 13.6) in which it corresponds to the 50th percentile where half the population by weight is coarser grained and half finer grained. This value is expressed in phi units as Φ_{50} and is obtained by locating the 50th percentile on the y-axis and finding the corresponding phi value on the x-axis (e.g., 0.6Φ in Figure 13.6).

The **mean** particle size ($\bar{x}$) is the average size of the particles. Ideally, mean particle size is determined by (1) measuring the mean diameter of every particle, (2) summing the particle diameters, and (3) dividing the sum by the number of particles in the population. Mathematically, this is expressed by the formula:

$$\bar{x} = (\Sigma x)/n$$

where Σx is the sum of all diameter values (x), n is the number of values and $\bar{x}$ is the mean value. In reality, measuring the diameter of each particle is impractical. Based on the cumulative curve (Figure 13.6), one can estimate the mean size using the **graphic geometric mean (GM)**, which is calculated using the following formula:

$$GM = (\Phi_{16} + \Phi_{50} + \Phi_{84})/3$$

where Φ_{16} is the size value of the 16th percentile – the size such that 16% of the population by weight is coarser; Φ_{50} is the value of the 50th percentile – the median such that 50% by weight is coarser; and Φ_{84} is the value such that 84% of the population by weight is coarser.

The statistical significance of the three phi values used is that they represent the median (Φ_{50}), one standard deviation coarser than the median (Φ_{16}) and one standard deviation finer than the median (Φ_{84}) for a normal probability distribution in which the mean and the median coincide. This permits the central 68% of the population to be used in estimating the mean grain size.

For a more detailed discussion of the statistics of normal probability distributions and their use in sediment studies, readers should consult Blatt et al. (2006), Prothero and Schwab (2004), Folk (1974) and Krumbein and Graybill (1966).

13.1.4 Sorting

Sorting is a measure of the degree of similarity of particle sizes in clastic sediment. As the range of particle sizes decreases, sorting increases. An everyday analogy is the sorting of eggs by size (medium, large, jumbo) before they are shipped to market. Sediment in which all particle grain diameters are of similar size, such as all fine sand, is well sorted. Poorly sorted deposits contain a wide range of particle diameters. A good example is glacial till, which commonly consists of particles that range in size from boulders to clay. Transportation media such as wind and water are effective at sorting epiclastic sediments by size during transportation and deposition. In contrast, deposits transported and deposited by glaciers and most mass flows are generally very poorly sorted. As discussed later in this chapter, sorting renders important information concerning the medium from which the sediment was deposited and the environmental conditions under which deposition may have occurred.

Many quantitative measures have been developed to express sorting. In statistics, one might use the standard deviation (σ) or variance (σ^2). Most studies of detrital sediments utilize the phi scale and a simple mathematical formula to calculate a **coefficient of sorting (So)**. Where quantitative information is available concerning the percentage by weight of grains in various size classes, a cumulative curve showing these distributions can be constructed. Steeper cumulative curves represent better sorted sediments with fewer size classes, whereas more gently sloped curves drawn to the same scale represent sediments with more size classes and therefore poorer sorting (Figure 13.7).

From a brief analysis of these curves, one can quickly determine the phi size value for any percentile (the size at which a specific percent of the grains by weight are larger). It is easy to find values for the 10th (Φ_{10}) percentile (the size at which 10% of the grains by weight are larger) and for the 90th (Φ_{90}) percentile (the size at which 90% of the grains by weight are larger). The coefficient of sorting, as given by Compton (1962), is then:

$$So = |\Phi_{10} - \Phi_{90}|$$

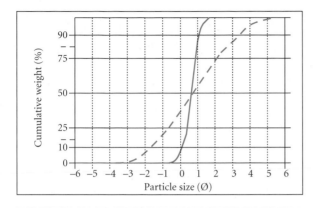

Figure 13.7 Cumulative curves showing the median (Φ_{50}) and different degrees of sorting expressed by the range of sizes in the distribution. The solid curve shows well-sorted sand with a narrow range of sizes (low variance), whereas the dashed curve depicts a relatively poorly sorted, gravelly, muddy sand with a similar median (Φ_{50}) value, but a large range of sizes (large variance).

Table 13.1 Descriptive terminology for sorting coefficients proposed by Compton (1962) and Folk (1974).

Sorting description	Compton sorting coefficient (So)	Inclusive graphic standard deviation
Very well sorted	<1 Φ	<0.35 Φ
Well sorted	1–3 Φ	0.35–0.50 Φ
Moderately well sorted	NA	0.50–0.71 Φ
Moderately sorted	3–5 Φ	0.71–1.00 Φ
Poorly sorted	5–7 Φ	1.00–2.00 Φ
Very poorly sorted	>7 Φ	>2.00 Φ

A more complicated measure of sorting, following Folk (1974), is the **inclusive graphic standard deviation**. It involves finding values that are one standard deviation (Φ_{84} and Φ_{16}) and two standard deviations (Φ_{95} and Φ_5) above and below the mean and is given by:

$$So = \frac{\Phi_{84} - \Phi_{16}}{4} + \frac{\Phi_{95} - \Phi_5}{6.6}$$

General descriptive terminology for sorting based on numerical calculations following Compton (1962) and Folk (1974) is summarized in Table 13.1.

A rapid visual estimation of sorting may be made by determining the size of a particle such that only 10% of the particles by volume are larger (roughly Φ_{10}), and the size of another particle such that only 10% by volume are smaller (roughly Φ_{90}), using the formula and terminology above. This is quite satisfactory for many general purposes. Sorting, using Compton's criteria, can also be estimated by visual comparison with charts showing different degrees of sorting (Figure 13.8).

Transportation and depositional media of relatively low viscosity, such as wind and water, tend to sort detrital sediments according to size. Wind deposits are typically very well sorted to well sorted. Sorting in water-laid sediment is more variable, generally ranging from very well sorted to moderately sorted, depending on specific flow conditions. Transport and depositional media of high viscosity such as glaciers and many mass flows such as landslides and debris flows are quite ineffective at sorting detrital sediments. Most poorly sorted and very poorly sorted sediments are deposited from glaciers or by mass flows. Sorting is extremely important in initial assessments of the likely medium of deposition for any sediment or sedimentary rock.

13.1.5   Particle shape

A rather complex set of quantitative terminology has been developed to describe particle shapes. Many of these are based on three essential measurements:

1   The longest dimension (a).
2   The shortest dimension perpendicular to it (c).
3   An intermediate dimension that is perpendicular to the other two (b).

In such descriptions then a ≥ b ≥ c. Figure 13.9, after Zingg (in Schulz et al., 1954), illustrates four common particle shapes based on the ratios b/a and c/b and the terminology associated with each one:

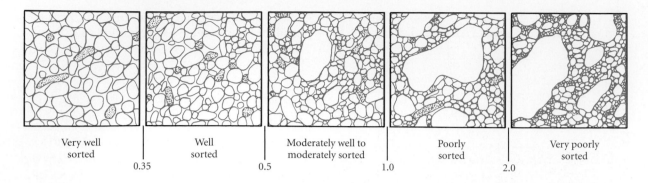

Very well sorted | Well sorted | Moderately well to moderately sorted | Poorly sorted | Very poorly sorted

0.35   0.5   1.0   2.0

Figure 13.8 Diagram for the determination of sorting by visual comparison, using Folk (1974) parameters. (After Compton, 1962.)

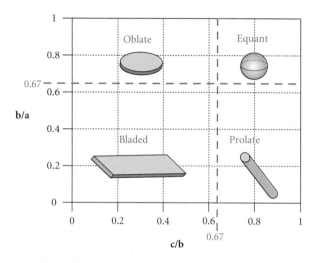

Figure 13.9 Grain shapes defined from a, b and c dimensions, where $a \geq b \geq c$. (After Zingg, in Schulz et al., 1954.)

1 **Equant** (spheroid) particles are those that have similar dimensions in all directions so that $a \approx b \approx c$.
2 **Prolate** (rod-shaped) particles have an elongate cylindrical or cigar shape such that one axis is much longer than the other two and $a \gg b \approx c$.
3 **Oblate** (disk-shaped) particles have a flattened cylindrical or disk-like shape such that one axis is much shorter than the other two with $a \approx b \ll c$.
4 **Bladed** particles have a shape that resembles a knife blade with one long axis, one short axis and one intermediate axis so that $a > b \gg c$.

Prolate and bladed grains are sometimes oriented with their long axes parallel to the flow of the current that deposited them. In such cases they can, used with other features where possible, allow geologists to infer paleoflow directions from ancient sedimentary rocks. Care must be exercised, however, as in some circumstances such particles come to rest with their long axes perpendicular to current flow. Large bladed and oblate clasts sometimes show **imbrication** in which they are more or less uniformly inclined in one direction. Such imbrication commonly dips in an up-current direction because large clasts come to rest on one another in a "piggy-back" manner when deposited (Figure 13.10).

Sphericity and roundness

Sphericity and roundness are two terms that at first glance appear synonymous but have different meanings. **Sphericity** is a measure of the degree to which a particle approaches the dimensions of a perfect sphere. A perfectly spherical particle would be both equant and round. An angular equant particle will have a much higher sphericity than a rounded but inequant particle. The particles in the lower row of Figure 13.11 have lower sphericity than the corresponding particles in the top row.

Roundness is measure of the degree of smoothing or curvature of grain edges. A visual comparison chart (Figure 13.11), first developed by Powers (1953), is generally used to estimate grain roundness. This chart allows one to visually compare any particle with visual standards that include relatively equant grains with high sphericity and relatively inequant grains with low sphericity so that one focuses on the degree of smoothing when estimating roundness. Powers recognized six roundness classes, sequenced in order of increasing roundness: (1) very angular, (2) angular, (3) subangular, (4) subrounded, (5) rounded, and (6) well rounded. Quantitative measures of sphericity and roundness are complex and rarely warrant the time and effort involved.

Most rounding of detrital grains results from abrasion caused by grain collisions during transport. Rounding therefore has the potential to yield information concerning transport history. Several factors affect the rounding of grains during transport; these include (1) size, (2) hardness and durability, (3) intensity of collisions, (4) duration of transport, and (5) number of sedimentary cycles. Larger grains round much more quickly than smaller ones because they impact other grains with much larger force. As a result, pebbles and cobbles tend to round quite rapidly during transport. Sand grains round much more slowly, and mud grains may not round at all. Soft particles round much more quickly than hard particles because they abrade much more rapidly. Therefore mudstone, limestone and marble round much more quickly than quartzite or granitic rock fragments. Grain rounding may be inherited from an earlier sedimentary cycle.

(a)

(b)

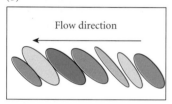

Flow direction

Figure 13.10 Imbricated clasts. (a) Cambrian, New Jersey. (Photo by John O'Brien.) (b) Imbrication diagram. The flow is from right to left, in both cases. (For color version, see Plate 13.10a, opposite p. 408.)

13.1.6 Porosity and permeability

Porosity and permeability are critical factors in the storage and transport of fluids, such as hydrocarbons and water, in sediments and rocks. They are defined and discussed in the

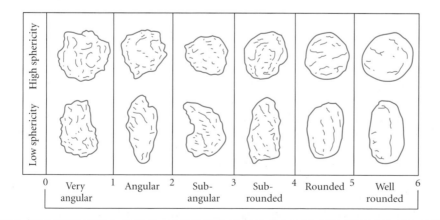

Figure 13.11 Diagram for the determination of rounding in grains of varying sphericity by visual estimation. (After Powers, 1953; Pettijohn et al., 1987; from Tucker, 2001.)

context of soils in Chapter 12. Here we focus on porosity and permeability in relationship to the textures of sediments and sedimentary rocks.

Porosity (Ps) is simply the volume percentage of void spaces, called pores, in a rock given by the simple formula, Ps = % pore space = (volume of pores/total rock volume) × 100. In general, porosity increases with sorting and with grain roundness. In poorly sorted sediments, the finer constituents tend to fill the pore spaces between the larger ones. Angular grains tend to fill pore spaces more effectively than round ones.

The primary importance of porosity is that it represents the total capacity of a rock to hold or store fluids such as groundwater, natural gas or liquid petroleum in the void spaces between solid grains. The search for groundwater or petroleum reservoirs often begins as a search for subsurface rocks with substantial amounts of porosity. Any process, such as cementation, that reduces porosity also reduces a rock's ability to store fluids. For this reason, much attention is paid to the distribution of mineral cementation patterns in the subsurface during petroleum and hydrological exploration surveys.

Permeability (K) expresses the rate of fluid flow through a rock. Permeability can be determined by using Darcy's law, $Q = A[K \times (h/L)]$, where A is cross-sectional area and h/L the hydraulic gradient, as discussed in Chapter 12. In this equation, K is the hydraulic conductivity, which is given by $K = \kappa \rho g / \mu$, where κ is effective permeability, ρ is density, g is acceleration due to gravity and μ is viscosity. The two formulas show that flow rates (Q) through porous media increase with increasing effective permeability. Permeability increases rapidly with pore size and pore interconnectedness, both of which tend to increase with particle size and sorting. Well-sorted sediments with large pore spaces (and large particles) tend to posses very high permeabilities; rocks with secondary porosity due to dissolution can possess even higher permeability; think caves! Permeability is extremely important in the petroleum industry because petroleum and natural gas must be able to flow into porous reservoir rocks in order to collect there and must be able to flow out of them as extraction proceeds. In addition, fluid migration is severely

Table 13.2 Typical hydraulic conductivity of common detrital sediments.

Sediment type	Hydraulic conductivity
Clays	Low ≈ 10^{-6} cm/s
Silts	Low to moderate ≈ 10^{-4} cm/s
Sands	Moderate to high ≈ 10^{-2} to 10^{-3} cm/s
Gravels	High to very high ≈ 10^{-1} to 10^{1} cm/s

restricted by relatively impermeable layers that serve as petrotards (cap rocks) that trap petroleum and natural gas in underlying permeable rocks. Permeability is also fundamentally important in groundwater aquifer studies and in determining contaminant flow behavior in the subsurface. Typical hydraulic conductivities for unconsolidated sediments are given in Table 13.2. Note the general increase in permeability with increasing particle size. Sands and gravels tend to be quite permeable and act as **aquifers**, which store and transmit water. Clay-rich layers tend to be relatively impermeable and to act as **aquitards**, which retard fluid flow.

13.1.7 Textural maturity

Plumley (1948) and Folk (1951) applied the concept of textural maturity to sands and sandstones. The concept does not apply to mudrocks and gravelstones. Textural maturity is based on the answers to three questions regarding (1) the presence or absence of significant (>5%) mud matrix, (2) the degree of sorting, and (3) the degree of grain rounding. In Table 13.3 the textural maturity increases from bottom to top as these three criteria are successively applied to the sediment or sedimentary rock in question. When applying these criteria, if the answer to any question is no, the remaining questions are not required as the textural maturity has been established.

Question 1 Is there more than 5% matrix in the sediment or sedimentary rock? Sediments and sedimentary rocks that contain more than 5% mud matrix between the sand and gravel grains are **texturally immature**. Note that question 1 is really asking

Table 13.3 Criteria for the textural maturity of sands and sandstones.

(1) Is more than 5% mud matrix present? If no:	(2) Is it well sorted? If no:	(3) Are the average grains rounded?	Textural maturity
		Yes	Supermature
	Yes	No	Mature
No	No		Submature
Yes			Immature

whether the mud particles were largely kept in suspension (or later winnowed out) while the sand grains accumulated or whether a significant amount of mud matrix accumulated with the coarse fraction. Since wind and many aqueous media transport mud in suspension while they deposit sand and gravel from the bed load (Chapter 11), wind- and water-laid deposits are rarely texturally immature. On the other hand, glaciers and many mass flows, such as landslides, debris flows and high concentration sediment gravity flows (Chapter 11), transport and deposit a wide variety of sizes together, and their deposits are commonly texturally immature. If less than 5% mud matrix occurs, one proceeds to question 2.

Question 2 Is the sediment or sedimentary rock well sorted? If the material is not well sorted, it is **texturally submature**. Question 2 assesses whether or not the medium of deposition, which was capable of keeping mud in suspension, was also capable of producing well-sorted sediment. If it is not well sorted, it was most likely deposited by rapidly decelerating or fluctuating flows that can deposit a range of sizes in a very short time. Such flows might include river systems (e.g., braided streams and levees), many storm deposits and low concentration sediment gravity flows such as turbidity currents. If the sediment is well sorted, then it was almost certainly deposited by either wind or aqueous media in which sediments were well washed during slow accumulation (e.g., beaches, shallow bars and some stream channels). If the material is well sorted or very well sorted, it is at least texturally mature and one proceeds to question 3.

Question 3 Are the average grains rounded? If they are rounded (≥4 on the Powers scale of roundness), the sediment is **texturally**

supermature; if they are not rounded (≤3 on the Powers scale), it is **texturally mature**. This is especially significant when working with medium and finer sands as such sizes are only rounded significantly during wind transport. Care must be exercised! Wind-rounded sands can blow into any number of aqueous environments. They can even be picked up by glaciers and are frequently entrained in mass flows. In the following sections, we will describe gravelstones, sandstones and mudstones and what we learn from their properties.

13.2 GRAVELSTONES

Gravelstones such as conglomerates and breccias constitute roughly 5% of detrital sediments and sedimentary rocks. Mudstones and sandstones are much more abundant. Still, the vast majority of detrital sediments deposited directly by glaciers and debris flows and a significant proportion of water-laid sediments contain enough gravel to qualify as gravelstones.

13.2.1 Gravelstone classification

A gravelstone is a solid sedimentary rock in which at least 30% of the detrital grains by volume are gravel. Gravelstones, also called **rudites**, are classified primarily on the basis of three criteria: (1) the nature of the support framework for the gravel, (2) the roundness of the gravel fragments or clasts, and (3) the composition of the clasts.

Many classifications for gravelstones exist. Table 13.4, modified from Raymond's (2002) classification, provides an excellent basis for discriminating between the major types of gravelstone. Clast size can be used as a modifier (e.g., pebble conglomerate or boulder diamictite), as can clast composition (e.g.,

Table 13.4   Classification of gravelstones: conglomerates, breccias and diamictites. (Modified from Raymond, 2002.)

Gravelstone shape	Matrix support	Gravelstone composition	Rock name
Subrounded to very rounded (conglomerate)	Gravel- or sand-supported framework	Single composition	Oligomictic conglomerate (e.g., quartz conglomerate)
		Multiple compositions	Polymictic conglomerate
Subangular to very angular (breccia)	Gravel- or sand-supported framework	Single composition	Oligomictic breccia (e.g., limestone breccia)
		Multiple compositions	Polymictic breccia
Any shape	Mud-supported framework (diamictite)	Single composition	Oligomictic diamictite
		Multiple compositions	Polymictic diamictite

quartz pebble conglomerate or limestone breccia).

Gravelstone clast shape

Gravelstones in which gravel clasts are generally rounded are called **conglomerates**. Since large clasts are rounded rapidly during transport, conglomerates (Figure 13.12a, b) are the most common type of gravelstone based on clast shape. Gravelstones that contain angular clasts are called **breccia**. The angular clasts of breccias (Figure 13.12c) suggest minimal transport and derivation from a nearby source area. Breccias occur in a number of environmental settings, including:

1   Fault zones where angular fragments are produced by fragmentation during slip.
2   Talus slopes at the base of cliffs and ridges due to the effects of rock weathering and mass wasting processes such as rockfall.
3   Areas where bolides have collided with Earth's surface, producing **impactites**.
4   Karst regions where breccias are produced by the dissolution and collapse of carbonate rocks (Chapter 12).
5   The upper portions of alluvial fans where debris flows are common.
6   Wherever cohesive mud clasts are eroded and transported a short distance by streams, currents or sediment gravity flows.

Gravelstone framework

Conglomerates and breccias occur in clast-supported and sand or mud matrix-supported varieties. Gravelstones with a **clast-supported framework** contain gravel clasts that are generally in contact and therefore support one another to form a clast-supported framework. A clast-supported conglomerate can be called an **orthoconglomerate**. Clast-supported gravelstones (Figure 13.12a, b) commonly accumulate in the following environments:

1   Aqueous environments such as stream channels, beaches and marine shoals in which sand and mud continue to be transported while gravel accumulates.
2   Alluvial fans and ephemeral braided stream systems where the fines are removed by surface water infiltrating into the ground between clasts, as **sieve deposits**.
3   Where winds entrain sand and mud from the surface, leaving the gravel behind as **lag gravels**.

Many gravelstones possess a **matrix-supported framework** in which the gravel clasts are generally not in contact and are supported by the finer sand and/or gravel matrix that separates them. If the matrix is mostly sand, the gravelstone possesses a **sand-supported framework** (Figure 13.12c). Gravelstones with sand-supported frameworks occur wherever significant amounts of sand

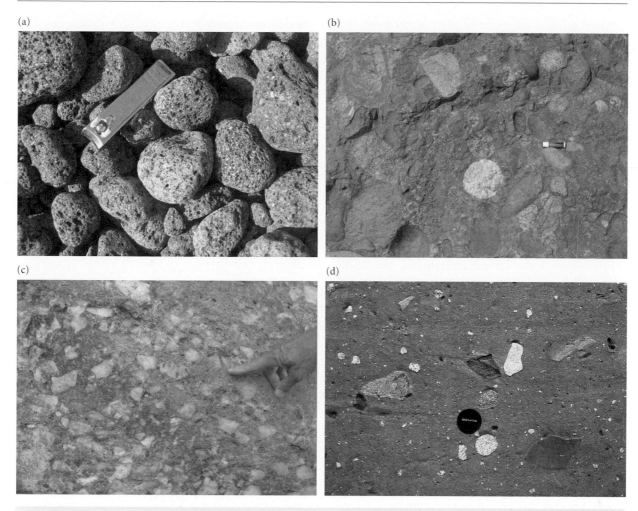

Figure 13.12 (a) Oligomictic conglomerate with rounded gravel and a clast-supported framework, Stromboli, Italy. (b) Polymictic conglomerate with a clast-supported framework, Kata Tjuta, northern Australia. (c) Oligomictic quartz breccia with sand-supported framework, Kakadu National Park, northern Australia. (d) Polymictic diamictite (a tillite) with a mud-supported framework, South Africa. (a–c, photos by John O'Brien; d, photo courtesy of Michael Hambrey, with permission from www.glaciers-online.net.) (For color version, see Plate 13.12, between pp. 408 and 409.)

and gravel accumulate together. These environments include alluvial fans, river channels, braid bars and point bars in stream systems, as well as beaches, marine shoals and storm deposits in marine environments. Some high concentration sediment gravity flows also produce sand-supported gravelstones. If the matrix is mostly mud, the gravelstone has a **mud-supported framework** (Figure 13.12d). A gravelstone with a mud-supported matrix is called a **diamictite** or a **paraconglomerate**. Because they contain significant amounts of both gravel and mud, diamictites are invariably poorly or very poorly sorted. Diamictites

form by a number of different processes, including the following:

1   Glaciers deposit gravel and mud-rich till that are lithified to form **tillites**.
2   Mudflows and many debris flows and lahars also produce diamictites.
3   Some gravel-bearing, sediment gravity flows produce diamictites, especially those that bring gravel-rich sediment into areas with extensive mud accumulations where the two are mixed.
4   Tectonic **mélanges** at convergent plate boundaries are diamictites produced by

the mixing of gravel clasts with muds by a combination of sedimentary and tectonic processes.

Gravelstone composition

Gravelstones are also classified according to clast composition. Gravelstones in which the clasts are largely of one composition are called **oligomictic** conglomerates, breccias or diamictites (see Figure 13.12a, c). Fault, collapse and rockfall breccias are typically oligomictic, as are compositionally mature, quartz-rich conglomerates.

Gravelstones that contain a variety of clast compositions are called **polymictic** conglomerates, breccias or diamictites (Figure 13.12b, d). Polymictic gravelstones can yield significant, easily accessible information concerning the nature of the rocks in the source area. Mélange gravelstones, produced by various combinations of sedimentary and tectonic processes that mix several rock types are typically **polymictic diamictites**, as are many glacial tills produced during continental glaciation where expanding glaciers erode rocks from a large region.

13.2.2   Gravelstone provenance

Gravelstones are by far the easiest detrital sedimentary rocks to use in hand-specimen provenance studies. Fragments of the original source rocks are often preserved intact in the pebble, cobble and boulder populations. Many clasts can be identified macroscopically, using fresh surfaces, and almost all can be identified using microscopic methods. Large clast size allows for the identification of possible source rocks and for making inferences about climate and relief, the duration and intensity of transport and the tectonic setting in which the gravelstones accumulated. Preservation of representative unaltered clasts is favored by factors that include:

1   High mechanical durability and chemical stability of clast components.
2   Low precipitation and temperatures, which inhibit chemical decomposition.
3   High relief and low vegetative cover, which promote rapid erosion and minimize the duration of chemical decomposition and clast disintegration.

4   Short transportation, which promotes survival of mechanically unstable clasts.

Gravelstones derived from volcanic source areas are often rich in volcanic rock fragments (VRF or Lv) that record whether the source region was a volcanic arc (rich in VRF of andesitic to rhyolitic composition), or a hotspot flood plateau (rich in VRF of basaltic composition) or a bimodal suite (rich in both basaltic and rhyolitic VRF). Most nonpyroclastic volcanic rock fragments are quite hard and can survive relatively long periods of transport. Basaltic rocks, rich in ferromagnesian minerals and calcic plagioclase, tend to decompose rapidly (Chapter 12) and survive weathering only in areas with relatively high relief, low rainfall and/or low temperatures.

Gravelstones derived from plutonic igneous source areas tend to disintegrate along grain boundaries during long or intense transport. An abundance of plutonic rock fragments (Lp) suggests a relatively nearby source area. Once again, mafic fragments tend to decompose rapidly; their occurrence implies some combination of high relief/rapid erosion, and/or low rainfall and/or temperature. Granitoid rock fragments (GRF), on the other hand, decompose less rapidly and are more commonly preserved in gravelstones. GRF are most commonly derived from magmatic arc and intracratonic rift settings in which GRF are shed from areas of reasonably high relief into nearby basins of deposition before constituent feldspars undergo significant decomposition

Gravelstones derived from metamorphic source areas may be rich in metamorphic rock fragments (MRF or Lm). MRF can be subdivided into low grade fragments (Lm_1) and higher grade rock fragments (Lm_2). Hard, fine-grained MRF such as metaquartzite survive long periods of transportation, as does the vein quartz common in metamorphic terrains, but softer MRF such as slate, phyllite and marble do not. The occurrence of such soft MRF in gravelstone implies derivation from a nearby source. The most varied assemblages of MRF are derived from the erosion or dissection of orogenic belts. Progressive dissection of such orogenic belts generally causes the exposure of progressively higher grade MRF over time, with increases in Lm_2 and decreases in Lm_1.

Gravelstones derived from sedimentary source areas can be rich in sedimentary rock fragments (SRF or Ls). Hard, fine-grained SRF such as chert survive long periods of transportation and multiple sedimentary cycles. Softer SRF such as shale, claystone and limestone do not commonly survive transportation, and their occurrence as clasts implies derivation from a nearby source. Readily soluble SRF such as limestone and evaporates and MRF such as marble do not survive weathering unless the source area has high relief and erosion rates, and/or low temperature and/or rainfall.

With the proviso that fresh material must be used, any of the techniques used by igneous petrologists to discriminate magma sources and tectonic settings (Chapter 10) and by metamorphic petrologists to determine protoliths, metamorphic facies and tectonic settings (Chapter 18) may be applied to inferring the history of the source area from which the gravelstone clasts have been derived (Box 13.1).

13.3 SANDSTONES

13.3.1 Sandstone classification

Sandstones are rocks that contain less than 30% gravel in their detrital fraction and contain more sand than mud (sand: mud > 1:1). Dozens of sandstone classifications have been proposed. They use related criteria to organize sandstones into groups, similar boundaries to separate different types and related terms to name different varieties of sandstone. One widely used sandstone classification is that proposed by Folk (1974). Folk's scheme (Figure 13.13) classifies sandstones on the proportions of three components:

Box 13.1   Using conglomerate clasts to document slip on faults

For more than half a century, since Crowell (1952) used Ridge Basin breccias to document strike-slip on the San Gabriel Fault in southern California, geologists have utilized synkinematic deposition of conglomerates to try to unravel the timing and movement history of faults. The trick is to recognize conglomerate clasts in formations adjacent to the fault that were derived from a unique source on the other side of the fault. This source may have been adjacent to the conglomerates at the time of deposition, but if subsequent strike-slip along the fault has occurred, the source will have been relocated relative to the conglomerate. The distance of relocation between the source rock and the conglomerate permits the horizontal component of slip to be documented. Knowledge of the age of the conglomerate permits the average slip rate to be calculated (e.g., in km/Ma).

Geologists in New Zealand have documented the movement history of the large Alpine Fault (Figure B13.1), which has been the site of several magnitude 8 earthquakes in the past 1000 years. Like the San Andreas Fault system in southern California, the Alpine Fault is part of a major transform plate boundary system, in this case between the Australian and Pacific Plates. Sutherland (1994) recognized unusual clasts derived from ultramafic rocks and greenshist–amphibolite grade schists in Pliocene (3.6 Ma) conglomerates deposited in Cascade Valley. Sutherland traced these clasts to source rocks in the Red Mountain ultramafic sequence and Haast schist that today are located 95 km to the southeast of the Miocene conglomerates. From these data, Sutherland concluded that the minimum right lateral slip on the Alpine Fault since 3.6 Ma has been 27 km/Ma and that up to 35 km/Ma was possible.

Previously Cutten (1979) had posited that Miocene (11.5 Ma) sandstone and metamorphic clast assemblages in the Maruia Basin on the northwest side of the fault had been derived from the Caples and Torlesse Terranes, which are now located 420 km to the southwest of the Maruia Basin (Figure B13.1). Recent geochemical work on both clasts and source rocks by Cutten et al. (2006) has confirmed this interpretation. The average dextral (right lateral) slip on the Alpine Fault responsible for the present offset between the source rocks and the clasts shed from them is roughly 37 km/Ma. These numbers correspond well with plate tectonic studies that imply approximately 35 km/Ma of

Continued

Box 13.1   *Continued*

dextral slip on the Alpine Fault system transform plate boundary. Synkinematic conglomerate clasts are useful in many other tectonic interpretations. Similar procedures are utilized to determine the docking history of terranes in accretionary collisional complexes (Chapter 1). If clasts from one terrane were deposited atop another terrane, the two terranes were in close proximity (docked) at the time the conglomerates accumulated. So much critical information from so few cobbles!

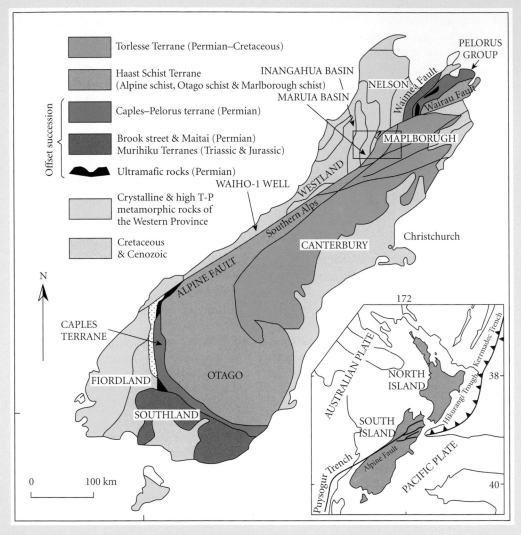

Figure B13.1   Geological map of South Island, New Zealand, showing the Alpine Fault, Caples Terrane and Maruia Basin. The inset map shows the plate tectonic setting. T-P, temperature–pressure. (From Cutten et al., 2006; with permission of the American Geophysical Union.)

1 The percentage of quartz grains by volume (%Q) in the sand fraction.
2 The percentage of feldspar grains by volume (%F) in the sand fraction.
3 The percentage of polycrystalline rock fragments, called lithic fragments by volume (%L), in the sand fraction.

All other components are ignored for the purposes of basic sandstone classification. A three-component (QFL) diagram is used and is subdivided into seven sectors. Each of the seven sectors represents a different type of sandstone with a specific name that reflects its composition (Figure 13.13).

The first criterion used to classify sandstones is the recalculated percentage of quartz grains ($\%Q = [\%Q/(\%Q + \%F + \%L)] \times 100$) in the sand fraction. Any sandstone that possesses more than 95%Q plots in the uppermost sector of the diagram and is called **quartzarenite**. Some classifications use the alternate spelling quartz arenite; others require only 90%Q. Quartzarenites are compositionally mature or supermature.

Sandstones that contain 75–95% recalculated quartz and have more feldspar than lithic fragments (F:L > 1:1) are called **subarkoses**; those with 75–95% quartz and less feldspar than lithic fragments (F:L < 1:1) are called **sublitharenites**. Subarkoses and sublitharenites are compositionally submature to mature.

All sandstones that contain less than 75%Q plot in one of the four sectors in the bottom portion of the QFL triangle. These sandstones tend to be compositionally submature or immature. Feldspar-rich sandstones that contain less than 75% quartz and have an F:L ratio > 3:1 plot in the lower left-hand sector of the QFL triangle and are called **arkoses**. **Litharenites** are rock fragment-rich sandstones that plot in the lower right-hand sector of QFL diagrams because they contain less than 75% quartz and have an F:L ratio < 1:3. Many types of sandstone contain substantial proportions of both feldspars and lithic fragments. These are called **lithic arkoses** when they contain less than 75% quartz and have F:L ratios between 1:1 and 3:1 and **feldspathic litharenites** when they contain less than 75% quartz and have F:L ratios between 1:1 and 1:3.

Another valuable set of classifications for sandstones is based on a combination of sandstone composition and sandstone texture. In addition to the QFL triangle, these classifications incorporate a fourth variable, the percentage of mud matrix (Figure 13.14), which increases from zero in the front triangle to 100% in the back triangle of the diagram. Sandstones with significant mud matrix are called wackes; those with negligible mud

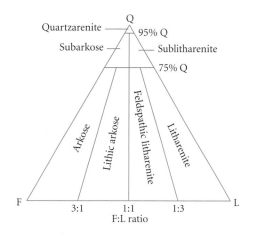

Figure 13.13 Sandstone classification of Folk (1974). F, feldspar; L, lithic fragments; Q, quartz.

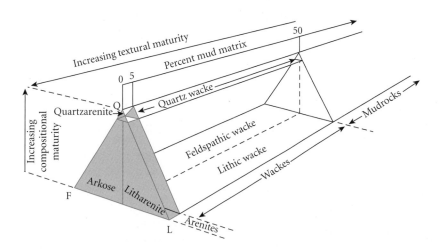

Figure 13.14 Four-component classification sandstones. F, feldspar; L, lithic fragments; Q, quartz. (After Blatt and Tracy, 1996.)

matrix are called arenites. Once again, boundaries and terminology vary between classifications. Blatt and Tracy (1996) put the boundary between wacke sandstones and arenites at 5% mud (Figure 13.14) so that texturally immature sandstones are wackes. They place the boundary between wackes and mudstones at 50% mud, a practice consistent with other classifications based on sand:mud ratios. Sandstones with less than 5% mud matrix are arenites of the types discussed previously (e.g., quartzarenites, arkoses and litharenites).

An older term, **graywacke** (or greywacke), is sometimes used as a general term for matrix-rich sandstones but has been used with so many different meanings that its value is questionable. Wackes can be subdivided into **quartz wackes/graywackes, feldspathic wackes/graywackes** or **lithic wackes/graywackes** depending on the percentage of quartz, feldspar and lithic fragments in the sand fraction percentage.

Detrital sedimentary rocks with more than 50% mud in their detrital fractions are mudrocks rather than sandstones and are classified using a different set of criteria, discussed later in this chapter.

13.3.2   Sandstone provenance

The composition of sandstones reflects several factors, including:

1   The rock types exposed in the source area.
2   The chemical stability and decomposition history of the rock types.
3   The climate, relief and rates of erosion in the source terrain.
4   The proximity of the source area and the intensity and duration of transport as expressed by mechanical stability.
5   The tectonic setting in which sedimentation occurred.

For decades, petrologists have used various types of **discrimination diagrams** to try to distinguish igneous rocks produced in different tectonic settings, in some cases even before the advent of the formal theory of plate tectonics. No one discrimination diagram makes clear distinctions in every case, but combinations of such diagrams have proved useful in

(a)

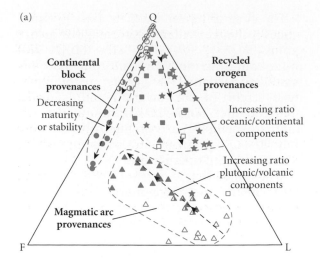

(b)

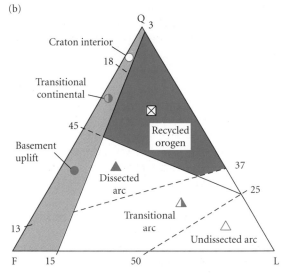

Figure 13.15 (a) QFL (quartz, feldspar, lithic fragments) diagram after Dickenson and Suczek (1979). (From Raymond, 2002; with permission of Waveland Press.) (b) Modified QFL diagram after Dickenson et al. (1983). (From Tucker, 2001; with permission of Blackwell.)

distinguishing igneous rocks formed in different tectonic settings (Chapter 10).

Dickenson (1974) and Dickenson and Suczek (1979) introduced the use of discrimination diagrams (Figure 13.15) for the determination of sandstone tectonic settings. They presented several different diagrams which, when used in combination and with other datasets (such as rock texture, field relations, geochemistry and geochronology), allow one to infer the likely tectonic setting in which a particular sandstone formed. The most frequently used discrimination diagram is a QFL triangular diagram, much like the ones used to classify sandstones by composition. In this case, however, the Q constituent includes all quartz grains, including monocrystalline quartz, polycrystalline quartz and chert. The latter two are considered to be rock fragments in Folk's classification. The F constituent includes all feldspars and the L constituent includes all lithic fragments except polycrystalline quartz and chert. Effective utilization of such diagrams requires a quantitative determination of the percentages of each constituent, e.g., from detailed examination of a thin section. In the diagrams of Dickenson and Suczek (1979) and Dickenson et al. (1983), three major provenances or tectonic settings, each with subdivisions, are recognized:

1 **Magmatic arcs** in which sediments are derived from volcanic–magmatic arcs formed over subduction zones along convergent plate boundaries.
2 **Recycled orogens** in which the sediments are derived from orogenic mountain belts that developed at collision plate boundaries.
3 **Continental blocks** in which sediments are derived from stable cratonic source rocks from shields and/or platforms or continental rift systems.

A basic knowledge of plate tectonics (Chapter 1) is essential to understanding patterns of sedimentation. Detailed petrography supplemented by precise geochemical analyses, radiometric ages and cooling ages (Box 13.2) of the constituents can allow one to test and refine hypotheses about provenance and/or tectonic setting, which can lead to more detailed and sophisticated portrayals of the provenance and tectonic settings of sandstone deposition (e.g., Armstrong-Altrin et al., 2004; Dickinson and Gehrels, 2003; Preston et al., 2002).

With few exceptions, sandstones from **volcanic–magmatic arc source areas** plot in an area in the lower part of the diagram and are mostly litharenites, feldspathic litharenites or lithic arkoses (Figure 13.15a). These rock compositions reflect the generally low compositional maturity of sediments shed rapidly

from primary source rocks in areas of high relief. Dickenson and Suczek (1979) also noted an important trend within the magmatic arc field. Progressive erosion of a volcanic–magmatic arc produces a general decrease in lithic rock fragments and an increase in both feldspars and quartz so that arc dissection produces the trend line from litharenites toward arkoses shown in Figure 13.15a (lower dashed arrow). This occurs largely because young, undissected volcanic arcs generate large numbers of fine-grained, largely aphanitic, rock fragments, whereas dissected plutonic arcs consist largely of coarser grained, phaneritic granitoids that generate larger amounts of sand-size feldspar and quartz grains.

Dickenson and Suczek (1979) subdivided sandstones from **recycled orogen source areas** into:

1 **Subduction complexes** that contain sedimentary rocks, metamorphic rocks and recycled volcanic–magmatic arc sediments derived from uplifted recycled accretionary complexes.
2 **Collision orogens** that contain sedimentary and metamorphic rocks, of both continental and oceanic origin, uplifted during the continental collisions that close ocean basins.
3 **Foreland uplifts** that consist of diverse sedimentary, metamorphic and plutonic igneous rock assemblages exposed in orogenic belts some distance from convergent plate boundaries.

With some exceptions, sandstones from orogenic belt source areas plot in an area in the central to right upper portion of QFL diagrams (Figure 13.15a). These sandstones are mostly litharenites and sublitharenites along with subordinate subarkoses and a few lithic arkoses. Orogenic belt rocks are mostly recycled from sedimentary or metasedimentary source rocks that have experienced at least one previous sedimentary cycle. As a result, these sediments tend to be more compositionally mature (mostly submature to mature) than those eroded from the primary source rocks in volcanic–magmatic arcs. Because recycled orogen sediments are shed relatively rapidly, they also tend to contain at least some rock fragments. One important trend in such areas is the tendency for compositions to become enriched in quartz and depleted in lithic grains and feldspars as the orogenic belt is worn down, making relief and erosion rates lower and decomposition longer.

Continental block source areas consist principally of stable cratonic areas that are subdivided into shields and platforms. **Shields** consist primarily of Precambrian plutonic and high-grade metamorphic rocks such as granitoids, gneisses and granulites. **Platforms** are characterized by a relatively thin veneer of largely mature detrital sedimentary rocks and/ or carbonate sedimentary rocks that overlie shield rocks. Both shields and platforms have long histories of relative tectonic stability.

Dickenson and Suczek (1979) divided continental block provenances into the following (Figure 13.15b):

1 **Craton interior** in which sediments are derived from pre-existing, generally mature, sedimentary rocks that overly plutonic basement rocks in a stable platform setting with very low relief.
2 **Transitional cratons** in which both platform sedimentary rocks and plutonic basement shield rocks are exposed as source rock types in settings of low to moderate relief.
3 **Uplifted basement** in which the basement plutonic igneous and metamorphic rocks are exposed in an area of high relief resulting from uplift, e.g., along faults in continental rift setting.

With very few exceptions, sandstones from cratonic settings plot in an area along the left side of QFL diagrams (Figure 13.15a). Depending upon mineralogical maturity, these sandstones range from less mature arkoses and subarkoses to more mature quartzarenites. Dickenson and Suczek noted a significant trend within continental block sandstones that extends from feldspar-rich arkoses toward quartzarenites. They explained this trend in terms of relief: the higher the relief in the source area, the more feldspar is present in the sandstones; the lower the relief, the more quartz is present. Uplifted basement sources provide more feldspar for dispersion than do stable craton interiors.

In the past two decades, sedimentologists have used a variety of geochemical techniques to match detrital sediments to source rocks in their efforts to more accurately determine their most likely provenance (Box 13.2).

13.4 MUDROCKS

Mudrocks dominate the sedimentary record, constituting 60–65% of all sedimentary rocks. This results primarily from the abundance of clay particles that are produced by decomposition in the source area and silt particles produced by the disintegration of larger particles in the source area and by abrasion during transportation. Despite their abundance, mudrocks generally are *not* well exposed at the surface. This is because most mudrocks are fairly soft and easily eroded and because they often support abundant vegetation in non-arid climates.

Mudrocks contain less than 30% gravel in their detrital fraction and have more mud than sand (sand:mud < 1:1). Those with more than 5% gravel are gravelly mudrocks. Mudrocks with more than 10% sand (sand:mud ratio > 1:9) are sandy mudrocks. The vast majority of mudrocks contain little or no gravel and sand. This is because the traction conditions under which most gravel and sand are transported and deposited by wind and water are significantly different from the suspension conditions under which most mud is transported and deposited. As a result, the two size populations are largely separated during transportation and deposition.

13.4.1 Mudrock textures and structures

The two major components of mudrocks are silt $(0.0625–0.004\,\text{mm} = 4–8\,\Phi)$ and clay $(<0.004\,\text{mm} = <4\,\mu\text{m} = >8\,\Phi)$. Three common types of mudrocks are classified and named (Table 13.5) on the basis their silt:sand ratios:

1 **Siltstones** are coarse mudrocks that contain more than two-thirds silt in the mud fraction (silt:clay > 2:1).
2 **Claystones** are fine mudrocks that contain more than two-thirds clay in the mud fraction (silt:clay < 1:2).
3 **Mudstones** contain substantial amounts of both clay and silt and possess silt:clay ratios of between 2:1 and 1:2.

Box 13.2   Appalachian sources for western sandstones

The Permian and Jurassic aeolian quartzarenites and subarkoses of the Colorado Plateau in the western United States have long inspired geologists with their impressive, large-scale cross-stratification (Figure B13.2). Now they have another claim to fame. The majority of the sand grains in these wind-blown deposits were derived from the Appalachian Mountains in the eastern United States.

The story began when Dickenson and Gehrels (2003) analyzed 468 detrital zircons from five Permian aeolian formations, using U/Pb dating techniques to establish their ages and delimit their provenance. Their analysis revealed six age peaks for the zircons. About half corresponded to ages of local source rocks in the ancestral Rocky Mountains (1800 to 1365 Ma) and the North American Craton (3015 to 1800 Ma). But the other half could not have been derived from local sources. These zircons (1315 to 1000 Ma, 750 to 515 Ma and 515 to 310 Ma) could only have been derived from sources located in the vicinity of the Grenville and Appalachian orogens, now located on the eastern margin of the North American Craton. Dickenson and Gehrels concluded that a large transcontinental river system, analogous to the modern Amazon River in South America, transported detritus from the Appalachian region to the western part of the craton in Permian time. Because this area was located just north of the equator at this time, northeast trade winds did the rest, depositing the vast seas of sand that became the Permian Coconino and Jurassic Navajo Formations and their equivalents.

Rahl et al. (2003) used a more sophisticated set of analyses to support and extend these contentions. They used U/Pb isotope methods to date zircon formation and He isotope analysis to determine zircon cooling ages, which approximates their time of exposure and erosion at Earth's surface. These

Figure B13.2   Large-scale cross-strata, Navajo sandstone, Jurassic, Utah. (Photo by John O'Brien.) (For color version, see Plate B13.2, between pp. 408 and 409.)

Continued

Box 13.2 *Continued*

so-called double dating methods revealed that most of the zircons formed during the Grenville orogeny (1200 to 950 Ma) but had cooling ages of 500 to 225 Ma that correspond to the formation and unroofing of the Appalachian orogen. Their work confirmed the Dickenson and Gehrels hypothesis and suggested that the transcontinental river system may have existed from the Permian into the Triassic.

Campbell et al. (2005) have now suggested that 75% of the zircons in the Navajo sandstone were derived from Appalachian–Grenville sources. By tying together detrital sediments and source areas, detrital zircons have the potential to test ancient continental reconstructions by identifying the presence of particular source regions, the timing of their erosion and subsequent dispersal patterns. What tools we have at our disposal!

Table 13.5 Textural classification of mudrocks based on silt:clay ratios.

Silt:clay ratio	Mudrock name
Predominantly silt (>2:1)	Siltstone
Abundant silt and clay (2:1 to 1:2)	Mudstone
Predominantly clay (<1:2)	Claystone

Determination of the percentages of silt and clay in outcrop or hand specimens is made difficult by their fine size, though coarse silt grains are still distinguishable to the unaided eye. One can use a taste test as a rough guide. As a rule of thumb, clay particles feel soft, smooth and pasty when placed between the tongue and teeth, whereas most silt particles feel rather hard and gritty. This is not recommended for hazardous waste sites! Wet mud can be rolled into a string using a shearing motion with the hands; clay forms a long cohesive string while silt, being less cohesive, tends to disaggregate. The presence of mudcracks formed by the dessication of wet mud indicates a degree of cohesiveness in the original sediment. For more accurate determinations, one can use laboratory methods such as settling tube analysis for unconsolidated muds and microscopic methods for consolidated rocks.

The feature called fissility is so common in mudstones and claystones that it requires discussion. **Fissility** is the tendency of certain mudstones and claystones to split into thin layers, roughly parallel to stratification. Fissile mudstones and claystones are called **shales**. Although the term shale is often used informally for all mudrocks, it should be reserved for those that display fissility. Fissility generally results from the sub-parallel alignment of clay and mica minerals with their characteristic sheet (phyllosilicate) structures. The rock splits parallel to the mica and clay sheets. Mudrocks, such as siltstones, that contain little or no clay minerals or mudrocks in which the phyllosilicate minerals are randomly oriented do not possess fissility and are not shales.

13.4.2 Mudrock composition and color

The two major components of mudrocks are quartz and clay minerals. It is important to point out that the term "clay" is used in two distinctly different ways. The size term "clay" is used for any clastic particle smaller than 0.004 mm. The mineralogical term "clay" is used for a group of aluminum-bearing phyllosilicates (discussed in detail in Chapter 12), as the major products of decomposition during weathering. The relationship between the two uses of the term is that most of the clay size fraction is composed of very small clay mineral particles.

In addition to clay minerals, the clay size fraction may contain very fine-grained quartz, feldspar, micas and other minerals. Silt-sized minerals are predominantly angular to subangular quartz, feldspars and mica flakes. Lithic fragments are uncommon because most polycrystalline rock fragments are too large to be included in the mud fraction. Other significant mudrock components include carbonate minerals, iron oxides and hydroxides, zeolite minerals, sulfide minerals and organic materials. The significance of organic material in mudrocks cannot be overstated. More than

95% of cellular organic material, whose diagenetic alteration produces petroleum and natural gas, initially accumulates in detrital or carbonate muds.

Because of their tiny particle size, clay mineral analysis is difficult by standard thin section techniques, but laboratory techniques such as X-ray diffraction can yield semiquantitative analyses. All the major clay minerals occur in modern muds and ancient mudrocks. The general distribution of clay minerals in modern soils (Chapter 12) depends primarily on the composition of the bedrock, the climate – especially rainfall and temperature – and the intensity of chemical decomposition. A brief review of the major clay minerals in mudrocks follows:

1   Chlorites are produced by the minimal decomposition of ferromagnesian minerals and commonly form in alkaline soils with impeded drainage, especially at high latitudes where precipitation and temperatures are low.
2   Smectites such as montmorillonite are the product of the weathering of ferromagnesian minerals plus plagioclase. Their formation is favored by impeded drainage, alkaline conditions and semiarid climates.
3   Illites are common products of the weathering of feldspars (especially K-spars) and occur most commonly in temperate region soils with near neutral pH.
4   Mixed layer illite–smectite clays are also common in mid-latitude semi-arid to temperate soils with slightly alkaline pH.
5   Under warm, humid, acidic soil conditions, such as those common in the subtropics, cations tend to be leached from interlayer sites, which gives rise to degraded illites and kandites such as kaolinite.
6   Under warm, humid conditions of unusually high acidity and low pH, intense decomposition allows silica to dissolve readily, giving rise to gibbsite and other minerals of the bauxite suite.

These patterns have been traced from the source areas to adjacent areas of sediment dispersal. Chlorites are most common at high latitudes, kaolinite is most common at low latitudes, and illite, smectites and mixed layer clays dominate sediments elsewhere. However, these patterns do not survive in the oceans or in the subsurface because clay minerals are reconstituted and transformed as they react with seawater and with pore water during diagenesis. Smectite clays, such as montmorillonite, and mixed layer smectite–illite clays dominate modern sediments and are common in younger Tertiary rocks. Older rocks are dominated by illite clays. The progressive decrease in the proportions of smectite and kandite (e.g., kaolinite) with age results from their alteration to illite and/or chlorite. These diagenetic processes are detailed in the next section.

A locally important clay mineral that has not yet been mentioned is **glauconite**. Glauconite is a generally green-colored, potassium-iron-rich illite that is produced in marine environments. Some glauconite is generated by slow precipitation in agitated, oxidizing, marine environments. However, glauconite also forms by the replacement of fecal pellets under marine conditions that are somewhat reducing. Although glauconite is a clay mineral, some composite glauconite grains are of sand size. Glauconite sand grains occur as disseminated grains in sands in areas where detrital influx is large, but they can be concentrated in areas where it is small. In a few cases, extensive deposits occur with large concentrations of glauconite, such as the Cretaceous greensands of the New Jersey coastal plain and passive margin. Glauconite, whether it occurs in mudrocks or as grains in sandstones, is an excellent, if not foolproof, indicator of marine sedimentation.

Mudrocks occur in a wide variety of colors that in many cases are closely related to their composition. Many mudrocks have white to pale gray to green colors that reflect the colors characteristic of clay minerals such as kaolinite (white), montmorillonite (pale gray to green), illite (pale green) and chlorite and glauconite (green). Other mudrocks possess colors that reflect the presence of non-clay coloring agents such as organic matter and iron oxides and hydroxides. Most red to purple mudrocks owe their color to the occurrence of the ferric iron (Fe^{+3}) oxide mineral **hematite** (Fe_2O_3). This records oxidizing conditions at the time the hematite was produced. Many red mudstones contain **reduction spots** (Figure 13.16). Reduction spots commonly form around decomposing organic matter

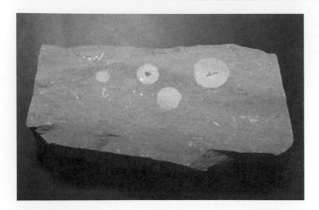

Figure 13.16 Reduction spots around organic particles in red–purple mudrock. (Photo courtesy of Steve Dutch.) (For color version, see Plate 13.16, between pp. 408 and 409.)

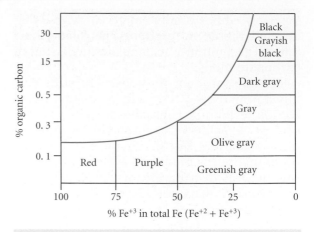

Figure 13.17 Diagram showing the general relationships between color, organic content and oxidation state of iron in mudrocks. (After Potter et al., 1980, in Raymond, 2002.)

that reduces the iron, enabling it to be removed in solution, which results in the removal of the red coloration.

Yellow to rusty-colored mudrocks contain the mineraloid **limonite** [~FeO(OH)·nH$_2$O], and some brown mudrocks have colors related to the presence of the closely related mineral **goethite** (FeOOH). Most medium to dark gray and black mudrocks owe their color to the presence of finely divided, carbon-rich organic material, although finely divided **pyrite** (FeS$_2$) with its black streak may also contribute. The diagram in Figure 13.17 summarizes some general relationships between color and mudrock composition.

13.4.3 Mudrock deposition

Mudrocks are the most widely distributed sedimentary rocks. Mud particles finer than coarse silt are generally transported in suspension by wind and water, whereas coarser particles are transported in the bed load (Chapter 11). This separation is maintained during deposition as small mud particles remain in suspension at very low velocities that are not capable of moving larger particles. Muds are deposited from suspension only in very calm environments where current velocities are exceedingly low. Mud-dominated sediment depositional environments include:

1   Generally calm marine environments such as passive margin shelves below wave base, continental slopes, rises, submarine fans, abyssal plains and elsewhere on deep ocean floors; active margin trenches, forearc basins and backarc basins; and the deeper portions of marine foreland basins.
2   Transitional (paralic) shorelines protected from significant wave and current activity such as lagoons, estuaries, fjords, bays and tidal flats.
3   Terrestrial environments such as deltas, river floodplains, swamps and the deeper portions of lakes characterized by weak waves and/or currents.
4   Wind-blown muds delivered to standing bodies of water that settle out of suspension to accumulate on the bottom.
5   Wind-blown loess (Chapter 11), a silt-dominated mud that is produced when winds generated by the temperature differences across continental glacier margins erode fine glacial rock flour from glacial deposits, transport it away from the glacier and then deposit it on land as the winds die down.

13.4.4 Distinctive mudrock varieties

Bentonites

Bentonites are smectite-rich claystones formed by the alteration of volcanic ash deposits generated by explosive eruptions (Chapter 9).

Figure 13.18   Bentonite, with its typical, lumpy popcorn-like appearance, Mowry Formation, Cretaceous, Wyoming. (Photo courtesy of Wayne Sutherland, with permission of Wyoming State Geological Survey.) (For color version, see Plate 13.18, between pp. 408 and 409.)

Figure 13.19   Oil shale with a dark color due to its oil content. (Photo courtesy of US Department of Energy, Argonne National Laboratory.) (For color version, see Plate 13.19, between pp. 408 and 409.)

The smectite, typically a variety of montmorillonite, forms by the alteration of glass shards and pumice fragments in contact with diagenetic fluids. In addition to shards, clay and silica, bentonites commonly contain the unaltered remains of euhedral crystals and angular mineral fragments such as sanidine, quartz and plagioclase from the partially crystallized melt and various lithic fragments ripped from the walls of the vent. Because they consist of expandable clays with large shrink–swell potentials, bentonites swell when wet. When dried, they contract into small pieces, giving dry bentonite deposits a distinctive, rubbly surface appearance that resembles popcorn (Figure 13.18). Their ability to swell when wet makes bentonites a potential geohazard (Chapter 12).

Bentonite layers up to 50 m thick are known. Their thickness and the size of the crystal and lithic fragments tends to decrease away from the explosive source, and fragment size decreases upward within the layer due to the more rapid settling of larger fragments. Distinctive bentonite layers can be distinguished on the basis of shard geometry, crystal composition, lithic fragment composition and bulk chemical composition. Wherever they occur, recognizable bentonites are of high stratigraphic value because each bentonite forms in a very short period of time, during short-lived explosive eruptions, yet can cover a large area. For this reason, bentonites are of great value as time horizons used to correlate geological events from one place to another.

Carbonaceous mudrocks

Carbonaceous mudrocks are mudstones and claystones that contain sufficient carbon-rich organic material to substantially influence their properties. A wide array of terms is used for these rocks. The terms reflect differences in properties but also reflect the context in which they are used. **Black shales** are characterized by a black color that is largely the result of their elevated content (generally >2%) of incompletely decomposed, carbon-rich organic matter. The term **sapropel** (sapro = organic; pel = mud) is also used for organic-rich mudstones and claystones, especially by oceanographers and paleo-oceanographers. **Oil shales** are characterized by organic material contents (generally 15–30%) that have the potential to yield profitable amounts of petroleum and/or natural gas when heated sufficiently. Most oil shales are black (Figure 13.19); some possess a dark brown color. Oil can be extracted from oil shales by the process of **pyrolysis**, which involves crushing the shale and heating it, in the absence of air, to 500°C. The costs of refining oil from shales, the costs of environmental protection associated with large amounts of solid waste material, and the low price of crude oil from traditional sources have inhibited large-scale production from oil shales. However, in light of the recent surge in petroleum prices, oil shales now represent a significant resource. A recent estimate from the American Association of Petroleum Geologists puts the potential supply at 2.6×10^{12} barrels, far more than the total reserves from traditional sources.

Carbonaceous mudrocks are also the **source rocks** for petroleum and natural gas. When a bituminous organic material in such mudrocks called **kerogen** is buried and heated to 100–140°C, it is converted into petroleum, and when heated to over 160°C, it is converted into natural gas. Because petroleum and natural gas are lighter than water, they migrate upward from source rocks to accumulate beneath impermeable "cap rocks" in porous, permeable rocks such as limestones and sandstones (Chapter 14).

Thick and thin accumulations of finely laminated or fissile black shales, sapropels and/or oil shales are widespread and locally abundant in Phanerozoic rocks. Very commonly they contain reduced iron sulfide ($Fe^{+2}S_2$) minerals in the form of the polymorphs pyrite or marcasite. These features suggest that black shales form when fine muds and organic matter accumulate together under low oxygen, reducing conditions. Several important factors interact to produce the conditions under which black shales, oil shales or sapropels accumulate. These include: (1) high biological productivity of organic material in surface waters, (2) sluggish circulation that impedes both the influx of coarse detritus and oxygen replenishment of bottom waters, and (3) the development of low oxygen conditions in bottom waters that impede bacterial decomposition of organic material, allowing it to accumulate in bottom sediments.

Biological productivity in near surface waters is a critical factor because high rates of biological productivity produce high rates of **biological oxygen demand** (**BOD**). This occurs because the higher the rate of biological productivity, the higher the demand for oxygen by the aerobic bacteria that decompose organic material. As organisms die and the cellular material drifts toward the bottom, elevated BOD begins to deplete dissolved oxygen, especially in areas of sluggish circulation where the replenishment of oxygen to bottom waters is impeded. Such anaerobic conditions create a dead zone in which aerobic bacteria and organisms dependent on oxygen for respiration cannot survive. As a result, organic matter settling from surface waters is decomposed at much reduced rates by anaerobic bacteria that do not require oxygen and tends to accumulate in bottom sediments. The sluggish circulation in these relatively stagnant waters, so important in impeding oxygen replenishment, drastically reduces the influx of coarse detrital sediments. Instead, relatively small amounts of mud accumulate with the organic matter without diluting it significantly. Because both the organic material and the mud particles are of very small size, they settle out of suspension and accumulate together to produce black, organic-rich muds that become black shales or oil shales during diagenesis. Pyrite and marcasite form when reduced iron (Fe^{+2}) combines with sulfide produced by sulfate-reducing anaerobic bacteria to form iron sulfide (FeS_2).

The ideal conditions for the formation of black shales occur when large quantities of organic matter are produced in surface waters creating a high BOD, the rate of influx of detrital sediments is restricted to small amounts of mud per unit of time, and the bottom waters are depleted in oxygen because of sluggish circulation so that bacterial decomposition is inhibited. These conditions can exist in any basin in which circulation is sufficiently restricted, whether it is a lake, a marginal sea, an epicontinental sea or a deep ocean basin. Examples of black shales, sapropels and/or oil shales from all of these environments are well known. Figure 13.20 illustrates a sapropel in Jurassic lake sediments from Connecticut.

13.5 DIAGENESIS OF DETRITAL SEDIMENTS

As discussed in Chapter 11, all sedimentary rocks experience diagenesis. **Diagenesis** includes all sub-metamorphic, post-depositional changes that affect sediment after its accumulation. Most of these changes

Figure 13.20 Organic-rich sapropel layer (middle) between lighter colored layers in lake sediments, Jurassic Berlin Formation, Connecticut. (Photo by John O'Brien.) (For color version, see Plate 13.20, between pp. 408 and 409.)

occur after the sediment is buried below the surface by the accumulation of additional sediments. Diagenesis occurs at relatively low temperatures (up to ~150 ± 50°C) and pressures (~0–3 kbar). These conditions result in the alteration of minerals and sediment textures under sub-metamorphic conditions. It is useful to distinguish the stages of diagenesis: (1) **eodiagenesis** – early, shallow diagenesis that occurs shortly after burial; (2) **mesodiagenesis** – later, deeper diagenesis; and (3) **telodiagenesis** – still later, shallow diagenesis that occurs as sedimentary rocks approach the surface due to erosion.

The major factors that control the diagenesis of detrital sediments include:

1 Temperature, which generally increases with depth.
2 Pressure, which increases with depth.
3 Fluid chemistry, including concentrations of dissolved solids, gases, pH and oxidation–reduction potential (Eh).
4 Aqueous fluid circulation rates and patterns, which depend on porosity, permeability, hydraulic head and rock structure.

Diagenesis includes all the **lithification** processes that convert unconsolidated sediments into sedimentary rocks but also involves many other changes, discussed below. Because many of these processes change the porosity and permeability of sediments and sedimentary rocks, they can profoundly affect the rock's ability to hold and transmit fluids such as water, oil and natural gas. As a result, diagenesis is of great interest to hydrologists, petroleum geologists and engineers. Because many diagenetic processes change the volume, thickness and density of sediments, they are also of great interest to scientists who study the evolution of sedimentary basins. The discussion of diagenesis that follows is necessarily a brief overview. For more comprehensive treatments, refer to books such as those by Wolf and Chingarian (1992, 1994) and McIlreath and Morrow (1990).

13.5.1 Compaction and pressure solution

Sediment **compaction** occurs as a result of increasing confining pressures as sediments are buried progressively deeper beneath the surface. The resulting increase in static pressure (1) forces the particles closer together, (2) expels pore fluids such as air and water, which (3) results in a progressive decrease in porosity (Figure 13.21). These processes also tend to progressively reduce the volume and thickness of stratigraphic units, while increasing their density.

Rigid grains such as quartz and feldspar remain largely undeformed during compaction. In the absence of other diagenetic processes, reasonably well-sorted quartzarenite and arkosic sands retain substantial porosity and permeability during burial (Figure 13.21). Such sand layers are excellent fluid transmitters and generally act as aquifers or water-bearing horizons.

Wet muds commonly possess initial porosities in the range of 60 to 80%. During progressive burial, as pore fluids are expelled, muds undergo rapid reductions in porosity (Figure 13.21). Silt and especially plastic clay particles are squeezed together, and the static charges on their surfaces cause the grains to be attracted to one another, increasing their cohesion and converting muds into mudrocks. Eventually permeability may approach zero, making well-compacted mudstone layers extremely impermeable. Such impermeable layers are very poor transmitters of fluids and

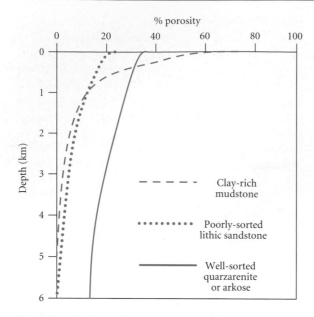

Figure 13.21 Idealized compaction and porosity curves for well-sorted quartzarenite or arkose, poorly sorted lithic sandstone and clay-rich mudstone.

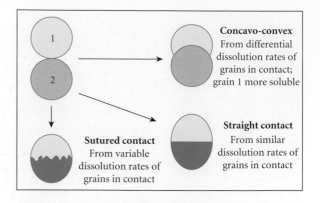

Figure 13.22 Major varieties of long grain contacts produced by diagenetic pressure solution.

act as aquicludes or aquitards. They form the confining layers that keep water from migrating upward from underlying **confined aquifers**. They also form **cap rocks** that keep petroleum and natural gas from migrating upward, thus effectively trapping them in underlying permeable layers.

In lithic sandstones, softer, more plastic grains such as mudrock, slate and phyllite fragments may be substantially deformed. In thin section, plastic grains can be seen to have been bent around more rigid grains and to have been squeezed between them to produce a diagenetic **psuedomatrix** that can be difficult to distinguish from detrital matrix constituents. The creation of psuedomatrix can drastically reduce the porosity of lithic sandstones and conglomerates, making them less efficient storers and transmitters of fluids.

As a general rule, the solubility of mineral crystals increases with increasing pressure or stress. During compaction, the highest stresses are transmitted at point contacts between rigid grains. When even a thin film of water is present, these grain contacts become the sites of dissolution, the rate of which increases with pressure. Such pressure-induced dissolu-

tion is called **pressure solution** or **pressolution**. Pressolution is most common between grains in quartz sands and gravels with little or no plastic matrix and occurs less frequently in feldspathic sands and gravels. It occurs over a range of depths and causes significant decreases in porosity. Dissolution is controlled by (1) the relative solubility of the two grains, (2) irregularities in the grain contacts so that not all parts of the grains are in contact, and (3) the presence of grain microfractures and other grain imperfections. As one or both grains dissolve, the length of their contacts tends to increase (Figure 13.22). As they evolve, such contacts may become relatively straight if both phases dissolve at similar rates, concavo-convex if one phase (concave) is more soluble than the other (convex) or sutured if microfractures or other imperfections lead to variable solubility between the two grains across their contact.

13.5.2   Dissolution and cementation

Dissolution occurs when crystalline solids partially or completely dissolve in pore fluids. These dissolved solids in aqueous solutions migrate through fractures and between grains where they may eventually be precipitated as veins and mineral cements. **Cementation** occurs when pore fluids precipitate intergranular mineral or mineraloid cements that bind the grains together. In order for pore fluids to precipitate mineral cements, they must first dissolve the constituents that are eventually precipitated. Determining the source of the dissolved solids that provide the material for mineral cements is an important research topic in sediment diagenesis.

Dissolution is an important process in the production of secondary porosity in sediments and sedimentary rocks, leaving a void space where the soluble solid previously existed. If such pore spaces are interconnected, dissolution can greatly increase permeability. Dissolution is selective; the most soluble constituents are preferentially dissolved.

Cementation, on the other hand, generally decreases porosity by partially or wholly filling the voids between grains. It is possible, of course, for an earlier generation of cement to later be dissolved, increasing both porosity and permeability and making the rock an excellent host for water, natural gas and/or petroleum.

Because coarse silts, sands and gravels commonly possess significant porosity and permeability, the precipitation of mineral cements that bind grains together is a significant process in the lithification of coarse siltstones, sandstones and conglomerates. Given the relatively low dissolved solid contents of most subsurface waters, extraordinarily large amounts of groundwater must flow through such sediments in order to produce the significant amounts of pore-filling cements that they commonly possess. The major cements in detrital sedimentary rocks include (1) silica minerals, (2) carbonate minerals, (3) iron oxides and hydroxides, (4) feldspars, and (5) clay minerals.

Silica cements

The most abundant type of **silica cement** is quartz. Quartz cement occurs chiefly in the form of **syntaxial quartz** overgrowths in which the silica that precipitates from pore solutions initially nucleates on a pre-existing detrital quartz grain. The quartz cement has the same crystallographic orientation as the detrital grain on which it nucleated and therefore the same orientation of its crystallographic axes; thus the term syntaxial. In thin section, both the syntaxial overgrowth and the detrital host grain go to extinction (Chapter 6) at the same time. This can make them difficult to distinguish. If the original

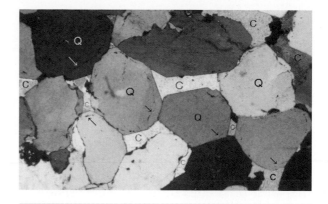

Figure 13.23   Syntaxial quartz overgrowths (arrows) and blocky calcite cement (C) in quartzarenite (Q). (Photo courtesy of L. Bruce Railsback.) (For color version, see Plate 13.23, between pp. 408 and 409.)

quartz grain had a coating of clay or hematite or if gas bubbles (vacuoles) adhered to its surface, the boundary between the detrital quartz and the syntaxial cement is easily seen. Figure 13.23 shows rounded, detrital quartz grains surrounded by syntaxial quartz overgrowths. The overgrowths are separated from the host detrital grains by a surface with dust and vacuoles. If no dust or vacuoles are present, the boundary may be invisible and the size and shape of the original grain indistinguishable. In such cases, cement generations may develop long contacts with each other that mimic those produced by pressolution. Laboratory studies using cathodoluminescence make such relationships easier to determine. Quartz cement is particularly common in quartz-rich sandstones and gravelstones that accumulated in beach, dune and shallow marine environments associated with intracratonic basins and passive margins.

The sources of the dissolved silica that is later precipitated during quartz cementation are poorly understood. Quartz is rather insoluble at low temperatures, but its solubility increases with temperature and pressure. Amorphous opal is more soluble at all temperatures, even at low temperatures. Proposed sources for silica cements include: (1) the dissolution of opalline silica shells of organisms such as diatoms, radiolarians and sponges, (2) the near surface conversion of volcanic ash to bentonites, which releases silica,

(3) pressolution of silicate minerals, (4) authigenic reactions that release silica (see below), and (5) metamorphic reactions that release silica to fluids, which cool as they rise through the overlying sediments.

Less common varieties of silica cement include opal, chalcedony, chert and non-syntaxial quartz. **Opal** ($SiO_2 \cdot nH_2O$) is an amorphous mineraloid that contains various amounts of water. Opal is a common cement in volcanoclastic sediments. It is thought that the silica originates during the conversion of glass shards and pumice fragments into smectites clays (bentonites), which yields dissolved silica as a byproduct. Because opal solubility rapidly increases with temperature and depth, it is likely that opalline cements are precipitated at lower temperatures and pressures close to the surface during eodiagenesis. Opal solubility also depends on pH; opal is more soluble in mildly alkaline waters than in mildly acidic waters. Therefore, a decrease in pH can cause opal cement precipitation, whereas an increase in pH can cause dissolution. **Chert** (SiO_2) and **chalcedony** are cryptocrystalline to microcrystalline varieties of silica. Chert is characterized by microscopic, equant crystals, whereas chalcedony crystals are bladed, commonly with a radiating to divergent habit when observed under a petrographic microscope.

Carbonate cements

Carbonate cements are the most abundant cements in sandstones and gravelstones. **Calcite** ($CaCO_3$) is by far the most abundant carbonate mineral cement. The solubility of calcite is strongly affected by the acidity–alkalinity (roughly pH) of subsurface fluids (Figure 13.24). Acidity is strongly influenced by the amount of dissolved carbon dioxide. Any process such as organic respiration or bacterial decomposition of organic matter that releases carbon dioxide to pore solutions consequently lowers their pH and increases their acidity. Any process that removes carbon dioxide from subsurface waters raises the pH, making them more alkaline.

As a result, the precipitation of calcite cements is favored by processes that increase the pH and alkalinity of subsurface solutions, which causes them to become supersaturated with respect to calcium carbonate. Con-

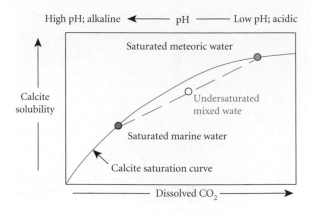

Figure 13.24   Solubility of calcite as a function of dissolved CO_2 content and acidity of natural pore waters; note that two saturated natural waters may mix to produce an undersaturated solution capable of dissolving calcite.

versely, the dissolution of calcite is favored by any process that decreases the pH and increases the acidity of subsurface waters. The most striking examples of subsurface dissolution of calcium carbonate, the formation of caves and other karst features, are discussed in Chapter 12.

Calcite cements in sandstones and gravelstones are commonly either blocky or poikiloptic. **Blocky cement** is composed of one or more calcite crystals that occupy small pore spaces between detrital grains (see Figure 13.23). **Poikiloptic cement** is composed of a single, large calcite crystal that nucleates and grows to fill multiple pore spaces so that it completely envelops several detrital grains (Figure 13.25), which appear as inclusions (called poikils) within a single calcite.

Rarer carbonate cements include aragonite ($CaCO_3$), the orthorhombic polymorph of calcium carbonate, dolomite [$CaMg(CO_3)_2$], ankerite [$CaFe(CO_3)_2$] and siderite [$Fe(CO_3)$]. Because these minerals are important constituents of biochemical sedimentary rocks, their chemistry is discussed in more detail in Chapter 14.

Iron-rich cements

The third most common cementing agents in sedimentary rocks are iron oxide and hydroxide minerals. Chief among these are hematite (Fe_2O_3) and a variety of water-bearing miner-

Figure 13.25 Photomicrograph of calcite-cemented sandstone with poikiloptic cement enclosing multiple quartz and feldspar grains (crossed polars). (Photo courtesy of Lee Phillips.) (For color version, see Plate 13.25, between pp. 408 and 409.)

als such as goethite (FeOOH) and mineraloids including limonite (complex, amorphous FeO(OH)·nH$_2$O). The dissolved iron required for these minerals originates primarily from the alteration of ferromagnesian silicate minerals during weathering and diagenesis. The dissolved iron is the form of reduced ferrous iron (Fe^{+2}) that is quite soluble in waters with low oxidation–reduction potentials (Eh). Oxidation of such waters converts ferrous iron into oxidized ferric iron, which is much less soluble. This leads to the precipitation of oxidized iron-bearing cements. Hematite is especially common as cement; its bright red color produces classic detrital red bed sequences. Hematite cementation is particularly common in sediments deposited in fairly arid terrestrial environments such as alluvial fans, braided streams, meandering stream channels and flood plains and deserts. In such environments, the paucity of vegetative cover and consequent bacterial decomposition lead to oxidizing groundwaters that precipitate amorphous hydrated iron oxide [FeO(OH)·nH$_2$O], which is rapidly dehydrated into hematite (Fe$_2$O$_3$).

Rarer cements

Feldspar cements occur in feldspar-rich detrital sedimentary rocks such as arkosic sandstones and gravelstones. Feldspar cements appear microscopically as optically clear overgrowths that nucleated on feldspar host grains. They include orthoclase overgrowths on potassium feldspar grains and albite overgrowths on both plagioclase and potassium feldspar grains.

Clay cements occur in some detrital sedimentary rocks. Clay mineral stabilities are strongly controlled by temperature and pH. **Kaolinite cement** generally occurs as stacks of platy layers called "books" that precipitate at fairly shallow depths from low potassium, acidic pore waters during eodiagenesis and telodiagenesis. Acidic pore waters are especially common in continental settings, which is where most kaolinite cements are precipitated. **Illite cement** generally forms at higher temperatures and depths from high potassium, alkaline pore waters, especially in marine settings. Illite cements form mostly during late eodiagenesis and mesodiagenesis.

13.5.3 Additional diagenetic processes

Mineral alteration or **replacement** occurs when one mineral crystal is altered to another during diagenesis. Chemically stable minerals such as quartz typically display little or no alteration. Many potassium feldspar crystals are altered to clays and to the fine-grained mica called sericite, which is closely related to muscovite. At higher temperatures, both calcium and potassium feldspars are commonly altered to albite. Typically such alteration begins at grain margins or along cleavage or fracture surfaces where minerals are in contact with pore solutions or aqueous films. Volcanic rock fragments commonly alter to zeolite minerals (Chapter 5) during progressive burial and diagenesis. The type of zeolite mineral yields insights into the temperature conditions and burial depth at which diagenesis occurred.

Clay mineral stability is strongly influenced by temperature. During burial, increasing temperatures lead to the transformation of lower temperature clay minerals into higher temperature mineral assemblages. In general,

- Smectites become unstable and are transformed into mixed layer clays above 100°C.
- Kaolinite is converted into illite or chlorite above 150°C.

- Mixed layer clays are transformed into more ordered illite above 200°C.
- All clay minerals are transformed into chlorite or micas such as muscovite above 300°C.

Many of these reactions also involve the zeolite minerals, with analcime and heulandite stable below 100°C, laumontite stable from 100 to 200°C and prehnite and pumpellyite stable above 200°C. In fact, very low grade metamorphic rocks are often ascribed to the zeolite or prehnite–pumpellyite facies, reflecting the transitional nature between diagenesis and low grade metamorphic processes (Chapter 18).

13.5.4   Diagenetic structures

Several diagenetic sedimentary structures occur in detrital sediments. The most frequently seen or significant of these include (1) concretions, (2) nodules, (3) geodes, and (4) liesegang rings or bands.

Concretions form by the precipitation of material around a nucleation surface, such as a fossil, sand grain or shale chip. Multiple periods of precipitation cause many concretions to have a concentric structure and a roughly spherical shape (Figure 13.26a). Continued growth may cause several small concretions to coalesce into larger concretions with more complex shapes. The precipitation of mineral cement such as limonite or hematite in oxidizing conditions often causes the concretion to be harder and more resistant to weathering than the rest of the rock. As a result concretions weather out of outcrops as cannon-ball-like structures (Figure 13.26b). Siderite, pyrite or marcasite concretions form under reducing conditions. Calcite is a common component of concretions formed under a variety of oxidizing–reducing conditions.

The terms nodule and concretion are sometimes used in an overlapping manner. Strictly speaking, **nodules** are similar to concretions but lack a well-defined nucleus and generally lack concentric growth rings. Chert nodules are especially abundant in limestones and dolostones. Recall that limestones are increasingly soluble as pH decreases and acidity increases. These same conditions often favor the precipitation of opalline silica, which is

(a)

(b)

Figure 13.26   Images of concretions. (a) A nucleus and concentric structure in sandstone. (Photo courtesy of John Merck.) (b) A concretion in shale, Devonian, Virginia. (Photo courtesy of Duncan Heron.) (For color version, see Plate 13.26, between pp. 408 and 409.)

less soluble in acidic waters. Under such conditions, opalline silica may replace the calcite in limestones to produce nodules that recrystallize into chert or flint (Figure 13.27a). A spectacular group of nodules are **septarian nodules** (Figure 13.27b) that are characterized by mineral-filled cracks whose origin remains controversial.

Geodes are nodules that were or are partially hollow. Many geodes have roughly spherical to more irregular outer layers of chalcedony and a cavity lined with crystals (Figure 13.28) that, in some cases, have grown to fill the cavity. Their origin remains controversial and may well involve more than one method of formation. One widely accepted explanation is that geodes originate as nodules

(a)

(b)

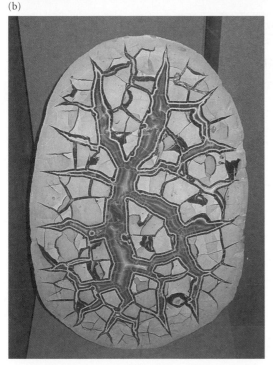

Figure 13.28 Geodes showing banded chalcedony rims and quartz (plus bladed gypsum) crystal linings. (For color version, see Plate 13.28, between pp. 408 and 409.)

Figure 13.27 Nodules. (a) Three flint nodules in limestone. (Photo by John O'Brien.) (b) Septarian nodule. (Photo courtesy of Keith Pomakis; nodule from the Pinch Collection, Canadian Museum of Nature, Ottawa, Ontario.) (For color version, see Plate 13.27, between pp. 408 and 409.)

Figure 13.29 Liesegang bands in a sandstone block; note the truncation against the joint surfaces. (Photo courtesy of Peter Adderley.) (For color version, see Plate 13.29, between pp. 408 and 409.)

or concretions of calcite or anhydrite. In acidic ground waters, their exteriors are replaced by chalcedony; the interiors eventually dissolve permitting the precipitation of the crystals that line or fill their interiors.

Liesegang bands (Figure 13.29) are common in many detrital sedimentary rocks, especially those with carbonate or iron-rich cements. Although the bands sometimes mimic stratification, they are clearly secondary features that are often truncated against prominent joint surfaces. They form by the precipitation of various iron oxide minerals moving through bodies of rock separated by fractures. Fluids moving into such bodies of rock from the outside–in often produce ring-like patterns of bands called Liesegang rings.

The diagenesis of sedimentary rocks provides insights into the subsurface processes that lead to lithification. Diagenetic processes significantly alter the porosity, permeability and density of sediments as they undergo compaction, pressolution, cementation and chemical alteration. This in turn has a profound effect on their ability to transmit fluids such as water, natural gas and petroleum.

Chapter 14

Biochemical sedimentary rocks

In the previous chapter, we discussed the rocks created when detrital sediments produced by weathering are dispersed and deposited by water, wind, glaciers and gravity and lithified into detrital sedimentary rocks. In this chapter we discuss the fate of the dissolved solids produced by decomposition, volcanism and surface and subsurface dissolution processes.

Organisms produce **organic (biogenic) sediments** via biochemical precipitation of skeletal materials such as shells, bones and teeth, and by organic synthesis of cellular materials to make organic tissues. In precipitating shells, bones and teeth, organisms extract dissolved ions from the environment and secrete skeletal materials composed of "biominerals" such as calcite, aragonite, silica or calcium phosphate that accumulate on Earth's surface. In synthesizing organic tissues, organisms remove dissolved carbon, hydrogen, oxygen, nitrogen, sulfur and phosphorous from solution to synthesize a variety of solid organic molecules. Once produced, organic sediments can accumulate in situ or be dispersed and deposited elsewhere on Earth's surface. **Chemical sediments** such as gypsum ($CaSO_4 \cdot 2H_2O$), halite ($NaCl$) and some cherts (SiO_2) are formed by inorganic precipitation of minerals from solution to form solid sediments, which then accumulate on Earth's surface.

The boundary between organic and chemical sediments is often fuzzy. Organic activities cause changes in solution geochemistry such as oxidation–reduction potential (Eh) and acidity–alkalinity (roughly pH) that in turn trigger chemical precipitation. Are the crystals thus formed of organic or inorganic origin, or both? Because the dividing line between organic and chemical sediments is sometimes unclear, it is convenient to combine the two into a larger group called **biochemical sediments**. This is also convenient because the solid crystals produced by organic precipitation and inorganic precipitation commonly have the same composition. A good example is provided by the calcium carbonate ($CaCO_3$) minerals calcite and aragonite. They are the principle components of most organically

Earth Materials, 1st edition. By K. Hefferan and J. O'Brien. Published 2010 by Blackwell Publishing Ltd.

precipitated shells but also are inorganically precipitated in caves, around springs, from lakes and as microcrystals from seawater. Many limestones are composed of calcium carbonate produced by both organic and inorganic processes. This chapter discusses a variety of biochemical sedimentary rocks, with emphasis on their occurrence, composition, classification, origins, diagenesis and uses. Biochemical sedimentary rocks have been the focus of intensive study because they are extremely important as hydrocarbon reservoirs, groundwater aquifer units and the sources of many critical industrial materials.

14.1   CARBONATE SEDIMENTARY ROCKS

Carbonate sedimentary rocks are by far the most abundant group of biochemical sedimentary rocks and constitute some 15% of all sedimentary rocks. They are composed chiefly of calcite and aragonite, the polymorphs of calcium carbonate ($CaCO_3$), and the calcium–magnesium carbonate mineral dolomite [$CaMg(CO_3)_2$]. Following common usage, we will employ the term minerals to describe these crystals whether or not they are of organic origin. The two most abundant carbonate sedimentary rocks are **limestone**, composed primarily of calcite and/or aragonite, and **dolostone**, composed principally of dolomite. Limestones are essential to the construction industry, as they provide the essential ingredient lime (CaO) used in the production of cement products.

14.1.1   Carbonate mineralogy

The most abundant mineral in carbonate sedimentary rocks is **calcite ($CaCO_{3(R)}$)**, the rhombohedral polymorph of calcium carbonate. Because limited amounts of smaller magnesium (Mg^{+2}) ions can substitute for calcium (Ca^{+2}) ions in the crystal lattice, calcite crystals can be classified according to their magnesium content. This substitution can be viewed as a limited substitution of **magnesite ($MgCO_3$)** for calcite (Chapter 2). Calcite crystals with less than 4% magnesium substitution for calcium are referred to as **low magnesium calcite**, whereas those with more than 4% magnesium substitution are referred to as **high magnesium calcite** (Figure 14.1).

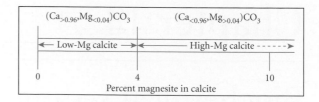

Figure 14.1   Compositions of low magnesium and high magnesium calcites.

The percentage of magnesium in calcite is significant as it plays a major role in the solubility of calcite in different pore waters and provides important clues that permit interpretation of the history of carbonate sediments and sedimentary rocks. High magnesium calcite is far more abundant and low magnesium calcite less abundant in modern carbonate sediments than in ancient carbonate sedimentary rocks, where their relative abundances are reversed.

The second common polymorph of calcium carbonate is the orthorhombic polymorph **aragonite ($CaCO_{3(O)}$)**. Together with calcite, it is an essential mineral in limestones. Because of its orthorhombic structure, small cations such as magnesium do not substitute for calcium to a significant degree. However, larger cations such as strontium (Sr^{+2}) substitute for calcium (Ca^{+2}) to a much larger degree than they do in the rhombohedral calcite structure. These large ion substitutions occur on a scale of only a few parts per thousand. Aragonite is far more abundant in modern carbonate sediments than in ancient carbonate sedimentary rocks.

In general, aragonite is most unstable during diagenesis, especially in meteoric water, whereas low magnesium calcite is stable, especially in meteoric water. This largely explains why aragonite and high magnesium calcite become progressively less common in older rocks. The composition of skeletal materials produced by organisms that secrete calcium carbonate varies between and within groups of organisms (Table 14.1).

The major mineral in dolostones and a subordinate constituent in many limestones is the double carbonate mineral **dolomite [$CaMg(CO_3)_2$]**. Because of their similar ionic radius, charge and availability, a significant substitution of ferrous iron (Fe^{+2}) for magne-

Table 14.1 Common compositions of carbonate minerals for major groups of organisms that secrete or secreted calcium carbonate skeletons.

Group	Subgroup	Aragonite	Calcite Low-Mg	Calcite High-Mg
Bacteria	Blue-green		✓	
Algae	Red			✓
	Green	✓		
	Coccoliths		✓	
Foramininera	Planktonic		✓	
	Benthic		✓	✓
Sponges				✓
Annelids	Serpulids	✓	✓	✓
Coelenterates	Stromotoporoids	✓	✓	
	Rugose corals		✓	
	Tabulate corals		✓	
	Scleractinids	✓		
	Alcyonarians			✓
Bryozoans		✓		
Brachiopods		✓		
Mollusks	Bivalves	✓	✓	
	Gastropods	✓	✓	
	Pteropods	✓		
	Cephalopods	✓		
	Belemonoids		✓	
Arthropods	Decapods			✓
	Ostracods		✓	✓
	Trilobites		✓	
Echinoderms			✓	✓

sium (Mg^{+2}) occurs in many dolomite crystals. This can be modeled as a solid–solution substitution between **ankerite [$CaFe(CO_3)_2$]** and dolomite. While dolomite is a rather rare component of modern carbonate sediments, its abundance increases with age in ancient carbonate sedimentary rocks, especially in the Precambrian. Once again, this raises significant questions about changes in the composition of carbonate rocks with age.

Siderite ($FeCO_3$) is a rock-forming carbonate mineral common in iron-rich sediments (Section 14.4). Table 14.2 summarizes the significant minerals of carbonate sediments and sedimentary rocks.

14.1.2 Conditions for carbonate accumulation

Although carbonate sedimentary rocks are widely distributed in the geological record, the production of thick carbonate sediment accumulations requires a specific set of favorable environmental conditions:

1 Large rates of carbonate production.
2 Large rates of carbonate preservation.
3 Small rates of detrital sediment influx.

For reasons elaborated below, these conditions occur primarily in warm, shallow, nutrient-rich marine environments in tropical and subtropical regions located in:

1 Low relief, tectonically stable areas such as passive margins or intracratonic seas where cratons are flooded during sea level high stands.
2 Relatively shallow areas far from continents such as oceanic platforms, ocean islands, seamounts and ocean ridges.

Table 14.2 Common minerals in carbonate rocks.

Mineral	Composition	Temporal distribution
Low-Mg calcite	$(Ca_{>0.96}, Mg_{<0.04})CO_3$	More abundant in ancient than in modern carbonate sequences
High-Mg calcite	$(Ca_{<0.96}, Mg_{>0.04})CO_3$	More abundant in modern than in ancient carbonate sequences
Aragonite	$(CaCO_3)$; limited substitution of Sr^{+2} for Ca^{+2}	More abundant in modern than in ancient carbonate sequences
Dolomite	$[CaMg(CO_3)_2]$; limited substitution of Fe^{+2} for Mg^{+2}	More abundant in ancient than in modern carbonate sequences

Because the majority of calcium carbonate ($CaCO_3$) sediment is produced by organic activities such as shell secretion and precipitation by algae and bacteria, any set of processes that increases the biomass of such organisms increases calcium carbonate production. Because sunlight is essential to photosynthesis and biological productivity, most carbonate sediment is produced in the shallow, sunlit waters of the photic zone. Because nutrients such as nitrate (NO_3^{-1}), phosphate (PO_4^{-2}) and sulfate (SO_4^{-2}) are essential to the synthesis of complex organic molecules, carbonate production is especially high in shallow, nutrient-rich waters. Oceanic conditions that favor the oversaturation and precipitation of calcium carbonate also include high temperature, wave agitation, low dissolved CO_2 content and elevated alkalinity (high pH).

Once calcium carbonate sediments are produced, one might expect them to eventually accumulate on Earth's surface. But this is not always the case. Calcium carbonate dissolution occurs wherever warm, CO_2-poor, alkaline surface waters are underlain by colder,

lower pH, CO_2-rich deeper waters. In most parts of the deep ocean, as carbonate shells sink to deeper levels, they begin to undergo dissolution. This dissolution becomes significant at a depth called the **lysocline** and becomes complete where bottom waters are sufficiently cold and acidic, so that below a certain depth all $CaCO_3$ is dissolved. This depth below which calcium carbonate sediments do not accumulate is called the **carbonate compensation depth (CCD)**. In the tropics, the CCD occurs at a depth of 4000–5000 m. In subpolar regions, where cold, CO_2-rich water occurs close to or at the surface, the CCD does so as well. This helps to explain why carbonate sediments are abundant in warm, tropical seas where alkaline waters favor $CaCO_3$ precipitation and preservation, and much less common in cold, subpolar seas where more acidic waters inhibit precipitation and aid in the dissolution of $CaCO_3$.

If $CaCO_3$ is produced and preserved, it will accumulate as carbonate sediment on Earth's surface. Whether carbonates dominate sediment accumulation depends on the rate at which other types of sediment accumulate. Most important, it depends on the rate at which detrital mud, sand and gravel are delivered to the site in question. If the amount of detrital sediment flowing into an area exceeds carbonate production and preservation, carbonate-bearing detrital sediment (e.g., a fossil-bearing sandstone) will form instead of a limestone with little or no detrital sediment content. Rates of detrital sediment influx increase with proximity to areas of (1) high relief, which increases erosion rates, (2) elevated precipitation, which generally increases erosion rates, and (3) proximity to large rivers that deliver significant volumes of detrital sediment to the coastline. This helps to explain why limestones and dolostones are rare in trench arc systems and in the portions of foreland basins adjacent to orogenic belts and are abundant along passive margins, in epicontinental seas, around islands and over platforms far from continents.

Smaller amounts of carbonate sediments are generated in terrestrial environments and include (1) stalactites, stalagmites and columns precipitated from groundwater in caves, (2) travertine and tufa deposits formed around springs, and (3) carbonate sediments precipitated from some lake waters.

14.1.3 Components of carbonate rocks

Most carbonate rocks consist of various combinations of three major groups of biochemical components: (1) sand- or gravel-size clastic particles called grains or allochemical constituents (allochems), (2) mud-sized particles called mud or micrite, and (3) mineral cements. Two other types of constituents are important contributors to some carbonate rocks: (4) organically bound accumulations of carbonate called boundstones or biolithites, and (5) an array of diagenetic products that record the dissolution, replacement and recrystallization of carbonate minerals during their low temperature alteration.

Allochems

Grains or **allochems** are sand- and/or gravel-size carbonate particles. If you have ever seen a "shell" beach in Florida, Bermuda, Hawaii or elsewhere (Figure 14.2a), you have seen a beach composed largely of carbonate grains or allochems. Allochems include:

1 Shells and other skeletal particles.
2 Spherical particles called ooids.
3 Clasts of carbonate sediment called limeclasts.
4 Smaller pellet-like particles called peloids.

Each grain type is an important constituent of carbonate sediments forming today, so that their occurrences and origins can be directly investigated. The properties and origins of each of these grain types are discussed in the sections that follow.

Skeletal particles

The most abundant sand- and/or gravel-size components of carbonate sedimentary rocks such as limestones and dolostones are **skeletal particles**, initially precipitated by organisms. Skeletal particles (Figure 14.2b) include (1) whole and disarticulated shells, (2) shell fragments, and (3) a variety of internal support structures. The vast majority of shelled macroorganisms, including mollusks, echinoderms, corals, bryozoans, brachiopods and arthropods, secrete calcium carbonate shells composed of low magnesium calcite, high magnesium calcite and/or aragonite. Many microorganisms, including foraminifera, coccoliths, pteropods and many algae, also secrete calcium carbonate shells or internal skeletal particles. Some microorganisms are large enough to contribute to carbonate grain populations; others contribute to carbonate mud populations. Skeletal particles of bottom-dwelling organisms may be preserved in the place where they were created. If their ecology is well known, such shells may record signifi-

(a) (b)

Figure 14.2 (a) Shell beach, Hinchinbrook Island, Queensland, Australia: an example of carbonate sediment composed largely of grains or allochems. (b) Fossil-bearing limestone, middle Devonian, New York, which records marine life of 435 million years ago. (Photos by John O'Brien.) (For color version, see Plate 14.2, between pp. 408 and 409.)

(a) 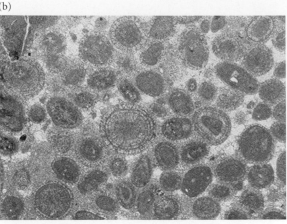 (b)

Figure 14.3 (a) Ooid-rich dolostone with ooid nuclei clearly visible. (Photo courtesy of Jack Morelock.) (b) Thin section showing various ooids with nuclei and radial and concentric laminations. (From Adams et al., 1984.) (For color version, see Plate 14.3, between pp. 408 and 409.)

cant information about the environment of sediment accumulation. More commonly, shells behave as bioclastic particles, transported from the place where they formed to the site of their final accumulation. For example, the mollusk shells in Figure 14.2a were transported to their present location by a storm surge. In either case, skeletal particles record significant information about the kinds of organisms that have inhabited Earth's surface over time; when they evolved and when they became extinct.

Ooids

Among the most intriguing components of carbonate grain populations are ooids (Figure 14.3). Ideally, **ooids** are roughly spherical, concentrically laminated, sand-size particles that possess a nucleus.

Modern ooid sands occur primarily in shallow, subtropical and tropical environments that are agitated by waves and/or tidal currents. Ooids form by the accretion of calcium carbonate laminae about a particle such as a shell fragment or sand grain that acts as a nucleus for precipitation. Ooids can display concentric and/or radial calcium carbonate structures (Figure 14.3b). The manner of such accretion remains controversial but seems in many cases to involve endolithic bacteria such as cyanophytes. These bacteria inhabit the surface of the particle and aid in the binding and/or precipitation of additional

calcium carbonate laminae as the ooid grows. The continual movement of ooids in the bed load of wave- and current-agitated environments ensures that the laminae will be of subequal thickness on all sides, thus producing approximately spherical particles. The size of the ooids is governed in part by the maximum size that can be entrained in the bed load and is generally in the sand range (<2 mm), although larger ooids are produced in exceptionally agitated environments. Ooids from less agitated environments tend to have fewer and less symmetrical coatings that reflect their less continuous motion. Although ooids can be dispersed from such environments into both deeper marine environments and even terrestrial environments, ooid-rich limestones and dolostones are an excellent indication of shallow, marine, wave- and/or tidal current-agitated environments of deposition in a tropical or subtropical setting.

Limeclasts

Another type of grain or allochem common in limestones includes gravel-sized clasts of cohesive carbonate sediment called **limeclasts**. These carbonate clasts are produced when clasts of cohesive carbonate sediments or sedimentary rocks are eroded, then transported to the site of deposition. Most are derived from nearby coeval deposits of cohesive carbonate muds, within the immediate area of deposition. Because these are derived from within

(a)   (b)

Figure 14.4   (a) Rounded intraclasts, Jurassic, Sundance Formation, Wyoming. (b) Angular intraclasts, Cambro-Ordovician, Allentown dolostone, New Jersey. (Photos by John O'Brien.) (For color version, see Plate 14.4, between pp. 408 and 409.)

the area of deposition, they are called **intraclasts** (Figure 14.4). Rarer clasts derived from the erosion of older source rocks outside the area of deposition are called **lithoclasts** or **extraclasts**.

Because penecontemporaneous cementation occurs early during carbonate diagenesis, cohesive carbonate sediments are common in environments where carbonate sediments occur. Many intraclasts are generated when tidal and/or storm surges move across mudcracked carbonate sediments, eroding the upturned edges of the surface. Others are generated by storms and tsunamis moving across partially lithified sediments in lagoonal and/or subtidal, below wave base environments, and still others are produced as mass flows moving down slopes into deeper water environments erode the bottom. In all these situations, intraclasts are a significant type of allochem in carbonate sediments. Some intraclasts are generated by the micritization (see below) of skeletal fragments or aggregates in which endolithic bacteria convert the components of gravel-size grains into micrite or lime mud. Another fascinating type of limeclast consists of cemented grains rather than cohesive muds; such limeclasts are called **aggregates**. These include **grapestones**, which consist of cemented grains such as ooids and so resemble bunches of grapes, and **botryoidal grains** that have a colloform coating of car-

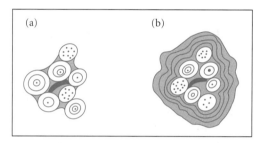

Figure 14.5   Idealized aggregate limeclasts: (a) grapestone, (b) botryoidal aggregate.

bonate laminations (Figure 14.5). Grapestones are produced when partially cemented grain clusters are eroded during storms. If these fragments are encrusted by cyanophytes, carbonate laminae may form, transforming them into botryoidal grains.

Peloids
Smaller, sand-size particles composed of carbonate mud are also common as carbonate grains. Such grains are called **peloids** because they commonly resemble the ellipsoidal fecal pellets excreted by many organisms. Many peloids, particularly those with elevated levels of organic matter, are fecal pellets excreted by sediment-feeding organisms for which the term **pellet** is appropriate. Other peloids are generated by various processes that produce sand-size particles composed of mud that

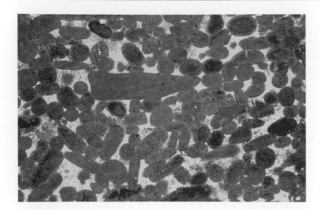

Figure 14.6 Peloids of various shapes in grain-supported framework, with interstitial cement. (Photo courtesy of C. G. St. C. Kendall.) (For color version, see Plate 14.6, between pp. 408 and 409.)

include (1) micritization of sand-size ooids and skeletal fragments, and (2) the production of sand-size clasts, which are otherwise similar to intraclasts, by the erosion of cohesive lime muds. That these particles are cohesive enough to be eroded and transported as clastic particles is indicated by their occurrence as grains within grain-supported frameworks, as seen in Figure 14.6.

Sand- and gravel-size particles called grains or allochems are common constituents of limestones and dolostones. Major types of allochems include a variety of skeletal grains, ooids, limeclasts and peloids – all of which yield clues concerning the depositional history of the carbonate rocks in which they occur. Mud-sized carbonate particles are discussed below.

Micrite

The silt- and clay-size carbonate particles in limestones and dolostones are called **mud** or **micrite** (Figure 14.7a). Because the term mud is also used for detrital mud particles, we prefer to use the terms carbonate mud or micrite for clarity. Most micrite consists of rather tiny, clay-size particles, with a diameter of less than 4 μm. Coarser carbonate mud particles are sometimes referred to as **microspar** (Figure 14.7b).

The origin of carbonate mud particles or micrite has engendered a significant amount of controversy, but it is now clear that such

(a)

(b)

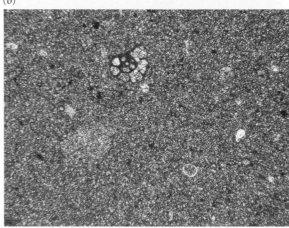

Figure 14.7 Carbonate mud in thin section, under plane light: (a) micrite, (b) microspar. (Photos courtesy of Frederic Boulvain.) (For color version, see Plate 14.7, between pp. 408 and 409.)

particles are produced by a variety of organic and inorganic processes, including:

1 Secretion by calcareous algae.
2 Micritization of pre-existing carbonate by microbes.
3 Mechanical abrasion of pre-existing carbonate grains.
4 Precipitation from solution.

A large percentage of modern carbonate mud is produced by calcareous green algae that secrete fine needles of aragonite. Upon death and decomposition, the needles are released, generating carbonate mud particles. Other carbonate mud particles are generated by

micritization, in which the micro-boring activity of blue-green, cyanophyte bacteria and algae, accompanied by precipitation of carbonate in micropores, converts original carbonate material into micrite. Still other micrite particles are generated by the mechanical abrasion of larger carbonate particles and accumulations, especially in agitated environments with vigorous wave and tidal current activity. Controversial at one time but now widely accepted is the hypothesis that calcium carbonate mud particles form by precipitation from seawater. In warm environments, calcium carbonate can become supersaturated with respect to alkaline seawater, especially if the salinity is slightly elevated. Under such conditions, widespread calcium carbonate precipitation can produce sufficient carbonate mud to turn the water white.

Clastic carbonate mud settles out of suspension under calm conditions with negligible flow velocities. Fine organic matter also settles out of suspension with micrite particles. This organic matter is an important food source for sediment-feeding organisms that extract the organic matter from the sediment and excrete fecal pellets, so that many micrites are peloidal.

Many carbonate deposits appear to consist of carbonate sediments that formed in place, as **organically bound accumulations,** rather than as separate grains and micrite. This important group of sediments includes organic reefs, stromatolites and some bioherms, all of which are discussed in the sections that follow.

14.1.4 Classification of carbonate rocks

This section introduces three classification systems. One is used primarily for rapid field work, while the other two work best in conjunction with microscopic analyses of etched surfaces, acetate peels or thin sections.

The field classification is based on (1) the average or modal size of the constituents, and (2) whether they are composed primarily of calcium carbonate ($CaCO_3$) or dolomite [$CaMg(CO_3)_2$]. The textural terms, originally introduced by Grabau (1913), are **rudite** for gravel, **arenite** for sand and **lutite** for mud. The compositional terms, used as modifiers, are calc(i) for limestones and dol(o) for dolostones. Combining the two sets of terms produces the general descriptive terms in

Table 14.3 Field classification of carbonate rocks.

Rock name	Modal particle size	Principal composition
Calcirudite or dolorudite	Gravel	Calcium carbonate or dolomite
Calcarenite or dolarenite	Sand	Calcium carbonate or dolomite
Calcilutite or dololutite	Mud	Calcium carbonate or dolomite

Table 14.3. Using this classification, it is generally easy to make identifications in the field using a hand lens, dilute HCl, which causes calcite but not dolomite to readily effervesce, and/or alizarin red-S, which stains calcite pink, but not dolomite.

Two more sophisticated classification systems were introduced by Folk (1959, 1962) and Dunham (1962) with the latter two appearing together in an American Association of Petroleum Geologists (AAPG) memoir (Ham, 1962) on the classification of carbonate rocks. Each has its advantages and disadvantages, its adherents and detractors, and each has been modified since its introduction.

Dunham's classification system

Dunham's classification system emphasizes the texture of carbonate rocks, utilizing rather simple terminology. Dunham's classification system recognized six major varieties of carbonate rocks (Figure 14.8). Four of these were based on rock texture, the fifth on the rock's inferred origin and the sixth on the notion that the original texture could not be recognized. The first criterion used to distinguish the first four rock types is whether the rock possessed a mud-supported or a grain-supported framework (Chapter 13). Carbonate rocks with mud-supported frameworks are subdivided into **mudstone**, which contains less than 10% grains, and **wackestone**, which contains more than 10% grains. Any grains in such rocks are mud supported and therefore appear to be suspended in a mud matrix.

Depositional texture recognizable					Depositional texture not recognizable
Constituents not bound together at time of deposition				Constituents bound together at time of deposition	
Contains mud			Lacks mud		
Mud-supported framework		Grain-supported framework			
<10% grains	>10% grains	Some mud matrix	No mud matrix		
Mudstone	Wackestone	Packstone	Grainstone	Boundstone	Crystalline

Figure 14.8   Dunham's classification of limestones. (After Dunham, 1962.)

The size of the grains is not used in the name, but the major type of grain may be used as a modifier, as in fossiliferous mudstone or peloidal wackestone. Both mudstones and wackestones accumulate in generally calm environments where significant amounts of mud settle from suspension. Carbonate rocks with grain-supported frameworks are subdivided into **packstone**, which contains mud matrix between the grains, and **grainstone**, which lacks significant mud matrix. Most grainstones contain interstitial diagenetic mineral cements that bind the grains together. As with mudstones and wackestones, the size of the grains is not used in the name, but the major type of grain may be used as a modifier, as in intraclastic packstone or ooid grainstone. Grainstones imply deposition in an environment in which currents kept mud in suspension. Packstones suggest situations in which mud periodically settled from suspension to infiltrate grain frameworks deposited under more agitated conditions.

Two other rock types are recognized. **Boundstone** is used for in situ carbonate accumulations, such as reefs and stromatolites, which were organically bound at the time of accumulation. As we will see in the section that follows, the primary textures of carbonate rocks can be largely obliterated by recrystallization during diagenesis. In addition, some carbonate rocks, such as travertine, consist solely of coarsely crystalline calcium carbonate precipitated directly from solution. Dunham suggested the term **crystalline carbonate** for such carbonate rocks in which no depositional texture is recognizable.

With a sawed slab, etched by HCl, an acetate peel and/or a thin section, Dunham's classification is easy to use. For epiclastic carbonates, it involves recognition of (1) the type of framework (mud supported or grain supported), (2) the percentage of grains in rocks with mud-supported frameworks, and (3) the presence or absence of interstitial mud in grain-supported frameworks. For organically bound carbonates, recognition of their nature is sufficient.

Two significant modifications to Dunham's classification are widely used. Embry and Klovan (1971) recommended subdividing boundstones on the basis of the process involved in binding the carbonate sediment at the time of accumulation (Figure 14.9). **Framestone** is produced by organisms that build rigid organic structures such as reefs by secreting the calcium carbonate. **Bindstone** is produced by organisms that build organic structures such as stromatolites and reefs by binding and/or encrusting pre-existing carbonate material. **Bafflestone** is generated by organisms that trap carbonate sediment by acting as baffles that hinder its movement across the bed, causing it to be trapped – an important process in reefs and bioherms. James (1984) introduced the terms **rudstone** for carbonate gravel-bearing rocks with a clast-supported framework and **floatstone** for carbonate gravel-bearing rocks with a matrix-supported framework.

(a)

(b)

Plate 13.3 Gravelstones. (a) A matrix-supported framework with gravel particles "floating" in and supported by finer particles; in this case a mud-supported matrix. (Photo courtesy of Mario Coniglio.) (b) A clast-supported framework with gravel particles in contact, from the Cretaceous, California. (Photo by John O'Brien.)

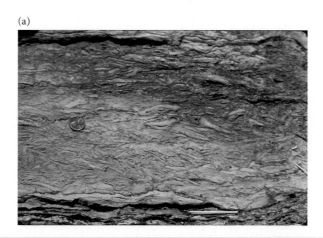

(a)

Plate 13.10a Imbricated clasts, Cambrian, New Jersey. (Photo by John O'Brien.) The flow is from right to left.

Plate 13.12 (a) Oligomictic conglomerate with rounded gravel and a clast-supported framework, Stromboli, Italy. (b) Polymictic conglomerate with a clast-supported framework, Kata Tjuta, northern Australia. (c) Oligomictic quartz breccia with sand-supported framework, Kakadu National Park, northern Australia. (d) Polymictic diamictite (a tillite) with a mud-supported framework, South Africa. (a–c, photos by John O'Brien; d, photo courtesy of Michael Hambrey, with permission from www.glaciers-online.net.)

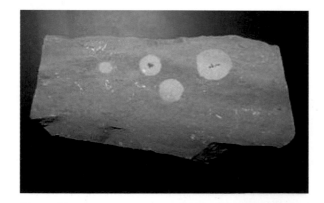

Plate 13.16 Reduction spots around organic particles in red–purple mudrock. (Photo courtesy of Steve Dutch.)

Plate 13.18 Bentonite, with its typical, lumpy popcorn-like appearance, Mowry Formation, Cretaceous, Wyoming. (Photo courtesy of Wayne Sutherland, with permission of Wyoming State Geological Survey.)

Plate 13.19 Oil shale with a dark color due to its oil content. (Photo courtesy of US Department of Energy, Argonne National Laboratory.)

Plate 13.20 Organic-rich sapropel layer (middle) between lighter colored layers in lake sediments, Jurassic Berlin Formation, Connecticut. (Photo by John O'Brien.)

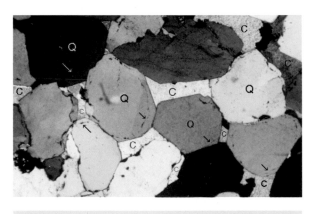

Plate 13.23 Syntaxial quartz overgrowths (arrows) and blocky calcite cement (C) in quartzarenite (Q). (Photo courtesy of L. Bruce Railsback.)

Plate 13.25 Photomicrograph of calcite-cemented sandstone with poikiloptic cement enclosing multiple quartz and feldspar grains (crossed polars). (Photo courtesy of Lee Phillips.)

(a)

(b)

Plate 13.26 Images of concretions. (a) A nucleus and concentric structure in sandstone. (Photo courtesy of John Merck.) (b) A concretion in shale, Devonian, Virginia. (Photo courtesy of Duncan Heron.)

(a)

(b)

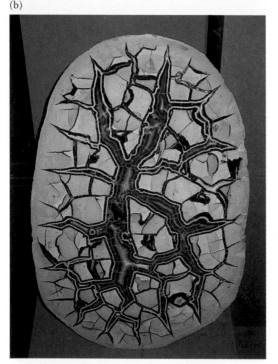

Plate 13.28 Geodes showing banded chalcedony rims and quartz (plus bladed gypsum) crystal linings.

Plate 13.27 Nodules. (a) Three flint nodules in limestone. (Photo by John O'Brien.) (b) Septarian nodule. (Photo courtesy of Keith Pomakis; nodule from the Pinch Collection, Canadian Museum of Nature, Ottawa, Ontario.)

Plate 13.29 Liesegang bands in a sandstone block; note the truncation against the joint surfaces. (Photo courtesy of Peter Adderley.)

(a)

(b)

Plate B13.2 Large-scale cross-strata, Navajo sandstone, Jurassic, Utah. (Photo by John O'Brien.)

Plate 14.2 (a) Shell beach, Hinchinbrook Island, Queensland, Australia: an example of carbonate sediment composed largely of grains or allochems. (b) Fossil-bearing limestone, middle Devonian, New York, which records marine life of 435 million years ago. (Photos by John O'Brien.)

(a)

(b)

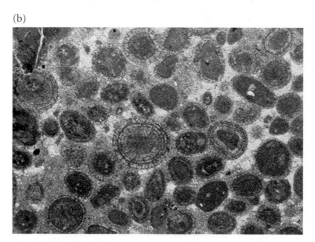

Plate 14.3 (a) Ooid-rich dolostone with ooid nuclei clearly visible. (Photo courtesy of Jack Morelock.) (b) Thin section showing various ooids with nuclei and radial and concentric laminations. (From Adams et al., 1984.)

(a)

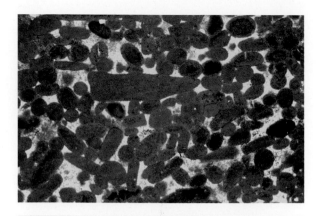

(b)

Plate 14.6 Peloids of various shapes in grain-supported framework, with interstitial cement. (Photo courtesy of C. G. St. C. Kendall.)

Plate 14.4 (a) Rounded intraclasts, Jurassic, Sundance Formation, Wyoming. (b) Angular intraclasts, Cambro-Ordovician, Allentown dolostone, New Jersey. (Photos by John O'Brien.)

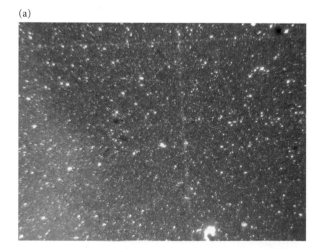

(a)

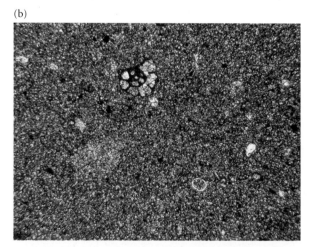

(b)

Plate 14.7 Carbonate mud in thin section, under plane light: (a) micrite, (b) microspar. (Photos courtesy of Frederic Boulvain.)

Plate 14.14 Stylolites with a "toothed" pattern (below the penny) and a thicker dissolution seam near the bottom. (Photo by John O'Brien.)

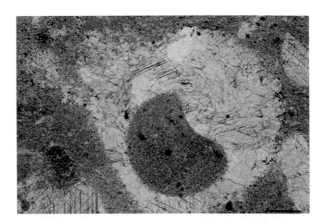

Plate 14.15 Moldic porosity, showing a dissolved gastropod shell, later filled with precipitated sparry cement. (Photo courtesy of C. M. Woo.)

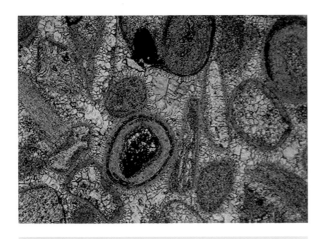

Plate 14.16 Marine, isopachous rim cement (brownish) on grains with pore spaces filled with second generation meteoric drusy calcite cement, Ouanamane Formation (Jurassic), Morocco. (From Adams et al., 1984.)

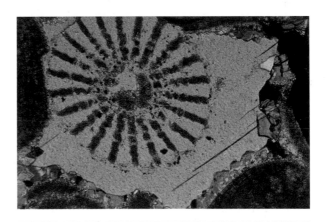

Plate 14.17 Syntaxial calcite in optical continuity on an echinoderm spine. (Photo courtesy of Maria Simon-Neuser, with permission of LUMIC.)

(a)

(b)

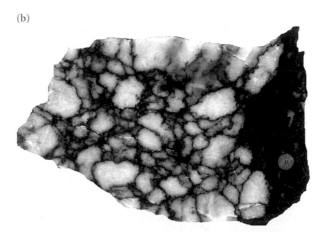

Plate 14.19 (a) Nodular gypsum, Triassic Mercia Group, Watchet Beach, England. (Photo courtesy of Nicola Scarselli.) (b) Chicken-wire anhydrite, Carboniferous, Belgium. (Photo courtesy of Frederic Boulvain.)

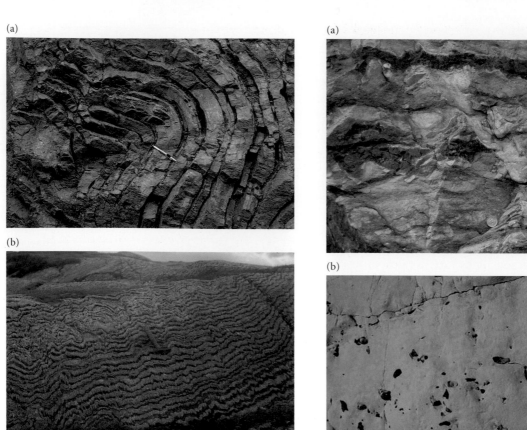

(a)

(b)

Plate 14.23 Bedded "ribbon" chert in outcrops. (a) Radiolarian chert, Cretaceous mélange, Marin County, California. (Photo courtesy of Steve Newton.) (b) Ribbon chert, Ordovician, Norway. (Photo courtesy of Roger Suthren.)

(a)

(b)

Plate 14.24 Chert nodules. (a) Flint nodules in Kalkberg limestone, Devonian, New York. (b) Flint in chalk cliffs, Cretaceous, Seven Sisters, East Sussex, England. (Photos by John O'Brien.)

(a)

(b)

Plate 14.25 (a) Siliceous sinter precipitated around hot springs, Yellowstone National Park, Wyoming. (Photo by Kevin Hefferan.) (b) Siliceous sinter terraces, Mammoth Hot Springs, Yellowstone National Park. (Photo by John O'Brien.)

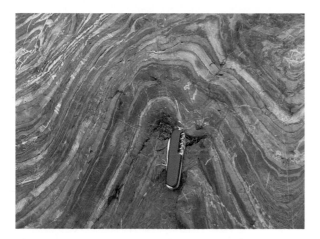

Plate 14.26 Banded iron formation, Vermillion Range, Minnesota. (Photo by Kevin Hefferan.)

(a)

Skarn Hornfels

Anthracite coal Marble Metaquartzite

(b)

Slate Pyllite Schist

Gneiss Migmatite Mylonite

Plate 15.3 Common metamorphic rocks: (a) non-foliated rocks, (b) foliated rocks. (Photos by Kevin Hefferan.)

Growth rings

1 cm

Plate 14.27 Manganese nodule. (Photo by John O'Brien.)

Plate 14.30 Peat and the major ranks of coal derived from it during coalification (left to right): peat, lignite, bituminous and anthracite. Note the increases in compactness, hardness and luster and change in color from brown to black (carbon rich) with increasing rank. (Photo by John O'Brien.)

Plate 17.1 (a) Photomicrograph of hornfels displaying equant crystals separated by biotite-rich layers that may represent relict sedimentary bedding. (Photo courtesy of Kent Ratajeski.)
(b) Hornfels collected from the base of the Palisades Sill, New Jersey. (Photo by Kevin Hefferan.)

Plate 17.2 Photomicrograph of granoblastic texture with equant calcite crystals. (Photo courtesy of Kent Ratajeski.)

Plate 17.4 Marble occurs in multiple colored varieties depending upon chemical substitutions in calcite and dolomite and the presence of accessory minerals. (Photo by Kevin Hefferan.)

Plate 17.7 Metabreccia containing oxidized hematitic basalt rock fragments and epidote cemented together with calcite. (Photo by Kevin Hefferan.)

Plate 17.13 In this garnet mica schist, viewed between crossed polarizers, the less competent mica minerals are thinned and bent around black garnet porphyroblasts. The field of view is 6 mm. (Photo courtesy of David Waters and Department of Earth Sciences, University of Oxford.)

Plate 18.24 Franciscan mélange with a mixture of blueschist and greenschist facies rocks. Individual blueschist clasts are boulder sized and are referred to as "knockers". (Photos by Kevin Hefferan.)

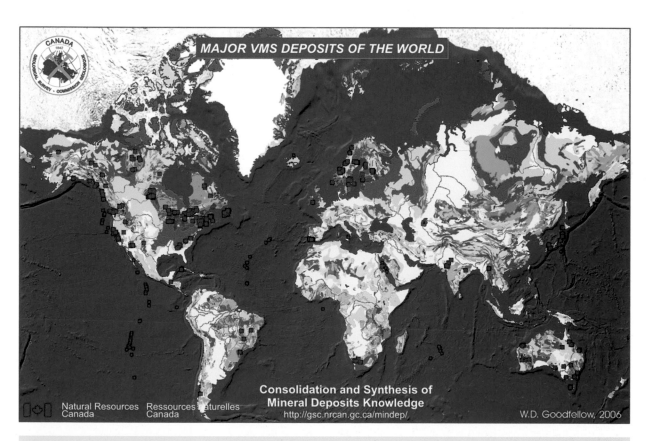

Plate 19.3 Global map of the major volcanogenic massive sulfide (VMS) deposits. Note that these deposits occur along ancient and modern divergent and convergent plate boundaries (From Galley et al., 2007; with permission of Natural Resources of Canada, 2009, courtesy of W. D. Goodfellow and the Geological Survey of Canada.)

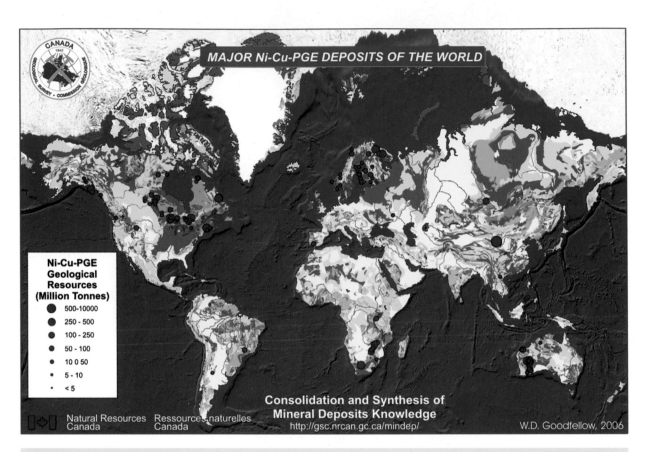

Plate 19.6 Global map of the major layered gabbroic intrusions containing platinum group elements, nickel and copper. Note that these deposits are concentrated in continental crust and are generally of Archean (>2.5 Ga) age. (From Eckstrand and Hulbert, 2007; with permission of Natural Resources of Canada, 2009, courtesy of W. D. Goodfellow and the Geological Survey of Canada.)

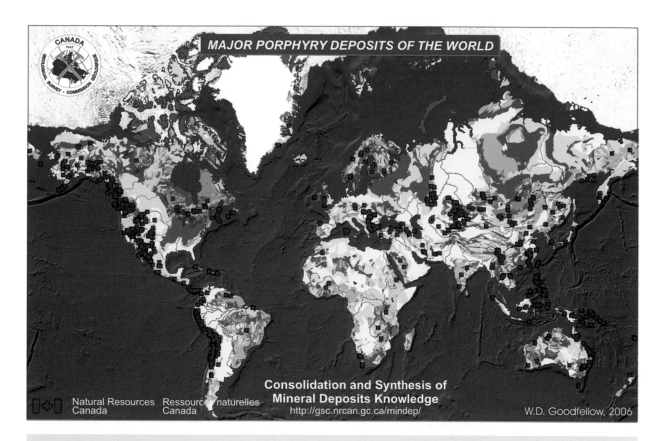

Plate 19.8   Global map of the major porphyry deposits. Note that these deposits are concentrated along modern and ancient convergent plate boundaries (From Sinclair, 2007; with permission of Natural Resources of Canada, 2009, courtesy of W. D. Goodfellow and the Geological Survey of Canada.)

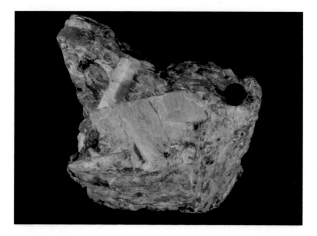

Plate 19.10   Granite pegmatite from the Black Hills of South Dakota containing large crystals of beryl, quartz and feldspar. Gold deposits are associated with the Black Hills pegmatite. (Photo by Kevin Hefferan.)

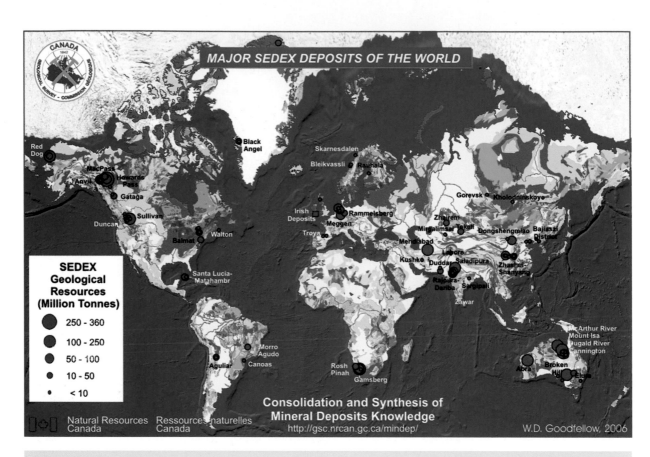

Plate 19.16 Global map of the major sedex deposits. Note that these deposits only occur in continental settings. Data are from the synthesis of sedex deposits by Goodfellow and Lydon (2007). (From Goodfellow and Lydon, 2007; with permission of Natural Resources of Canada, 2009, courtesy of W. D. Goodfellow and the Geological Survey of Canada.)

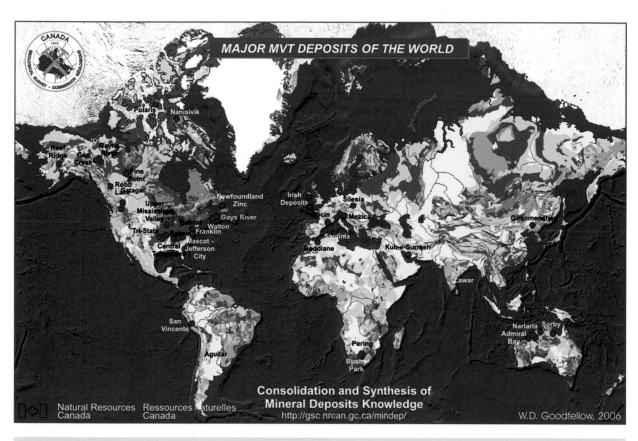

Plate 19.18 Global map of the major Mississippi Valley-type (MVT) deposits. Note that these deposits only occur in continental settings. (From Paradis et al., 2007; with permission of Natural Resources of Canada, 2009, courtesy of W. D. Goodfellow and the Geological Survey of Canada.)

Plate 19.22 (a) Sandstone quarry, Wisconsin. (b) Cut and polished granite, Wisconsin. (c) Esker containing glaciofluvial sands, Wisconsin. (d) Crushed aggregate. (e) Limestone quarry Wisconsin. (f) Kiln for baking limestone. (g) Moroccan adobe house. (h) Brick and concrete stadium in Iowa. (Photos by Kevin Hefferan.)

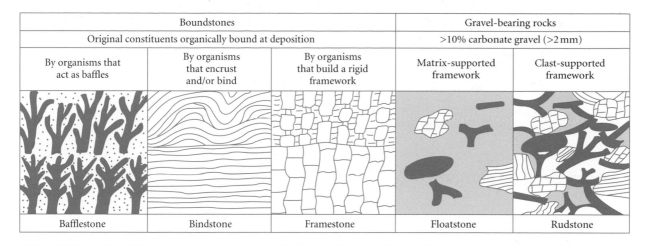

Boundstones			Gravel-bearing rocks	
Original constituents organically bound at deposition			>10% carbonate gravel (>2 mm)	
By organisms that act as baffles	By organisms that encrust and/or bind	By organisms that build a rigid framework	Matrix-supported framework	Clast-supported framework
Bafflestone	Bindstone	Framestone	Floatstone	Rudstone

Figure 14.9   Modifications of Dunham's classification by Embry and Klovan (1971) and James (1984).

Because of its emphasis on primary textures, Dunham's classification has been widely adopted in the petroleum industry. Rocks such as grainstones and rudstones have sufficient primary porosity and permeability to make excellent reservoir rocks, whereas mud-rich rocks such as mudstones, wackestones and floatstones may be sufficiently impermeable to trap petroleum, and impede its flow.

Folk's classification system

Folk's classification revolutionized our understanding of carbonate rocks. It encouraged workers to pay careful attention to the components of carbonate rocks that lie at the heart of the classification scheme (Figure 14.10). Because our understanding of carbonate rock components has evolved substantially since 1959, the following discussion emphasizes those aspects of Folk's classification that are still widely used.

Folk's classification (1959) recognized four major types of grains or allochems: (1) intraclasts (limeclasts), (2) oolites (ooids), (3) fossils (skeletal fragments), and (4) pellets (peloids). The term micrite, short for microcrystalline calcite, was used for carbonate mud – a valuable and widely used term that clearly distinguishes carbonate mud from siliciclastic mud. Intergranular cements (see next section) were lumped together under the rubric of sparry cement or spar. Allochem-bearing carbonate rocks were classified according to the percentages of various allochems and the ratio of sparry calcite cement to micrite (Figure 14.11). Carbonate rocks with significant micrite accumulate chiefly in periodically calm environments where mud settles from suspension. Carbonate rocks with sparry cement accumulate largely in continuously agitated environments where muds stay in suspension and only allochemical grains are deposited. Sparry cements are precipitated in the pore spaces between allochems during diagenesis. Rocks that lack allochems were classified separately (Figure 14.10).

In Folk's classification, allochemical rocks that contain a minimum of 25% intraclasts in the allochem population are given the prefix "intra". If the interstices between intraclasts are largely filled with diagenetic cement, the rock is an **intrasparite**; if filled with mud, it is an **intramicrite**. If fewer than 25% intraclasts occur, but the rock contains more than 25% ooids, the names **oosparite** and **oomicrite** are used. If neither intraclasts nor ooids exceed 25% of the allochems, then fossils or pellets (peloids in current usage) will be the dominant allochemical constituents. Where fossils dominate, the rocks are **biosparite** or **biomicrite**; where pellets dominate, they are **pelsparite** or **pelmicrite**. Rocks in which allochemical grains are sparse to absent are classified separately. These include **micrite** for rocks composed primarily of carbonate mud, **dismicrite** for micrites that contain small, spar-filled voids produced during diagenesis

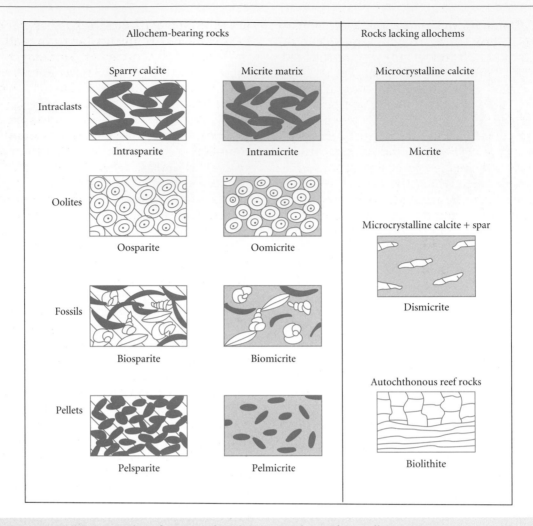

Figure 14.10   Folk's basic classification of carbonate rocks. (After Folk, 1959.)

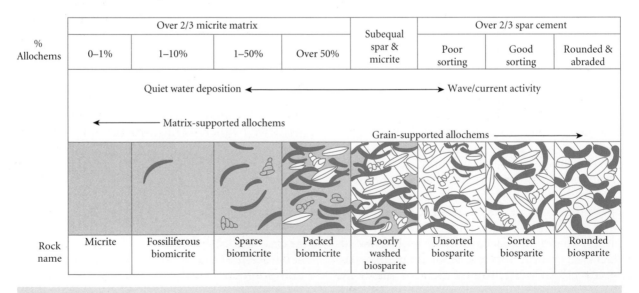

Figure 14.11   Folk's (1962) textural classification of carbonates.

and **biolithite** for in situ carbonate accumulations (see Box 14.1) roughly equivalent to Dunham's boundstones. Folk also introduced the prefix "dolo" for carbonate rocks composed largely of dolomite as in dolomicrite.

Folk (1962) also developed a textural classification system (Figure 14.11) based on the proportions of micrite and sparry cement in carbonate rocks. This textural scheme permits inferences to be made concerning depositional conditions under which the carbonate particles accumulated. Carbonate rocks are subdivided into three major textural groups: (1) rocks with more than two-thirds mud matrix between the allochems (if any), called micrites, (2) rocks with more than two-thirds sparry cement between the allochems, called sparites, and (3) rocks with at least one-third of both spar and micrite between allochems, called poorly washed sparites. Micrites are subdivided according to the percentage of allochems into micrite (<1% allochems), allochemical micrite (1–10% allochems), sparse allochemical micrite (10–50% allochems) and packed allochemical micrite (>50% allochems). The latter two are roughly equivalent to the distinction between wackestone and packstone in Dunham's classification. The name of the allochem is used, as in Figure 14.11 in which the major allochem is assumed to be fossil grains. Sparites are subdivided according to sorting and rounding into unsorted allochemical sparites, sorted allochemical sparites and rounded allochemical sparites. These terms echo Folk's terminology for the textural maturity of detrital sedimentary rocks (Chapter 12).

The classification systems developed by Folk, Dunham and others have provided us with a valuable conceptual framework in which to carefully describe and interpret the components of carbonate sedimentary rocks.

14.1.5   Environments of carbonate deposition

The conditions that favor the production of carbonate sediments include warm, fairly shallow, nutrient-rich water, so that most carbonate rocks originally accumulated in shallow seas in the tropics and subtropics. Carbonate sedimentary rock formation also requires minimal influx of detrital sediments, so that most carbonate rocks initially accumulated in areas of low relief such as on intracratonic platforms or passive margins or in oceanic environments far from land.

Tectonic–physiographic settings for the accumulation of carbonate sediments include (1) carbonate ramps, (2) rimmed carbonate platforms, (3) epeiric sea platforms, (4) isolated platforms, and (5) submerged platforms or "deep" water environments. Many environments, subenvironments and facies are shared between settings, but their spatial distributions depend on the tectonic–physiographic setting, local factors and how both change with time. Within these five tectonic–physiographic settings, several environments of carbonate sediment accumulation are recognized. These include (1) the supratidal zone, (2) the intertidal zone, (3) the subtidal zone above the normal wave base, (4) the subtidal zone between the normal and storm wave bases, (5) barrier-fringing reefs, (6) sand shoals, (7) lagoons, and (8) deeper shelf and basin environments.

Research over the past half century, driven in part by the fact that most of the world's petroleum and natural gas reserves occur in carbonate rocks, has developed criteria that permit geologists to recognize the processes and conditions that produced particular carbonate rocks, including their tectonic–physiographic setting and the specific environment within that setting in which they accumulated. Table 14.4 summarizes features that permit carbonate sedimentary rocks to be ascribed to specific environmental settings. A more detailed discussion of carbonate depositional environments and the settings in which they occur is beyond the scope of this text. Readers interested in investigating this topic in more detail are referred to Flugal (2004), Tucker and Wright (1990), Friedman (1981) and Bathurst (1975).

The simplest distribution of carbonate depositional environments occurs on **carbonate ramps** where, in the ideal model, the bottom slopes gently seaward over distances of 10^1–10^3 km (Figure 14.12a). Carbonate ramps typically develop on the margins of shallow subtropical seas. Environments occupy roughly shore-parallel bands characterized by water depths that gradually increase seaward. The ideal shore to deep ocean sequence of environments on carbonate ramps is (1) supratidal, (2) intertidal, (3) subtidal above normal wave base, (4) subtidal above storm

Table 14.4 General distribution of carbonate lithologies, sedimentary structures and fauna in major carbonate depositional environments.

Environment	Lithologies	Structures	Fauna
Supratidal: region above normal high tide line, periodically flooded by storm surges and tsunami	Micrite; some peloidal; intraclastic/skeletal floatstone/packstone from storm tides; nodular gypsum/chicken-wire anhydrite in arid climates; aeolian grainstones	Dessication cracks; teepee structure; soils and calcrete; vadose fenestrae and dissolution collapse breccias; cross-strata in aeolian grainstones	Generally sparse and restricted; flat laminated stromatolites; some terrestrial forms; rootlets where not arid
Intertidal: area between high tide and low tide line flooded and exposed daily; extensive intertidal flats with ponds and tidal channels	Micrite; some peloidal and skeletal wackestone and packstone in protected areas; grainstones and occasional rudstones on exposed coasts and in tidal channels	Plane beds and cross-strata on exposed coasts and in tidal channels; oscillation and combined flow ripples; flaser and lenticular bedding	Laterally-linked hemispheroid and stacked hemispheroid stromatolites; robust skeletal fragments on exposed coasts; delicate forms in protected areas
Subtidal shoal: below low tide line, above normal wave base; constantly swept by waves and currents; mud stays in suspension	Skeletal and ooid grainstones dominate; skeletal rudstones; no micrite	Abundant cross-stratification; plane beds; vertical dwelling and feeding traces	Oncolite stromatolites; robust shallow water invertebrates including mollusks, brachiopods, corals bryozoans and echinoderms
Subtidal: below normal wave base, above storm wave base; bottom periodically swept by storm waves	Micrites deposited from suspension; skeletal wackestones; interlayered with skeletal/ intraclastic packstone and rudstone storm deposits	Hummocky cross-strata and graded bedding common in storm units; resting, feeding, grazing and escape traces	Less robust forms including mollusks, brachiopods, bryozoans, trilobites and echinoderms
Subtidal: below storm wave base	Micrites; skeletal wackestones	Nodular bedding common; extensive bioturbation, resting, feeding and grazing traces	Many sessile forms such as brachiopods, bryozoans, and mollusks in growth positions
Reef: wave-resistant organic accumulations that rise above the surrounding sea floor; banks and mounds are organic accumulations lacking reef properties	Boundstones including framestones, bindstones and bafflestones in reef core; skeletal rudstones and floatstones on steeply inclined forereef flanks; skeletal rudstones associated with packstones and grainstones in backreef; bafflestones in banks and mounds	Robust reef-building and -binding organisms in massive reef core; steeply inclined thick beds in forereef; cross-strata and rubbly beds in backreef areas	Corals, coralline algae, stromatoporoids, mollusks, sponges, bryozoans, brachiopods and echinoderms; skeletal fragments and micrite in banks and mounds
Lagoon: calm, shallow water areas on landward side of rimmed platform reefs and sand shoals	Peloidal micrites dominate; skeletal and aggregate wackestones; patch reef boundstones; grainstones in tidal deltas; packstones in washover fans; evaporites in restricted lagoons	Laminated unless, as is common, extensively bioturbated; cross-strata in grainstones	Variable, depending on salinity; restricted where salinity abnormal; ostracods common

Table 14.4   *Continued*

Environment	Lithologies	Structures	Fauna
"Deep" pelagic: deposits that settle from suspension in offshore environments	Micrites; chalk; various amounts of detrital mud	Lamination; thin beds; interlayered with mass flow deposits adjacent to platform margins	Foraminifera, coccoliths and pteropods since Mesozoic; plantic and invertebrates and deeper water benthic fauna
"Deep" mass flow: deposits from rockfall, slides, debris flows and turbidity currents that flow into deeper water	Mega-rudstones and floatstones in rockfalls, rock slides and debris flows from platform margin; packstones, grainstones and micrites from turbidites	Massive bedding in rock slides and debris flows; distorted bedding in slumps; graded bedding in turbidites; olistoliths from rock fall and slides	Shallow water fauna in rocks displaced from platforms; deeper water fauna in interlayered pelagic units

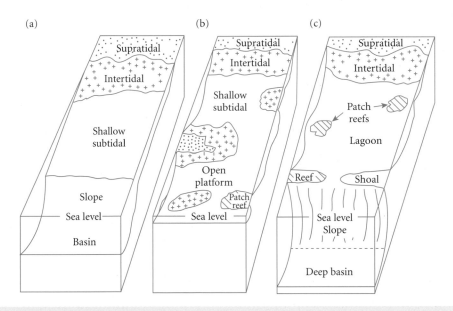

Figure 14.12   Carbonate environment models showing tectonic–physiographic settings and the distribution of carbonate depositional environments in: (a) a carbonate ramp, (b) an epeiric sea platform, and (c) a rimmed platform. By extension, common environments in isolated platforms and submerged platforms are also illustrated.

wave base, and (5) subtidal below wave base. The gentle slopes preclude the development of significant mass flow deposits in the latter environments. Complications involve the local development of small patch reefs that grow upward into the subtidal zone and the local development of sand shoals in the subti-dal zone above the normal wave base that may lead to the development of small islands (cays) and even shallow lagoons.

Rimmed platforms, typically tens to thou-sands of kilometers wide, develop near shelf or platform margins where nutrient-rich waters upwelling from depth encourage the development of carbonate buildups such as reefs and/or sand shoals (Figure 14.12c). The development of carbonate buildups creates a quiet-water lagoon on the landward side. On the seaward side of the shelf edge, buildups are steeply dipping forereef deposits produced by rockfall, slides and mass flows from the adjacent reef. For rimmed carbonate platforms the ideal landward to seaward sequence of environments is (1) supratidal, (2) intertidal, (3) lagoonal, (4) reef or subtidal shoal/island with local tidal channels, (5) reef flank/platform slope, and (6) deep water mass flow/pelagic (Figure 14.12c).

Epeiric sea platforms develop during sea level high stands when oceans flood large portions of the craton to form shallow seas with widths of $10^2–10^4$ km (Figure 14.12b). Flooded cratons typically contain a patchwork of basins with somewhat deeper water, extensive platforms covered by shallow water and slightly elevated areas, some of which are above sea level. As a result, carbonate depositional environments show a similar patchwork pattern that depends significantly on local slope. Typically, slopes are gentle and the environments and facies that develop are those typical of carbonate ramps. However, their patterns are such that they wrap around upwarped high areas and encircle subsiding basins. As a result, environments around elevated areas are arranged in a bull's-eye pattern from terrestrial through (1) supratidal, (2) intertidal, to (3) subtidal. Environments around subsiding basins are arranged in a bull's-eye pattern sequence: (1) supratidal, (2) intertidal, (3) subtidal above the normal wave base, (4) subtidal above the storm wave base, and (5) subtidal below the wave base.

Brief mention should be made of **isolated carbonate platforms** with widths from one to thousands of kilometers, which are very abundant in modern oceans. Such platforms develop, often on volcanic seamounts or platforms, separated from land by a considerable distance. Typically carbonate buildups develop on the platform margins and enclose a lagoon between them. Reefs (Box 14.1) typically develop on the windward side of platforms, sand shoals develop on the leeward side with lagoons, and bays develop between them. Complications include the presence of patch reefs and islands with the lagoon. Adjacent to islands, supratidal to intertidal environments

may occur. **Submerged carbonate platforms** occur in deeper water settings below the wave base and above the carbonate compensation depth. Submerged platforms located far from land masses, including most oceanic ridges and plateaus, are characterized by pelagic sediment deposition; those adjacent to land masses are sites of mass flow deposits as well.

14.1.6 Carbonate diagenesis

Carbonate rocks generally undergo extensive changes during diagenesis. One important set of controls on diagenetic reactions is pore fluid geochemistry, especially its (1) alkalinity–acidity, (2) temperature, (3) total dissolved solids, (4) dissolved Mg^{+2}/Ca^{+2} ratios, and (5) dissolved sulfate ions (SO_4^{-2}). The solubility of carbonate minerals is extremely sensitive to alkalinity–acidity (roughly pH). Small increases in dissolved CO_2 cause pore waters to become more acidic as their pH decreases, which leads to extensive dissolution of carbonate minerals. In contrast, small decreases in dissolved CO_2 cause pore waters to become more alkaline as their pH increases, which leads to extensive precipitation of carbonate minerals. A second set of factors important in diagenesis is the mineralogy of the carbonate material, which includes aragonite, high magnesium calcite, low magnesium calcite and dolomite in a variety of crystal habits. These mineralogical and pore fluid factors interact during diagenesis to play significant roles in whether a particular species is dissolved, precipitated or otherwise altered.

Three principle carbonate diagenetic environments (Figure 14.13) are defined by the type of water that penetrates the pore spaces: (1) **marine connate water**, mostly beneath the sea floor, (2) **meteoric water**, groundwater from surface infiltration, located mostly below land surfaces, and (3) a **zone of mixing**, typically near the boundary between less dense meteoric water and a wedge of denser marine water. The meteoric zone can be subdivided into a non-saturated meteoric **vadose zone**, above the water table, and a saturated **phreatic zone** below it. The phreatic zone, or zone of saturation, can be subdivided into a near surface zone and a deep zone.

The processes that occur during the diagenesis of carbonate rocks include (1) microbial micritization, (2) compaction, (3) dissolution,

Box 14.1   Reefs through time

Carbonate buildups are local accumulations of carbonate sediment that possess significant relief above the surrounding ocean floors. They include a range of features from mud mounds through sand shoals to organic reefs. The terms **bioherm** or **biostrome** distinguish buildups in which in situ construction by organic activity is important from buildups that originate by mechanical processes such as waves and tidal currents. Important organic activities involved in biostrome buildup include the production of (1) framestone, as large numbers of organisms secrete skeletal material, (2) bindstone, as organisms encrust and bind together carbonate fragments, and (3) bafflestone, as organisms trap carbonate grains or mud.

Reefs are biostromes in which organisms have built relatively rigid, wave-resistant structures over substantial periods of time. Every reef starts as a small organic accumulation, often developed on a local shoal. Baffling and binding by these organisms eventually produces a firmer substrate on which framework-building organisms can begin to build a wave resistant structure. As organisms die and new ones build additional framework, the structure grows into the wave zone. The wave resistance results from the production of a rigid framework of framestone and bindstone during reef construction by organic activity. Large blocks of reef material are eroded from reefs, especially during storms, but many blocks are reincorporated into the reef by later encrustation to produce bindstone and by additional generation of framestone. Typical reef cores contain only 10–20% framestone. Spaces within the framework are filled with carbonate grains and mud whose eventual cementation serves to strengthen the reef structure. Auxiliary organisms that inhabit the reef add to its growth and encrusting organisms bind much of this material into bindstone. Major types of reefs include (1) **fringing reefs** that develop adjacent to and fringe shorelines, (2) **barrier reefs** that are separated from shorelines by lagoons, (3) roughly circular **atolls** that enclose lagoons without major land masses, (4) small, isolated, high relief **pinnacle reefs**, (5) small, isolated, low-relief **patch reefs**, and (6) **submerged reefs**, formerly active reefs drowned during subsidence or sea level rise.

Many groups of organisms have contributed substantially to the formation of reefs and other types of biostromes since the lower Paleozoic (Figure B14.1a).

Reef-building organisms may record significant oscillations in ocean water chemistry over time. Figure B14.1a summarizes the major periods of reef building and the organisms that contributed substantially to reef growth. Stanley and Hardie (1999) pointed out that organisms with calcite skeletons such as stromatoporoids, rugose and tabulate corals and receptaculitid algae were the dominant reef-building organisms in extensive Ordovician, Silurian and Devonian reefs. This corresponds to earlier work by Sandberg (1983), who suggested that during this time **calcite seas** favored low magnesium calcite precipitation over high magnesium calcite and aragonite (Figure B14.1b). Sandberg also suggested that from the Carboniferous through the mid-Jurassic, low calcium **aragonite seas** favored the precipitation of high magnesium calcite and aragonite. Stanley and Hardie (1999) pointed out that this period was characterized by a paucity of true reefs and that biostromal buildups were dominated by organisms with aragonite skeletons such as scleractinian corals, most sponges and some algae and by algae with high magnesium calcite skeletons. In the late Jurassic and throughout the Cretaceous, when calcite sea conditions returned, widespread construction of large reefs resumed with calcitic rudistid mollusks playing a dominant role. However, few correlations are perfect and scleractinian corals with aragonite skeletons were also important. Aragonite seas returned in the Oligocene–Miocene and coral reefs have since been built primarily by scleractinian corals and high magnesium coralline algae. Could it be that fluctuations in major reef-building organisms are controlled by and record fluctuations in seawater chemistry?

What might cause such changes in seawater chemistry and reef-building organisms? Sandberg (1983) argued that calcite seas correlated with major global warming events ("greenhouse" conditions) generated by elevated atmospheric CO_2 levels and aragonite seas with periods of global cooling ("icehouse" conditions) and major glaciations associated with depressed atmospheric CO_2. He proposed that during greenhouse periods elevated CO_2 increased the acidity of ocean water to the point

Continued

Box 14.1   *Continued*

where aragonite precipitation was inhibited so that calcite precipitation was dominant. On the other hand, during cool periods lower CO_2 concentration produced lower acidity that permitted aragonite to precipitate. Stanley and Hardie (1999) argued that Sandberg's mechanism was unlikely. Pointing to a correlation with plate tectonics, they suggested that the cyclic variations in ocean chemistry and skeletal composition are driven by variations in the rate of sea floor spreading. These periods of elevated sea level correspond to periods of global warming produced in part by the CO_2 generated by additional volcanic activity along the ridge system. On the other hand, during periods of decelerated sea floor spreading, such as those associated with the existence of Pangea (Carboniferous–

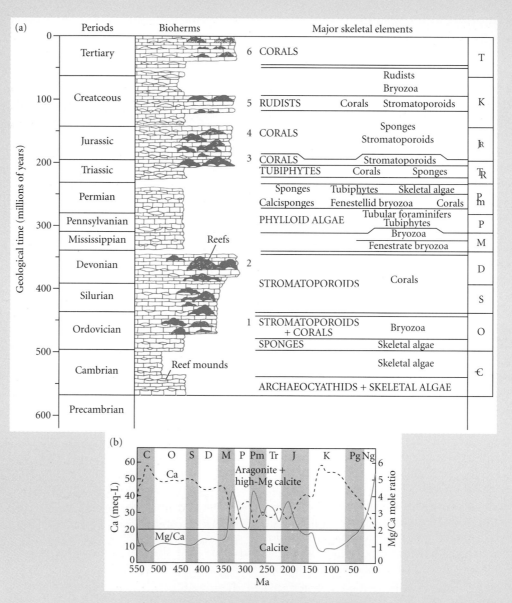

Figure B14.1   (a) Major reef and carbonate buildup organisms through time. Major framework builders are given in capitals, and lesser contributors in lower case letters. (After James, 1983; from Raymond, 2002.) (b) Changes in major reef and carbonate buildup organisms and their composition through time, compared with periods of calcite and aragonite seas. (After Stanley and Hardie, 1999; from Prothero and Schwab, 2004.)

Box 14.1   *Continued*

Triassic) and since the collision of India with Asia (Oligocene), (1) the global rate of sea floor spreading slowed, (2) the sea level fell, (3) less magnesium was removed from seawater as hydrothermal metamorphism decreased, (4) the Mg/Ca ratio of seawater rose, and (5) aragonite seas were produced. These periods of sea level low stand correspond to periods of global cooling and glaciation produced by a combination of less CO_2 generated by volcanic activity along the ridge system and increased albedo. On the other hand, Kiessling et al. (2008) argued that changes in skeletal compositions are driven by a complex set of processes, of which mass extinctions are the most important. Groups that recover well from such extinction events tend to dominate skeletal compositions in the succeeding period. Of course, all these ideas are somewhat speculative, but they are very interesting indeed!

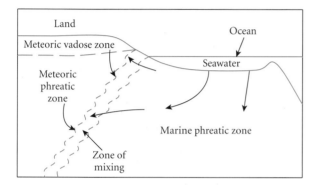

Figure 14.13 Distribution of the major zones in which carbonate diagenesis occurs.

(4) dissolution including pressolution, (5) cementation, (6) recrystallization or neomorphism, and (7) replacement by new minerals. Each of these processes is discussed in relationship to pore water chemistry, carbonate mineralogy and diagenetic setting in the sections that follow.

Limestone diagenesis

Microbial micritization occurs largely in marine surface and near surface environments where the micro-boring activity of blue-green endolithic bacteria and algae, accompanied by precipitation of micrite into the micropores generated by such boring, converts original carbonate materials into micrite. Micritization of ooids, skeletal particles and aggregate grains is common. All stages in this process have been observed, from partial micritization where the original material is recognizable to complete micritization where it is not. As discussed earlier, micritization is an important source of carbonate mud. In the context of this section it is an important source of mud generated during eodiagenesis by endolithic microbes that inhabit carbonate sediments.

As carbonate sediments are buried, progressive increases in confining pressure cause them to undergo **compaction**. If grainstones are largely uncemented, significant breaking and crushing of rigid, but delicate, skeletal grains can occur and soft grains may undergo significant plastic deformation. In any case, the porosity and permeability of carbonate sediments tend to decrease during compaction in ways similar to that which occurs during the compaction of detrital sedimentary rocks (Chapter 12).

Burial also leads to significant amounts of **pressolution** in which the portions of grains under maximum stress, especially at grain contacts, undergo selective dissolution. As in detrital sediments (Chapter 12), pressolution leads to larger grain contacts where grains interpenetrate along concavo-convex and sutured contacts. Extensive pressolution causes interpenetration of carbonate grains producing a texture called a **fitted fabric**.

Dissolution of impure carbonate rocks leaves an insoluble residue of constituents such as clays and other siliclastic minerals, iron oxides and organic matter. **Stylolites** (Figure 14.14) are insoluble residue seams that cross-cut partially dissolved grains and commonly have a toothed pattern when viewed in outcrop or microscopically. **Dissolution seams** are thicker seams of insoluble residue that often anastomose to produce a braided pattern (Figure 14.14). Many contacts between carbonate layers are marked by

Figure 14.14 Stylolites with a "toothed" pattern (below the penny) and a thicker dissolution seam near the bottom. (Photo by John O'Brien.) (For color version, see Plate 14.14, between pp. 408 and 409.)

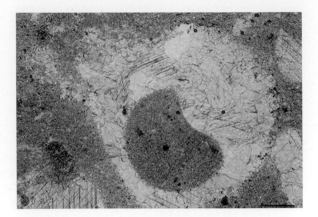

Figure 14.15 Moldic porosity, showing a dissolved gastropod shell, later filled with precipitated sparry cement. (Photo courtesy of C. M. Woo.) (For color version, see Plate 14.15, between pp. 408 and 409.)

stylolites or dissolution seams along which substantial dissolution has occurred. It is important to remember that these are not depositional contacts and that dissolution has likely thinned the strata in question, often by a considerable amount. Especially soluble strata may disappear completely.

The existence of caves, sinkholes and other karst features (Chapter 12) implies that carbonate rocks are extremely soluble under the appropriate conditions. Dissolution of carbonate sediments is relatively rare in warm, marine pore waters because they are approximately saturated with respect to calcium carbonate. By analogy with the carbonate compensation depth (CCD) discussed earlier, colder, more acidic, marine pore waters are undersaturated with respect to calcium carbonate and have the ability to dissolve calcium carbonate sediments during diagenesis. Nonetheless, most calcium carbonate dissolution occurs in the presence of meteoric water, which is generally much more acidic than marine pore waters and generally has a lower Mg/Ca ratio as well. Although carbonates of any composition may be dissolved by meteoric groundwater, aragonite and high magnesium calcite are the most susceptible. This helps to explain why high magnesium calcite and, especially, aragonite become increasingly rare in older carbonate rocks. The dissolution of grains composed of aragonite or high magnesium calcite produces **moldic porosity** in

which the form of ooids and fossils is preserved as a cavity of similar shape. Aragonitic ooids and mollusk shells (Figure 14.15) are especially susceptible to dissolution. Later precipitation of pore-filling cements produces a cast of the original grain. Dissolution produces a variety of other void spaces that range from tiny voids to large caverns and that represent storage capacity for fluids such as petroleum and natural gas, as well as for aqueous fluids capable of precipitating of mineral cements.

As implied by the "sparry calcite" in Folk's classification, mineral cements are common products of carbonate diagenesis. Their composition depends largely on the pore water composition and their form depends primarily on whether cementation occurs above or below the water table, i.e., within the vadose zone or within the phreatic zone. **Marine cements** are almost exclusively aragonite and high magnesium calcite. Because grains in ocean floor sediments are completely bathed in pore fluid, cements that nucleate on grains grow at similar rates to produce coatings of nearly constant thickness called **isopachous rim cements**. They occur primarily as coatings of radial, acicular to fibrous crystals and as micritic coatings of aragonite or high magnesium calcite (Figure 14.16).

On the other hand, **meteoric cements** are almost exclusively composed of low magnesium calcite. In the meteoric phreatic zone

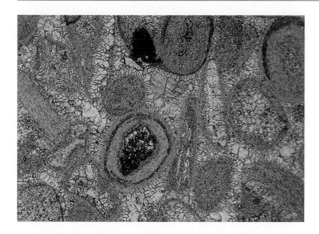

Figure 14.16 Marine, isopachous rim cement (brownish) on grains with pore spaces filled with second generation meteoric drusy calcite cement, Ouanamane Formation (Jurassic), Morocco. (From Adams et al., 1984.) (For color version, see Plate 14.16, between pp. 408 and 409.)

Figure 14.17 Syntaxial calcite in optical continuity on an echinoderm spine. (Photo courtesy of Maria Simon-Neuser, with permission of LUMIC.) (For color version, see Plate 14.17, between pp. 408 and 409.)

where pore spaces are saturated with groundwater, three types of cement are common. The precipitation of **drusy calcite** cements (Figure 14.16) involves the nucleation on host grains of multiple crystals that grow outward into pore spaces to produce a fringe of crystals with relatively straight boundaries whose size increases away from the host grains. **Syntaxial calcite** cements (Figure 14.17) involve the precipitation of low magnesium calcite that nucleates in optical continuity with a low

magnesium calcite grain. These cements are especially common around single-crystal echinoderm skeletal particles. As in detrital rocks, **poikiloptic calcite** cement consists of a single crystal large enough to incorporate multiple grains during its growth.

Vadose zone cements display a very different geometry. As groundwater moves downward through the vadose zone toward the water table, its high surface tension causes it to adhere to the underside of grains and to the constricted spaces between grains. Continued precipitation on grain bottoms leads to the development of **pendant cements** that hang downward from the grain like a pendant on a necklace. Vadose zone water infiltrating between grains forms a meniscus, analogous to the one in a capillary tube or burette, from which precipitation of low magnesium calcite produces **meniscus cement** which may resemble an hour-glass suspended between adjacent grains.

In reality, the position of the water table varies both seasonally and over the long term so that carbonate rocks may spend multiple periods of time in both the vadose and phreatic zones. In addition, the position of the boundary or zone of mixing between marine and meteoric waters varies, especially between sea level high stands and low stands. By paying careful attention to the form and composition of calcium carbonate cements, carbonate workers can infer a complex set of processes that involve changes in pore water composition, nucleation history and cement growth in marine environments and in meteoric vadose and phreatic zones over long periods of time. For more details on calcium carbonate diagenesis, the reader is referred to works by Scholle and Ulmer-Scholle (2003), Tucker and Wright (1990), Friedman (1981), Longman (1980) and Bathurst (1975).

Carbonate sediments are also subject to recrystallization during diagenesis in which older crystals are dissolved and reprecipitated as new crystals of similar composition. In carbonate rocks the term **neomorphism** is used for processes in which new crystals form without a significant change in composition. If the crystals become larger, the neomorphism is called aggrading neomorphism; if they become smaller, it is called degrading neomorphism. The most frequently encountered neomorphism involves the progressive recrystallization of micrite into microspar and of microspar into sparry calcite by aggrading neomorphism. Such neomorphic sparry calcite is called **pseudospar** to distinguish it from directly precipitated calcite cements. Pseudospar can be recognized where neomorphism is incomplete because patches of dusty micrite and microspar remain. In original wackestones and floatstones, pseudospar can often be inferred because grains formerly supported by mud matrix are widely separated and seem to float unsupported in sparry calcite (pseudospar).

Dolomitization

Dolomite is far less common in modern carbonate sediments than it is in older carbonate sequences. This fact, along with abundant evidence for calcium carbonate replacement by dolomite, supports the hypothesis that most, if not all, dolostones are of diagenetic origin – produced by a process called **dolomitization**. Dolomite is especially abundant in rocks formed during sea level high stands such as those during the Ordovician–Devonian and Jurassic–Cretaceous periods. They are also abundant in the Precambrian. Only a few occurrences of dolomite represent primary dolomite precipitated in specific modern environments.

The virtual absence of primary dolomite is a long-standing conundrum. Because ocean water is oversaturated with respect to dolomite, its rarity in modern sediments suggests that something must inhibit its nucleation and block its precipitation from solution. Research suggests that dolomite precipitation from marine waters requires either elevated Mg/Ca ratios or decreases in dissolved sulfate (SO_4^{-2}). **Primary dolomite** forms in modern intertidal to supratidal zones in the subtropics where evaporation and subsequent precipitation of aragonite ($CaCO_3$) and gypsum ($CaSO_4 \cdot 2H_2O$) produce the necessary increase in Mg/Ca ratio and decrease in sulfate to initiate the precipitation of dolomite. It also forms where magnesium-rich groundwater discharges into lakes, raising the Mg/Ca ratio, and where sulfate-reducing bacteria lower sulfate concentrations in saline lakes (Wright and Wacey, 2004).

The origin of abundant diagenetic dolomites has been the subject of discussion and fervent controversy for more than a century (Machel, 2004). The abundance of secondary dolomite requires that large volumes of water move through limestones over time. Here we will discuss four types of dolomitization processes that have had a significant number of adherents in recent decades. One should remember that hybrids involving multiple processes may be significant.

1 It is well known that in warm, arid areas with high evaporation rates, evaporation of surface and/or groundwater leads to increased salinity and the precipitation of calcium sulfate minerals such as gypsum and anhydrite (see Section 14.3). The simultaneous removal of both calcium and sulfate from such waters raises their Mg/Ca ratios and lowers their dissolved sulfate concentrations, while the increased salinity increases the density of such waters. Various models show that such dense brines will percolate downward through carbonate sequences by a process known as **reflux** or **evaporative drawdown**. Dense brines form beneath shallow lagoons, sabkhas and larger evaporite basins and have the requisite composition to replace calcium carbonate minerals with dolomite. The reflux mechanism gains support from the occurrence of relatively young, diagenetic dolostones in areas where evaporative drawdown is known to occur.

2 In the constantly shifting zone of mixing (see Figure 14.13) that lies between marine and meteoric pore waters, pore waters with the appropriate composition for dolomitization can theoretically be produced. This zone would, over time, occupy a large area as it shifts during periods of prolonged sea level rise during which dolomitization is known to have been particularly common. Mixing meteoric water with seawater in proportions between 1:2 and 1:3 is sufficient to cause calcite to become soluble, while the Mg/Ca ratio remains high enough for dolomite to form. However, this process is severely limited by the fact that the dissolution of calcite rapidly lowers the Mg/Ca ratio so that dolomite is unlikely to precipitate. This process, popular in the 1970s and 1980s, is no longer considered as important in dolomitization as it once was.

3 Isotopic studies suggest that many dolostones have equilibrated with elevated temperatures (50–100°C) commensurate with depths of 500–2000 m (Tucker, 2001; Machel, 2004). This suggests deep circulation of dolomitizing fluids by some type of convective mechanism such as **Kohout convection** (Kohout, 1965; Tucker, 2001). The most common model for this process involves the pumping of cold seawater, undersaturated with respect to calcium carbonate and oversaturated with respect to dolomite, into limestones by some combination of tidal and ocean currents. Convection through the limestones is driven by some combination of heating from below and/or sinking of evaporative brines. Perhaps aided by sulfate-reducing bacteria (see below) and elevated temperature, this convection is believed by many to be the major agent in dolostone formation. Limestone permeability is essential to the effectiveness of convection-driven dolomitization. Because dolomite occupies a smaller volume than calcite, it is easy to envision a positive feedback mechanism in which permeability increases as dolomitization proceeds, assuring continued dolomitization.

4 Many subsurface diagenetic environments are reducing environments and some are anoxic. **Bacterial reduction of sulfate** tends to occur in such environments as bacteria utilize the oxygen in dissolved sulfate ions to decompose buried organic material. Such sulfate-reducing bacteria lower the dissolved sulfate contents of subsurface water, which should favor the replacement of calcite by dolomite. In conjunction with reflux or some type of deep convection, sulfate-reducing bacteria may play a very significant role in dolomitization.

The origin of secondary dolomite remains controversial. Currently some combination of evaporative reflux, deep convection and sulfate reduction by bacteria seems to provide the most promising explanation. Stay tuned!

14.2 EVAPORITES

Evaporites are sedimentary rocks that form by chemical precipitation from highly saline waters (brines) that have become oversaturated with respect to one or more dissolved solids as the result of evaporation. They are important cap rocks for petroleum reservoirs; in fact the US Strategic Petroleum Reserve is capped by evaporite deposits. The essential steps in the formation of evaporites are:

1 The presence of surface or shallow groundwater that contains dissolved solids.
2 Warm, dry conditions that permit net evaporation in which the progressive removal of water by evaporation exceeds the replenishment of water over time.
3 Sufficient net evaporation to progressively concentrate dissolved solids to the point where the water becomes oversaturated with respect to one or more dissolved solids.
4 Leading to the precipitation of one or more dissolved solids from saline brines in the form of evaporite minerals.

Because of their high solubility, evaporites are far more abundant in the subsurface (they underlie roughly 30% of the United States) than they are in outcrop, although the less soluble minerals are exposed at the surface in areas with arid climates. Still, the conditions under which evaporite minerals form are sufficiently limited in time and space that evaporite rocks constitute less than 1% of the sedimentary rock record. Evaporite rocks form from saline groundwater, saline lakes and highly saline, restricted seas. Evaporite rocks are climate sensitive in that the vast majority formed under warm, arid climatic conditions that promote net evaporation. These conditions most commonly occur at subtropical latitudes between 10° and 30° from the equator and in rain shadows at somewhat higher latitudes. Most evaporites have formed in enclosed or restricted basins in which replenishment of water from outside sources, such as the open ocean, rainfall, groundwater inflow and river runoff is restricted. Evaporites dominate sedimentary sequences only where the influx of detrital sediment is low relative to evaporite precipitation.

14.2.1 Marine evaporites

Marine evaporites form where marine water undergoes extensive net evaporation to become hypersaline brine. Continued evapo-

Table 14.5 Major marine evaporite minerals and their relative abundance.

Mineral	Chemical composition	Abundance
Anhydrite	$CaSO_4$	Abundant
Bischofite	$MgCl_2 \cdot 6H_2O$	Scarce
Carnellite	$KMgCl_3 \cdot 6H_2O$	Common
Gypsum	$CaSO_4 \cdot 2H_2O$	Abundant
Halite	$NaCl$	Abundant
Kainite	$KMg(SO_4)Cl \cdot 3H_2O$	Common
Keiserite	$MgSO_4 \cdot H_2O$	Common
Langbenite	$KMg_2(SO_4)_3$	Scarce
Polyhalite	$K_2Ca_2Mg(SO_4)_4 \cdot 2H_2O$	Common
Sylvite	KCl	Common

Table 14.6 Precipitation sequence of common evaporite minerals from seawater.

Mineral	Evaporation (%)	Salinity (ppt)	Density (g/cm^3)
Seawater	~0	~35	1.04
Calcite	>50	>70	1.08
Gypsum	>75	>135	1.14
Halite	>90	>350	1.21
Potassium and magnesium minerals	>96	>750	1.27

ration leads to progressive concentration of dissolved solids in the remaining water. Under such conditions, hypersaline brines can become oversaturated with respect to one or more minerals, which leads to their precipitation from solution as evaporite minerals. Literally hundreds of minerals, both primary precipitates and secondary products of diagenetic replacement and recrystallization have been reported from marine (and non-marine) evaporites. Most are relatively rare. The major halide and sulfate minerals in marine evaporites are listed in Table 14.5, along with their relative abundance in the context of marine evaporite rocks.

All minerals possess different degrees of solubility that depend on the nature of the solution and other environmental conditions. When a body of water undergoes evaporation, increasing water salinity and density, the least soluble minerals are precipitated first and progressively more soluble minerals are precipitated later in a specific sequence. The conditions and sequence of crystallization for average seawater are well known (Table 14.6). As seawater evaporates the following sequence occurs: (1) a small amount of calcite ($CaCO_3$) is precipitated when about 50% of the water has evaporated; (2) gypsum ($CaSO_4 \cdot 2H_2O$) begins to precipitated at 75% evaporation and precipitates as evaporation continues; (3) halite ($NaCl$) precipitation is initiated at 90% evaporation; and (4) a variety of potassium and magnesium sulfates and halides (e.g., carnellite) are precipitated at >96% evaporation from very dense brines.

Additions of saline terrestrial water in various proportions can change both the minerals that precipitate and their sequence of crystallization, which helps to explain why lacustrine evaporite mineralogy is quite different from that in marine evaporites.

Modern marine evaporates precipitate in warm, arid, shallow, marine lagoons and/or sabkhas on a relatively small scale. **Sabkha evaporites** form in areas of very low relief called sabkhas that occur along arid coastal plains in the transition zone between marine and non-marine environments. The best known sabkha occurs along the Trucial Coast of the Persian Gulf, but many other examples are known from the subtropics. Where the influx of siliciclastic detritus is relatively small, carbonate rocks dominate and evaporites form in the upper intertidal and supratidal zones (Figure 14.18). The essential processes involve the evaporation of pore waters that are of either marine or mixed marine–terrestrial origin. As the water evaporates during extensive dry periods, groundwater is transformed into hypersaline brines from which gypsum and/or anhydrite are precipitated.

Much of the gypsum is precipitated at the top of the water table as bladed gypsum or as gypsum rosettes; some is precipitated in the vadose zone as nodules (Figure 14.19a). Higher temperatures and salinities favor the dehydration of gypsum to anhydrite. These conditions typically exist in the middle–upper portions of the supratidal zone, farther from normal marine influences. Here much of the gypsum is replaced by anhydrite or polyhalite, a process that involves a substantial decrease in volume. This leads to the formation of

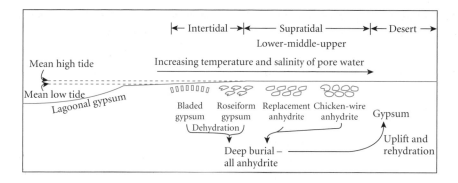

Figure 14.18 Environments of evaporite formation in modern sabkhas and the conversions between gypsum and anhydrite.

(a)   (b)

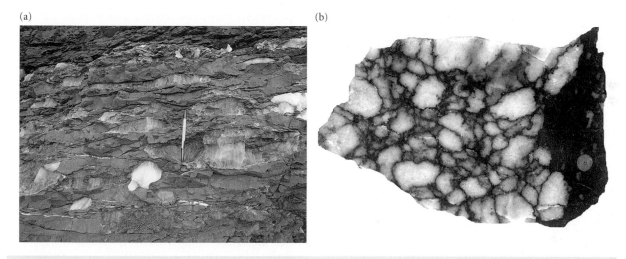

Figure 14.19 (a) Nodular gypsum, Triassic Mercia Group, Watchet Beach, England. (Photo courtesy of Nicola Scarselli.) (b) Chicken-wire anhydrite, Carboniferous, Belgium. (Photo courtesy of Frederic Boulvain.) (For color version, see Plate 14.19, between pp. 408 and 409.)

nodular or **chicken-wire anhydrite** (Figure 14.19b), so-called because the anhydrite nodules are enclosed in stringers of calcareous or siliciclastic sediment that resembles chicken-wire fencing. During burial, anhydrite generally remains the stable calcium sulfate mineral, but as rocks approach the surface in areas with sufficient rainfall, anhydrite is commonly hydrated into gypsum.

Very **large marine evaporite sequences** with thicknesses of 10^3–10^4 m and aerial extents of 10^3–10^4 km^2 are abundant in the geological record. Large evaporite sequences have been deposited in shallow, subsiding cratonic basins in which water depths were generally less than a few tens of meters. Examples include the Paleozoic Williston Basin centered

in North Dakota and the Silurian Michigan Basin. Other large evaporites have been deposited in foreland basins of variable depth, including the Silurian Salina Group of New York and Pennsylvania in the Taconic foreland basin and the Pennsylvanian–Permian Paradox Group in a foreland basin of the Ancestral Rocky Mountains. Still other large evaporite sequences were deposited in rift basins, e.g., the Gulf Coast Basin and the proto-Atlantic Ocean Basin, of Jurassic to Cretaceous age. A few large evaporite sequences such as those in the Permian Delaware Basin in west Texas were deposited in deep basins associated with the irregular closing of ocean basins at convergent plate boundaries (Box 14.2).

Box 14.2 When the Mediterranean dried up!

In the 1960s, oceanographers using seismic reflection profilers detected a mysterious reflecting horizon known as the "M" layer in the sediments under the deeper parts (>2000–4000 m) of the Mediterranean Basin. In the summer of 1970, the Deep Sea Drilling Project sampled sedimentary rocks from the same area and a startling discovery was made (Hsu, 1983). The "M" layer contains sequences up to 2000 m thick of 5.3–6.0 Ma Miocene evaporite rocks that contain anhydrite, halite, gypsum and rarer evaporite minerals. These evaporites accumulated as isolated sequences in the deeper parts of the Mediterranean Basin (Figure B14.2). In addition, these evaporites are intimately associated with stromatolite-bearing carbonate rocks that were clearly deposited in shallow water. The deep parts of the Mediterranean Basin were occupied by shallow water so saline that evaporites were precipitated. For nearly 700,000 years, the Mediterranean Sea had, at least periodically, dried up! What caused this to occur?

Prior to 6.0 Ma, one or more connections existed between the Atlantic Ocean and the Mediterranean Sea, in the area now occupied by the Straits of Gibraltar between Spain and Morocco. As long as a connection existed, any Mediterranean water removed by evaporation was replenished by the inflow of normal salinity water from the Atlantic Ocean. Just after 6.0 Ma, as Africa and Spain converged, causing uplift, the Atlantic–Mediterranean connection was apparently closed and a barrier formed to produce a restricted sea. Barrier formation was possibly aided by a 50 m fall in sea level caused by Miocene glaciation in Antarctica. Once the connection was closed, the results were inevitable. Because of the warm, arid climate, high evaporation rates removed large amounts of water from the sea surface. The low rainfall and river runoff in the region and leakage through the barrier were frequently insufficient to replenish the water lost by evaporation. Several separate saline seas formed in the deepest parts of the Mediterranean Basin while terrestrial sediments were spread across the former sea floor. Gypsum/anhydrite were precipitated from saline brines over large areas when net evaporation exceeded 70% and halite and rarer evaporites were precipitated over smaller areas when net evaporation exceeded 90%. The result is the bull's-eye pattern of evaporite minerals seen in Figure B14.2. The Atlantic–Mediterranean connection was re-established through the Straits of Gibraltar around 5.3 Ma; a giant waterfall formed and the Mediterranean basin filled rapidly.

The volume of evaporites formed in this brief interlude (5.3–6 Ma) requires the evaporation of at least 40 times the volume of the modern Mediterranean Sea. Scientists now have a recent example of the large marine evaporite basins that have formed periodically during geological history.

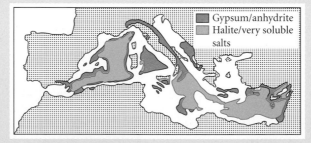

Figure B14.2 Distribution of Miocene evaporates in the Mediterranean Basin.

There are no modern analogs for these large ancient marine evaporite deposits. Most appear to have been formed in large barred basins (Figure 14.20). Some of these basins were shallow (<100 m deep) and some were very deep (>1000 m). The depth of water in the deep basins has been the subject of considerable debate with some supporting generally deep water/deep basin models (Figure 14.20b) and others supporting shallow water/deep basin models (Figure 14.20c).

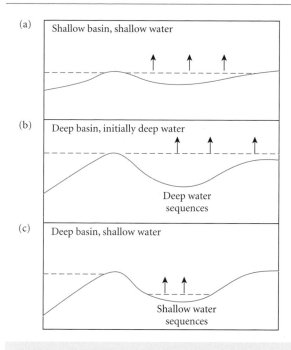

Figure 14.20   Models for large barred basin evaporite formation: (a) shallow basin, (b) deep basin with deep water, (c) deep basin with shallow water.

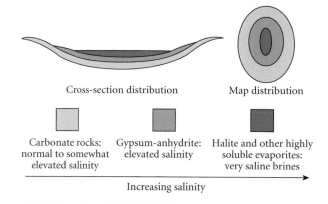

Figure 14.21   Cross-section and map view of an idealized evaporite basin. The bull's-eye map pattern shows that marine carbonates surround gypsum deposits, which in turn surround a halite bull's-eye, reflecting increased salinity and decreased aerial coverage as seas underwent increasing isolation and evaporation.

The volume of evaporites in these basins is far too large to be explained by the simple evaporation of a body of water that filled them at one time. In many cases, the volume of evaporites implies the evaporation of volumes of seawater that exceeded maximum basin volume by factors of 10–100. This can only have been accomplished by repeated dessication, and flooding of such basins or by gradual replenishment of seawater through "leaky" barriers and/or the inflow of saline continental groundwater.

Thick sequences of anhydrite and gypsum indicate prolonged periods when salinities hovered in the appropriate range for continued precipitation of calcium sulfate. This suggests a balance between evaporation and inflow, perhaps through leaky barriers, that is consistent with the large volumes of evaporites produced. Thick sequences of halite, carnellite and other highly soluble minerals imply severe dessication concentration of dissolved solids in dense brines. As is to be expected, these minerals are deposited from waters that covered smaller areas than those that precipitated gypsum and anhydrite so that the depos-

its form a bull's-eye pattern (Figure 14.21) with normal salinity carbonates and sabkha evaporites surrounding elevated salinity laminated sulfates, which in turn surround higher salinity deposits of halite and other highly soluble evaporites (Box 14.2).

Because evaporite minerals are highly soluble in most groundwater systems, evaporites are rare in Precambrian sequences. In those few instances where they do occur, they are represented by the least soluble evaporite minerals such as anhydrite and gypsum. Most evaporite rocks are Phanerozoic and halides and other highly soluble evaporites are known only from rocks of this age. As might be expected in rocks composed of highly soluble minerals, the diagenesis of evaporite rocks is exceedingly complex and can involve multiple replacement and neomorphism events. During burial, gypsum commonly loses water and dehydrates into anhydrite during eodiagenesis. The reverse process, the hydration of anhydrite to gypsum, commonly occurs during telodiagenesis as evaporite sequences approach the surface. Another very important aspect of evaporite diagenesis involves the tendency of halite and other soluble halides to become extremely mobile when subjected to elevated temperatures and confining stresses during diagenesis. Under such conditions, large masses of halide minerals may begin to flow upward by plastic flow as buoyant masses called salt **diapirs** or **salt domes**. A well-known set of examples occurs in the Gulf of Mexico and is associated with the largest petroleum reserves in the United States.

14.2.2   Lacustrine evaporites

Lakes are filled with water that contains dissolved solids. Lakes located in interior basins and rift valleys with warm, arid climates suffer a fate similar to that of restricted seas. Small, playa lakes are seasonal, filling during rainy periods and evaporating during the dry season to leave thin crusts of evaporite minerals. Larger lakes can, over longer periods of times, produce more extensive evaporite deposits. In either case, lacustrine evaporite deposits commonly reveal a bull's-eye pattern in which extensive deposits of the least soluble minerals precipitated first when the lake occupied a larger area, with ring deposits of progressively more soluble minerals deposited as the lake shrinks to a progressively smaller area occupied by progressively more saline waters from which more soluble minerals are finally precipitated.

Because their water is derived from local interior drainage via surface water and groundwater, lakes tend to have more variable dissolved solid compositions, which reflect local source rock compositions, compared to marine waters. As a result, it is more difficult to generalize about lacustrine evaporite mineralogy than about marine evaporites. Depending on local conditions, lacustrine evaporites may be dominated by carbonates, sulfates, borates or even nitrates. Halides, such as halite ($NaCl$), although present locally, tend to be less abundant than they are in marine evaporite sequences. Common evaporite minerals include the carbonate minerals calcite ($CaCO_3$), aragonite ($CaCO_3$), magnesite ($MgCO_3$) and dolomite [$CaMg(CO_3)_2$]. Common sulfate minerals include gypsum ($CaSO_4 \cdot 2H_2O$) and anhydrite ($CaSO_4$). In addition to these minerals, the major non-marine carbonate, sulfate and borate evaporite minerals that form in lacustrine evaporite sequences are those listed in Table 14.7. The minerals shown with asterisks are excellent indicators of lacustrine evaporites.

Table 14.7 Listing of some important lacustrine evaporite minerals.

Mineral	Chemical composition
Bloedite	$Na_2SO_4 \cdot MgSO_4 \cdot 4H_2O$
Borax*	$Na_2B_4O_5(OH)_4 \cdot 8H_2O$
Colemanite	$CaB_5O_4(OH)_3 \cdot H_2O$
Epsomite*	$MgSO_4 \cdot 7H_2O$
Gaylussite*	$CaCO_3 \cdot Na_2CO_3 \cdot 5H_2O$
Glauberite*	$CaSO_4 \cdot Na_2SO_4$
Kernite	$Ca_2B_4O_6(OH)_2 \cdot 3H_2O$
Mirabilite	$Na_2SO_4 \cdot 10H_2O$
Natron*	$NaCO_3 \cdot 10H_2O$
Nahcolite	$NaHCO_3$
Thenardite	Na_2SO_4
Trona*	$NaHCO_3 Na_2CO_3 \cdot 2H_2O$
Ulexite*	$Na_2B_4O_5(OH)_4 \cdot 3H_2O$

* Excellent indicators of lacustrine evaporates.

Older North American examples of lacustrine evaporites include the sodium-rich trona–halite deposits in the Eocene Wilkins Peak Member of the Green River Formation in southwestern Wyoming and the Jurassic Lockatong and New Berlin Formations deposited in rift graben lakes in New Jersey and Connecticut.

14.3   SILICEOUS SEDIMENTARY ROCKS

Siliceous sedimentary rocks occur in a wide variety of forms created in many different settings over the past 3.8 Ga. They consist primarily of silica (SiO_2) minerals of the chert family (Chapter 5), such as (1) cryptocrystalline–microcrystalline–macrocrystalline **quartz**, (2) cryptocrystalline, radial-fibrous **chalcedony**, and (3) amorphous **opal**. These silica minerals combine to form hard rocks with conchoidal fracture, collectively referred to as **chert**.

Modern siliceous sediments are composed of the remains of diatoms, radiolaria and/or silica flagellates that secrete microscopic shells of amorphous opaline silica ($SiO_2 \cdot nH_2O$). Such deposits commonly form on modern sea floors beneath areas of upwelling, below the carbonate compensation depth (CCD) and where the influx of detrital sediments is minimal (Figure 14.22). Upwelling brings dissolved silica into surface waters where microorganisms extract it to secrete opaline silica

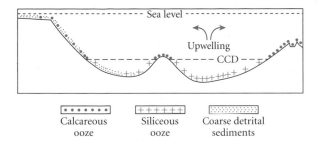

Calcareous ooze   Siliceous ooze   Coarse detrital sediments

Figure 14.22 Conditions under which modern siliceous oozes accumulate below the carbonate compensation depth (CCD), beneath the productive areas of upwelling, where detrital influx is minimal.

(a)

(b)

Figure 14.23 Bedded "ribbon" chert in outcrops. (a) Radiolarian chert, Cretaceous mélange, Marin County, California. (Photo courtesy of Steve Newton.) (b) Ribbon chert, Ordovician, Norway. (Photo courtesy of Roger Suthren.) (For color version, see Plate 14.23, between pp. 408 and 409.)

shells that sink to form a significant portion of the sediment in areas where little detrital and carbonate sediment accumulates. During diagenesis, opaline silica shells are converted to a partially crystalline form of silica called opal-CT and perhaps eventually to chert. Despite the abundance of recent siliceous sediments on the sea floor, no incontrovertible chert equivalents have been recovered by deep sea drilling.

Two contrasting groups of chert occurrences dominate siliceous sedimentary rocks: bedded cherts and nodular cherts. **Bedded cherts** occur as layers in sequences up to several hundred meters thick in which chert is most commonly interstratified with mudrocks to form **ribbon rocks** (Figure 14.23). Most bedded cherts formed in shelf to deeper water marine environments, below the wave base along active margins and in pull-apart basins associated with transform margins. Most bedded cherts are interlayered with mudrocks, but a few are interlayered with turbidite sandstones and even carbonates. Bedded cherts are also abundant in association with Precambrian banded iron formations (BIF) and with phosphorites, as discussed later in this chapter. They are also a significant part of the sedimentary portion of many ophiolite and alpine mélange sequences (Chapters 10 and 18).

Nodular cherts occur as ellipsoidal to bulbous to irregular masses less than 1 m in length that tend to be elongate, parallel to and concentrated in certain strata (Figure 14.24). Although they occur in many different lithol-

ogies, the vast majority of nodular cherts occur in carbonate rocks such as limestone and dolostone that formed in supratidal to subtidal marine environments. Some nodules are concentrically zoned, others are massive and still others preserve the structures of the enclosing rocks. In thin section, primary carbonate grains such as ooids, fossils and pellets are observed to have been replaced by silica minerals. The preservation of primary structures and carbonate grains as silica in nodular cherts strongly suggests that most nodular cherts are secondary and form by the replacement of pre-existing, shallow water carbonates during diagenesis. For this reason, many

(a)

(b)

Figure 14.24 Chert nodules. (a) Flint nodules in Kalkberg limestone, Devonian, New York. (b) Flint in chalk cliffs, Cretaceous, Seven Sisters, East Sussex, England. (Photos by John O'Brien.) (For color version, see Plate 14.24, between pp. 408 and 409.)

nodular cherts are referred to as **secondary** or **replacement cherts**.

Knauth (1979) pointed out that the formation of nodular cherts by limestone replacement requires pore waters supersaturated with respect to silica and undersaturated with respect to calcium carbonate. These conditions would favor the dissolution of calcium carbonate and the simultaneous precipitation of silica, especially where the pH is decreasing. Because alkaline marine surface waters are highly undersaturated with respect to silica and nearly saturated with respect to calcium carbonate, it is unlikely that chert

nodules form from marine pore waters. Such conditions might occur under several sets of diagenetic conditions that involve meteoric pore waters enriched in silica from the dissolution of silicate minerals or silica skeletal particles. Wherever silica-rich, calcium carbonate-undersaturated meteoric or mixed continental–marine waters occur the necessary conditions of reduced pH and undersaturation with respect to calcium carbonate can be created. Most models for the formation of chert nodules involve carbonate replacement by opal or cryptocrystalline quartz during eodiagenesis in the meteoric zone or the zone of mixing.

Siliceous sediments also form in smaller amounts in alkaline lakes, because the solubility of silica rapidly decreases at elevated pH (>9), and as siliceous sinter precipitated around hot springs (Figure 14.25), because the solubility of silica rapidly decreases as spring water cools.

14.4 IRON-RICH SEDIMENTARY ROCKS

Iron-rich sedimentary rocks contain more than 15% iron by weight and occur in rocks that range in age from 3.8 Ga to the present. As the major source for iron ore, used in the manufacture of all iron and steel products, they represent a significant economic resource. Several different types of iron-rich sedimentary rocks occur including (1) Precambrian iron formations, (2) Phanerozoic ironstones, and (3) smaller accumulations of bog iron, iron-rich manganese nodules and pyrite-rich black shales. Table 14.8 summarizes the significant differences between Precambrian iron formations and Phanerozoic ironstones.

14.4.1 Precambrian iron formations

Most, but not all, Precambrian iron formations contain iron-bearing minerals interlayered with siliceous sediments from the chert family in a way that gives them a banded appearance in outcrop (Figure 14.26). These distinctive laminated to thinly bedded iron formations are called **banded iron formations (BIF)** and contain 60% of the world's iron ores (Wenk and Bulakh, 2004).

The major iron-bearing minerals (Table 14.8) vary and include: (1) iron oxides such as hematite and magnetite, (2) iron carbon-

(a)

(b)

Figure 14.25   (a) Siliceous sinter precipitated around hot springs, Yellowstone National Park, Wyoming. (Photo by Kevin Hefferan.) (b) Siliceous sinter terraces, Mammoth Hot Springs, Yellowstone National Park. (Photo by John O'Brien.) (For color version, see Plate 14.25, between pp. 408 and 409.)

Table 14.8   Common minerals in Precambrian and Phanerozoic iron-rich sedimentary rocks.

Mineral group	Mineral	Formula	Iron formations	Ironstones
Iron oxides and hydroxides	Hematite	Fe_2O_3	Abundant	Common
	Magnetite	$FeFe_2O_4$	Abundant	Rare
	Goethite	$FeOOH$		Abundant
Iron silicates	Stilpnomelane	$K(Fe,Mg,Fe)_8(Si,Al)_{12}(O,OH)_{27}\cdot nH_2O$	Common	
	Greenalite	$(Fe,Mg)_6Si_4O_{10}(OH)_8$	Common	
	Minnesotaite	$(Fe,Mg)_3Si_4O_{10}(OH)_2$	Common	
	Riebeckite	$Na_2Fe_5Si_8O_{22}(OH)_2$	Scarce	
	Chamosite	$(Fe,Mg,Fe)5AlAlSi_3O_{10}(O,OH)_8$		Common
Iron carbonates	Siderite	$FeCO_3$	Common	Common
	Ankerite	$FeMg(CO_3)_2$		
Iron sulfides	Pyrite	FeS_2	Common	Common
	Pyrrhotite	$Fe_{1-x}S$	Scarce	
Common associated minerals	Chert group	Microcrystalline SiO_2	Abundant	Rare
	Calcite	$CaCO_3$	Rare	Common
	Dolomite	$CaMg(CO_3)_2$	Scarce	Common
	Collophane	$Ca_5(PO_4)_3(F,OH)$		Common

ates such as siderite and ankerite, and (3) iron silicates such as the iron-rich serpentine mineral greenalite, the iron-rich chlorite group mineral stilpnomelane and iron-rich talc group mineral minnesotaite.

Banded iron formations are common only in Archean and early Proterozoic rocks that range in age from 3.8 to 1.8 Ga, with a peak in abundance between 2.5 and 2.2 Ga. After a hiatus of 1.0 Ga, a few examples occur in late Proterozoic rocks formed from 0.8 to 0.5 Ga. Because there are no known modern analogs for these Precambrian rocks, their origin remains both controversial and enigmatic. What were the conditions that permitted widespread iron formations to accumulate during the Precambrian, but not during the Phanerozoic? Why did they reappear briefly after a hiatus of a billion years?

Two major types of iron formation, each dominant during different parts of the Precambrian, are commonly recognized. **Algoma-type banded iron formations** (after deposits in the Algoma District, Ontario, Canada)

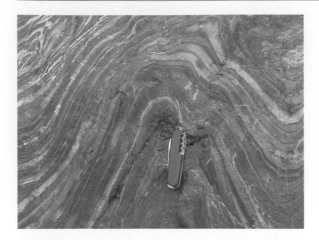

Figure 14.26   Banded iron formation, Vermillion Range, Minnesota. (Photo by Kevin Hefferan.) (For color version, see Plate 14.26, between pp. 408 and 409.)

dominate Archean iron-rich sedimentary rocks formed between 3.8 and 2.6 Ga. These iron formations tend to occur as fairly thin (<10–100 m), elongate lenses of limited lateral extent (<1–10 km) that occur within Archean greenstone belts. The iron-rich bands are composed almost exclusively of extremely fine-grained rocks called **femicrites**, by analogy with fine-grained carbonate micrites. In addition to interlayered chert group minerals, Algoma-type BIF are associated with submarine ultramafic–mafic volcanic rocks, mudrocks and sparse volcanoclastic "greywacke" sandstones. These are inferred to have been formed in fairly deep water marine environments, perhaps in forearc or backarc basin or advanced intracratonic rift settings (Kappler et al., 2005), well below the wave base, and with minimal influx of detrital sediments.

Superior-type iron formations (after deposits near Lake Superior, Minnesota) dominate Proterozoic iron-rich sedimentary rocks formed between 2.6 and 1.8 Ga and again from 0.8 to 0.5 Ga. Although their mineralogy is similar to that of Algoma-type formations, their size, textures, associations and inferred depositional environments are quite different. Lake Superior-type iron formations tend to be much larger than Algoma-type formations. Their thickness is commonly 100–1000 m, they occur in broad belts up to

100–1000 km wide with aerial extents of as much as 10^5 km². Although they contain BIF femicrites similar to those in Algoma-type sequences, they also contain **granular iron formations (GIF)** with ooids, pisoliths, intraclasts and pellets similar to those in shallow water carbonate sequences (Blatt et al., 2006). Stromatolites and abundant traction current features such as ripple marks and cross-strata further attest to shallow water conditions under which many GIF were formed. Lake Superior-type GIF occur in a regional context that includes quartzarentites, limestones, dolostones and mudrocks. Taken together, the features suggest that GIF formed in fairly shallow water environments on passive margin continental shelves, with limited influx of detrital sediments. Their abundance increases in younger Superior-type sequences and they appear to be associated with transgressive episodes associated with rising sea level. A substantial change in the conditions of iron formation production seems to have occurred during the transition from the Archean into the Proterozoic. What changes occurred during this transition to account for these differences?

Perhaps because of the lack of modern analogs, many intriguing hypotheses have been put forth for the origin of both Algoma-type and Superior-type iron formations. Such hypotheses must explain the origin of the iron-rich minerals, the origin of the interlayered chert, their banded nature, and the physical and temporal contrasts between the two types of iron formation. They must also explain why their occurrence is largely restricted to rocks older than 1.8 Ga and why they formed in the period 0.8–0.5 Ga after a hiatus of a billion years. Over the past two decades a broad consensus has developed concerning some aspects of their formation, but the details, especially concerning the role of microbes, remains controversial (Nealson and Myers, 1990; Klein and Buekes, 1992; Simonson, 2003; Kappler et al., 2005; Klein, 2005).

Several lines of evidence suggest that the Achaean–Proterozoic atmosphere was depleted in free oxygen (O_2 gas), which should have depleted the oceans of dissolved oxygen as well. Oxidized hematite-bearing soils and red hematite-cemented sandstones do not occur until 1.8 Ga, at about the same time that reduced detrital pyrite and uraninite (unstable in the presence of free oxygen) disappear from the geological record. Although correlation is not cause, it is likely that the scarcity of oxygen in the pre-1.8 Ga atmosphere played some role in the conditions under which iron formations were produced.

Early workers (e.g., James, 1954) suggested a continental source for the iron. They argued that an early atmosphere, depleted in O_2 and enriched in CO_2, provided the ideal conditions for iron enrichment in ocean water. Under such conditions, more iron would have been leached from the continents and, because ferrous iron is highly soluble in acidic water if free oxygen is not present, large quantities of dissolved iron would have resulted. More recent work (Simonson, 2003) suggests that surface waters at this time contained too much oxygen for ferrous iron to be transported in solution in sufficient quantities. A new consensus has developed, supported by trace chemical and isotopic signatures, that the iron originated largely as dense brines from hydrothermal vents on the sea floor (Klein and Buekes, 1992; Simonson, 2003; Klein, 2005). This certainly makes sense for the deep water Algoma-type formations whose elongate character might well be related to hydrothermal systems associated with elongate fracture systems in the sea floor. The source of silica for the cherts presents a separate problem. Silica solubility [as $Si(OH)_4$] is remarkably independent of acidity for pH ranges between 2 and 10. However, in colloidal form, the solubility of silica increases rapidly at pH levels from 4 to 5. In an acidic atmosphere silica would be more strongly leached from continental source rocks producing acidic oceans with far higher dissolved silica contents than modern oceans. Such oceans might on occasion become oversaturated with respect to dissolved silica, leading to its precipitation. The stage is set for banded iron formation; the remaining questions concern the mechanisms by which they originated.

Klein and Buekes (1992) proposed a widely accepted model in which Archean–early Proterozoic oceans were chemically stratified, with dense masses of anoxic, acidic, deep, ferrous iron-rich waters separated by a **chemocline** from less dense, more oxidized surface waters with little dissolved iron. Support for a stratified ocean comes from carbon isotope data that strongly suggest that Archean–early Proterozoic oceans were stratified with respect to carbon in both carbonates and organic matter (Klein, 2005). A variation on this model (Cameron, 1983; Simonson and Hassler, 1996, cited in Simonson, 2003) suggests that maximum dissolved iron concentrations occurred well above the bottom, just below the chemocline. This would certainly help to explain the origin of Superior-type iron formations in shallower water environments.

Periodic mixing of higher Eh and pH surface water with lower Eh and pH deeper water – caused by changes in the position of the chemocline, overturning or upwelling of deep, iron-rich waters onto continental margins – could account for the cyclic precipitation of iron-bearing minerals and chert. Iron oxide minerals would develop under the highest Eh conditions, iron carbonates under the highest pH conditions and iron sulfides under the lowest Eh conditions. Hamade et al. (2003) proposed that BIF resulted from alternations between fluxes of silica derived from the continents and ferrous iron derived periodically from hydrothermal vents. Nealson and Myers (1990) suggested that bacteria played a significant role in the precipitation of both silica and iron oxides. The presence of stromatolites in many GIF-type iron formations suggests that cyanophyte, photosynthetic bacteria may have produced the oxygen necessary to cause precipitation of iron in certain shallow water Superior-type deposits. Kappler et al. (2005) claimed that anoxic, photosynthetic bacteria, living in deeper water than cyanophytes, produced the oxygen that caused the precipitation of iron in BIF-type deposits. Posth et al. (2008) have also argued for a fundamental role for cyanophytes in producing the oxygen that led to the huge BIF deposits of the early Proterozoic. They have argued that microbial-induced precipitation of iron occurred during periods of high ocean temperatures and alternated with non-microbial precipitation of silica at lower temperatures. Yet as Klein (2005) noted, direct evidence for a microbial role in the origin of BIF remains elusive. Researchers continue to address these issues.

The reappearance of iron formations in 0.8 to 0.5 Ga rocks corresponds roughly to the time of the controversial Neoproterozoic "snowball Earth" (0.75 to 0.58 Ga), when it has been suggested (Hoffman et al., 1998) that the world's oceans were frozen over for a long period of time and glaciers reached within 10° of the equator. A surface layer of ice on the oceans would have impeded the exchange of oxygen with the atmosphere, leading to the development of an anoxic ocean. The occurrence of what appear to be glacial dropstones in some BIF of this age are certainly intriguing. A snowball Earth would explain the reappearance of BIF at this time, but the theory remains highly controversial.

14.4.2 Phanerozoic ironstones and other iron-rich rocks

Phanerozoic iron-rich rocks are quite different from their Precambrian iron-rich counterparts in terms of mineralogy, texture, scale, associated rock types and the environment and processes under which they accumulated. The most widespread type of Phanerozoic iron-rich sedimentary rock is called **ironstone**. Most ironstone deposits are quite thin (<20–30 m thick) and are composed predominantly of goethite and hematite, with smaller amounts of the iron-bearing chlorite mineral chamosite. Siderite, magnetite or pyrite occur in some deposits. Ironstones are commonly associated with, even gradational into, carbonate rocks such as limestone or dolostone and in many cases the iron minerals appear to replace carbonate grains. Quartz-rich sandstone, phosphorites and/or cherts are less common associates. Sedimentary textures and structures, such as cross-strata, ripples, erosional scours and abraded fossils, indicate accumulation in shallow water (<100 m), well-oxidized, marine environments.

Ironstones range in age from Cambrian through mid-Tertiary. The vast majority of them formed during two periods of peak ironstone formation, Ordovician–Devonian and Jurassic–Cretaceous, which correspond to periods of maximum global warming and sea level high stands. Periods of ironstone formation are associated with significant marine transgressions. The clear association of Phanerozoic ironstones with periods of global warming supports the idea that the iron originated from the erosion of lateritic soils formed by weathering in warm, humid,

Figure 14.27 Manganese nodule. (Photo by John O'Brien.) (For color version, see Plate 14.27, between pp. 408 and 409.)

tropical climates that promoted the leaching of iron by acidic soil waters. To avoid oxidation, the iron would have to be delivered as colloidal ferrous iron attached to tiny clay particles. Once in the marine environment, the iron could then be mobilized to replace carbonate grains and/or to be precipitated as primary goethite, as hematite in near shore environments or as chamosite farther offshore.

Additional types of Phanerozoic iron formations occur, all of which have modern analogs. **Bog iron deposits** consist largely of goethite, siderite and the manganese oxide minerals psilomelane and pyrolusite. They form where acidic groundwater delivers ferrous iron into swamps and lakes where it is oxidized and precipitated as surface crusts. Polyminerallic **manganese nodules** (Figure 14.27) and encrustations form under oxidizing conditions on the sea floor where iron and manganese oxide minerals precipitate along with copper, cobalt, nickel and other metals in the form of concentric nodules precipitated about a nucleus or as layered encrustations on rock outcrops. Some black shale deposited under anoxic conditions (Chapter 13) contains sufficient pyrite to be classified as iron-rich sediment. Hydrothermally precipitated chimneys around "black smoker" hot springs on the sea floor commonly contain significant amounts of iron-bearing sulfide minerals. The black color of such hot springs results from finely divided, opaque sulfide minerals, such as pyrite (FeS_2) and chalcopyrite ($CuFeS_2$).

Although minor rock types compared to detrital sedimentary rocks, carbonates and even evaporites, iron-rich sedimentary rocks continue to intrigue with their variety, the mysteries surrounding their formation and their importance as significant sources of ore for the manufacture of steel.

14.5   SEDIMENTARY PHOSPHATES

Phosphatic materials, in the form detrital **apatite** [$Ca_5(PO_4)_3(OH,F,Cl)$], bones and teeth composed of **hydroxyapatite** [$Ca_5(PO_4)_3(OH)$], cryptocrystalline **fluorapatite** [$Ca_5(PO_4)_3(F)$] and amorphous **collophane** are minor components of many sedimentary rocks. Much rarer are phosphate-rich rocks called **phosphorites** that contain more than 50% phosphate minerals and/or 20% phosphate by weight. Because phosphate is an important nutrient element for organic synthesis, these deposits have been extensively mined as sources of phosphate fertilizers.

The major occurrence of phosphorites is in the form of laminae and beds of cryptocrystalline fluorapatite and amorphous collophane interlayered with carbonates, siliceous sediments and detrital mudrocks (Figure 14.28). The phosphatic strata appear nearly black in outcrop; collophane is medium to dark brown and isotropic in thin section. The constituents of these phosphorites are similar

to those of carbonate rocks and include mud particles and grains such as ooids, fossils, peloids and clasts. In some cases the phosphate partially or totally replaces earlier carbonate grains, whereas in other cases its origin is less clear. Like carbonates, phosphorites tend to accumulate in areas where the influx of detrital sediment is minimal, thus ensuring that phosphate content is high.

Modern examples of such phosphorites occur mostly as crusts near the sediment–water interface, in fairly shallow (30–500 m) tropical to subtropical waters (latitude <40°), beneath areas of upwelling. The upwelling of cold deep waters brings substantial nutrients such as phosphate to near surface waters, which increases biological productivity so that phosphate is incorporated into organic matter. As these organisms die, the organic matter drifts toward the ocean floor where bacterial decomposition releases phosphate to the bottom and pore waters while utilizing sufficient oxygen to lower their oxidation–reduction potential (Eh). The details of phosphate precipitation and replacement remain uncertain, but most current work focuses on the role of various bacteria on and within the sediment in creating the geochemical conditions that allow phosphate to form by some combination of eodiagenetic replacement and/or precipitation.

Another significant occurrence of phosphorites is in so-called bone beds where various forms of apatite have been concentrated in **placer deposits** of bones, teeth and other phosphatic material formed as lag deposits, where finer constituents have been removed by currents, often over long periods of time. Yet another occurrence of phosphorites, rapidly vanishing due to mining, is as **guano deposits**, formed where bird colonies have generated thick accumulations of phosphate-rich fecal material over long periods of time.

Figure 14.28 Laminated phosphorites from the Miocene, Monterrey Formation, California. (Photo courtesy of Esther Arning.)

14.6   CARBON-RICH SEDIMENTARY ROCKS AND MATERIALS

Sedimentary rocks typically contain very little carbonaceous organic matter. This is clearly indicated by the average amounts of carbonaceous material in sandstones (0.05%), limestones (0.3%) and mudrocks (2.0%), cited by Tucker (2001). Carbon-rich sediments contain elevated amounts of organic carbon derived from the preservation of organic tissues. As pointed out in the discussions of humus-rich organic soils such as histosols (Chapter 12) and black and oil shales (Chapter 13), the ideal conditions for the preservation of organic matter occur when large quantities of organic matter are produced in surface environments that are depleted in oxygen so that bacterial decomposition is inhibited. Carbon-rich sedimentary materials that preserve significant organic matter include the coal family and solid, liquid and gaseous hydrocarbons of the petroleum family such as asphalt, crude oil and natural gas. These are the major sources for the carbon-based fossil fuels that provide most of the energy required by modern societies. Their use has significant side effects, including atmospheric pollution and global warming with its potential for significant long-range consequences.

14.6.1 Coal

Coal consists primarily of plant materials (humus) that have been buried, compacted, heated and biochemically altered during diagenesis. Additional constituents include small amounts of siliclastic and/or carbonate sediment, pyrite and variable amounts of moisture. The vast majority of coal consists mostly of altered organic materials derived from wood and leaves and smaller amounts of mosses, grasses and phytoplankton that initially accumulated under anoxic conditions in swamps, bogs and marshes. Anoxic swamps typically develop (1) in paralic shoreline environments, especially in parts of deltaic systems bypassed by distributary channels; (2) on river floodplains in the ultra-low gradient parts of meandering stream systems; (3) along the shores of shallow lakes, in cratonic or continental rift valley settings; and (4) in areas of poor drainage, such as those recently covered by glaciers. In such environments, thick, dense deposits of organic **peat** may accumulate over long periods of time. Peat is converted into coal by a variety of biochemical transformations that are driven primarily by increasing temperature and, to a lesser extent, pressure during progressively deeper and longer burial. The progressive transformations that occur during coal formation are collectively referred to as **coalification**.

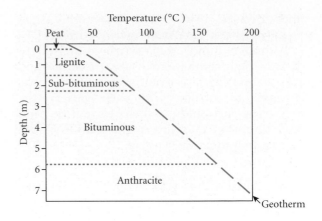

Figure 14.29 Major ranks of coal produced during progressive coalification, with approximate burial depths and temperature ranges that produce the major ranks of coal.

Because peat forms primarily from woody materials that accumulated in terrestrial environments and because woody plants did not inhabit terrestrial environments until the middle Devonian, coal deposits do not occur in rocks older than Devonian and do not become abundant until the Carboniferous. Coals are commonly associated with (1) soils, especially vertisols and histosols, (2) mudrocks such as tonsteins and bentonites (Chapter 13), and (3) a variety of marine–terrestrial transitional and lacustrine facies. Most coals belong to a main series that is subdivided into a number of different coal ranks. **Coal ranks** are based on the progressive changes in coal composition, texture and appearance that occur during coalification (Figure 14.29). The sequence of coal rank varieties during progressive burial and heating of peat is (1) lignite, (2) sub-bituminous, (3) bituminous, and (4) anthracite (Figure 14.30). Other ranks, such as sub-anthracite and meta-anthracite are sometimes recognized. Figure 14.29 shows the approximate maximum burial depths and temperatures at which each coal rank develops over sufficient periods of time.

Progressive coalification is characterized by a number of significant trends, including (1) increasing carbon content, accompanied by (2) gradual color change from brownish to black, (3) decreasing content of moisture and other volatiles, (4) increasing hardness and

Figure 14.30 Peat and the major ranks of coal derived from it during coalification (left to right): peat, lignite, bituminous and anthracite. Note the increases in compactness, hardness and luster and change in color from brown to black (carbon rich) with increasing rank. (Photo by John O'Brien.) (For color version, see Plate 14.30, between pp. 408 and 409.)

compactness, and (5) increasing reflectivity (Figure 14.30).

Lignite is a coal produced when peat is buried, compacted and biochemically transformed by bacterial activity. Most lignite is quite soft, somewhat porous, and possesses a brownish color and rather dull luster (Figure 14.30). Carbon content in the organic fraction is relatively low (50–70%) and volatile content (including moisture) is high (45–55%). Common volatiles include water (H_2O), carbon dioxide (CO_2), hydrogen (H_2) and methane (CH_4); less common are sulfur and nitrogen compounds. **Sub-bituminous coal** is produced by the burial and heating of lignite, which increases carbon content, drives off volatile components and transforms some of the woody material into reflective organic compounds collectively called **vitrinite**. Sub-bituminous coal is still soft, but less porous and slightly more reflective than lignite, and generally possesses a brownish black color. Carbon content in the organic fraction is slightly higher (70–80%) and volatile contents are slightly lower (40–50%) than for lignite. Because of their relatively low carbon content and hardness, lignite and sub-bituminous coals are considered **low rank coals** or

soft coals. Such coals constitute almost half of the available coal reserves worldwide and are used primarily as fuel in electric power-generating stations. Their high content of volatile substances means that when burned they have an especially large potential to produce significant amounts of airborne pollutants. If these are emitted into the atmosphere, rather than being collected in the power-generating plant, they can be significant contributors to acid rain and global warming.

Increased coalification produces harder coals that are valued for their high carbon content. **Bituminous coal** is a harder coal produced by additional burial and heating of sub-bituminous coal, accompanied by additional formation of vitrinite, increased carbon content and loss of volatiles. Bituminous coal is typically compact, hard, black and somewhat reflective (Figure 14.30). Carbon content in the organic fraction is higher (80–90%) and volatile contents are lower (25–40%) than for sub-bituminous coals. High rank bituminous coal called **coking coal** is essential in the manufacture of steel. Relatively pure bituminous coal is first burned in an oven to drive off volatile constituents. What remains is solid coke, mostly carbon with small amounts of impurities or ash consisting largely of fused siliciclastic materials. The coke is then used as the primary fuel in the blast furnaces that smelt iron for the production of various steel products. The development of coke-fired blast furnaces for the smelting of iron in 18th century England was a critical event in the development of the industrial revolution.

With unusually deep burial and heating, bituminous coal is transformed into the relatively rare, vitrinite-rich coal called **anthracite**. Anthracite is very compact, hard, black and reflective (Figure 14.30). Carbon content in the organic fraction is higher (>90%) and volatile contents are lower (5–15%) than for bituminous coals. Because of its high carbon and low volatile contents, anthracite is highly valued. Its high carbon content permits anthracite to produce very large amounts of energy when burned and the low volatile content minimizes both smoke and pollution. Unfortunately, relatively clean-burning anthracite comprises less than 1% of coal deposits worldwide.

14.6.2 Petroleum: crude oil and natural gas

Petroleum is the general name for carbon-rich fluids that accumulate in the pore spaces of rock bodies, most commonly limestones, sandstone and dolostones. Petroleum is composed principally of different kinds of organic hydrocarbons with various amounts of chemical impurities. As the name suggests, **hydrocarbons** are organic molecules that contain hydrogen and carbon as essential constituents. Hydrocarbons may also contain significant amounts of sulfur, nitrogen, oxygen and phosphorous, as well as many less common constituents. Hydrocarbons are classified according to their structure, composition and mass or density. Natural gas consists of lighter, more volatile hydrocarbons, whereas crude oil consists primarily of heavier, less volatile hydrocarbons.

Natural gas is subdivided into dry gas and wet gas components. **Dry gas** consists of very light molecules, dominated by chain-structured alkanes such as methane (CH_4), ethane (C_2H_6) and propane (C_3H_8) with the general formula C_nH_{2n+2}. **Wet gas** consists of somewhat heavier molecules, including alkanes such as butane (C_4H_{10}) and ring-structured cycloalkanes such as cyclobutane (C_4H_6) and cyclohexane (C_6H_{12}) with the general formula C_nH_n. **Crude oil** is composed of heavier, generally more complex hydrocarbon molecules, many with substantial amounts of sulfur, nitrogen, oxygen and phosphorous. Most natural petroleum deposits that contain crude oil also contain wet gas and dry gas in various proportions which can be separated by distillation in an oil refinery.

The source material for petroleum is organic material that is deposited and preserved in sediments by incomplete decomposition, especially under low oxygen disoxic and/or anoxic conditions. The source materials are varied and incompletely understood, but rather than the dinosaurs suggested by certain TV advertisements, they are principally sapropels derived from marine plankton, microbial mats and other fine organic material from marine and terrestrial sources. Most of this fine organic material accumulates in fine-grained sediments, including potential shales, mudstones and micrites. During burial of these **source rocks**, progressive heating over time causes sapropels to undergo conversion to hydrocarbons by a process called **maturation**. Maturation is marked by three transitional stages: (1) diagenesis, (2) catagenesis, and (3) metagenesis. During shallow burial and **diagenesis**, at temperatures less than 50°C, bacteria convert some of the sapropels to methane while increases in temperature produce thermocatalytic conversion of sapropels to kerogens, which are composed of very heavy, insoluble organic molecules. As discussed in Chapter 13, kerogens are important constituents of oil shales. The significant action begins as burial depths exceed 1.0–1.5 km and burial temperatures exceed 50°C. Under such condition, **catagenesis** begins to convert kerogens into crude oil and natural gas (Figure 14.31). Conversion to crude oil, rich in heavy hydrocarbons, continues to increase until temperatures approach 100°C, the maximum oil production temperature. In areas within the normal range of geothermal gradients, this temperature is reached at depths of 2.0–4.5 km. At higher temperatures, increasing amounts of natural gas are produced by the conversion of crude oil into lighter molecules. Conversion of petroleum to wet gas reaches a maximum near 125°C and dry gas conversion peaks near 150°C. At still higher temperatures and/or longer duration, only dry gas (methane) continues to be

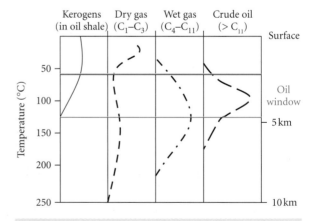

Figure 14.31 Formation of crude oil, wet gas, dry gas and kerogens during petroleum maturation as a function of temperature and approximate burial depth in sedimentary basins. C_n refers to the number of carbon atoms per organic molecule, a rough guide to the molecular density of petroleum molecules.

Structural traps

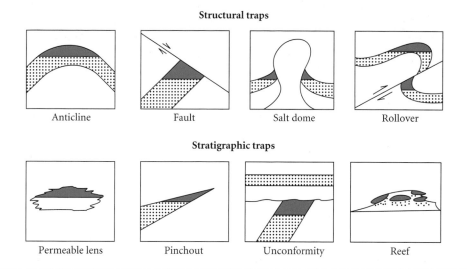

Anticline   Fault   Salt dome   Rollover

Stratigraphic traps

Permeable lens   Pinchout   Unconformity   Reef

Figure 14.32 Major types of petroleum traps: (top) structural traps produced by deformation, and (below) stratigraphic traps produced by deposition and/or diagenesis. White, impermeable rocks; stippled, permeable rocks; black, oil and/or gas.

generated and it is gradually lost from source rocks during **metagenesis** (Figure 14.31).

Only when organic material is cooked to temperatures of 60–120°C, the so-called **oil window** at depths of ~2.0–4.5 km, are significant amounts of crude oil produced. At lower and higher temperatures, only kerogens and natural gas are produced and, at still higher temperatures (>200°C), very few hydrocarbons remain.

Once crude oil and natural gas are produced, their fluidity permits them to migrate from the source rocks where they were produced into rocks with sufficient porosity and permeability to contain and transmit them. Their relatively low density permits fluid hydrocarbons to rise toward the surface. The rocks in which crude oil and natural gas eventually accumulate are called **reservoir rocks**. Such rocks are typically aquifers that contain significant water in addition to petroleum. Just as relatively impermeable aquitards

(Chapter 12) and aquicludes trap water in confined aquifers, relatively impermeable rocks surrounding and overlying reservoir rocks trap oil in reservoirs. Where the impermeable rocks, typically mudrocks or evaporites, overly the oil trap or reservoir and keep oil from rising toward the surface, they are called **cap rocks**. Typically, lighter crude oil rises above water and natural gas sits atop both in reservoir pore spaces. Figure 14.32 depicts several major types of petroleum traps. Petroleum traps that involve geological structures produced by deformation such as folds and faults are called **structural traps**; those produced by depositional patterns that trap petroleum in reservoir rocks are called **stratigraphic traps**.

For more detailed discussions of biochemical rocks than can be encompassed in a book on Earth materials, the reader is referred to Flugal (2004), Tucker and Wright (1990), Friedman (1981) and Bathurst (1975).

Chapter 15

Metamorphism

15.1   METAMORPHISM AND ITS AGENTS

Metamorphism refers to predominantly solid state mineral and/or textural changes to a pre-existing "parent" rock, or **protolith**. These changes are due to high temperatures, high pressures or the action of hot fluids. Solid state implies that the rock does not melt to form magma. While it is true that the rock maintains at least a semi-solid form, liquids and gases play important roles in metamorphism and serve as catalysts in chemical reactions. Although metamorphic rocks can develop from any pre-existing rock, we will consider protoliths as the original igneous or sedimentary source rock, prior to one or more metamorphic events. Metamorphic rock mineralogy is highly dependent upon the chemical composition of the protolith. For example, a quartz arenite can be metamorphosed into a metaquartzite and a limestone can be transformed into a marble. A quartz arenite cannot be changed into a marble because the chemical components of the protolith and daughter rock are not the same.

Metamorphic rocks contain minerals stable at the temperature and pressure conditions of metamorphism. Changes in temperature and pressure require that existing minerals either:

- Are stable with respect to these new conditions; or
- Alter to produce new minerals stable at new temperature/pressure conditions.

Metamorphism is a response to changes in temperature and/or pressure over time from some initial state. Geologists refer to pressure–temperature–time relationships, which are abbreviated as P-T-t. **Prograde metamorphism** results from increasing temperature and/or pressure conditions over time. Sufficient time is required so that mineral assemblages may form and equilibrate to the conditions of metamorphism. Mineral replacement and transformation occur due to processes such as recrystallization, neocrystallization and other processes induced by increasing pressure and/or temperature. For example, increasing temperatures can result

Earth Materials, 1st edition. By K. Hefferan and J. O'Brien. Published 2010 by Blackwell Publishing Ltd.

in devolatilization, which means that minerals such as amphibole lose highly mobile volatile components such as water and transform to anhydrous minerals. Devolatilization processes are particularly important in convergent margins where the loss of volatiles results in:

1 The loss of highly mobile volatile components that migrate out of the system.
2 Enhanced melting by an influx of volatiles into the surrounding system.
3 The generation of high temperature anhydrous minerals such as pyroxene.

Retrograde metamorphism results from decreasing temperature and/or pressure so that lower temperature/pressure mineral assemblages develop that overprint earlier peak temperature/pressure mineral assemblages. Volatile components serve as catalysts in driving retrograde metamorphic reactions. Without the addition of volatiles from an external source, retrograde metamorphic conditions are difficult to attain because previously occurring prograde metamorphism has already depleted the rock in volatile components.

Geothermobarometry involves the use of mineral assemblages or deformation characteristics of specific minerals to infer peak temperature and/or pressure conditions of metamorphism. For example, minerals such as glaucophane occur within specific low temperature and high pressure conditions of subduction zones. Such index minerals provide information related to the conditions of metamorphism. In response to changes in the metamorphic environment, unstable minerals break down to form new stable minerals. The stable mineral suite that occurs within a metamorphic rock represents an "equilibrium assemblage" of minerals stable at the temperature and pressure conditions of metamorphism.

Metamorphic rocks are classified based on composition and texture. Rock composition is largely determined by protolith chemistry, and modified by fluid-driven processes that add or remove constituents. Metamorphic textures are produced in response to temperature and pressure. For example, increasing temperature can result in mineral growth producing larger crystals; unequal pressures can change grain shape. Let us consider the role played by each of these factors during metamorphism.

15.1.1 Heat

The onset of metamorphism begins at ~150–200°C. Diagenesis is a set of sedimentary processes that occur at temperatures less than ~150–200°C and at relatively low pressures (<3 kbar or 10 km depth). The higher temperature/pressure range of diagenesis marks the transition to low grade metamorphism (Figure 15.1a). Progressively higher temperatures and/or pressures result in higher grades of metamorphism. At temperatures of ~600–800°C, igneous processes can be initiated due to partial melting (anatexis). This high temperature limit for metamorphism marks the transition from metamorphic to igneous processes. Temperature plays a critical role in determining which minerals occur within a metamorphic rock (Figure 15.1b).

15.1.2 Hydrothermal alteration

Hydrothermal alteration occurs when the bulk composition of rocks changes as a result of chemical reactions with hot fluids of variable origin. Hydrothermal fluids can dissolve and leach elements from rock, transport dissolved ions and precipitate ions, in some cases producing valuable ore deposits.

Fluids and vapors containing H_2O, CO_2, CH_4, K, Na, B, S and Cl serve as catalysts in metamorphic reactions. Volatile sources consist of:

1 "Juvenile" fluids derived from magma, principally at ocean spreading ridges, magmatic arcs and over hotspots.
2 Seawater, which infiltrates fractures and pore spaces at ocean spreading ridges and subduction zones.
3 Fluids derived by devolatilization reactions, principally in subduction zones where heating of the subducted ocean lithosphere releases:
   • H_2O from amphibole, mica and serpentine minerals.
   • CO_2 from limestone and dolostone.
4 Meteoritic fluids associated with precipitation, surface water and groundwater.
5 Connate fluids, also known as formation pore fluids, stored in spaces between grains or crystals.

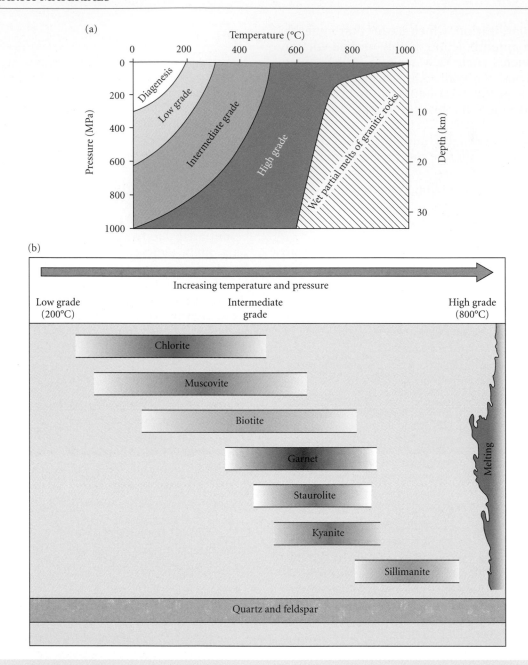

Figure 15.1 (a) Diagram showing temperature and pressure conditions of diagenesis, different grades of metamorphic conditions, and the high temperature magmatic field. (Courtesy of Stephen Nelson.) (b) Common metamorphic mineral temperature stability fields. (From Levin, 2006; with permission of John Wiley & Sons.)

Metamorphism via chemically active fluids is indicated by the presence of:

1   Volatile-rich, secondary minerals such as calcite, epidote, muscovite, serpentine, zeolite and amphibole minerals.

2   Secondary minerals in fluid inclusions and fracture-filling veins.

While hydrothermal fluids play a significant role in most metamorphic settings, they are particularly important around hydrothermal

Table 15.1 Major types of hydrothermal alteration.

Type	Major minerals generated	Description
Albitic	Albite, paragonite (Na-rich mica)	High temperature alteration resulting in Na enrichment
Alunitic	Alunite and sulfate minerals	Occurs in hot spring environments and gold and copper porphyry deposits by the oxidation of sulfide minerals
Argillic	Kaolinite, smectite, illite	Low temperature decomposition of feldspars in acidic (low pH) conditions; occurs in gold deposits hosted by sedimentary rocks
Carbonatization	Carbonate minerals such as calcite, dolomite and ankerite and accessory minerals chlorite, sericite and albite	Replacement by carbonate minerals at variable temperatures
Phyllic	Sericite, quartz, pyrite	Decomposition of silicic rocks; associated with porphyry copper deposits
Potassic	Biotite, K-feldspar, adularia	High temperature alteration of silicic magma resulting in K enrichment; commonly underlies phyllic zones
Propylitic	Chlorite, epidote, actinolite, tremolite	Low to moderate temperature decomposition of basic and ultrabasic rocks enriched in pyroxene, amphibole, biotite and plagioclase; also occurs in gold and copper porphyry deposits
Sericitic	Sericite (fine-grained, white mica)	Alteration of feldspars
Serpentinization	Serpentine, talc	Low temperature alteration of basic and ultrabasic rocks
Silicification	Quartz, chert	Replacement by silica minerals at variable temperatures
Spilitization	Albite	Low temperature alteration of Ca-plagioclase to albite
Zeolite	Zeolite minerals	Low temperature replacement of glass in volcanic rocks

vents, magma intrusions (contact metamorphism) and in fault and shear zones. The major types of hydrothermal alteration are summarized in Table 15.1.

Hydrothermal alteration occurs via deuteric reactions and metasomatic reactions. **Deuteric reactions** involve reactions in which igneous rocks are "stewing in their own juices". Hot, vapor-rich fluids are commonly associated with igneous intrusions that provide the heat, fluid and corrosive compounds to chemically alter minerals. Minerals produced by deuteric reactions include albite, calcite, epidote, sericite, chlorite, serpentine and talc. For example, in hydrothermal reactions affecting volcanic tuff protoliths, clays are altered to sericite, plagioclase to epidote and micas to chlorite.

Metasomatism involves changes in solid rock composition resulting from hydrothermal fluids exchanging constituents with an outside source. Metasomatism can involve leaching, whereby elements are removed from the rock, as well as precipitation, whereby elements are introduced by hydrothermal fluids into the rock from an outside source. Metasomatism is an important process in submarine volcanic settings such as the oceanic ridge system where basaltic magma interacts with seawater. Magma containing calcic plagioclase (e.g., labradorite) reacts with sodium-rich seawater; ionic exchanges occur that convert the calcic plagioclase into sodic plagioclase (e.g., albite) in a process called spilitization. In addition, ore metals such as cobalt, copper and manganese may be leached from the magma, incorporated into the hot fluids and precipitated along the ocean floor. Thus, oceanic crust and ophiolites are sites of metallic mineral deposits due to metasomatic processes.

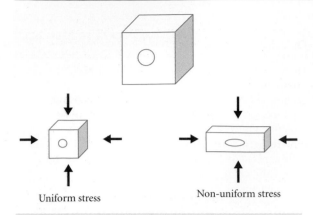

Figure 15.2 An undeformed cube (above) subjected to uniform stress changes volume but not shape; when subjected to non-uniform stress the cube changes shape, which can produce foliations.

15.1.3   Pressure

A third major metamorphic agent is **pressure**, which is a type of **stress**. We will discuss rock deformation in response to stress in greater detail in Chapter 16. Two major types of stress exist: uniform stress and non-uniform stress (Figure 15.2).

Uniform (isotropic) stress

Uniform stress, also known as isotropic stress, is equal in all directions. This is also referred to as confining stress because it is associated with burial depth in which equal compressive forces are directed towards a common central point. As confining stresses are equal in all directions, they produce volume changes, but not changes in shape. Uniform stresses tend to produce metamorphic rocks with:

1   Equant grains in which crystals have similar lengths, widths and height.
2   Non-foliated textures that lack well-defined metamorphic layering.

Non-uniform (anisotropic) stress

Non-uniform stress is not equal in all directions and tends to produce metamorphic rocks containing:

1   Inequant grains in which mineral crystals are flattened or elongated so that at least

one direction is longer or shorter than the other directions.
2   Foliated textures in which a metamorphic layering has developed due to the preferred aligned arrangement of inequant grains.

Changes in rock texture determined by uniform or non-uniform stress are the primary basis on which metamorphic rocks are classified.

15.2   CLASSIFICATION OF COMMON METAMORPHIC ROCKS

Metamorphic rocks are broadly classified into two groups based on texture (Table 15.2): common metamorphic rocks with equant grains and the absence of metamorphic foliations (Figure 15.3a) and common metamorphic rocks with inequant grains and the presence of metamorphic foliations (Figure 15.3b).

In the following section, we will consider how protolith composition influences metamorphic rock composition. In Chapter 17, we will expand upon metamorphic rock classification and address complexities within the foliated/non-foliated classification approach.

15.3   COMMON PROTOLITHS

In this section we will consider five common groups of protoliths and describe their general chemical and mineral components.

15.3.1   Pelites

Pelite protoliths include aluminum-rich rocks such as shale, mudstone and altered volcanic tuff (bentonite). As a result of protolith chemical composition, pelitic metamorphic rocks commonly include minerals enriched in SiO_2, Al_2O_3 and K_2O. With increasing temperature and pressure, clay minerals such as kaolinite, smectite and illite become unstable and are transformed into the aluminosilicate minerals listed in Table 15.3. Together with non-uniform stress, these newly formed minerals enhance the development of a metamorphic layering (foliation). Thus, slates, phyllites and mica schists commonly develop from mudstone protoliths. Contact metamorphism

Table 15.2 Common metamorphic rock types.

Metamorphic rock	Protolith	Major mineral
With equant grains and absence of foliations		
Anthracite coal	Bituminous coal	None
Hornfels	Shale	Mica, quartz, feldspar, andalusite, graphite, tourmaline, hematite
Marble	Limestone	Carbonates
Metaquartzite	Quartz arenite	Quartz
Skarn	Limestone	Carbonates
With inequant grains and foliations		
Migmatite	Gneiss	Quartz, feldspar, pyroxene
Mylonite	Granite, gneiss	Quartz, feldspar
Gneiss	Granite, diorite, shale	Quartz, feldspar, pyroxene, biotite
Schist	Shale, tuff	Mica, quartz, feldspar, kyanite, garnet
Phyllite	Shale, tuff	Chlorite, quartz, talc, mica
Slate	Mudstone, shale, tuff	Mica, quartz, feldspar, andalusite, graphite, hematite, chlorite

Figure 15.3 Common metamorphic rocks: (a) non-foliated rocks, (b) foliated rocks. (Photos by Kevin Hefferan.) (For color version, see Plate 15.3, between pp. 408 and 409.)

of pelitic protoliths produces non-foliated hornfels.

The aluminous polymorph minerals (Al_2SiO_5) kyanite, andalusite and sillimanite are useful temperature and pressure indicators – **geothermobarometers** – based on their respective mineral stability fields. For example, andalusite is a low pressure polymorph stable at pressures less than ~4 kbar (~11 km depth). Sillimanite is a high temperature polymorph that becomes increasingly more common at temperatures above ~525°C. Kyanite is a high pressure polymorph that may occur with the high pressure mica mineral paragonite. The stability field for each of the three Al_2SiO_5 polymorph minerals is illustrated in Figure 15.4. Any two of the three polymorphs can coexist along a line separating two fields. The only condition where all three polymorphs can coexist defines the "triple point", where all three lines intersect. The triple point on Figure 15.4 occurs at a temperature of ~510°C and a pressure of ~3.8 kbar.

Other common peraluminous minerals include corundum, pyrophyllite, cordierite, staurolite, tourmaline, chloritoid and alman-

Table 15.3 Aluminosilicate minerals.

Common rocks	Common protoliths	Common minerals
Slate, phyllite, mica schist, hornfels	Shale, siltstone, mudstone, greywacke, tuff	Quartz, graphite, muscovite, chlorite, chloritoid, biotite, plagioclase, K-feldspar, pyrophyllite, corundum, paragonite, andalusite, kyanite, sillimanite, staurolite, garnet, cordierite, paragonite, tourmaline, pyrite

Table 15.4 Quartzofeldspathic protoliths.

Common rocks	Common protoliths	Common minerals
Metaquartzite, gneiss, migmatite, granulite	Quartz arenite, arkose, graywacke, quartz conglomerate, granite, granodiorite, rhyolite	Quartz, plagioclase, alkali feldspar, muscovite, chlorite, biotite, pyrophyllite, sillimanite, kyanite, andalusite, cordierite, aegerine, crossite, stilpnomelane, garnet, coesite

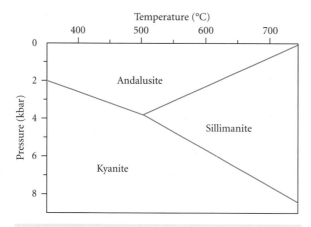

Figure 15.4 The kyanite–andalusite–sillimanite stability field.

dine garnet. Cordierite commonly occurs with andalusite in low pressure assemblages. Staurolite and chloritoid occur in moderate to high pressure environments and almandine garnet is a moderate temperature mineral.

15.3.2 Quartzofeldspathic rocks

Quartzofeldspathic or **psammitic protoliths** include quartz sandstone, arkosic sandstone and intermediate to silicic igneous rocks such as granite, granodiorite and their volcanic equivalents. Quartzofeldspathic protoliths contain high concentrations of SiO_2, Na_2O and K_2O and relatively low concentrations of FeO and MgO. As with the metapelites, the quartzofeldspathic rocks are enriched in quartz and feldspars, such as the alkali feldspars and plagioclase feldspar minerals. Unlike the peraluminous pelitic rocks, the Al_2O_3 concentrations are more variable depending upon the protolith. The most common metamorphic rocks produced from quartzofeldspathic protoliths include metaquartzite (if sandstone was the protolith) and silicic gneisses (Table 15.4).

Aluminous minerals, common in pelitic rocks, may occur in quartzofeldspathic rocks, especially if the protolith is a graywacke or micaceous granite. With increasing FeO concentrations, chlorite, biotite, cordierite, stilpnomelane, crossite, garnet and aegerine occur. Under very high pressure conditions, the quartz polymorph coesite can form.

15.3.3 Calcareous rocks

Calcareous protoliths include carbonate sedimentary rocks such as limestone and dolostone (Table 15.5). Calcareous rocks are enriched in CaO, MgO, CO_2 and FeO and may also contain SiO_2 and Al_2O_3 where chert, siliciclastic sand or mud contents are significant.

Calcareous protoliths are dominated by carbonate minerals such as calcite, aragonite and dolomite. Chert, siliciclastic sand and mud are common in some carbonate protoliths. Calcareous protoliths generate marbles enriched in calcite and/or dolomite. Minor

Table 15.5 Calcareous protoliths.

Common rocks	Common protoliths	Common minerals
Marble, skarn	Limestone, dolostone	Calcite, dolomite, aragonite, brucite, magnesite, quartz, graphite, pyrite, periclase, idocrase, (vesuvianite), anorthite olivine, talc, phlogopite, garnet, tremolite, wollastonite, diopside, coesite, lawsonite

Table 15.6 Intermediate to basic rocks.

Common rocks	Common protoliths	Common minerals
Amphibolite, eclogite, blueschist, greenschist, greenstone, serpentinite	Basalt, gabbro, norite, diorite, andesite	Plagioclase, epidote, zeolite, prehnite, pumpellyite, magnesite, serpentine, talc, apatite, magnetite, ilmenite, sphene, biotite, actinolite, hornblende, garnet, crossite, riebeckite, jadeite, glaucophane, augite, omphacite, hypersthene, diopside, lawsonite, quartz, calcite

minerals may include brucite, diopside, talc, phlogopite, periclase, grossular garnet, idocrase and olivine. Under high pressure conditions, lawsonite and the SiO_2 polymorph coesite may form. Contact metamorphism of carbonate rocks produces skarn deposits containing minerals such as wollastonite, tremolite and grossular garnet, spessartine garnet and andradite garnet.

15.3.4 Intermediate to basic rocks

Intermediate to basic protoliths include igneous rocks such as basalt, andesite, diorite and gabbro. These rocks are enriched in MgO, FeO and CaO and relatively depleted in SiO_2, Na_2O and K_2O. As a result, amphibole, pyroxene, calcium plagioclase, mafic phyllosilicate and iron oxide minerals commonly occur, as indicated in Table 15.6.

Minerals such as zeolite, pumpellyite, chlorite, chrysotile serpentine and epidote occur in low temperature/low pressure environments. Glaucophane, lawsonite, jadeite, crossite and riebeckite occur in low temperature/high pressure rocks.

15.3.5 Ultrabasic rocks

Ultrabasic protoliths include pyroxenites and peridotites. These rocks are enriched in MgO, FeO, CaO, Ni and Cr and strongly depleted in SiO_2, Na_2O and K_2O. Common ultrabasic minerals include pyroxene, olivine, plagioclase, amphibole, biotite and iron oxide minerals (Table 15.7). Epidote, prehnite, pumpellyite, zeolite, talc and serpentine group

Table 15.7 Ultrabasic rocks.

Common rocks	Common protoliths	Common minerals
Amphibolite, serpentinite, soapstone, eclogite	Peridotite, pyroxenite, hornblendite	Olivine, epidote, prehnite, pumpellyite, zeolite, calcite, chlorite, talc, antigorite, chrysotile, amosite, crocidolite, periclase, brucite, phlogopite, magnetite, chromite, anthophyllite, actinolite, cummingtonite, grunerite, tremolite, augite, diopside, hypersthene, garnet, omphacite

minerals commonly occur in low temperature/low pressure conditions. In contrast, the pyroxene minerals indicate high temperature conditions. The garnet minerals pyrope and uvarovite indicate moderate to high temperatures and pressures. Omphacite is a particularly noteworthy pyroxene mineral that forms in high temperature and very high pressure environments in association with garnet.

As with their igneous protoliths, ultrabasic metamorphic assemblages are highly valued for metallic ore deposits – particularly nickel and chromium. Ultrabasic metamorphic rocks also contain amphibole and serpentine asbestiform minerals that have both resource and hazard considerations (Box 15.1).

15.4   METAMORPHIC PROCESSES

Metamorphism can involve a number of processes that cause chemical and/or structural changes to minerals. These processes, which may act alone or in unison, include the following.

Box 15.1   Asbestos

Asbestos has been widely used for over 100 years in fireproof clothing, roofing, insulation, cement, pipes, steel, brake linings, flooring, gaskets, coatings, plastics, textile and paper. Asbestos has been valued for a combination of properties such as resistance to heat and friction, durability, flexibility, resistance to acids and the potential for asbestos fibers to be woven into a strong interlocking fabric. The US Occupational Safety and Health Administration (OSHA) defines asbestiform minerals as filiform minerals with lengths greater than $5\,\mu m$, diameters less than $5\,\mu m$ and length:width ratios $> 3:1$. These minerals are considered fibrous if their length:width ratio is greater than $10:1$. Two major groups of asbestiform minerals exist: the serpentine mineral group and the amphibole mineral group.

1 **Chrysotile** $[Mg_3Si_2O_5(OH)_4]$, also known as "white asbestos" is a widely used serpentine asbestos mineral. Chrysotile consists of soft, curly, flexible fibers and constitutes ~95% of all the asbestos used in industry.
2 **Amphibole** asbestos varieties are hard and brittle. The hard, brittle amphibole minerals are less suitable for common industrial uses and represent <5% of all asbestos used in industry. Amphibole asbestos includes five different minerals, listed in decreasing order of use:
    - Crocidolite $[Na_2(Fe^{2+},Mg)_3Fe_2^{3+}Si_8O_{22}(OH)_2]$, also known as "blue asbestos", is a variety of the mineral riebeckite.
    - Amosite $[(Fe^{2+})_2(Fe^{2+}Mg)_5Si_8O_{22}(OH)_2]$, also known as "brown asbestos", is a variety of the mineral grunerite.
    - Anthophyllite $[Mg_7Si_8O_{22}(OH)_2]$.
    - Actinolite $[Ca_2(Fe,Mg)_5Si_8O_{22}(OH)_2]$.
    - Tremolite $[Ca_2Mg_5Si_8O_{22}(OH)_2]$.

Why are asbestiform minerals hazardous? Lightweight acicular to fibrous asbestiform minerals are easily inhaled or ingested. For those individuals heavily exposed to asbestos, particularly asbestos miners and refiners, severe health problems can develop which include:

1 **Asbestosis:** a disease whereby lung tissue encapsulates asbestos particles. As a result, lung tissue hardens and decreases essential O_2/CO_2 exchange. This effect weakens the heart and destroys the lungs.
2 **Mesothelioma:** a rare disease of the lining of the lung and stomach caused by asbestos. Amazingly, mesothelioma has a 35–40-year latency between exposure and disease onset.
3 **Lung cancer:** the third major disease associated with asbestos.

Are all asbestos minerals equally hazardous? No. In fact, the most widely used chrysotile fibers decompose naturally within the lungs over a period of 9 months. As a result, the three diseases listed above are not generally associated with chrysotile asbestos. In contrast, the less widely used amphibole asbestos minerals do not decompose over the course of decades and are serious health hazards, particularly to amphibole asbestos miners in locations such as Montana, Western Australia and South Africa (Gunter, 1994; Gibbons, 2000).

15.4.1 Cataclasis

Cataclasis is a low temperature, brittle grain-fracturing process that involves grain size reduction through the mechanical grinding, rotation and crushing of rock. Cataclasis produces small, angular grains that infill between larger grains as in cement mortar. As a result, cataclasis is said to produce a **mortar texture**. Cataclasis occurs in high strain rate fault zones within the upper crust as well as in rare meteorite impacts.

15.4.2 Mylonitization

Mylonitization is a ductile grain reduction process that produces oriented grains of smaller diameter. Deformation occurs via a combination of grain fracturing, plastic bending and internal deformation of grains and rotation, which produce visible foliations in response to non-uniform stress. Although mylonitization appears to be a ductile process to the eye, microscopic grain fracturing occurs that facilitates plastic deformation. Mylonitization does not involve mineralogical change but commonly occurs with other processes discussed below. Mylonitization occurs in high temperature, high strain shear zones of the lower crust and upper mantle.

15.4.3 Diffusion

Diffusion occurs in metamorphic rocks whereby individual atoms or molecules can migrate in gaseous, liquid or solid phases from one location in a rock body to a new location. Diffusion is most efficient in gaseous states, where molecules are in motion and encounter minimal resistance. In the liquid state, diffusion is highly variable depending upon fluid viscosity: high viscosity inhibits migration whereas low viscosity enhances molecule migration. Diffusion is most limited in the solid state, where chemical bonding restricts the migration of atoms from one site to another. In the solid state, which characterizes most metamorphic reactions, diffusion is facilitated by crystal dislocations and by the presence of small amounts of intergranular fluids that catalyze chemical reactions. This is particularly important in high temperature metamorphic reactions where volatiles such as H_2O and CO_2 effectively stimulate ion transfer. In addition to high temperature environments, diffusion also plays a critical role in the pressure solution reactions described below.

15.4.4 Pressure solution

Pressure solution (pressolution) involves the dissolution of solid grains under high compressive stress conditions. High, localized compressive stress regimes result in alterations to the crystal lattice structure so that an aqueous phase is produced at high stress sites. Dissolution occurs so that one grain impinges upon another grain, initiating a soluble phase in the indented grain. Continued grain dissolution along a boundary produces a sutured contact between the grains. Local precipitation can occur in regions of lower stress. Pressure solution is enhanced by high concentrations of crystal defects within the lattice structures, which promotes dissolution (Passchier and Trouw, 2005).

Pressure solution is well demonstrated in calcareous, quartzofeldspathic and pelitic rocks, where it can result in up to 50% volume loss (Wright and Platt, 1982). As soluble minerals dissolve, insoluble minerals (iron oxides, micas, graphite) accumulate as an insoluble seam called a **stylolite** (Chapter 14). Pressure solution is very important in the development of cleavage, folds and gneissic layering where the quartz and carbonate minerals dissolve in preference to clays and micas. Pressure solution involves not only dissolution and volume loss but also recrystallization.

15.4.5 Recrystallization

Recrystallization occurs when existing minerals are transformed under higher temperature and/or pressure conditions, without experiencing a significant change in chemical composition. Recrystallization results from crystal lattice reorganization without breakage:

- Extracrystalline recrystallization occurs by grain rotation, grain boundary migration or the bulging of one grain into another. Extracrystalline recrystallization involves the addition or removal of mineral material by diffusion.
- Intracrystalline deformation occurs within individual grains due to microscopic

movements related to defects, vacancies or dislocations within a crystal. Recovery is a means by which crystal defects are minimized by the movement of ions within a grain. Intracrystalline deformation and recovery are permanent, plastic processes that do not involve visible breakage.

The net effect of these processes is to reduce the state of stress within a crystal in response to elevated temperature/pressure conditions. Both extracrystalline and intracrystalline recrystallization processes are enhanced by high volatile content, temperatures and/or pressures.

15.4.6   Neocrystallization

Neocrystallization refers to the nucleation and growth of new minerals as pre-existing minerals become unstable due to temperature/ pressure changes. Diffusion enhances the nucleation and growth of new minerals. Newly formed minerals distinctly larger than the minerals in the surrounding matrix are referred to as **porphyroblasts**. Continued neocrystallization is dependent upon the continuation of favorable temperature/pressure conditions and the availability of ions necessary to permit crystal growth.

15.4.7   Differentiation

Differentiation refers to the segregation of minerals in an initially homogeneous rock due to different physical or chemical characteristics such as solubility, ductility, mineral growth or crystallization temperature. For example, during intense folding more soluble minerals (e.g., calcite, quartz and feldspars) are concentrated in fold hinges, while less soluble minerals (e.g., ferromagnesium minerals and clays) occur in fold limbs. Continued deformation can produce transposed layering and mineral segregation as the hinges and limbs are disconnected.

Through differentiation, more competent (rigid) porphyroblast minerals such as garnet, quartz or feldspars can extend and grow, producing elongated forms (Figure 15.5). Growth of these rigid grains produces pressure shadows about which more ductile minerals such as micas can accumulate and grow. As a result of their ductility contrast, segrega-

Figure 15.5   Microphotograph of a quartz porphyroblast that has experienced elongation and growth parallel to foliations in a mylonite. Field of view ~2 mm. (Photo by Kevin Hefferan.)

tion can occur, resulting in metamorphic differentiation.

Under high temperature conditions, minerals with different crystallization temperatures migrate into distinct horizons or bands. Color-banded gneisses and migmatites can form in this manner by developing light-colored (quartz, feldspar) layers and dark-colored (amphibole) layers. Let us now consider different types of metamorphism under which some of these mechanisms occur.

15.5   METAMORPHISM TYPES

Metamorphism occurs on a variety of scales. Local metamorphism affects areas of less than 100 km². Types of local metamorphism include impact metamorphism, dynamic metamorphism along fault and shear zones and contact metamorphism. Regional metamorphism affects areas greater than 100 km² and may encompass thousands of square kilometers. Regional metamorphism is most commonly associated with convergent and divergent plate boundaries. In the following sections we will briefly describe different types of local and regional metamorphism.

15.5.1   Impact (shock) metamorphism

Impact metamorphism is generated by explosive volcanic eruptions or relatively rare collisions of extraterrestrial objects with Earth. The high strain rate associated with impact metamorphism produces breccias, shocked quartz lamellae, pseudotachylites produced by impact melting and ultra-high pressure (UHP) minerals such as the silica (SiO_2) polymorphs coesite and stishovite. These features collectively produce rocks called "impactites". Large numbers of meteorites bombarded Earth around 4.3 Ga resulting in a pock-marked surface similar to that which still exists on the moon. Over time, weathering, erosion and sedimentation have largely resculpted Earth's surface removing the pock marks.

The most famous meteorite impact example is the ~65 Ma Chicxulub Crater near the Yucatan Peninsula. Breccias, shocked quartz and UHP minerals occur in association with this impact feature. While the Chicxulub meteorite had been proposed to be responsible for the extinction of dinosaurs, the impact is now known to predate the Cretaceous–Tertiary extinction by 300,000 years (Keller et al., 2004). Meteor Crater (Figure 15.6), located in the Arizona Desert of the southwestern USA, is ~1.2 km in diameter and 180 m deep. Meteor Crater's rim has been uplifted 30–60 m due to debris associated with the impact that collected around the depression (Shoemaker, 1979). Originally interpreted as a volcanic crater, an impact origin was proposed by D. M. Barringer early in the 20th century. On the basis of his work, Meteor Crater – also known as Barringer Crater – is now thought to have formed 25,000–50,000 years ago by a meteor of

Figure 15.6 Meteor Crater, Arizona. (Photo courtesy of D. Roddy, with permission of NASA and the Lunar Planetary Institute.)

~50 m diameter. Interestingly, when astronauts were preparing for their moon walks in the late 1960s, geologists trained the astronauts in Meteor Crater as it resembles impact craters common on the moon. Debate continues regarding the distinction between magmatic and impact features on Earth – for example, the 1.85 Ga Sudbury Complex of Ontario, long considered a layered intrusive igneous complex, is now also recognized as a meteorite impact structure.

15.5.2 Dynamic metamorphism

Dynamic metamorphism is induced primarily by non-uniform stress in fault zones and shear zones (Figure 15.7). These high strain rate environments are mostly of local extent, but may be of regional extent in large fault or shear zones. Dynamic metamorphism can occur over relatively short time intervals during which faults and shear zones are in motion. Dynamic metamorphism has a high likelihood of recurring in the same fault or shear zone because of the inherent instability of these zones in response to stress.

The rock texture generated through dynamic metamorphism depends largely on the temperature, pressure and strain rate. Relatively cold rocks in Earth's low pressure, upper crust rocks are brittle and fracture in response to stress producing cataclasites. **Cataclasites** are fragmented rocks that have experienced the breakage of brittle rock into smaller sized fragments. Within the upper 5 km of Earth's surface, the brittle crushing and grinding of rocks produces fault breccia or a mélange. At depths of 5–10 km, temperatures increase so that the frictional heat and geothermal gradient combine to heat rocks within the fault zone closer to their melting point. As a result of localized rock melting, pseudotachylites develop within the most intensely deformed zones. **Pseudotachylites** are partially melted rocks that form by quenching under high strain rates in shear zone fractures. Ductile shear zones are zones of ductile deformation at depths greater than ~10–15 km. Ductile shear zones deform plastically producing **mylonite** rocks. Mylonites are characterized by grain size reduction via microscopic scale cataclasis, plastic stretching and thinning associated with ductile deformation.

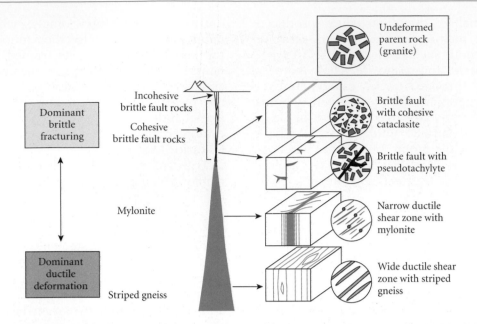

Figure 15.7 Brittle faulting produces cataclasites and high strain rate pseudotachylites; ductile shearing produces mylonitic rocks. (From Passchier and Trouw, 2005; with permission of Springer Science+Business Media.)

15.5.3 Contact metamorphism

Contact metamorphism develops locally where hot magma intrudes relatively cold, upper crustal (<10 km) country rock. Contact metamorphism is a variable temperature, low pressure metamorphism. Rocks produced by contact metamorphism typically display altered mineral assemblages with non-foliated textures. Temperature variations are largely related to the size and temperature of the pluton, as well as the distance from the intrusion.

The heat from the igneous intrusion produces a metamorphic **aureole** in the contact zone that surrounds it (Figure 15.8). Metamorphic aureoles range from centimeters to hundreds of meters in diameter. Extensional forces associated with the intrusion can fracture the country rock, resulting in secondary vein development and precipitation of minerals by hydrothermal fluids. Many metallic ore minerals of significant economic importance precipitate in metamorphic aureoles.

Contact metamorphism produces non-foliated rock types such as hornfels, metaquartzite and skarns. **Hornfels** is a general term for fine-grained, contact metamorphic rock rich in silicate minerals. The protolith is commonly a mudstone, but may also be a basic or ultrabasic rock. Metaquartzites can develop by contact metamorphism of quartz sandstones. **Skarns** develop when igneous plutons intrude carbonate rocks (Figure 15.9). Chemical reactions with magmatic fluids result in enrichment of carbonate rocks with SiO_2, producing a calcium-, magnesium-, iron-rich silicate rock. Common calc-silicate minerals produced by contact metamorphism of carbonate include wollastonite, tremolite, grossular garnet, spessartine garnet and andradite garnet. Skarn deposits are strongly affected by hydrothermal alteration, as discussed below.

While contact metamorphism is usually considered as a local type of metamorphism, it is also prevalent in combination with other types of metamorphism at convergent and divergent boundaries discussed below.

15.5.4 Ocean floor metamorphism

Hydrothermal alteration is particularly pervasive at ocean spreading ridges that experience tension, thinning and uplift. Forces within the rift valley produce extensional fracture and normal fault systems that serve as two-way

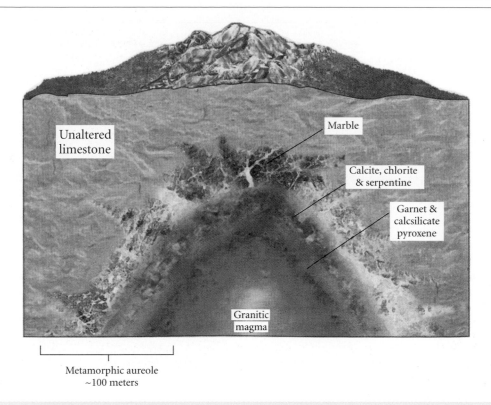

Figure 15.8   Simplified map view of contact metamorphism with an aureole. (Courtesy of Murck et al. (2010); with permission of John Wiley & Sons.)

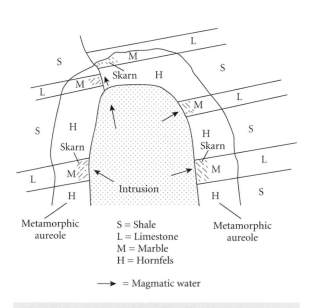

S = Shale
L = Limestone
M = Marble
H = Hornfels

⟶ = Magmatic water

Figure 15.9   Skarn deposit formed by the intrusion of magma into limestone country rock. (From Evans, 1993; with permission of Blackwell Publishers.)

conduits. Hot magma rises upward towards the ocean floor to cool and crystallize as basic igneous rocks. Reducing conditions within the upwelling magma result in extensive metal sulfide deposits. At the same time cold, oxygenated seawater descends through the fissures and reacts with the basaltic crust. Hydrothermal fluid from seawater enriches the basalt in sodium and magnesium. Hot magmatic plumes release helium, manganese oxide, hydrogen sulfate, methane, iron sulfide, chromium, phosphate and trace metals (Figure 15.10).

Metasomatism at ocean spreading ridges involves a number of chemical reactions that include alunitic, propylitic and sericitic reactions, silicification, carbonitization, serpentinization and spilitization (see Table 15.1). In shallow oxidizing environments, olivine and pyroxene minerals are altered to iron oxide minerals such as magnetite and hematite. Minerals such as zeolites, epidote, chlorite, talc, brucite and magnesite are produced by similar reactions involving hydroxyl ions. **Serpentinization** occurs when magnesium-rich

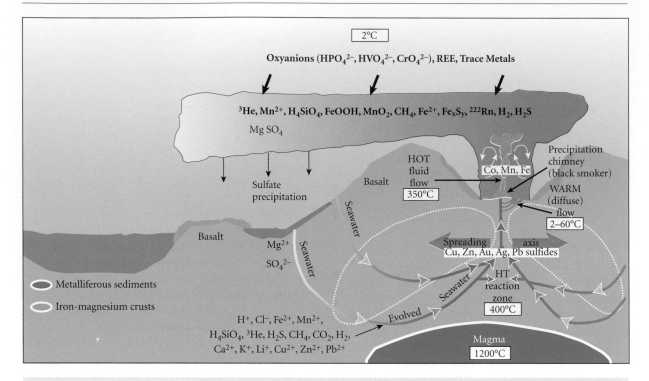

Figure 15.10   Cross-section depicting the chemical reactions that occur at mid-ocean ridges. HT, high temperature; REE, rare Earth elements. (Courtesy of NOAA.)

olivine or pyroxene minerals are altered to serpentinite by seawater-derived hydrothermal fluids. Typical reactions include the following:

$$2Mg_2SiO_4 + 3H_2O$$
$$\underset{\text{Forsterite}}{} \quad \underset{\text{Water}}{}$$
$$= Mg_3Si_2O_5(OH)_4 + Mg(OH)_2$$
$$\underset{\text{Serpentine}}{} \quad \underset{\text{Brucite}}{}$$

(Hyndman, 1985; Best, 2003)

$$6MgSiO_3 + 3H_2O$$
$$\underset{\text{Enstatite}}{} \quad \underset{\text{Water}}{}$$
$$= Mg_3Si_2O_5(OH)_4 + Mg_3Si_4O_{10}(OH)_2.$$
$$\underset{\text{Serpentine}}{} \quad \underset{\text{Talc}}{}$$

(Barker, 1983)

Spilitization occurs as a result of the exchange of sodium from seawater for calcium in plagioclase, which converts the plagioclase into albite. Spilite is the name for sodium-rich basalts that form along ocean ridges and volcanic arcs. Spilites commonly occur in Precambrian greenstone belts and in ophiolites (Chapter 18) with minerals such as chlorite, epidote and actinolite that are also produced

by hydrothermal alteration. Perhaps most importantly, metallic ore deposits precipitate from black smokers around ocean vents. As essentially all ocean crust is generated at ocean ridges, the entire ocean crust is affected by low grade metamorphic reactions involving aspects of contact metamorphism and hydrothermal alteration.

15.5.5   Burial (static) metamorphism

Burial metamorphism results from increases in lithostatic stress induced by deep burial of rock and produces non-foliated textures (Coombs, 1961). Burial metamorphism affects regional subsiding basins that accumulate thick sequences of sediments and volcanic debris. Sediment-rich, subsiding basins occur in a variety of environments which include: rifts, foreland basins, passive marine basins adjacent to continental margins, forearc and backarc basins at convergent margins, and thick sedimentary sequences that form in pull-apart basins associated with transform faults.

Sediment burial initially results in diagenetic processes such as compaction, cementa-

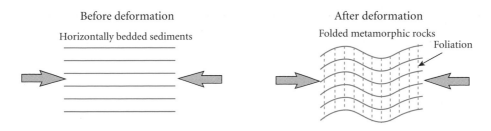

Figure 15.11   Compressive stress produces foliations typically at a high angle to bedding. (Courtesy of Stephen Nelson.)

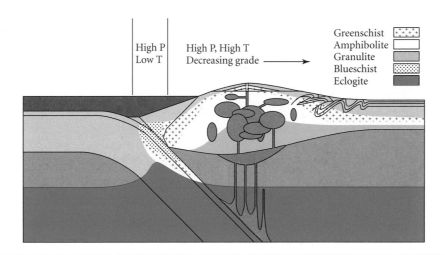

Figure 15.12   Dynamothermal metamorphism occurs at convergent plate boundaries. P, pressure; T, temperature. (Courtesy of Steve Dutch.)

tion and lithification. The onset of burial metamorphism, at temperatures of ~150°C, is gradational with diagenesis. Burial metamorphism temperatures generally range from ~150° to 350°C, producing relatively low temperature mineral assemblages that include zeolite and prehnite group minerals, pumpellyite, chlorite and micas.

Notable modern-day examples of burial metamorphism include the Bay of Bengal and the Gulf of Mexico and pull-apart basins in southern California. Curray (1991) suggests that the lower part of the 22 km thick sedimentary deposits in the Bay of Bengal have experienced temperatures in excess of 350°C. In addition to the pressure and heat associated with deep burial, warm formation pore fluids induce relatively low temperature hydrothermal alteration. Perhaps the most significant economic aspect of burial meta-

morphism is in the generation of hydrocarbon deposits of oil, gas and coal (see Section 14.6).

15.5.6   Dynamothermal metamorphism

Dynamothermal metamorphism is a regional metamorphism induced by increases in both pressure and temperature. Non-uniform stress commonly produces rocks with inequant grain shapes and foliated textures (Figure 15.11). With the exception of ocean spreading ridge hydrothermal alteration, dynamothermal metamorphism is the most aerially extensive type of metamorphism on Earth. Dynamothermal metamorphism dominates convergent margins and associated fold and thrust belts. Together, ocean floor metamorphism and dynamothermal metamorphism are by far the most volumetrically and economically important types of metamorphism on Earth, responsible for the concentration of most crustal metal deposits and many hydrocarbon traps.

Dynamothermal metamorphism combines a complex set of processes that involve crustal shortening due to the convergence of two lithospheric plates. As a result of crustal shortening, immense fold and thrust belts develop that characterize convergent margins throughout the world. All of the great mountain belts – Himalayas, Alps, Atlas, Urals, Pyrenees, Carpathian, Cordilleran, Zagros, Andes, Appalachians – formed as a result of crustal shortening and igneous activity and are the sites of intense dynamothermal metamorphism (Figure 15.12). Dynamothermal metamorphism produces a succession of pressure and temperature conditions that will be discussed in greater detail in Chapter 18.

Metamorphic rocks are the dominant rock type in Earth's lower crust and mantle, due to the elevated temperatures and pressures deep within Earth. Metamorphic rocks are also widespread at convergent and divergent plate boundaries. This chapter serves as a brief introduction to metamorphism. In succeeding chapters we will learn more about rock deformation, metamorphic rock classification and our understanding of the relationship between tectonics and metamorphism based upon mineral assemblages.

Chapter 16

Metamorphism: stress, deformation and structures

16.1 STRESS

An understanding of metamorphic rock textures requires a basic understanding of stress and deformation – the foundation of structural geology. **Stress** is a directed force of some magnitude applied over an area. **Deformation** is a change induced by stress. We discuss stress and deformation in the context of metamorphic rocks because of their importance in forming metamorphic textures. Stress and deformation are also critically important to sedimentary and igneous rocks, particularly with respect to their resource potential (e.g., as hydrocarbon traps and ore deposits) and their role in geohazards such as earthquakes and mass wasting.

Because stress is a directed force with magnitude it can be depicted as a vector. A vector is a line whose length is proportional to its magnitude and which has an arrow indicating direction. In Figure 16.1, force F is applied vertically downward to an inclined plane.

Earth Materials, 1st edition. By K. Hefferan and J. O'Brien. Published 2010 by Blackwell Publishing Ltd.

Force F can be resolved into two component vectors: Fn is a normal force oriented perpendicular to the inclined plane and Fs is a shear force oriented parallel to the inclined plane.

Stress can be applied to a rock body in three fundamental ways: compression, tension and shear (Figure 16.2).

1 **Compression** occurs where forces are directed towards a point or a plane.
2 **Tension** occurs where forces are directed away from a point or plane.
3 **Shear** occurs where forces are oriented parallel to a plane.

The state of stress on three-dimensional rock bodies can be described in relation to three principal stress axes (Figure 16.3). The three **principal stress axes** are mutually perpendicular vectors, representing normal forces of equal (uniform) or unequal (non-uniform) magnitude that intersect a principal plane at right angles. Each of the three **principal planes** is parallel to two principal stress axes and normal (perpendicular) to the third stress axis. Because principal planes are perpendicular to principal stress axes, no shear stresses

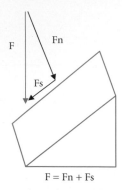

Figure 16.1 Stress acting on a two-dimensional plane can be depicted as a vector force, F, which can be resolved into Fn (normal force) and Fs (shear force) components.

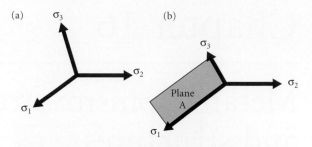

Figure 16.3 (a) Uniform stress in which the three principal stress axes are of equal magnitude. (b) Non-uniform stress occurs in which the three principal stress axes are not of equal magnitude as indicated by three stress axes of different lengths. Plane A is parallel to σ_1 and σ_3, and is perpendicular to σ_2.

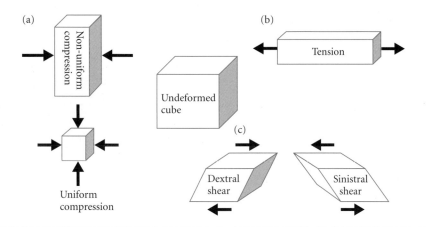

Figure 16.2 The three main forms of stress: (a) compression, (b) tension, and (c) shear stress.

occur along these surfaces. As no shear stresses exist on principal stress axes, these axes are described in terms of compressive and tensional stresses (Box 16.1).

As compressive stresses dominate within the Earth due to the inward directed forces of surrounding rock mass, principal stresses are usually described in relation to compressive stresses. The Greek symbol sigma (σ) is the standard notation for stress. The three principal stress axes are described as follows:

- σ_1 = maximum compressive stress (or minimal tensional stress).
- σ_2 = intermediate compressive stress.
- σ_3 = minimum compressive stress (or maximum tensional stress).

Geologists consider compressive stresses to be positive and tensional stresses are assigned negative values (engineers, on the other hand, consider tensional stresses to be positive and compressive stresses to be negative). σ_1 may be greater or equal to σ_2, which may be greater or equal to σ_3. So the following relationship exists: $\sigma_1 \geq \sigma_2 \geq \sigma_3$.

Different states of stress can be described in relation to the three principal stress axes. These include uniform and non-uniform stresses (Figure 16.3).

16.1.1   Uniform (isotropic) stress

Uniform (isotropic) stress occurs when all three principal stress axes (Figure 16.3a) are

Box 16.1   Principal stress planes and the teeter-totter analogy

Does the concept of principal stress axes and principal planes sound confusing? Malcolm Hill suggests a playground analogy to help visualize this concept. If two friends orient a teeter-totter (see-saw) so that the board is exactly parallel to Earth's surface (Figure B16.1a), compressive stress is directed perpendicular to the board and there will be no component of shear stress on the board surface, which is to say you will not slide down the teeter-totter. Note that the vertical support on which the teeter-totter rests is a principal stress axis perpendicular to a principal plane. If your friend gradually tilts one end of the teeter-totter upwards (Figure B16.1b), the board is no longer a principal plane because it is not perpendicular to the maximum compressive stress direction extending from your head, through your feet, to Earth's center of mass. The tilted board has a component of shear directed "downhill" along the sloping plane. If the shear stress exceeds the frictional stresses holding you in place, you may slide down the teeter-totter board.

(a)   (b)

Figure B16.1   Teeter-totter analogy of: (a) a principal plane oriented perpendicular to a principal stress axis, and (b) a plane on which shear stress exists. (Photos by Kevin Hefferan.)

of equal magnitude ($\sigma_1 = \sigma_2 = \sigma_3$). In uniform stress the following conditions exist:

1   The three perpendicular, principal stress axes can have any orientation as stress is equal in all directions.
2   No shear stresses occur.
3   No change in shape occurs.
4   Volume change can occur.

Earth materials are overlain by rocks, sediment or water which exert a compressive uniform stress referred to as confining stress. **Hydrostatic stress** refers to the uniform compressive force directed radially inward by the surrounding mass of water. We have a 12-ounce Styrofoam coffee cup that went to the bottom of the Mid-Atlantic Ridge with the

submersible Alvin. When the coffee cup returned to the ocean surface it had retained its original shape but was the size of a 3-ounce cup as a result of hydrostatic stress. **Lithostatic stress** refers to a uniform compressive force exerted radially inward due to the mass of surrounding rock. Rock stresses are commonly expressed in kilobars (kbar), megapascals (MPa) or gigapascals (GPa). Most rocks have a density of ~2.6 g/cm^3; therefore, a relationship exists between depth of burial and lithostatic stress in which 1 kbar (0.1 GPa or 100 MPa) pressure = 3.3 km depth of burial, so that for each 10 km of depth, the lithostatic pressure increases by ~3 kbar. As discussed in Chapter 15, uniform stresses tend to produce equant grains arranged in a non-foliated orientation.

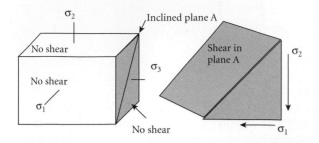

Figure 16.4 Deformation of a cube. Note that shear stresses do not exist on the principal stress planes but shear stresses do exist on other planes such as plane A.

16.1.2   Non-uniform (anisotropic) stress

Non-uniform (anisotropic or deviatoric) stress occurs when at least one principal stress has a magnitude not equal to the other principal stresses (Figure 16.3b). In non-uniform stresses the following conditions may exist:

1   Stress axes are not equal in all directions.
2   Shear stresses can occur on rock bodies, but not on a principal plane (Figure 16.4).
3   Shape changes can occur.
4   Volume change can occur with corresponding changes in density.

Non-uniform stress results in the deformation of a spherical ball into an ellipsoid egg shape as indicated in Figure 16.5. Non-uniform stress also promotes the development of inequant grain growth and foliations. We will discuss different types of deformation in the following section.

16.2   DEFORMATION

Stresses of sufficient magnitude produce rock deformation. **Deformation** is a physical change in the rock due to an applied stress. Deformation consists of four components: distortion, dilation, translation and rotation (Figure 16.6).

Distortion, also known as **strain**, indicates a change in shape from some initial form. **Homogeneous strain** (Figure 16.7a) occurs when strain is equal throughout the rock body so that parallel lines remain parallel, perpendicular lines remain perpendicular and

circles flatten to become ellipses. **Heterogeneous strain** occurs when strain intensity varies within a rock body. Heterogeneous strain (Figure 16.7b) produces angular changes so that lines that were once parallel or perpendicular to one another are no longer parallel or perpendicular, and circles do not deform to ellipses. In the real world, heterogeneous strain predominates.

Dilation indicates a change in volume. Two commonplace examples of dilation in our lives are described here. If you get tested for glaucoma during an eye exam, a fluid is dropped into your eyes which makes your pupils expand. As you exit the optometrist's office, you are blinded by the brilliance of light that floods through your dilated pupils. Any woman who has given birth to a baby is also very familiar with the importance of 10 cm dilation prior to the delivery of a baby. In both of these physiological examples, dilation refers to a volume increase. How does dilation affect rocks? Rocks can also experience an increase in volume in response to tensional stresses such as in ocean spreading and continental rift environments. But in most cases, rock dilation involves a volume decrease because buried rocks experience compressive lithostatic stress. Compression reduces pore space, increases density and decreases rock volume.

Translation, also known as **displacement**, means that an object has moved from one point to another point. In a game such as golf, one of the measures of player skill is the length of their drive. Stress is initiated when the golf club hits the ball. The golf ball has not permanently changed shape or volume, but has been displaced. We can determine the amount of displacement by measuring the distance from the tee spot, where the golf club hit the ball, to the final resting point of the golf ball. Rocks are moved from one location to another via folding, faulting and shearing mechanisms. Unfortunately, except in rare cases we do not usually know the distance the rock particles have been displaced. An extreme case to consider would be the dismembered tectonic fragments of the Avalon Terrane, which extends from Scandinavia through Ireland, Newfoundland, Massachusetts and southward to Florida.

Rotation infers that an object has moved in a circular arc about an axis, sort of like tight-

(b)

(a)

(c)

σ_1

Grain growth

σ_3 ← → σ_3

Grain shortening

σ_1

Figure 16.5 Non-uniform stresses transform an undeformed sphere (a) into an ellipsoid (b) and promote inequant grain growth and the development of foliations perpendicular to the σ_1 direction (c).

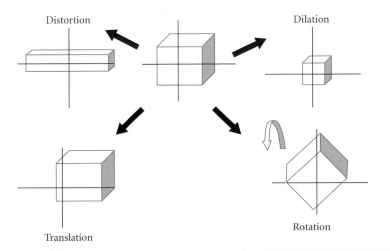

Distortion

Dilation

Translation

Rotation

Figure 16.6 Summary diagram of the four aspects of rock deformation.

(a)

(b)

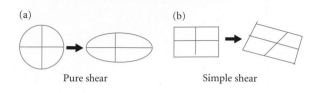

Pure shear   Simple shear

Figure 16.7   (a) Homogeneous strain in which parallel lines remain parallel, perpendicular lines remain perpendicular and circles deform to ellipse shapes. (b) Heterogeneous strain in which parallel lines do not remain parallel, perpendicular lines do not remain perpendicular but circles do deform to ellipses.

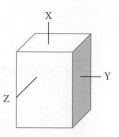

Figure 16.9   Illustration depicting the X, Y and Z strain axes.

Figure 16.8   Photomicrograph of a ~5 mm snowball garnet that has been rotated counterclockwise during crystal growth. (Photo courtesy of Bruce Yardley.)

ening a screw into a wall or the way in which a wheel rotates around an axle. Rock particles can move in a circular motion as they are displaced. One of the best rock examples is the "snowball" garnet in which the garnet porphyroblast has grown during rotation, resulting in a swirling growth pattern that indicates direction of motion (Figure 16.8).

16.2.1   Principal strain axes

Deformation is sometimes referred to as strain. Strictly speaking, this is incorrect because strain is only one component of deformation; but get used to it as it is commonly done, particularly with respect to strain axes. **Principal strain axes**, like stress axes, are imaginary lines that are perpendicular to each

other and that intersect planes of zero shear strain. Strain axes (Figure 16.9) are denoted by the letters X, Y and Z as follows:

X = maximum direction of extension or minimal compressive strain
Y = intermediate strain axis
Z = maximum direction of shortening or minimum extension

Stress induces strain. In the simplest scenario, it would be quite easy to correspond the three principal stress axes (σ_3, σ_2, σ_1) with three principal strain axes X, Y and Z. σ_3 is the minimum compressive or maximum tensional force that produces maximum extension parallel to X. σ_2 is the intermediate stress that produces intermediate deformation par-

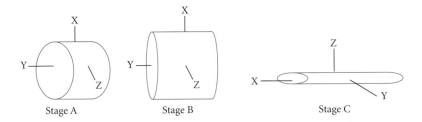

Stage A   Stage B   Stage C

Figure 16.10   The equidimensional cylinder in stage A represents an undeformed state. Stress produces a deformed intermediate stage B in which the cylinder has been compressed laterally. Further stress produces a deformed cylinder in stage C that has been stretched horizontally and flattened vertically. Note that significant changes in strain orientations have occurred from stages A to C. This strain history for stages A and B is rarely preserved in rocks.

allel to Y. σ_1 is the maximum compressive or minimum tensional stress that produces maximum shortening parallel to Z. The correspondence of stress axes with strain axes works in the simplest cases, such as taking a ball of clay and squeezing it in one direction. However, in most cases we cannot confidently correlate principal stress axes with principal strain axes in rocks.

16.2.2   Strain kinematics

In studying deformed rocks, scientists attempt to determine the strain path. The strain path describes the **kinematic strain** development from an initial to a final state. A kinematic study describes a series of strain events resulting in a final strain state. **Incremental strain** refers to one or more intermediate strain steps describing separate strain conditions. Strain analysis requires knowledge of the original undeformed state of the material (rare in nature) as well as the incremental stress and strain steps. In most cases, we can only study the final state of the rock and make generalizations about the state of stress that may have produced the observed deformation features. Figure 16.10 illustrates a very simple three-stage strain history.

Rarely do rocks in the field preserve evidence of their initial state, intermediate states or the stress and strain history that produced the final rock state. So one has to ask:

1   Was the deformation a singular event?
2   Did the stress and strain axes remain the same throughout the deformation history?

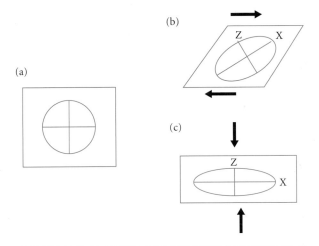

Figure 16.11   Diagrams illustrating: (a) the initial undeformed state, (b) simple shear, and (c) pure shear.

3   Did the stress and strain axes rotate and change during the deformation history?

We usually do not have sufficient information related to stress and strain history to answer these questions. Nevertheless, later in this chapter we will be discussing rock textures that allow us to make some observations about the state of stress and strain. Let us consider two end member examples of non-rotational and rotational strain (Figure 16.11).

Coaxial strain infers that no rotation of the incremental strain axes occurred from an initial to final strain state. Coaxial strain produces **pure shear (irrotational strain)** structures in which the X, Y and Z axes do not rotate during progressive strain. Pure shear is analogous to standing on a stationary ball so that a circular object approaches that of an ellipse (Figure 16.11c). Pure shear requires that:

- Uniform elongation occurs in only one direction.
- Uniform contraction occurs in a perpendicular direction.
- Strain axes are parallel to principal stress axes (σ_1 = Z, σ_2 = Y, σ_3 = X).
- Principal strain axes correspond to principal stress axes throughout deformation.
- No change in volume occurs.

Non-coaxial strain is a rotational strain in which the strain axes rotate through time, so that the stationary ball described above is now rolling and rotating. During incremental strain steps, the principal strain axes do not remain the same. Instead, different principal strain axes occur at each incremental step as a result of axis rotation. Non-coaxial strain results in simple shear (rotational shear) structures (Figure 16.11b). In simple shear, the following conditions exist:

- Strain axes do not remain parallel during progressive deformation.
- X, Y and Z strain axes rotate during progressive deformation for a fixed single stress orientation.
- As a result of rotation, the direction of maximum elongation is not parallel to the direction of minimum compressive stress or maximum tensional stress.
- The direction of maximum shortening (minimum extension) is not parallel to the direction of minimum tension or maximum compressive stress.

Consider a card deck with circles drawn on the side of the card deck. In simple shear, as you slide the cards one on top of the other, circles on the side of the card deck deform. The total distance perpendicular to the shear plane remains constant; in other words, the thickness of the card deck remains the same. All points move parallel to a fixed direction with an amount of displacement proportional to a distance from some defined plane (e.g., parallel to the faces of cards). As slip occurs

on the card deck planes, the strain axes rotate in the direction of shear.

Pure shear and simple shear are two idealized end members. In reality, rocks commonly experience general shear. **General shear** is a combination of pure shear and simple shear. Let us now consider different means by which strain is accommodated in rocks.

16.3   DEFORMATION BEHAVIOR

The type of deformation that occurs is related to rock rheology. **Rheology** refers to how materials respond to stress. We will focus upon three rock behavior types: elastic, plastic and brittle. Plastic and brittle behavior are most common in rocks; however, elastic behavior also occurs at low stress levels and is important in earthquake studies.

1 **Elastic deformation**   occurs when a body is deformed in response to a stress, but returns to its original shape when the stress is removed. Deformation produced by stress is totally and instantaneously reversible or recoverable.
2 **Plastic deformation** is an irreversible strain without visible fractures, although microfracturing can occur. Stress results in permanent, irreversible deformation.
3 **Rupture deformation** creates visible fractures in response to stress. Rupture deformation results in loss of cohesion of rock particles producing permanent, irreversible deformation.

16.3.1   Elastic behavior

Many materials we use have elastic properties, such as undergarments, hair accessories (such as pony-tail holders) and rubber bands. When we apply tensional stress by pulling them apart, these items deform by stretching or changing shape so that:

- Even a small amount of tension results in some noticeable amount of stretching.
- The amount of stretching is directly related to the amount of tension you apply.
- If you release the stress, the material returns to its initial shape.

Elastic behavior is temporary, reversible strain in which a linear relationship exists between stress and strain (Box 16.2). Upon the application of stress, bonds temporarily stretch but return to their original position following stress removal. Rock behavior in response to stress can be depicted on a graph where stress and strain (or strain rate) are depicted on x- and y-axes. Figure 16.12 illustrates a series of idealized stress versus strain graphs for elastic behavior. Note that in all cases a linear relationship exists between stress and strain so that the strain is proportional to the amount of stress. As this relationship is known as **Hooke's law**, elastic behavior is also referred to as **Hookean behavior**. Note that this rela-

Box 16.2 Elastic behavior

Elastic behavior can be described in terms of **length change** (translation), shape change (strain or distortion) and volume change (dilation). Length change, represented by the symbol e, refers to elongation of a linear feature. Length change is determined by dividing the change in line length (L – Lo) by the original line length (Lo), whereby:

$$e = \frac{(L - Lo)}{Lo}$$

Strain refers to shape changes or distortion. Shear strain (γ) is a measure of the angular changes to linear features. Rigidity (G), also known as shear modulus, is a measure of resistance to change in shape. For elastic materials, rigidity is a ratio of shear stress (σ_s) to shear strain such that:

$$G = \sigma_s / \gamma$$

Dilation, represented by the symbol D, refers to a change in volume such that:

$$D = K \frac{(V - Vo)}{Vo}$$

where Vo is the original volume, V the final volume and K the bulk modulus or incompressibility. The **bulk modulus** is a measure of the resistance to a change in shape. Bulk modulus, also known as incompressibility (Table B16.2a), can be expressed as $K = \Delta p / \Delta v$ where K is equal to the change in pressure divided by the change in volume.

Table B16.2 (a) Incompressibility and rigidity values at 1 atm and 25°C. Note that iron and copper are highly incompressible and that halite and ice are very compressible. (From Poirier, 1985; cited in van der Pluijm and Marshak, 2004.)

Mineral	Incompressibility (K)	Rigidity (G)
Iron	1.7	0.8
Copper	1.33	0.5
Olivine	1.29	0.81
Silicon	0.98	0.7
Calcite	0.69	0.4
Quartz	0.3	0.5
Halite	0.14	0.3
Ice	0.07	0.03

Decreasing incompressibility ↓

Box 16.2 *Continued*

Bulk modulus is related to **Poisson's ratio**, which is a measure of material "fattening" compared to its "lengthening" in response to compressive stress (Figure B16.2). Poisson's ratio (υ) is the change in object diameter divided by change in object length.

Earth materials have Poisson's ratios that range from 0 to 0.5, which means that all Earth materials increase in diameter and decrease in length in response to compressive stress. Poisson's ratios for some common Earth materials are listed in Table B16.2b.

Incompressibility and Poisson's ratio are very important factors in determining Earth material strength. In using Earth materials for construction, engineers need information relating to material strength and rheology so that roads, bridges, dams, tunnels, buildings and other structures do not collapse.

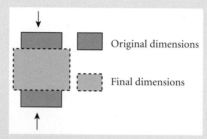

Figure B16.2 Illustration of Poisson's ratio in which material "fattening" is compared to its "lengthening" in response to vertical compressive stress.

Table B16.2 (b) Poisson's ratio values for Earth materials and processed materials. (Data from Van der Pluijm and Marshak, 2004; Hatcher, 1995.)

Material	Poisson's ratio
Fine-grained soil	0.45
Gold	0.44
Lead	0.43
Crushed stone	0.40
Concrete	0.35
Gabbro	0.33
Aluminum	0.33
Limestone	0.32
Slate	0.30
Steel	0.29
Gneiss, peridotite	0.27
Shale, sandstone	0.26
Basalt, granite	0.25
Glass	0.22
Metaquartzite	0.10

Increasing stiffness and reducing "thickening"

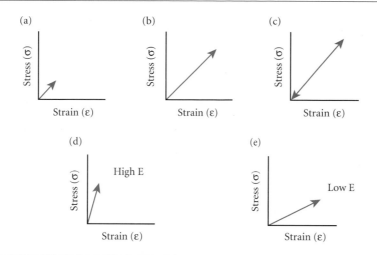

Figure 16.12 Elastic behavior depicted on idealized stress–strain graphs. (a) A small amount of stress produces a small amount of strain. (b) A larger amount of stress produces a larger amount of strain. (c) Stress is proportional to strain and is temporary; as stress is removed, strain diminishes. (d) Stiff rocks require high stress values to achieve a given strain value. (e) Rocks that are not stiff deform more with a given amount of stress.

tionship is not time dependent; no time lag occurs, so that strain begins when stress is first applied; and strain ceases immediately upon the removal of stress.

The slope of the stress–strain line is referred to as **Young's modulus of elasticity** (E), where E = stress/strain. Young's modulus of elasticity is a constant of proportionality that describes the slope of the line. The slope steepness of line E is a measure of resistance to elastic distortion. The E slope is dependent upon the stiffness or rigidity of the material. A rigid, stiff rock (high E) such as granite requires greater stress to achieve a given strain than a soft, pliable shale (low E). Young's modulus has an average value of $\sim 10^{-11}$ Pa for crustal rocks.

16.3.2 Plastic behavior

Plastic behavior is an irreversible strain that occurs without visible (mesoscopic) fractures. Microscopic fracturing may occur, but is not observable without the use of a petrographic microscope. Plastic behavior occurs through reorientation of the crystal structures. These processes (Chapter 4) include grain boundary sliding, dissolution, intracrystalline slip, twinning, kinking, bond elongation, dislocation and crystal defect processes that occur in

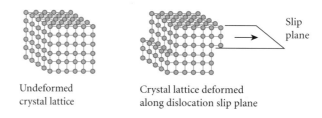

Undeformed crystal lattice

Crystal lattice deformed along dislocation slip plane

Figure 16.13 Plastic deformation proceeds through microscopic mechanisms such as slip along dislocation planes or other intracrystalline processes discussed in Chapter 4.

response to stress (Figure 16.13). Plastic behavior is favored by high temperature, high pressure, high stress and low strain rate. Other factors relate to grain size, rock composition and fluid pore pressure.

Plastic behavior occurs through the following mechanisms:

1 **Cataclastic flow**, in which mesoscopic ductile behavior is facilitated by microscopic fracturing and frictional sliding. Cataclastic flow occurs at low lithostatic pressures in the shallow crust.

2 **Diffusional mass transfer**, a high temperature and high pressure process that involves the flow of material through the

crystals. This is a temperature-dependent process, as thermal energy is necessary to break bonds by vibrating the crystal lattice structure. Diffusional mass transfer processes include pressure solution and solid state diffusion:

- **Pressure solution** is a high pressure diffusional mass transfer process (Chapter 15). In pressure solution, grain boundaries are compressed and dissolved, resulting in the generation of a fluid phase (Figure 16.14). As each mineral has different dissolution tendencies, pressure solution results in mineral differentiation whereby more soluble minerals are removed and less soluble minerals are concentrated (quartz vs. mica-rich zones; calcite vs. clays). Pressure solution can involve substantial volume loss and is particularly important in marble, metaquartzite, slate, limestone, dolostone, shale and quartz sandstone. Pressure solution in carbonate rocks is commonly indicated by stylolites, which are jagged seams of insoluble mineral residue that accumulate and concentrate along a dissolution seam. Stylolites (Figure 16.15) are commonly black – due to enrichment in carbon and iron oxides – and have the appearance of wound sutures stitched together. Pressure solution in clay-rich rocks produces cleavage in slates, as well as embayed grains and

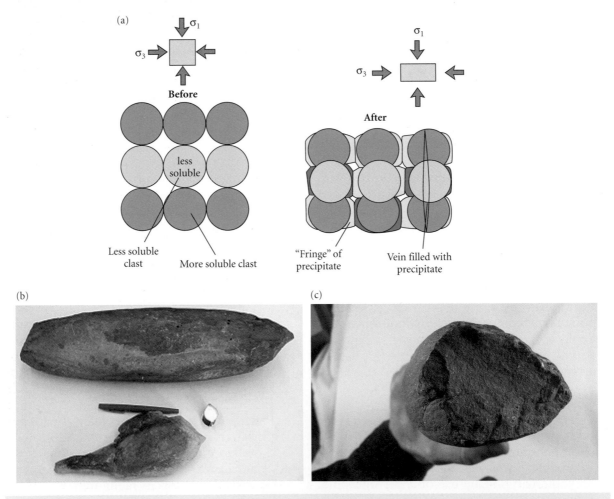

Figure 16.14 (a) Diffusional mass transfer where pressure solution occurs as a result of compression. (Courtesy of Robert Butler.) (b) Stretched quartz cobbles from Narraganset Basin, Rhode Island. The quartz pebbles experienced dissolution resulting in volume removal in the center of the cobble and precipitation on either ends, resulting in quartz beards or tails. Elongation is parallel to the pen. (c) Cross-section view of a quartz cobble. (Photos by Kevin Hefferan.)

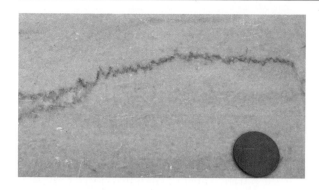

Figure 16.15 Stylolites in a marble slab formed by the concentration of insoluble residue along suture-like black seams. (Photo by Kevin Hefferan.)

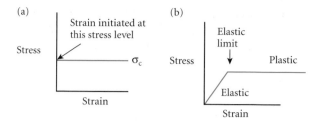

Figure 16.16 (a) Idealized plastic deformation initiates after a critical stress or yield stress value is attained. (b) Plastic behavior preceded by an elastic response at lower stress levels.

grain overgrowths in quartz-rich rocks such as sandstone and metaquartzite.

- **Solid state diffusion**, also known as grain boundary or volume diffusion is a high temperature and high pressure process by which solid particles experience translation within a mineral. Crystal lattice vacancies migrate to sites of greatest stress and atoms relocate to sites of minimal stress.

3 **Crystal defect** processes create adjustments in deformed crystal lattice structures. These deformities include point defects, line defects, edge dislocations and screw dislocations (Chapter 4).

4 **Crystal plasticity** is achieved by bending the lattice through gliding along weak planes within crystalline structures. **Mechanical twinning** is common in calcite and feldspar minerals. **Kinking** is common in micas and other platy minerals such as clays.

Diffusion, crystal defects and crystal plasticity are important plastic mechanisms in recrystallization and neocrystallization processes.

Plastic behavior is preceded by reversible elastic deformation prior to the onset of permanent plastic strain. Ideal plastic behavior initiates after some critical stress (σ_c) value, also known as yield stress, is achieved and maintained over time (Figure 16.16). The critical stress is analogous to trying to blow up a balloon for the first time. An initial stress level is necessary to inflate the balloon; once

achieved, that constant air pressure results in continued balloon expansion.

Rocks in the lower crust typically deform through plastic behavior as evidenced in folds and shear zones on a variety of scales ranging from millimeters to kilometers. Some rocks, such as rock salt, also deform plastically in the upper crust. Rock salt's ability to flow at shallow depths results in unusual qualities that make rock salt highly suitable for the storage of oil as part of the strategic petroleum reserve of the US government. Unlike brittle rocks, rock salt flows plastically and serves as an impermeable boundary surrounding a reservoir. On the basis of these properties, rock salt has also been proposed as a suitable rock for a nuclear waste repository.

16.3.3 Brittle behavior

Brittle behavior is permanent, irreversible deformation characterized by the development of visible fractures and loss of cohesion between rock particles. Brittle behavior is commonly preceded at lower stress levels by elastic behavior, and can also be preceded by significant plastic deformation (Figure 16.17). Rocks experience elastic deformation until a rupture point (rupture strength) is attained, whereat the rock loses cohesion and ruptures. The ultimate strength of the material is the maximum stress level that can be achieved prior to the onset of brittle failure. Brittle behavior characterizes most rocks in the upper crust resulting in widespread fracturing and fault displacement of rocks.

The **brittle–ductile boundary** (or brittle–ductile transition) is the depth within Earth where rock behavior changes from brittle to ductile behavior (Figure 16.18). Brittle rocks exhibit elastic behavior followed by rupture. Ductile rocks experience large amounts of plastic deformation before rupturing. Ductile rocks may deform elastically to a point called the **elastic limit**. Beyond the elastic limit, plastic deformation ensues with increasing stress. With the exception of subduction zones, earthquakes occur in the brittle portion of the lithosphere above the brittle–ductile boundary. The brittle–ductile boundary depth varies based on rock type, geothermal gradient and tectonic location. However, this boundary zone can be generalized as existing at depths ~10–20 km and temperatures of ~300°C.

16.3.4 Mineral deformation behavior

The rheology of Earth materials is fundamentally important in addressing earthquake behavior, engineering studies and in understanding deep crustal and mantle processes within Earth. Factors such as depth, temperature, stress conditions, mineral composition, rock texture, rock competency and strain rate affect rheology. Strain rate refers to the rate at which a rock is pulled apart, compressed or sheared. Generally, Earth materials display the following behaviors:

- Brittle behavior at shallow depths, low temperatures and high strain rate conditions.
- Ductile behavior at greater depths, higher temperatures and low strain rate conditions.

In many rocks, minerals such as quartz, feldspars, amphiboles, garnet and biotite are useful geothermometers because these common minerals change strain behavior with increasing temperatures (Passchier and Trouw, 2005). Table 16.1 indicates approximate temperatures at which these minerals change from brittle to ductile behavior. Actual temperatures may vary based upon other

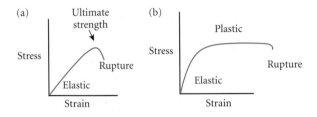

Figure 16.17 (a) Idealized rock response to stress in which low stress levels produce elastic behavior. As stress increases, the rock reaches a maximum stress level (ultimate strength) prior to the initiation of failure. The delay between ultimate strength and visible rupture marks the development of microfracturing processes that precede visible rupture. (b) Idealized rock response in which elastic behavior is followed by permanent plastic deformation prior to rupture.

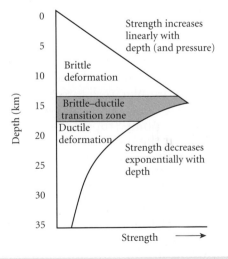
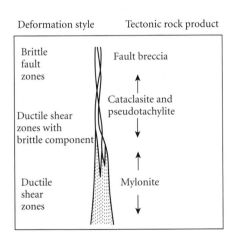

Figure 16.18 Brittle–ductile transition within Earth. (Courtesy of Mike Norton; with permission of Wikipedia.)

Table 16.1 Approximate temperatures at which some major minerals change from brittle to ductile behavior. (After Passchier and Trouw, 2005.)

Mineral	Brittle–ductile transition
Biotite	~250°C
Quartz	~300°C
Feldspar	~400°C
Amphibole	~650–700°C
Garnet	600–800°C

important factors such as pressure, volatile content and strain rate. By observing the deformation characteristics of minerals within a rock, geologists can make some educated guesses as to the metamorphic conditions.

Quartz and feldspars deform by brittle processes at temperature <300°C. At <250°C quartz exhibits greater strength than feldspars due to higher hardness and absence of cleavage. However, quartz progressively weakens as temperatures exceed 250°C. At temperatures greater than 300°C, quartz deforms via ductile processes such as dislocation creep and glide. Ductile processes produce undulose extinction, deformation lamellae and recrystallization. In contrast, feldspars continue to deform by brittle processes at temperatures up to 400°C, producing fractures, twinning and deformed cleavage planes. At temperatures >400°C, ductile processes dominate through dislocation glide and recrystallization. So, in assessing the deformation conditions in a granite containing quartz, feldspar, amphibole, biotite and garnet, the following conditions can be identified:

- <250°C: all minerals exhibit brittle deformation.
- 250–300°C: biotite exhibits ductile deformation; quartz, feldspars, amphibole and garnet remain brittle.
- 300–400°C: quartz becomes ductile but feldspars, amphibole and garnet remain brittle.
- >400°C: quartz and feldspars exhibit ductile behavior; amphibole and garnet exhibit brittle behavior.
- 600–650°C: garnet may begin to exhibit ductile deformation.

- >650°C: garnet and amphibole exhibit ductile deformation.

Let us consider other ways to describe the behavior of rocks under stress.

16.3.5 Rock competency

Competency is a term that describes the resistance of rocks to flow. Rocks that flow easily are less competent, or incompetent.

1 **Incompetent** rocks commonly display ductile behavior and include rock salt, shale, siltstone, slate, phyllite and schist. These rocks contain clays, micas, evaporates, talc, chlorite and other relatively soft minerals with Mohr's hardness <3.
2 **Competent** rocks commonly display brittle behavior and include metaquartzite, granite, gneiss, quartz sandstone, basalt, gabbro and diorite. These rocks contain minerals with Mohr's hardness >3 such as quartz, feldspars and ferromagnesian minerals.

Rock competency is dependent upon changes in pressure and temperature. In general, rock competency increases with higher pressure but decreases with higher temperature. Table 16.2 lists the relative competency of some common rocks.

Competency is related to strength. **Strength** refers to the amount of stress necessary to induce failure. The values listed in Table 16.3 are generalized averages of strength under compression and tension. Note that Earth materials are relatively strong under compressive stress and relatively weak under tensional stress. Other factors such as temperature, pressure, planes of weakness, specific rock composition, amount of weathering and fluids present also affect material strength.

A rock outcrop can display both brittle and ductile behavior based upon competency contrasts between rock layers. For example, competent rock such as a quartz sandstone or metaquartzite will rupture, while shale or phyllite will flow and fold in a ductile fashion (Figure 16.19a). The rupture of competent layers produces "French bread" or sausage-shaped structures called **boudins** (Figure 16.19b). Boudins are isolated rem-

Table 16.2 Relative competency of some common rocks at 1 atm and 25°C. (After Davis and Reynolds, 1996.)

Rock	Major minerals
Metaquartzite	Quartz
Granite	Quartz, feldspar, amphibole
Gneiss	Quartz, feldspar, amphibole
Quartz sandstone	Quartz
Basalt, gabbro	Pyroxene, plagioclase, amphibole
Dolostone	Dolomite
Limestone	Calcite
Schist	Mica, quartz, clay
Marble	Calcite
Shale	Clay
Rock gypsum	Gypsum, anhydrite
Rock salt	Halite

(Decreasing competency — indicated by downward arrow on left)

Table 16.3 Mineral and rock strength under compressive and tensile conditions at 1 atm and 25°C. (Data from Handin, 1966; Middleton and Wilcox, 1994; cited in Davis and Reynolds, 1996.)

Mineral or rock	Compressive strength (MPa)	Tensile strength (MPa)
Metaquartzite	360	–
Granite	160	14
Basalt	100	10
Limestone	80	10
Sandstone	50	10
Shale	30	8
Calcite	27	–
Halite	14	–
Clay	10	–

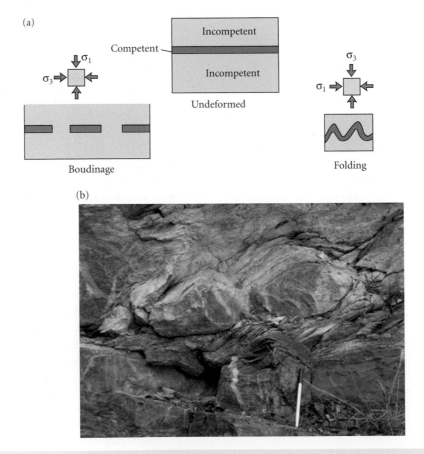

Figure 16.19 (a) Competence contrast results in both brittle and ductile behavior in response to the same stress conditions. (Courtesy of Robert Butler.) (b) Boudinage develops within competent Baraboo metaquartzite while the incompetent phyllite flows around the brittle boudin structures. (Photo by Kevin Hefferan.)

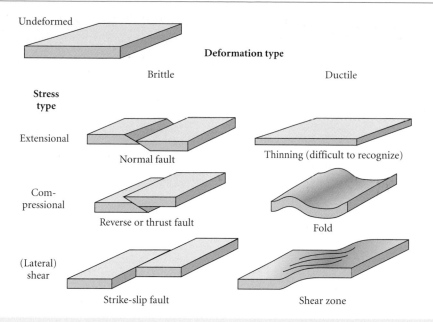

Figure 16.20 General types of structures produced by brittle and ductile deformation. (Courtesy of Bruce Railsback.)

nants of competent rock that once formed a continuous bed surrounded by less competent rocks.

Let us take a look at some common brittle and ductile rock structures. Brittle deformation results in the development of fractures such as joints and faults. Ductile deformation produces crustal thinning, thickening, folds and shear zones (Figure 16.20).

16.4 BRITTLE STRUCTURES

16.4.1 Fractures

Brittle behavior commonly occurs at depths less than 10 km because of upper crustal low temperature/low lithostatic pressure conditions, which allows for the development of fractures. **Fractures** are brittle structures that develop by rupturing either (1) previously intact rock, or (2) pre-existing weak surfaces in rock. Pre-existing weak surfaces include sedimentary bedding, volcanic flow contacts, weathering surfaces or metamorphic foliations. Fractures initiate from a point and migrate outward to form a discontinuity surface, which grows with time. **Joints** are fractures with minimal displacement.

Faults are fractures that involve measureable displacement. Two main types of faults

exist: (1) dip-slip faults, and (2) strike-slip faults. Dip-slip faults involve vertical displacement parallel to the dip of the fault. Dip-slip faults include normal faults, in which the hanging wall moves down relative to the footwall, and reverse faults, in which the hanging wall moves up relative to the footwall. Reverse faults in which the fault angle dip is less than 45° are referred to as thrust faults. Strike-slip faults involve horizontal displacement parallel to the strike of the fault. Relative to the observer, if the rock block on the far side of the fault moves to the left, it is a left lateral or sinistral strike-slip fault. If the block on the far side of the fault moves to the right, it is a right lateral or dextral strike-slip fault. Figure 16.21 illustrates the general orientation of the principal stress axes that produce normal, reverse and strike-slip faults.

16.4.2 Veins

Veins are the products of fluids flowing within fractures, producing one or more secondary minerals that precipitate from solution. Common secondary vein minerals include quartz, calcite, zeolite and chlorite. Ore minerals containing copper, silver, gold and other valuable metallic elements also occur in veins. Veins originate due to tensile or shear stresses.

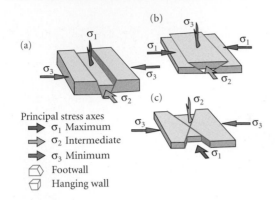

Principal stress axes
→ σ₁ Maximum
→ σ₂ Intermediate
→ σ₃ Minimum
⬡ Footwall
⬡ Hanging wall

Figure 16.21 (a) Two conjugate normal dip-slip faults. (b) Two conjugate reverse dip-slip faults. (c) Two conjugate strike-slip faults. The principal stress axes directions are indicated. (Courtesy of Robert Butler.)

Figure 16.22 Non-systematic, random stockwork veins in metaquartzite breccia, Baraboo, Wisconsin (USA). (Photo by Kevin Hefferan.)

While veins are highly associated with brittle, fractured rock, they also occur in ductile shear zones. Veins can be found as isolated features or in a vein array, in which groups of veins occur in a rock body. As with fractures, they can occur as non-systematic vein arrays in which rocks randomly fracture and infill with secondary mineral material. **Non-systematic vein arrays** are produced by high strain rate events that blast rock apart due to high pressures. Igneous intrusions with high volatile contents are capable of hydrofracturing rock producing random breakage. Rare meteorite impacts are also high strain events that produce massive disruption of rock. High strain rates can produce stockwork veins, which are a cluster of irregularly shaped veins of variable orientation that occur in a pervasively fractured rock body (Figure 16.22).

Systematic vein arrays consist of veins that display orientations suggesting a common origin in response to directed stress. For example, Figure 16.23 illustrates metaquartzite beds bounded by phyllite layers. In response to nearly vertical compressive stress, the metaquartzite experiences horizontal tension, resulting in brittle fracturing and the generation of extensional joints. Silica-rich fluids precipitate in the joints creating tension veins parallel to the maximum compressive stress direction. Figure 16.24 illustrates an **en echelon quartz vein array** that consists of a series of offset, parallel veins that formed in response to sinistral shear within metaquartzite.

16.4.3 Vein filling

Fluids that precipitate as secondary minerals in veins develop blocky or fibrous textures. **Blocky** or **sparry** minerals are equant and may display euhedral crystal faces indicating growth within an unimpeded open space (Figure 16.25).

Fibrous veins displays a linear, acicular character (Figure 16.26) suggesting that vein growth was incremental in response to fracture width increases. Fibrous veins develop by repeated cycles of a "crack and seal" mechanism whereby elevated fluid pore pressures crack a vein, followed by sealing from mineralized solution. Fiber growth provides information regarding displacement sense as well as progressive vein growth in both brittle and ductile environments.

16.5 DUCTILE STRUCTURES

Ductile deformation includes folds and shear zones. Folds form by the plastic bending of rock layers. Folded rock layers can consist of sedimentary beds, igneous intrusions or flows, or metamorphic layers or foliations. Shear zones are essentially ductile faults in which displacement is dominated by plastic deformation processes rather than brittle rupture.

(a) (b)

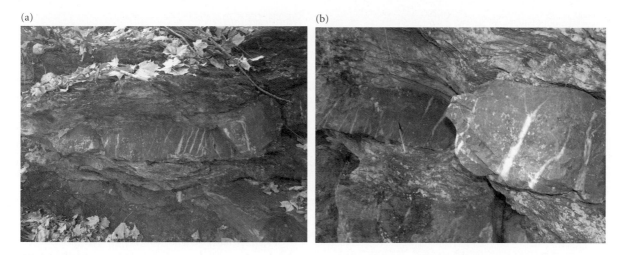

Figure 16.23 (a) Parallel, regularly spaced, systematic veins in competent metaquartzite boudins encased within less competent phyllite, Wisconsin (USA), with close up (b). (Photos by Kevin Hefferan.)

Figure 16.24 En echelon quartz vein array that formed in response to sinistral shear in metaquartzite, Wisconsin (USA). (Photo by Kevin Hefferan.)

Some structures, such as boudins and veins (previously discussed in brittle structures above), occur in ductile settings. Let us first consider folds and then move on to ductile shear zones.

16.5.1 Folds

Folds form by the plastic bending of rock layers without displaying mesoscopic brittle

Figure 16.25 Blocky spar quartz and feldspar vein filling within a granite. (Photo by Kevin Hefferan.)

Figure 16.26 Fibrous calcite veins indicating sinistral shear within metavolcanic rock, Bou Azzer, Morocco. (Photo by Kevin Hefferan.)

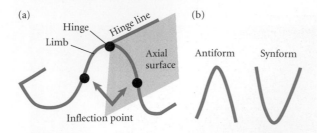

Figure 16.27 (a) Major components of folds. (b) Terms used to describe the form of elongate folds.

behavior. Most rocks fold due to compressive stress but folds can also be produced by tensional and shear stress. Folds consist of limbs and hinges (Figure 16.27). **Limbs** are relatively straight (low curvature) layers separated by a high curvature region of the hinge. The **hinge** is a point of maximum curvature separating two limbs. A **hinge line** is an imaginary line connecting a series of hinge points along the strike of the fold. An **axial surface** (**axial plane**) is an imaginary plane connecting a series of hinge lines. The **inflection point** is the point at which the sense of curvature changes from one fold to another. Elongate folds characterized by a convex-upward structure are called **antiforms** whereas those structures with concave-upward shapes are referred to as **synforms**.

Fold structures in which the relative ages of rock units are known include synclines, anticlines, domes and basins (Figure 16.28):

1 **Synclines** consist of two limbs that dip towards the hinge. Synclines contain young rock in the hinge and progressively older rock away from the hinge.
2 **Anticlines** consist of two limbs that dip away from the hinge. Anticlines contain old rocks in the hinge and progressively younger rock further away from the hinge.
3 **Domes** are circular to oval structures in which rock layers dip away from the center and the oldest rocks are exposed in the center of the structure.
4 **Basins** are circular to oval structures in which rock layers dip towards the center and the youngest rocks are contained in the center.

Refolded folds

Rocks may experience multiple episodes or generations of folding during one orogenic cycle or multiple orogenic pulses. As a result, complex fold patterns develop that represent a culmination from a number of different fold cycles. These can produce fold interference patterns due to the successive refolding of pre-existing structures. Because younger fold structures are superimposed upon earlier fold structures, refolded folds are referred to as **superposed folds** or superimposed folds (Figure 16.29).

Transposed folds consist of folds in which the limbs and hinges have been pulled apart due to extension (Figure 16.30). Transposed folds occur with multiple fold generations, involving the replacement of an earlier tectonic fabric (S_1) by a more recent tectonic fabric (S_2) by ductile mechanisms such as recrystallization and pressure solution. Transposed folds are associated with high temperature and high pressure metamorphism.

Parasitic folds are small folds occurring in the limbs and hinges of larger scale folds. The form of the parasitic fold can provide information as to the limb or hinge location on a large fold structure. Bedding plane slip commonly produces rotation on fold limbs. The upper bed surface is displaced towards the hinge, while the lower bed surface moves away from the hinge. Rotation to the right produces clockwise bed rotation (producing a Z-shape) while bedding plane rotation to the left produces counterclockwise (or S-shape) parasitic folds. Parasitic folds in the hinge experience minimal rotation. Figure 16.31 illustrates how parasitic fold orientation can be used in the analysis of a map-scale antiform. Geologists can determine position within a map-scale fold structure by looking down the plunge direction of the parasitic folds.

Faults and folds in rocks are prime targets for oil and gas exploration as well as ore mineral deposits. Faults and folds create structural traps that encase and concentrate hydrocarbons in sedimentary rock material. The high heat and pressures of metamorphism generally exceed the conditions at which hydrocarbons can be preserved. However, high fluid flow rates through deformed crustal rocks can concentrate metallic ore deposits in

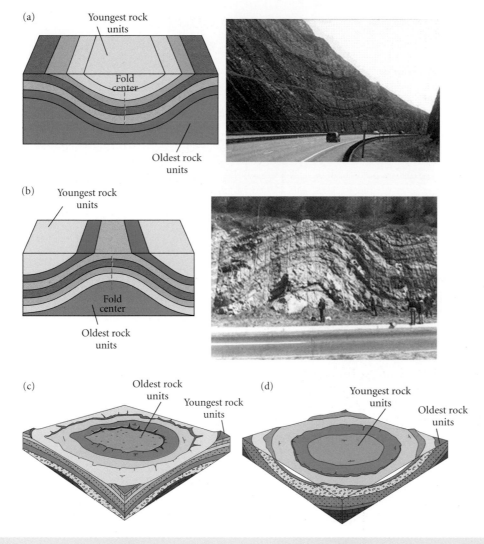

Figure 16.28   Idealized block diagrams illustrating: (a) a syncline, (b) an anticline, (c) a dome, and (d) a basin. (Photos by Kevin Hefferan.)

sedimentary, igneous and metamorphic rocks. In summary, fold and fault structures are studied intensely for their potential as economically viable exploration targets.

16.5.2   Shear zones

Shear zones are essentially ductile fault zones that accommodate displacement. Shear zones develop in ductile lower crustal rocks. In the field, ductile shear zones commonly occur in tectonic mélanges (Figure 16.32), mylonites and pseudotachylites (Figure 16.33). Mélange consists of larger rock blocks encased within a scaly, clay-rich matrix. The matrix develops a crude layering and internal structures can

indicate the sense of motion within the shear zone. The larger rock blocks may change shape or orientation and also provide information regarding sense of shear.

16.6   PLANAR AND LINEAR STRUCTURES

We have summarized many of the common brittle and ductile structures that occur within deformed rocks. These structural features are all part of a rock's fabric. **Fabric**, also known as texture, refers to the geometric arrangement of grains within a rock. Metamorphic rock fabric is a critical component in rock

(a)

(b)

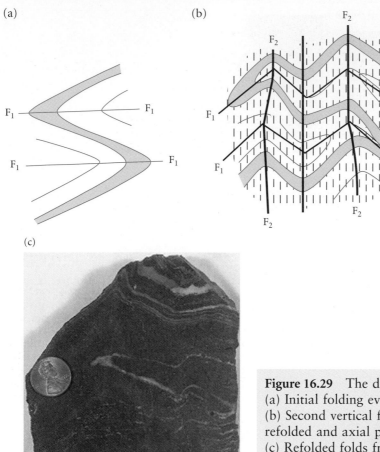

(c)

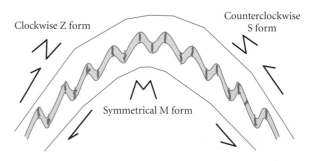

Figure 16.29 The development of superposed folds.
(a) Initial folding event with F_1 recumbent folds.
(b) Second vertical folding event in which F_1 folds are refolded and axial planar F_2 cleavage develops.
(c) Refolded folds from the Penokean Range of Wisconsin. (a, b, after Hobbs et al. (1976); reproduced by permission of Win Means. c, photo by Kevin Hefferan.)

Figure 16.30 Transposed folds in metamorphosed ironstone in which "rootless" fold hinges and limbs have become disassociated as a result of intense deformation. (Photo by Kevin Hefferan.)

Figure 16.31 Parasitic folds produce Z (clockwise rotation), M (symmetrical form) or S (counterclockwise rotation) shapes indicating shear sense.

Figure 16.32 Tectonic mélange displaying metabasalt encased within a muddy matrix, Bou Azzer, Morocco. (Photo by Kevin Hefferan.)

Figure 16.33 Intense deformation can result in localized melting and the development of pseudotachylite rock within a mylonite, such as in the Siroua region of Morocco. (Photo by Kevin Hefferan.)

identification (Chapter 17). Fabric can be described in a number of ways, such as:

1 When did the fabric develop? A **primary fabric** develops during lithification processes that produce the rock. A **tectonic fabric** is produced by deformation processes after the initial lithification of the rock. Metamorphic rocks are dominated by tectonic fabrics although some primary

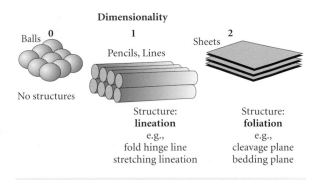

Figure 16.34 Random, linear and planar fabric elements. (Courtesy of Robert Butler.)

fabrics (e.g., sedimentary bedding) may be preserved.

2 What is the spatial arrangement of the fabric?

• A **continuous fabric** infers that the structures are continuous on a millimeter scale so that no undeformed parts of the rock remain. A **spaced fabric** means that visible spacing exists between fabric elements so that both deformed and undeformed parts of the rock are visible.

• **Random fabric** infers that no preferred orientation of component elements exists; instead, rock components are arranged in a random fashion. **Preferred fabric** means that rock elements are aligned in a predictable manner. Two main classes of preferred fabrics exist: **foliations** are planar fabrics and **lineations** are linear fabrics (Figure 16.34).

16.6.1 Planar fabrics

Planar features are sheet-like structures that include joints, veins, faults, axial surfaces of folds, shear zones and cleavage. We have already provided brief overviews of many planar rock objects in our discussion of brittle and ductile structures. In this section, we will focus upon cleavage.

Cleavage

Cleavage refers to the tendency of rocks to break along sub-parallel surfaces. Cleavage is a tectonic planar fabric produced by non-uniform stresses under relatively low temperature metamorphic conditions. Processes such

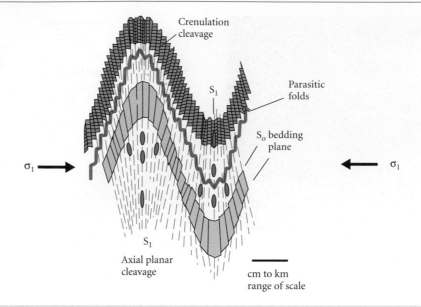

Figure 16.35   Cleavage bedding relationships. Note that the axial planar cleavage (S1) dips at a steeper angle than the bedding (So). Also note the association of fracture cleavage, crenulation cleavage, slaty cleavage and parasitic folds with axial planar cleavage. These structures commonly occur together in folded rock. (Courtesy of Patrice Rey.)

as pressure solution and recrystallization result in the alignment of inequant minerals so that rocks tend to split or cleave along planes. Cleavage commonly develops in response to compressive stresses associated with dynamic or dynamothermal metamorphism. Cleavage is described on the basis of orientation to geological structures, origin and spacing.

Foliations are particularly well developed in folded rock layers, where they form an axial planar cleavage. **Axial planar cleavage** consists of parallel foliations oriented nearly perpendicular to the maximum compressive stress, and converging towards the inner arc of the fold hinge area. Figure 16.35 illustrates rock beds (So) that have been folded into an antiform and synform pair. Note that the axial plane cleavage (S1) is steeper than the bedding angle in upright folds. In cases where the axial planar cleavage is dipping less steeply than the bedding, complex folding patterns have occurred such as refolded folds or overturned folds wherein the rock stratigraphy is inverted.

Other types of planar structures include slaty and phyllitic cleavage as well as schistosity, gneissic banding and mylonitic folia-

tions. These will be presented in our discussion of foliated textures in Chapter 17. Let us now take a brief look at linear fabrics.

16.6.2   Linear fabrics

Linear structures contain one long axis and two short axes, producing needle-like structures. Linear features include intersection lineations, form lineations, crenulation lineations, stretching lineations and surface or slip lineations (Figure 16.36).

Intersection lineations form by the intersection of two planar fabrics (Figure 16.36a). **Pencil cleavage** is an intersection lineation marked by the development of elongate, pencil-like shards (Figure 16.37). Pencil cleavage commonly forms due to the intersection of two fracture sets or a fracture set and bedding. The "pencils" are ~5–10 cm long and generally less than 2 cm in width and height. The pencil-like shape is facilitated by the internal alignment of inequant clay minerals. Pencil cleavage develops in weakly deformed shales or mudstone and is an early stage in the development of slaty cleavage.

Form lineations are linear features produced by geological structures.

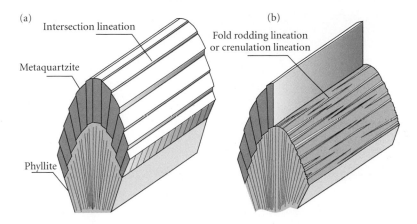

Figure 16.36 (a) Intersection lineations formed by the intersection of cleavage and bedding. (b) Form lineations such as the development of rodding structures. (Courtesy of Patrice Rey.)

Figure 16.37 Pencil cleavage from slate within the Anti-Atlas Mountains, Morocco. (Photo by Kevin Hefferan.)

Figure 16.38 Crenulation lineations within metavolcanic rock, Colorado (USA). (Photo by Kevin Hefferan.)

Crenulation lineations are linear features that occur as a result of a secondary cleavage imposed upon a fine-grained rock (slate or phyllite) that experienced an earlier cleavage (Figure 16.36b). Crenulation lineations represent hinge lines formed by the intersection of two planar surfaces (Figure 16.38).

Stretching lineations develop as elongated mineral or rock grains that define a linear fabric (Figure 16.39). Stretching lineations commonly form on metamorphic foliations, on shear surfaces or on mylonitic planes. These lineations may be due to either (1) growth of a crystal in a preferred orientation, or (2) rotation of crystals toward a principal strain direction.

Slip or **fiber lineations** refer to vein mineral fibers that precipitate on rock surfaces via crack–seal processes. Slickenlines are fiber lineations produced during displacement in faults and shear zones (Figure 16.40).

This chapter serves as an introduction to stress and deformation structures encountered in Earth materials. Metamorphic rock textures, and therefore the rocks themselves, are primarily defined by rock structures produced by stress. An understanding of metamorphic rock textures requires knowledge of stress and deformation. For more detailed information on structural geology, the reader

Figure 16.39 Stretching lineations in hornblende in metamorphosed granite, Big Thompson Canyon, Colorado (USA). (Photo by Kevin Hefferan.)

Figure 16.40 Slickenlines on an exposed portion of the San Andreas Fault, San Francisco (USA). (Photo by Kevin Hefferan.)

is referred to excellent textbooks by Twiss and Moores (2006), Passchier and Trouw (2005), van der Pluijm and Marshak (2004), Davis and Reynolds (1996), Hatcher (1995), Ramsay and Huber (1984, 1987), Suppe (1984), and Hobbs et al. (1976) among others.

In Chapters 17 and 18 we will discuss metamorphic textures, rocks, zones, facies and facies series. Hopefully the stress and deformation features discussed in this chapter will provide a basis for understanding metamorphic rock structures.

Chapter 17

Texture and classification of metamorphic rocks

All rocks are classified on the basis of composition and texture. As discussed in Chapter 15, metamorphic rock composition is largely determined by protolith chemistry. Despite the importance of composition, the primary name for many metamorphic rocks is based on their texture or rock structure.

Texture, also known as **fabric**, refers to the size, shape, orientation and intergranular relationships of the rock's constituents. Metamorphic textures may be due to the metamorphic changes induced by the temperature and pressure conditions of metamorphism or **relict textures** inherited from the protolith. Sedimentary protoliths contain a variety of stratification features such as ripple marks, mudcracks, graded bedding, fossils and tool marks that can be preserved in lower grade metamorphic rocks. Relict features derived from igneous rocks include volcanic flow contacts, intrusive contacts, vesicles, phenocrysts and other features that survive the temperatures and pressures of metamorphism.

Earth Materials, 1st edition. By K. Hefferan and J. O'Brien. Published 2010 by Blackwell Publishing Ltd.

17.1   GRAIN TEXTURE

17.1.1   Grain shape

Metamorphic grains are dominated by minerals formed in response to high heat and pressures. Grain shape is subdivided into two broad groups: equant grains and inequant grains.

1   **Equant grains** possess nearly equal diameters in all directions and assume forms approximated by spheres or cubes. They are described by three mutually perpendicular axes of roughly equal length.
2   **Inequant grains** contain at least one direction in which the grain diameter is not equal to the other grain diameters. The sub-parallel alignment of inequant grains produces metamorphic foliations. Inequant forms include:
   • Tabular, disc-shaped (pancake or paper-like) grains in which one axis is significantly shorter than the other two axes.
   • Bladed (prismatic) grains in which one axis is significantly longer than the

Table 17.1 The tendency of common metamorphic minerals to develop complete crystal forms. (After Winter, 2001.)

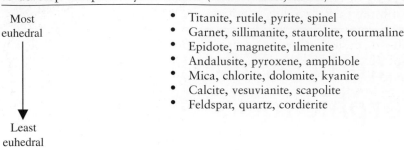

Most euhedral

- Titanite, rutile, pyrite, spinel
- Garnet, sillimanite, staurolite, tourmaline
- Epidote, magnetite, ilmenite
- Andalusite, pyroxene, amphibole
- Mica, chlorite, dolomite, kyanite
- Calcite, vesuvianite, scapolite
- Feldspar, quartz, cordierite

Least euhedral

other two axes, which are not equal to one another.

- Acicular (needle-like) or prolate (cigar-shaped) grains in which one axis is significantly longer than the other two axes, which are of equal length.

17.1.2 Grain size

Metamorphic grain size classification is similar to that used for igneous rocks. Aphanitic grains possess diameters of less than 1 mm and phaneritic grains have diameters of 1 mm or greater. In Chapter 7, we discussed porphyritic igneous textures in which larger phenocryst crystals occur amid smaller diameter groundmass minerals. Metamorphic rocks can also display similar textures with bimodal size distributions in which larger crystals are called porphyroclasts or porphyroblasts.

Porphyroclasts are large relict grains from the protolith that have experienced deformation but have retained their original composition. Their large size relative to surrounding minerals is due to significant crushing or stretching of the surrounding minerals and/or the growth of new, smaller crystals. Common porphyroclast minerals include quartz and feldspars. **Augen** are oval-shaped feldspar porphyroclasts that resemble the shape of an eye and are particularly common in gneisses, whereas **flaser** are composed of quartz.

Porphyroblasts are large grains that have experienced neocrystallization and growth in response to favorable temperature and pressure conditions during metamorphism. As described in Chapter 15, neocrystallization refers to the growth of new minerals stable at the temperature and pressure conditions of metamorphism. Common porphyroblast minerals include garnet, staurolite and cordierite.

You will recall that in our discussion of igneous textures (Chapter 7) the terms euhedral, subhedral and anhedral were used to describe the degree to which crystals are bounded by crystal faces; they are also appropriate terms for metamorphic crystals. Whether porphyroblasts are euhedral, subhedral or anhedral depends upon many factors related to the conditions of metamorphism. However, as indicated in Table 17.1, metamorphic minerals possess different tendencies to develop crystal faces during solid state growth.

17.1.3 Grain orientation

Metamorphic grains can be arranged in a random or a preferred orientation. A **random grain orientation** occurs when no preferred orientation of inequant grains is visible. Random grain orientation results in a non-foliated texture. **Preferred grain orientation** occurs when inequant grains are oriented sub-parallel to one another and can produce lineations and foliations. Preferred orientation of acicular, bladed or rod-like (e.g., inosilicate minerals) grains with sub-parallel long axes produces lineations. **Lineations** are line-like features similar to pencils all pointing in a common direction. Preferred orientation of tabular grains, especially phyllosilicate minerals, with sub-parallel long axes produces foliations. **Foliations** appear as metamorphic layers oriented parallel to one another like pages in a book. As discussed in

Chapter 16, uniform stresses commonly produce random grain orientations; lineations and foliations commonly develop due to non-uniform stresses.

We have briefly discussed the shape, size and orientation of metamorphic crystals and their relationship to lineations and foliations. On the basis of grain orientation, metamorphic rocks can be divided into two major groups:

1 **Rocks with non-foliated textures** in which crystals lack a preferred orientation;
2 **Rocks with foliated textures** in which crystals possess a preferred orientation.

Metamorphic rock classification is largely based on the presence or absence of foliations. As we shall see, this simple approach poses some difficulties because:

- Gradations exist between weakly foliated and non-foliated textures.
- Rocks can retain relict layering derived from the protolith.
- Some rocks, especially those defined primarily on the basis of composition, occur in both foliated and non-foliated forms depending upon the stress conditions.

Despite these limitations, texture is an appropriate first approach to metamorphic rock classification. Mineralogy, chemical composition and origin are also important factors in naming metamorphic rocks. In the next two sections, we consider the common non-foliated and foliated metamorphic rocks.

17.2 NON-FOLIATED METAMORPHIC ROCKS

Non-foliated textures lack metamorphic layering as defined by preferred mineral orientations. Equant (e.g., quartz) and inequant (e.g., clay, mica, amphibole) grains can occur in non-foliated rocks; however, inequant grains are not aligned sub-parallel to one another. Non-foliated textures are commonly produced by dynamic metamorphism, contact metamorphism or burial metamorphism (Chapter 15), in which uniform lithostatic stress produces equant or randomly arranged minerals so that prominent foliated textures are not produced.

(a)

0.5 mm

(b)

Figure 17.1 (a) Photomicrograph of hornfels displaying equant crystals separated by biotite-rich layers that may represent relict sedimentary bedding. (Photo courtesy of Kent Ratajeski.) (b) Hornfels collected from the base of the Palisades Sill, New Jersey. (Photo by Kevin Hefferan.) (For color version, see Plate 17.1, between pp. 408 and 409.)

17.2.1 Metamorphic rocks with hornfelsic or granoblastic textures

Hornfels

Hornfelsic texture is a fine-grained (<1.0 mm diameter), non-foliated fabric that develops by contact metamorphism, producing a rock called **hornfels** (Figure 17.1). Hornfels is derived from fine-grained protolith rocks such as shale, mudstone, tuff or basalt. The crystals in hornfelsic textures are predominantly equant. Where present, tabular to acicular crystals lack preferred alignment. Relict sedimentary fabrics such as bedding may be preserved.

Hornfels develops in metamorphic aureoles, adjacent to igneous intrusions. Hornfels minerals include muscovite, biotite, andalusite, cordierite, plagioclase, potassium feldspar, epidote, amphibole and pyroxene. Mineral assemblages are strongly influenced by protolith composition and the temperatures achieved during metamorphism. Recrystallization causes hornfels to be somewhat harder and more brittle than the mudstone protoliths from which it commonly forms. Although hornfels is composed of fine-grained crystals, spotted hornfels with larger porphyroblasts are common.

Granoblastic rocks

Granoblastic textures are characterized by large (>1.0 mm diameter) equant grains or large inequant crystals that lack preferred orientation (Figure 17.2). Granoblastic textures also occur in high grade rocks known as granulites, that form at elevated temperature and pressure conditions associated with deep burial. Granoblastic textures develop during metamorphism of a wide range of protoliths under uniform stress conditions. Crystals in granoblastic rocks commonly display anhedral, sutured boundaries that reflect a combination of pressure solution, recrystallization and annealing. Granoblastic textures most commonly develop with minerals such as quartz, feldspar and calcite that have low euhedral form potential (see Table 17.1) and tend to crystallize in subequant forms. Phyllosilicate or other tabular to prismatic minerals that form foliations are commonly absent. Granoblastic texture characterizes many nonfoliated rocks such as metaquartzite and marble and contact metamorphosed skarn deposits. The most common rocks with granoblastic textures are described below.

Metaquartzite

Metaquartzites contain >90% quartz and are derived from quartz-rich sandstone or chert protoliths (Figure 17.3). Accessory minerals commonly include hematite and feldspars. In the transformation of quartz sandstone or chert to metaquartzite, recrystallization, pressure solution and intracrystalline plastic deformation mechanisms produce interlocking quartz grains that commonly display granoblastic texture. Because the quartz grains interlock, rupture occurs through

0.25 mm

Figure 17.2 Photomicrograph of granoblastic texture with equant calcite crystals. (Photo courtesy of Kent Ratajeski.) (For color version, see Plate 17.2, between pp. 408 and 409.)

Figure 17.3 The Proterozoic age Baraboo metaquartzite in Wisconsin contains relict features that include ripple marks that formed during the deposition of quartz sand in a coastal environment over 1 billion years ago. (Photo by Kevin Hefferan.)

grains, rather than around grain boundaries. This gives metaquartzites a smooth, glazed appearance as opposed to the granular appearance of many quartzarenites. Because quartz is stable over a wide range of metamorphic temperatures and pressures, metaquartzite, also referred to as quartzite, is a common metamorphic rock. Metaquartzites are hard, durable rocks that produce angular surfaces when fractured. As a result of the sharp edges, metaquartzites are not ideal for road construction because of the tendency to tear rubber tires. Metaquartzites are used for rock walls, railroad ballast and drainage culverts.

Marble

Marbles are granoblastic metamorphic rocks rich in calcite and/or dolomite. Marbles are derived by recrystallization of limestone or dolostone protoliths via dynamothermal, deep burial or contact metamorphism. Because both calcite and dolomite are stable over a broad range of metamorphic temperatures and pressures, both calcitic and dolomitic marbles are common. Common accessory minerals include graphite and calcium- and/or magnesium-bearing minerals such as brucite, diopside, forsterite, wollastonite, epidote, serpentine, idocrase (vesuvianite), tremolite and grossular garnet. Accessory minerals provide distinctive hues that allow marble to assume a wide variety of colors. For example, iron oxides can impart a reddish or yellow color to marble, graphite colors marble gray to black, while serpentine produces green marble (Figure 17.4). Marbles can either retain some relict sedimentary structures and fossils or exhibit total recrystallization so that all prior sedimentary textures and structures are obliterated.

Marbles are relatively soft and are easy to cut. Because of their great color diversity and softness, marbles have been extensively used for sculptures and building stone for thousands of years (Figure 17.5). Artisans preferred marble for creating classic sculptures such as Venus de Milo and Michelangelo's David, sculptured from Greek and Italian marble, respectively.

Skarn

Skarns, also known as tactites, are granoblastic calc-silicate rocks formed by contact metamorphism of carbonate country rocks such as limestone or dolostone. The release of silica and volatiles from the magma results in extensive metasomatism, generating calc-silicate mineral assemblages and/or metallic ore deposits. Skarns commonly contain carbonate minerals such as calcite, dolomite and

Figure 17.4 Marble occurs in multiple colored varieties depending upon chemical substitutions in calcite and dolomite and the presence of accessory minerals. (Photo by Kevin Hefferan.) (For color version, see Plate 17.4, between pp. 408 and 409.)

Figure 17.5 Italian marble is used in construction and sculpture as in the Doge's Palace in Venice. (Photo courtesy of Tony Crowe.)

ankerite and silica group minerals such as quartz. Common calc-silicate skarn minerals (calcium–magnesium silicates) include wollastonite, tremolite, diopside, talc, epidote, grossular garnet, phlogopite and idocrase (vesuvianite). A number of different skarn types occur; their classification is largely related to the associated ore minerals such as gold, silver, tungsten, molybdenum or iron.

17.2.2   Metamorphic rocks with cataclastic or non-crystalline textures

Cataclastic rocks

Metamorphic rocks with **cataclastic textures** are composed of fractured, angular particles that form in response to the brittle crushing of grains during deformation in upper crustal fault zones (Figure 17.6). Cataclastic textures develop during dynamic metamorphism in a wide variety of rocks within 15 km of Earth's surface.

Cataclastic textures are described with respect to the relative percentages of larger clasts and finer matrix, and their degree of cohesion (Higgins, 1971; Sibson, 1977). Cataclastic rocks that lack cohesion are called **breccia** if they contain coarse (>2 mm diameter), angular fragments and **gouge** if they are composed of finer sized fragments.

Metabreccia
Metabreccias are derived from metamorphism of sedimentary or igneous breccias. Metabreccias commonly develop during dynamic or dynamothermal metamorphism. Metabreccias contain subangular to angular clasts with diameters >2 mm (Figure 17.7). Extensively recrystallized matrix grains interlock so tightly that rupture occurs through grains, rather than around grain boundaries. Metabreccias can be composed of a single clast type or many different clast types.

Cataclasite
Cataclasites are cohesive rocks with cataclastic textures produced by brittle deformation. Cataclasites form under low temperature, high strain, dynamic metamorphic conditions such as upper crustal fault zones. Cataclasites retain primary cohesion, which refers to their ability to remain as a cohesive mass during deformation. Primary cohesion distinguishes cataclasites from non-cohesive fault zone rocks such as gouge and breccia. Cataclasites are commonly non-foliated; however, phyllosilicate-rich cataclasites can exhibit foliations. These rocks are classified on the basis of percent matrix (Table 17.2).

While cataclasites are not volumetrically significant components of Earth's crust, they do serve as impermeable seals in fault zones,

Figure 17.6 Photomicrograph of cataclastically deformed sand from Miocene sands in an accretionary prism off southern Mexico. (From Lucas and Moore, 1986; courtesy of J. Casey Moore.)

Figure 17.7 Metabreccia containing oxidized hematitic basalt rock fragments and epidote cemented together with calcite. (Photo by Kevin Hefferan.) (For color version, see Plate 17.7, between pp. 408 and 409.)

impeding the escape of resources such as oil and gas. In addition, when rocks are subjected to new stresses, cataclasites tend to rupture more easily than the surrounding host rock, so that faults are reactivated. As a result, cataclasites are of interest in both resource exploration and seismology.

Pseudotachylite

Pseudotachylites are glassy rocks produced by high strain rates generating localized melting in fault zones (Figure 17.8). Near instantaneous solidification of melts produced by pressure release during fracturing produces very dark-colored, vitreous to flinty rocks. These rocks commonly occur as vein material in cataclastic rocks such as fault breccias and cataclasites.

Impactite

Impactites are high strain rate cataclastic rocks created by the tremendous short-term stresses associated with extraterrestrial rock bodies impacting Earth (Chapter 15). Four major components of impactites are recognized:

Table 17.2 Cataclasite series rock terms developed by Sibson (1977).

Percent matrix	Cataclasite series rock name
10–50	Protocataclasite
50–90	Cataclasite
90–100	Ultracataclasite

1 Impact breccia produced by the fragmentation of rock upon impact.
2 Glassy spherules called tektites that form as rocks are locally melted due to impact. The impact of extraterrestrial rock bodies produces showers of droplets that cool very rapidly as tektites.
3 Deformation lamellae that form due to the intense stresses that deform crystal structures (Figure 17.9). These are especially common in "shocked" quartz crystals;
4 Ultra-high pressure mineral assemblages such as the high pressure polymorphs of silica including coesite or stishovite. Although produced synthetically in the laboratory by Loring Coes in 1953, the

Figure 17.8 Pseudotachylite formed in a ductile shear zone within leucogabbro. (Photo by Kevin Hefferan.)

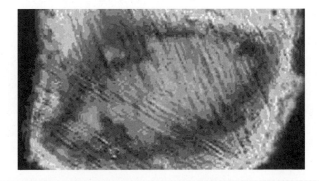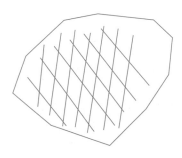

Figure 17.9 Photomicrograph of deformation lamellae in shocked quartz from the Chesapeake Bay impact site (USA). The sample was collected from 250 m underneath NASA's Langley Research Center in Hampton, Virginia, which sits on the crater rim. The line drawing illustrates two sets of shock lamellae. (Courtesy of the US Geological Survey and Glen A. Izett.)

high pressure mineral coesite was first discovered in the field at Meteor Crater (Chao et al., 1960).

Anthracite coal

Anthracites are non-crystalline, high grade coals that form by heating, compressing and chemically altering bituminous coal. Anthracite coals form by burial and dynamothermal metamorphism and represent a very important energy resource. Anthracite coals are vitreous, light weight and jet black in color and commonly display conchoidal fractures. As discussed in Chapter 14, they are harder, more vitreous and have substantially higher thermal capacity than bituminous coals due to the loss of volatiles and the concentration of carbon. Unfortunately, the mining and burning of coal have serious environmental consequences. Surface or strip mining is highly destructive to the landscape both in the removal of rock material and subsequent dumping of tailings in valleys and streams. Underground coal mining poses great dangers to the miners excavating the rock. Abandoned underground mines experience roof collapse, causing subsidence and sinkhole development on Earth's surface. Coal burning is also a major source of smog and greenhouse gases (SO, SO_2, NO, NO_2, CO, CO_2, CH_4), particulate pollution and heavy metals such as mercury and uranium. Of all the fossil fuels, coal is the most detrimental to our environment. Nevertheless, because of its abundance and ease of use, we will continue to utilize coal for the foreseeable future.

17.2.3 Non-foliated to foliated metamorphic rocks

The rocks discussed in the section above constitute the more common and/or interesting non-foliated metamorphic rocks. The distinction between non-foliated and foliated textures is not always well defined either in the field or in hand-specimens. In a given rock exposure, a non-foliated rock may grade into poorly foliated or even a well-foliated rock. Rocks such as metaconglomerates, serpentinites, soapstones, greenstones, granulites, eclogites and amphibolites can occur with non-foliated or foliated textures as discussed below.

Metaconglomerate

Metaconglomerates are derived from conglomerate protoliths and contain sub-rounded to rounded relict clasts with diameters >2 mm. Metaconglomerates can form by deep burial, dynamothermal or contact metamorphism. Most metaconglomerates display non-foliated textures. Extensively recrystallized matrix grains interlock so tightly that rupture occurs through grains, rather than around grain boundaries. The rock and mineral clast assemblage of metaconglomerates varies widely as a result of the wide range of protolith clast compositions, as well as the temperature and pressure conditions of metamorphism. Rock clasts may be derived from sedimentary, igneous or metamorphic rocks. As with sedimentary conglomerates (Chapter 13), metaconglomerates may be oligomictic (one clast type) or polymictic (many clast types).

Stretched pebble metaconglomerates (Figure 17.10) form by the metamorphism of conglomerates and/or breccias in response to strong non-uniform stress during dynamothermal or dynamic metamorphism. During metamorphism, pebbles and cobbles are shortened and/or flattened parallel to the Z-strain direction and relatively elongated parallel to the X-strain direction (Chapter 16). Pebble alignment may define a metamorphic foliation or lineation fabric. Excellent

Figure 17.10 Stretched pebble conglomerate in which the clasts have been flattened parallel to the direction of the least compressive stress. (Photo by Kevin Hefferan.)

examples of lineations occur in Pennsylvanian Purgatory Formation metaconglomerate in the Narraganset Basin, Rhode Island (USA) where stretched pebbles are arranged parallel to one another like loaves of French bread.

Serpentinite

Serpentinites are serpentine-rich metamorphic rocks that occur in non-foliated or foliated forms. Serpentinite forms by hydrothermal alteration of ultrabasic rocks at temperatures below ~500°C. During metamorphism, olivines and pyroxenes are hydrated to form serpentine group minerals in a process aptly named serpentinization. Serpentine group minerals (Chapter 5) include lizardite $[Mg_3Si_2O_5(OH)_4]$, chrysotile $[Mg_3Si_2O_5(OH)_4]$ and antigorite $[(Mg,Fe)_3Si_2O_5(OH)_4]$. Lizardite and chrysotile are low temperature minerals and antigorite is the higher temperature mineral. Serpentinization occurs (1) at ocean spreading ridges, where hydrothermal fluids circulate through fracture systems that penetrate the ultrabasic mantle, and (2) at subduction zones where ocean water interacts with ultrabasic rocks of the ocean lithosphere. Serpentinite slivers are commonly imbricated with other rocks in tectonic mélanges within subduction complexes. These include metamorphic rocks discussed later in this chapter, such as amphibolite, greenstones and eclogites, as well as other rocks found in ophiolite assemblages. Serpentines are used for building stone and as industrial and residential asbestos. Serpentinites commonly occur with soapstones, discussed below.

Soapstone

Soapstones are fine-grained rocks that form through the alteration of ultrabasic rocks, or magnesium-rich sedimentary rocks such as dolostone, by low temperature and low pressure hydrothermal fluids. Soapstones contain talc in combination with other minerals such as magnesite, serpentine and/or tremolite. These minerals impart a low hardness and white to green color. Talc contributes a soapy feel and low hardness that allows it to be cut easily. Soapstones are the material of choice for many sculptures, especially Native American sculptors in the Arctic rim. They are also used for ornaments, kitchen sinks and coun-

tertops. Soapstone's low porosity prevents staining or seepage. Soapstones are used in woodstoves and fireplaces – even smoking pipes – due to their ability to withstand and evenly dissipate heat.

Greenstone

Greenstones are green-colored rocks rich in silicate minerals that commonly include chlorite, epidote, prehnite, pumpellyite, talc, serpentine, actinolite and albite. Greenstones form by low to moderate (200–500°C) temperature alteration of basic and, to a lesser extent, ultrabasic igneous rocks. During metamorphism, plagioclase and primary ferromagnesian silicates such as olivine, pyroxene and amphibole are converted into the greenstone minerals listed above. Greenstones are commonly produced by hydrothermal metamorphism of basalts and gabbros in oceanic crust near divergent plate boundaries. Many **spilites** (sodium-rich basalt) and **keratophyres** (sodium-rich andesite) occur in greenstones. These greenstones can later be incorporated into mélanges and orogenic belts along convergent plate boundaries.

Greenstones also occur on a very large scale in **greenstone belts,** which are abundant in Precambrian rocks (Box 17.1). Greenstone belts occur in large synclinal structures. In a typical greenstone belt sequence, ultrabasic metavolcanic rocks (**komatiites**) and metabasalt form the basal layers and are overlain successively by intermediate and silicic metavolcanic and metavolcaniclastic sequences, which are in turn capped by graywackes and chert. Greenstone belts commonly parallel granulite belts containing rocks of granitic to dioritic composition metamorphosed at high temperatures and pressures. Greenstone belts are tens of kilometers wide and hundreds of kilometers long and occur in Archean (>2.5 Ga) and Proterozoic (2.5 Ga to 544 Ma) cratons around the world. Phanerozoic greenstone belts are rare. The best known greenstone localities include the Barberton belt in South Africa, the eastern goldfields of Western Australia, the Superior and Slave Provinces in North America, and the Sao Francisco Craton in Brazil. Greenstone belts yield valuable metallic ore deposits containing copper, gold, silver, nickel, zinc and lead (Hoatson et al., 2006).

Box 17.1   Greenstone belts

Greenstone belts dominate the history of continents in the Archean and Early Proterozoic Eons (>2 Ga). An estimated 260 greenstone belts occur throughout the world (de Wit and Ashwal, 1997). Greenstone belts are so named because they contain green-colored minerals such as actinolite, chlorite, epidote, prehnite, pumpellyite, serpentine and talc. These minerals formed by extensive metasomatic alteration of basic and ultrabasic rocks through chemical reactions with H_2O and CO_2. Greenstone belts are synclinal to tabular rock assemblages that contain peridotite and gabbroic intrusive rocks and ultrabasic to basic volcanic rocks called komatiites. Komatiites are overlain successively by basalt and rhyolite layers. The igneous layers are capped by metasedimentary layers that include chert, graywacke, banded iron formations and conglomerates. Archean and Proterozoic greenstones occur in association with granitic gneisses in Precambrian cratons (Figure B17.1). Two major questions remain regarding greenstone belts:

1   In what tectonic setting did Precambrian greenstone belts form? The origin of Precambrian greenstone belts remains controversial (Bickle et al., 1994; de Wit, 1998). Proposed environments for greenstone belt formation include ocean ridges, continental rifts (Grove et al., 1978), accreted subduction zone/island arc fragments (de Wit and Ashwal, 1997), oceanic plateaus formed above deep mantle plume hotspots, or extensive lava plains unrelated to plate tectonics (Hamilton, 1998).

2   Why are greenstone belts uncommon in the Phanerozoic Eon? This question also continues to perplex geologists and is the subject of current research. The answer is perhaps related to lower geothermal gradients in the Phanerozoic as well as changes in plate tectonics through time. A higher Precambrian geothermal gradient may have resulted in higher grade metamorphism at shallow depths and prevented deep subduction along steeply inclined Benioff zones. These conditions may have favored the development of shallow greenstones and prevented the development of the deep, high pressure assemblages that characterize modern subduction zones.

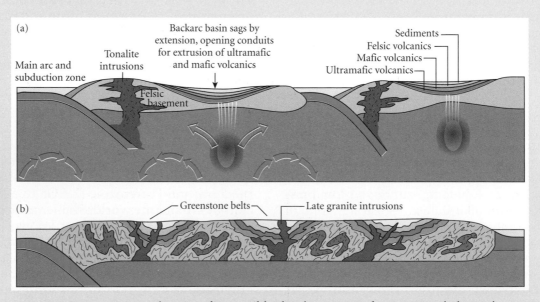

Figure B17.1   Cross-sections showing the possible developement of greenstone belts and associated granulite rocks. (a) Synkinematic Precambrian tectonic activity. (b) Modern erosional setting. (From Levin, 2006; with permission of John Wiley & Sons.)

Amphibolite

Amphibolites are dark-colored rocks composed largely of amphiboles, such as hornblende, and plagioclase. Garnet is also a common major mineral. Accessory minerals include quartz, potassium feldspar, tremolite, biotite, magnetite, chlorite, calcite, sphene and epidote. Amphibolites form by the medium to high temperature (>550°C) regional metamorphism of basic igneous rocks or sedimentary rocks such as calcareous mudrocks and graywackes. Amphibolites derived from basic igneous rocks such as basalt and gabbro are called **ortho-amphibolites**, whereas amphibolites produced from sedimentary protoliths are called **para-amphibolites**. Amphibolites may or may not be foliated.

Granulite

Granulites are medium- to coarse-grained rocks that contain granoblastic to foliated textures. Granulites form by high temperature (>800°C) and high pressure (>10 kbar; ~33 km depth) metamorphism. High temperature granulite conditions trigger dehydration reactions resulting in the transformation of hydrous amphibole and mica minerals into anhydrous minerals such as pyroxene, potassium feldspar, kyanite and garnet. The high pressure and very low water content prevents melting, and preserves the rock's metamorphic fabric. Granulites from the type locality of Saxony, Germany are granoblastic rocks of granitic composition that contain quartz, feldspars and minor amounts of kyanite and garnet. Modern usage of the term granulite refers to rocks of the granulite facies (Chapter 18) and may be derived from a wide variety of protoliths that include granitoids, diorite, gabbro, peridotite and pelitic rocks such as mudstone and graywacke. Common granulite minerals include orthopyroxene, clinopyroxene, calcic plagioclase, garnet and quartz. With increasing temperature and pressure within granulites, orthopyroxene and plagioclase transform into clinopyroxene, garnet and quartz (O'Brien and Rötzler, 2003). Granulites form in high temperature and high pressure conditions of the lower continental crust, upper mantle and as a result of subduction of crust at convergent margins. Granulites are also common in Precambrian greenstone belts and in association with eclogites.

Eclogite

Eclogites are very high pressure, high temperature rocks that develop principally from basalt/gabbro protoliths. Eclogites may be the major rock type in Earth's lower crust because they are stable at temperatures that exceed 400°C and pressures that exceed 1.2 GPa (>40 km depth). Eclogites are commonly red and green because they contain green jadeite pyroxene, omphacite (sodium/calcium pyroxene) and red garnet as major minerals. Garnets in eclogite include pyrope, almandine and grossular garnet. Accessory minerals consist of quartz, rutile, phengite mica, coesite, lawsonite, kyanite, corundum, zoisite, glaucophane, epidote, allanite and diamond (Enami and Banno, 2000).

Eclogite can form by a number of processes, which include (1) high pressure recrystallization of deep continental crustal rocks during thickening at continent–continent collisions, (2) partial melting of the mantle followed by deep crystallization as high pressure eclogite, or (3) high pressure metamorphism of subducted oceanic lithosphere deep within Earth. In fact, eclogite's high density (3.5–4.0 g/cm^3) may be one of the driving forces for plate motion. The slab-pull effect generated by eclogite in subducted slabs is thought to drive mantle convection and plate motion.

Let us now consider some of the more common foliated metamorphic rocks.

17.3 METAMORPHIC ROCKS CONTAINING FOLIATED TEXTURES

Foliated textures are marked by the planar alignment of inequant crystals such as tabular phyllosilicates (micas, clays) and prismatic/acicular inosilicates (amphiboles, pyroxenes). Foliated textures are commonly associated with dynamic or dynamothermal metamorphism, in which rocks change shape in response to non-uniform stresses, producing minerals aligned in a planar fabric. Foliated textures include slaty cleavage, phyllitic cleavage, schistosity, gneissic layering, migmatitic layering and mylonitic foliation. Rocks containing these textures are briefly described below.

Figure 17.11   Slate quarry in Arvonia, VA (USA). (Photo courtesy of Duncan Heron.)

Figure 17.12   Note the wavy foliation and the overall fine grain size of this photomicrograph of a phyllite. (Photo courtesy of Kent Ratajeski.)

Slate

Slates are fine-grained, aluminum-rich, pelitic rocks that possess flat, planar cleavage. Slaty cleavage consists of closely spaced layers along which the rock breaks or cleaves readily to produce flat surfaces with a dull luster. The layering is defined by the sub-parallel orientation of microscopic phyllosilicate mineral grains. Slaty cleavage commonly forms during the metamorphism of clay-rich sedimentary protoliths such as shale, siltstone and mudstone or altered pyroclastic volcanic rocks such as tuff. Slaty cleavage develops during metamorphism under non-uniform stress at relatively low temperatures (~150–250°C) and low pressures. At these relatively low temperatures, relict sedimentary textures and structures may be preserved. During metamorphism, inequant phyllosilicate minerals such as clays and micas are reoriented into planes that are roughly perpendicular to the direction of maximum shortening, creating foliations that are typically at a high angle to the original bedding. With increasing metamorphic temperatures clay minerals such as kaolinite and smectite and zeolite minerals decrease in abundance, while illite, chlorite and mica concentrations progressively increase. Accessory minerals include quartz, graphite, pyrite, ilmenite, chlorite, plagioclase, muscovite and iron oxide minerals such as hematite.

Slates, like the mudstones from which they commonly form, possess many colors including red, green, gray and black, which is largely inherited from the protolith. Oxidized iron imparts a brick red color; reduced iron produces green colors. Organic matter deposited in anoxic environments and commonly recrystallized into graphite, produces dark gray and black-colored slates. Slates are actively quarried for many construction applications (Figure 17.11). The flat, planar nature of slaty foliations allows it to be split into thin sheets that make it suitable for sidewalks, roof shingles, school blackboards, pool tables, patios and floor tiles. In rugged topography, slates present a significant geohazard when steeply dipping foliations are inclined downslope as this can result in unstable weathered rock faces that slide downslope by mass wasting.

As temperatures rise progressively above 250°C, slates are transformed into higher grade metamorphic rocks such as phyllite, schist and gneiss.

Phyllite

Phyllites display phyllitic cleavage, characterized by larger crystals and more wavy surfaces than slaty cleavage (Figure 17.12). The larger, more reflective microscopic grains in phyllitic cleavage induce a silky or glossy sheen. Phyllites commonly develop by the recrystallization of slate and therefore from the same protoliths as slates. At temperatures of ~250–

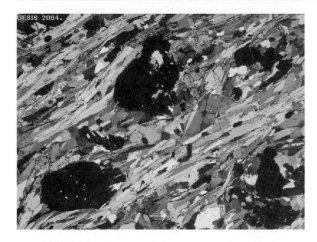

Figure 17.13 In this garnet mica schist, viewed between crossed polarizers, the less competent mica minerals are thinned and bent around black garnet porphyroblasts. The field of view is 6 mm. (Photo courtesy of David Waters and Department of Earth Sciences, University of Oxford.) (For color version, see Plate 17.13, between pp. 408 and 409.)

300°C, smectite and illite clays metamorphose to slightly coarser grained minerals such as sericite, muscovite, talc and chlorite that align parallel to each other and define foliations. Other common metamorphic minerals include quartz, feldspar, biotite, graphite, prehnite and pumpellyite. Phyllitic cleavage develops in response to non-uniform stresses at temperatures and pressures that exceed those that produce slaty cleavage. Phyllitic cleavage forms perpendicular to the direction of maximum shortening. Many phyllites display a crenulation cleavage that cross-cuts an earlier generation of cleavage.

Despite higher temperature conditions and levels of deformation, relict sedimentary structures are preserved in many phyllites. As temperatures increase, progressively greater amounts of recrystallization and neocrystallization occur so that grain size increases, visible porphyroblasts are produced and relict textures disappear. At temperatures above ~300°C, phyllites are transformed gradationally into schist (Miyashiro, 1973). Like slates, phyllites present a mass wasting hazard when steeply dipping foliations are inclined downslope.

Schist

Schists are characterized by a foliation called schistosity. Schistosity is a very common foliation defined by the sub-parallel arrangement of macroscopic platy minerals such as phyllosilicates in closely spaced metamorphic layers. This foliation is commonly less regular than that of either slates or phyllites. Light-reflecting, macroscopic crystals generally impart a high sheen or sparkle to the schistosity. Even relatively rigid minerals such as quartz and feldspars are commonly stretched or flattened so as to be parallel to the foliation direction. Less competent minerals such as micas are commonly stretched and folded around more rigid porphyroclasts and porphyroblasts (Figure 17.13). Major minerals are used in the rock name as a preceding adjective; for example, sillimanite-rich schists are called sillimanite schists.

Mineral assemblages in schists depend on both metamorphic conditions and protolith composition. Schists, like slates and phyllites, can develop from pelitic (shale, mudstone, graywacke) or altered tuff protoliths. Pelitic schists are enriched in aluminosilicate minerals such as andalusite, kyanite, sillimanite, micas and other minerals listed in Table 15.3. Schists can also be derived from many other protoliths including a wide variety of igneous rocks. For example, basalt protoliths commonly contain hornblende, garnet, actinolite and glaucophane. Thus, visible mineral grains provide key information related to rock chemistry, as well as the specific temperature and pressure range under which peak metamorphism occurred.

Schists are produced by dynamothermal metamorphism at convergent plate boundaries with temperatures >300°C (Miyashiro, 1973). Under such conditions, both brittle and ductile deformation may occur. For example, quartz begins to deform plastically at ~300°C, while feldspar minerals remain rigid. Moderate to high temperatures and non-uniform stresses induce the development of visible porphyroclasts (e.g., feldspars) or porphyroblasts (e.g., garnet and/or staurolite) in a strongly foliated, schistose fabric (Figure 17.13). As a result of extensive deformation, relict textures are rarely preserved in schists.

Gemstones and industrial minerals such as garnet and corundum provide moderate resource value for schists. As with other foliated phyllosilicate-rich metamorphic rocks, schists present a mass wasting hazard where foliations are inclined downslope.

Hydrous minerals are stable in schists, which begin to form at temperatures of ~300°C. As temperatures exceed ~400°C, dehydration processes initiate so that hydrous minerals become unstable (e.g., talc and chlorite). As a result, anhydrous minerals increasingly predominate in high grade rocks such as gneisses, migmatites and granulites.

Gneiss

Gneisses are characterized by gneissic layering, a foliation characterized by the arrangement of minerals into distinct color bands. The color bands are most commonly due to alternating light-colored quartz and/or feldspar-rich layers and dark-colored layers rich in biotite, amphibole and – at increasing temperatures – pyroxenes. In addition to the major non-ferromagnesian and ferromagnesian minerals listed above, accessory minerals include garnet, sillimanite, cordierite and corundum. Rigid feldspar grains may have been sheared into oval eye-shaped (augen) crystals producing augen gneiss (Figure 17.14). Quartz grains are commonly been sheared into lens-shaped grains called flaser. Both are testimony to the development of gneissic banding in response to non-uniform stresses.

Gneisses develop from lower grade metamorphic rocks as well as from a variety of protoliths including granites, diorites, gabbros, mudrocks, tuffs and graywackes. The protolith dictates the mineral components in gneiss (Chapter 15). Gneisses that develop from an igneous protoliths are called **orthogneiss** whereas those that develop from sedimentary protoliths are called **paragneiss**. Many gneisses with a granitic composition are referred to as granite gneiss. Gneisses derived from gabbro can produce mafic gneiss.

Gneissic foliations develop due to extensive layer transposition, recrystallization and neocrystallization processes that result in the segregation of minerals into separate layers. Several processes have been suggested for the development of gneissic banding:

1   The vast majority of gneissic bands originate by layer transposition. Transposition results from the pulling apart of earlier folded layers resulting in the separation of hinges and limbs (Figure 17.15). Insoluble phyllosilicate minerals (e.g., micas) and inosilicate minerals (e.g., amphibole and pyroxene) are concentrated in fold limbs; soluble quartz and feldspars are concentrated in hinge areas of tightly compressed folds resulting in alternating light and dark color bands.

Figure 17.14   Augen gneiss with oval feldspar porphyroblasts. (Photo by Kevin Hefferan.)

Figure 17.15   Gneissic color banding with light-colored quartz and feldspar layers and dark pyroxene layers. Note the fold hinge within the mafic layer in the left side of image, which suggests transposition processes. (Photo by Kevin Hefferan.)

2 Differentiation through partial melting (anatexis) and recrystallization. Ferromagnesian minerals have higher melting temperatures than quartzofeldspathic minerals. As a result, anatexis produces igneous layers enriched in quartz and feldspar between unmelted, refractory layers enriched in dark-colored ferromagnesian minerals. Together, they create gneissic banding.

3 "Lit par lit intrusion", which refers to the thin, sill-like intrusion of magma into parallel country rock layers. Lit par lit intrusions can occur to a limited extent when granitic magma intrudes mafic country rock producing alternating light and dark color bands.

Gneisses are used as dimension stones for tiles and countertops or for the construction of tombstones, crypts and buildings. Because their hardness and crude foliation impede easy splitting, gneisses do not pose a mass wasting geohazard.

Gneisses form in dynamothermal settings at temperatures that commonly exceed ~600°C. All relict sedimentary textures are erased due to high temperature neocrystallization processes and layer transposition. However, original igneous textures and structures can be preserved in orthogneisses. At temperatures above ~700°C, extensive melting of gneiss can lead to the disassembly of gneissic banding and the formation of migmatites as discussed below.

Migmatite

Migmatites are "mixed" rocks (Figure 17.16) that possess textural and structural characteristics of both igneous and metamorphic rocks. Whereas gneisses have well-defined color bands, migmatites commonly display an irregular, swirling mix of colors. Migmatites typically contain zones of rock with the outward appearance of granitoid igneous rocks mixed with zones of rock that resemble typical gneiss. Light-colored segments are enriched in quartz and feldspars; dark-colored components are enriched in mafic minerals. With additional melting, migmatites melt sufficiently so as to produce magma.

Migmatites develop under high temperature (>800°C) conditions in the lower crust.

Figure 17.16 Proterozoic migmatite that formed in a convergent margin zone in Morocco during the Pan African orogeny. (Photo by Kevin Hefferan.)

They form via dynamothermal metamorphism at convergent plate boundaries by a number of processes that involve some combination of (1) partial melting (anatexis), (2) magma injection, and/or (3) ductile deformation and plastic flow of rocks in the lower crust. Migmatites have no resource or hazard considerations.

Ironstone

Ironstones, discussed in more detail in Chapter 14, are silica- and iron-rich rocks that formed primarily in the Early Proterozoic and Archean. Ironstones commonly occur as banded iron formations, in which alternating hematite, magnetite and chert layers form red and black color bands. Metamorphosed ironstones can also have gneissic banding with alternating quartz-rich layers and magnetite-rich layers, which may or may not be related to the original compositional layering (Figure 17.17). Deposits in the Lake Superior region of North America and in Western Australia are among the richest iron deposits on Earth.

Figure 17.17 Lake Superior taconite rock sample displaying transposed folds. To the right, note the processed taconite pebbles produced in the mining process. (Photo by Kevin Hefferan.)

Table 17.3 Mylonite series as defined by Sibson (1977). Note that this terminology follows that used for the cataclasite series in Table 17.2.

Percent matrix	Mylonite series rock name
10–50	Protomylonite
50–90	Mylonite
90–100	Ultramylonite

(a)

(b)

Figure 17.18 (a) Protomylonite sample containing asymmetrical porphyroclasts. (b) Photomicrograph (~4 mm across) of a mylonite with an asymmetrical feldspar porphyroclast. (Photos by Kevin Hefferan.)

Taconites are metaquartzites that contain 20–30% iron and are also commonly banded. Ironstones have been major sources of iron ore since the mid 1800s. Taconites, once considered waste rocks because of their lower iron content, are now the principal ore rocks mined in the Lake Superior region due to depletion of richer ore deposits.

Mylonite

Mylonites are fine-grained, foliated rocks produced in the ductile shear zones of the lower crust and mantle (Chapter 16). In ductile shear zones, rocks undergo intense ductile deformation that may involve both crushing and grinding (cataclasis) and plastic flow. Grains become extensively elongated parallel to the shear zone. High temperatures and pressures associated with dynamic metamorphism in the lower crust and mantle result in

intensely sheared and recrystallized porphyroclasts and/or neocrystallized porphyroblasts (Figure 17.18).

Mylonites can occur in a variety of rocks such as schist, gneiss or granite. Mylonites are classified on the basis of the percentage of matrix material to porphyroclasts (Table 17.3). Higher percentages of matrix generally record greater intensity of grain size reduction and more intense shear strains during deformation.

Mylonites provide important information regarding the sense of displacement in shear zones. In the following section, we will address some shear sense indicators used in the study of mylonitic rocks. Mylonites have no significant resource or geohazard considerations.

Tectonite

Metamorphic **tectonites** are pervasively deformed rocks so that their original composition and texture are largely obliterated. Tectonites are defined by their solid state flow fabric generated through intense ductile or brittle–ductile deformation. Foliated tectonites are called **S tectonites**. Tectonites with a pronounced lineation, but no foliation, are called **L tectonites**. Tectonites with both foliation and a lineation are referred to as **L-S tectonites**.

17.4   SHEAR SENSE INDICATORS

Shear sense indicators are rock structures or textural elements that provide information concerning the relative sense of displacement of rock components within fault zones or shear zones (Simpson, 1986). Shear sense indicators include a number of microscopic and megascopic structures discussed below.

Offset markers

Offset markers indicate the sense of displacement in shear zones and may also provide information on displacement distance. Linear features (e.g., fold axes) cut by planar faults or shear zones provide "**piercing points**" that allow determination of the absolute displacement distance between offset lines. When a linear feature, such as a fold axis, intersects that plane a point is produced. If the linear feature is cut into two separate objects due to displacement, then two offset points are produced. The distance between those two offset piercing points, measured in the shear plane, records absolute displacement distance. Unfortunately, piercing points are exceedingly rare in the field. More commonly, offset grains or foliations provide a means to determine relative sense of displacement (Figure 17.19).

Grain tail complexes

Grain tail complexes are asymmetrical porphyroclasts or poryphyroblasts with mineral tails that "point" in the direction of shear. Tail complexes form commonly about minerals such as feldspars and quartz. Mineral tails may form by a combination of:

Figure 17.19   The dextral offset of a gneissic felsic layer indicates the sense of displacement. (Photo by Kevin Hefferan.)

- Plastic flattening of pre-existing mineral grains (commonly feldspars or quartz).
- Pressure solution of material from the grain center.
- Dynamic recrystallization of material at the rim of the grain.
- Neocrystallization in pressure shadows around the grain.

Grain tail complexes are particularly well developed in schists, gneisses and mylonitic rocks. Mylonitic rocks experience intense ductile shearing, which favors the plastic deformation processes required to create grain tail complexes. Sigma and delta grain tail complexes are distinguished by the relationship between the grain tails and an imaginary reference plane of shear that passes through the grain center.

Sigma grain tail complex
Sigma (σ) grain tail complexes consist of wedge-shaped tails that do not cross the reference plane of shear (Figure 17.20). Through pressure solution and rotation, mineral material is removed from the outer part of the grain and is precipitated in pressure shadows parallel to the minimum stress direction. Incremental tail growth occurs in the direction of minimum compressive stress (σ_3), located at 45° to the reference plane (and

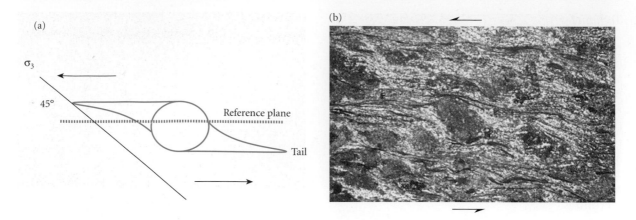

Figure 17.20 (a) Cross-section illustrating a σ grain tail complex in a sinistral shear zone. (b) Photomicrograph of σ grain tail complexes within mylonitic rock subjected to sinistral shear, Bou Azzer, Morocco. (Photo by Kevin Hefferan.)

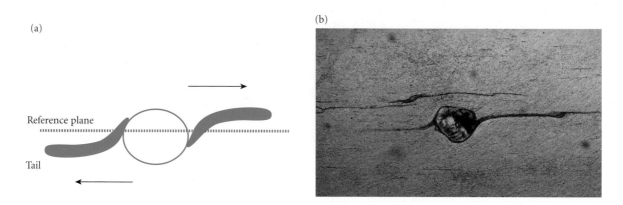

Figure 17.21 (a) Cross-section illustrating dextral shear as indicated by a δ grain tail complex that crosses the imaginary reference plane. (b) A δ grain tail complex in an ultramylonite that has experienced dextral shear. (Photo courtesy of Cees Passchier and Rudolph Trouw; with permission of Springer Science+Business Media.)

shear plane), during rotational shear. σ grain tail complexes develop by slow grain rotation relative to tail growth.

Delta grain tail complex
Delta (δ) grain tail complexes (Figure 17.21) are produced by relatively rapid grain rotation relative to tail growth rate. Rapid grain rotation results in a significant bending of the earlier formed parts of the tail so that it crosses the reference plane.

Fracture patterns

Fracture patterns can develop preferred orientations that indicate sense of shear. **Synthetic fractures** are preferentially oriented in the direction of shear. The fractures are inclined at low angles (<45°) to foliations. Synthetic fractures show displacement consistent with the overall sense of shear (Figure 17.22). **Antithetic fractures** are preferentially oriented in a direction opposite to the sense of shear. Antithetic fractures are inclined at high angles (>45°) to the foliation plane and display an opposite sense of movement to overall sense of shear (Figure 17.23).

S-C foliations

S-C foliations develop in mylonitic, schistose and gneissic rocks subjected to ductile shear.

Figure 17.22   A set of synthetic fractures, all of which are inclined to the right and parallel to the sense of movement, that indicate dextral shear in metamorphosed cherts, Bou Azzer, Morocco. (Photo by Kevin Hefferan.)

Figure 17.23   Outcrop-scale en echelon, antithetic fractures in metavolcanic rock recording dextral shear in Bou Azzer, Morocco. (Photo by Kevin Hefferan.)

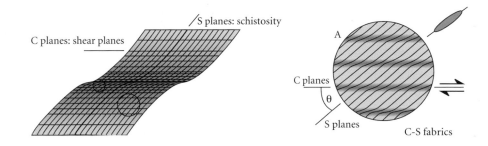

Figure 17.24   S-C structures in a dextral shear zone. (Courtesy of Patrice Rey.)

The letter S represents schistosity (foliation) and the letter C is for "cisaillement", a French term for shear direction, which lies in the C plane (Figure 17.24).

Figure 17.25 illustrates S-C structures developed within granite mylonite subjected to dextral shear. This granite mylonite is used as floor tiles in the Denver International Airport, Colorado (USA). Figure 17.26 illustrates a stunning photomicrograph of S-C structures in which phyllosilicates form the schistosity and dissolution seams occur within the cisaillment zone.

All of these different shear sense indicators can be useful in determining sense of shear in deformed rocks. Only under relatively rare circumstances can displacement be accurately

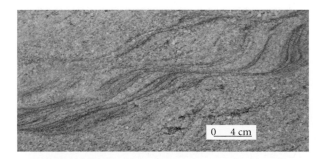

Figure 17.25   S-C structures in Proterozoic mylonitic granite from Colorado. This rock is used as floor tiles in the Denver Airport. (Photo by Kevin Hefferan.)

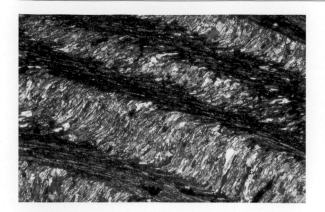

Figure 17.26   S-C structures in phyllosilicate-rich mylonite. Note the metamorphic segregation that has produced gneissic layers, wherein ferromagnesian minerals are concentrated in the C planes, which represent fold limbs. The light-colored mineral bands of the S plane occur in the S fold hinges. (Photo courtesy of Cees Passchier and Rudolph Trouw; with permission of Springer Science+Business Media.)

measured. For additional information regarding shear sense indicators, the reader is referred to Simpson (1986) and Passchier and Trouw (2005). In the following chapter we will focus upon mineral and rock assemblages that develop in response to changes in temperature and pressure.

Chapter 18

Metamorphic zones, facies and facies series

A major objective in the study of metamorphic rocks is the recognition of geothermobarometers that permit us to accurately infer metamorphic temperature and pressure conditions during metamorphism. As we shall see, key minerals and mineral assemblages and subtle variations in mineral chemistry effectively serve this vital role. Ideally metamorphic reactions produce minerals that are in equilibrium with the temperature and pressure conditions of metamorphism. To review:

- Temperatures of metamorphism vary from ~150°C to over 1200°C. This broad range can be subdivided into low temperature (~150–400°C), moderate temperature (400–600°C) and high temperature (>600°C) conditions (Fyfe et al., 1958).
- Pressures of metamorphism vary from near atmospheric conditions in some contact metamorphic reactions to extremely high pressures in the lower crust, mantle and subduction zone environments. This broad pressure range can be subdivided into

low pressure (0–2 kbar ≈ 0–6 km depth), moderate pressure (2–6 kbar) or high pressure (>6 kbar ≈ >20 km depth). Well into the second half of the 20th century, metamorphic petrologists considered the highest metamorphic pressures to be ~15 kbar (Fyfe et al., 1958). Remember that 3.3 kbar corresponds to ~10 km depth within the Earth, so the maximum depth of formation of metamorphic rocks now exposed at the surface was considered to be 45–50 km. With a greater understanding of convergent margins and the discovery of ultra-high pressure rocks, geologists now recognize that metamorphic pressures may exceed 30 kbar, which is equivalent to depths >100 km. In fact, some ultra-high pressure xenoliths are derived from depths of 400 km (Pearson et al., 2003). Ultra-high pressure metamorphism is a new and fascinating field of discovery.

- Volatiles serve as catalysts in hydrothermal and metasomatic reactions, thereby altering mineral composition and creating ore deposits. Without volatiles, metamorphic reaction kinetics slow or cease. This explains why anhydrous high temperature

Earth Materials, 1st edition. By K. Hefferan and J. O'Brien. Published 2010 by Blackwell Publishing Ltd.

and pressure assemblages return to Earth's surface relatively unchanged. The absence of volatile catalysts inhibits retrograde metamorphic reactions.

Despite the wide variety of possible mineral components, most metamorphic rocks contain from one to six essential minerals. The mineral assemblage provides critical information regarding the protolith and/or the conditions of metamorphism. In the following discussion we will consider key individual minerals and then address mineral assemblages that serve to provide constraints on the conditions of metamorphism. We will also introduce some chemical reactions by which unstable minerals are converted into minerals stable at the new temperature/pressure conditions. It is on the basis of our knowledge of these reactions that we can infer the increasing temperature and/or pressure conditions of progressive metamorphism or, less commonly, the decreasing temperature and/or pressure of retrograde metamorphism. A historical perspective of metamorphic petrology involves the concept of metamorphic zones, a discussion of which follows.

18.1 METAMORPHIC ZONES

Since the late 19th century, geologists have recognized that specific groups of minerals effectively define different metamorphic temperature and pressure conditions. Minerals such as quartz and calcite, stable over a wide range of temperature and pressure, are of little value in this regard. However, other minerals form within a more limited temperature and/or pressure range. These minerals, referred to as **index minerals**, provide critical information because they effectively indicate the temperature/pressure conditions of metamorphism. Using index minerals, George Barrow (1912) and C. E. Tilley (1925) made some remarkable discoveries regarding regional metamorphism while mapping pelitic (mudstone) regional metamorphic rocks in the Scottish Highlands. The index minerals used by Barrow and Tilley were, in order of increasing temperature: chlorite, biotite, almandine, staurolite, kyanite and sillimanite. On the basis of mapping the locations of these six index minerals in the field, George Barrow

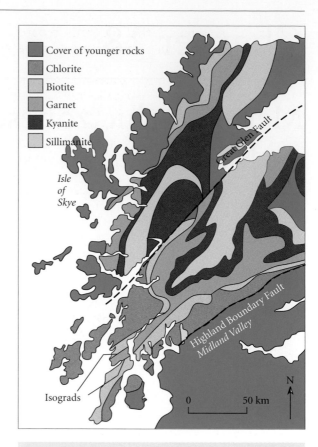

Figure 18.1 Barrovian zones of metamorphism based on index minerals from pelitic rocks of the Highlands of Scotland. Note that the staurolite zone is not shown in this figure. (After Tilley, 1925; from Murck et al., 2010; with permission of John Wiley & Sons.)

defined isograds and metamorphic zones (Figure 18.1). **Isograds** are lines drawn on geological maps that mark the first appearance of a particular index mineral. For example, the chlorite isograd marks the first appearance of chlorite and the biotite isograd marks the first appearance of biotite.

A **metamorphic zone** consists of the region bounded by two isograd lines. For example, the chlorite zone (Figure 18.1) is bounded on one side by the chlorite isograd, which marks where chlorite first appears, and on the other side by the biotite isograd, which marks the higher temperature conditions where biotite first appears. The biotite isograd marks the transition to the higher grade biotite zone (in which chlorite may still occur). Barrow, Tilley and others recognized that isograds record important changes in temperature during metamorphism (Figure 18.1).

The six metamorphic zones based on the six index mineral isograds are called **Barrovian zones** in honor of Barrow's pioneering work. In essence, Barrow discovered a key to interpreting progressive metamorphism whereby progressively higher grades of metamorphic minerals and rocks are produced with increasing temperature in an evolving orogenic belt. The Barrovian zones developed by Barrow (1912), Tilley (1925) and Harker (1932) are discussed below from lowest to highest grades of metamorphism.

18.1.1 Chlorite zone

The **chlorite zone** is bounded by the chlorite and biotite isograds. The chlorite isograd records metamorphic grades sufficient for the first appearance of chlorite group minerals. Common rocks in the chlorite zone are chlorite-bearing slate, chlorite-sericite phyllite and chlorite-sericite schist. Other minerals common to the chlorite zone include quartz, muscovite, albite (sodium plagioclase) and pyrophyllite. Chlorite becomes unstable and begins to be replaced by biotite at the high temperature limit of the chlorite zone.

18.1.2 Biotite zone

The **biotite zone** occurs between the biotite isograd – marking the first appearance of biotite – and the almandine (garnet) isograd. Common rocks in the biotite zone include sericite-biotite phyllite and biotite schist. Other minerals common to pelitic rocks in the biotite zone include quartz, sodium plagioclase, chlorite and muscovite. Biotite forms by chemical reactions that involve minerals such as chlorite, muscovite, quartz, magnetite and rutile. Note that lower temperature index minerals are not confined to the Barrovian zone that bears their name. Index minerals such as chlorite, biotite and almandine (which we will discuss next) persist in higher temperature zones. Why is this so? The lower grade minerals persist because the transformation from chlorite to biotite, or biotite to almandine, is not attained at a single temperature/pressure, nor by a simple transformation of one index mineral into another, but instead

occurs incrementally over a temperature range by reactions that involve several coexisting minerals.

18.1.3 Almandine (garnet) zone

The **almandine zone** is bounded by the almandine garnet isograd – marking the first appearance of almandine garnet – and the staurolite isograd. Common rocks in the almandine zone include garnet schist or garnet-mica schist. Garnet commonly occurs as porphyroblasts. Other minerals common in pelitic rocks of the almandine zone include biotite, muscovite, magnetite, quartz and sodium plagioclase minerals such as albite or oligoclase. Almandine forms through the chemical transformation of chlorite and magnetite.

18.1.4 Staurolite zone

The **staurolite zone** lies between the staurolite isograd – marking the first appearance of the higher temperature mineral staurolite – and the kyanite isograd. Staurolite-mica schist and staurolite-garnet-mica schist are common rock types in this zone. Other minerals common to the staurolite zone include quartz, almandine, potassium feldspar, biotite and muscovite. Potassium feldspar forms through the breakdown of muscovite. Staurolite forms through the chemical transformation of minerals such as almandine, chlorite and muscovite.

Unlike the lower temperature index minerals – chlorite, biotite and almandine, which exist in multiple zones – staurolite exists only within the staurolite zone. As we shall see, the higher grade index minerals – kyanite and sillimanite – also only occur in their namesake zones.

18.1.5 Kyanite zone

The **kyanite zone** occurs between the kyanite and sillimanite isograds that mark the first appearances of kyanite and sillimanite, respectively. Kyanite is a high pressure member of the Al_2SiO_5 (otherwise written as $AlAlOSiO_4$) polymorph series of andalusite, kyanite and sillimanite. Common rock types include kyanite schist and kyanite-mica schist. Other minerals commonly occurring in pelitic rocks of the kyanite zone include biotite, muscovite, almandine garnet, cordierite and quartz. Kyanite forms by transformation of alumino-silicate minerals such as andalusite (a low pressure polymorph) or through a dehydration reaction involving staurolite or pyrophyllite. For example:

$$Al_2Si_4O_{10}(OH)_2 \rightarrow Al_2SiO_5 + 3SiO_2 + H_2O$$
$$\text{Pyrophyllite} \qquad \text{Kyanite} \quad \text{Quartz} \quad \text{Water}$$

(Haas and Holdaway, 1973)

18.1.6 Sillimanite zone

The **sillimanite zone** occurs inside the sillimanite isograd and marks the highest temperature zone defined by Barrow and Tilley (see Figure 18.1). Common rock types include sillimanite schist, sillimanite gneiss and cordierite gneiss. Sillimanite is the high temperature Al_2SiO_5 polymorph mineral. Other minerals that commonly occur in the sillimanite zone include biotite, muscovite, cordierite, quartz, oligoclase and orthoclase. Sillimanite and potassium feldspar can also develop by dehydration of muscovite in the presence of quartz, as in the reaction:

$$KAl_2(AlSi_3O_{10})(OH)_2 + SiO_2$$
$$\text{Muscovite} \qquad\qquad \text{Quartz}$$
$$\rightarrow K(AlSi_3O_8) + Al_2SiO_5 + H_2O$$
$$\text{K-feldspar} \quad \text{Sillimanite} \quad \text{Water}$$

(Philpotts and Ague, 2009)

The early mapping of Barrow (1893, 1912), Tilley (1925) and others focused on the pelitic rocks of Scotland. Subsequently, Barrovian zones have been recognized in many orogenic belts that contain both pelitic and non-pelitic assemblages (Harker, 1932; Wiseman, 1934; Kennedy, 1949). Geologists now recognize approximate equivalents of Barrovian zones in rocks that do not have pelitic compositions. Table 18.1 presents a list of both pelitic and non-pelitic metamorphic protoliths and common index minerals associated with Barrovian zones.

Barrovian zones and metamorphic isograds remain in use today by geologists studying pelitic metamorphic rocks in the field. However, the zones and isograds are less useful in non-pelitic rocks, or in rocks that form in subduction zone or contact metamorphic environments. How can we infer the conditions that produced metamorphic rocks of all compositions and environments? Eskola (1915, 1920, 1939) introduced the concept of metamorphic facies – a more comprehensive approach to assessing the conditions recorded by metamorphic rocks. Mapping in Finland, Eskola noticed that certain metamorphic rock and mineral assemblages of various compositions recurred together in time and space, which he called "facies". Metamorphic facies are defined by a group or assemblage of critical minerals, rather than a single index mineral as used for Barrovian zones. It is therefore theoretically possible to place the metamorphic rocks of any area, regardless of their composition, into a metamorphic facies.

18.2 METAMORPHIC FACIES

Metamorphic facies are distinctive mineral assemblages in metamorphic rocks that form in response to a particular range of temperature and/or pressure conditions. Note that the definition of a metamorphic facies does not specifically imply its genetic origin. However, the assemblage of minerals within a facies closely constrains the temperature and/or pressure conditions. Most of the metamorphic facies were named for common basic rocks or minerals that occur within it. Eskola (1920) initially identified five metamorphic facies: sanidinite, hornfels, greenschist, amphibolite and eclogite. Additional facies proposed since Eskola's early work included the blueschist and granulite facies (Eskola, 1939; Turner, 1958) as well as the zeolite (Turner, 1958) and prehnite-pumpellyite facies (Coombs, 1961). The original hornfels facies proposed by Eskola (1920) was expanded by Eskola (1939) and Fyfe et al. (1958) into four facies that include, in order of increasing temperature: (1) albite-epidote hornfels, (2) hornblende hornfels, (3) pyroxene hornfels, and (4) sanidinite hornfels. In addition to the four hornfels facies, the other seven widely recognized metamorphic facies include: (1) zeolite, (2) prehnite-pumpellyite, (3) greenschist, (4) amphibolite, (5) granulite, (6) blueschist, and (7) eclogite (Figure 18.2).

We will begin our discussion with metamorphic facies that contain low pressure mineral assemblages associated with contact metamorphism and shallow dynamothermal metamorphism. Then we will consider higher pressure and temperature mineral assem-

Table 18.1 Barrovian zones in pelitic, calcareous and basic protoliths. (After Harker, 1932; Turner, 1958.)

Zone	Pelitic protolith	Calcareous protolith	Basic protolith
Chlorite	Chlorite, muscovite, quartz, pyrophyllite, albite, graphite	Calcite, albite, biotite, zoisite, quartz	Chlorite, albite, epidote, sphene, calcite, actinolite
Biotite	Biotite, muscovite, chlorite, quartz		
Almandine	Almandine, biotite, magnetite, muscovite, quartz	Garnet, andesine, zoisite, biotite	
Staurolite	Staurolite, biotite, muscovite, almandine, quartz	Bytownite, anorthite, hornblende, garnet	Hornblende, sodium–calcium plagioclase, epidote, almandine, diopside
Kyanite	Kyanite, biotite, muscovite, almandine, quartz		
Sillimanite	Sillimanite, biotite, cordierite, muscovite, almandine, quartz, oligoclase, orthoclase	Bytownite, anorthite, diopside, garnet	

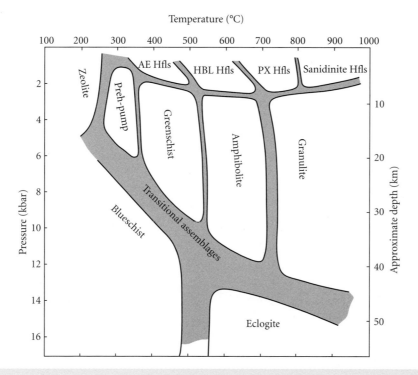

Figure 18.2 Common metamorphic facies depicted on a temperature and depth/pressure graph. Hfls, hornfels; Preh-pump, prehnite-pumpellyite. (After Yardley, 1989; courtesy of Bruce Yardley.)

blages associated with burial and dynamothermal metamorphism.

18.2.1 Hornfels facies

Hornfels facies include non-foliated, fine-grained hornfels rocks and coarser grained rocks with granoblastic textures. Hornfels facies form by heat-induced metamorphism in aureoles surrounding igneous intrusions. The metamorphic aureoles are localized around the intrusion, commonly having widths of 100 m or less. Ocean spreading centers represent regionally extensive zones of hydrothermal metamorphism that produce hornfels facies rocks on a large scale, effectively altering entire ocean basins.

The four types of hornfels facies are named for characteristic minerals that occur. Note that these critical minerals are not unique to

Table 18.2 Mineral assemblages and rocks in albite-epidote hornfels facies related to protolith composition (Turner and Verhoogen, 1951; Turner, 1958).

Protolith	Common minerals	Common rocks
Pelitic	Chloritoid, albite, epidote, muscovite, chlorite, biotite, andalusite, quartz	Hornfels
Basic	Epidote, albite, chlorite, actinolite, biotite, talc, sphene	Hornfels
Ultrabasic	Antigorite (serpentine), actinolite, tremolite, talc, chlorite, biotite, albite, magnesite, brucite, dolomite	Hornfels
Quartz-feldspathic	Albite, epidote, quartz, microcline, muscovite, biotite	Hornfels, metaquartzite
Calcareous	Calcite, dolomite, epidote, tremolite, idocrase (vesuvianite), magnesite, brucite	Marble, skarn

the hornfels facies. Later in this chapter we will see that these minerals also occur in greenschist, amphibolite and granulite facies rocks of the equivalent temperature but higher pressure conditions of dynamothermal metamorphism, which tends to produce foliated metamorphic rocks. It is the combination of non-foliated textures and the mineral assemblages that defines each particular hornfels facies.

Albite-epidote hornfels facies

The **albite-epidote hornfels facies** (Figure 18.2) is the low temperature hornfels facies, with temperatures generally <450°C and pressures <2 kbar (depth <6 km). Albite-epidote hornfels minerals develop in the outer fringes of many metamorphic aureoles. The characteristic minerals of this assemblage are albite and epidote, which most commonly occur in basaltic tuffs and lavas thermally metamorphosed at ocean ridges, hotspots and in volcanic–magmatic arcs. The minerals and rocks associated with the albite-epidote hornfels facies depend upon the protolith composition as indicated in Table 18.2. The albite-epidote hornfels facies is roughly the low pressure equivalent of the greenschist facies, which will be discussed in Section 18.2.4.

Hornblende hornfels facies

Hornblende hornfels facies rocks (Figure 18.2) compose the bulk of many metamorphic aureoles, forming at temperatures generally between 450 and 600°C and at pressures

<2.5 kbar (<8 km). The minerals and rocks associated with the hornblende hornfels facies depend upon the protolith composition as indicated in Table 18.3. Chlorite, albite, epidote and actinolite – common in albite-epidote hornfels facies – are notable by their absence in the hornblende hornfels assemblage. At temperatures above 450°C, dolomite is replaced by diopside via the following chemical reaction:

$$CaMg(CO_3)_2 + 2SiO_2$$
$$\text{Dolomite} \qquad \text{Quartz}$$
$$\rightarrow CaMgSi_2O_6 + 2CO_2$$
$$\text{Diopside} \qquad \text{Carbon dioxide}$$

(Turner, 1981)

The hornblende hornfels facies is roughly the low pressure equivalent of the amphibolite facies, which will be discussed in Section 18.2.5.

Pyroxene hornfels facies

Because its development requires very high temperatures, not commonly attained by granitoid magmas, the **pyroxene hornfels facies** is less common than the lower temperature hornfels facies discussed above. Where present, commonly adjacent to higher temperature basic intrusions, pyroxene hornfels facies rocks (Figure 18.2) develop at temperatures of 600–800°C and at pressures <2.5 kbar (<8 km). The minerals and rocks associated with the pyroxene hornfels facies depend upon the protolith composition as indicated in Table 18.4. Note that with the exception of biotite, hydrous minerals do not occur. At

Table 18.3 Mineral assemblages and rocks in hornblende hornfels facies related to protolith composition (Turner and Verhoogen, 1951; Turner, 1958).

Protolith	Common minerals	Common rocks
Pelitic	Andalusite, cordierite, anthophyllite, plagioclase, muscovite, biotite, quartz, microcline	Hornfels
Basic	Hornblende, plagioclase, biotite, diopside	Hornfels
Ultrabasic	Talc, forsterite, cummingtonite, grunerite, hornblende, lizardite, anthophyllite, plagioclase, diopside, tremolite, clinochlore, brucite, spinel, bronzite	Hornfels
Quartz-feldspathic	Hornblende, microcline, quartz, plagioclase, muscovite, biotite	Hornfels, metaquartzite
Calcareous	Calcite, diopside, tremolite, calcium plagioclase, wollastonite, grossular garnet, idocrase (vesuvianite), brucite	Marble, skarn

Table 18.4 Mineral assemblages and rocks in pyroxene hornfels facies related to protolith composition (Turner and Verhoogen, 1951; Turner, 1958).

Protolith	Common minerals	Common rocks
Pelitic	Cordierite, andalusite or sillimanite, plagioclase, orthoclase, quartz, biotite, corundum, spinel, hypersthene	Hornfels
Basic	Hypersthene, diopside, cordierite, plagioclase, biotite, augite	Hornfels
Ultrabasic	Forsterite, hypersthene, diopside, augite, grossular garnet, enstatite	Hornfels
Quartz-feldspathic	Augite, hypersthene, diopside, orthoclase, andalusite or sillimanite, quartz, plagioclase, biotite	Hornfels, metaquartzite
Calcareous	Diopside, wollastonite, calcium plagioclase (anorthite, bytownite), grossular garnet, forsterite, periclase	Marble, skarn

the higher temperatures at which pyroxene hornfels form, dehydration reactions produce a largely anhydrous suite of minerals. The pyroxene hornfels facies is roughly the low pressure equivalent of the granulite facies, which will be discussed in Section 18.2.6.

Sanidinite hornfels facies

Sanidinite hornfels facies rocks (Figure 18.2) are very rare, forming in very high temperature (>800°C) and low pressure (<2.5 kbar ≈ <8 km) conditions in association with basic and ultrabasic intrusions. Sanidinite hornfels develop where the country rock is in contact with the intrusion or in country rock inclusions (xenoliths) within the intrusion. The minerals and rocks associated with the sanidinite hornfels facies depend upon the protolith composition as indicated in Table 18.5. Note the absence of hydrous minerals, the result of reactions such as the following:

$$KMg_3AlSi_3O_{10}(OH)_2 + 3SiO_2$$
$$\text{Phlogopite} \quad \text{Quartz}$$
$$\rightarrow K(AlSi_3O_8) + 3MgSiO_3 + H_2O$$
$$\text{Sanidine} \quad \text{Enstatite} \quad \text{Water}$$

(Wones and Dodge, 1968)

The contact metamorphic facies described above record temperature-driven mineral changes as a result of igneous intrusions into country rock, or hydrothermal metamorphism related to shallow magmatic activity, for example along the oceanic ridge system. Contact metamorphism can occur anywhere within Earth's crust, but predominates in the upper 10 km. Contact metamorphic facies provide a means to place hydrothermal metamorphic rocks and ocean ridge alterations into the facies concept.

The next two facies we will discuss – zeolite and prehnite-pumpellyite – contain low to moderate temperature and low to moderate pressure mineral assemblages. Zeolite and

Table 18.5 Mineral assemblages and rocks in sanidinite hornfels facies related to protolith composition (Turner and Verhoogen, 1951; Turner, 1958).

Protolith	Common minerals	Common rocks
Pelitic	Cordierite, sillimanite, corundum, sanidine, calcium plagioclase (anorthite, bytownite), tridymite	Hornfels
Basic	Hypersthene, pigeonite, augite, enstatite, forsterite, magnetite, ilmenite, hematite, diopside, plagioclase, corundum, spinel	Hornfels
Ultrabasic	Forsterite, diopside, pigeonite, enstatite, hypersthene, cordierite, augite, spinel, ilmenite, magnetite	Hornfels
Quartz-feldspathic	Sanidine, tridymite, calcium plagioclase (anorthite, bytownite), diopside, hypersthene, pigeonite	Hornfels, metaquartzite
Calcareous	Wollastonite, diopside, calcium plagioclase (anorthite, bytownite)	Marble, skarn

prehnite-pumpellyite minerals form by burial metamorphism in sedimentary basins as well as at ocean ridges, hotspots and volcanic arcs.

18.2.2 Zeolite facies

Turner (1958) defined the **zeolite facies** (see Figure 18.2) as a low grade metamorphic facies produced by temperatures between ~150 and 300°C and pressures less than 5 kbar (~15 km depth). The zeolite facies is named after the zeolite mineral group (Table 18.6). Zeolites are a hydrous sodium and calcium aluminum tectosilicate mineral group formed by diagenetic or low temperature metamorphic reactions.

Critical zeolite facies minerals, which commonly coexist with quartz, include analcime, laumontite, heulandite and wairakite. Accessory minerals in the zeolite facies may include albite, kaolinite, vermiculite, adularia, pumpellyite, sphene, epidote, prehnite, montmorillonite, smectite, muscovite, chlorite and calcite (Turner, 1958; Coombs, 1961). The minerals and rocks associated with the zeolite facies depend upon the protolith composition as indicated in Table 18.7.

Zeolite facies minerals originate from the hydrothermal alteration of volcanic protoliths such as basalt and andesite, the devitrification of basaltic glass and tuff, and the reaction of pelites and graywackes with saline waters. Zeolite minerals occur in isolated vesicles and veins or as pervasively disseminated minerals within the metamorphic rock. The low temperature and low pressure conditions of zeolite facies metamorphism commonly preserve relict structures inherited from the protolith.

Turner (1958) formalized the zeolite facies based upon Coombs' (1954) geological mapping of metamorphosed volcanic graywackes in Southland, New Zealand. There, metagraywackes that contain andesitic volcanic lapilli and glass particles were affected by low temperature diagenesis and zeolite facies metamorphism. Metagraywacke rocks subjected to low temperature diagenetic alteration were enriched in stilbite and heulandite. As temperatures and pressures increased into the zeolite facies, stilbite and heulandite were replaced by laumontite and wairakite:

$$\text{Stilbite} \rightarrow \text{heulandite}$$
$$\rightarrow \text{laumontite} \rightarrow \text{wairakite}$$

(Miyashiro and Shido, 1970)

$$\underset{\text{Heulandite}}{CaAl_2Si_7O_{18} \cdot 6H_2O}$$
$$\rightarrow \underset{\text{Laumontite}}{Ca(Al_2Si_4)O_{12} \cdot 5H_2O} + \underset{\text{Quartz}}{3SiO_2} + \underset{\text{Water}}{H_2O}$$

(Coombs et al., 1959)

$$\underset{\text{Laumontite}}{Ca(Al_2Si_4)O_{12} \cdot 4H_2O}$$
$$\rightarrow \underset{\text{Wairakite}}{Ca(Al_2Si_4)O_{12} \cdot 2H_2O} + \underset{\text{Water}}{2H_2O}$$

(Bird and Helgeson, 1981)

Zeolite facies metamorphism develops by hydrothermal alteration at divergent margins, hotspots and convergent margins or during burial metamorphism at depths less than

Table 18.6 Zeolite minerals and chemical formulas.

Zeolite mineral	Chemical formula	Occurrence
Natrolite	$Na_2Al_2Si_3O_{10} \cdot 2H_2O$	Diagenesis and zeolite facies
Stilbite	$NaCa_2Al_5Si_{13}O_{36} \cdot 14H_2O$	Diagenesis and zeolite facies
Chabazite	$Ca_2Al_2Si_4O_{12} \cdot 14H_2O$	Diagenesis and zeolite facies
Heulandite	$CaAl_2Si_7O_{18} \cdot 6H_2O$	Diagenesis and zeolite facies
Thompsonite	$NaCa_2Al_5Si_5O_{20} \cdot 6H_2O$	Zeolite facies
Analcime	$NaAlSi_2O_6 \cdot H_2O$	Zeolite facies
Laumontite	$CaAl_2Si_4O_{12} \cdot 4H_2O$	Zeolite facies
Wairakite	$CaAl_2Si_4O_{12} \cdot 2H_2O$	Zeolite facies

Table 18.7 Mineral assemblages and rocks in zeolite facies related to protolith composition.

Protolith	Common minerals	Common rocks
Pelitic	Kaolinite, zeolite, quartz, montmorillonite, vermiculite, phengite, epidote, muscovite	Metapelite, argillite, slate
Basic	Zeolite, albite, quartz, phengite, sphene, epidote, chlorite, prehnite, pumpellyite	Metabasite or greenstone
Ultrabasic	Lizardite serpentine, talc, olivine, chlorite, prehnite, pumpellyite	Serpentinite or greenstone
Quartz-feldspathic	Quartz, zeolite, albite, sphene, epidote, quartz, muscovite	Metaquartzite or metagraywacke
Calcareous	Zeolite, calcite, quartz, epidote, dolomite, lawsonite, talc, muscovite	Marble

5 km. Laumontite and heulandite are particularly common in the zeolite facies. At temperatures approaching 250°C and depths of 3–5 km, zeolite facies minerals begin to alter to prehnite and pumpellyite facies minerals, which occur in the upper zeolite facies but are the dominant minerals in the prehnite-pumpellyite facies, discussed below.

18.2.3 Prehnite-pumpellyite facies

Coombs (1961) defined the **prehnite-pumpellyite facies** (see Figure 18.2) based on metasedimentary basin deposits in New Zealand. Prehnite-pumpellyite facies minerals are produced by hydrothermal alteration and burial metamorphism at temperatures and pressures that exceed zeolite facies conditions. Zeolite facies and prehnite-pumpellyite facies rocks formed by hydrothermal alteration are widespread at oceanic ridges and therefore affect substantial portions of the oceanic crust generated at spreading ridges. Although any protolith can be metamorphosed to prehnite-pumpellyite facies, the most common protoliths include basalt, graywackes and mudstones (pelites).

The prehnite-pumpellyite facies generally forms under low temperature (250–350°C) and fairly low pressure (<6 kbar, ~20 km depth) conditions. In addition to prehnite and pumpellyite, other common minerals include quartz, albite, chlorite, muscovite, illite, phengite, smectite, sphene, titanite, epidote, lawsonite and stilpnomelane. Although some zeolite minerals occur in the prehnite-pumpellyite facies, laumontite and heulandite are restricted to the zeolite facies. The minerals and rocks associated with the prehnite-pumpellyite facies depend upon the protolith composition as indicated in Table 18.8.

Due to the relatively low temperature and low pressure conditions, prehnite-pumpellyite facies rocks commonly retain relict textures and structures. As with zeolite facies rocks, prehnite-pumpellyite facies rocks are commonly referred to by referencing the protolith as in metabasites, metagraywackes and metapelites. Higher temperature alteration of

Table 18.8 Mineral assemblages and rocks in prehnite-pumpellyite facies related to protolith composition.

Protolith	Common minerals	Common rocks
Pelitic	Chlorite, quartz, illite, phengite, smectite, sphene, titanite, epidote, stilpnomelane	Slate or phyllite
Basic	Prehnite, pumpellyite, albite, quartz, phengite, sphene, titanite, epidote, chlorite, actinolite, zeolite	Metabasite or greenstone
Ultrabasic	Lizardite serpentine, talc, forsterite, tremolite, chlorite, zeolite	Serpentinite, soapstone or greenstone
Quartz-feldspathic	Quartz, epidote, feldspar	Metaquartzite or metagraywacke
Calcareous	Prehnite, calcite, quartz, epidote, dolomite, lawsonite, talc, muscovite, zeolite	Marble

prehnite and pumpellyite results in the neocrystallization of actinolite and epidote, two minerals that mark the transition to the higher grade albite-epidote hornfels facies discussed previously and the greenschist facies discussed below. The higher temperature assemblage containing pumpellyite and actinolite has been called the transitional pumpellyite-actinolite facies (Hashimoto, 1966; Turner, 1981).

The facies discussed in the following sections largely form in response to higher temperatures and pressures than those that produce the zeolite and prehnite-pumpellyite facies. They are typically produced by dynamothermal metamorphism under conditions of non-uniform stress, particularly at convergent plate boundaries.

18.2.4 Greenschist facies

Greenschist facies rocks (see Figure 18.2) generally form under medium temperature (350–550°C) and pressure (3–10 kbar ≈ 10–30 km depth) conditions associated with dynamothermal metamorphism at convergent plate boundaries. At these higher temperature and pressure conditions, pervasive recrystallization and/or neocrystallization commonly results in the obliteration of relict textures. Although greenschist facies metamorphism can affect rocks of any composition, the most common protoliths include basic and ultrabasic igneous rocks, tuff, sandstones, mudrocks and limestone. Key minerals in the greenschist facies include epidote, chlorite and actinolite (green amphibole). These minerals impart the green color to both foliated (schistose)

greenschist rocks and non-foliated greenstone rocks in this facies. Minerals and rocks common to the greenschist facies are listed in Table 18.9.

Increasing temperatures associated with greenschist facies metamorphism result in the liberation of volatile components such as CO_2 and H_2O, as indicated in the following reactions:

$$\underset{\text{Kaolinite}}{Al_4(Si_4O_{10})(OH)_8} + \underset{\text{Quartz}}{4SiO_2}$$
$$\rightarrow \underset{\text{Pyrophyllite}}{2Al_2Si_4O_{10}(OH)_2} + \underset{\text{Water}}{2H_2O}$$

(Hower et al., 1976)

$$\underset{\text{Dolomite}}{3CaMg(CO_3)_2} + \underset{\text{Quartz}}{4SiO_2} + \underset{\text{Water}}{H_2O}$$
$$\rightarrow \underset{\text{Talc}}{Mg_3(Si_4O_{10})(OH)_2}$$
$$+ \underset{\text{Calcite}}{3CaCO_3} + \underset{\text{Carbon dioxide}}{3CO_2}$$

(Metz and Winkler, 1963)

Greenschist facies rocks are abundant in orogenic fold and thrust belts, where they record regional, moderate temperature/pressure metamorphic conditions at convergent plate boundaries. Classic greenschist facies rocks are extensively exposed in orogenic belts such as the Appalachians, the Alps and the Otago fold and thrust belt of southern New Zealand.

Earlier in this chapter, we discussed index minerals in pelitic rocks used to map Barrovian zones. Where metapelites occur, the greenschist facies can be subdivided into three Barrovian zones:

Table 18.9 Mineral assemblages and rocks in greenschist facies related to protolith composition. (Turner and Verhoogen, 1951; Turner, 1958).

Protolith	Common minerals	Common rocks
Pelitic	Pyrophyllite, andalusite, kyanite, quartz, albite, epidote, chlorite, chloritoid, biotite, phengite muscovite, garnet (almandine, spessartine), chloritoid, sphene, tourmaline	Slate, phyllite or schist
Basic	Actinolite, albite, epidote, chlorite, quartz, biotite, magnetite, sphene, stilpnomelane, calcite	Greenschist or greenstone
Ultrabasic	Antigorite serpentine, talc, epidote, chlorite, actinolite, tremolite, magnetite, almandine or spessartine garnet, sphene	Serpentinite, soapstone or greenstone
Quartz-feldspathic	Quartz, albite, muscovite, epidote, chlorite	Metaquartzite
Calcareous	Calcite, talc, tremolite, dolomite, stilpnomelane, scapolite, tremolite, magnesite, quartz, pumpellyite	Marble

1 The chlorite zone corresponds to lower greenschist facies conditions with minerals such as chlorite, dolomite, stilpnomelane and calcite.
2 The biotite zone corresponds to upper greenschist facies conditions and contains biotite and tremolite.
3 The lower part of the almandine garnet zone corresponds to the uppermost greenschist to epidote-amphibolite facies.

Figure 18.3 illustrates how Barrovian zones relate to the metamorphic facies common in orogenic belts.

As pressures decrease, the low pressure field of greenschist metamorphism grades into the albite-epidote hornfels facies. As progressively higher temperature conditions develop within orogenic belts, the greenschist facies transforms into higher grade amphibolite facies metamorphism, discussed below.

18.2.5 Amphibolite facies

Amphibolite facies rocks (see Figure 18.2) form at high temperatures (~550–750°C) and moderate to high pressures (4–12 kbar ≈ 12–40 km depth) in regional orogenic belts at convergent margins. Amphibolite facies metamorphism can affect rocks of any composition, although the most common protoliths include basic and ultrabasic igneous rocks, tuff, sandstone, mudstone and limestone. Minerals and rocks common to the amphibo-

lite facies are listed in Table 18.10. The transition from greenschist to amphibolite facies is marked by an increase in hornblende, garnet and anthophyllite and a decrease in actinolite, chlorite, biotite and talc in basic and ultrabasic rocks. In addition, plagioclase minerals become less sodic and more calcic. The amphibolite facies also marks the appearance of staurolite in pelitic rocks, where staurolite may occur with kyanite. Increasing temperature in amphibolite facies rocks results in the transformation from kyanite to sillimanite in pelitic rocks where sillimanite commonly occurs with muscovite. In calcareous rocks, dolomite breaks down to diopside and releases carbon dioxide via the following chemical reactions:

$$\underset{\text{Dolomite}}{CaMg(CO_3)_2} + \underset{\text{Quartz}}{2SiO_2}$$
$$\rightarrow \underset{\text{Diopside}}{CaMgSi_2O_6} + \underset{\text{Carbon dioxide}}{2CO_2}$$

(Turner, 1981)

Increasing intensity of metamorphism also results in devolatization reactions resulting in the release of $H_2O + CO_2$ as shown by the two examples below:

$$\underset{\text{Talc}}{7Mg_3Si_4O_{10}(OH)_2}$$
$$\rightarrow \underset{\text{Anthophyllite}}{3Mg_7Si_8O_{22}(OH)_2} + \underset{\text{Quartz}}{4SiO_2} + \underset{\text{Water}}{4H_2O}$$

(Chernosky, 1976)

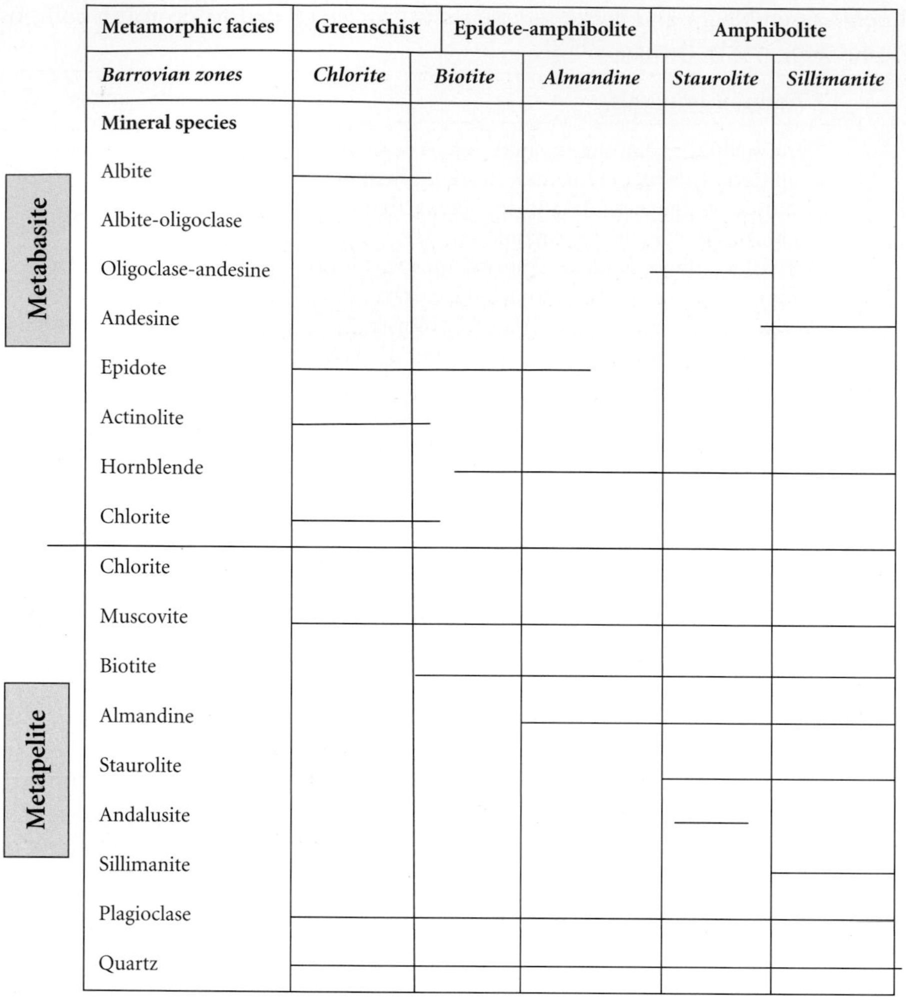

Metamorphic facies	Greenschist	Epidote-amphibolite		Amphibolite	
Barrovian zones	*Chlorite*	*Biotite*	*Almandine*	*Staurolite*	*Sillimanite*
Mineral species					
Albite					
Albite-oligoclase					
Oligoclase-andesine					
Andesine					
Epidote					
Actinolite					
Hornblende					
Chlorite					
Chlorite					
Muscovite					
Biotite					
Almandine					
Staurolite					
Andalusite					
Sillimanite					
Plagioclase					
Quartz					

Figure 18.3   This figure illustrates the relationship between Barrovian zones and metamorphic facies. (After Yardley, 1989; courtesy of Bruce Yardley.)

$$Ca_2Mg_5Si_8O_{22}(OH)_2 + 3CaCO_3 + 2SiO_2$$

Tremolite   Calcite   Quartz

$$\rightarrow 5CaMgSi_2O_6 + H_2O + 3CO_2$$

Diopside   Water   Carbon dioxide

(Turner, 1981)

The amphibolite facies encompasses a number of different Barrovian zones (see Figure 18.3) that, with increasing temperature, include the upper part of the almandine zone, all of the staurolite and the lower part of the sillimanite zone. The low temperature part of the amphibolite facies that corresponds with the almandine zone is also known as the epidote-amphibolite facies ("transitional" facies) because both epidote and hornblende can coexist under these conditions.

With decreasing pressure, the lowest pressure field of the amphibolite facies metamorphism grades into the hornblende hornfels field. As temperatures increase, the amphibolite facies grades into the higher temperature granulite facies discussed below.

18.2.6   Granulite facies

The **granulite facies** (see Figure 18.2) consists of high temperature (~700–900°C) and moderate to high pressure (3–15 kbar ≈ 10–50 km depth) mineral assemblages. Granulite facies metamorphism can affect rocks of any composition, although the most common protoliths include granitic to ultrabasic igneous

Table 18.10 Mineral assemblages and rocks in amphibolite facies related to protolith composition. (Turner and Verhoogen, 1951; Turner, 1958).

Protolith	Common minerals	Common rocks
Pelitic	Kyanite, sillimanite, staurolite, muscovite, almandine garnet, quartz, plagioclase, biotite, epidote	Schist, gneiss
Basic	Hornblende, calcium plagioclase (oligoclase-anorthite), epidote, almandine garnet, diopside, cummingtonite, grunerite, augite, quartz, ilmenite, magnetite	Amphibolite, schist
Ultrabasic	Talc, anthophyllite, olivine, diopside, cummingtonite, serpentine, spinel, calcium plagioclase (oligoclase-anorthite), enstatite, phlogopite, tremolite	Schist
Quartz-feldspathic	Quartz, microcline, hornblende, biotite, muscovite, almandine garnet	Metaquartzite, gneiss
Calcareous	Calcite, dolomite, tremolite, anorthite plagioclase, almandine to grossular garnet, diopside, augite phlogopite	Marble

Table 18.11 Mineral assemblages and rocks in granulite facies related to protolith composition. (Turner and Verhoogen, 1951; Turner, 1958).

Protolith	Common minerals	Common rocks
Pelitic	Sillimanite, kyanite, orthoclase, quartz, almandine garnet, calcium plagioclase (anorthite, bytownite, andesine), cordierite, sapphire, magnetite, ilmenite, rutile	Gneiss, granulite
Basic	Hypersthene, calcium plagioclase (anorthite, bytownite, andesine), diopside, garnet, cummingtonite, grunerite, magnetite, ilmenite	Gneiss, granulite
Ultrabasic	Enstatite, hypersthene, diopside, olivine, calcium plagioclase, corundum, spinel	Gneiss, granulite
Quartz-feldspathic	Orthoclase, quartz, hypersthene, almandine garnet, plagioclase	Gneiss, charnockite, granulite
Calcareous	Calcite, diopside, almandine garnet, forsterite, scapolite, corundum	Marble

rocks, schists, gneisses, pelites, sandstones and limestones. The minerals and rocks associated with the granulite facies largely depend upon the protolith composition as indicated in Table 18.11.

Granulite facies rocks are commonly, but not always, dominated by rocks with granoblastic textures arranged in a non-foliated to foliated fabric. Foliated granulite rocks are less common because many of the inequant, hydrous phyllosilicate minerals (e.g., muscovite) have been largely transformed to minerals (e.g., potassium feldspar) with equant forms. Granulite facies minerals are predominantly anhydrous (Table 18.11), due to dehydration reactions at high temperatures. Hydrous minerals hornblende and biotite,

but not muscovite, can occur in the lower part of the granulite facies, sometimes referred to as granulite I. The upper part of the granulite facies, sometimes referred to as granulite II, is characterized entirely by anhydrous minerals.

Amphibole minerals (tremolite, anthophyllite, hornblende) dehydrate to pyroxene minerals (enstatite, diopside, hypersthene), and phyllosilicate minerals (such as muscovite) dehydrate to anhydrous minerals (orthoclase) in response to high temperatures. Such reactions are completed either in the transition from amphibolite to granulite facies or in the transition from granulite I to granulite II. Examples of these dehydration reactions include:

$$2Ca_2Mg_5Si_8O_{22}(OH)_2$$

Tremolite

$$\rightarrow 4CaMgSi_2O_6 + 6MgSiO_3$$

Diopside   Enstatite

$$+ 2SiO_2 + 2H_2O$$

Quartz   Water

(Boyd, 1954)

$$Mg_7Si_8O_{22}(OH)_2 \rightarrow 7MgSiO_3 + SiO_2 + H_2O$$

Anthophyllite   Enstatite   Quartz   Water

(Greenwood, 1963; Chernosky, 1979)

$$Ca_2Mg_4Al_2Si_7O_{22}(OH)_2$$

Hornblende

$$\Rightarrow CaMgSi_2O_6 + 3MgSiO_3$$

Diopside   Hypersthene

$$+ CaAl_2Si_2O_8 + H_2O$$

Anorthite   Water

$$KAl_2AlSi_3O_{10}(OH)_2 + SiO_2$$

Muscovite   Quartz

$$\rightarrow KAlSi_3O_8 + Al_2SiO_5 + H_2O$$

Orthoclase   Sillimanite   Water

(Philpotts and Ague, 2009)

Quartz and orthoclase are common granulite facies minerals in pelitic and quartz feldspathic rocks. Calcareous rocks are marked by the appearance of wollastonite and the absence of hydrous minerals such as phlogopite. Meta-basic and meta-ultrabasic granulites commonly contain both orthopyroxene (hypersthene) and clinopyroxene (diopside). This two-pyroxene association that characterizes many granulites does not occur in amphibolites or other lower temperature facies.

Granulite facies rocks include gneisses, charnockites and migmatites. **Charnockites** are hypersthene-bearing granitic gneisses. **Migmatites** are mixed rocks, meaning they display the textural relations of both metamorphic and plutonic igneous rocks due to varying degrees of partial melting.

Granulite facies metamorphism occurs in the highest temperature dynamothermal metamorphism region at (1) convergent plate boundaries, (2) at the base of thick continental crust, and (3) in the uppermost part of the mantle. Some basic granulites may represent the refractory residual rock material following partial melting at the base of continental lithosphere. Granulites occur in rocks of all ages, but are especially common in Precam-

brian shields and associated anorthosite complexes (Chapter 10) where long-term erosion has exposed rock formed deep below the surface.

The granulite facies corresponds with the upper parts of the Barrovian sillimanite zone and – at still higher temperature – the cordierite-garnet zone. With decreasing pressure, the low pressure field of granulite facies metamorphism grades into the pyroxene and sanidinite hornfels facies. As progressively higher temperature conditions develop within orogenic belts, migmatites of the granulite facies are produced. So far, we have focused largely on facies changes affected primarily by temperature conditions. The facies described below are defined largely on changes in pressure.

18.2.7 Blueschist facies

The **blueschist facies** (see Figure 18.2) consists of moderate to high pressure (4–20 kbar ≈ 13–66 km depth), low temperature (150–500°C) mineral assemblages. The conditions that produce this facies are characterized by unusually high pressure/temperature ratios reflecting the existence of unusually low geothermal gradients. Although blueschist facies metamorphism affects rocks of any composition, common protoliths include basic to ultrabasic igneous rock and sedimentary graywackes and mudstones. The blue amphibole mineral glaucophane provides the distinctive color for which this facies is named. In addition to glaucophane, other common minerals include magnesio-riebeckite, lawsonite, jadeite pyroxene, aegirine, crossite and kyanite. Minerals and rocks occurring in the blueschist facies depend upon the protolith composition as indicated in Table 18.12.

Blueschists form in subduction zones where oceanic lithosphere is forced downward to great depths at geologically rapid rates. As cold oceanic lithosphere is dragged downward into Earth's interior it absorbs heat from the surrounding asthenosphere very slowly, reaching great depths while remaining relatively cool. The subduction of cold lithosphere lowers the temperature in the subduction zone environment to temperatures not found anywhere else on Earth at comparable depths. This has the effect of depressing the isotherms (lines of equal temperature) in the

Table 18.12 Mineral assemblages and rocks in blueschist facies related to protolith composition. (Turner and Verhoogen, 1951; Turner, 1958).

Protolith	Common minerals	Common rocks
Pelitic	Kyanite, chlorite, epidote, titanite, almandine garnet, quartz, phengite, coesite	Schist
Basic	Glaucophane, jadeite, crossite, lawsonite, albite plagioclase, chlorite, epidote, titanite, magnesio-riebeckite, magnesio-arfvedsonite, aegerine, augite, pumpellyite, stilpnomelane	Schist
Ultrabasic	Lizardite, glaucophane, magnesio-riebeckite, magnesio-arfvedsonite, jadeite, aegerine, lawsonite, pyrope garnet, acmite	Schist
Quartz-feldspathic	Kyanite, quartz, coesite, jadeite, magnesio-riebeckite, stilpnomelane	Kyanite metaquartzite
Calcareous	Calcite, epidote, chlorite	Marble

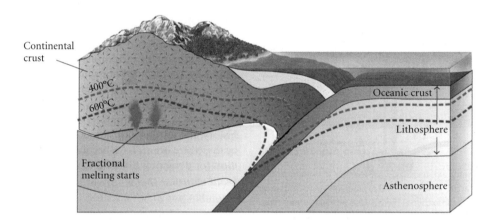

Figure 18.4 Generalized cross-section indicating depressed temperatures in the subduction zone. (After Furukama, 1993; from Murck et al., 2010; with permission of John Wiley & Sons.)

vicinity of the subducted slab (Figure 18.4). Subduction of oceanic lithosphere serves as a proverbial icicle in Earth's interior. The result is the creation of a unique set of high pressure, low temperature conditions that produces blueschist facies metamorphism. In fact, blueschist facies rocks are a critical indicator for subduction zones in convergent margins.

Blueschist facies rocks present several very interesting scientific dilemmas. Two critical questions are:

1 How do blueschist mineral assemblages produced at great depth return to the Earth's surface? Suggested mechanisms include:
   - Extreme extension perpendicular to the arc that allows deep-seated rocks to rise toward the surface (Lister et al., 1984; Platt, 1986; Shermer, 1990).
   - Extension associated with oblique subduction (Avé Lallemant and Guth, 1990).
   - Uplift associated with rising, buoyant serpentinites (Pilchin, 2003).
   - A reverse loop within the base of an accretionary wedge that regurgitates deeply buried (underplated) subducted rock fragments upwards to Earth's surface (Platt, 1986).
   - Uplift along thrust fault systems combined with hanging wall erosion (Unruh et al., 1995).

2 Why are blueschists relatively common in the Phanerozoic but relatively rare in the Precambrian? Explanations include:

- Their precarious positions at the leading edges of subduction zones makes them vulnerable to further subduction and crustal recycling (Möller et al., 1995).
- Uplift and removal by erosion during continental collisions.
- Retrograde metamorphism to greenschist facies assemblages during later tectonic events with more normal pressure:termperature ratio conditions (Zwart, 1967; Ernst, 1988).
- Elevated Precambrian geotherms that lowered pressure:temperature ratios by raising temperatures during metamorphism (De Roever, 1956; Burke et al., 1977).
- More limited upper mantle convection and thinner lithosphere in the Precambrian, which caused shallower subduction (Liou et al., 1990).

Research regarding the scarcity of Precambrian blueschists and the uplift of blueschists continues.

At temperatures between 350 and 450°C and pressures of ~8 kbar, a blueschist and greenschist facies transition occurs (Maruyama et al., 1986). With increasing temperature and pressures exceeding 12 kbar, blueschist facies converts to the eclogite facies.

18.2.8 Eclogite facies

The **eclogite facies** (see Figure 18.2) typically develops at high temperatures (400–900°C) and very high pressures (12–25 kbar ≈ 40–82 km). Ultra-high pressure rocks, also considered part of the eclogite facies, form at even higher pressures and greater depths, as discussed in the next section. True eclogites are chemically similar to a silica-undersaturated, anhydrous basalt and generally develop from basic protoliths. As discussed in Chapter 17, eclogite rocks are fine- to coarse-grained, dense, dark green rocks, commonly with reddish brown garnet porphyroblasts. The two key minerals are garnet and omphacite, a sodium-rich jadeitic clinopyroxene. Garnet group minerals stable at high pressure eclogite conditions include pyrope, majorite, and to a lesser degree, grossular; at lower pressures almandine can also occur. While the rock eclogite does derive from basic protoliths, the

facies includes rocks of other compositions. For example, kyanite is common in pelitic rocks. Other minerals that can occur within the eclogite facies include enstatite, jadeite, rutile, zoisite, coesite, phengite, lawsonite, corundum and diamond. The eclogite facies constitutes a high pressure facies in which plagioclase is unstable as shown by the reaction:

$$\underset{\text{Albite}}{Na(AlSi_3O_8)} \rightarrow \underset{\text{Jadeite}}{NaAlSi_2O_6} + \underset{\text{Quartz}}{SiO_2}$$

(Philpotts and Ague, 2009)

Minerals and rocks occurring in the eclogite facies depend upon the protolith composition as indicated in Table 18.13.

Eclogite facies rocks occur in three major environments:

1 In the lower continental crust and mantle (>40 km depth) and later exposed on Earth's surface in deeply eroded fold and thrust belts.
2 At convergent margins in ophiolite complexes and subduction zone mélanges.
3 As xenoliths in diamond-bearing kimberlite pipes.

Baldwin et al. (2004) studied the youngest documented eclogites in the world, produced along the Papua New Guinea oblique subduction zone. Oblique subduction involves components of both thrusting and strike-slip faulting. The 4.33 ± 0.36 Ma eclogites were metamorphosed at depths of 75 km. Geothermobarometric analysis indicates that the minimum rate of uplift was ~1.7 cm/yr, comparable to spreading rates at some ocean ridges. The 1.7 cm/yr estimate assumes vertical uplift that is unlikely in the oblique shear zones of Papua New Guinea. Likely transport rates associated with the strike-slip component of motion ranged between 3.5 and 5.1 cm/yr. We will now consider ultra-high pressure minerals, which geologists consider as part of the eclogite facies.

Ultra-high pressure minerals

Ultra-high pressure (UHP) minerals occur within the eclogite facies (Figure 18.5) at pressures >25 kbar (>80 km depth) and temperatures >600°C. UHP conditions are indicated by critical minerals such as:

Table 18.13 Mineral assemblages and rocks in eclogite facies related to protolith composition. (Turner and Verhoogen, 1951; Turner, 1958).

Protolith	Common minerals	Common rocks
Pelitic	Kyanite, almandine garnet, phengite, coesite, diamond	Eclogite
Basic	Omphacite, pyrope garnet, glaucophane, jadeite, crossite, lawsonite, diamond	Eclogite
Ultrabasic	Omphacite, pyrope garnet, glaucophane, jadeite, aegerine, lawsonite, diamond	Eclogite
Quartz-feldspathic	Kyanite, coesite, jadeite, diamond	Kyanite metaquartzite
Calcareous	Calcite, lawsonite, pyrope-almandine garnet	Marble

Figure 18.5 Eurasian–African view of ultra-high pressure mineral occurrences of the minerals coesite, diamond and majorite. (Courtesy of Bradley Hacker.)

- Coesite, a high pressure polymorph of silica.
- Diamond, the high pressure polymorph of carbon.
- Majorite, a high pressure mineral ($Mg_2Al_3Si_2O_{12}$–$MgSiO_3$).

Coesite was first observed in a laboratory, where it was synthetically created by Loring Coes, Jr., in 1953. Naturally occurring coesite was first noted at Meteor (Barringer) Crater, Arizona where it formed by high pressure impact metamorphism (Chao et al., 1960). Christian Chopin (1984) discovered the first dynamothermally produced coesite in Alpine rocks. Since that time, researchers have found UHP assemblages in many orogenic belts (Figure 18.5). In addition to coesite, diamond and majorite, other minerals include potassium-rich clinopyroxene, magnesium-rich garnet (pyrope), lawsonite, aluminum rutile, glaucophane, phengite, diopside, kyanite, jadeite and ellenbergite (Ernst and Liou, 2000). UHP minerals can occur in a number of environments, which include:

- Meteorite impact sites such as at Meteor Crater, Arizona.
- Kimberlite pipes in which UHP minerals such as coesite and diamond occur in brecciated ultrabasic host rock.
- Deepest levels of ocean lithosphere subduction where quartz transforms to coesite.
- Convergent margins involving two continental lithospheric plates that collide and partially subduct. Although prevailing thought suggests that continental crust is too thick and buoyant to subduct, UHP rocks indicate that continental lithosphere can indeed be subducted to depths of 100–150 km (Ernst and Liou, 2000). UHP rocks are distributed along continent–continent collision zones such as in the Alpine–Himalayan belt, the Ural Mountains, the Western Gneiss region of Norway and in pan-African belt rocks (Figure 18.5).

It is a matter of current debate as to whether UHP minerals define a separate metamorphic facies or represent the high pressure end of the eclogite facies. Given that increasingly higher pressure suites are being recognized, it is likely that a separate UHP facies will be formally recognized in the future. UHP research is one of the dynamic aspects of current research.

On the basis of the discussion above, it should be clear that temperature and pressure conditions produce distinct metamorphic facies. The evolving concept of metamorphic facies – together with experimental laboratory work on mineral stability fields – provides powerful tools for inferring metamorphic conditions. The individual facies do occur alone, but commonly are associated in time and space with other facies. These multiple, associated facies form as a result of temperature and pressure conditions that vary laterally and vertically in an area during metamorphism. In recognition of the existence of patterns of multiple metamorphic facies within a given region, Miyashiro (1961, 1975, 1994) proposed the concept of metamorphic facies series, which we will now address.

18.3   METAMORPHIC FACIES SERIES

A **metamorphic facies series** is a sequence of facies that occurs across a metamorphic terrane due to differences in pressure and temperature (P/T) conditions. Variations in P/T conditions are related to both space and time. In order to describe a sequence of changing metamorphic conditions, geologists refer to **pressure–temperature–time (P-T-t)** relations in which the history of pressure and temperature changes over some period of time are inferred from the rock record. Each facies series is characterized by the development of a particular sequence of individual facies, with each facies stable at a specific range of temperature and pressure conditions. Why is the concept of a facies series so important? Facies series provide key information concerning the progressive P-T-t conditions as well as the tectonic setting in which metamorphism occurred.

Metamorphic facies series were defined (Miyashiro, 1994) on the basis of pressure and temperature gradients, both of which are related to the conditions of metamorphism and tectonic setting (Figure 18.6). Five metamorphic facies series, assigned to three major groups, are recognized.

1 **Low P/T series group:** two low pressure and high temperature facies series are recognized: (1) the very low P/T contact facies series, and (2) the somewhat higher P/T Buchan facies series. These are represented by zone A in Figure 18.6 and by separate lines in Figure 18.7.
2 **Moderate P/T series group:** moderate P/T gradients characterize the Barrovian facies series, represented by zone B in Figure 18.6 and an inclined line in Figure 18.7.
3 **High P/T series group:** the high P/T group includes the Sanbagawa facies series and Franciscan facies series, represented by zone C in Figure 18.6 and as two separate lines in Figure 18.7.

18.3.1   Contact facies series

Contact facies series consists of relatively low pressure (<2.5 kbar ≈ 8 km depth) but moderate to high temperature mineral assemblages. The geothermal gradient for the contact facies

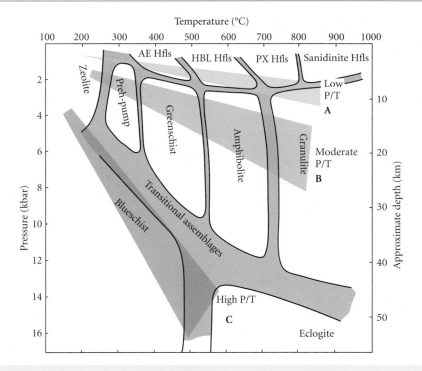

Figure 18.6 Metamorphic facies series trends as indicated by a graph depicting temperature (T) and pressure (P), metamorphic facies and metamorphic facies series. See text for explanation of zone A–C. AE, albite-epidote; HBL, hornblende; Hfls, hornfels; PX, pyroxene. (Courtesy of Bruce Yardley.)

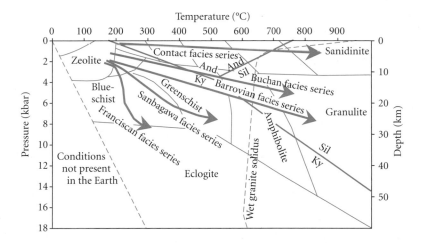

Figure 18.7 The five metamorphic facies series as defined by Miyashiro (1961) and the prehnite-pumpellyite and contact facies series. And, andalusite; Ky, kyanite; Sil, sillimanite. (Courtesy of Stephen Nelson.)

series is >80°/km, implying a significantly higher than average heat input due to magmatic activity. This facies series develops by contact metamorphism in aureoles adjacent to igneous intrusions. While contact meta-morphism commonly occurs at shallow depths, it may also develop in moderate pressure conditions of the lower crust. Low pressure contact metamorphism produces hornfelsic and/or granoblastic rather than foliated textures. However, the rocks may display (1) relict foliated fabrics due to deformation that preceded contact metamorphism, or (2) overprinted foliated fabrics from deformation that occur after contact metamorphism.

Contact facies series metamorphism is recorded by the series of hornfels facies; each of the individual facies is defined based on a low pressure assemblage of minerals stable at specific temperature ranges (see Section 18.2.1). For example, pelitic rocks commonly contain the low temperature polymorph andalusite in low temperature contact facies and the high temperature polymorph sillimanite in high temperature contact facies. The high pressure aluminum silicate polymorph kyanite does not occur (see Figure 18.7).

With increasing temperature the contact facies series progresses through the sequence (1) zeolite facies, (2) albite-epidote hornfels facies, (3) hornblende hornfels facies, (4) the rarer pyroxene hornfels facies, and (5) at very high temperatures, sanidinite hornfels facies. On a temperature–pressure diagram, a line marking the trajectory of progressive contact metamorphism (see Figure 18.6) has a gentle slope, which reflects the progressive increase in temperature. Higher temperature facies occur in close proximity to the intrusion with progressively lower temperature facies toward the outer margins of the contact metamorphic aureole.

The contact facies series can occur anywhere hot plutons intrude rock, such as at convergent margins, ocean spreading ridges, hotspots and localized aureoles around intrusions. In contrast, the Buchan, Barrovian, Sanbagawa and Franciscan facies series are associated with orogenic belts at convergent plate boundaries.

18.3.2   Buchan facies series

The **Buchan facies series** record high geothermal gradients ranging from 40 to 80°C/km. This facies series is named after the Buchan area of northeastern Scotland (Figure 18.8). The Buchan facies series is also known as the Abukuma facies series, named after the Abukuma region in Japan (Miyashiro, 1961).

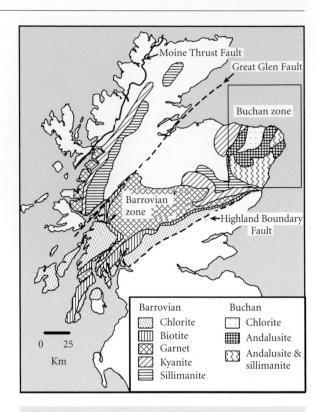

Figure 18.8   Generalized geological map of northern Scotland showing the Barrovian and Buchan zones (after Miyashiro, 1961). (From Best, 2003; with permission of Wiley-Blackwell Publishers and Oxford University Press.)

The individual facies within the Buchan facies series are defined by low to moderate P/T mineral assemblages. As with the contact facies series, pelitic rocks may contain the low temperature polymorph andalusite or the high temperature polymorph sillimanite; the high pressure polymorph mineral kyanite is commonly absent. However, because of its higher P/T ratio, the trajectory of progressive metamorphism followed by the Buchan series on a P/T diagram is somewhat steeper than that of the contact facies series (Figure 18.7).

The Buchan facies series progresses, with increasing temperature and pressure, through (1) zeolite, (2) prehnite-pumpellyite, (3) greenschist, (4) amphibolite, to (5) the high temperature, moderate pressure granulite facies. Buchan facies series metamorphism reflects higher temperatures, but only moderate increases in pressure. Buchan facies series develop by regional metamorphism and magmatic arc activity at convergent margins. Because of the non-uniform stress states produced in orogenic belts, rocks of the Buchan facies series are commonly, but not always, foliated. Non-foliated Buchan facies series rocks can occur in regions that experience crustal thinning and heating, which causes low to moderate P/T ratio metamorphism. Higher P/T ratios result in the development of Barrovian facies series assemblages, which are described below.

18.3.3 Barrovian facies series

The **Barrovian facies series** develop in response to geothermal gradients of ~20–40°C/km, reflecting the progressive increase in both temperature and pressure during regional metamorphism. The Barrovian facies series is named for George Barrow, the geologist who first mapped isograd zones in the Scottish highlands (see Section 18.1). With increasing temperature, the Barrovian facies series progresses through the same facies sequence as the Buchan facies series, from (1) zeolite, (2) prehnite-pumpellyite, (3) greenschist, (4) amphibolite, to (5) the high temperature, moderate to high pressure granulite facies. Because of its higher P/T ratio, the trajectory of progressive metamorphism followed by

Barrovian series metamorphism on a P/T diagram is somewhat steeper than that of the Buchan series (see Figure 18.7). However, the equilibrium mineral assemblage differs from the Buchan system because of the higher P/T ratios during metamorphism. For example, pelitic rocks may contain not only the low temperature polymorph andalusite and the high temperature polymorph sillimanite, but also the high pressure polymorph kyanite. The Barrovian facies series develops in response to increasing temperature and pressure in thickening orogenic belts at convergent plate boundaries, especially collisional orogens. As a result, Barrovian facies series rocks commonly display foliated textures as a result of non-uniform stresses.

18.3.4 Sanbagawa facies series

The **Sanbagawa facies series** are produced under geothermal gradients in the range of 10 to 20°C/km. Miyashiro named this facies after the Sanbagawa belt of Japan (Figure 18.9). The Sanbagawa facies series progression includes (1) zeolite, (2) prehnite-pumpellyite, (3) blueschist facies, followed in some cases by (4) greenschist, and/or (5) amphibolite facies. Sanbagawa facies series metamorphism reflects the rapid increase of pressure

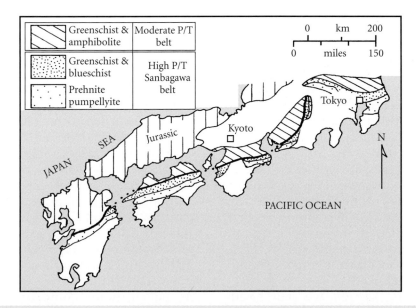

Figure 18.9 Generalized map indicating the location of Japan's Sanbagawa belt (after Miyashiro, 1994). (From Best, 2003; with permission of Wiley-Blackwell Publishers and Oxford University Press.)

relative to temperature during progressive regional metamorphism at convergent plate boundaries. As a result of non-uniform stress, rocks within this facies series commonly are foliated and highly deformed. Because of its higher P/T ratio, the trajectory of progressive metamorphism during Sanbagawa series metamorphism on a temperature–pressure diagram is somewhat steeper than that of the Barrovian series (see Figure 18.7). Compared to the high pressure/very low temperature Franciscan facies series described below, the Sanbagawa facies series is characterized by slightly higher temperatures. This may result from (1) slower subduction giving the rocks more time to heat up as pressures increase, or (2) higher geothermal gradients during subduction.

18.3.5   Franciscan facies series

The **Franciscan facies series** develop where geothermal gradients are <10°C/km. Franciscan facies rocks are characterized by unusually high P/T ratios during progressive metamorphism. Because of its very high P/T ratio, the trajectory of progressive metamorphism during Franciscan series metamorphism on a temperature–pressure diagram is very steep (see Figure 18.7). Miyashiro named this facies series after the Franciscan Complex of California (USA) (Figure 18.10). The Franciscan facies series progresses from (1) zeolite, (2) prehnite-pumpellyite, (3) blueschist, possibly to (4) the eclogite facies (see Figure 18.7). Franciscan series metamorphism reflects the progressive rapid increase in pressure relative to slow increases in temperature during regional metamorphism as rocks are rapidly dragged downward in subduction zones. The high pressure minerals jadeite, glaucophane and lawsonite are particularly important indicators of the high pressure, low temperature conditions. Kyanite, the high pressure polymorph of aluminum silicate, and phengite are common in pelitic rocks. Franciscan facies series rocks are commonly highly deformed and foliated.

In the following section we will briefly introduce the study of equilibrium suites of minerals through the use of ternary diagrams, a powerful tool in the interpretation of metamorphic mineral assemblages.

18.4   THE PHASE RULE, CHEMICAL REACTIONS AND THREE-COMPONENT DIAGRAMS

In this section, we pose two major questions:

1   What are the factors that determine the equilibrium assemblage of metamorphic minerals that develop during closed system metamorphism?
2   How can these mineral assemblages be graphically represented in a useful manner?

When rocks experience high temperatures and/or high pressures over long periods of time, in the presence of appropriate catalysts that permit chemical reactions to proceed, pre-existing mineral assemblages (the reactants) are converted into new mineral assemblages (the products). **Equilibrium conditions** occur whenever chemical reactions have gone to completion – that is, chemical reactions in a system have proceeded to the point where no further net change occurs. When a system has reached equilibrium, no further changes in the mineral assemblage will occur without further changes in temperature and/or pressure. The system is said to be in chemical equilibrium with the existing conditions. Of course, if temperature and/or pressure conditions or any other variable in the system should change, additional reactions will occur in order to re-establish chemical equilibrium with respect to the new conditions. The concept of equilibrium mineral assemblages is the basis for the Barrovian zones, metamorphic facies and metamorphic facies series described earlier in this chapter.

How can we determine whether a mineral assemblage records equilibrium or disequilibrium conditions? Microscopic analysis of thin sections can provide information. In thin section, disequilibrium conditions can be indicated by reaction rims on minerals, coexisting minerals that cannot exist in equilibrium, and incomplete replacement of minerals. Equilibrium conditions are indicated by the absence of the features stated above as well as planar grain contacts between mineral crystals. If equilibrium conditions appear to have been achieved based on thin section or hand-sample analysis of rocks, metamorphic petrologists can make some inferences regard-

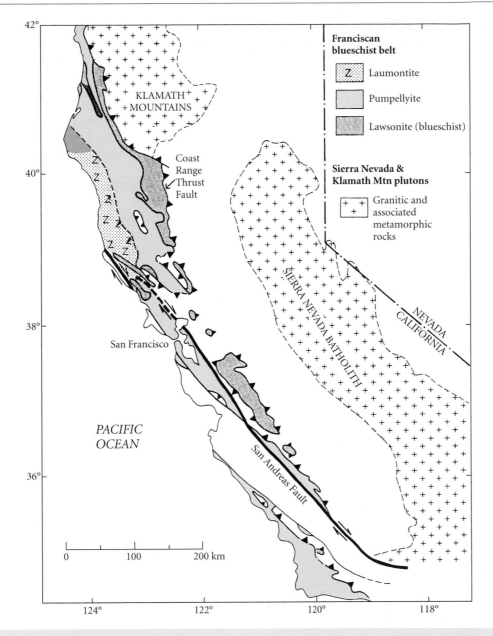

Figure 18.10 Generalized map of the Franciscan belt (after Ernst, 1975; Miyashiro, 1994). (From Best, 2003; with permission of Wiley-Blackwell Publishers and Oxford University Press.)

ing the conditions that produced the equilibrium mineral assemblages using the phase rule.

18.4.1 The phase rule and metamorphic minerals

The phase rule (Gibbs, 1928) is a means by which petrologists predict equilibrium assemblages of minerals in many systems, including both igneous and metamorphic rocks. Recall the phase rule (see Section 3.2.1), which states that

$$P = C + 2 - F$$

where P represents the number of phases present in a system, C designates the minimum number of chemical components and F refers

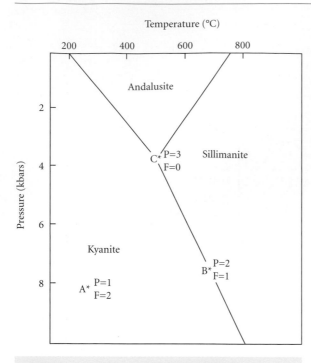

Figure 18.11 Phase rule as applied to the Al$_2$SiO$_5$ polymorph mineral group.

to the number of degrees of freedom or variance.

As discussed in Chapter 3, a **phase diagram** is a visual means by which mineral stability fields can be displayed. Let us examine a very simple and familiar example of how the phase rule can be applied to phase diagrams in metamorphic petrology (Figure 18.11). Consider the one chemical component (C = 1) system Al$_2$SiO$_5$, which can exist as three different mineral phases (P): (1) andalusite (low temperature polymorph), (2) kyanite (high pressure polymorph), and (3) sillimanite (high temperature polymorph). Temperature–pressure stability fields for different minerals are separated by lines. At any point within the kyanite stability field (e.g., point A), Al$_2$SiO$_5$ exists as kyanite, a single phase (P = 1). Using the phase rule P = C + 2 − F, since 1 = 1 + 2 − 2, F = 2. This means that both temperature and pressure can be varied independently of one another while the system remains within the kyanite stability field. This stability field is called a **divariant field** because the two variables can change independently of one another without changing the phase composition of the system. As suggested by Figure 18.11, similar large divariant fields occur at higher

temperature for sillimanite and at lower temperature for andalusite and are separated from one another by boundary lines on the phase diagram that separate the stability fields for each phase.

However, on the boundary line between the kyanite and sillimanite stability fields, such as at point B, kyanite and sillimanite coexist (P = 2). From the phase rule it is clear that along this line only one independent variable exists (F = 1) in this one component system. An independent change in either the temperature or pressure requires a change in the other dependent variable in order for the system to remain on the kyanite–sillimanite boundary line. All phase boundary lines in this system are **univariant lines** because only one variable can change independently. Any change in one variable requires a compensating change in the second variable in order to maintain the two phases in equilibrium. Univariant lines are fundamentally important in a one-component system because they indicate:

- Boundaries between divariant fields in which only one mineral species is stable.
- Conditions under which two phases can coexist.
- Conditions across which changes in equilibrium mineral assemblages occur.

Point C, the only point on the diagram where three univariant phase boundaries intersect, defines the unique temperature–pressure conditions under which all three Al$_2$SiO$_5$ mineral phases coexist (P = 3) (see Figure 18.11). From the phase rule, since 3 = 1 + 2 − 0, the coexistence of three phases at a single point implies that F = 0. Only at this fixed temperature and pressure, called an **invariant point**, can all three phases coexist. In the Al$_2$SiO$_5$ system, kyanite, andalusite and sillimanite all coexist in equilibrium only at a temperature of 510°C and pressure of 3.8 kbar. Note that univariant lines intersect at invariant points. Invariant points are critically important because they indicate:

- Unique points in three divariant fields where three univariant lines intersect.
- Specific temperature and pressure conditions at which three stable phases coexist in a one-component system.

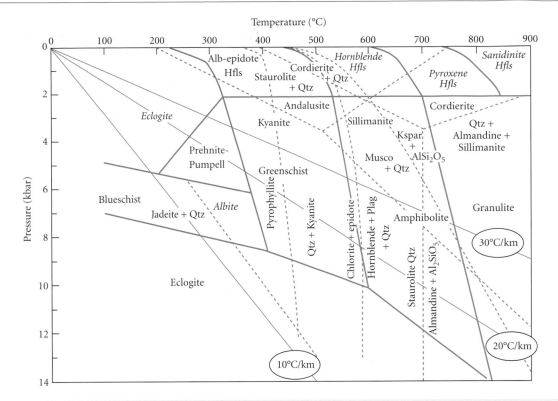

Figure 18.12   Equilibrium assemblage grid for minerals derived from pelitic protoliths. See text for further explanation of figure. Hfls, hornfels. (Courtesy of Stephen Nelson.)

Most metamorphic mineral assemblages have compositions that require more than one component in order to adequately describe phase compositions. These are discussed in the section that follows.

18.4.2   Equilibrium mineral assemblage grids

The phase rule can be applied in metamorphic systems that involve more than one component. By displaying a number of phase stability relationships on a temperature–pressure diagram, we can graphically demonstrate the temperature and pressure conditions at which specific mineral transformations occur, and the minerals likely to occur within specific temperature and pressure ranges. By overlaying such information onto the temperature–pressure fields for metamorphic facies, we can demonstrate which facies and facies series involve specific reactions and mineral stability fields (Figure 18.12). Such equilibrium mineral reaction and assemblage diagrams are referred to as **petrogenetic** or **paragenetic grids**. Petrogenetic refers to the conditions under which the rock originated, whereas the term para-

genesis refers to the formation sequence of an equilibrium set of minerals that formed at different times, e.g., along a metamorphic temperature–pressure trajectory. Both are important in understanding the formation and evolution of metamorphic rocks.

Figure 18.12 illustrates a paragenetic grid for minerals derived from a pelitic protolith. Note the many divariant fields, univariant lines and invariant points for a number of different aluminosilicate mineral transformations. Phase transformations across univariant lines (dashed lines) that separate mineral stability fields are labeled with the mineral reactions that occur. This information is overlain on the metamorphic facies stability fields (solid lines). Progressive metamorphic paths or metamorphic trajectories for three different geothermal gradients that approximately range from Franciscan through Barrovian facies series (10 to 30°C/km) are also shown. This grid provides an example of the various types of data that can be succinctly summarized in a single diagram. By following any metamorphic trajectory, one can see metamorphic transformations that can occur

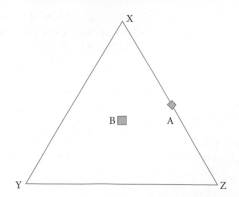

Figure 18.13   Generalized ternary diagram with X, Y and Z representing end member oxide compounds. Point A has a normalized composition of 50% X and 50% Z. Point B has a normalized concentration of 33% X, 33% Y and 33% Z.

during progressive metamorphism, compare it with the different reactions that characterize other trajectories, and gain insights into the variations in mineral assemblages between metamorphic facies.

18.4.3   Ternary diagrams

Equilibrium assemblages of minerals can be graphically displayed on **three-component (ternary) diagrams**. Ternary plots are widely used in the study of metamorphic rocks to illustrate the minerals that develop under various conditions for specific rock compositions. The major chemical components in metamorphic reactions include SiO_2, Al_2O_3, FeO, Fe_2O_3, MgO, CaO, K_2O, Na_2O, H_2O and CO_2. Other chemical components, such as TiO_2, MnO_2 and P_2O_5, are also important in metamorphic rocks but our focus will be on the major chemical compounds. Ternary diagrams display three sets of oxide compounds at the X, Y and Z apices of the triangular diagram, respectively (Figure 18.13). Common X, Y and Z oxide compounds selected include Na_2O, K_2O, CaO, Al_2O_3, FeO and MgO. Chemically similar components, such as ferrous iron and magnesium oxides, are commonly grouped as a single component (FeO + MgO) for simplification. The three major chemical components combine to produce up to five major coexisting minerals in a given metamorphic rock.

Ternary diagrams depict (1) stable mineral assemblages that exist under specific temperature and pressure conditions, and (2) how mineral assemblages change in response to changes in temperature and pressure. As such, ternary diagrams are a powerful tool in understanding the evolution of metamorphic mineral assemblages, rocks and series. Pentti Eskola (1915) used ternary diagrams to display the composition of common metamorphic mineral assemblages in terms of their concentration of common oxide components. Mineral concentrations are normalized so that the concentrations of the three oxide components in a given mineral sum to a total of 100% (e.g., X + Y + Z = 100%). As discussed in Chapter 3 and shown in Figure 18.13, a mineral that plots at point A has a composition expressed as 50% X, 50% Z and 0% Y, whereas a mineral represented by point B has a composition expressed as 33.3% X, 33.3% Y and 33.3% Z.

ACF ternary diagrams

The **ACF ternary diagram** proposed by Eskola (1915) is based on the molecular amounts of three components, ACF, where A = (Al_2O_3 + Fe_2O_3) − (Na_2O + K_2O), C = (CaO − $3.33P_2O_5$) and F = FeO + MgO + MnO. The ACF diagram can be used to depict average compositional differences between the five major compositional groups of metamorphic rocks: (1) aluminum-rich pelitic rocks, (2) calcium/magnesium-rich, aluminum-poor calcareous rocks, (3) magnesium/iron-rich ultrabasic rocks, (4) iron/magnesium/calcium-rich basic rocks, and (5) quartz feldspathic rocks that contain on average similar amounts of all three components. The ACF diagram is particularly useful for displaying common equilibrium mineral assemblages that occur in rocks derived from the quartzo-feldspathic, basic, calcareous and pelitic protoliths (Figure 18.14).

Figure 18.15 illustrates common metamorphic minerals whose compositions can be plotted on an ACF diagram. Only a small subset of these minerals will occur under any given temperature and pressure conditions, depending upon mineral stability ranges. Minerals with fairly constant compositions are shown by black squares. Minerals with extensive substitution solid solution can have

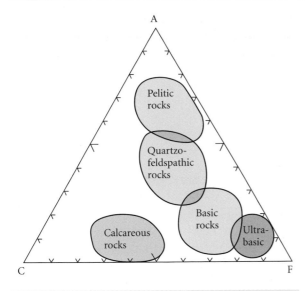

Figure 18.14 Diagrammatic illustration of general chemical composition in terms of A, C and F for different protoliths where A = (Al$_2$O$_3$ + Fe$_2$O$_3$) − (Na$_2$O+K$_2$O), C = CaO − 3.33P$_2$O$_5$ and F = FeO + MgO + MnO. (Courtesy of Stephen Nelson.)

variable compositions (e.g., chlorite, hornblende, anthophyllite) that occupy larger areas within ternary diagrams.

A′KF ternary diagrams

The **A′KF diagram** proposed by Eskola (1915) is used to discriminate equilibrium mineral assemblages derived from pelitic and quartzofeldspathic protoliths with excess Al$_2$O$_3$ and SiO$_2$ (Figure 18.16). In this diagram the three apices indicate the following: A′ = (Al$_2$O$_3$ + Fe$_2$O$_3$) − (K$_2$O + CaO + Na$_2$O), K = K$_2$O and F = (FeO + MgO).

While ACF and A′KF diagrams combine MgO, FeO ± MnO as F, Thompson (1957) utilized an **AFM diagram** for metamorphic rocks where A = Al$_2$O$_3$, F = FeO and M = MgO. The AFM diagram is particularly useful in discriminating mineral compositions in ferromagnesian-rich basic and ultrabasic rocks as well as many pelitic rocks. Figure 18.17 illustrates the compositions of aluminosilicate minerals associated with pelitic rocks

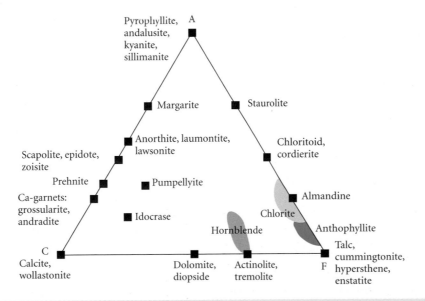

Figure 18.15 ACF diagram illustrating the approximate location of common metamorphic minerals on a ternary diagram in terms of three components: A = (Al$_2$O$_3$ + Fe$_2$O$_3$) − (Na$_2$O+K$_2$O), C = CaO − 3.33P$_2$O$_5$ and F = FeO + MgO + MnO. The black boxes indicate a fairly constant mineral composition. Shaded zones indicate a variable mineral composition due to solid solution or variable purity. (Courtesy of Steve Dutch.)

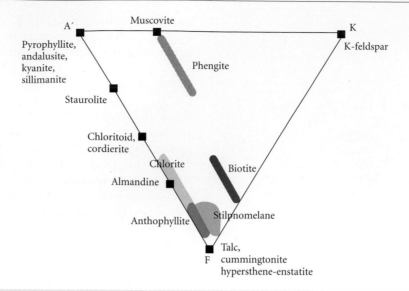

Figure 18.16 A′KF diagram illustrating the approximate location of common pelitic and quartzo-feldspathic metamorphic minerals on a ternary diagram where the end points are: A′ = Al_2O_3 – $(Na_2O + K_2O + CaO)$, K = K_2O and F = $(FeO + MgO + MnO)$. (Courtesy of Steve Dutch.)

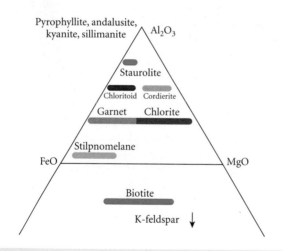

Figure 18.17 AFM diagram illustrating some common silicate minerals. (Courtesy of Steve Dutch.)

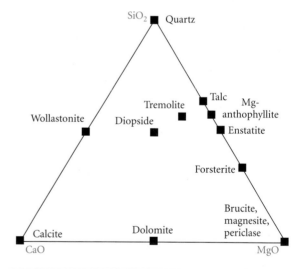

Figure 18.18 Ternary CMS diagram illustrating common metamorphic minerals that occur in rocks enriched in CaO, SiO_2 or MgO. (Courtesy of Steve Dutch.)

such as biotite, cordierite, andalusite, sillimanite, kyanite, chloritoid, stilpnomelane, garnet and staurolite on the Thompson AFM diagram.

CMS ternary diagrams

CMS ternary diagrams have also been developed for calcareous rocks as illustrated in

Figure 18.18, in which the three components are CaO, MgO and SiO_2. The CMS diagram also has powerful applications in the depiction of mineral assemblages in calc-silicate rocks (containing minerals such as calcite, dolomite, wollastonite and diopside) and

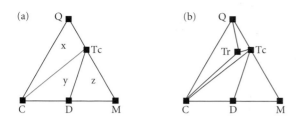

Figure 18.19 A simplified CMS ternary diagram showing at two temperatures: (a) ~400°C mineral assemblage; (b) >500°C mineral assemblage. C, calcite; D, dolomite; M, magnesite; Q, quartz; Tc, talc; Tr, tremolite. (Courtesy of Steve Dutch.)

ultrabasic (containing minerals such as tremolite, anthophyllite and forsterite) rocks. It is assumed that sufficient CO_2 is present to make the carbonate minerals such as calcite and dolomite and sufficient H_2O to make hydrous minerals such as talc, tremolite, antigorite and brucite.

Ternary diagrams such as these depict the equilibrium assemblage of minerals that may occur. However the constraints imposed by the phase rule limit to no more than five ($P = C + 2 - F$), the number of minerals that coexist in equilibrium at a given set of temperature and pressure conditions ($F = 0$) in a three-component system.

Tie lines

Tie lines are drawn on ternary diagrams between equilibrium minerals that coexist under specific temperature and pressure conditions. Compositions located on tie lines contain the two minerals joined by the tie line while points on either end of a tie line represent only one stable mineral. Mineral stability fields are defined by three tie lines. Figure 18.19 depicts a simplified model of a CMS ternary diagram with CaO (C), MgO (M) and SiO_2 (S, here represented by Q for quartz) components. Figure 18.19a illustrates the stable minerals in this system at ~400°C, in which tie lines connect mineral pairs. Some tie lines occur along the margins of the triangle (e.g., Q–C), others cross the interior (e.g., D–Tc). The tie lines divide the ternary diagram in Figure 18.19a into three separate fields,

each of which is completely bounded by tie lines between stable mineral phases. Mineral assemblages within any stability field contain the minerals located on the tie lines that bound the stability field. For example, compositions in field x contain quartz, calcite and talc. Field y contains calcite, dolomite and talc, and field z contains talc, dolomite and magnesite. This permits prediction of minerals that can coexist at 400°C for any composition expressed by the three end member components.

Figure 18.19b illustrates the same ternary diagram at >500°C. Note the occurrence of tremolite, which documents higher temperature conditions. The Tr–Tc and Tr–Q tie lines cause the ternary diagram to be divided into five distinct fields. Compositions that lie within the x field on this diagram are composed of quartz, calcite and talc at 400°C, but at 500°C are composed of (1) quartz-calcite-tremolite in the field bounded by the Q-C-Tr tie lines, (2) quartz-talc-tremolite in the field bounded by the Q-Tc-Tr tie lines, and (3) tremolite-talc-calcite in the field bounded by the Tr-Tc-C tie lines. Tremolite forms at ~500°C by the chemical reaction of quartz, talc and calcite, until one of these phases disappears and a new chemical equilibrium is achieved. The equilibrium mineral assemblages whose compositions plot in stability fields y and z do not change over this temperature range. Figure 18.19 illustrates a simple example of how mineral assemblages change as a result of chemical reactions during progressive metamorphism.

As noted earlier, many minerals have compositions that vary considerably as a result of solid solution. These include plagioclase, garnet, chlorite and amphibole minerals such as hornblende and actinolite-tremolite. In many of the ternary diagrams presented in this chapter, bold black lines are used to indicate the compositional range of minerals such as chlorite with variable composition due to solid solution (Figure 18.20a). In scientific journals and other textbooks, you may see a series of sub-parallel, diverging lines directed towards the solid solution mineral (Figure 18.20b). These are tie lines between other minerals and various compositions of the mineral in question. In Figure 18.20b, multiple tie lines are shown between epidote and different compositions of chlorite.

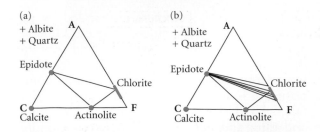

Figure 18.20 (a) ACF diagram in which the different chemical compositions of the chlorite group minerals are represented by a blue bar line. (b) ACF diagram in which a series of sub-parallel divergent (radiating) tie lines connect various compositions of chlorite group minerals. (Courtesy of Stephen Nelson.)

Ternary diagrams are used as a predictive model for mineral assemblages in rocks. In the real world, where the number of components is generally more than three, mineral assemblages are more complex than the simple diagrams presented above. Nevertheless, three-component models serve as a very useful starting point for determining the likely assemblage of major metamorphic minerals produced from different protoliths. For these reasons, ternary diagrams continue to be widely used in the study of metamorphic rocks.

Figure 18.21 presents a series of ACF, A′KF and AFM diagrams for the greenschist, amphibolite, granulite, blueschist and eclogite facies for metabasic/calc-silicate (left column) or pelitic (right column) protoliths. Note that several of the diagrams indicate other minerals that occur such as quartz, muscovite and potassium feldspar, which require additional components. One can take any composition on one of these sets of diagrams and trace the changes in equilibrium mineral assemblages that occur as temperature and/or pressure conditions change during progressive or retrograde metamorphism. Of course different diagrams will be used for Franciscan facies series and Barrovian facies series depending on the sequence in which metamorphic assemblages and facies develop along the metamorphic trajectory. The above discussion is an abbreviated introduction to ternary diagrams. Readers who wish to learn more of the construction, use and assumptions of ternary diagrams are referred to Winter (2009), Philpotts

and Ague (2009), Raymond (2007), Blatt et al. (2003), Best (2003), Yardley (1989), Hyndman (1985) or Miyashiro (1973).

While the pioneering work on Barrovian zones, metamorphic facies, metamorphic facies series, mineral assemblage grids and ternary diagrams occurred primarily in the early 20th century, the second half of the 20th century provided a context by which we were able to incorporate these mineral and rock assemblages into a global tectonic framework. In the following section, we will relate metamorphic rocks, facies and facies series to the context of plate tectonics.

18.5   METAMORPHIC ROCKS AND PLATE TECTONICS

We introduced plate tectonics in Chapter 1 and have continued to incorporate the concepts of plate tectonics throughout our discussion of igneous, sedimentary and metamorphic rocks. Various types of metamorphism – including meteorite impacts, dynamic metamorphism, contact metamorphism and burial metamorphism – can occur anywhere on Earth under the appropriate conditions. Most large-scale metamorphism is, however, inextricably linked to plate tectonics (Figure 18.22). Regional metamorphism, which produces the vast majority of metamorphic rocks, occurs primarily at:

1   Divergent plate boundaries, associated with continental and oceanic rifts and sea floor spreading, where hydrothermal processes dominate.

Figure 18.21 ACF, A′KF and AFM diagrams for rocks derived from basites, calc-silicates (A′KF and AFM diagrams on right) and pelitic rocks (middle ACF diagrams): (a) greenschist facies; (b) amphibolite facies; (c) granulite facies; (d) blueschist facies, and (e) eclogite facies. Note the ACF inset diagram to the left of each part indicating a generalized composition of protolith rock type on an ACF diagram. (Courtesy of Stephen Nelson.)

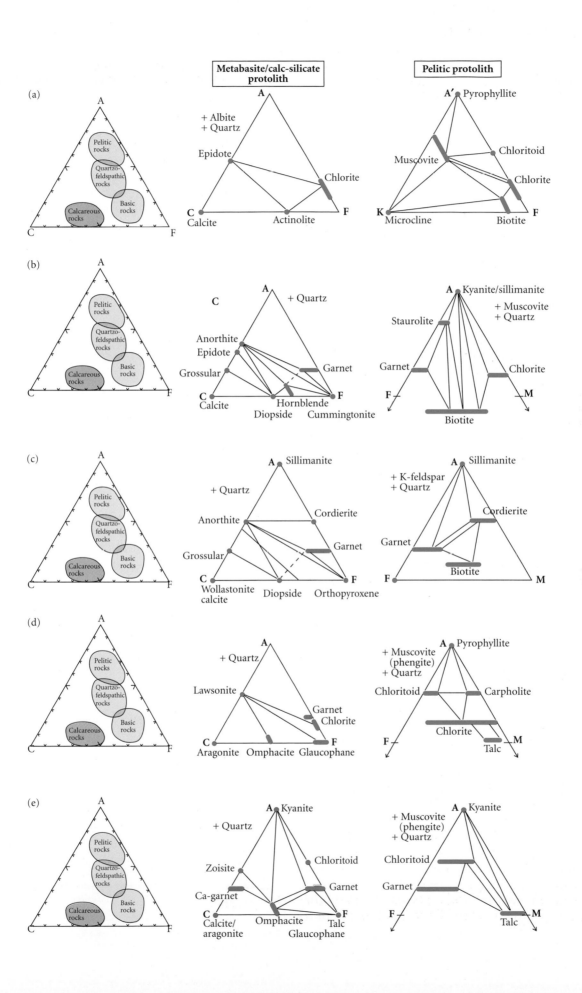

(a)

A

+ Albite
+ Quartz

Epidote

C — Calcite — Actinolite — Chlorite — F

A′ Pyrophyllite

Muscovite — Chloritoid — Chlorite

K — Microcline — Biotite — F

(b)

A

C — + Quartz

Anorthite
Epidote
Grossular

C — Calcite — Diopside — Hornblende — Cummingtonite — F — Garnet

A Kyanite/sillimanite

+ Muscovite
+ Quartz

Staurolite

Garnet — Chlorite

F — Biotite — M

(c)

A Sillimanite

+ Quartz

Anorthite — Cordierite

Grossular — Garnet

C — Wollastonite calcite — Diopside — Orthopyroxene — F

A Sillimanite

+ K-feldspar
+ Quartz

Garnet — Cordierite

F — Biotite — M

(d)

A

+ Quartz

Lawsonite

Garnet
Chlorite

C — Aragonite — Omphacite — Glaucophane — F

A Pyrophyllite

+ Muscovite
(phengite)
+ Quartz

Chloritoid — Carpholite

F — Chlorite — M
Talc

(e)

A Kyanite

+ Quartz

Zoisite — Chloritoid

Ca-garnet — Garnet

C — Calcite/ aragonite — Omphacite — Talc — F
Glaucophane

A Kyanite

+ Muscovite
(phengite)
+ Quartz

Chloritoid

Garnet

F — Talc — M

2 Convergent plate boundaries, associated with subduction, magmatic arcs and continental collisions, where dynamothermal metamorphic processes dominate.

Because most metamorphism occurs in intimate association with plate boundaries it is appropriate to weave a story that integrates the topics discussed in Chapters 15–18 with their plate tectonic settings. In the section that follows we will relate metamorphic processes, mineral assemblages, rocks, facies and facies series to the tectonic processes that occur at divergent and convergent plate boundaries.

18.5.1   Divergent plate boundaries

Divergent plate boundaries are the site of widespread metamorphism. As discussed in Chapter 1, divergent margins include continental rifts and ocean ridge systems. Tensional stresses and extensional deformation result in the uplift, thinning and fracturing of continental lithosphere. Buoyant, hot basaltic magma rises upward through extensional fissures. Continued extension and magmatism generates rift valleys that can evolve to ocean spreading ridges.

Continental rifts

Continental rift basin metamorphism can occur by a number of processes:

- Contact metamorphism from the intrusion of shallow dikes, sills and flood basalts.
- Hydrothermal alteration associated with hot magmatic and wall rock volatile fluids.
- Dynamic metamorphism due to brittle extensional faulting in the upper crust and ductile shearing in the lower crust.
- Burial metamorphism producing zeolite and prehnite-pumpellyite facies assemblages due to the deposition of thick, non-marine detrital sediment sequences in rift basins.

Ocean ridges

As discussed in Chapter 1, continued rifting of continents creates narrow seaways and eventually wider ocean basins by sea floor spreading along an oceanic ridge system. In such marine environments, cold, saline seawater flows down through fractures and interacts with hot basic and ultrabasic rock of the recently formed ocean lithosphere (Figure 18.22a). As a result, extensive hydrothermal metamorphism of basalt, gabbro and peridotite occurs at ocean ridges resulting in albite-epidote hornfels, zeolite and prehnite-pumpellyite facies mineral assemblages. At deeper levels within the oceanic crust and upper mantle, gabbro and peridotite are altered to higher temperature hornblende hornfels facies assemblages (see Section 18.2). As a result, minerals such as zeolites, prehnite, pumpellyite, chlorite, calcite-epidote, actinolite, albite, talc, serpentine, brucite, magnesite, biotite and hornblende are formed. Because hydrothermal metamorphism of the sea floor occurs in an environment with minimal compressive stress and low confining pressures, foliations and high pressure minerals such as garnet are not formed. Metasomatism via the reaction of cold seawater infiltrating into fractured ocean lithosphere, results in the addition of Na, Cl, Br, CO_3, SO_4 and O_2 to the oceanic crust. As a result, the ocean crust experiences extensive hydrothermal alteration and contains spilitized (sodium-rich) basalts. At the same time, elements such as Mn, Ni, Zn, Cu, Co, Au, Ag and Fe are leached from magma and precipitate onto the sea floor via black smokers producing metallic ore deposits (Chapters 9 and 19). Eventually, hydrothermally altered oceanic crust, which covers ~70% of Earth's surface, moves away from the spreading ridge and encounters a convergent plate boundary, discussed below.

18.5.2   Convergent plate boundaries

As noted earlier, the concepts of metamorphic zones, facies and facies series predated our understanding of plate tectonics involving divergent and convergent plate boundaries. Miyashiro (1961) noted the existence of paired metamorphic belts, consisting of high P/T terranes adjacent to low P/T terranes, in young mountains of Japan and other regions within the circum-Pacific region. The question arose as to why young paired metamorphic belts exist in the circum-Pacific, Caribbean,

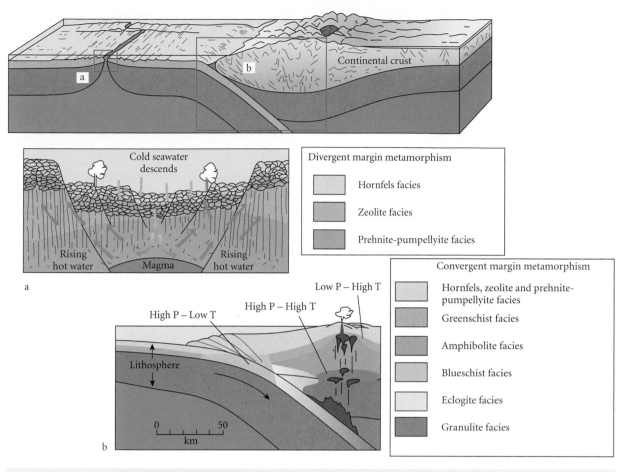

Figure 18.22   Divergent and convergent plate boundaries and metamorphic facies. P, pressure; T, temperature.

Scotia and eastern Indian Ocean region and not elsewhere? Miyashiro's observation was particularly timely as 1960s research revolutionized our knowledge of Earth's dynamic nature. Within a few years, the paired metamorphic belts that encircle the Pacific Ocean were recognized as subduction zone (outer metamorphic belt) and magmatic arc (inner metamorphic belt) assemblages that developed in response to lithospheric plate convergence and subduction.

The **paired metamorphic belts** correspond with the two major components of convergent margins (Figure 18.22b):

1   The outer metamorphic belt occurs on the ocean or trench side and consists of Sanbagawa or Franciscan facies series rocks, characterized lower thermal gradi-

ents and high P/T to very high P/T mineral assemblages (see Section 18.3). Rapid, steep subduction favors the development of very high P/T ratios that characterize the Franciscan facies series, whereas slower, shallower subduction favors the development of the moderate to high P/T ratios that characterize the Sanbagawa facies series. The outer metamorphic belt also contains hornfels, zeolite and prehnite-pumpellyite facies metamorphism from early ocean ridge alteration. However, these rocks are commonly overprinted by more recent greenschist facies and high pressure assemblages such as blueschist and eclogite facies rocks due to burial in the subduction zone. Notably absent are the high temperature assemblages of the amphibolite and granulite facies. The outer metamorphic belt is exposed in subduction zone complexes, accretionary wedges and mélanges.

2 The inner metamorphic belt occurs on the continent or arc side and consists of Buchan or Barrovian facies series rocks, characterized by moderate to high temperature gradients and by moderate P/T to low P/T mineral assemblages (see Section 18.3). The inner metamorphic belt is marked by the occurrence of hornfels, zeolite, prehnite-pumpellyite, greenschist, amphibolite and granulite metamorphic facies produced by arc magmatism and compressive stresses. Notably absent are high pressure assemblages of the blueschist and eclogite facies. The inner metamorphic belt occurs along the magmatic arc complex.

Convergent plate boundaries are sites where two lithospheric plates move toward one another, generating immense compressive stresses and producing widespread, regional belts of dynamothermally metamorphosed rock. Three different convergent margin settings occur based on the type of lithosphere at the plate leading edges: ocean–ocean convergence, ocean–continent convergence and continent–continent collision. In the following discussion we will briefly present some of the major components of these convergent margin settings and describe the metamorphic processes and rocks that occur in each.

Ocean–ocean convergence

Trenches and subduction zones
Trenches (Figure 18.23) are deep linear troughs in ocean basins that mark the surface expression of inclined subduction zones. Subduction zones occur along the Pacific rim, the eastern Indian Ocean, the Caribbean, the Mediterranean Sea and around the Scotia Plate. The subduction of relatively cold lithosphere to depths as great as 700 km produces high P/T Sanbagawa and Franciscan facies that occur only in association with ocean plate subduction. While these high P/T blueschist rocks are generated in the subduction zone, they can later be incorporated into forearc accretionary wedges and mélanges, discussed below.

Forearc accretionary wedges and mélanges
Forearc regions, located on the overriding plate between the trench and the volcanic arc, include the accretionary wedge (prism), forearc basin and forearc basement. The forearc accretionary wedge (Figure 18.23) contains a diverse assemblage of deformed rocks that are tectonically dismembered through faulting, folding and shearing. The oceanic lithosphere that undergoes subduction contains ocean plateaus, seamounts, guyots, ocean ridges and sediment layers of varying thickness. While much of the ocean lithosphere is subducted, some is offscraped, underplated, folded and sheared into tectonically deformed rock material that accumulates as an imbricated accretionary mélange package. As a result of subduction zone tectonism and offscraping, a diverse suite of rock assemblages are incorporated into the accretionary wedge and include the following:

- Basalt, gabbro and peridotite from the downgoing ocean lithosphere, ocean plateaus, ocean islands and seamounts.
- Limestone and chert from pelagic sediment that accumulated on the ocean floor.
- Muds, mudrocks and coarser volcano–detrital sediment derived from the arc.
- Intermediate to silicic volcanic and plutonic rocks derived from the adjacent volcanic arc complex or from far-traveled arcs that have migrated towards the subduction zone.
- Metamorphic suites of rocks highly variable in composition, origin and facies. The metamorphic suite includes zeolite, prehnite-pumpellyite, greenschist, amphibolite and granulite.
- High P/T assemblages such as blueschist – and in rare instances eclogite – facies rocks derived from the underlying subduction zone.

Tectonic mélanges are produced within the forearc accretionary prism. Mélange, from the French word for chaotic mixture, commonly contains fragments of rocks that include basalt, gabbro, peridotite, chert, limestone, sandstone, serpentinite, phyllite, zeolite, prehnite-pumpellyite, greenschist, amphibolite, blueschist and eclogite rock fragments encased in a scaly mudrock matrix. Some of these rock materials are offscraped at the

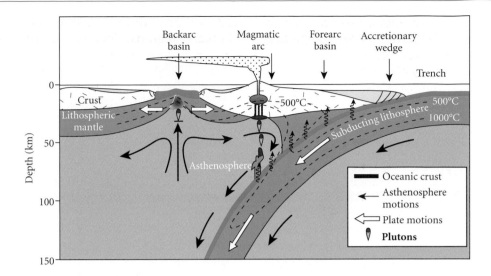

Figure 18.23   Convergent margin setting with a trench, accretionary wedge (prism), forearc basin and underlying basement, magmatic arc and backarc basin. (Courtesy of Wikipedia.)

inner trench wall where the downgoing lithospheric slab is stripped of overlying rock material. Other rock material is transferred from the subducted slab to the base of the overriding slab accretionary wedge by underplating, and then rises upward as regurgitated rock components metamorphosed at greater depths by enigmatic processes involving faulting and uplift. The accretionary mélange matrix is enriched in pore fluids that are expelled upon further compression to produce a scaley matrix (Figure 18.24). The release of heated pore fluids enhances hydrothermal alteration and metasomatic reactions. Widespread shortening produces fold and thrust belts in the accretionary wedge whereby rock slices are stacked vertically. Intense folding, thrusting and local mass wasting juxtaposes variably metamorphosed rock fragments in the mélange.

Highly faulted ophiolite complexes (Chapter 10) commonly occur in accretionary wedges. Because of their location above subduction zones (Figure 18.25), these are referred to as supra-subduction zone (SSZ) ophiolites. SSZ ophiolites form by the off-scraping and offslicing of oceanic and/or arc lithosphere resulting in the emplacement of oceanic/arc lithosphere fragments onto the overriding hanging wall plate – a process called **obduction**.

Blueschist facies mineral assemblages, the diagnostic facies of Phanerozoic subduction zones, are exposed in subduction mélanges within the accretionary wedge of most young convergent margins. The subduction mélange is characterized by a diverse suite of metamorphic facies including high P/T blueschist and eclogite assemblages of the Franciscan and/or Sanbagawa metamorphic facies series (Figure 18.26). In contrast, the forearc basin is largely affected by burial metamorphism and the arc complex is characterized by low to moderate P/T assemblages as discussed below.

Forearc basins and basements
Forearc basins (see Figure 18.23) develop between the uplifted accretionary wedge/subduction zone complex and the volcanic arc. The forearc basins occur between the high P/T rocks of the accretionary prism and the low P/T paired metamorphic belts of the magmatic arc (see Figure 18.22b). Forearc basins are commonly affected by burial metamorphism due to deposition of volcaniclastic sediment derived from the adjacent volcanic arc. Total basin-fill thicknesses can exceed 10 km, resulting in zeolite to prehnite-pumpellyite facies assemblages. Coombs' research in forearc basins of New Zealand served as the pioneering work in understanding the role of burial metamorphism at sub-greenschist facies

Figure 18.24 Franciscan mélange with a mixture of blueschist and greenschist facies rocks. Individual blueschist clasts are boulder sized and are referred to as "knockers". (Photos by Kevin Hefferan.) (For color version, see Plate 18.24, between pp. 408 and 409.)

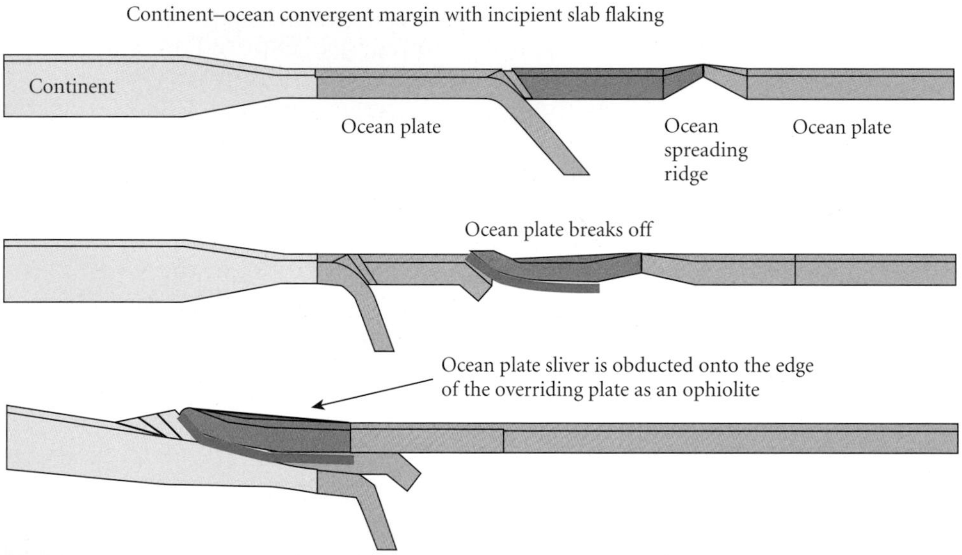

Continent–ocean convergent margin with incipient slab flaking

Continent

Ocean plate

Ocean spreading ridge

Ocean plate

Ocean plate breaks off

Ocean plate sliver is obducted onto the edge of the overriding plate as an ophiolite

Figure 18.25 Simplified development of an ophiolite by obduction of ocean lithosphere above a subduction zone. (Courtesy of Bradley Hacker.)

conditions (Coombs et al., 1959; Coombs, 1961). Hot fluids and magma generated above the subducted slab can result in hydrothermal alteration and contact metamorphism, respectively. Forearc basins, such as the Great Valley Basin in California (USA), provide for excellent agriculture due to the rapid deposition of nutrient-rich sediment derived from nearby volcanic arc complexes (Figure 18.27).

The forearc basin is underlain by forearc basement, which consists of overriding plate rocks that existed prior to subduction as well as younger volcanic–magmatic arc material that formed after subduction began. The

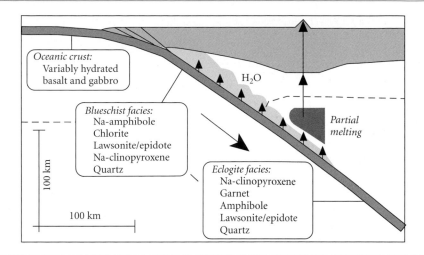

Figure 18.26 High pressure mineral assemblages within the subducting slab. (Courtesy of Simon Peacock.)

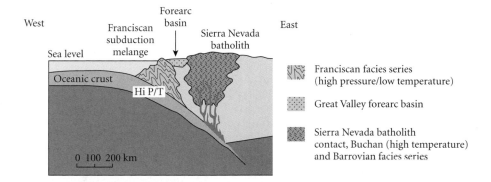

Figure 18.27 Cross-section showing Mesozoic subduction, a Franciscan accretionary wedge, and the Great Valley forearc basin and volcanic arc in California (USA). Note the paired metamorphic belts occurring in the Franciscan complex and the Sierra Nevada. P, pressure; T, temperature. (From Levin, 2006; with permission of John Wiley & Sons.)

forearc basement may experience subduction erosion as the arc system migrates towards the trench. In subduction erosion, eroded forearc basement rock is incorporated into the accretionary wedge or cannibalized into deeper levels of the subduction zone. Some SSZ ophiolites develop by tectonic slicing of the forearc basement. The forearc basement can experience burial metamorphism due to the thick sedimentary deposits in the overlying basin. Dynamothermal metamorphism can also produce either high P/T facies associated with the underlying subduction zone or moderate P/T due to the adjacent magmatic arc.

Magmatic arc complexes
The magmatic arc complex (see Figure 18.23) consists of deep intermediate to silicic plutons that are overlain by the composite volcanoes of the volcanic arc. The duration and aerial extent of igneous activity leads to long-term, intense heating of rocks in the overlying plate. This heating creates higher than normal geothermal gradients ($\geq 40°C/km$) generating contact, Buchan and, to a less degree, Barro-

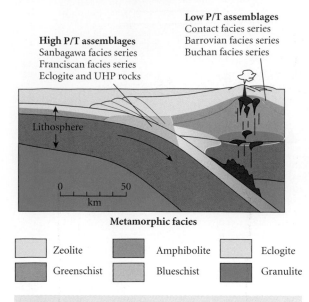

Figure 18.28   The metamorphic facies that develop at convergent plate boundaries. P, pressure; T, temperature; UHP, ultra-high pressure.

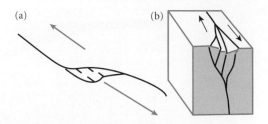

Figure 18.29   (a) Map view of a pull-apart basin developing by tension along a bend in a sinistral strike-slip fault system. (b) Block diagram of a pull-apart basin.

vian facies series assemblages. Progressive Buchan facies series metamorphism generates regionally developed zeolite, greenschist, amphibolite and granulite facies metamorphism. High geothermal gradients occur in close proximity to arc plutons where low P/T assemblages are developed in contact metamorphic aureoles (Figure 18.28). Modern magmatic arc complexes with extensive moderate to low P/T Buchan and Barrovian facies series rocks are well developed throughout the world, including the Pacific rim, eastern Indian Ocean, Caribbean and Scotia region. Ancient examples occur in the Caledonian, Appalachian, Ural and Alpine–Himalayan fold and thrust belts. Magmatic underplating and the tectonic underplating of subducted rock slices results in long-term thickening of arc complex lithosphere. In extreme instances, increasing thickness can lead to the development of the high pressure, high temperature conditions required for eclogite and granulite to develop at the base of the lithosphere.

Backarc basins and pull-apart basins

In some convergent margin settings, backarc basins (see Figure 18.23) and pull-apart basins develop in response to tensional stresses within the arc complex (Chapter 10). Backarc extension in an ocean–ocean convergent margin, such as in the western Pacific Ocean, results in the generation of actively spreading marine backarc basins that can eventually evolve into marginal seas. Backarc basins can also form in ocean–continent convergent margins, as for example in the Tyrrhenian Sea. Metamorphism in backarc basins is similar to the processes that occur in ocean ridges because both systems involve rifting and metasomatism involving cold seawater and hot magma reacting in fractured basic rock. Zeolite and prehnite-pumpellyite facies as well as hornfels facies metamorphism occur at relatively shallow depths. Greenschist facies temperatures can be attained at greater depths.

Pull-apart basins form as a result of transtensional stress in oblique-slip environments or along bends in faults (Figure 18.29). Local extension produces a down-dropped block which infills with sediments with thicknesses capable of producing burial metamorphism. Pull-apart basins can occur in transform, divergent or convergent plate boundaries as well as within plate settings. As with backarc basins and forearc basins, pull-apart basins are generally subjected to sub-greenschist conditions in the zeolite and prehnite-pumpellyite facies. Pull-apart basins occur notably along California's San Andreas Fault.

Ocean–continent convergence: fold and thrust belts and foreland basins

Fold and thrust belts and foreland basins are among the most distal tectonic features created

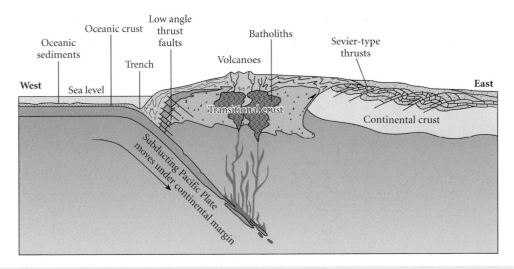

Figure 18.30   Mesozoic active convergent margin setting in the western United States ~100 Ma. (From Levin, 2006; with permission of John Wiley & Sons.)

in the overlying plates of many convergent margins. Formerly gently dipping to horizontal continental slope and continental shelf sedimentary rocks – such as shales, limestones, dolostones and sandstones deposited in passive marine settings – are subjected to intense subhorizontal compressive stress. The horizontal shortening is accommodated by folds and thrust faults that result in the telescoping or "piggybacking" of thrust slices. The net effect is that the lithosphere is thickened by vertically stacked thrust sheets which produces an isostatically depressed lithosphere in front of the fold and thrust belt. The heat for metamorphism comes largely from the higher temperatures associated with rocks being buried progressively deeper as the orogenic belt is thickened by folding, thrusting and telescoping. For example, the Sevier fold and thrust belt and Rocky Mountain foreland basin (Figure 18.30) developed in the Mesozoic and Early Cenozoic in response to subduction activity. In the fold and thrust belt, Precambrian, Paleozoic and Mesozoic sedimentary rocks were folded and displaced eastward by thrust faults. Syn- to post-orogenic magmatism intruded the fold and thrust belt. Lithospheric loading associated with the thickened pile of folded and faulted terranes produced a subsiding foreland basin eastward of the thrust sheets into which sediments were deposited. The combination of intense com-

pression, thickened lithosphere and late silicic intrusions produced regional Buchan to Barrovian facies series assemblages in fold and thrust belts.

Orogenic thickening of the lithosphere causes an isostatically depressed lithosphere in front of the fold and thrust belt. In the adjoining foreland basin, an initially deep basin fills with marine deposits producing alternating shale, sandstone, chert and carbonate layers producing what is referred to as **flysch** deposits. With continued thrusting and infilling, fine-grained marine rocks are succeeded by sandstones and conglomerates in what are referred to as **molasse** deposits. Total layer thicknesses in foreland basins can range from a few to tens of kilometers, resulting in burial metamorphism in the zeolite or prehnite-pumpellyite facies. Fold and thrust belts and foreland basins are exceedingly important for the generation of oil and gas deposits as well as extensive lignite, bituminous and anthracite coal deposits. For these reasons, fold and thrust belt and foreland basin environments are of great interest with respect to fossil fuel energy production.

Continent–continent collision

The end result of continued subduction is the elimination of ocean lithosphere, the accretion of far-traveled arc complexes and microcontinents, the closing of an ocean basin, and the cessation of subduction as two continental lithospheric plates collide. A suture zone develops – marking the former site of the subduction trench – within an accretionary mélange containing intensely deformed rocks which can include ophiolite, forearc fragments, blueschist, ocean plateaus and ultra-high pressure (UHP) assemblages discussed earlier in this chapter. Continued horizontal shortening produces vertical uplift and intense lithospheric thickening accommodated by folding and thrust faulting, producing widespread classic Barrovian facies series assemblages. The elevated temperatures and pressures produce greenschist, amphibolite and granulite facies reflecting geothermal gradients ranging from ~20 to 40°C/km in the Barrovian facies series. This situation is well displayed in the Himalayan belt, which contains the highest mountain peaks on Earth. The regionally elevated terrain is due to a double lithosphere thickness because of the partial subduction and thrust telescoping of the Indian lithosphere and significant thickening of the Eurasian Plate. The Himalayan Mountains contain an uplifted Tibetan Plateau, suture zone and fold and thrust belt that formed as a result of continent–continent collision. The Himalayan Mountains are an important UHP study area.

Metamorphism, igneous activity and sedimentation are intricately related so that numerous themes recur when discussing Earth materials. Hopefully, this chapter has elucidated key connections from the early work of Barrow to the modern study of UHP assemblages. Many questions remain unresolved such as the origin of greenstone belts and the processes by which blueschists and UHP rocks return to Earth's surface. Geology is a journey without end. We hope you have enjoyed the ride thus far.

Chapter 19

Mineral resources and hazards

Minerals have many applications that can be broadly grouped into metallic ores, industrial minerals and gemstones. An **ore** is a naturally occurring mineral or rock deposit sufficiently enriched in metal that it may be economically mined, processed and utilized for a profit. Metallic ores have been so important in human history that civilizations are largely classified based upon the metals used during that time such as the Bronze Age (~3000 to 2000 BC) and the Iron Age (~2000 BC to 0 AD). In what may now be termed the plastic or silicon age, metals continue to play critical roles in manufacturing, agriculture and technology largely due to the special properties associated with metallic bonds.

The use of industrial minerals has increased greatly in the 20th and 21st centuries as Earth's human population increases exponentially, urban areas expand and farmland is required to provide greater yields per acre. Based on physical and/or chemical properties, **industrial minerals** serve as fertilizers, construction material, abrasives, fillers, refrac-

Earth Materials, 1st edition. By K. Hefferan and J. O'Brien. Published 2010 by Blackwell Publishing Ltd.

tory materials and various other uses in manufacturing.

Throughout the course of human history, **gem minerals** have been valued for their physical appearance and used for ornamentation and as jewelry. While gem minerals are the least useful from an application viewpoint, they are among the most highly prized in human societies. In the following discussion we will provide an overview of how these three areas impact every person on Earth.

19.1   ORE MINERALS

Over 3500 minerals are known to exist. Of these, less than 100 minerals are considered to be important metallic ore minerals. In the discussion that follows we will address some of the environments in which ore minerals form as well as the uses of these ore minerals in our global economy. Metallic minerals are widely distributed in Earth in low concentrations. However, enriched mineral deposits occur that constitute ore bodies. Ores are metallic deposits that can be mined, processed and sold for a profit. Profitability is dependent upon ore price, refining cost, transportation, demand and competition. Enriched ore bodies commonly occur with sub-ore concentrations of **gangue minerals**, which are not economically valuable. Common gangue minerals include quartz, calcite, dolomite, barite, gypsum, feldspar, garnet, chlorite, clay, fluorite, apatite, pyrite, marcasite, pyrrhotite and arsenopyrite.

An ore is deemed valuable due to a combination of availability, physical or chemical properties and demand. When metal prices are low, rock that is less enriched in ore, called **protore**, would be discarded in waste piles because the cost associated with processing exceeds the value of the metal contained in the rock. With an increase in price and demand, waste rock can be reclassified as ore grade material. Ore minerals contain metallic elements such as platinum, gold, silver, nickel, cobalt, iron, lead, palladium and copper. As cations, metals commonly bond with anions such as oxygen and sulfur. As a result most metallic ore minerals are oxides, sulfides or native elements from which the metallic components are easily separated during smelting and refining processes releasing oxygen or sulfur-bearing waste material. Sulfur-bearing waste material is a particular hazard because, in reacting with water, sulfuric acid is produced which contaminates groundwater and surface water.

Relatively few ores are silicate minerals, due in part to the difficulty in separating metallic elements from the tightly bonded silica compound and the large amounts of waste material (referred to as slag) generated. For example, aluminum is one of the most common elements in Earth's crust and is an essential element in many feldspar minerals – the most common minerals in Earth's crust. However, it is not economically feasible to extract aluminum ore from feldspars at present – or foreseeable – market prices.

Ores concentrate in igneous, metamorphic or sedimentary rocks either during the time the rock forms or due to subsequent enrichment. **Syngenetic** ore deposits represent primary mineralization wherein ores form at the same time – synchronous – as the rock. **Epigenetic** ore deposits form by secondary mineralization wherein ore concentrates some time after the rock has formed. Epigenetic mineralization can postdate the host rock by hundreds of millions of years. Most ore deposits originate as dissolved metallic ions that precipitate from magma. These primary ore deposits are commonly later altered within Earth to create metamorphic ore deposits or reworked by near surface processes to produce sedimentary ore deposits. In all of these ore-enhancing processes, hot silicate fluids, water and other volatile fluids play critical roles in dissolving, transporting and precipitating metals in concentrated deposits. Water tends to leach and concentrate metallic ores in two fundamental ways. **Supergene enrichment** occurs as surface waters percolate downward, in so doing leaching near surface metals and concentrating them at deeper levels within Earth's crust. Supergene enrichment is an epigenetic process that commonly develops in oxidizing zones above the water table. **Hypogene enrichment** is a primary or syngenetic process that occurs as deep, upwelling magmatic fluids concentrate ore synchronous with rock development.

Depending upon the rock characteristics and the mode of emplacement, ores may assume tabular, cylindrical or irregular forms. Ores can accumulate in narrow, concentrated zones or be broadly disseminated in low concentrations over large areas. **Tabular** ore bodies are considered to be veins or layered, stratiform bodies. Tabular ore bodies commonly form along fracture systems, igneous layers, metamorphic foliations or sedimentary beds. **Cylindrical** ore bodies, appropriately referred to as pipes or chimneys, commonly form in response to ore-enriched magma or hydrothermal solutions that rise buoyantly towards Earth's surface. Lensoid ore deposits are referred to as **podiform**, meaning that they have a foot-like shape. **Irregular** or broadly disseminated ore deposits typically develop in close proximity to large igneous intrusions in which ore-bearing fluids infiltrate – and in many cases locally metamorphose – surrounding rock. Intense forces associated with igneous intrusions result in elevated fluid pressures. These fluid pressures are directed outward into the rock body resulting in extensive hydrofracturing of the surrounding country or host rock.

In the following discussion we will briefly address some of the major ore-forming environments, many of which are associated with tectonic plate boundaries.

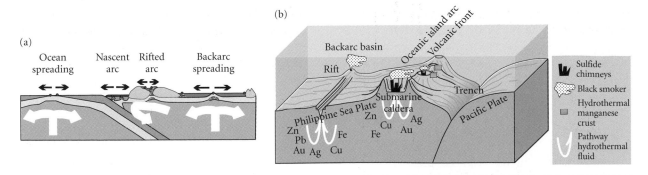

Figure 19.1 (a) Tectonic environments in which volcanogenic massive sulfide deposits form. (From Galley et al., 2007; with permission of Natural Resources of Canada, 2009, courtesy of the Geological Survey of Canada.) (b) Metal enrichment due to hydrothermal interaction between upwelling hypogene magmatic fluids and descending supergene ocean fluids. (Courtesy of the National Institute of Advanced Industrial Science and Technology of Japan.)

19.1.1 Igneous ore-forming environments

Igneous processes create large-scale ore deposits at convergent and divergent plate boundaries as well as within intraplate settings. Magmatic ore-forming processes develop from eruptions of lava on Earth's surface – as in volcanogenic massive sulfides – as well as the intrusion of magma within Earth.

Volcanogenic massive sulfide deposits

Volcanogenic massive sulfide (VMS) deposits are copper–zinc–lead sulfide deposits concentrated on the ocean floor at divergent and convergent plate boundaries. Other VMS ores include silver, gold, cobalt, nickel, iron, tin, selenium, manganese, cadmium, bismuth, germanium, gallium, indium and tellurium. While plate margin magmatism at ocean spreading ridges or volcanic arcs is the driving force behind the development of VMS deposits, extensive hydrothermal alteration and metamorphism play critical roles in altering the igneous rocks and concentrating ores through both hypogene and supergene enrichment. How does this occur? Tensional forces at ocean ridges, backarc basins and intra-arc basins produce extensional fractures through which hot magmatic fluids rise upward and interact with descending cold seawater (Figure 19.1). On the basis of their genetic origin, VMS deposits have been classified as indicated in Table 19.1.

As first documented by Francheteau et al. (1979), upwelling plumes of black, metal-laden "smoke" from ocean ridge vents release hydrothermal fluids. The chimney-like structures produced by >360°C black smokers are enriched in chalcopyrite, sphalerite, pyrite and anhydrite. Metasomatic mixing of seawater and black smoker hydrothermal fluids also produce "white smokers", erupting <300°C plumes that precipitate quartz, calcite, anhydrite, pyrite and barite on the sea floor. Other common sulfide and oxide ore minerals at VMS include galena, chalcocite, bornite, cobaltite, magnetite, hematite, pyrolusite and enargite (Herzig and Hannington, 1995; Humphris et al., 1995; Galley et al., 2007).

VMS deposits accumulate by the growth and subsequent collapse of black smoker chimneys resulting in layered (stratiform) or lens-shaped (podiform) metal deposits and collapse breccias overlying stockwork sulphide–silicate dike structures (Figure 19.2). The black smoker environments produce mounds and nodules enriched in manganese, zinc, iron, cobalt, copper and nickel. Beneath the overlying mound, hydrothermal fluid flow and steeply inclined chemical and thermal gradients commonly produce cylindrical pipe zones with a higher temperature, chalcopyrite-rich inner core and lower temperature, sphalerite–galena-rich outer zones (Galley et al., 2007). While ancient VMS deposits on land have been mined for thousands of years,

Table 19.1 Three different varieties of volcanogenic massive sulfide (VMS) deposits (Cox and Singer, 1986).

VMS name	Origin	Description
Cyprus type	Form in ocean ridge (East Pacific Rise) or backarc basin	Named after the Troodos ophiolite (Cyprus) in the Mediterranean Sea. Cyprus-type VMS are basalt-dominated deposits associated with ophiolites and enriched in copper, zinc, nickel, chromium and manganese and with minor amounts of silver and gold
Besshi type	Juvenile, nascent volcanic arc	Named after the Besshi copper mine in Japan. Besshi-type VMS are early formed convergent margin deposits containing basalt, rhyolite and greywacke rocks. Besshi deposits are notable for their ore concentrations of copper and cobalt and only minor concentrations of zinc
Kuroko type	Mature volcanic arc or backarc basin (Okinawa Trough)	Named after mature convergent margin deposits in the Japanese convergent arc system. Kuroko-type VMS are dominated by silicic rocks such as rhyolite. Kuroko deposits are enriched in copper, zinc and lead and may also contain substantial gold and silver

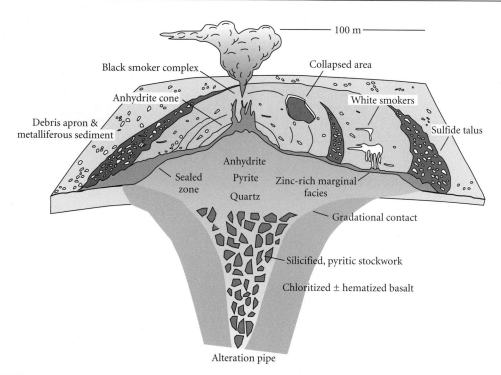

Figure 19.2   Schematic cross-section of volcanogenic massive sulfide deposits, which commonly display a lower temperature stratiform cap rock enriched in anhydrite, pyrite and quartz. Below the overlying rock cap, higher temperature sulfide and oxide ore assemblages occur within a highly altered pipe vent system (Hannington et al., 1991; Franklin et al., 2005; Galley et al., 2007). (From Galley et al., 2007; with permission of Natural Resources of Canada, 2009, courtesy of the Geological Survey of Canada.)

it was not until the discovery of actively forming VMS deposits at the 21°N latitude of the East Pacific Rise that the genesis of these metal deposits was well understood (Herzig and Hannington, 1995). Figure 19.2 illus-trates an idealized cross-section of a VMS deposit. Ancient VMS deposits occur globally and are recognized at ancient and modern divergent and convergent boundaries (Figure 19.3).

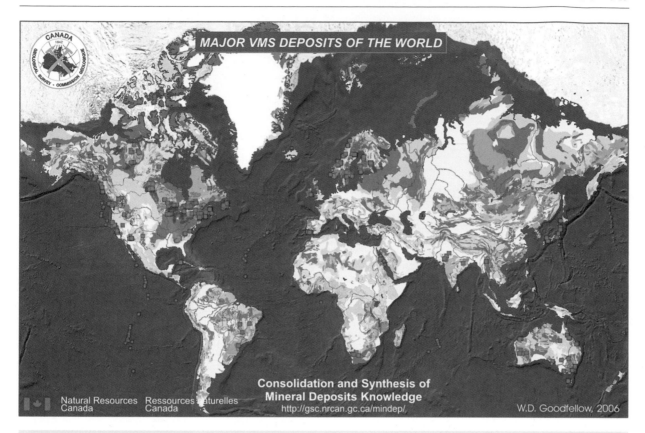

Figure 19.3   Global map of the major volcanogenic massive sulfide (VMS) deposits. Note that these deposits occur along ancient and modern divergent and convergent plate boundaries (From Galley et al., 2007; with permission of Natural Resources of Canada, 2009, courtesy of W. D. Goodfellow and the Geological Survey of Canada.) (For color version, see Plate 19.3, between (pp. 408 and 409.)

Rift deposits

Within intracontinental rift basins, ore deposits form by the initial precipitation of metals from magma. Basalt and rhyolite lavas erupt in continental rift basins producing vesicular textures associated with high gas content. In response to hydrothermal fluids and oxidation processes, metals such as iron, copper, zinc, nickel and platinum group elements precipitate in gas vesicles forming ore deposits. In some locations, rift basin ores are leached from their igneous host rocks and concentrated in overlying sedimentary rocks. The Bethlehem, Pennsylvania iron deposits and the Keweenaw, Michigan (USA) iron–copper–nickel deposits represent two important **rift deposits** involving sedimentary deposits bearing ore minerals derived from underlying rift basalts.

The Keweenaw Basin in the Lake Superior region of Michigan is the most famous example of a rift basin deposit (Figure 19.4). Keweenaw ore, principally copper, was mined from the 1840s to the 1950s. The Keweenaw belt represents a 1.1 Ga continental rift that formed synchronous with the Grenville Orogeny in eastern North America. Continental rifting produced a down-dropped basin which infilled with 2.5–5 km thick basaltic lava flows and underlying gabbro, troctolite and granite intrusions. Copper, nickel, iron, titanium, platinum group elements (platinum and palladium) and silver precipitated in the igneous rocks. Weathering resulted in dissolution of metals and subsequent secondary enrichment in overlying conglomerate, sandstone and shale layers (Figure 19.4).

While volcanic activity is important in producing surficial ore deposits, deeper levels

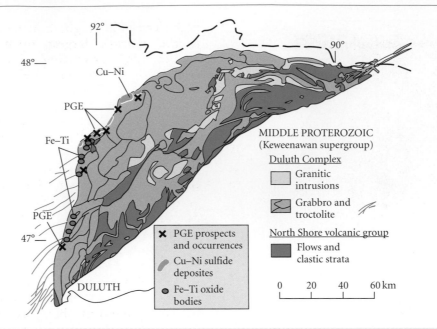

Figure 19.4 Generalized geological map of the Duluth Complex and Keweenaw rift illustrating platinum group elements (PGE), copper, nickel, titanium and iron deposits associated with rifting and magmatism. (From Jirsa and Southwick, 2003; with permission of the Minnesota Geological Survey.)

of igneous activity are also significant. For example, the Keweenaw rift is underlain by the Duluth gabbro complex, which formed from basaltic magma that crystallized at deeper levels within Earth's interior. As magma fractures and intrudes the pre-existing host rock, dissolved metal ions infiltrate, cool and precipitate via magmatic crystallization processes or magma segregation processes. The Duluth gabbro complex is recognized as an important platinum group element ore deposit yet to be mined.

Immiscible sulfide deposits and layered igneous intrusions

At depth within Earth's interior, **immiscible metal sulfide** liquids separate from metal-rich, silicate magma. Evidence for immiscible liquid separation includes immiscible fluid inclusions in a host material (Figure 19.5) such as occurs with oil and water in salad dressing. Immiscible liquid separation in silicate magma results in the concentration of metallic sulfide deposits containing copper, iron, nickel, chromium, vanadium, palladium and platinum. Common ore minerals include chalcocite, bornite, chalcopyrite, chromite, pentlandite,

nickelline, magnetite and hematite. These valuable ores occur in layered igneous intrusions (Chapter 10), which consist of stratiform gabbroic and pyroxenite rocks that crystallize in tabular plutons. Exceptional examples of ore-bearing layered igneous intrusions include the Sudbury Mine in Ontario (Canada), the Duluth Complex and Stillwater Mine in Montana (USA), the Skaergaard Intrusion of Greenland and the Bushveld Complex of South Africa. Layered igneous intrusions occur in all of the continents, as illustrated in Figure 19.6.

Porphyry deposits

Porphyry deposits – the principal source of copper – form as silica-rich magma intrudes and fractures the host rock and subsequently crystallizes. Forces associated with magma injection, coupled with hydrothermal fluid pressures, result in the diffuse infiltration of ore-bearing fluids into complex network of fractures and pore spaces of the surrounding rock at temperatures >500°C. Cooling and crystallization results in massive, low concentration (<2%) deposits of copper, molybdenum, gold, zinc, mercury, silver, lead, lithium

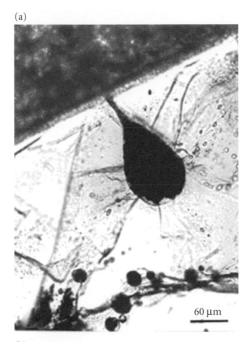

Figure 19.5 (a) Scanning electron microscope image of an hourglass inclusion from the Pine Grove porphyry, Utah. A small tadpole-shaped capillary connects the bulk of the inclusion with the outside of the host crystal. In a similar way, immiscible metals can separate from silicate magmas to produce ore deposits in plutonic igneous rocks. (Photo courtesy of Jake Lowenstern, with permission of the US Geological Survey.) (b) Platinum group elements enriched through a combination of immiscibility and crystallization processes at the Stillwater Mine, Montana. (Photo by Kevin Hefferan.)

and tin disseminated throughout a zone of alteration within the host rock and in more concentrated veins surrounding the intrusion (Figure 19.7a). For example, the 0.8 km deep

and 4 km wide Bingham Mine in Utah (USA) is the world's largest open-pit porphyry copper mine, producing 12 million tons of copper since open-pit operations began in 1906 (Figure 19.7b).

Common ore minerals in porphyry deposits include chalcocite, chalcopyrite, bornite, molybdenite, sphalerite, cinnabar, enargite, spodumene, cassiterite and galena. Porphyry deposits commonly occur in association with silicic to intermediate intrusions, such as granite or granodiorite, at convergent plate boundaries. Porphyry deposits of copper, molybdenum, mercury, silver and gold occur throughout the Cordilleran fold belt extending from Alaska southwards through the Andes Mountains. Similar deposits also occur in Indonesia and throughout the Alpine–Himalayan belt as illustrated in Figure 19.8 (Matthes, 1987; Kesler, 1994; Wenk and Bulakh, 2004).

Vein deposits

Vein deposits are commonly associated with magma intrusions due to extensional hydro-fractures that develop as hydrothermal fluids escape upward from the magma, and later cool and precipitate as vein deposits. Vein deposits occur at plate boundaries as well as intraplate settings. Because of the high concentration of metals, vein deposits are referred to as **lode deposits** and can contain gold, silver, copper or metal sulfides that occur in association with gangue minerals such as quartz or calcite (Figure 19.9a). Metals such as gold are dissolved in volatilized gases that subsequently cool and crystallize producing highly concentrated ore vein deposits (Figure 19.9b). Familiar examples are the lode veins that precipitated the 1849 California gold rush and the Klondike gold rush of Canada. Both of these famous gold vein deposits originated as vein-filling fractures produced by granitic intrusions at convergent margin systems. Figure 19.9a illustrates copper vein deposits associated with the gabbro and basalt of the Keweenaw Duluth Complex.

Pegmatite deposits

Pegmatites are coarse-grained igneous textures that develop in plutons of granitic composition (Chapter 7). Coarse-grained crystals

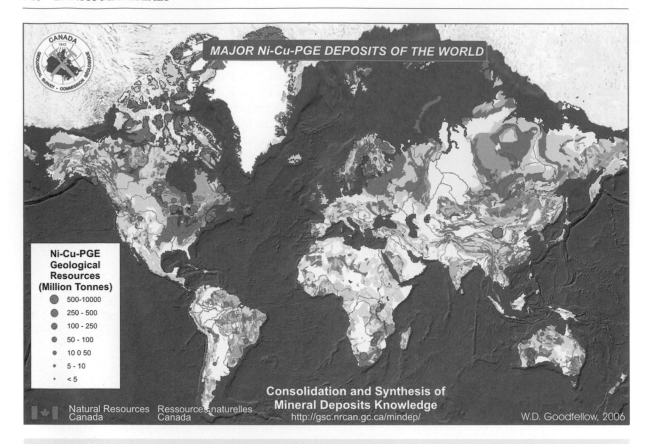

Figure 19.6   Global map of the major layered gabbroic intrusions containing platinum group elements, nickel and copper. Note that these deposits are concentrated in continental crust and are generally of Archean (>2.5 Ga) age. (From Eckstrand and Hulbert, 2007; with permission of Natural Resources of Canada, 2009, courtesy of W. D. Goodfellow and the Geological Survey of Canada.) (For color version, see Plate 19.6, between pp. 408 and 409.)

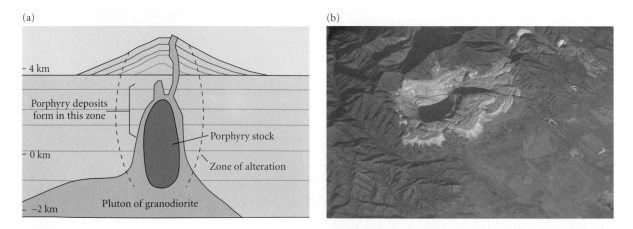

Figure 19.7   (a) Cross-section of porphyry ore deposits created by igneous intrusion into a host rock. (After Sillitoe, 1973.) (b) Bingham Canyon Mine, Utah, is featured in this NASA image photographed by an Expedition 15 crew member on the International Space Station. The Bingham Canyon Mine (center) is one of the largest open pit porphyry copper mines in the world, measuring over 4 km wide and 1200 m deep. (Reproduced with permission of NASA.)

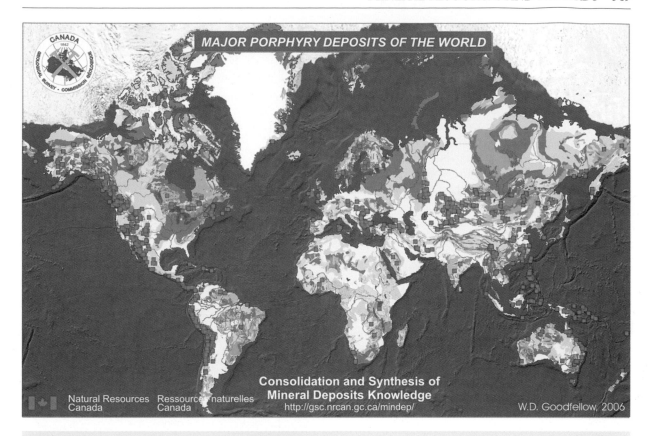

Figure 19.8   Global map of the major porphyry deposits. Note that these deposits are concentrated along modern and ancient convergent plate boundaries (From Sinclair, 2007; with permission of Natural Resources of Canada, 2009, courtesy of W. D. Goodfellow and the Geological Survey of Canada.) (For color version, see Plate 19.8, between pp. 408 and 409.)

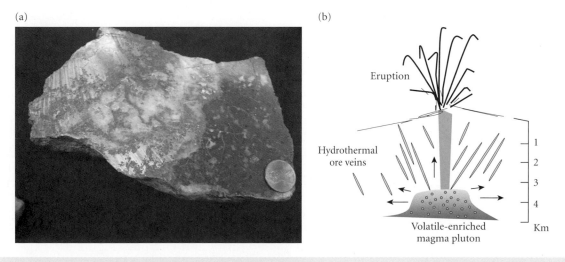

Figure 19.9   (a) A copper and quartz vein within metamorphosed basalt. (Photo by Kevin Hefferan.) (b) Gold vein deposits forming from volatile, gas-rich magma.

develop due to slow cooling of a low viscosity magma characterized by high volatile content. Volatiles such as OH, fluorine, boron and H_2O promote ion diffusion and the development of large crystals (Chapter 8). **Pegmatite deposits** are closely associated with the vein deposits described above. Granitic pegmatites – containing quartz, feldspars, amphiboles and micas as major minerals – occur within continental plates and at convergent plate boundaries. Minor and accessory minerals in pegmatites include beryl, apatite, lepidolite, spodumene, cassiterite, wulfenite, molybdenite, scheelite, tourmaline, topaz, uraninite, lithiophillite, columbite, tantalite, gold and silver. Pegmatites are important sources of tin, molybdenum, gold and silver and the primary source for beryllium, lithium, tantalum, niobium and rare Earth element ores. Figure 19.10 illustrates a beryl-bearing granite pegmatite from the Black Hills of South Dakota (USA). Let us now briefly consider metamorphic processes that occur in regions surrounding igneous intrusions.

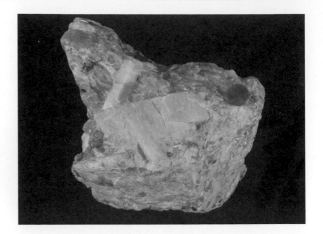

Figure 19.10   Granite pegmatite from the Black Hills of South Dakota containing large crystals of beryl, quartz and feldspar. Gold deposits are associated with the Black Hills pegmatite. (Photo by Kevin Hefferan.) (For color version, see Plate 19.10, between pp. 408 and 409.)

19.1.2   Metamorphic ore-forming environments

Metamorphic processes concentrate metallic elements into ore deposits through solid state changes as well as hydrothermal fluid reactions. Hydrothermal fluids are critically important in driving metamorphic reactions and in concentrating ore metals.

When we think of hydrothermal processes, Yellowstone or Iceland may come to mind where hot magma reacts with subsurface fluids to create geysers or hot springs. In fact, hydrothermal deposits are formed by a number of different means that may originate as meteoric (surface) waters, seawater, groundwater, formation pore fluids or deep magmatic fluids.

Greenstone belts

Greenstone belts (Chapter 18) are hydrothermally altered assemblages that contain thick sequences of volcanic suites and interbedded sedimentary layers. They are called greenstones because of green-colored metamorphic minerals such as chlorite, epidote and serpentine. Greenstone belts are particularly abundant in Precambrian cratonic belts where they

Figure 19.11   Cobalt mine in the greenstone belt in Bou Azzer, Morocco. (Photo by Kevin Hefferan.)

contain among the world's greatest concentrations of copper, chromium, nickel, gold, cobalt and silver deposits. Common ore minerals include chalcocite, chalcopyrite, chromite, nickelline, pentlandite, cobaltite, erythrite and enargite (Figure 19.11). Greenstone belts consist of down-warped basinal deposits in which peridotite rocks are overlain successively by basaltic layers, silicic igneous rocks and marine sediments. Many of these greenstone ore deposits are derived from altered komatiite, basalt and peridotite rocks (Chapter 10) that form at very high temperatures (>1400°C) near the base of the greenstone assemblage. Greenstone belts are particularly extensive in the Precambrian cratons of Africa, Canada and Australia.

Skarns

Skarns are contact metamorphosed rocks enriched in calc-silicate minerals. Skarns form through the high temperature alteration of country rocks, usually carbonate rocks, by the intrusion of silicate magmas. Hot magma intruding carbonate rock produces ion exchange via hydrothermal solutions. Minerals such as calcite and dolomite release CO_2 and obtain SiO_2 from the magma; as a result, a distinctive suite of calc-silicate minerals form that include calcium pyroxenoid (e.g., wollastonite), calcium amphibole (e.g., tremolite), calcium pyroxene (e.g., diopside) and calcium garnet (grossular, andradite or uvarovite). Other associated minerals include quartz, calcite, phlogopite, brucite, talc, serpentine and periclase. In addition, dissolved volatiles such as H_2O, CO_2, SO_2, H_2S and HCl serve as catalysts in promoting the dissolution of metallic ions and transporting them in solution (Chapter 15). Ion exchange commonly results in the concentration of copper, lead, zinc, iron, molybdenum, tin, tungsten, cobalt, manganese, silver and gold ore deposits. Common ore minerals include chalcopyrite, chalcocite, sphalerite, galena, hematite, magnetite, siderite, molybdenite, cassiterite, wulfenite, wolframite, cobaltite, rhodochrosite, rhodonite and enargite. Not all skarns form in carbonate rock. **Exoskarns** are skarns that develop in any sedimentary country rock whereas **endoskarns** occur in igneous country rock. Modern skarns form in geothermal systems, hot springs, hydrothermal vents on the sea floor and at convergent and divergent plate boundaries. In these environments, skarns are commonly associated with porphyry, pegmatite, vein and VMS deposits (Figure 19.12).

19.1.3 Sedimentary ore-forming environments

Sedimentary ore-forming environments develop as a result of hydrothermal, depositional and weathering processes. In all cases, water plays a role in leaching and/or concentrating ores in sedimentary deposits. Let us begin by considering the role of hot fluids in leaching, transporting and concentrating ore minerals in sedimentary deposits.

Banded iron deposits

The largest and most important iron deposits (Chapter 14) formed as a result of chemical precipitation in shallow marine environments 1.8–2.5 billion years ago. These deposits are well developed in the Lake Superior region of North America, where they have represented over 80% of US production since 1900, and are referred to as Superior-type deposits. Similar massive deposits also occur in the Hamersley Basin of Australia, the Minas Gerais deposits of Brazil and the Kursk region of Russia. **Superior-type banded iron formations** (BIF) consist of alternating iron-rich and silica-rich layers. The iron-rich layers contain both ferrous and ferric iron. Ferrous iron minerals include magnetite and siderite, whereas ferric minerals include hematite and goethite. These shallow marine deposits formed in the Early Proterozoic when iron-rich, deep seawater mixed with shallow oxygenated shelves (Figure 19.13). As a result, BIF layers several hundred meters thick and encompassing >100 km² in area formed the greatest iron deposits on Earth.

Algoma-type deposits

Algoma-type deposits contain iron ore concentrations that occur in metasedimentary deposits, most of which are Archean (>2.5 Ga) in age. The Algoma ironstones contain hematite and magnetite interbedded with volcanic rocks, graywackes, turbidites and pelagic sedimentary rocks. Algoma ironstones form in deep abyssal basins heated by submarine volcanic activity (Figure 19.14). Volcanic hot springs release heated waters containing dissolved iron which precipitate in down-warped sedimentary basins. Algoma iron deposits form concentrated iron-rich layers ~30–100 m thick and extending a few square kilometers in area. Algoma-type deposits are associated with VMS deposits and likely form at ocean ridges or volcanic arc environments. These deposits have been extensively mined in South Africa and in the Algoma region of Canada,

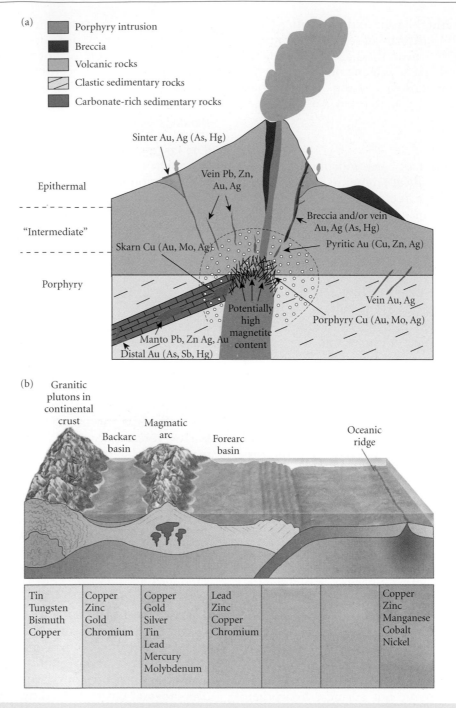

(a)

- Porphyry intrusion
- Breccia
- Volcanic rocks
- Clastic sedimentary rocks
- Carbonate-rich sedimentary rocks

Sinter Au, Ag (As, Hg)

Vein Pb, Zn, Au, Ag

Epithermal

"Intermediate"

Breccia and/or vein Au, Ag (As, Hg)

Skarn Cu (Au, Mo, Ag)

Pyritic Au (Cu, Zn, Ag)

Porphyry

Vein Au, Ag

Potentially high magnetite content

Porphyry Cu (Au, Mo, Ag)

Manto Pb, Zn Ag, Au

Distal Au (As, Sb, Hg)

(b) Granitic plutons in continental crust

Backarc basin

Magmatic arc

Forearc basin

Oceanic ridge

Tin Tungsten Bismuth Copper	Copper Zinc Gold Chromium	Copper Gold Silver Tin Lead Mercury Molybdenum	Lead Zinc Copper Chromium			Copper Zinc Manganese Cobalt Nickel

Figure 19.12 (a) Cross-section illustrating skarns developing in association with vein deposits and porphyry deposits. These enrichment processes are not isolated but commonly occur together, particularly at convergent and divergent plate boundaries (From Sinclair, 2007; with permission of Natural Resources of Canada, 2009, courtesy of the Geological Survey of Canada.) (b) Cross-section illustrating common ores at convergent and divergent margins. (After Murck et al., 2010; with permission of John Wiley & Sons.)

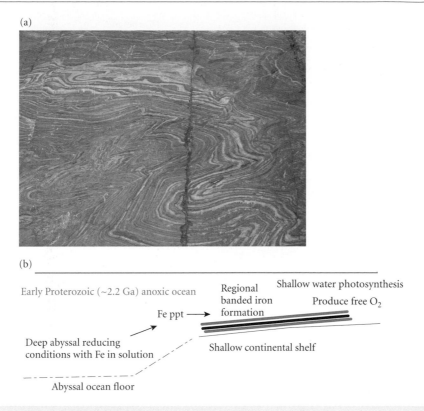

Figure 19.13 (a) Banded iron formation rock. (Photo by Kevin Hefferan.) (b) Anoxic ocean model for banded iron formation development.

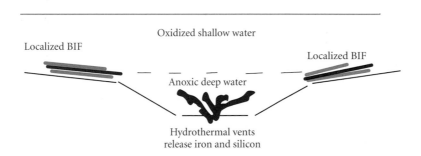

Figure 19.14 Algoma-type deposits that form due to hydrothermal fluids interacting in deep sedimentary basins near the oxic–anoxic boundary. BIF, banded iron formation.

for which these deposits are named (Kesler, 1994).

Sedimentary exhalative deposits

Sedimentary exhalative (sedex) deposits are similar to Algoma-type deposits in that hydrothermal fluids leach and concentrate metallic ore. Whereas Algoma deposits are enriched in iron, sedex deposits contain lead–zinc–iron

sulfides precipitated by submarine hot springs. Hydrothermal fluids containing dissolved metals rise upward and are "exhaled" into clastic sedimentary basins releasing metal-rich brine solutions into the surrounding country rock (Figure 19.15). The brines precipitate as hot spring sedimentary deposits creating massive lead, zinc and iron sulfides in sedimentary rocks such as shale, chert and carbonate rocks. Common minerals include

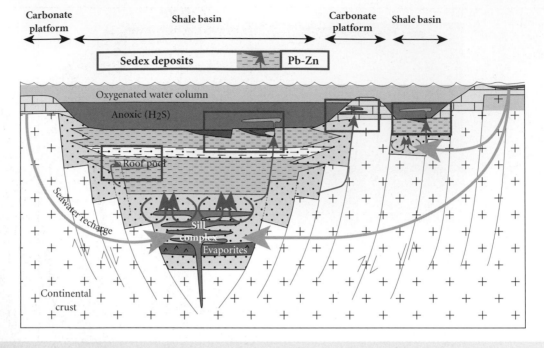

Figure 19.15 Cross-section of sedex deposits formed by an external, magmatic heat source "exhaling" into a sedimentary basin (From Goodfellow and Lydon, 2007; with permission of Natural Resources of Canada, 2009, courtesy of the Geological Survey of Canada.)

galena, sphalerite, pyrite and pyrrhotite. Sedex deposits are mined for valuable lead, zinc and silver in regions such as Australia, British Columbia, northern Europe and the Yukon (Kesler, 1994). The global distribution of sedex deposits is illustrated in Figure 19.16.

Mississippi Valley-type deposits

Unlike the Algoma and sedex deposits that have an external heat source, **Mississippi Valley-type (MVT) deposits** form from warm (<300°C) saline solutions that flow in the pore spaces within permeable carbonate or sandstone rocks in deep sedimentary basins. MVT deposits precipitate lead and zinc in thick limestone, dolostone or sandstone deposits (Figure 19.17). These deposits most commonly occur in distal foreland basins where brine flow produces secondary, epigenetic deposits. In addition to lead and zinc minerals such as galena and sphalerite, the brine formation waters precipitate halite, sylvite, gypsum, calcium chloride, barite and minor amounts of gold, silver, copper, mercury and molybdenum. MVT deposits are named for

deposits occurring from Missouri to Wisconsin in the central USA. Notable MVT deposits also occur in Poland, Spain, Ireland, the Alps and the Northwest Territories of Canada (Kesler, 1994). The global distribution of MVT deposits is illustrated in Figure 19.18.

Placer deposits

As most metallic elements have a high specific gravity, they tend to be concentrated as **placer deposits** (Figure 19.19) in high energy stream or beach environments. Placer deposits contain heavy metals that may have originated in igneous intrusions or metamorphic belts. Upon uplift and weathering, the metals are eroded and transported by streams, waves or gravity to a high energy depositional site. Strong water action transports lower density minerals down-current and tends to deposit moderate to high specific gravity minerals that are somewhat stable at Earth's surface.

Placer mining usually is conducted by dredging or panning methods by which sediments are disturbed and sorted by density using water. Common high specific gravity

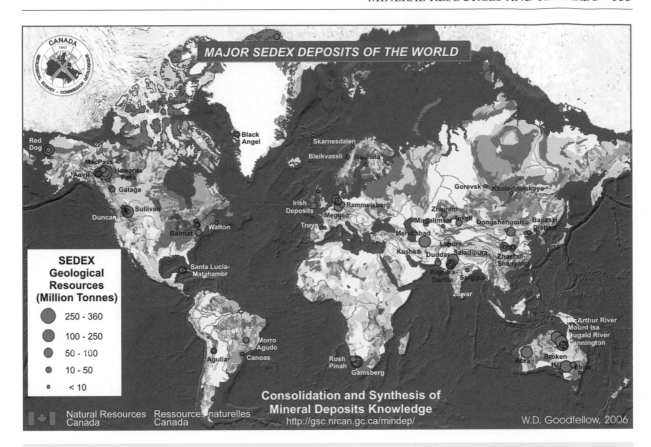

Figure 19.16 Global map of the major sedex deposits. Note that these deposits only occur in continental settings. Data are from the synthesis of sedex deposits by Goodfellow and Lydon (2007). (From Goodfellow and Lydon, 2007; with permission of Natural Resources of Canada, 2009, courtesy of W. D. Goodfellow and the Geological Survey of Canada.) (For color version, see Plate 19.16, between pp. 408 and 409.)

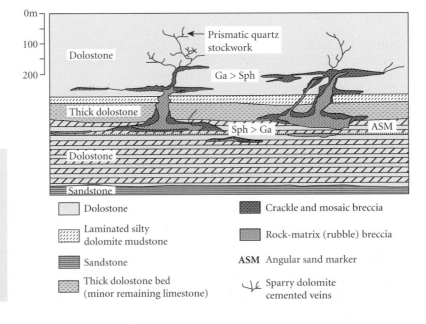

Figure 19.17 Schematic cross-section of the Robb Lake (Canada) Zn–Pb sulphide bodies in brecciated MVT deposits Ga, galena; Sph, sphalerite. (From Paradis et al., 2007; with permission of Natural Resources of Canada, courtesy of the Geological Survey of Canada.)

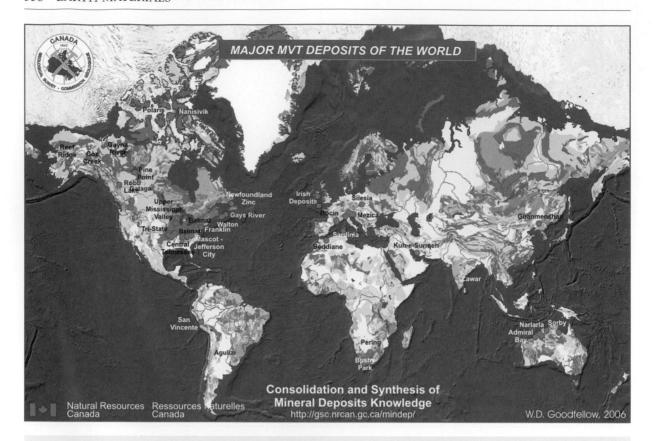

Figure 19.18 Global map of the major Mississippi Valley-type (MVT) deposits. Note that these deposits only occur in continental settings. (From Paradis et al., 2007; with permission of Natural Resources of Canada, 2009, courtesy of W. D. Goodfellow and the Geological Survey of Canada.) (For color version, see Plate 19.18, between pp. 408 and 409.)

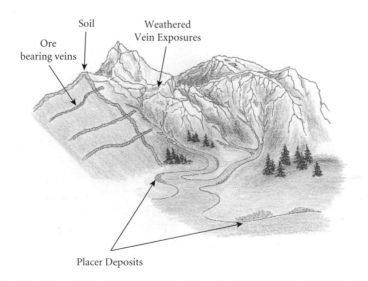

Figure 19.19 Erosion tends to transport and concentrate moderate to heavy metallic ores in stream or beach placer deposits. The metals may have originally been precipitated in veins as a result of an igneous intrusion. (Artwork created by Rebecca A. Gregory and used by permission.)

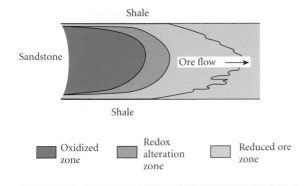

Shale

Sandstone

Ore flow →

Shale

Oxidized zone   Redox alteration zone   Reduced ore zone

Figure 19.20 Roll-front uranium deposits form in stream channels. (After Reynolds and Goldhaber, 1978.)

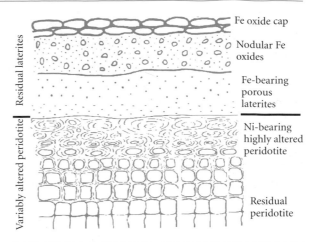

Fe oxide cap

Nodular Fe oxides

Fe-bearing porous laterites

Ni-bearing highly altered peridotite

Residual peridotite

Residual laterites

Variably altered peridotite

Figure 19.21 Intense weathering of peridotite igneous rock concentrates iron and nickel deposits in overlying sediments. (After Chetelat, 1947.)

minerals in placer deposits include gold (20), platinum (14–19), silver (10), uraninite (7.5–10), cassiterite (6.8–7.1), columbite (5.2–7.3), ilmenite and zircon (4.7), chromite (4.6), rutile (4.2) and diamond (3.5). For example, uranium occurs in placer deposits (e.g., the Athabasca Basin) within sandstones deposited in stream and beach environments. Canada and Australia are the two largest uranium exporters in the world and many of these deposits are associated either with placer deposits or with major unconformity deposits.

Unconformity deposits

Unconformities are erosional surfaces representing time gaps between depositional cycles. In Precambrian (>1 Ga) cratons consisting of gneisses and granites, 100–300°C hydrothermal fluids leach metals from underlying rocks and deposit them as ores in tabular vein deposits along unconformity surfaces. Major unconformity-derived deposits of uranium were discovered in Canada's Rabbit Lake deposit in 1968 and in Australia's East Alligator River field in 1970. In addition to uranium, other ores that occur in these unconformity deposits include nickel, copper, arsenic, silver, molybdenum and selenium (Evans, 1993).

Roll-front deposits

Sedimentary deposits of uranium occur in tongue-shaped, **roll-front deposits** (Figure 19.20). Roll-front deposits form in fluvial sandstones as dissolved uranium is transported in stream channels. The uranium is derived from granitic rocks and silicic tuffs. In the presence of oxygen, uranium is soluble and moves downstream as a dissolved phase. Under reducing conditions, uranium precipitates within the pores of sandstone. Variations in oxygen concentrations result in cycles of dissolution and precipitation producing irregular, tongue-shaped deposits. In addition to uranium, other roll-front ores include vanadium, copper, silver and selenium (Evans, 1993).

Weathering processes

Weathering processes, involving the chemical breakdown of pre-existing rock material, can produce residual ores by concentrating aluminum, nickel, manganese and iron. Through the action of surface waters and shallow groundwater, supergene processes result in the decomposition and oxidation of igneous, metamorphic and sedimentary rocks on Earth's surface. Generally, decomposition rates correlate with precipitation and temperature so that tropical environments – regions of high temperature and high precipitation rates – are sites of intense chemical weathering capable of producing laterite soils (Figure 19.21). **Laterite** soils are intensely leached soils in which soluble components have been removed from the decomposed rock materials and insoluble residues have been concentrated. Laterites are commonly enriched in clay minerals – produced by the breakdown of feldspars – and can contain concentrations of metal ores. The specific ore type is determined by the chemical composition of the residual rock from which the laterite is derived. Laterite soils enriched in aluminum hydroxide minerals are collectively referred to as bauxite. **Bauxite**, which is the primary source of aluminum, is produced by the intense leaching of granitic rocks containing aluminum-rich feldspar minerals and, in some cases, gold ore. Laterites derived from the breakdown of igneous rocks such as gabbro, pyroxenite or peridotite (Chapter 7) produce iron, nickel, cobalt, chromium, titanium or manganese ores. Laterites are extensively mined in countries such as China, Guyana, Australia, Jamaica, New Caledonia, Brazil, India and Surinam.

Let us now consider some of the important metals – produced by igneous, metamorphic and sedimentary processes – and their uses in our society.

19.1.4 Metals and alloys

Metals are widely used because of their physical and chemical properties. For over 8000 years, people have been using metals for cooking utensils, weapons, tools, buildings, gems and as currency. Initially people used pure base metals such as copper, zinc or tin. Over time it was discovered that tools increase in strength and durability when two metals are alloyed or forged together using heat. As a result, the Bronze Age developed around 3000 BC when copper and tin were alloyed together. Subsequently brass was created by combining copper and zinc. The use of **metal alloys** continues to the present because alloys serve three primary functions: (1) increasing hardness, (2) reducing corrosion, and (3) increasing high temperature strength. Metals continue to be critically important in all aspects of our existence, from bodily function and health, to our homes, industry, agriculture, medicine and technology. Metals are non-renewable resources that impart a significant financial and environmental cost in mining. Conservation and recycling of these materials is increasingly recognized as a requirement. Metals are commonly grouped into non-ferrous and ferrous metals. **Non-ferrous metals** are classified as (1) precious metals, (2) light metals, and (3) base metals. **Ferrous metals** are alloyed with other elements – such as manganese, cobalt, nickel, chromium, silicon, molybdenum and tungsten – to make steel. We will begin by considering non-ferrous metals.

Precious metals

The **precious metals** – including the platinum group elements, gold and silver – are in great demand. They are primarily used for the following: (1) as catalysts in chemical reactions, (2) as conductors of electricity, (3) for the production of nitric acid, (4) in the fabrication of laboratory equipment, (5) as fillings and caps in dental restoration, (6) as currency, (7) in electronic equipment such as cell phones, and (8) for jewelry. **Platinum group elements (PGE)** occur largely in their native state and include platinum, palladium, rhodium, ruthenium, iridium and osmium. PGE have similar chemical characteristics, serving primarily as catalysts in chemical reactions. Platinum and palladium are the "most abundant" of the rare PGE, occurring in concentrations of ~5 ppb in Earth's crust. PGE are obtained primarily from layered gabbroic intrusions (Chapter 10) such as the Bushveld Complex in South Africa and the Stillwater and Duluth Complexes in the USA where PGE occur with chromite and nickel deposits. PGE also occur in modern ocean ridge black smokers, ancient VMS deposits and in placer deposits. PGE are widely used as the catalysts in automotive catalytic converters, in oil refining and in converting hydrogen and oxygen to electricity in fuel cells. In catalytic converters, oxidation reactions convert hydrocarbons, nitrous oxides and carbon monoxide into less harmful carbon dioxide, nitrogen and water. Given the increasing demand for automobiles and fuel cells, PGE prices will continue to rise. Other PGE such as palladium, rhodium and iridium can be used as a substitute for platinum. These extremely rare metals are mined in only nine countries – South Africa, Russia, the United States, Zimbabwe, Canada, Australia, Finland, Columbia and Ethiopia –

making them a rare and exclusive group of metals (Kesler, 1994).

Gold occurs in a number of different forms, which include (1) a native element state, (2) bonding with silver to form electrum, or (3) bonding with tellurium to form telluride minerals such as calaverite. Gold is largely used in jewelry because of its ductility, softness and resistance to tarnish. Gold is also used in dentistry, as building ornamentation, as currency and – because of its very high conductivity – is widely used in the electronics industry. **Silver** occurs in its native state, with gold in electrum and with base metals in sulfide minerals such as argentite, tennantite and tetrahedrite. Like gold, silver is widely used in jewelry, coins and dentistry due to its ductility and softness and in electronics because of its high conductivity. However, unlike gold, silver readily tarnishes due to oxidation reactions. Gold and silver originate via magmatic processes and concentrate through hydrothermal reactions in a number of igneous environments. The largest gold deposits on Earth – the Witwatersrand deposits of South Africa – occur in Archean age (>2.5 Ga) placer deposits that formed by weathering granite source rocks and depositing high density gold and uranium in a clastic sedimentary basin. The Witwatersrand deposit was discovered in 1886 and mining continues to this day. Notable gold rushes in the 19th century, which included the California and Klondike gold rushes, also involved placer deposits in which high specific gravity gold flakes and nuggets weathered from vein deposits, settled to the base of streams and were later extracted by panning techniques.

The gold and silver found in placer deposits typically originate via granitic intrusions, Precambrian greenstone belts, deep turbidite sedimentary basins (Chapter 11) and as finely disseminated grains within porphyry copper deposits and VMS deposits. Porphyry deposits such as Utah's Bingham copper mine and the Grasberg copper mine in Indonesia produce significant tonnage of both gold and silver (Kesler, 1994). The leading producers of gold include South Africa, Australia, the United States, China, Peru, Russia, Indonesia and Canada. The leading producers of silver include Peru, Mexico, China, Australia, Chile, Poland, the United States and Canada.

Light metals

Light metals consist of low density elements such as magnesium ($1.7\,g/cm^3$), beryllium ($1.85\,g/cm^3$), aluminum ($2.7\,g/cm^3$) and titanium ($4.5\,g/cm^3$). Because of their light weight and relatively high strength, demand for light metals has increased greatly, particularly for use in the aerospace and transportation industries (rockets, aircraft, trains, cars and trucks) as well as for common uses such as aluminum foil and baseball bats. Aluminum baseball bats are alloyed with scandium and cored with light-weight materials such as graphite. Compared to wooden bats, aluminum bats have a greater elastic response upon impact with a baseball, which results in the ball traveling greater distances.

Aluminum and magnesium are the fourth and seventh most abundant elements in Earth's crust, respectively. Despite the fact that **aluminum** constitutes ~8% of Earth's crust by weight, it is a difficult metal to obtain and process. Aluminum is obtained primarily from laterite soils that have experienced extreme leaching in tropical environments. Laterite soils derived from the weathering of granitic or clay-rich rocks commonly contain the bauxite group minerals diaspore, gibbsite and boehmite. Although bauxite is the only commercial source of aluminum, this light metal could be processed from clay or feldspar group minerals at a significantly higher cost. Whether derived from bauxite, clay or feldspars, aluminum production requires an enormous amount of energy. Bauxite is strip mined in over 20 countries, with the largest deposits occurring in Australia, Brazil, China, India, Guyana and Jamaica (USGS, 2007).

Earth's crust contains about 2% **magnesium** by weight, which occurs in minerals such as olivine, pyroxene, amphibole, dolomite, magnesite and brucite. Minerals containing magnesium are mined in greenstone belts, ophiolites, hydrothermal environments, seawater and evaporate brine basins. Magnesium is used in the automotive industry and as an alloy with aluminum to increase aluminum's hardness and resistance to corrosion. Magnesium is also used as a refractory material for furnaces and molds in the steel industry. Leading producers of magnesium include China, Turkey, North Korea, Russia and Slovakia (USGS, 2007).

Beryllium is a very expensive, relatively rare element that occurs in the minerals beryl and beryllonite. Ore deposits of these minerals occur in silicic pegmatite intrusions and in related hydrothermal veins. Beryllium is alloyed with copper to increase hardness and is widely used in computer, telecommunications, aerospace, military and automotive electronics industries due to its high conductivity, light weight, stability at high temperature and resistance to corrosion. Beryllium is also used in the medical industry, for X-ray windows and treatments, as well as in nuclear fusion reactor experiments. Beryllium is a known carcinogen and inhalation of beryllium dust is particularly lethal, which is the reason why beryllium is no longer used in fluourscent light fixtures. The leading producers of beryllium are the United States, China and Mozambique (USGS, 2007).

Titanium represents about 1% by weight of Earth's crust and occurs in ore minerals such as ilmenite, rutile and anatase. Titanium ore occurs in layered gabbroic intrusions and in coastal placer deposits derived from the weathering of gabbroic intrusions. Titanium is a light-weight metal with a very high melting temperature (1678°C) and is widely used in aircraft engines and high speed turbines. Kesler (1994) reports that titanium represents up to 30% of modern aircraft weight. Ocean research submersibles such as the Alvin also consist largely of titanium. Titanium is widely used in paint, plastic, paper, space stations and recreational sports equipment such as bicycles, tennis racquets and golf clubs. Because of its high strength, low weight, low toxicity and hypoallergic properties, titanium is also widely used for joint replacements and prostheses in the medical industry. For people with allergic reactions to gold or silver metal jewelry such as earrings, titanium is also the metal of choice. Significant producers of titanium include Australia, South Africa, Canada, China, Norway and the Unites States.

Base metals and rare Earth metals

Base metals are non-ferrous metals that oxidize easily. Base metals include copper, zinc, lead, tin, lithium, uranium, mercury, arsenic, cadmium, antimony, germanium, rhenium, tantalum, zirconium, hafnium, indium, selenium, bismuth, tellurium and thallium.

Copper occurs in the native state as well as in sulfide, oxide, hydroxide and carbonate minerals. Copper ore minerals include chalcopyrite, chalcocite, bornite, cuprite, enargite, tetrahedrite, malachite and azurite. Copper ores form in many different environments which include porphyry copper deposits, VMS deposits, rift basin deposits, hydrothermal vein deposits and in oxidized secondary supergene enrichment environments. Copper is perhaps the earliest metal used by humans and continues to be among the most widely used, particularly in electronics, due to its high conductivity. The demand for and value of copper is rising dramatically in response to the industrialization of China and growth in India. Major producers of copper include Chile, the United States, Peru, Indonesia, Australia, Russia, China, Canada, Mexico, Kazakhstan and Zambia (USGS, 2007).

Zinc occurs primarily in the sulfide mineral sphalerite, and to a lesser degree in the zinc silicate mineral willemite. Zinc oxide minerals zincite and franklinite were exclusively mined at the Sterling Mine in Franklin, New Jersey, which ceased operations in the 1980s. Sphalerite occurs with copper minerals listed above in VMS deposits, skarns and other hydrothermal environments. Sphalerite occurs with galena in MVT and sedex deposits. Zinc is widely used as a metal alloy to prevent oxidation and corrosion in galvanized sheet metal, nails, bolts and other construction equipment. Zinc is also used in the production of brass, bronze and in ammunition. Like copper, zinc is greatly in demand due to economic expansion in China and India. The leading producers of zinc include China, Australia, Peru, the United States, Canada, Mexico and Kazakhstan (Kesler, 1994; USGS, 2007).

Lead occurs primarily in the sulfide mineral galena. Minor minerals include anglesite, cerussite and crocoite. As noted previously, lead occurs with zinc in MVT and sedex environments. Lead has been used since ancient times in lead crystal glassware, lead glass and as a sweetener for wine. Physical and mental debilitation due to ingestion of lead has been cited as one of the possible causes for the fall of the Roman Empire. Until recently, lead was commonly used in paints, gasoline and in fishing equipment, among other uses. Since the 1970s restrictions have been emplaced on lead use due to severe health risks. Lead is now used primarily in lead-acid batteries for industrial equipment, aircraft and the automotive industry. Lead is also used in ammunition, radioactive shields, solder, pipes, glass, pigments and ceramics. Lead remains a very significant health risk in our environment as will be discussed later in this chapter. The principal producers of lead include China, Australia, Peru, the United States and Mexico (Kesler, 1994; USGS, 2007).

Tin ore is derived primarily from the mineral cassiterite and is associated with silicic igneous intrusions in sedimentary rock. Tin is closely associated with tungsten and molybdenum ore deposits in granite plutons and associated hydrothermal vein networks in continental crust overlying subduction zones. Tin also occurs in VMS, MVT and placer deposits. In the past, tin was widely used for canned food and beverages. However, its market share has been dramatically replaced by glass, aluminum and plastic. The major producers of tin include China, Indonesia, Peru, Bolivia, Brazil, Russia, Malaysia and Australia.

Uranium ores are obtained from oxide and phosphate minerals such as uraninite and carnotite in unconformity and placer deposits derived by weathering granite source rocks. Uranium is the primary fuel for nuclear reactors and is a major energy source. While the production of nuclear energy does not emit the contaminants or greenhouse gases associated with fossil fuels, the mining and disposal of uranium material poses major environmental problems. Uranium mining is associated with a high lung cancer incidence due to inhalation of radioactive radon gas, produced as a breakdown product of uranium. Radium, another uranium daughter product, is a major contaminant of groundwater. As radium is chemically similar to calcium, it is readily absorbed into bones causing cancer. Uranium waste from nuclear reactors poses two long-standing problems: (1) uranium ore can be enriched to weapons grade material that can be used in nuclear weapons, and (2) permanent nuclear waste repositories for safe disposal do not yet exist. In the United States, Yucca Mountain, Nevada, had been proposed as a permanent nuclear waste repository but its future use as of this writing is uncertain.

While Yucca Mountain is geographically isolated from major populations and has a very low water table, it is located in a seismically and volcanically active region, has active faults in the proposed mine and the host rock is porous volcanic tuff. Yucca Mountain was chosen largely based on political, rather than geological, considerations. Despite the issues related to the safe disposal of uranium, energy demands will result in increased use of uranium for nuclear reactors. Major producers of uranium include Canada, the United States, Australia, Russia, Namibia and France.

The origins and uses of other base metals are listed in Table 19.2.

Rare Earth metals include scandium, yttrium and the 15 lanthanide elements (Table 19.3). Rare Earth metals are found in minerals such as monazite, which occurs in granite pegmatites, hydrothermal veins and placer deposits. Rare Earth metals are widely used as catalysts in oil refining, in chemical synthesis, as catalytic converters in automobiles, as glass additives, in glass polishing, in fiber optic lasers, phosphors for fluorescent lighting, in color televisions, cell phones, electronic thermometers and X-ray screens, and as pigments, superconductors, dopants and more. A **phosphor** exhibits the phenomenon of phosphorescence and is utilized in electrical equipment such as fluorescent lights and cathode ray tubes. A **dopant** is an impurity that alters the optical and electrical properties of semiconductors. Major sources of rare Earth metals include China, India and Malaysia (Kesler, 1994; USGS, 2007).

Ferrous metals and ferrous alloys

The term "ferrous" refers to **iron**. Iron is derived from: (1) oxide minerals such as magnetite and hematite, (2) hydroxide minerals such as goethite and limonite, and (3) carbonate minerals such as siderite. Iron is largely mined from Superior-type or Algoma-type deposits. Iron also occurs in magmatic deposits and skarns. Demand for iron is steadily increasing largely due to economic growth in China. Major producers of iron include China, Australia, Brazil, India, Russia, Ukraine and the United States. Iron is used principally in the production of steel through combining iron with various metal alloys. **Steel** is an alloy consisting mostly of iron with

Table 19.2   Some additional base metal occurrences and uses (Kesler, 1994; USGS, 2007).

Element and minerals	Occurrence	Uses and/or hazards
Antimony: in stibnite, tetrahedrite and jamesonite	Hydrothermal vein deposits, MVT deposits and Kuroko-type VMS deposits	Fire-retardant materials, batteries, ceramics and glass. While antimony use and production is declining, China, Bolivia, Mexico, Russia, South Africa, Tajikistan and Guatemala continue to mine antimony in association with Pb, Zn, Ag, Sn and Wo
Arsenic: in realgar, orpiment, enargite, arsenopyrite and tennantite	Hydrothermal veins with Cu, Ni, Ag and Au and in Cu porphyry deposits	Toxic aspect used in copper chromate arsenic (CCA) wood preservatives, herbicides, insecticides and ammunition. Arsenic use continues to decline due to adverse health effects that include breathing and heart rhythm problems and increased risk of bladder, lung and skin cancer. Producers include China, Chile, Morocco, Peru, Russia, Mexico and Kazakhstan
Bismuth: in bismuthinite	Byproduct of Wo, Mo and Pb mining in porphyry deposits	Over-the-counter stomach remedies (Pepto-Bismol), foundry equipment and pigments. As a non-toxic replacement for lead, Bi is increasingly being used in plumbing, fishing weights, ammunition, lubricating grease and soldering alloys. Because of its low melting temperature, Bi is used as an impermeable low temperature coating on fire sprinklers. Bi is mined in China, Peru, Mexico, Canada, Kazakhstan and Bolivia
Cadmium: in greenockite; primarily derived from sphalerite	Byproduct of Zn mining in VMS and MVT-type deposits	NiCd rechargeable batteries for alarm systems, cordless power tools, medical equipment, electric cars and semiconductor industry, and steel and PVC pipes for corrosion resistance and durability. Previously used as a yellow, orange, red and maroon pigment. Unfortunately Cd interferes with Ca, Cu and Fe metabolism resulting in softening of the bones and vitamin D deficiency. Because of adverse health effects, Cd use is declining. Producers include China, Canada, South Korea, Kazakhstan, Mexico, the United States, Russia, Germany, India, Australia and Peru
Germanium: rarely forms its own mineral but occurs with Zn and Cu	Byproduct from MVT deposits and in Cu ore deposits	Fiberoptic cables, where it has replaced Cu in wireless communication, solar panels, semiconductors, microscope lenses and infrared devices for night-vision applications in luxury cars, military security and surveillance equipment. Ge is also used as a catalyst in the production of polyethylene terephthalate (PET) plastic containers and has potential for killing harmful bacteria. The USA is the leading producer of Ge
Hafnium: in ilmenite and rutile	Placer deposits	Used in the construction of nuclear rods because Hf does not transmit neutrons. Major producers of zirconium and hafnium include Australia and South Africa
Indium is in sphalerite, cassiterite and wolframite	VMS, MVT and hydrothermal veins with Sn, Wo and Zn	Indium–tin oxide (ITO) is used in the production of flat panel displays and other LCD products. ITO is also used in windshield glass, semiconductors, breathalyzers and dental crowns. Major producers of indium include China, Canada, Belgium and Russia

Table 19.2 *Continued*

Element and minerals	Occurrence	Uses and/or hazards
Lithium: in spodumene, lepidolite and lithiophilite	Granite pegmatites and alkali brines from playa basins	Used in glass, ceramics, greases and batteries. Because of the adverse health issues associated with Cd and Pb, Li batteries are increasingly being used in power tools, calculators, cameras, computers, electronic games, cellphones, watches and other electronic devices. The primary producers include Chile, Australia, Russia, China, Argentina and Canada
Mercury: in cinnabar	Low temperature (<200°C) hydrothermal vein deposits	Only metal that is liquid at room temperature for which it is commonly referred to as quicksilver. Previously used for automotive switches, cosmetics, pigments, gold processing and hat making. Hg is highly soluble and readily enters the bloodstream where it attacks the nervous system causing psychotic behavior ("mad as a hatter"), irritability, tremors and can lead to death. Hg continues to be used in thermometers, batteries, electrical fixtures and dental amalgam fillings and is used as a catalyst in paper production. Hg use in these and other capacities will continue to decline because of its adverse health effects. Hg is obtained from mines in China or Kyrgyzstan or obtained as a secondary ore in copper, zinc, lead and gold mines throughout the world
Rhenium substitutes for molybdenum: in molybdenite	Cu porphyry deposits	Used with Pt as a catalyst in oil refining and the generation of high octane, lead-free gasoline. Re is a superalloy used in high temperature turbine engine components, crucibles, electrical contacts, electromagnets, electron tubes and targets, heating elements, ionization gauges, mass spectrographs, semiconductors, thermocouples, vacuum tubes and other uses. Producers of Re include Chile, the United States, Kazakhstan, Peru, Canada, Russia and Armenia
Selenium: in selenite gypsum; more commonly occurs with sulfide minerals such as FeS and CuS minerals	Byproduct of Cu ore deposits	Decolorized green tints are caused by iron impurities in glass; reduces solar heat transmission in architectural plate glass. Together with Cd, Se is used to generate ruby red colors in traffic lights, plastics, ceramics and glass. Selenium's photoelectrical properties were used as photoreceptors in replacement drums for older plain paper photocopiers. In the digital age, Se enables the conversion of X-ray data to digital form. Se serves as a catalyst in oxidation reactions, and is used in blasting caps, rubber compounds, brass alloys and in dandruff shampoos. Se is also used in fertilizers and as a dietary supplement for livestock. Se has both positive and negative health aspects: deficiencies increase the incidence of stroke while excess selenium is related to deformities. Major producers of Se include Japan, Canada and Belgium
Tantalum: in microlite, pyrochlore and tantalite	Pegmatites, hydrothermal veins and placer deposits	Used in electrical capacitors, automotive electronics, pagers, personal computers and cell phones. Major producers of Ta include Australia, Brazil, Mozambique, Canada, Ethiopia, Congo (Kinshasa) and Rwanda

Continued

Table 19.2 *Continued*

Element and minerals	Occurrence	Uses and/or hazards
Tellurium: in calaverite	Pegmatites, veins.	Used as a semiconductor and as a metal alloy for ductility and strength. Te is among the rarest elements in Earth's crust
Thallium: in association with sphalerite	VMS and MVT deposits	Previously used in rat and ant poison until its toxicity to humans became apparent. Tl interferes with the metabolism of K and can cause death. Tl-201, a radioactive isotope, is used in the medical industry for cardiovascular imaging. Tl is also used an activator in gamma radiation detection equipment, infrared detectors and in high temperature superconductors used for wireless communication. Major producers include Canada, European countries and the USA
Zirconium: in rutile and ilmenite	Placer deposits	Used as a fuel in nuclear reactors due to the ease with which it transmits neutrons; also used in ceramics and as an abrasive, metal alloy and refractory material due to its very high melting temperature (2550°C). Major producers include Australia and South Africa

LCD, liquid crystal display; MVT, Mississippi Valley-type; VMS, volcanogenic massive sulfide.

Table 19.3 Common uses of rare Earth metals. Data from J. B. Hendrick as cited by Kesler (1994).

Rare Earth metals	Use
Cerium	Polishing compounds, radiation shield, glass, ammonia synthesis
Dysprosium	Permanent magnets
Erbium	Fiber optic amplifier, glass additive
Europium	Phosphors in cathode ray tubes
Gadolinium	Phosphor and laser crystals
Holmiun	Dopant in laser crystals
Lanthanum	Catalyst in petroleum refining, glass additive, rechargeable batteries
Lutetium	Phosphor
Neodynium	Permanent magnets, glass additive, dopant in laser crystals
Praseodymium	Yellow pigment in ceramics
Promethium	Fluorescent lighting starter
Samarium	Permanent magnets
Scandium	Metal halide lamps
Terbium	Phosphors
Thulium	Isotope used in medicine
Ytterbium	X-ray source, glass and laser additives
Yttrium	Phosphor, synthetic gems, superalloy

0.02% to ~2.0% carbon content by weight. Carbon is the most cost-effective alloy, serving as a hardening agent and preventing dislocations in the iron crystal lattice structure. In addition to carbon, other ferrous alloys include aluminum, chromium, cobalt, magnesium, manganese, molybdenum, nickel, silicon, titanium, tungsten and vanadium. Important aspects of these alloys are briefly described in Table 19.4.

19.2 INDUSTRIAL MINERALS AND ROCKS

Industrial minerals and rocks constitute Earth materials other than metals or fuel that have economic value. Industrial minerals and rocks serve a number of vital roles in the global economy and are widely used (1) as fertilizers and chemicals, (2) in construction, and (3) in manufacturing.

Fertilizers are widely used to increase crop production. Fertilizers consist of primary nutrients such as nitrogen, phosphorous and potassium, and secondary nutrients such as calcium, magnesium and sulfur. In addition to serving as nutrients, industrial minerals containing arsenic (realgar and orpiment) and mercury (cinnabar) have been widely used as pesticides. While the use of arsenic and mercury compounds has diminished somewhat due to the detrimental health effects on humans and other animals, arsenic continues to be used in pressure-treated lumber used for outdoor recreation purposes.

The chemical industry relies on minerals to produce mass quantities of sulfur, fluorine and chlorine for diverse purposes such as pharmaceuticals, cosmetics and purifiers and in various manufacturing processes. The construction of roadways, buildings, dams and other structures requires aggregate minerals that include quartz, calcite, gypsum, dimension stone, bricks, cement, clay, sand and gravel. Manufacturing includes a wide array of uses such as abrasives, absorbents, fillers, filters and more. In the following section we will address the three major categories of industrial minerals and demonstrate their significance to the global economy.

19.2.1 Fertilizers and chemicals

Fertilizers and industrial chemicals are derived from a number of sources, which include evaporite minerals, marine sediments and organic deposits representing the remains of animals and bacterial alteration. **Fertilizers** are used to improve nutrient level within soils and serve as soil conditioners to alter pH. Fertilizers include nitrogen compounds, phosphates, potash, limestone, dolostone, sulfur and salts. Nitrogen is a key fertilizer element widely used in agriculture. Although nitrates can be extracted from the evaporite mineral niter, most nitrogen is obtained from chemical processes and not from the mining of nitrates. In addition, bat excrement, known as guano has been used as a minor nitrogen source.

Phosphate is derived from the calcium phosphate mineral apatite and represents a critical fertilizer element. Phosphate is a component of DNA and other critical organic compounds, forming the skeletal material of many vertebrates. Approximately 95% of world phosphate production is used as fertilizer and ~5% is used in products such as detergents, fire retardants and toothpaste. Phosphate occurs in three major settings: (1) continental shelves rich in organic material such as bones and teeth, (2) igneous intrusions (syenite, carbonatite) at continental rifts, and (3) guano deposits from bat and bird excrement in caves and tropical islands.

Halide elements such as the evaporite minerals halite, sylvite and polyhalite are used for potash fertilizer (sylvite) as well as for food additives (salt), water purification (chlorine) and in the production of sodium (caustic soda) and chlorine gas. In the days before refrigeration, salting meat was essential to prevent food from going rancid. Salt continues to be used as a very common additive in food and still serves as a preservative, as lovers of jerked beef and salted cod will attest. Sodium and lime are used to adjust the pH of soils. Lime is produced from limestone that forms in marine environments, beach systems, continental hot springs and freshwater lakes. Lime is produced by heating limestone to 700–1000°C, effectively driving off CO_2:

$$\underset{\text{Calcite}}{CaCO_3} + heat \rightarrow \underset{\text{Lime}}{CaO} + \underset{\text{Carbon dioxide}}{CO_2}$$

Lime is used to neutralize soil acidity, purify water and to control the pH of natural lake systems. For example, the liming of lakes

Table 19.4 Common steel alloys (Kesler, 1994; USGS, 2007).

Alloy element and minerals	Origin	Attributes and major producers
Aluminum: diaspore, boehmite, gibbsite	Bauxite in laterite deposits derived from weathering silicic igneous rocks	Light-weight and high strength metal. Major producers include Australia, Papua New Guinea, Jamaica, Brazil, India and Guyana
Beryllium: beryl	Granite pegmatites, hydrothermal veins around silicic igneous rocks	Light-weight metal stable with high temperature strength. Major producers include the USA, China and Mozambique
Carbon: graphite	Organic or inorganic sources	Alloyed for hardness. Producers are widespread globally
Chromium: chromite	Layered gabbroic intrusions, ophiolites, VMS, laterites, placer deposits	Corrosion resistance; important alloy in "stainless steel". Major producers include South Africa, Kazakhstan and India
Cobalt: cobaltite	Layered gabbroic intrusions, hydrothermal veins, evaporite brines in desert basins	Corrosion and abrasion resistance; used in industrial and gas turbine engines. Major producers include Congo (Kinshasa), Zambia, Australia, Russia, Canada, Cuba, Morocco, New Caledonia and Brazil
Magnesium: magnesite, olivine	Marine carbonate rocks and layered igneous intrusions.	Hardens Al and corrosion resistance. Magnesium producers are widespread globally
Manganese: pyrolusite, psilomelane, rhodonite, rhodochrosite	Supergene weathering of Mn rich rocks in laterites; Mn nodules in VMS; marine limestones and shales	Corrosion resistance in Al; alloyed with Cu for strength; replaced lead in fuel. Major producers include South Africa, Gabon, Australia, Brazil and China
Molybdenum: molybdenite	Porphyry type deposits and granite pegmatites	Hardness; corrosion resistance, high temperature strength. Producers include the USA, Chile, China and Peru
Nickel: nickelline, pentlandite, millerite	Layered gabbroic intrusions ophiolites, komatiites and laterites derived from ultrabasic rocks	High temperature strength; corrosion resistance; used in jet engines and turbines. Major producers include Russia, Canada, Australia, Indonesia, New Caledonia, Columbia and Brazil
Silicon: quartz	Clastic sedimentary rocks, pegmatites and hydrothermal veins	Deoxidant in steel; increases strength and corrosion resistance. Silicon producers are widespread globally
Titanium: ilmenite, rutile	Placer sand deposits (beach dunes) or from layered gabbroic igneous intrusions	Light weight and strong; strategic mineral with limited reserves; six times stronger than steel; aircraft commonly contain one-third Ti. Producers include South Africa, Australia, China and India
Tungsten: scheelite, wolframite	Pegmatites and hydrothermal veins derived from granitic intrusions. Also skarns, hot springs	High temperature strength; high speed drills, armor plating, light filaments. Major producers are China, Russia, Kazakhstan, Austria, Portugal and Canada
Vanadium: vanadinite, uraninite	Pegmatites, veins, layered igneous intrusions, phosphate sedimentary rock.	High strength, ductility, toughness; used in high rises, oil platforms. Major producers are South Africa, China and Russia
Zinc: sphalerite, zincite	VMS or MVT	Corrosion resistance. Major producers include China, Peru, Australia and Ireland

MVT, Mississippi Valley-type; VMS, volcanogenic massive sulfide.

increases the pH (less acidic, more basic) and reduces acidification. Industrial processes involving lime include making cement and for use in whitewashing houses to increase reflectivity in many parts of the world.

Sulfur is obtained from native sulfur and to a lesser degree pyrite. Sulfur forms by bacterial reduction of gypsum and other sulfate minerals and by sublimation from volcanic gases around vents. Sulfur is used for the production of sulfuric acid, which is a key component in the generation of superphosphate fertilizers. Sulfur is also used as a catalyst in the production of high octane gasoline. Fluorite is an important mineral used in the production of hydrofluoric acid and fluorine. Fluorite occurs in low to medium temperature hydrothermal veins and is associated with granitic plutons. Fluorite is used in the enrichment of uranium, the production of chlorofluorocarbons (CFC) for refrigerants and propellants, for cavity protection, and as a flux in steel making, glass and ceramics.

Other **chemicals** are widely used in pharmaceuticals, health products and cosmetics. For example, barite is used to enhance X-ray images of organs and blood vessels; compounds containing calcium are used to prevent bone loss; and clays (kaopectate), magnesium and calcium are widely used for digestion. Other dietary supplements contain zinc, iron, selenium, cobalt and sodium. Shampoos and soaps contain nitrogen compounds, zinc, potassium, chlorine and titanium. Sunscreens are composed largely of zinc oxide and titanium oxide. Cosmetics are largely composed of clays, talc and micas with color additives containing oxides of iron, titanium, zinc, chrome, aluminum and copper. Micas provide the sheen in lipsticks. The list of chemicals derived from minerals in our everyday life goes on and on.

19.2.2   Construction material

Common materials used in **construction** include dimension stone, aggregate, concrete, plaster and other clay building materials (Figure 19.22). **Dimension stone** consists of rocks such as limestone, marble, sandstone, granite, gneiss and slate. Dimension stone has been used since at least the construction of the pyramids over 4000 years ago and, until recently, dimension stone remained the

primary material used in sidewalks, roads, bridges, dams, building foundations, walls and roofs, as well as in interior uses such as countertops and fireplaces. Dimension stone has been acquired by quarrying operations throughout the world. Michelangelo used marble quarried locally in Carrara, Italy. Vermont slate was the rock of choice for sidewalks, blackboards and roof tiles in the eastern United States; slate from Wales provided an equal service in Wales and England. Homes in Ireland and England were largely constructed of limestone. The Palisades Sill basalt studied by Norman Bowen (Chapter 8) was the "trap" rock commonly used for road construction in the New York metropolitan area. The rock of choice in any given location was determined by the underlying geology at that site. Dimension stone continues to be used in high end construction such as in granite kitchen countertops. However, in most large construction projects it has largely been displaced by less costly aggregate and cement materials.

Aggregate consists of sand, gravel and crushed stone. Aggregate material is most commonly derived from unconsolidated sedimentary layers deposited in beaches, deltas, deserts and stream systems or by glaciers. In the absence of available sand and gravel, aggregate is produced by mechanically crushing rock. The equipment and energy required to crush rock makes this a more expensive operation. Due to their low hardness and abundance, limestone and dolostone are rocks of choice for crushed aggregate. Rocks such as granite, basalt, sandstone and metaquartzite are also readily available in many locations but their greater hardness results in higher costs. Aggregate quality is lowered by the presence of pyrite, which oxidizes to produce rust, and shale, which weathers easily.

Concrete, developed by the Romans 2000 years ago, is the dominant construction material used in world today. Concrete consists of Portland cement mixed with an aggregate of sand or gravel. Portland cement is made by heating limestone and clay in a kiln to produce clinker. Clinker is then ground to a powder and mixed with gypsum. Portland cement is the most common hydraulic cement, meaning that it hardens with water. In many construction projects cement has replaced dimension stone because of its ease of

Figure 19.22 (a) Sandstone quarry, Wisconsin. (b) Cut and polished granite, Wisconsin. (c) Esker containing glaciofluvial sands, Wisconsin. (d) Crushed aggregate. (e) Limestone quarry, Wisconsin. (f) Kiln for baking limestone. (g) Moroccan adobe house. (h) Brick and concrete stadium in Iowa. (Photos by Kevin Hefferan.) (For color version, see Plate 19.22, between pp. 408 and 409.)

transport, workability and the ability to create forms to fit site-specific construction projects. Cement quality is affected by the presence of chert (promotes cracks in concrete), magnesium content (MgO must be <5%) and pyrite concentration (produces SO_2 gas).

Gypsum is an evaporite mineral that commonly forms in sabkhas or intertidal flats of marine systems. In addition to its use in cement, gypsum is widely used in plaster and wallboard (sheetrock) for building construction. Plaster of Paris (gypsum cement) is produced by heating and dehydrating gypsum at temperatures of ~150°C. Other minerals such as clay, quartz, hematite and calcite are major components of common construction materials such as bricks, adobe, stucco, roof tiles, drains and sewer pipes. Bricks, adobe and stucco are widely used where dimension stone and lumber are not readily available. Again, in the absence of ostentatious wealth, construction utilizes the resources at hand, largely due to the cost associated with transporting rock material great distances. In these times of rapidly expanding populations and urbanization, construction represents the single greatest growth area for non-fuel mineral resources.

19.2.3 Manufacturing minerals

Manufacturing materials include a wide array of mineral applications such as abrasives, absorbents, ceramics, detergents, drilling fluids, fillers, filters, fluxes, insulators, lubricants, pigments, refractory materials and glass. All of these applications are used in manufacturing and many of these are utilized in our daily lives throughout the world. Many minerals are adept at multitasking. For example, clay minerals serve as absorbents and fillers, and in drilling mud and in construction.

Abrasive minerals

Abrasives are used to grind, polish, abrade, scour and clean surfaces. Abrasive minerals display high hardness and rigidity, retain their grain shape and size, and do not contain softer impurities. Abrasives are used for saws, drills, sandpaper, grinding wheels and abrasive cleansers. Common abrasive minerals include diamond, emery (corundum), garnet and quartz. Large, high quality mineral samples of diamond and corundum are considered gemstones but the varieties used as abrasives have imperfect, flawed and/or small crystals so that they are considered to be of industrial grade rather than gem quality. Abrasives are used on industrial scales for cutting and polishing manufactured goods. We also use these for cleaning kitchen utensils, finishing wood products, in emery boards for fingernails and in toothpaste for brushing teeth.

Absorbents

Absorbents are used to remediate spills or to remove waste products from liquids. Absorbents are used for industrial-scale removal of wastes and contaminants as well as in household items such as baby's diapers, feminine hygiene pads, antiperspirants and cat litter. Silica is also used as an absorbent in packaging electronics and footwear. Minerals commonly used as absorbants include the hydrous aluminosilicate zeolite minerals and phyllosilicate minerals such as the expansive smectite clays and vermiculite. These hydrous aluminum silicate minerals have crystal structures that allow them to incorporate molecules such as water between silica tetrahedral sheets. Vermiculite is used to increase soil aeration and retain moisture and chemicals in agricultural soils.

Ceramics

Ceramics utilize clays and other minerals such as feldspars, hematite, bauxite, bentonite, pyrophyllite, borax, wollastonite, barite, lepidolite and spodumene. Clay forms the basis of pottery and other ceramic materials. Feldspars, borax and wollastonite are used to produce ceramic glazes on pottery that increase hardness, provide a vitreous luster and preserve color. Hematite serves as a pigment. Barite hardens ceramics and serves to preserve pigments. Bauxite and pyrophyllite provide high temperature strength. Lithium minerals lepidolite and spodumene are added to ceramics to prevent volume change with temperature variations. Each of these minerals serves to improve the quality of clay-based ceramic materials (Chang, 2002).

Detergents

Minerals used as cleansing **detergents** include evaporate minerals such as borax, halite and trona. Together with baking soda (HNO_3) or ammonia, evaporite minerals are very effective in cleansing soiled items. Depending on the application, detergents can be used in combination with abrasive materials for effective scrubbing action. Phosphate minerals were used in the past but have largely been removed from detergents. Phosphates promote eutrophication of water bodies by deteriorating water quality due to excessive nutrient input.

Drilling fluids

Drilling fluids are used in well drilling for oil, gas and groundwater, and in geophysical studies such as earthquake studies. Inert minerals with high specific gravities such as barite are combined with water to produce drilling muds, which allow rock chips to float to the surface for analysis of borehole geology. Drilling muds also control well pressure, maintain borehole stability, lubricate the drilling apparatus and protect the drill target from contamination. Barite, which has a specific gravity of 4.5–5, is combined with bentonite (smectite) clay to produce an expansive mud capable of supporting rock chips and cuttings. After constructing a borehole well using drilling fluids, swelling bentonite pellets are used to pack the top of the well to prevent surface fluids from flowing into the borehole and contaminating the well. Upon exposure to water, swelling clays can expand up to 20 times their original volume. In this capacity, the bentonite pellets serve as fillers discussed below.

Fillers

Fillers are inert, inexpensive materials that extend the volume of material at low cost. Fillers include barite and phyllosilicate minerals such as kaolinite, smectite, mica, talc and pyrophyllite. The ability of fillers to expand volume without reducing material quality can substantially reduce manufacturing costs. The most effective fillers may in fact improve material quality. For example, in paper production fillers such as lime and kaolin improve print and appearance of paper. Fillers are used in the production of food, rubber, paint, wallpaper, cosmetics, plastics, roofing, textiles, soaps, ceramics and paper. For example, vermiculite is used as a light-weight filler in cement, concrete and plaster. Barite is used as a filler in chocolate and kaolinite is used as a filler in ice creams and shakes. In a sense water can be considered as a filler when it is used to dilute juices or other drinks to reduce their sugar content or simply to extend the volume.

Filters

Filters are used to purify, clarify and clean liquids by removing solids or other contaminants. Materials commonly used as filters include zeolites, graphite, clays, charcoal and diatomite. Diatomite is a siliceous rock composed of microscopic marine organisms called diatoms. Filters are used in a wide variety of industries, which include water softeners in homes, beer filters in breweries and hydrocarbon filters in oil and gas refining. Filters are also used in the production of medicines and fruit juices and in the remediation of contaminated groundwater and surface water. Zeolites are multipurpose industrial minerals in that they serve as absorbents as well as filters. Zeolites filter lead, cadmium and other potentially hazardous elements from potable water. Zeolites also serve as filters in oil refining and the petrochemical industry.

Fluxes

A **flux** breaks chemical bonds, thereby lowering the melting temperature of materials. A common analogy is the fluxing or melting of ice by adding salt. In manufacturing, a lower melting temperature drastically reduces energy demands. Common industrial fluxes include fluorite, limestone and borate minerals. Limestone is widely used as flux in the steel industry so that metals melt at a lower temperature, requiring less heat input. Fluorite and boron act as fluxes in the manufacture of glass, glass fibers and insulation. Cryolite deposits from

Greenland have been used as a flux in aluminum refining.

Insulation

Insulators are materials that reduce the transmission of heat or sound. Common insulating materials include asbestos, micas, silica and vermiculite. Insulators are commonly used in flooring, walls and ceiling tiles, brake linings, roof shingles, around pipes and in the production of fire-resistant clothing. Lead is also an insulator in preventing the transmission of radioactive energy as anyone who has had an X-ray can attest. Chrysotile asbestos was one of the most popular insulating materials in homes and buildings due to its ability to retain heat, its ease of use and its fireproof nature. However, since the 1980s chrysotile asbestos has largely been replaced due to health concerns about mesothelomia, asbestosis and lung cancer – despite the fact that these illnesses are generally caused by amphibole asbestos minerals, which are less commonly used. Asbestos has been replaced by other materials, such as fiberglass and vermiculite. Although commonly used as rolled insulation, use of fiberglass is a health concern because the silica glass fibers lacerate the lung tissue when inhaled. Vermiculite is blown into walls, crawl spaces and attics as an unconsolidated fluffy material. Although vermiculite is inert, it commonly contains amphibole (tremolite) asbestos which is a known carcinogen as well as a known contributor to mesothelomia and asbestosis. Change is not always progress.

Lubricants

Lubricants are used for a variety of purposes ranging from petroleum jelly use, to automotive engines in transportation, to heavy machinery in agricultural and industrial applications. The very low hardness of minerals such as molybdenite and graphite permits their use as dry lubricants. In all cases, lubricants reduce friction and abrasion, minimize heat generation and extend the usability of materials. Molybdenite is widely used in manufacturing oils and greases to reduce friction and abrasion of machinery. Graphite is widely used in grease, brake linings and pencils. The hardness of a pencil (no. 1 being soft and no. 4 being hard) relates to the ratio of graphite to clay. Soft pencil lead contains more graphite and hard pencil lead contains more clay. Other uses for these soft, light-weight and strong minerals include sports equipment such as bicycles, tennis rackets and golf clubs, batteries and military purposes such as in Stealth aircraft.

Pigments

Pigments provide color variations in paints and other materials and have been used for thousands of years. Minerals used as pigments commonly include metallic elements that provide vibrant hues to Earth materials. Table 19.5 lists common mineral pigments used in art and construction materials. Some of the metallic elements, such as arsenic, cadmium, uranium, mercury and lead, are harmful and potentially lethal.

Refractory minerals

Refractory minerals are heat-resistant materials that have very high melting temperatures and are chemically resistant to breakdown. Because refractory minerals do not melt readily they can be used for high temperature industrial applications such as in furnaces or crucibles. Refractory minerals are also used in jet engines, turbines and high speed drills that generate extremely high temperatures. Refractory minerals include magnesite, wolframite, magnesium chromite, olivine, zircon, bauxite, kyanite, sillimanite and andalusite. Many of these minerals are key alloys in the production of high strength steel discussed earlier.

Glass

Quartz is the primary mineral used in the production of **glass**. The purest, highest quality quartz sand occurs in beach or desert dune deposits. Freshwater deposits are preferable as salts corrode metal. High quality quartz must lack color (no iron) and lack refractory elements (Fe, Al, Sn, Zr, Cr). High quality quartz sand, such as from St Peters sandstone, is used in glass making, as foundry sand for making metal molds, and as an abra-

Table 19.5   Common mineral pigments used for providing color to materials.

Pigment color	Pigment minerals or elements
Yellow	Limonite, goethite, spinel, orpiment cadmium, uranium
Blue	Azurite, lazurite, zircon, spinel, cobalt
Green	Malachite, glauconite, epidote, celadonite, atacamite [$Cu_2Cl(OH)_3$], barite, chromium
Red	Hematite, cinnabar, iron, mercury, strontium
Orange	Realgar
Turquoise	Azurite, chrysocolla, turquoise
Copper	Malachite, dioptase, gaspeite [$(Ni,Mg)CO_3$], copper
Blue gray	Vivianite [$Fe_3(PO_4)_28(H_2O)$], aegerine
Purple	Hematite, pyrolusite, cuprite, purpurite (Mn_3PO_4)
Lapis lazuli	Lazurite [$(Na,Ca)_8(AlSiO_4)_6(SO_4,S,Cl)_2$]
Pink	Montmorillonite, lepidolite, rhodonite, rhodochrosite, titanium, manganese
Gray–black	Graphite, magnetite, pyrolusite

sive. While quartz is the dominant glass component, other minerals and elements include feldspar (reduces breakage), trona soda ash (glaze and pigment), fluorite (pigment), barite (scratch resistance), selenium (reduces tint and passage of infrared rays), spodumene and zircon (reduce thermal expansion), celestite (for television glass), borax (for optical lenses) and lead (used in artisian glass works).

19.3   GEMS

Gems form an important part of our lives, culture and economy. Although over 3500 minerals are known to exist, fewer than 100 minerals are considered to be gems. Exactly what are gems? **Gems** are durable, beautiful, somewhat rare, solid substances that, with proper cut and polish, may be used as jewels or for ornamentation. Gems include naturally occurring, hard, crystalline minerals that are the most valuable precious stones. Gems also include mineraloids and manufactured gems produced in a laboratory. **Mineraloid gems** are naturally occurring solids that lack one of the critical aspects of minerals such as crystal form or inorganic origin. The gems amber, ivory and pearls are mineraloids. Amber is both non-crystalline and organic, having formed from plant material. Ivory is non-crystalline and organic, forming the tusks of large animals slaughtered for these gem materials. Pearls are organic substances produced by oysters in the marine environment. **Manufactured gems** consist of treated gems, synthetic gems and imitation gems as well as doublets and triplets. **Treated gems** constitute natural stones "improved" by artificial means such as artificial coloring or staining, heat treatment, irradiation or special mountings. **Synthetic gems** are solid substances produced in a laboratory. One well-known synthetic gem is the isometric zirconium oxide crystal called cubic zirconia. **Imitation gems** resemble gemstones but are composed of inferior materials such as glass or plastics and are commonly referred to as "costume jewelry". **Doublets** and **triplets** consist, respectively, of two or three layers fused together to resemble a single coherent crystal. The doublet or triplet may consist of natural gemstone material or a less expensive imitation substance.

Why are gems deemed so valuable? First and foremost, gems are treasured for their beauty. However, gem value varies somewhat with fashion. Minerals valued today as precious stones may have at one time been considered a nuisance! Platinum, the most precious of metals, was in earlier times cursed for its difficult-to-isolate occurrence with the more highly valued gold and silver. We should all be so cursed.

Gem minerals may be classified based upon relative value. "Precious" gems are those perceived to have the greatest value; these include the non-metals diamond, emerald, ruby, sapphire and alexandrite. Precious metals such as platinum, gold and silver may also be considered precious gemstones. These metallic minerals are commonly used for gem mountings. However, as these metals bear hardnesses ranging from 2.5 to 4.5, they must be alloyed with other metals to increase their hardness and wearability. Pure **platinum** is soft and flexible, with a hardness of 4–4.5. Platinum hardness increases substantially when alloyed with 5–10% iridium. Pure 24 karat **gold** is relatively soft (2.5–3). Alloying 18 parts gold with 6 parts metal alloy yields 18 karat gold, which is significantly harder than 24 karat gold. Alloyed gold results in a color change. Yellow gold, alloyed with silver and copper, becomes progressively lighter in color with increasing silver to copper ratio and becomes darker in color with increasing copper to silver content. White gold is produced by alloying 18 karat gold with nickel, copper and zinc. Green gold is generated by alloying 18 karat gold with silver, nickel and copper. Rolled gold or gold plated refers to a thin veneer of gold overlying a base metal (Kraus and Slawson, 1947). Silver is another soft, flexible precious metal with a hardness of 2.5. Silver must be alloyed with 7.5–10% copper to produce a harder "sterling silver" or "coin silver" (Kraus and Slawson, 1947).

Table 19.6 lists some of the more common non-metal gemstones.

19.3.1   Physical properties of gems

Gem beauty and value results from the specimen's color, play of color, brilliance, purity, transparency, luster, wearability, cut and rarity; many of these properties are interrelated such that when one aspect is diminished, a cascading effect occurs reducing the gem's beauty and value.

Color refers to the degree of hue, tone and intensity of light within a substance. **Hue** is a function of the frequency of light and is described as red, orange, yellow, blue, green, indigo and violet. **Tone** varies from very light to very dark. **Intensity** refers to the saturation or purity of a color. It is generally true that for colored gems, the greater the hue, tone and intensity, the greater the value of the stone. Notwithstanding this general rule, some colorless gems are more prized for exactly that attribute – the absence of color. This is particularly true with diamonds.

Gems consist of both metallic and non-metallic minerals. Metallic minerals include gold, platinum, silver, titanium, copper, brass (copper and zinc) and bronze (copper and tin). Even though most gems consist of non-metals, metals play a vital role in imparting color to nearly all gems. Metallic elements behave as brilliant dyes in transforming colorless minerals into minerals exhibiting hue, tone and saturation. For example, the metals listed in Table 19.7 are responsible for different gemstone colors. Let us consider how optical properties affect gem quality.

The **refractive index** (RI) refers to the ratio between the velocity of light through air and the velocity of light through a mineral substance. Minerals with a high refractive index (diamond RI = ~2.4) tend to be more brilliant than minerals with lower refractive indices (quartz RI = ~1.5). **Brilliance** refers to the tendency of a mineral to sparkle or radiate transmitted and reflected light outward in all directions, producing a brilliant display of color and light. Diamonds, with their adamantine luster, are most noted for brilliance. Brilliance is dependent upon naturally occurring physical properties such as dispersion. **Dispersion** refers to the ability to split light into its component colors based upon the varying wavelengths and velocities of light. The rainbow effect produced by white light penetrating a prism is an example of dispersion.

Gems may be isotropic or anisotropic. Isotropic minerals, which crystallize in the isometric system, have the same refractive index in different directions and tend to have a constant color. Anisotropic minerals crystallize in non-isometric crystal systems and are characterized by having different refractive indices in different directions. Anisotropic gems tend to display **pleochroism**, characterized by changes in color in different directions. The optical properties of isotropic and anisotropic minerals are discussed in more detail in Chapter 6.

Play of color refers to the ability of some anisotropic gems to display multiple changing colors when subjected to the transmission (light passing through a mineral) and reflection (light bouncing off a surface) of light. Gems such as pearls and opals display a "mother of pearl" opalescence – a play of color characterized by an intermingled rainbow-like assortment of green, yellow, orange and blue. Acicular-fibrous gems such as tiger's

Table 19.6   Some of the more common gems.

Mineral	Chemical formula	Gem variety
Aragonite	$CaCO_3$	Pearls
Beryl	$Be_3Al_2(Si_6O_{18})$	Emerald (chromium–green), aquamarine (blue), helliodore (yellow)
Chrysoberyl	$BeAl_2O_4$	Alexandrite (green by day, red by night), cat's eye (yellow–green–brown)
Chrysocolla	$Cu_4H_4O_{10}(OH)_8$	Chrysocolla (green)
Corundum	Al_2O_3	Ruby (red), sapphire (blue)
Diamond	C	Diamond
Garnet	$A_3B_2(SiO_4)_3$ $A = Ca, Mn, Mg, Fe$ $B = Al, Fe, Cr$	Demantoid (andradite garnet: green–yellow), hessonite (grossular garnet: green–yellow), uvarovite (chromium–green), almandine, pyrope, spessartine (red)
Jadeite	$NaAlSi_2O_6$	Jade
Lazurite	$(Na,Ca)_8(AlSiO_4)_6(SO_4,S,Cl)_2$	Lapis lazuli (lazurite with calcite, pyroxene)
Malachite	$Cu_2CO_3(OH)_2$	Malachite (green)
Microcline	$KalSi_3O_8$	Amazonite
Olivine	$(Fe,Mg)_2SiO_4$	Peridot
Plagioclase	$(Na,Ca)AlSi_3O_8$	Labradorite (iridescent), moonstone (opalesque), sunstone (golden shine)
Quartz	SiO_2	Agate (banded chalcedony), amethyst (purple), aventurine (mint green), jasper (red chert), chrysoprase (green chalcedony), citrine (yellow–orange), moss agate (mottled gray), onyx (layered chalcedony), rock crystal quartz, rose quartz, smoky quartz, tiger's eye
Rhodochrosite	$MnCO_3$	Rhodochrosie (pink)
Sodalite	$Na_4(AlSiO_4)_3Cl$	Sodalite (violet-blue)
Spinel	$MgAl_2O_4$	Spinel (white, red, all colors)
Topaz	$Al_2SiO_4(F,OH)_2$	Topaz (yellow, clear, pink, red, green, blue)
Tourmaline	$(Na,Ca)(Li,Mg,Al)(Al,Fe,Mn)_6$ $(BO_3)_3(Si_6O_{18})(OH)_4$	Tourmaline (various colors)
Tremolite-actinolite	$Ca_2Mg_5Si_8O_{22}(OH)_2$	Nephrite jade (amphibole)
Turquoise	$CuAl_6(PO_4)_4(OH)_8 \cdot 4H_2O$	Turquoise
Zircon	$ZrSiO_4$	Zircon (diamond like)
Zoisite	$Ca_2Al_3(Si_3O_{12})(OH)$	Tanzanite (violet blue)

eye (yellow-orange quartz) and cat's eye (chrysoberyl) display **chatoyancy**, which is a translucent play of yellow, orange, brown and gray colors. Another precious variety of chrysoberyl, alexandrite, displays **photochroism** whereby colors change under different lighting conditions; for example, alexandrite appears green in natural light (sunlight) but turns ruby red in artificial (incandescent) light. Why? Incandescent light is enriched in red wavelengths of light and is relatively depleted in blue–green wavelengths.

Purity is a measure of the lack of impurities (other minerals or substances) and flaws

Table 19.7   Metals that commonly provide vibrant color to gemstones.

Element	Gem
Chromium	Ruby, emerald, pyrope, grossular, uvarovite, tourmaline
Copper	Dioptase, malachite, azurite
Iron	Sapphire, aquamarine, olivine, almandine garnet, amethyst quartz
Manganese	Morganite, pink tourmaline, spessartite garnet
Vanadium	Green beryl, blue zoisite, garnet
Titanium	Blue sapphire

in gemstones. Flaws include fractures, inclusions, notched corners, cleavage or earthy, granular, rough or splintery textures. Impure, flawed gems are characterized by reduced transparency and brilliance. Fractures, bubbles and inclusions of other substances reduce the transmission of light through the gemstone, thus reducing their **transparency**. Impurities and flaws also reduce **luster**, the appearance of a mineral in reflected light, by reducing the quality and intensity of reflected light. In contrast, pure, unflawed, transparent minerals with high refractive indices display an adamantine or highly vitreous luster (Chapter 5). Even the most valuable gems have flaws. The most valuable gems, however, are marked by extraordinary color, clarity, transparency, luster, cut, brilliance, hardness and/or rarity.

Wearability refers to the ability of a gem to be worn and exposed to "wear and tear". A mineral with high wearability retains its luster, brilliance and polish and is not chemically or physically diminished through prolonged use. Wearability is dependent upon the **durability** of a gem. Durable gems are hard and tough. **Hardness** is defined as the ability to resist abrasion or scratching. **Toughness**, the resistance to breakage, is distinctly different to hardness. Very hard minerals may be deficient in toughness due to properties such as cleavage. For example, diamonds have a hardness of 10, but also contain four sets of cleavage planes that reduce their toughness. Toughness may also be reduced by susceptibility to chemical weathering, such as the tendency of some minerals to break down in acidic solutions and others to oxidize when exposed to the atmosphere.

No story of gem properties and value would be complete without a description of the modifying role played by humans. Cut, brilliance and rarity are at least partially due to the impacts of humans. **Cut** refers to the manner in which a jeweler has sliced the mineral specimen to enhance its natural beauty and increase its transparency, luster and brilliance. The two most common types of cuts are the cabochon and the faceted (brilliant) cuts. The **cabochon cut** features a smooth, rounded top and a flat base, resembling the form of a contact lens. A **faceted cut**, which is highly prized in diamonds, consists of flat, polished, planar surfaces (facets) bounded by sharp angles that result in a sparkling, radiating, adamantine luster. The upper flat surface is called the **top, bezel** or **crown** with a flat **table** in the center of the crown. The base of the faceted cut is referred to as the **pavilion** or **back**. The **girdle** constitutes those faces between the crown and the base. Gem brilliance is closely allayed with the quality of the cut provided by the jeweler (Box 19.1).

Rarity is a measure of the lack of availability of gem-quality specimens. Certainly most gemstones are rare in nature. The most precious gems – diamond, emerald, ruby and sapphire as well as precious metals such as platinum – are deemed exceedingly rare. However, in some cases rarity is due to highly effective marketing (Box 19.1).

19.4   MINERALS AND HEALTH

Just as minerals play pivotal roles in the health of the global economy, minerals also serve vital roles in human health. How are minerals incorporated into our bodies? Rocks containing minerals break down on Earth's surface in the presence of water to produce soils (Chapter 12). Soils contain organic material, weathered rock material and dissolved ions in pore waters. With continued weathering, dissolved ions are distributed throughout the soil profile and are absorbed by plants. Animals such as cows, sheep, fowl and humans eat plant materials – or other animals – and incorporate these dissolved nutrient ions in their bodies. Diet and location determines nutrient intake. Climate and the underlying geology play a critical role in determining the concentration of soil nutrients. Plants grown in limestone soils will have more calcium but less potassium than those grown in granite. Leafy vegetables provide elements such as potassium, iron, calcium, magnesium, phosphorous and copper. Cereal grains provide manganese, iron, zinc and chromium to our diet.

Nutritionists classify "minerals" – we would consider them elements – with regard to the amount needed for physiological function. "Macrominerals" are required in relatively large amounts while only trace amounts (parts per million) of "microminerals" are required to satisfy physiological needs within

Box 19.1 Diamonds

Diamonds, which crystallize in the isometric system, are the hardest of minerals. The most common crystal forms are the octahedron and the dodecahedron. Diamonds have octahedral (four-sided) cleavage. Diamonds vary in size from microscopic grains to over 3000 karats in weight. Approximately 75% of all diamonds are of lower quality industrial grade and the remaining 25% are gem-quality stones. Gem-quality diamonds are rated and sold according to the **four Cs**: carat (karat), color, clarity and cut.

1 **Carat** refers to the weight of the diamond. One carat is equivalent to 200 mg. A carat is defined of consisting of "100 points" so that a 300 mg diamond is 1.5 carats or 150 points.
2 **Color** is measured from a D (perfectly colorless) to Z (markedly colored) scale where the following ranges exist:
   - D–F: colorless range; these are area and extremely expensive.
   - G–J: nearly colorless, wherein color is only detected under magnification
   - K–M: faintly yellow, wherein slight yellow hues can be noted by the unaided eye.
   - N–R: very light yellow, where a light yellow color is obvious.
   - S–Z: light yellow, where a yellow tinge is distinctly visible.
3 **Clarity** refers to the degree of flaws within a gem. Flaws such as inclusions, blemishes or fractures diminish the gem's transparency and brilliance. Clarity is measured according to six different classifications, with 1–3 ranges within each classification:
   - **Flawless** (Fl) refers to perfect clarity with no detectable imperfections either internally or on the surface of the crystal. Flawless crystals are rare and exorbitantly priced.
   - **Internally flawless** (IF) crystals are, as the name suggests, internally flawless but contain surface blemishes.
   - **Very, very small inclusions** (VVS_1, VVS_2) refer to crystals with near perfect clarity. A VVS_1 is slightly less flawed than a VVS_2. Very, very small inclusion crystals appear flawless to the naked eye but may be observed by trained gemologists with a 10× magnification hand lens.
   - **Very small inclusions** (VS_1, VS_2) refer to crystals with flaws that may be visible with a 10× magnification hand lens.
   - **Small inclusions** (S_1, S_2) contain inclusions visible to most individuals with a 10× magnification hand lens. However, to the naked eye, the inclusions are not noticeable.
   - **Imperfect** (I_1, I_2, I_3) diamonds contain flaws visible to the naked eye with the flaws progressively increasing from I_1 to I_3.
4 **Cut.** Six major diamond cuts exist with the **brilliant** cut being the most common and as, the name suggests, the most brilliant. The deep, brilliant cut allows for maximum light penetration and dispersion while increasing the adamantine effect cherished in diamonds. The **marquise** cut is a football-shaped cut with a length : width ratio of ~2 : 1. The **emerald** cut is a rectangular form with beveled corner edges. The **pear, oval** and **heart** cuts resemble the forms for which they are named (Figure B19.1).

Buying a diamond?

If you are buying a diamond, first read the preceding paragraphs. If you are still buying a diamond, consider an appropriately affordable karat size (0.5–1 karat) and select a nearly colorless G–J color range and slightly flawed S_1 clarity range. And remember, diamonds are forever – once you buy it you cannot sell it as the value of a previously purchased diamond drops like a rock.

Value due to rarity or effective marketing?

Diamonds have been highly prized for thousands of years. This trend continues. Consider such worn clichés as "diamonds are a girl's best friend" or "diamonds are forever"; these statements –

Box 19.1 *Continued*

embedded in western culture at least – represent the pinnacle of salesmanship. Diamonds are perceived to be extremely rare. However, diamond "rarity" has been due to supply restrictions by a diamond cartel which, by limiting the supply of gem-quality diamonds, has generated artificially high prices. The cartel has been highly effective at limiting the availability of diamonds on the world market and artificially maintaining exorbitant prices since the mid 1800s. Presently the cartel controls ~75% of the global diamond market. To keep the supply artificially low, diamonds from a recently discovered Russian diamond source are largely purchased by the cartel to keep the Russian diamond supply from reaching the market. The supply restrictions, coupled with an unprecedented advertising campaign, have effectively convinced the world that diamonds are the most valued and rarest of gems. Both claims are in reality unfounded. Nevertheless, almost every guy that goes to a jeweler for an engagement ring falls for this marketing trap, present company included.

Recently, the diamond cartel has been challenged by diamond suppliers in the Congo, Angola and Australia. Some of these diamond suppliers are even more unscrupulous, leading to bloody diamond wars in Africa fueled by greed. Returning to one of our earlier questions in this chapter: are diamonds really forever? The answer is "yes" based on two reasons: (1) diamond marketers have convinced consumers that they should never sell their diamonds, and (2) no market value exists for used diamonds. In short, diamonds are a very bad investment; if you buy one, you had better love it (or the person to whom you are giving the diamond) because you will own it forever. In that sense, yes, diamonds are forever.

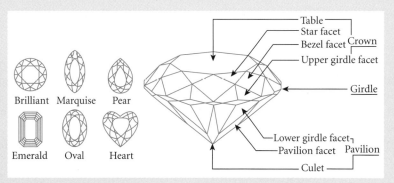

Figure B19.1 Popular diamond cuts.

the body (Table 19.8). Elements provide nutrients and serve as catalysts in chemical reactions within our bodies. For example, Cobalt is found in vitamin B_{12}; zinc is a vital component of enzymes, DNA, RNA, protein synthesis, carbohydrate metabolism and cell growth; copper allows for our bodies to process iron; iron allows our blood to carry oxygen; and magnesium promotes growth and stabilizes human behavior as well as providing a vital role in chlorophyll production in plants (Kesler, 1994). Without these essential ingredients, life as we know it would cease to exist.

On the other hand, excessive exposure to some elements can cause bodily harm. Exposure to amphibole asbestos causes potentially lethal diseases such as lung cancer, mesothelioma and asbestosis due to the ability of asbestos fibers to encapsulate and contami-

Table 19.8 Key elements, referred to by nutritionists as "macro-" and "microminerals", required for human health. Data from Dunn (1983), as cited by Wenk and Bulakh (2004).

Elements	Physiological function
"Macromineral"	
Calcium	Bones, teeth, cartilage, neural transmission, muscle function
Chlorine	Water and electrolyte balance, digestive acid
Magnesium	Regulates chemical reactions, nerve transmission, blood vessels, chlorophyll production in plants. Magnesium sulfate helps prevent cerebral palsy in infants
Phosphorous	Bones, teeth, cartilage, cell function, blood supply
Potassium	Growth, body fluids, muscle contractions, neural transmissions
Sodium	Regulates acid–base balance, neural transmission, blood pressure
Sulfur	Proteins, thiamine, hair, skin
"Micromineral"	
Chromium	Glucose metabolism
Cobalt	Vitamin B_{12}, red blood cells
Copper	Red blood cells, nervous system, metabolism, prevents anemia
Fluorine	Tooth decay, strengthens bones
Iron	Hemoglobin, immune system
Iodine	Thyroid hormones, reproduction
Manganese	Tendon and bones, central nervous system, enzyme reactions
Molybdenum	Growth and enzyme reactions
Selenium	Antioxidant, detoxifies pollutants, prevents cardiovascular disease, cancer
Zinc	Enzymes, red blood cells, sense of taste and smell, immune system, metabolizes alcohol, protects liver

nate lung tissue. Radioactive elements such as uranium, radium, radon, polonium and thorium are known carcinogens. Overexposure to copper can be indicated by green hair, intense overexposure can lead to death. Excess aluminum in drinking water has been correlated with Alzheimer's disease. Metals such as arsenic, mercury, cadmium and chromium are known carcinogens that also attack the nervous system. Prolonged inhalation of beryllium dust causes a chronic lung disease called berylliosis. Prolonged exposure to silicon dust causes the lung disease silicosis. Lead causes both physiological and neurological damage such as nausea, anemia, decreased nerve function, encephalitis, loss of coordination, coma, brain damage and death. Lead causes anemia by substituting for iron in blood and causes bone loss by substituting for calcium and blocking zinc enzymes. While lead has been removed from paint and gasoline since the 1970s in the USA, many products still contain lead. In 2007, toys manufactured in China were noted for their high lead concentration and many were belatedly pulled from shelves. Perhaps most importantly, the burning of fossil fuels such as coal, oil and gas release gases (carbon oxides, nitric oxides, sulfur oxides), metals (mercury, uranium, lead) and particulate materials that contaminate our air, water and soils. Table 19.9 summarizes the detrimental health aspects of elements based on whether they cause cancer (carcinogenic), birth defects (teratogenic) or are radioactive.

In addition to the ill effects due to the use and overexposure of potentially dangerous minerals, the mining of minerals poses significant risks. Mining can be conducted in deep open pits, shallow strip mines or underground mining. Underground mining poses a significant hazard due to underground explosions and mine collapse as well as subsidence on Earth's surface. Strip mines and mountain top mining remove vast quantities of near surface rock resulting in loss of topography and arable land use. Waste material is commonly dumped in low lying valleys and river systems.

Table 19.9 Naturally occurring toxic elements segregated based on whether they cause cancer, birth defects or are radioactive. (After Smith and Huyck, 1999; Wenk and Bulakh, 2004.)

Toxic effect	Element
Carcinogenic	Arsenic, beryllium, cadmium, chromium, cobalt, copper, iron, lead, nickel, tin, zirconium
Teratogenic	Aluminum, arsenic, beryllium, cadmium, copper, indium, lead, lithium, manganese, mercury, molybdenum, selenium, tellurium, thallium, zinc
Radioactive	Radium, radon, thorium, uranium

Open pit mining involves the excavation of holes several kilometers wide and over 1 km deep. Each of these types of mining operation results in immense waste piles. In metal mining and processing, rocks are exposed to cyanide leaching to segregate metallic ores. Pulverized rock contains sulfur which reacts with groundwater and surface water to produce sulfuric acid, resulting in the contamination of water resources. High temperature smelters vaporize metals such as lead, nickel, mercury and copper during the processing, resulting in release to the atmosphere and precipitation into soils and waters.

Prior to the 1970s, mining operations in most countries were largely unregulated with respect to the disposal of hazardous waste materials and impact to land and water. Beginning in the 1970s, environmental regulations were instituted that required mining operations to minimize contamination to the environment. However, these regulations are not universally applied and are challenged by mining companies. As a result, mining has increased in those countries where environmental regulations are relatively lax, resulting in contamination in developing countries. This trend continues – what are the answers? We need a steady supply of mineral resources for economic growth. However, we also need to utilize safe mining techniques to minimize impact to the environment, protect resources and to save lives. Recycling of existing mineral resources is also a major component in this solution.

Minerals have been critically important to human endeavors from ancient times to the present for their use as ores, industrial minerals and gems. Gems are used for personal wear or the adornment of buildings and, while aesthetically beautiful, have little practical impact upon our daily lives. In contrast, ore minerals and industrial minerals are vital to our survival. Wherever you may be at this moment, and in all moments of our lives, we are utilizing industrial minerals in our homes, offices, farms, industrial complexes, public spaces and on and within our bodies. From the moment you get up in the morning and flush a toilet you begin a daily routine that relies upon the availability of minerals: clay and feldspars in porcelain toilets, stainless steel in plumbing, copper in electrical wiring, water stored in rock and sand aquifers, cement sidewalks, gravel roads, metallic bicycles and automobiles, cement and stone buildings and homes – the list goes on and on. We hope this chapter has provided a framework on the global importance of minerals and rocks, as well as their daily role in our lives. This blue planet, with its dynamic tectonism, vital Earth materials and wonderful life forms, is nothing short of paradise.

References

Adams, A.E., Mackenzie, W.S., and Guilford, C. (1984) *Atlas of Sedimentary Rocks Under the Microscope*. Longman Publishers, Harlow, UK, 180 pp.

Aitken, B.G., and Echeverria, L.M. (1984) Petrology and geochemistry of komatiites and tholeiites from Gorgona Island, Columbia. *Contributions to Mineralogy and Petrology*, *86*, 94–105.

Allegre C.J. (1982) Genesis of Archaean komatiites in a wet ultramafic subducted plate. In: Arndt N.T., and Nisbet E.G. (eds) *Komatiites*. Springer-Verlag, Berlin, pp. 495–500.

Altinok, E. (2006) Soil formation beneath the world's oldest known (3.46 Ga) unconformity. *Geological Society of America Annual Meeting Abstracts with Programs*, *38*, 533.

Anderson, C.J., and Jessey, D.R. (2005) Geochemical analysis of the Ricardo volcanics, Southern El Paso Mountains, California. *Geological Society of America Abstracts with Programs*, *37*, 47. http://gsa.confex.com/gsa/2005CD/finalprogram/abstract_84281.htm.

Anderson, D.L. (1989) *Theory of the Earth*. Blackwell Scientific Publications, Oxford, UK, 366 pp.

Anderson, D.L., Sammis, C., and Jordan, T. (1971) Composition and evolution of the mantle and core. *Science*, *171*, 1103–1112.

Anderson, J.L. (1983) Proterozoic anorogenic granite plutonism of North America. In: Medaris, L.G., Byers, C.W., Michelson, D.M., and Shanks, W.C. (eds) *Proterozoic Geology*. Geological Society of America Memoir *161*, pp. 133–154.

Armstrong-Altrin, J.S., Lee, Y.I., Verma, S.P., and Ramasamy, S. (2004) Geochemistry of sandstones from the upper Miocene Kudankulam Formation, southern India: implications for provenance, weathering, and tectonic setting. *Journal of Sedimentary Research*, *74*, 285–297.

Arndt, N.T. (1976) Ultramafic lavas in Munro Township: economic and tectonic implications. In: Strong, D.F. (ed.) *Metallogeny and Plate Tectonics*. Geological Association of Canada, Ottawa, pp. 617–658.

Arndt, N.T. (1994) Komatiites. In: Condie, K.C. (ed.) *Archean Crustal Evolution*. Elsevier Publishers, Amsterdam, pp. 11–44.

Arndt, N.T., and Nesbitt, R.W. (1982) Geochemistry of Munro Township basalts. In: Arndt, N.T., and Nisbet, E.G. (eds) *Komatiites*. Allen and Unwin Publishers, London, pp. 309–330.

Avé Lallemant, H.G., and Guth, L.R. (1990) Role of extensional tectonics in exhumation of eclogites and blueschists in an oblique subduction setting, northeastern Venezuala. *Geology*, *18*, 950–953.

Bailey, D.L. (1974) Continental rifting and alkaline mamatism. In: Sorenson, H. (ed.) *The Alkaline Rocks*. John Wiley and Sons, New York, pp. 148–159.

Bailey, E.B., Clough, C.T., Wright, W.B., Richey, J.E., and Wilson, G.V. (1924) *Tertiary and Post-Tertiary Geology of Mull, Loch Aline and Oban*. Memoirs of the Geological Survey of Scotland, 445 pp.

Baker, D., Dalpe, C., and Poirier, G. (2004) The viscosities of food as analogs for silicate melts. *Journal of Geoscience Education*, *52*, 363–367.

Baldwin, S.L., Monteleone, B.D., Webb, L.E., Fitzgerald, P.G., Grove, M., and Hill, E.J. (2004) Pliocene eclogite exhumation at plate tectonic rates in eastern Papua New Guinea. *Nature*, *431*, 263–267.

Barker, F. (1979) Trondhjemite: definition, environment and hypothesis of origin. In: Barker, F. (ed.) *Trondhjemites, Dacites and Related Rocks*. Elsevier Publishers, New York, pp. 1–11.

Barker, F., Wones, D.R., Sharp, W.N., and Desborough, G.A. (1975) The Pikes Peak Batholith, Colorado, Front Range, and a model for the origin of the gabbro–anorthosite–syenite–potassic granite suite. *Precambrian Research*, *2*, 97–160.

Barker, S.D. (1983) *Igneous Rocks*. Prentice Hall Publishers, Englewood Cliffs, NJ, 417 pp.

Barrow, G. (1893) On an intrusion of muscovite-biotite gneiss in the southeast Highlands of Scotland, and its accompanying metamorphism. *Geological Society of London Quarterly Journal*, *49*, 330–358.

Barrow, G. (1912) On the geology of the lower Deeside and the southern highland border. *Proceedings of the Geologist's Association*, 23, 268–284.

Bathurst, R.G.C. (1975) *Carbonate Sediments and their Diagenesis*, 2nd edn. Elsevier Publishers, Amsterdam, 658 pp.

Beckinsale, R.D. (1979) Granite magmatism in the tin belt of southeast Asia. In: Atherton, M.P., and Tarney, J. (eds) *Origin of Granite Batholiths, Geochemical Evidence*. Shiva Publishing Ltd, Orpington, Kent, UK, pp. 34–44.

Benn, D.I. and Evans, D.J.A. (1998) *Glaciers and Glaciation*. Hodder Arnold Publishers, London, 744 pp.

Best, M.G. (2003) *Igneous and Metamorphic Petrology*, 2nd edn. Blackwell Publishing, Oxford, UK, 752 pp.

Bickle, M.J., Nisbet, E.G., and Martin, A. (1994) Archean greenstone belts are not oceanic crust. *Journal of Geology*, 102, 121–138.

Bird, D.K., and Helgeson, H.C. (1981) Chemical interaction of aqueous solutions with epidote-feldspar mineral assemblages in geologic systems: II. Equilibrium constraints in metamorphic/geothermal processes. *American Journal of Science*, 281, 576–614.

Blatt, H., and Tracey R. (1996) *Petrology: Igneous, Sedimentary, and Metamorphic*, 2nd edn. W.H. Freeman Publishers, New York, 497 pp.

Blatt, H., Tracey, R., and Owens, B.R. (2006) *Petrology: Igneous, Sedimentary and Metamorphic*. W.H. Freeman Publishers, New York, 530 pp.

Blichert-Toft, J., Frey, F.A., and Albarede, F. (1999) Hf isotope evidence for pelagic sediments in the source of Hawaiian basalts. *Science*, 285, 879–882.

Bloomer, S.H., and Hawkins, J.W. (1987) Petrology and geochemistry of boninite series volcanic rocks from the Marianas Trench. *Contributions to Mineralogy and Petrology*, 97, 361–377.

Boggs, S. (2005) *Principles of Sedimentology and Stratigraphy*. Prentice-Hall Publishers, Englewood Cliffs, NJ, 688 pp.

Bottinga, Y., and Weill, D.F. (1970) Densities of liquid silicate systems calculated from partial molar volumes of oxide components. *American Journal of Science*, 269, 169–182.

Boudreau, A.E., Mathez, E.A., and McCallum, I.S. (1986) Halogen geochemistry of the Stillwater and Bushveld Complex: evidence for transport of the platinum-group elements by Cl-rich fluids. *Journal of Petrology*, 27, 967–986.

Boudreau, A.E., Stewart, M.A., and Spivack, A.J. (1997) Stable Cl isotopes and origin of high-Cl magmas of the Stillwater Complex, Montana. *Geology*, 25, 791–794.

Bouma, A.H. (1962) *Sedimentology of Some Flysch Deposits: a Graphic Approach to Facies Interpretation*. Elsevier Publishers, Amsterdam, 168 pp.

Bowen, N.L. (1928) *The Evolution of Igneous Rocks*. Princeton University Press, Princeton, NJ, 332 pp.

Boyd, F.R. (1954) *Amphiboles*. Annual Report of the Director. Geophysical Laboratory, Carnegie Institution of Washington Year Book 53, pp. 108–111.

Brandeis, G., Jaupart, C., and Allegre, C.J. (1984) Nucleation, crystal growth and the thermal regime of cooling magmas. *Journal of Geophysical Research*, 89(B12), 10,161–10,177.

Bravais, A. (1850) Mémoire sur les systèmes formés par les points distribués régulièrement sur un plan ou dans l'espace. *Journal de L'Ecole Polytechnique*, 19, 1–128.

Bryan, S.E., Riley, T.R., Jerram, D.A., Stephens, C.J., and Leat, P.T. (2002) Silicic volcanism: an undervalued component of large igneous provinces and volcanic rifted margins. In: Menzies, M.A., Klemperer, S.L., Ebinger, C.J., and Baker, J. (eds) *Volcanic Rifted Margins*. Geological Society of America Special Paper 362, pp. 99–120.

Buddington, A.F. (1959) Granite emplacement with special reference to North America. *Geological Society of America Bulletin*, 70, 671–747.

Buddington, A.F., and Chapin, T. (1929) *Geology and Mineral Deposits of Southeastern Alaska*. US Geological Survey Bulletin No. 800, 398 pp.

Burbank, D., and Anderson, R. S. (2000) *Tectonic Geomorphology*. Blackwell Science, Oxford, UK, 274 pp.

Burke, K., Dewey, J.F., and Kidd, W.S.F. (1977) World distribution of sutures – the sites of former oceans. *Tectonophysics*, 40, 69–99.

Burke, K., Ashwal, L.D., and Webb, S.J. (2003) New way to map old sutures using deformed alkaline rocks and carbonatites. *Geology*, 31, 391–394.

Cameron, E.M. (1983) Evidence from early Proterozoic anhydrite for sulfur isotopic partitioning in Precambrian oceans. *Nature*, 304, 54–56.

Campbell, I.H., Reiners, P.W., Allen, C., Nicolescu, S., and Upadhyay, R. (2005) He-Pb double dating of detrital zircons from the Ganges and Indus Rivers: implications for quantifying sediment recycling studies. *Earth and Planetary Science Letters*, 237, 402–432.

Cann, J.R. (1970) New model for the structure of the ocean crust. *Nature*, 226, 928–930.

Cann, J.R. (1971) Major element variations in ocean floor basalts. *Philosophical Transactions of the Royal Society of London, Series A*, 268, 495–506.

Cashman, K.V., Sturtevant, B., Papale, P., and Navon, O. (2000) Magmatic fragmentation: In: Sigurdsson, H. (ed.) *Encyclopedia of Volcanoes*. Academic Press, New York, pp. 421–430.

Castillo, P.R. (2006) An overview of Adakite petrogenesis. *Chinese Science Bulletin*, 51(3), 257–268.

Cawthorn, R.G. (1999) Geological models for platinum-group metal mineralization in the Bushveld Complex. *South African Journal of Sciences*, 95, 490–498.

Chang, L.L.Y. (2002) *Industrial Mineralogy: Materials, Processes and Uses*. Prentice Hall Publishers, Englewood Cliffs, NJ, 472 pp.

Chao, E.C.T., Shoemaker, E.M., and Madsen, B.M. (1960) First natural occurrence of coesite from Meteor Crater, Arizona. *Science*, *132*, 220–222.

Chappell, B.W., and Stephens, W.E. (1988) Origin of intracrustal (T-type) granite magmas. *Royal Society of Edinburgh Transactions*, *79*, 71–86.

Chappell, B.W., and White, A.J.R. (1974) Two contrasting granite types. *Pacific Geology*, *8*, 173–174.

Chase, C.G. (1978) Extension behind island-arcs and motions relative to hot spots. *Journal of Geophysical Research*, *83*, 5385–5387.

Chernosky, J.V. (1976) The stability of anthophyllite – a reevaluation based on new experimental data. *American Mineralogist*, *61*, 1145–1155.

Chernosky, J.V. (1979) The stability of anthophyllite in the presence of quartz. *American Mineralogist*, *64*, 294–303.

Chesner, C.A., and Rose, W.I. (1990) Stratigraphy of the Toba Tuffs and the evolution of the Toba Caldera Complex, Sumatra, Indonesia. *Bulletin of Volcanology*, *53*, 343–356.

Chetelat, E. de (1947) La Genese et l'evolution des gisements de nickel de la Nouvelle-Caledonie. *Bulletin de Societie Geologie France, Series 5*, *17*, 105–160.

Chopin, C. (1984) Coesite and pure pyrope in high-grade blueschists of the western Alps: a first record and some consequences. *Contributions to Mineralogy and Petrology*, *86*, 107–118.

Christiansen, R.L. (2001) *The Quaternary and Pliocene Yellowstone Plateau Volcanic Field of Wyoming, Idaho, and Montana*. US Geological Survey Professional Paper No. 729-G, 145 pp.

Chung, S.L., Liu, D., Ji, J., Chu, M.F., Lee, H.Y., Wen, D.J., Lo, C.H., Lee, T.Y., Qian, Q., and Zhang, Q. (2003) Adakites from continental collision zones: melting of thickened lower crust beneath southern Tibet. *Geology*, *31*, 1021–1024.

Coffin, M.F., and Eldholm, O. (1994) Large igneous provinces: crustal structure, dimensions, and external consequences. *Reviews of Geophysics*, *32*(1), 1–36.

Coffin, M.F., Duncan, R.A., Eldholm, O., Fitton, J.G., Grey, F.A., Larsen, H.C., Mahoney, J.J., Saunders, A.D., Schlich, R., and Wallance, P.J. (2006) Large igneous province and scientific ocean drilling status quo and a look ahead. *Oceanography*, *19*, 150–160.

Coleman, D.S., Gray, W., and Glazner, A.F. (2004) Rethinking the emplacement and evolution of zoned plutons: geochronologic evidence for incremental assembly of the Tuolumne Intrusive Suite, California. *Geology*, *32*, 433–436.

Collins, W.J., Beans, S.D., White, A.J.R., and Chappell, B.W. (1982) Nature and origin of A-type granites with particular reference to southeastern Australia. *Contributions to Mineralogy and Petrology*, *80*, 189–200.

Compton, R.S. (1962) *Manual of Field Geology*. John Wiley, New York, 378 pp.

Condie, K.C. (1982) *Plate Tectonics and Crustal Evolution*. Pergamon Press, Oxford, UK, 476 pp.

Coombs, D.S. (1954) The nature and alteration of some Triassic sediments from Southland, New Zealand. *Royal Society of New Zealand Transactions*, *82*(I), 65–109.

Coombs, D.S. (1961) Some recent work on the lower grades of metamorphism. *Australian Journal of Science*, *24*, 203–215.

Coombs, D.S., Ellis, A.J., Fyfe, W.S., and Taylor, A.H. (1959) The zeolite facies with comments on the interpretation of hydrosynthesis. *Geochemica et Cosmochimica Acta*, *17*, 53–107.

Coombs, M.L., and Gardner, J.E. (2004) Reaction rim growth on olivine in silicic melts: implications for magma mixing. *American Mineralogist*, *89*, 748–759.

Courtillot, V., Feraud, G., Maluski, H., Vandamme, D., Moreau, M.G., and Besse, J. (1988) Deccan flood basalts and the Cretaceous/Tertiary boundary. *Nature*, *333*, 843–845.

Cox, D.P., and Singer, D.A. (1986) *Mineral Deposit Models*. US Geological Survey Bulletin No. 1693, 379 pp.

Creaser, R.A., Price, R.C., and Wormald, R.J. (1991) A-type granites revisited: assessment of a residual source model. *Geology*, *19*, 163–166.

Cross, W., Iddings, J.P., Pirsson, L.V., and Washington, H.S. (1902) A quantitative chemicominerological classification and nomenclature of igneous rocks. *Journal of Geology*, *10*, 555–690.

Crowell, J.C. (1952) Probable large lateral slip faulting on San Gabriel Fault. *American Association of Petroleum Geologists Bulletin*, *36*, 2026–2035.

Curray, J.R. (1991) Possible greenschist facies metamorphism at the base of a 22-km sedimentary section, Bay of Bengal. *Geology*, *19*, 1097–1100.

Cutten, H.N.C. (1979) Rappahannock Group: Late Cenozoic sedimentation and tectonics contemporaneous with Alpine Fault movement. *New Zealand Journal of Geology and Geophysics*, *22*, 535–553.

Cutten, H.N.C., Korsch, R.J., and Roser, B.P. (2006) Using geochemical fingerprinting to determine transpressive fault movement history: application to the New Zealand Alpine Fault. *Tectonics*, *25*, TC4014. doi:10.1029/2005TC001842.

Daly, R.A. (1928) Bushveld igneous complex of the Transvaal. *Geological Society of America Bulletin*, *39*, 703–768.

Daly, R.A. (1933) *Igneous Rocks and the Depth of the Earth*. McGraw Hill, New York, 598 pp.

Dalziel, I.W.D., Lawver, L.A., and Murphy, J.B. (2000) Plumes, orogenesis and supercontinental fragmentation. *Earth and Planetary Sciences*, *178*, 1–11.

Davies, G.F. (1992) On the emergence of plate tectonics. *Geology*, *20*, 963–966.

Davies, W.E., and LeGrand, H.E. (1972) Karst of the United States. In: Herak, M., and Stringfield, V.T. (eds) *Karst, Important Karst Regions of the Northern Hemisphere*. Elsevier Publishing, Amsterdam, pp. 467–505.

Davis, G.H. and Reynolds, S.J. (1996) *Structural Geology of Rocks and Regions*, 2nd edn. John Wiley and Sons, New York, 776 pp.

Dawson, J.B. (1962) Sodium carbonate lavas from Oldoinyou Lengai, Tanganyika. *Nature*, *195*, 1065–1066.

Dawson, J.B. (1980) *Kimberlites and their Xenoliths*. Springer-Verlag Publishers, Berlin, 250 pp.

De, A. (1974) Silicate liquid immiscibility in the Deccan traps and its petrogenetic significance. *Geological Society of America Bulletin*, *85*, 471–474.

De Roever, W. (1956) Some differences between post-Paleozoic and older regional metamorphism. *Geologie en Mijnbouw*, *18*, 123–127.

de Wit, M.J. (1998) On Archean granites, greenstones, cratons and tectonics: does the evidence demand a verdict? *Precambrian Research*, *91*, 181–226.

de Wit, M.J., and Ashwal, L.D. (1997) *Greenstone Belts*. Clarendon Publishers, Oxford, 836 pp.

Decelles, P.G. (2006) Geodynamic implications of supersols in the stratigraphic record. *Geological Society of America Abstracts with Programs*, *38*, 533.

Decker, R., and Decker, B. (2006) *Volcanoes*, 4th edn. W.H. Freeman Publishing, New York, 326 pp.

Defant, M.J., and Drummond, M.S. (1990) Derivation of some modern arc magmas by melting of young subducted lithosphere. *Nature*, *347*, 662–665.

D'Lemos, R. (1992) Magma mingling and melt modification between granitic pipes and host diorite, Guernsey, Channel Islands. *Journal of the Geological Society of London*, *149*, 709–720.

Detrick, R.S., Mutter, J.C., Buhl, P., and Kim, I.I. (1990) No evidence from multichannel reflection data for a crustal magma chamber in the MARK area on the Mid Atlantic Ridge. *Nature*, *347*, 61–64.

Dewey, J.F., and Bird, J.M. (1970) Mountain belts and new global tectonics. *Journal of Geophysical Research*, *75*, 2625–2647.

Dick, H.J.B., and Bullen, T. (1984) Chromium spinel as a petrogenetic indicator in abyssal and alpine-type peridotites and spatially associated lavas. *Contributions to Mineralogy and Petrology*, *86*(1), 54–76.

Dickenson, W.R., Beard, L.S., Brakenridge, G.R., Erjavec, J.L., Ferguson, R.C., Inman, K.F., Knepp, R.A., Lindberg, F.A., and Ryberg, P.T. (1983) Provenance of North American Phanerozoic sandstones in relation to tectonic setting. *Geological Society of America Bulletin*, *94*, 222–235.

Dickinson, W.R., and Gehrels, G.E. (2003) U-Pb ages of detrital zircons from Permian and Jurassic eolian sandstones of the Colorado Plateau, USA: paleogeo-graphic implications. *Sedimentary Geology*, *163*, 29–66.

Dickinson, W.R., and Suczek, C.A. (1979) Plate tectonics and sandstone compositions. *American Association of Petroleum Geologists Bulletin*, *63*, 2164–2182.

Dietz, R. (1961) Continental and ocean basin evolution by spreading of the sea floor. *Nature*, *190*, 854–857.

Drummond, M., and Defant, M. (1990) A model for trondjhemite-tonalite-dacite genesis and crystal growth via slab melting: Archean to modern comparisons. *Journal of Geophysical Research*, *95*, 21505–21521.

Ducea, M. (2001) The California arc: thick granitic batholiths, eclogitic residues, lithospheric-scale thrusting, and magmatic flare-ups. *GSA Today*, *11*(11), 4–10.

Dunham, R.J. (1962) Classification of carbonate rocks according to depositional texture. In: Ham, J. (ed.) *Classification of Carbonate Rocks*. American Association of Petroleum Geologists Memoir No. 1, pp. 108–121.

Dunn, M.D. (1983) *Fundamentals of Nutrition*. CBI Publishers, Boston, MA, 581 pp.

Durand, S.R., and Sen, G. (2004) Preeruption history of the Grande Ronde Formation lavas, Columbia River Basalt Group, American Northwest: evidence from phenocrysts. *Geology*, *32*, 293–296.

Dyer, M.D., Gunter, M.E., and Tasa, D. (2008) *Mineralogy and Optical Mineralogy*. Mineralogical Society of America, Chantilly, VA, 708 pp.

Echeverria, L. (1980) Tertiary or Mesozoic komatiites from Gorgona Island, Columbia: field relations and geochemistry. *Contributions to Mineralogy and Petrology*, *73*, 253–266.

Echeverria, L., and Aitken, B.G. (1986) Pyroclastic rocks: another manifestation of ultramafic volcanism on Gorgona Island, Columbia. *Contributions to Mineralogy and Petrology*, *92*, 428–436.

Eckstrand, O.R., and Hulbert, L. (2007) Magmatic nickel-copper-platinum group element deposits. In: Goodfellow, W.D. (ed.) *Mineral Deposits of Canada: a Synthesis of Major Deposit Types, District Metallogeny, the Evolution of Geological Provinces, and Exploration Methods*. Geological Association of Canada, Mineral Deposits Division, Special Publication No. 5, pp. 205–222.

Eiler, J.M., Schiano, P., Kitchen, N., and Stolper, E.M. (2000) Oxygen-isotope evidence for recycled crust in the sources of mid-ocean-ridge basalts. *Nature*, *403*, 530–534.

Embry, A.F., and Klovan, J.E. (1971) A late Devonian reef tract on northeastern Banks Island Northwest Territories. *Bulletin Canadian Petroleum Geologists*, *19*, 730–781.

Enami, M., and Banno, S. (2000) Major rock-forming minerals in UHP metamorphic rocks. In: Ernst, W.G., and Liou, J.G. (eds) *Ultra-high Pressure Meta-*

morphism and Geodynamics in Collision-type Orogenic Belts. Final report of the task group III-6 of the International Lithosphere Project, International Book Series, Vol. *4*, pp. 207–215.

Ernst, W.G. (1975) *Metamorphism and Plate Tectonic Regimes: Benchmark Papers in Geology.* Halsted Press, New York, 440 pp.

Ernst, W.G. (1988) Tectonic history of subduction zones inferred from retrograde blueschist P-T paths. *Geology, 16,* 1081–1084.

Ernst, W.G. (2007) Speculations on evolution of the terrestrial lithosphere–asthenosphere system – plumes and plates. *Gondwana Research, 11,* 38–49.

Ernst, W.G., and Liou, J.G. (2000) Overview of UHP metamorphism and tectonics in well-studied collisional orogens. In: Ernst, W.G., and Liou, J.G. (eds) *Ultra-high Pressure Metamorphism and Geodynamics in Collision-type Orogenic Belts.* Final report of the task group III-6 of the International Lithosphere Project, International Book Series, Vol. *4,* pp. 3–19.

Eskola, P. (1915) On the relations between the chemical and mineralogical composition in the metamorphic rocks of the Orijarvi region. *Bulletin of the Commission of Geology Finlande, 44,* 109–143.

Eskola, P. (1920) The mineral facies of rocks. *Norsk Geologisk Tidskrift, 6,* 143–194.

Eskola, P. (1939) Die metamorphen Gesteine. In: Barth, T.F.W., Correns, C.W., and Eskola, P. (eds) *Die Entstehung der Gesteine.* Julius Spring Publishers, Berlin, pp. 263–407.

Evans, A.M. (1993) *Ore Geology and Industrial Minerals,* 3rd edn. Blackwell Publishing, Oxford, 389 pp.

Evans, J.A., and Benn, D.I. (2004) *A Practical Guide to the Study of Glacial Sediments.* Hodder Arnold, London, 270 pp.

Faure, G. (1977) *Principles of Isotope Geology.* Wiley, New York, 464 pp.

Fedorenko, A., Lightfoot, P.C., Naldrett, A.J., Czamanske, G.K., Hawkesworth, C.J., Wooden, J.L., and Ebel, D.S. (1996) Petrogenesis of the Siberian floodbasalt sequence at Noril'sk. *International Geological Review, 38,* 99–135.

Fisher, R.V. (1961) Proposed classification of volcaniclastic sediments and rocks. *Geological Society of America Bulletin, 72,* 1409–1414.

Fisher, R.V. (1966) Rocks composed of volcanic fragments. *Earth Science Reviews, 1,* 287–298.

Fisher, R.V., and Schmincke, H.U. (1984) *Pyroclastic Rocks.* Springer-Verlag, New York, 472 pp.

Fitton, J.G., and Godard, M. (2004) Origin and evolution of magmas on the Ontong Java Plateau. In: Fitton, J.G., Mahoney, J.J., Wallace, P.J., and Saunders, A.D. (eds) *Origin and Evolution of the Ontong Java Plateau.* Geological Society of London Special Publications No. 229, pp. 151–178.

Fliedner, M.M., Klemperer, S.L., and Christensen, N.I. (2000) Three-dimensional seismic model of the Sierra Nevada arc, California, and its implications for crustal and upper mantle compositions. *Journal of Geophysical Research, 105,* 10,899–10,921.

Flugel, E. (2004) *Microfacies of Carbonate Rocks: Analysis, Interpretation and Application.* Springer Publishers, New York, 976 pp.

Folk, R.L. (1951) Stages of textural maturity in sedimentary rocks. *Journal of Sedimentary Petrology, 21,* 127–130.

Folk, R.L. (1959) Practical petrographic classification of limestones. *American Association of Petroleum Geologists Bulletin, 43,* 1–38.

Folk, R.L. (1962) Spectral subdivisions of limestone types. In: Ham, J. (ed.) *Classification of Carbonate Rocks.* American Association of Petroleum Geologists Memoir No. 1, pp. 62–85.

Folk, R.L. (1974) *Petrology of Sedimentary Rocks.* Hemphill Publishing, Austin, TX, 182 pp.

Folk, R.L., Andrews, P.B., and Lewis, D.W. (1970) Detrital sedimentary rock classification and nomenclature for use in New Zealand. *New Zealand Journal of Geology and Geophysics, 13,* 937–968.

Foulger, G.R., Natland, J.H., Presnall, D.C., and Anderson, D.L. (eds) (2005) *Plates, Plumes, and Paradigms.* Geological Society of America Special Vol. *388,* 861 pp.

Fowler, A.D., Berger, B., Shore, M., Jones, M.I., and Ropchan, J. (2002) Supercooled rocks: development and significance of varioles, spherulites, dendrites and spinifex in Archaean volcanic rocks, Abitibi Greenstone belt, Canada. *Precambrian Research, 115,* 311–328.

Francheteau, J., Needham, H.D., Choukroune, P., Juteau, T., Seguret, M., Ballard, R.D., Fox, P.J., Normark, W., Caranza, A., Cordoba, D., Guerrero, J., Rangin, C., Bougault, H., Cambon, P., and Hekinian, R. (1979) Massive deep-sea sulfide ore deposits discovered by submersible on the East Pacific Rise: Project RITA, 21°N. *Nature, 277,* 523–528.

Francis, P. (1993) *Volcanoes: A Planetary Perspective.* New York, Oxford, 443 pp.

Franklin, J.M., Gibson, H.L., Galley, A.G., and Jonasson, I.R. (2005) Volcanogenic massive sulfide deposits. In: Hedenquist, J.W., Thompson, J.F.H., Goldfarb, R.J., and Richards, J.P. (eds) *Economic Geology 100th Anniversary Volume.* Society of Economic Geologists, Littleton, CO, pp. 523–560.

Fretzdorff, S., Livermore, R.A., Devey, C.W., Leat, P.T., and Stoffers, P. (2002) Petrogenesis of the back-arc East Scotia Ridge, South Atlantic Ocean. *Journal of Petrology, 43,* 1435–1467.

Frey, F.A., and Haskins, L.A. (1964) Rare earths in oceanic basalts. *Journal of Geophysical Research, 69,* 775–779.

Frey, F.A., Green, D.H., and Roy, S.D. (1979) Integrated models of basalt petrogenesis: A study of quartz tholeiites to olivine melilitities from southeastern Australia utilizing geochemical and experimental petrological data. *Journal of Petrology, 19*, 463–513.

Frey, F.A., Wise, W.S., Garcia, M.O., West, H., Kwon, S.T., and Kennedy, A. (1990) Evolution of Mauna Kea volcano, Hawaii: the transition from shield building to the alkalic cap stage. *Journal of Geophysical Research, 95*, 1271–1300.

Friedman, G.M. (ed.) (1981) *Diagenesis of Carbonate Rocks: Cement–Porosity Relationships*. Society for Economic Paleontologists and Mineralogists, Tulsa, OK, 295 pp.

Friedman, G.M., and Sanders, J.E. (1978) *Principles of Sedimentology*. John Wiley and Sons, New York, 792 pp.

Fruh-Green, G.L., Kelley, D.S., Bernasconi, S.M., Karson, J.A., Ludwig, K.A., Butterfield, D.A., Boschi, C., and Proskurowski, G. (2003) 30,000 years of hydrothermal activity at the Lost City Vent Field. *Science, 301*, 495–498.

Fryer, P., Sinton, J.M., and Philpotts, J.A. (1981) Basaltic glasses from the Marina Trough. In: Hussong, D.M., Uyeda, S., Knapp, S.R., Ellis, H., Kling, S., and Natland, J. (eds) *Initial Reports of the Deep Sea Drilling Project*, Vol. 60. US Government Printing Office, Washington, DC, pp. 601–609.

Furukama, Y. (1993) Magmatic processes under arcs and formation of the volcanic front. *Journal of Geophysical Research, 98*, 8309–8319.

Furnes, H., de Wit M., Staudigel, H., Rosing, M., and Muehlenbachs, K. (2007) A vestige of Earth's oldest ophiotlite. *Science, 315*, 1704–1707.

Fyfe, W.S., Turner, F.J., and Verhoogen, J. (1958) *Metamorphic Reactions and Metamorphic Facies*. Geological Society of America Memoir 73, 259 pp.

Galley, A.G., Hannington, M.D., and Jonasson, I.R. (2007) Volcanogenic massive sulphide deposits. In: Goodfellow, W.D. (ed.) *Mineral Deposits of Canada: a Synthesis of Major Deposi Types, District Metallogeny, the Evolution of Geological Provinces, and Exploration Methods*. Geological Association of Canada, Mineral Deposits Division, Special Publication No. 5, pp. 141–161.

Gast, P.W. (1968) Trace element fractionation and the origin of tholeiitic and alkaline magma types. *Geochemica et Cosmochimica Acta, 32*, 1057–1086.

Ghose, N.C. (1976) Composition and origin of the Deccan basalts. *Lithos, 9*, 65–73.

Gibbons, W. (2000) Amphibole asbestos in Africa and Australia: geology, health hazard and mining legacy. *Journal of the Geological Society of London, 157*, 851–858.

Gibbs, J.W. (1928) *The Collected Works of J. Willard Gibbs, Vol. I. Thermodynamics*: Yale University Press, New Haven, CT, 438 pp.

Gilbert, G.K. (1877) *Report on the Geology of the Henry Mountains*. Department of the Interior, US Geographical and Geological Survey of the Rocky Mountain Region, Washington, DC, 170 pp.

Gill, J.B. (1981) *Orogenic Andesites and Plate Tectonics*. Springer-Verlag, New York, 390 pp.

Glazner, A.F., Bartley, M.M., Coleman, D.S., Gray, W., and Taylor, R.Z. (2004) Arc plutons assembled over millions of years by amalgamation from small magma chambers? *GSA Today, 14*, 4–11.

Goldich, S.S. (1938) A study in rock-weathering. *Journal of Geology, 46*, 17–58.

Gomez-Tuena, A., Langmuir, C.H., Goldstein, S.L., Straub, S.M., and Ortega-Gutierrez, F. (2007) Geochemical evidence for slab-melting in the Trans-Mexican volcanic belt. *Journal of Petrology, 48*, 537–562.

Goodfellow, W.D., and Lydon, J. (2007) Sedimentary-exhalative (SEDEX) deposits. In: Goodfellow, W.D. (ed.) *Mineral Deposits of Canada: a Synthesis of Major Deposit Types, District Metallogeny, the Evolution of Geological Provinces, and Exploration Methods*. Geological Association of Canada, Mineral Deposits Division, Special Publication No. 5, pp. 163–183.

Grabau, A.W. (1913) *Principles of Stratigraphy*. A.G. Seiler, New York, 1185 pp.

Green, D.H., and Ringwood, A.J. (1967) An experimental investigation of the gabbro to eclogite transformation and its petrologic application. *Geochemica et Cosmochimica Acta, 31*, 767–833.

Green, D.H., Nicholls, I.A., Viljoen, R., and Viljoen, M. (1975) Experimental demonstration of the existence of peridotitic liquids in earliest Archean magmatism. *Geology, 3*, 11–14.

Greenwood, H.J. (1963) The synthesis and stability of anthophyllite. *Journal of Petrology, l4*, 317–351.

Grim, R.E. (1968) *Clay Mineralogy*, 2nd edn. McGraw-Hill, New York, 596 pp.

Grove, D.I., Archibald, N.J., Bettenay, L.F., and Binns, R.A. (1978) Greenstone belts as ancient marginal basins or ensialic rift zones. *Nature, 273*, 460–461.

Grove, T.L., and Kinzler, R.J. (1986) Petrogenesis of andesites. *Annual Review of Earth and Planetary Sciences, 14*, 417–454.

Grove, T.L., de Wit, M.J., Dann, J.C. (1997) Komatiites from the Komatii type section, Barberton, South Africa. In: de Wit, M.J., and Ashwal, L.D. (eds) *Greenstone Belts*. Oxford University Press, Oxford, pp. 436–450.

Gunter, M.E. (1994) Asbestos as a metaphor for teaching risk perception. *Journal of Geological Education, 42*, 17–24.

Haas, H., and Holdaway, M.J. (1973) Equilibria in the system Al_2O_3–SiO_2–H_2O involving the stability limits of pyrophyllite. *American Journal of Science, 273*, 449–464.

Ham, W.E. (ed.) (1962) *Classification of Carbonate Rocks*. American Association of Petroleum Geologists Memoir No. 1, 270 pp.

Hamade, T., Konhauser, K.O., Raiswell, R., Goldsmith, S., and Morris, R.C. (2003) Using Ge/Si ratios to decouple iron and silica fluxes in Precambrian banded iron formations. *Geology, 31*, 35–38.

Hambrey, M., and Alean, J. (2004) *Glaciers.* Cambridge University Press, Cambridge, UK, 394 pp.

Hamilton, W. (1998) Archean magmatism and deformation were not products of plate tectonics. *Precambrian Research, 91*, 143–179.

Hamilton, W. (2003) An alternative Earth. *GSA Today, 13*(11), 4–12.

Hamilton, W. (2007) Comment on a vestige of Earth's oldest ophiolite. *Science, 318*, 746.

Handin, J. (1966) Strength and ductility. In: Clark, S.P., Jr (ed.) *Handbook of Physical Constraints.* Geological Society of America Memoir 97, pp. 223–289.

Hannington, M.D., Herzig, P.M., Scott, S.D., Thompson, G., and Rona, P.A. (1991) Comparative mineralogy and geochemistry of gold-bearing sulfide deposits on the mid-ocean ridges. *Marine Geology, 101*, 217–248.

Harker, A. (1909) *The Natural History of Igneous Rocks.* McMillan Publishers, New York, 384 pp.

Harker, A. (1932) *Metamorphism: a Study of the Transformation of Rock Masses.* Methuen Publishers, London, 376 pp.

Harms, J.C., Southard, J.B., and Walker, R.G. (1982) *Structures and Sequences in Clastic Rocks.* Society of Economic Paleontologists and Mineralogists Short Course No. 9. Society of Economic Paleontologists and Mineralogists, Tulsa, OK, 851 pp.

Hashimoto, M. (1966) On the prehnite-pumpellyite metagraywacke facies. *Geological Society of Japan Journal, 72*, 253–265.

Hatcher, R.D. (1995) *Structural Geology: Principles, Concepts and Problems*, 2nd edn. Prentice Hall, Englewood Cliffs, NJ, 528 pp.

Hawkesworth, C.J., and Kemp, A.I.S. (2006) Evolution of the continental crust. *Nature, 443*, 811–817.

Hawkesworth, C.J., Gallagher, K., Hergt, J.M., and Keynes, M. (1993) Mantle and slab contributions in arc magmas. *Annual Review of Earth and Planetary Sciences, 21*, 175–204.

Hawkins, J., Bloomer, S.H., Evans, C.A., and Melchior, J.T. (1984) Evolution of intra-oceanic arc-trench systems. *Tectonophysics, 102*, 175–205.

Heinrich, E.Wm. (1965) *Microscopic Identification of Minerals.* McGraw Hill, New York, 427 pp.

Herzberg, C.T. (1992) Depth and degree of melting of komatiites. *Journal of Geophysical Research, 97*, 4521–4540.

Herzberg, C.T. (1993) Lithosphere peridotites of the Kaapvaal craton. *Earth and Planetary Science Letters, 120*, 13–29.

Herzig, P.M., and Hannington, M.D. (1995) Polymetallic massive sulfides at the modern seafloor: a review. *Ore Geology Reviews, 10*, 95–115.

Hess, H.H. (1962) History of the ocean basins. In: Engel, A.D.J., James, H.L., and Leonard, B.F. (eds) *Petrologic Studies: a Volume in Honor of A. F. Buddington.* Geological Society of America Publications, pp. 599–620.

Hess, P.C. (1989) *Origins of Igneous Rocks.* Harvard University Press, Cambridge, MA, 336 pp.

Higgins, M.W. (1971) *Cataclastic Rocks.* United States Geological Survey Professional Paper No. 687.

Hirschmann, M.C., Kogiso, T., Baker, M.B., and Stolper, E.M. (2003) Alkalic magmas generated by partial melting of garnet pyroxenite. *Geology, 31*, 481–484.

Hjulstrom, F. (1939) Transportation of detritus by moving water. In: Trask, P.B. (ed.) *Recent Marine Sediment.* American Association of Petroleum Geologists, Tulsa OK, pp. 5–31.

Hoatson, D.M., Jaireth, S., and Jazues, A.L. (2006) Nickel sulfide deposits in Australia: characteristics, resources and potential. *Ore Geology Reviews, 29*, 177–241.

Hobbs, B.E., Means W.D., and Williams, P.E. (1976) *An Outline of Structural Geology.* John Wiley and Sons, New York, 571 pp.

Hoffman, P.F., Kaufman, A.J., Halverson, G., and Schrag, D.P. (1998) A Neoproterozoic snowball Earth. *Science, 281*, 1342–1346.

Hofmann, A.W. (1997) Mantle geochemistry: the message from oceanic volcanism. *Nature, 385*, 219–229.

Hofmann, A.W., and White, W.M. (1982) Mantle plumes from ancient ocean crust. *Earth and Planetary Science Letters, 57*, 421–436.

Holder, M.T. (1979) An emplacement mechanism for post-tectonic granites and its implications for their geochemical features. In: Atherton, M.P., and Tarney, J. (eds) *Origin of Granite Batholiths: Geochemical Evidence.* Shiva Publishing, Orpington, Kent, UK, pp. 116–128.

Hooper, P.R. (1982) The Columbia River basalts. *Science, 215*, 1463–1468.

Hower, J.E., Eslinger, M.E., Hower, M.E., and Perry, E.A. (1976) Mechanism of burial metamorphism of argillaeous sediments: I. Mineralogical and chemical evidence. *Geological Society of America Bulletin, 87*, 725–737.

Hsu, K.J. (1983) *The Mediterranean as a Desert: a Voyage of the Glomar Challenger.* Princeton University Press, Princeton, NJ, 197 pp.

Hugget, R. (2002) Fundamentals of Geomorphology. Routeledge Publishers, London, 336 pp.

Humphris, S.E., Herzig, P.M., Miller, D.J., Alt, J.C., Becker, K., Brown, D., Brugmann, G., Chiba, H., Fouquet, Y., Gemmell, J.B., Guerin, G., Hannington, M.D., Holm, N.G., Honnorez, J.J., Iturrino, G.J., Knott, R., Ludwig, R., Nakumura, K., Petersen, S., Reysenbach, A.L., Rona, P.A., Smith, S., Sturz, A.A., Tivey, M.K., and Zhao, X. (1995) The internal structure of an active sea-floor massive sulfide deposit. *Nature, 377*, 713–716.

Huppert, H.E., and Sparks, R.S.J. (1984) Double-diffusive convection due to crystallization in magmas. *Annual Reviews in Earth Planetary Sciences*, *12*, 11–37.

Hurlbut, C.S., Jr, and Sharp, W.E. (1998) *Dana's Minerals and How to Study Them*, 4th edn. Wiley Publishers, New York, 328 pp.

Hutko, A.R., Lay, T., Garnero, D.J., and Revenaugh, J. (2006) Seismic detection of folded, subducted lithosphere at the core–mantle boundary. *Nature*, *441*, 333–336.

Hyndman, D.W. (1985) *Petrology of Igneous and Metamorphic Rocks*, 2nd edn. McGraw Hill, New York, 786 pp.

Irvine, T.N. (1959) The ultramafic complex and related rocks of Duke Island, Southeastern Alaska. PhD dissertation, California Institute of Technology, Pasadena, CA, 337 pp. http://etd.caltech.edu/etd/available/etd-03102006-161603/unrestricted/Irvine_tn_1959.pdf.

Irvine, T.N. (1982) Terminology of layered intrusions. *Journal of Petrology*, *23*, 127–162.

Irvine, T.N., and Barager, W.R.A. (1971) A guide to the chemical classification of the common volcanic rocks. *Canadian Journal of Earth Sciences*, *8*, 523–548.

Irvine, T.N., Anderson, J.C., and Brooks, C.K. (1998) Included blocks (and blocks within blocks) in the Skaergaard intrusion: geologic relations and the origins of rhythmic modally graded layers. *Geological Society of America Bulletin*, *110*, 1398–1447.

Isacks, B., Oliver, J., and Sykes, L.R. (1968) Seismology and the new global tectonics. *Journal of Geophysical Research*, *73*, 5855–5899.

Ishiwatari, A., and Ichiyama, Y. (2004) Alaskan-type plutons and ultramafic lavas in Far East Russia, Northeast China and Japan. *International Geology Review*, *46*, 316–331.

Jakes, P., and White, A.J.R. (1972) Major and trace element abundances in volcanic rocks of orogenic areas. *Bulletin of the Geological Society of America*, *83*, 29–40.

James H.L. (1954) Sedimentary facies of iron-formation. *Economic Geology*, *9*, 235–293.

James, N.P. (1983) Reefs. In: Scholle, P.A., Bebout, D.G., and Moore, C.H. (eds) *Carbonate Depositional Models*. American Association of Petroleum Geologists Memoir No. 33, pp. 213–228, pp. 345–462.

James, N.P. (1984) Shallowing-upward cycles in carbonates. In: Walker, R.G. (ed.) *Facies Models*. Geological Association of Canada, Geoscience Canada Reprint Series 1, pp. 213–228.

Jeanloz, R. (1993) The mantle in sharper focus. *Nature*, *365*, 110–111.

Jirsa, M., and Southwick, D. (2003) *Mineral Potential and Geology of Minnesota*. Minnesota Geological Survey, University of Minnesota. http://www.geo.umn.edu/mgs/mnpot/MnpotGlg.html (last accessed November 20, 2009).

Kamo, S.L., Davis, D.W., Trofimov, V.R., Czamanske, G.K., Amelin, Y., and Fedorenko, V.A. (2003) Rapid eruption of Siberian flood-volcanic rocks and evidence for coincidence with the Permian–Triassic boundary and mass extinction at 251 Ma. *Earth and Planetary Science Letters*, *214*(1–2), 75–91.

Kappler, A., Pasquero, C., Konhauser, K.O., and Newman, D.K. (2005) Deposition of banded iron formations by anoxygenic phototrophic Fe(II)-oxidizing bacteria. *Geology*, *33*, 865–868.

Karson, J.A. (2002) Geologic structure of the uppermost oceanic crust created at fast- to intermediate-rate spreading centers. *Annual Review of Earth and Planetary Science*, *30*, 347–384.

Kay, R.W. (1978) Aleutian magnesian andesites: melts from subducted Pacific Ocean crust. *Journal of Volcanology and Geothermal Research*, *4*, 117–132.

Keller, G., Adatte, T., Stinnesbeck, W., Stuben, D., Berner, Z., Kramar, U., and Harting, M. (2004) More evidence that the Chicxulub impact predates the K/T mass extinction. *Meteoritics and Planetary Science*, *39*, 1127–1144.

Kelsey, C.H. (1965) Calculation of the C.I.P.W. norm. *Mineralogical Magazine*, *34*, 276–282.

Kennedy, G.C., and Walton, W.S. (1946) Geology and associated mineral deposits of some ultrabasic rock bodies in southeastern Alaska. *United States Geological Survey Bulletin*, *947-D*, 65–84.

Kennedy, W.Q. (1949) Zones of regional metamorphism in the Moine schists of the western Highlands of Scotland. *Geological Magazine*, *86*, 43–56.

Kennet, J.P., and Stott, L.D. (1991) Abrupt deep-sea warming, paleooceanographic changes and benthic extinctions at the end of the Palaeocene. *Nature*, *353*, 225–229.

Kerr, P.F. (1977) *Optical Mineralogy*, 4th edn. McGraw Hill, New York, 452 pp.

Kesler, S.E. (1994) *Mineral Resources, Economics and the Environment*. Macmillan Publishers, New York, 391 pp.

Kiessling, W., Aberhan, M., and Villier, L. (2008) Phanerozoic trends in skeletal mineralogy driven by mass extinctions. *Nature Geoscience*, *1*, 527–530.

Klein, C. (2005) Some Precambrian banded iron-formations (BIFs) from around the world: their age, geologic setting, mineralogy, metamorphism, geochemistry, and origins. *American Mineralogist*, *90*, 1473–1499.

Klein, C., and Buekes, N.J. (1992) Time distribution, stratigraphy, sedimentologic settings and geochemistry of Precambrian iron formations. *Science*, *275*, 136–146.

Klein, C., and Dutrow, B. (2007) *Manual of Mineral Science (Manual of Mineralogy)*, 23rd edn. John Wiley and Sons, New York, 704 pp.

Klein, C., and Hurlbut, C.S., Jr (1985) *Manual of Mineralogy*, 20th edn. Wiley, New York, 644 pp.

Knauth, L.P. (1979) A model for the origin of chert in limestone. *Geology*, 7, 274–277.

Kogiso, T., Hirose, K., and Takahashi, E. (1998) Melting experiments on homogeneous mixtures of peridotite and basalt: application to the genesis of ocean island basalts. *Earth and Planetary Science Letters*, 162, 45–61.

Kogiso, T., Hirschmann, M.M., and Frost, D.J. (2003) High pressure partial melting of garnet pyroxenite: possible mafic lithologies in the source of ocean island basalts. *Earth and Planetary Science Letters*, 216, 603–617.

Kohout, F.A. (1965) A hypothesis concerning cyclic flow of salt water related to geothermal heating in the Floridan aquifer. *New York Academy of Science Transactions, Series 2*, 28, 249–271.

Koppers, A.A.P., Morgan, J.P., Morgan, J.W., and Staudiget, H. (2001) Testing the fixed hotspot hypothesis using $^{40}Ar/^{39}Ar$ age progressions along seamount trails. *Earth and Planetary Science Letters*, 185, 237–252.

Kraus, E.H., and Slawson, C.B. (1947) *Gems and Gem Materials*. McGraw Hill Publishers, New York, 332 pp.

Krumbein, W.C. (1934) Size frequency distribution of sediments. *Journal of Sedimentary Petrology*, 4, 65–77.

Krumbein, W.C., and Graybill, F.A. (1966) *Introduction to Statistical Methods in Geology*. McGraw-Hill, New York, 475 pp.

Kusky, T.M., Li, J.H., and Tucker, R.D. (2001) The Archean Dongwanzi ophiolite complex, North China craton: 2.505-billion-year-old oceanic crust and mantle. *Science*, 292, 1142–1145.

Lange, R. (2002) Constraints on the preeruptive volatile concentrations in the Columbia River flood basalts. *Geology*, 30, 179–182.

Larson, R.V. (1991) Geological consequences of superplumes. *Geology*, 19, 963–966.

Le Pichon, X. (1968) Sea-floor spreading and continental drift. *Journal of Geophysical Research*, 73, 3661–3697.

LeBas, M.J., and Streckeisen, A.L. (1991) The IUGS systematics of igneous rocks. *Journal of the Geological Society of London*, 148, 825–833.

LeBas, M.J., LeMaitre, R.W., Streckeisen, A., and Zanettin, B. (1986) A chemical classification of volcanic rocks based on the total alkali–silica diagram. *Journal of Petrology*, 27, 745–750.

LeMaitre, R.W. (1984) A proposal by the IUGS subcommission on the systematics of igneous rocks for a chemical classification of volcanic rocks based on the total alkali silica (TAS) diagram. *Australian Journal of Earth Science*, 31, 243–255.

LeMaitre, R.W. (2002) *Igneous Rocks: a Classification and Glossary of Terms*, 2nd edn. Cambridge University Press, New York, 236 pp.

Levin, S. (2006) *The Earth Through Time*, 8th edn. John Wiley Publishers, New York, 547 pp.

Lewis, D.W. (1984) *Practical Sedimentology*. Hutchinson Ross, Stroudsburg, PA, 229 pp.

Liou, J.G., Maruyama, S., Wang, X., and Graham, S. (1990) Precambrian blueschist terranes of the world. *Tectonophysics*, 181, 97–111.

Lipman, P.W., and Mullineaux, D.R. (eds) (1981) The 1980 eruption of Mount St Helens, Washington. US Geological Surveys Professional Paper No. 1250, 844 pp.

Lister, G.S., Banga, B., and Feenstra, A. (1984) Metamorphic core complexes of Cordilleran type in the Cyclades, Aegean Sea, Greece. *Geology*, 12, 221–225.

Lofgren, G. (1980) Experimental studies on the dynamic crystallization of silicate melts. In: Hargraves, R.B. (ed.) *Physics of Magmatic Processes*. Princeton University, Press, Princeton, NJ, pp. 487–551.

Lofgren, G. (1983) Effect of heterogeneous nucleation on basaltic textures: a dynamic crystallization study. *Journal of Petrology*, 24, 229–255.

Loiselle, M.C., and Wones, D.R. (1979) Characteristics and origin of anorogenic granites. *Geological Society of America Abstracts with Programs*, 11, 468.

Longman, M.W. (1980) Carbonate diagenetic textures from nearsurface diagenetic environments. *American Association Petroleum Geologists Bulletin*, 64, 461–487.

Lovett, J.V. (1973) *The Environmental, Economic and Social Significance of Drought*. Angus and Robertson Publishers, Melbourne, 318 pp.

Lowe, D.R. and Guy, M. (2000) Slurry-flow deposits in the Britannia Formation (Lower Cretaceous) North Sea: a new perspective on the turbidity current and debris flow problem. *Sedimentology*, 47, 31–70.

Lucas, S.E., and Moore, J.C. (1986) Cataclastic deformation in accretionary wedges: DSDP Leg 66, and onland examples from Barbados and Kodiak Islands. In: Moore J.C. (ed.) *Structural Fabrics in Deep Sea Drilling Project Cores from Forearcs*. Geological Society of America Memoir 166, pp. 89–103.

Luhr, J., and Simkin, T. (1993) *Parícutin: a Volcano Born in a Mexican Cornfield*. Phoenix, Geoscience Press, 427 pp.

Machel, H.G. (2004) Concepts and models of dolomitization. In: Geological Society of London Special Publications No. 235, pp. 7–63.

MacKenzie, W.S., Donaldson, C.H., and Guilford, C. (1984) *Atlas of Igneous Rocks and their Textures*. John Wiley and Sons, New York, 148 pp.

Mahoney, J.J., and Coffin, M.F. (1997) *Large Igneous Provinces: Continental, Oceanic, and Planetary Flood Volcanism*. American Geophysical Union Monograph No. 100, 438 pp.

Mahoney, J.J., Sheth, H.C., Chandrasekharam, and Peng, Z.X. (2000) Geochemistry of flood basalts of the Toranmal section, northern Deccan Traps,

India: implications for regional Deccan stratigraphy. *Journal of Petrology*, 41, 1099–1120.

Marshak, S., and van der Pluijm, B.A. (2003) *Earth Structure: an Introduction to Structural Geology and Tectonics*, 2nd edn. W.W. Norton and Company, New York, 672 pp.

Martin, H. (1986) Effect of steeper Archean geothermal gradient on geochemistry of subduction-zone magmas. *Geology*, 14, 753–756.

Maruyama, S., Cho, M., and Liou, J.G. (1986) Experimental investigations of blue schist–green schist transition equilibria: pressure dependence of Al$_2$O$_3$ contents in sodic amphiboles – a new geobarometer. *Geological Society of America Memoir*, 164, 1–16.

Marzoli, A., Renne, P.R., Piccirillo, E.M., Ernesto, M., Bellieni, G., and DeMin, A. (1999) Extensive 200 million-year-old continental flood basalts of the Central Atlantic Magmatic Province. *Science*, 284, 616–618.

Matthes, S. (1987) *Minerolgie, Eine Einfuhrung in die spezielle Mineralogie, Petrologie, and Lagerstattenkunde*, 2nd edn. Springer-Verlag Publishers, Berlin, 444 pp.

McBirney, A.R. (2007) *Igneous Petrology*, 3rd edn. Jones and Bartlett Publishers, Boston, 550 pp.

McCallum, I.S., Raedeke, L.D., and Mathez, E.A. (1980) Investigations in the Stillwater Complex, Part I. Stratigraphy and structure of the Banded zone. *American Journal of Science*, 280-A, 59–87.

McCallum, I.S., Thurber, M.W., O'Brien, H.E., and Nelson, B.K. (1999) Lead isotopes in sulfides from the Stillwater Complex, Montana: evidence for subsolidus remobilization. *Contributions to Mineralogy and Petrology*, 137, 206–219.

McDougall, I. (1976) Geochemistry and origin of basalt of the Columbia River group, Oregon and Washington. *Geological Society of America Bulletin*, 87, 777–792.

McIlreath, I.A., and Morrow, D.W. (eds) (1990) *Diagenesis*. Geological Association of Canada, Geoscience Canada Reprint Series No. 4, 338 pp.

McKenzie, D.P., and Morgan, W.J. (1969) Evolution of triple junctions. *Nature*, 224, 125–133.

Menzies, J. (2002) *Modern and Past Glacial Environments*, revised student edition. Butterworth-Heinemann Publishers, Amsterdam, 352 pp.

Metz, P., and Winkler, H.G.F. (1963) Experimentelle gesteinsmetamorphose – VII. Die bildung von talc aus kieseligem dolomite. *Geochemica et Cosmochimica Acta*, 27, 431–457.

Meurer, W.P., Willmore, C.C., and Boudreau, A.E. (1999) Metal redistribution during fluid exsolution and migration in the Middle Banded series of the Stillwater Complex, Montana. *Lithos*, 47, 143–156.

Mickelson, D.M., and Attig, J.W. (eds) (1999) *Glacial Processes Past and Present*. Geological Society of America Special Publication No. 337, Denver, CO, 200 pp.

Middleton, G.V., and Wilcox, P.R. (1994) *Mechanics in the Earth and Environmental Science*. Cambridge University Press, Cambridge, UK, 459 pp.

Mitchell, R.H. (1986) *Kimberlites: Mineralogy, Geochemistry and Petrology*. Plenum Press, New York, 442 pp.

Miyashiro, A. (1961) Evolution of metamorphic belts. *Journal of Petrology*, 2, 277–311.

Miyashiro, A. (1973) *Metamorphism and Metamorphic Belts*. John Wiley and Sons, New York, 479 pp.

Miyashiro, A. (1974) Volcanic rock series in island arcs and active continental margins. *American Journal of Science*, 274, 321–355.

Miyashiro, A. (1975) Volcanic rock series and tectonic setting. *Annual Review of Earth and Planetary Sciences*, 3, 251–269.

Miyashiro, A. (1994) *Metamorphic Petrology*. Oxford University Press, New York, 416 pp.

Miyashiro, A., and Shido, F. (1970) Progressive metamorphism in zeolite assemblages. *Lithos*, 6, 13–20.

Möller, A., Appel, P., Mezger, K., and Schenk, V. (1995) Evidence for a 2 Ga subduction zone eclogites in the Usagaran belt of Tanzania. *Geology*, 23, 1067–1070.

Morgan, W.J. (1971) Convection plumes in the lower mantle. *Nature*, 230, 42–43.

Mullen, E.D. (1983) Mn/TiO$_2$/P$_2$O$_5$: a minor element discriminant for basaltic rocks of oceanic environments and its implications for petrogenesis. *Earth and Planetary Science Letters*, 62, 53–62.

Muller, R.D., Royer, J.Y., and Lawyer, L.A. (1993) Revised plate motions relative to the hotspots from combined Atlantic and Indian Ocean hotspot tracks. *Geology*, 21, 275–278.

Muller, R.D., Roest, W.R., Royer, J.Y., Gahagan, L.M., and Sclater, J.G. (1997) A digital age map of the oceanic floor. *Journal of Geophysical Research*, 102, 3211–3214.

Murck, B.W., Skinner, B.J., and Mackenzie, D. (2010) *Visualizing Geology*, 2nd edn. Wiley Publishers/National Geographic, New York, 532 pp.

Nataf, H.C. (2000) Seismic imaging of mantle plumes. *Annual Review of Earth and Planetary Science Letters*, 28, 391–417.

Natland, M.L., and Kuenen, Ph.H. (1951) *Sedimentary History of the Ventura Basin, California, and the Action of Turbidity Currents*. Society for Economic Paleontologists and Mineralogists Special Publication No. 2, pp. 76–104.

Naumann, T.R., and Geist, D.J. (1999) Generation alkalic basalt by crystal fractionation of tholeiitic magma. *Geology*, 27, 423–426.

Nealson, K.H., and Myers, C.R. (1990) Iron reduction by bacteria: a potential role in the genesis of banded iron formations. *American Journal of Science*, 290-A, 35–45.

Nesse, W.D. (2000) *Introduction to Mineralogy*. Oxford University Press, New York, 442 pp.

Nesse, W.D. (2004) *An Introduction to Optical Mineralogy*, 3rd edn. Oxford University Press, New York, 370 pp.

Nolet, S.-I. Karato, S.I., and Montelli, R. (2006) Plume fluxes from seismic tomography. *Earth and Planetary Science Letters*, 248, 685–699.

Nutman, A.P., and Friend, C.R.L. (2007) Comment on a vestige of Earth's oldest ophiolite. *Science*, 318, 746.

O'Brien, P.J., and Rötzler, J. (2003) High-pressure granulites: formation recovery of peak conditions and implications for tectonics. *Journal of Metamorphic Geology*, 21, 3–20.

O'Neil, J., Carlson, R.W., Francis, D., and Stevenson, R.K. (2008) Neodynium-142 evidence for Hadean mafic crust. *Science*, 321, 1828–1831.

Paradis, S., Hannigan, P., and Dewing, K. (2007) Mississippi Valley-type lead-zinc deposits. In: Goodfellow, W.D. (ed.) *Mineral Deposits of Canada: a Synthesis of Major Deposit Types, District Metallogeny, the Evolution of Geological Provinces, and Exploration Methods*. Geological Association of Canada, Mineral Deposits Division, Special Publication No. 5, pp. 185–203.

Parman, S.W., Grove, T.L., and Dann, J.C. (2001) The production of Barberton komatiites in an Archean subduction zone. *Geophysical Research Letters*, 28, 2513–2516.

Passchier, C.W., and Trouw, R.A.J. (2005) *Microtectonics*, 2nd edn. Springer-Verlag, New York, 366 pp.

Paterson, W.S.B. (1999) *Physics of Glaciers*, 3rd edn. Butterworth-Heinemann Publishers, Amsterdam, 496 pp.

Pauling, L. (1929) The principles determining the structure of complex ionic structures. *Journal American Chemical Society*, 51, 1010–1026.

Pearce, J.A., and Cann, J.R. (1973) Tectonic setting of basic volcanic rocks determined using trace chemical analyses. *Earth and Planetary Science Letters*, 19, 290–300.

Pearce, T.H., Gorman, B.E., and Birkett, T.C. (1975) The TiO_2–K_2O–P_2O_5 diagram: a method of discriminating between oceanic and non-oceanic basalts. *Earth and Planetary Science Letters*, 24, 419–426.

Pearce, T.H., Gorman, B.E., and Birkett, T.C. (1977) The relationship between major element chemistry and tectonic environment of basic and intermediate volcanic rocks. *Earth and Planetary Science Letters*, 36, 121–132.

Pearce, J.A., and Peate, D.W. (1995) Tectonic implications of the composition of volcanic arc magmas. *Annual Review of Earth and Planetary Sciences*, 23, 251–285.

Pearson, D.G., Canil, D., and Shirey, S.B. (2003) Mantle samples included in volcanic rocks: xenoliths and diamonds. In: Carlson, R.W. (ed.) (Holland, H.D., and Turekian, K.K., executive eds) *Treatise on Geochemistry*, Vol. 2. Elsevier, New York, pp. 171–275.

Perfit, M.R., Gust, D.A., Bence, A.E., Arculus, R.J., and Taylor, S.R. (1980) Chemical characteristics of island-arc basalts: implications for mantle sources. *Chemical Geology*, 30, 227–256.

Pettijohn, F.J., Potter, P.E., and Siever, R. (1987) *Sands and Sandstones*. Springer-Verlag, New York, 586 pp.

Philpotts, A.R. (1979) Silicate liquid immiscibility in tholeiitic basalts. *Journal of Petrology*, 20, 99–118.

Philpotts, A.R. (1989) *Petrography of Igneous and Metamorphic Rocks*. Prentice Hall, New York, 178 pp.

Philpotts, A.R. (1990) *Principles of Igneous and Metamorphic Petrology*. Prentice Hall, New York, 498 pp.

Philpotts, A.R., and Ague, J.J. (2009) *Principles of Igneous and Metamorphic Petrology*, 2nd edn. Cambridge University Press, New York, 667 pp.

Pierson, T.C., and Costa, J.E. (1987) A rheological classification of subaerial sediment-water flows. In: Costa, J.E., and Wieczorek, G.F. (eds) *Debris Flows/Avalanches: Processes, Recognition and Mitigation*. Geological Society of America Reviews in Engineering No. 7, pp. 1–12.

Pilchin, A. (2003) The role of serpentinization in exhumation of high- to ultra-high-pressure metamorphic rocks. *Earth and Planetary Science Letters*, 237, 815–828.

Pinkster, L.M. (2002) Sudden warming in the past. *Geotimes*, 30, 77.

Pitcher, W.S. (1979) The nature, ascent and emplacement of granite magmas. *Geological Society of London Journal*, 136, 627–662.

Pitcher, W.S. (1982) Granite type and tectonic environment. In: Hsu, K. (ed.) *Mountain Building Processes*. Academic Press, London, pp. 19–40.

Platt, J.P. (1986) Dynamics of orogenic wedges and the uplift of high pressure metamorphic rocks. *Geological Society of America Bulletin*, 97, 1037–1053.

Plumley, W.J. (1948) Black Hills terrace gravels: a study in sediment transport. *Journal of Geology*, 56, 526–577.

Poirier, J.P. (1985) *Creep of Crystals: High-temperature Deformation Processes in Metals, Ceramics and Minerals*. Cambridge Earth Science Series. Cambridge University Press, Cambridge, UK, 285 pp.

Posth, N.R., Hegler, F., Konhauser, K.O., and Kappler, A. (2008) Alternating Si and Fe deposition caused by temperature fluctuations in Precambrian oceans. *Nature Geoscience*, 1, 703–708.

Potter, P.E., Maynard, J.B., and Pryor, W.A. (1980) *Sedimentology of Shale*. Springer-Verlag, New York, 306 pp.

Powers, M.C. (1953) A new roundness scale for sedimentary particles. *Journal of Sedimentary Petrology*, 23, 117–119.

Premo, W.R., Helz, R.T., Zientek, M.L., and Langston, R.B. (1990) U-Pb and Sm-Nd ages for the Stillwater Complex and its associated sills and dikes, Beartooth Mountains, Montana: identification of a parent magma? *Geology*, 18, 1065–1068.

Preston, J., Hartley, A., Mange-Rajetzky, M., Hole, M., May, G., Buck, S., and Vaughan, L. (2002) The provenance of Triassic continental sandstones from the Beryl Field, northern North Sea: mineralogical, geochemical, and sedimentological constraints. *Journal of Sedimentary Research*, 72, 18–29.

Prothero, D.R., and Schwab, F. (2004) *Sedimentary Geology: an Introduction to Sedimentary Rocks and Stratigraphy*, 2nd edn. W.H. Freeman, New York, 557 pp.

Ragland, P.C. (1989) *Basic Analytical Petrology*. Oxford University Press, Oxford, UK, 296 pp.

Rahl, J.M., Reiners, P.W., Campbell, J.H., Nicolescu, S., and Allen, C.M. (2003) Combined single-grain (U-Th)/He and U/Pb dating of detrital zircons from the Navajo Sandstone, Utah. *Geology*, 31, 761–764.

Railsback, L.B. (2003) An earth scientist's periodic table of the elements and their ions. *Geology*, 31, 737–740.

Ramsay, J.G., and Huber, M.I. (1984) *The Techniques of Modern Structural Geology, Vol. 1. Strain Analysis*. Academic Press, Burlington, MA, 307 pp.

Ramsay, J.G., and Huber, M.I. (1987) *The Techniques of Modern Structural Geology, Vol. 2. Folds and Fractures*. Academic Press, Burlington, MA, 391 pp.

Raymond, L.A. (2002) *The Study of Igneous, Sedimentary and Metamorphic Rocks*. McGraw Hill, Boston, 720 pp.

Raymond, L.A. (2007) *Petrology: the Study of Igneous, Sedimentary and Metamorphic Rocks*. Waveland Press, New York, 736 pp.

Reading, H.G. (1996) *Sedimentary Environments: Processes, Facies and Stratigraphy*, 3rd edn. Blackwell Science, Oxford, UK, 704 pp.

Reay, A., and Parkinson, D. (1997) Adakites from Solander Island, New Zealand. *New Zealand Journal of Geology and Geophysics*, 40, 121–126.

Reichow, M.K., Saunders, A.D., White, R.V., Pringle, M.S., Al'Mukhamedov, A.I., and Kirda, N.P. (2002) Ar^{40}/Ar^{39} dates from the West Siberian Basin: Siberian flood basalt province doubled. *Science*, 296, 1846–1849.

Reidell, S.P. (1983) Stratigraphy and petrogenesis of the Grande Rhonde Basalt from the deep canyon country of Washington, Oregon and Idaho. *Geological Society of America Bulletin*, 94, 519–542.

Renne, P.R. (2002) Flood basalts – bigger and badder. *Science*, 296, 1812–1813.

Retallack, G.J. (2001a) Cenozoic expansion of grasslands and climatic cooling. *Journal of Geology*, 109, 407–426.

Retallack, G.J. (2001b) *Soils of the Past: an Introduction to Paleopedology*, 2nd edn. Blackwell Science, Oxford, UK, 404 pp.

Reynolds, R.L., and Goldhaber, M.B. (1978) Origin of a south Texas roll-type uranium deposit: I. Alteration of iron titanium oxide minerals. *Economic Geology*, 73, 1677–1689.

Ringwood, A.E. (1975) *Composition and Petrology of the Earth's Mantle*. McGraw-Hill, New York, 618 pp.

Ritter, D.F., Kochel, R.C., and Miller J.R. (2006) *Process Geomorphology*. Waveland Press, Long Grove, IL, 560 pp.

Rohl, V., Bralower, T.J., and Norris, R.D. (2000) New chronology for the late Paleocene thermal maximum and its environmental implications. *Geology*, 28, 927–930.

Rose, W.I., and Chesner, C.A. (1987) Dispersal of ash in the great Toba eruption, 75 ka. *Geology*, 15, 913–917.

Rubin, D.M., and Carter, C.L. (2006) *Bedforms and Cross-bedding in Animation*. Society for Economic Paleontologists and Mineralogists Society for Sedimentary Geology Atlas Series No. 2 (DVD).

Sandberg, P.A. (1983) An oscillating trend in Phanerozoic non-skeletal carbonate mineralogy. *Nature*, 305, 19–22.

Sano, T., Fujii, T., Deshmukh, S.S., Fukuoka, T., and Aramaki, S. (2001) Differentiation processes of Deccan trap basalts: contributions from geochemistry and experimental petrology. *Journal of Petrology*, 42, 2175–2195.

Schilling, J.-G., Thompson, G., Zajac, M., Evans, R., Johnson, T., White, W., Devine, J.D., and Kingsley, R. (1983) Petrologic and geochemical variations along the Mid-Atlantic Ridge from 27°N to 73°N. *American Journal of Science*, 283, 510–586.

Schirnick, C., van den Bogaard, P., and Schmincke, H.U. (1999) Cone sheet formation and intrusive growth of an oceanic island – the Miocene Tejeda complex on Fran Caniara (Canary Islands). *Geology*, 27, 207–210.

Schmid, R. (1981) Descriptive nomenclature and classification of pyroclastic deposits and fragments: recommendations of the IUGS Subcommision on the Systematics of Igneous Rocks. *Geology*, 9, 41–43.

Scholle, P.A., and Ulmer-Scholle, D.S. (2003) *A Color Guide to the Petrography of Carbonate Rocks*. American Association of Petroleum Geologists Memoir No. 77, 474 pp.

Schulz, E.F., Wilde, R.H., and Albert, M.L. (1954) *Influence of Shape on the Fall Velocity of Sedimentary Particles*. Sedimentation Series Report No. 5. US Army Corps of Engineers, Omaha, NB.

Selley, R.C. (1988) *Ancient Sedimentary Environments*, 3rd edn. Springer-Verlag, New York, 336 pp.

Sha, L.-K. (1995) Genesis of zoned hydrous ultramafic/mafic-silicic intrusive complexes: an MHFC hypothesis. *Earth Science Reviews, 39*, 59–90.

Shand, S.J. (1951) *Eruptive Rocks: Genesis, Composition, Classification and Relation to Ore Deposits*. Hafner Publishing, New York, 488 pp.

Sheridan, R. (1987) Pulsation tectonics as the control of long-term stratigraphic cycles. *Paleoceanography, 2*, 97–118.

Shermer, E.R. (1990) Mechanisms of blueschist facies metamorphism and preservation in an A-type subduction zone, Mount Olympos region, Greece. *Geology, 18*, 1130–1133.

Shervais, J.W. (1982) Ti-V plots and the petrogenesis of modern and ophiolitic lavas. *Earth and Planetary Science Letters, 59*, 101–118.

Shoemaker, E.M. (1979) Synopsis of the geology of Meteor Crater. In: Shoemaker, E.M., and Kieffer, S.W. (eds) *Geology of Meteor Crater, Arizona*. Arizona State University Laboratory for Meteorite Studies No. 17, pp. 1–11.

Sibson, R.H. (1977) Fault rocks and fault mechanisms. *Geological Society of London Journal, 133*, 191–213.

Sillitoe, R.H. (1973) The tops and bottoms of porphyry copper deposits. *Economic Geology, 68*, 799–815.

Simkin, T., Siebert, L., McClelland, L., Bridge, D., Newhall, C., and Latter, J.H. (1981) *Volcanoes of the World*. Smithsonian Institution, Washington, 232 pp.

Simonson, B. (2003) Origin and evolution of large Precambrian iron formations, in extreme depositional conditions: mega end members in geologic time: In: Chan, M.A., and Archer, A.A. (eds) *Sedimentary Giants – Extreme Depositional Environments: Mega End Members in Geologic Time*. Geological Society of America Special Paper 370, pp. 231–244.

Simpson, C. (1986) Determination of movement sense in mylonites. *Journal of Geological Education, 34*, 246–261.

Sinclair, W.D. (2007) Porphyry deposits. In: Goodfellow, W.D. (ed.) *Mineral Deposits of Canada: a Synthesis of Major Deposit Types, District Metallogeny, the Evolution of Geological Provinces, and Exploration Methods*. Geological Association of Canada, Mineral Deposits Division, Special Publication No. 5, pp. 223–243.

Sinton, J.M., Ford, L.L., Chappell, B., and McCulloch, M.T. (2003) Magma genesis and mantle heterogeneity in the Manus back arc basin, Papau New Guinea. *Journal of Petrology, 44*, 159–195.

Sisson, T.W., and Bacon, C.R. (1999) Gas-driven filter pressing in magmas. *Geology, 27*, 613–616.

Skinner, E.M.W. (1989) Contrasting group-1 and group-2 kimberlite petrology: towards a genetic model for kimberlites. In: Ross, J., Jaques, A.L., Ferguson, J., and Green, D.H. (eds) *Proceedings of the 4th International Kimberlite Conference, Perth, Australia*. Geological Society of Australia Special Publication No. 14, pp. 528–544.

Sleep, N.H. (1988) Tapping of melt by veins and dikes. *Journal of Geophysical Research, 93*, 10, 255–10, 272.

Smiley, T.L. (1958) The geology and dating of Sunset Crater, Flagstaff, Arizona. In: Anderson, R.Y., and Harshbarger, J.W. (eds) *Guidebook of the Black Mesa Basin, Northeastern Arizona, New Mexico*. Geological Society Guidebook, 9th Field Conference, pp. 186–190.

Smith, K.S., and Huyck, H.L.O. (1999) An overview of the abundance, relative mobility, bioavailability and human toxicity of metals. In: Plumlee, G.S., and Logsdon, M.J. (eds) *The Environmental Geochemistry of Mineral Deposits*. Society of Economic Geologists, Vol. A, pp. 29–70. Society of Economic Geologists, Littleton, CO.

Smithies, R.H., Champion, D.C., and Cassidy, K.F. (2003) Formation of Earth's Early Archean continental crust. *Precambrian Research, 127*, 89–111.

Sobolev, A.V., Hofmann, A.W., Sobolev, S.V., and Nikogosian, I.K. (2005) An olivine-free mantle source of Hawaiian shield basalts. *Nature, 434*, 590–597.

Soil Survey Staff (1999) *Soil Taxonomy, a Basic System of Soil Classification for Making and Interpreting Soil Surveys*, 2nd edn. United States Department of Agriculture, Natural Resources Conservation Service Agricultural Handbook No. 436, 870 pp.

Sorensen, H. (1974) *The Alkaline Rocks*. John Wiley and Sons, New York, 622 pp.

Sparks, R.S.J. (1978) The dynamics of bubble formation and growth in magma: A review and analysis. *Journal of Volcanology and Geothermal Research, 3*, 1–37.

Sparks, R.S.J. (1986) The dimensions and dynamics of volcanic eruption columns. *Bulletin of Volcanology, 48*, 3–15.

Sparks, R.S.J., and Wilson, L. (1976) A model for the formation of ignimbrite by gravitational column collapse. *Journal of the Geological Society of London, 132*, 441–451.

Sparks, R.S.J., Baker, L., Brown, R.J., Field, M., Schumacker, J., Stripp, G., and Walters, A. (2006) Dynamical constraints on kimberlite volcanism. *Journal of Volcanology and Geothermal Research, 155*, 18–48.

Spera, F. (2000) Physical properties of magma. In: Sigurdsson, H. (ed.) *Encyclopedia of Volcanoes*. Academic Press, New York, pp. 171–190.

Stanley, S.M., and Hardie, L.A. (1999) Hypercalcification: paleontology links plate tectonics and geochemistry to sedimentology. *GSA Today, 9*, 1–7

Steinmann, G. (1905) Geologische Beobachtungen in den Alpen. II. Die Schardtsche Ueberfaltungstheorie und die geologische Bedeutung der Tiefseeabsatze und der ophiolitischen Massengesteine. *Berichte der*

Naturfoschenden Gesellschaft zu Freiberg, iB, 16, 18–67.

Stern, C.R., and Killian, R. (1996) Role of the subducted slab, mantle wedge and continental crust in the generation of adakites from the Andean Austral volcanic zone. *Contributions to Mineralogy and Petrology, 123,* 263–281.

Stern, R.A., and Bleeker, W. (1998) Age of the world's oldest rocks refined using Canada's SHRIMP: the Acasta Gneiss Complex, Northwest Territories, Canada. *Geoscience Canada, 25,* 27–31.

Stern, R.J. (2005) Evidence from ophiolites, blueschists and ultrahigh-pressure metamorphic terranes that the modern episode of subduction tectonics began in Neoproterozoic time. *Geology, 33,* 557–560.

Stern, R.J. (2008) Modern-style plate tectonics began in Neoproterozoic time: an alternative interpretation of Earth's tectonic history. In: Condie, K.C., and Pease, V. (eds) *When Did Plate Tectonics Begin on Planet Earth?* Geological Society of America Special Paper 440, pp. 265–280.

Storey, M., Mahoney, J.J., Kroenke, L.W., and Saunders, A.D. (1991) Are oceanic plateaus the site of komatiite formation? *Geology, 19,* 376–379.

Streckeisen, A. (1973) Plutonic rocks: classification and nomenclature recommended by the IUGS Subcommission on the Systematics of Igneous Rocks. *Geotimes, 18*(10), 26–30.

Streckeisen, A. (1976) To each plutonic rock its proper name. *Earth Science Reviews, 12,* 1–33.

Sun, S., and McDonough, W.F. (1989) Chemical and isotopic systematics of oceanic basalts: implications for mantle composition and processes. In: Saunders, A.D., and Norry, M.J. (eds) *Magmatism in Ocean Basins.* Blackwell Scientific, Boston, pp. 313–345.

Sun, S.S., and Nesbitt, R.W. (1978) Petrogenesis of Archaean ultrabasic and basic volcanics: evidence from rare earth elements. *Contributions to Mineralogy and Petrology, 65,* 301–325.

Suppe, J. (1984) *Principles of Structural Geology.* Prentice Hall Publishers, Upper Saddle River, NJ, 560 pp.

Sutherland, R. (1994) Displacement since the Pliocene along the southern section of the Alpine Fault, New Zealand. *Geology, 22,* 327–330.

Swanson, D.A., and Wright, T.L. (1980) The regional approach to studying the Columbia River Basalt Group. *Geological Society of India Memoir, 3,* 58–80.

Swanson, D.A., and Wright, T.L. (1981) Guide to geologic field trip between Lewiston, Idaho, and Kimberly, Oregon, emphasizing the Columbia River Basalt Group. In: Johnston, D.A., and Donnelly, N.J. (eds) *Guides to Some Volcanic Terranes in Washington, Idaho, Oregon, and northern California.* US Geological Survey Circular No. 838, 189 pp.

Swanson, S.E. (1977) Relation of nucleation and crystal growth rate to the development of granitic textures. *American Mineralogist, 62,* 966–978.

Takahashi, E., Nakajima, K., and Wright, T.L. (1998) Origin of the Columbia River basalts: melting model of a heterogeneous plume head. *Earth and Planetary Science Letters, 162,* 63–80.

Tarduno, J., Bunge, H.P., Sleep, N., and Hansen, U. (2009) The bent Hawaiian–Emperor hotspot track: inheriting the mantle wind. *Science, 324,* 50–53.

Tatsumi, Y., and Eggins, S. (1995) *Subduction Zone Magmas.* Blackwell Publishers, Oxford, UK, 211 pp.

Taylor, H.P., Jr (1967) The zoned ultramafic complexes of southeastern Alaska. In: Wyllie, P.J. (ed.) *Ultramafic and Related Rocks.* John Wiley and Sons, New York, pp. 96–116.

Tejada, M.L.G., Mahoney, J.J., Castillo, P.R., Ingle, S.P., Sheth, H.C., and Weis, D. (2004) Pin-pricking the elephant: evidence on the origin of the Ontong Java Plateau from Pb-Sr-Hf-Nd isotopic characterizations of ODP Leg 192 basalts. In: Fitton, J.G., Mahoney, J.J., Wallace, P.J., and Saunders, A.D. (eds) *Origin and Evolution of the Ontong Java Plateau.* Geological Society of London Special Publications No. 229, pp. 133–150.

Thompson, J.B., Jr (1957) The graphical analysis of mineral assemblages in pelitic schists. *American Mineralogist, 42,* 842–858.

Tilley, C.E. (1925) A preliminary survey of metamorphic zones in the southern Highlands of Scotland. *Quarterly Journal of the Geological Society of London, 81,* 100–112.

Tilling, R.I., Heliker, C., and Wright, T.L. (1987) *Eruptions of Hawaiian Volcanoes: Past, Present and Future.* US Geological Survey General Interest Publications. US Geological Survey, Denver, CO, 54 pp.

Tistl, M., Burgath, K.P., Hohndorf, A., Kreuzer, H., Munoz, R., and Salinas, R. (1994) Origin and emplacement of Tertiary ultramafic complexes in northwest Columbia: evidence from geochemistry and K-Ar, Sm-Nd and Rb-Sr isotopes. *Earth and Planetary Science Letters, 126,* 41–59.

Todd, S.G., Keith, D.W., Le Roy, L.W., Schissel, D.J., Mann, E.L., and Irvine, T.N. (1982) The J-M platinum-palladium reef of the Stillwater Complex, Montana: I. Stratigraphy and petrology. *Economic Geology, 77,* 1454–1480.

Tolan, T.L., Reidel, S.P., Beeson, M.H., Anderson, J.L., Fecht, K.R., and Swanson, D.A. (1989) Revisions to the estimates of the aerial extent and volume of the Columbia River Basalt Group. In: Reidel, S.P., and Hooper, P.R. (eds) *Volcanism and Tectonism in the Columbia River Flood-basalt Province.* Geological Society of America Special Paper 239, pp. 1–20.

Tucker, M. (2001) *Sedimentary Petrology,* 3rd edn. Blackwell Science, Oxford, UK, 272 pp.

Tucker, M., and Wright, V.P. (1990) *Carbonate Sedimentology.* Blackwell Science, Oxford, UK, 496 pp.

Turner, F.J. (1958) Mineral assemblages of individual metamorphic facies. In: Fyfe, W.S., Turner, F.J., and Verhoogen, J. (eds) *Metamorphic Reactions and*

Metamorphic Facies. Geological Society of America Memoir 73, pp. 199–239.

Turner, F.J. (1981) *Metamorphic Petrology: Mineralogical, Field and Tectonic Aspects*. McGraw-Hill, New York, 524 pp.

Turner, F.J., and Verhoogen, J. (1951) *Igneous and Metamorphic Petrology*. McGraw-Hill, New York, 602 pp.

Turner, S., Tonarini, S., Bindeman, I., Leeman, W.P., and Schaefer, B.F. (2007) Boron and oxygen isotope evidence for recycling subducted components over the past 2.5 Ga. *Nature*, 447, 702–705.

Twiss, R.J., and Moores, E.M. (2006) *Structural Geology*, 2nd edn. W.H. Freeman Publishers, New York, 532 pp.

Unruh, J.R., Loewen, B.A., and Moores, E.M. (1995) Progressive arcward contraction of a Mesozoic–Tertiary fore-arc basin, southwestern Sacramento Valley, California. *Geological Society of America Bulletin*, 107, 38–53.

USDA-NRCS (US Department of Agriculture, Natural Resources Conservation Service) (1999) *Soil Taxonomy: a Basic System of Soil Classification for Making and Interpreting Soil Surveys*, 2nd edn. Agriculture Handbook No. 436. US Government Printing Office, Washington, 871 pp.

USGS (United States Geological Survey) (2007) *Mineral Commodity Summaries 2007*. http://minerals.usgs.gov/minerals/pubs/mcs/2007/mcs2007.pdf (last accessed November 7, 2007), 195 pp.

Van der Laan, S.R., Flower, M.F.J., and Van Groos, A.F.K. (1989) Experimental evidence for the origin of boninites: near-liquidus phase relations to 7.5 kbar. In: Crawford, A.J. (ed.) *Boninites and Related Rocks*. Unwin Hyman, London, pp. 112–147.

van der Pluijm, B.A., and Marshak, S. (2004) *Earth Structure: an Introduction to Structural Geology and Tectonics*, 2nd edn. W.W. Norton and Company, New York, 656 pp.

Vance, J.A. (1961) Zoned granite intrusions – an alternative hypothesis of origin. *Geological Society of America Bulletin*, 72, 1723–1728.

Vearncombe, S., and Vearncombe, J.R. (2002) Tectonic controls on kimberlite location, southern Africa. *Journal of Structural Geology*, 24, 1619–1625.

Vermaak, C.F. (1976) The Merensky Reef – thoughts on its environment and genesis. *Economic Geology*, 71, 1270–1298.

Viljoen, M.J., and Viljoen, R.P. (1969) Archean vulcanicity and continental evolution in the Barberton Region, Transvaal. In: Clifford, T.N., and Gass, I.G. (eds) *African Magmatism and Tectonics*. Oliver and Boyd, Edinburgh, pp. 27–39.

Vine, F.J., and Matthews, D.H. (1963) Magnetic anomolies over ocean ridges. *Nature*, 199, 947–949.

Wager, L.R., and Deer, W.A. (1939) Geological investigations in East Greenland. III. The petrology of the Skaergaard intrusion, Kangerdlugssuaq, East Greenland. *Medd Gronland*, 105(4), 1–352.

Walker, G.P.L. (1973) Explosive volcanic eruptions – a new classification scheme. *Geologie Rundschau*, 62, 431–446.

Wallace, P., and Anderson, A.T., Jr (2000) Volatiles in magma. In: Sigurdsson, H. (ed.) *Encyclopedia of Volcanoes*. Academic Press, New York, pp. 149–170.

Walter, M.J. (1998) Melting of garnet peridotite and the origin of komatiite and depleted lithosphere. *Journal of Petrology*, 39, 29–60.

Wei, J.F., Tronnes, R.G., and Scarfe, C.M. (1990) Phase relations of alumina-undepleted and alumina-depleted komatiites at pressures of 4–12 GPa. *Journal of Geophysical Research*, 95, 15,817–15,828.

Wenk, H.-R., and Bulakh, A. (2004) *Minerals: their Constitution and Origin*. Cambridge University Press, Cambridge, UK, 646 pp.

Whalen, J.B., Currie, K.L., and Chappell, B.W. (1987) A-type granites: geochemical characteristics, discrimination and petrogenesis. *Contributions to Mineralogy and Petrology*, 95, 407–419.

White, S.H., de Boorder, H., and Smith, C.B. (1995) Structural controls of kimberlite and lamproite emplacement. *Journal of Geochemical Exploration*, 51, 245–264.

Wilde, S., Valley, J., Peck W., and Graham, C. (2001) Evidence from detrital zircons for the existence of continental crust and oceans on the Earth at 4.4 Gyr ago. *Nature*, 409, 175–178.

Williams, Q., and Garnero, E.J. (1996) Seismic evidence for partial melt at the base of Earth's mantle. *Science*, 273, 1528–1530.

Wilson, J.T. (1963) Evidence from islands on the spreading of ocean floors. *Nature*, 197, 536–538.

Wilson, J.T. (1965) A new class of faults and their bearing on continental drift. *Nature*, 207, 343–347.

Wilson, L. (1980) Relationships between pressure, volatile content and ejecta velocity in three types of volcanic explosion. *Journal of Volcanology and Geothermal Research*, 8, 297–313.

Winter, J.D. (2001) *An Introduction to Igneous and Metamorphic Petrology*. Prentice Hall, Upper Saddle River, NJ, 697 pp.

Winter, J.D. (2009) *Principles of Igneous and Metamorphic Petrology*, 2nd edn. Prentice Hall, Upper Saddle River, NJ, 720 pp.

Wiseman, J.D.H. (1934) The Central and South-west Highland epidiorites: a study in progressive metamorphism. *Quaterly Journal of the Geological Society of London*, 90, 354–417.

Wolf, K.H., and Chingarian, G.V. (eds) (1992) *Diagenesis III*. Elsevier, Amsterdam, 674 pp.

Wolf, K.H., and Chingarian, G.V. (eds) (1994) *Diagenesis IV*. Elsevier, Amsterdam, 546 pp.

Wones, D.R., and Dodge, F.C.W. (1968) *On the Stability of Phlogopite*. Geological Society of America Special Paper 101, 242 pp.

Wright, D.T., and Wacey, D. (2004) Sedimentary dolomite: a reality check. In: Geological Society of London Special Publication No. 235, pp. 65–74.

Wright, J.V., Smith, A.L., and Self, S. (1980) A working terminology of pyroclastic deposits. *Journal of Volcanology and Geothermal Research*, 8, 315–336.

Wright, T.L., and Pierson, T.C. (1992) *Living with Volcanoes: the US Geological Survey's Volcano Hazards Program*. US Geological Survey Circular No. 1073, 57 pp.

Wright, T.O., and Platt, L.B. (1982) Pressure dissolution and cleavage in the Martinsburg Shale. *American Journal of Science*, 282, 122–135.

Wyman, D.A. (1999) Paleoproterozoic boninites in an ophiolite-like setting, trans-Hudson orogen, Canada. *Geology*, 27, 455–458.

Yardley, B.W.D. (1989) *An Introduction to Metamorphic Petrology*. Longman Publishers, London, UK, 248 pp.

Zachos, J.C., Lohmann, K.C., Walker, J.C.G., and Wise, S.W. (1993) Abrupt climate change and transient climates during the Paleogene – a marine perspective. *Journal of Geology*, *101*, 191–213.

Zdanowicz, C.M.; Zielinski, G.A., and Germani, M.S. (1999) Mount Mazama eruption; calendrical age verified and atmospheric impact assessed. *Geology*, *27*, 621–624.

Zolotukhin, V.V., and Al'Mukhamedov, A.I. (1988) Traps of the Siberian Platform. In: Macdougall, J.D. (ed.) *Continental Flood Basalts*. Kluwer Academic Publishers, New York, pp. 273–310.

Zoltai, T., and Stoudt, J.H. (1984) *Mineralogy: Concepts and Principles*. Burgess Publishing, Minneapolis, 506 pp.

Zwart, H.J. (1967) The duality of orogenic belts. *Geologie en Mijnbouw*, *46*, 283–309.

Index

Page numbers in **bold** refer to tables, page numbers in *italic* refer to figures.

Periodic table of the elements

IA 1	IIA 2	IIIB 3	IVB 4	VB 5	VIB 6	VIIB 7	8	----VIIIB---- 9	10
1 2.20 **H** 1.008 (1s₁) (±1)									
3 0.98 **Li** 6.941 (He + 2s₁) (+1)	**4** 1.57 **Be** 9.012 (He + 2s₂) (+2)								
11 0.93 **Na** 22.990 (Ne + 3s₁) (+1)	**12** 1.31 **Mg** 24.305 (Ne + 3s₂) (+2)								
19 0.82 **K** 39.098 (Ar + 4s₁) (+1)	**20** 1.00 **Ca** 40.080 (Ar + 4s₂) (+2)	**21** 1.36 **Sc** 44.956 (Ar + 4s₂ + 3d₁) (+3)	**22** 1.54 **Ti** 47.900 (Ar + 4s₂ + 3d₂) (+4,+2)	**23** 1.63 **V** 50.942 (Ar + 4s₂ + 3d₃) (many)	**24** 1.66 **Cr** 51.996 (Ar + 4s₁ + 3d₅) (+3,+6)	**25** 1.55 **Mn** 54.938 (Ar + 4s₂ + 3d₅) (many)	**26** 1.63 **Fe** 55.847 (Ar + 4s₂ + 3d₆) (+2,+3)	**27** 1.85 **Co** 58.933 (Ar + 4s₂ + 3d₇) (+2,+3)	**28** 1.91 **Ni** 58.700 (Ar + 4s₂ + 3d₈) (+2)
37 0.82 **Rb** 85.468 (Kr + 5s₁) (+1)	**38** 0.95 **Sr** 87.620 (Kr + 5s₂) (+2)	**39** 1.22 **Y** 88.906 (Kr + 5s₂ + 4d₁) (+3)	**40** 1.33 **Zr** 91.220 (Kr + 5s₂ + 4d₂) (+4,+3)	**41** 1.60 **Nb** 92.906 (Kr + 5s₁ + 4d₄) (many)	**42** 2.16 **Mo** 95.940 (Kr + 5s₁ + 4d₅) (many)	**43** 1.90 **Tc** (98) (Kr + 5s₂ + 4d₅) (many)	**44** 2.20 **Ru** 101.07 (Kr + 5s₁ + 4d₇) (many)	**45** 2.28 **Rh** 102.91 (Kr + 5s₁ + 4d₈) (many)	**46** 2.20 **Pd** 106.40 (Kr + 5s₂ + 4d₈) (+2,+4)
55 0.79 **Cs** 132.91 (Xe + 6s₁) (+1)	**56** 0.89 **Ba** 137.33 (Xe + 6s₂) (+2)	**57** 1.10 **La** 138.91 (Xe + 6s₂ + 5d₁) (+3)	**72** 1.30 **Hf** 178.48 (Xe + 6s₂ + 4f₁₄5d₂) (+4)	**73** 1.50 **Ta** 180.95 (Xe + 6s₂ + 4f₁₄5d₃) (+5)	**74** 2.36 **W** 183.85 (Xe + 6s₂ + 4f₁₄5d₄) (many)	**75** 1.90 **Re** 186.21 (Xe + 6s₂ + 4f₁₄5d₅) (many)	**76** 2.12 **Os** 190.20 (Xe + 6s₂ + 4f₁₄5d₆) (many)	**77** 2.20 **Ir** 192.22 (Xe + 6s₂ + 4f₁₄5d₇) (many)	**78** 2.28 **Pt** 195.09 (Xe + 6s₁ + 4f₁₄5d₉) (+2,+4)
87 0.70 **Fr** (223) (Rn + 7s₁) (+1)	**88** 0.87 **Ra** 226.03 (Rn + 7s₂) (+2)	**89** 1.10 **Ac** 227.03 (Rn + 7s₂ + 6d₁) (+3)							

Lanthanides	**58** 1.12 **Ce** 140.12 (Xe + 6s₂ + 5d₁4f₁) (+3,+4)	**59** 1.13 **Pr** 140.91 (Xe + 6s₂ + 5d₁4f₂) (+3,+4)	**60** 1.14 **Nd** 144.24 (Xe + 6s₂ + 5d₁4f₃) (+3)	**61** 1.13 **Pm** (145) (Xe + 6s₂ + 5d₁4f₄) (+3)	**62** 1.17 **Sm** 150.40 (Xe + 6s₂ + 5d₁4f₅) (+3,+2)	**63** 1.20 **Eu** 151.96 (Xe + 6s₂ + 5d₁4f₆) (+3,+2)	**64** 1.20 **Gd** 157.25 (Xe + 6s₂ + 5d₁4f₇) (+3)	**65** 1.20 **Tb** 158.93 (Xe + 6s₂ + 5d₁4f₈) (+3,+4)
Actinides	**90** 1.30 **Th** 232.04 (Rn + 7s₂ + 6d₁5f₁) (+4)	**91** 1.30 **Pa** 231.04 (Rn + 7s₂ + 5d₁4f₂) (+5,+4)	**92** 1.38 **U** 238.03 (Rn + 7s₂ + 5d₁4f₃) (many)					

IB 11	IIB 12	IIIA 13	IVA 14	VA 15	VIA 16	VIIA 17	VIIIA 18

Periodic Table (Groups IB–VIIIA) — each entry: Atomic number, Electronegativity, Symbol, Average mass, Electron configuration, Common valence state.

Z	Sym	EN	Mass	Configuration	Valence
2	He	----	4.003	$(1s_2)$	(0)
5	B	2.04	10.810	$(He + 2s_2 2p_1)$	(+3)
6	C	2.55	12.011	$(He + 2s_2 2p_2)$	(+4, 0)
7	N	3.04	14.007	$(He + 2s_2 2p_3)$	(many)
8	O	3.44	15.999	$(He + 2s_2 2p_4)$	(-2)
9	F	3.95	18.998	$(He + 2s_2 2p_5)$	(-1)
10	Ne	----	20.179	$(He + 2s_2 2p_6)$	(0)
13	Al		26.962	$(Ne + 3s_2 3p_1)$	(+3)
14	Si		28.086	$(Ne + 3s_2 3p_2)$	(+4)
15	P		30.974	$(Ne + 3s_2 3p_3)$	(many)
16	S		32.060	$(Ne + 3s_2 2p_4)$	(-2, +6)
17	Cl		35.453	$(Ne + 3s_1 3p_5)$	(-1)
18	Ar		39.948	$(Ne + 3s_2 3p_6)$	(0)
29	Cu	1.90	63.546	$(Ar + 4s_1 + 3d_{10})$	(+1,+2)
30	Zn	1.65	65.380	$(Ar + 4s_2 + 3d_{10})$	(+2)
31	Ga	1.81	69.720	$(Ar + 4s_2 4p_1 + 3d_{10})$	(+3)
32	Ge	2.01	72.590	$(Ar + 4s_2, 4p_2 + 3d_{10})$	(+4)
33	As	2.18	74.922	$(Ar + 4s_2, 4p_3 + 3d_{10})$	(many)
34	Se	2.55	78.960	$(Ar + 4s_2, 4p_4 + 3d_{10})$	(-2,+6)
35	Br	2.96	79.904	$(Ar + 4s_2, 4p_5 + 3d_{10})$	(-1)
36	Kr	----	83.800	$(Ar + 4s_2, 4p_6 + 3d_{10})$	(0)
47	Ag	1.93	107.87	$(Kr + 5s_1 + 4d_{10})$	(+1)
48	Cd	1.69	112.41	$(Kr + 5s_2 + 4d_{10})$	(+2)
49	In	1.78	114.82	$(Kr + 5s_2 5p_1 + 4d_{10})$	(+3)
50	Sn	1.96	118.69	$(Kr + 5s_2 5p_2 + 4d_{10})$	(+4,+2)
51	Sb	2.05	121.75	$(Kr + 5s_2 5p_3 + 4d_{10})$	(+5,+3)
52	Te	2.10	127.60	$(Kr + 5s_2 5p_4 + 4d_{10})$	(-2,+6)
53	I	2.66	126.90	$(Kr + 5s_2 5p_5 + 4d_{10})$	(-1)
54	Xe	----	131.30	$(Kr + 5s_2 5p_6 + 4d_{10})$	(0)
79	Au	2.54	196.97	$(Xe + 6s_1 + 4f_{14} 5d_{10})$	(+1,+3)
80	Hg	2.00	200.59	$(Xe + 6s_2 + 4f_{14} 5d_{10})$	(+2,+1)
81	Tl	2.04	204.37	$(Xe + 6s_2 6p_1 + 4f_{14} 5d_{10})$	(+3,+1)
82	Pb	2.33	208.98	$(Xe + 6s_2 6p_2 + 4f_{14} 5d_{10})$	(+4,+2)
83	Bi	2.02	208.98	$(Xe + 6s_2 6p_3 + 4f_{14} 5d_{10})$	(+5,+3)
84	Po	2.00	(209)	$(Xe + 6s_2 6p_4 + 4f_{14} 5d_{10})$	(+4,+2)
85	At	2.20	(210)	$(Xe + 6s_2 6p_5 + 4f_{14} 5d_{10})$	(many)
86	Rn	----	(222)	$(Xe + 6s_2 6p_6 + 4f_{14} 5d_{10})$	(0)

Lanthanides

Z	Sym	EN	Mass	Configuration	Valence
66	Dy	1.22	162.50	$(Xe + 6s_2 + 5d_1 4f_9)$	(+3)
67	Ho	1.23	164.93	$(Xe + 6s_2 + 5d_1 4f_{10})$	(+3)
68	Er	1.24	167.29	$(Xe + 6s_2 + 5d_1 4f_{11})$	(+3)
69	Tm	1.25	168.94	$(Xe + 6s_2 + 5d_1 4f_{12})$	(+3,+2)
70	Yb	1.10	173.04	$(Xe + 6s_2 + 5d_1 4f_{13})$	(+3,+2)
71	Lu	1.27	174.97	$(Xe + 6s_2 + 5d_1 4f_{14})$	(+3)

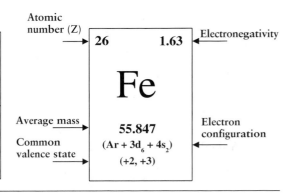

Legend:
- Atomic number (Z): 26
- Electronegativity: 1.63
- Fe
- Average mass: 55.847
- Electron configuration: $(Ar + 3d_6 + 4s_2)$
- Common valence state: (+2, +3)

Table of chemical elements

Symbol	Name	Symbol	Name
Ac	Actinium	Mn	Manganese
Ag	Silver (argentums)	Mo	Molybdenum
Al	Aluminum	N	Nitrogen
Am	Americium	Na	Sodium (natrium)
Ar	Argon	Nb	Niobium (columbium)
As	Arsenic	Nd	Neodymium
At	Astatine	Ne	Neon
Au	Gold (aurum)	Ni	Nickel
B	Boron	No	Nobelium
Ba	Barium	Np	Neptunium
Be	Beryllium (glucinum)	O	Oxygen
Bi	Bismuth	Os	Osmium
Bk	Berkelium	P	Phosphorus
Br	Bromine	Pa	Protactinium
C	Carbon	Pb	Lead (plumbum)
Ca	Calcium	Pd	Palladium
Cd	Cadmium	Pm	Promethium (illinium)
Ce	Cerium	Po	Polonium
Cf	Californium	Pr	Praseodymium
Cl	Chlorine	Pt	Platinum
Cm	Curium	Pu	Plutonium
Co	Cobalt	Ra	Radium
Cr	Chromium	Rb	Rubidium
Cs	Caesium	Re	Rhenium
Cu	Copper (cuprum)	Rf	Rutherfordium (kurchatovium)
Dy	Dysprosium	Rh	Rhodium
Er	Erbium	Rn	Radon (niton)
Es	Einsteinium	Ru	Ruthenium
Eu	Europium	S	Sulphur
F	Fluorine	Sb	Antimony (stibium)
Fe	Iron (ferrum)	Sc	Scandium
Fm	Fermium	Se	Selenium
Fr	Francium	Si	Silicon
Ga	Gallium	Sm	Samarium
Gd	Gadolinium	Sn	Tin (stannum)
Ge	Germanium	Sr	Strontium
H	Hydrogen	Ta	Tantalum
Ha	Hahnium	Tb	Terbium
He	Helium	Tc	Technetium (masurium)
Hf	Hafnium	Te	Tellurium
Hg	Mercury (hydragyrum)	Th	Thorium
Ho	Holmium	Ti	Titanium
I	Iodine	Tl	Thallium
In	Indium	Tm	Thulium
Ir	Iridium	U	Uranium
K	Potassium (kalium)	V	Vanadium
Kr	Krypton	W	Tungsten (wolfram)
La	Lanthanum	Xe	Xenon
Li	Lithium	Y	Yttrium
Lu	Lutetium	Yb	Ytterbium
Lr	Lawrencium	Zn	Zinc
Md	Mendelevium	Zr	Zirconium
Mg	Magnesium		